T0223498

Lecture Notes in Computer Science 9158

Commenced Publication in 1973
Founding and Former Series Editors:
Gerhard Goos, Juris Hartmanis, and Jan van Leeuwen

More information about this series at http://www.springer.com/series/7407

Osvaldo Gervasi · Beniamino Murgante
Sanjay Misra · Marina L. Gavrilova
Ana Maria Alves Coutinho Rocha · Carmelo Torre
David Taniar · Bernady O. Apduhan (Eds.)

Computational Science and Its Applications – ICCSA 2015

15th International Conference
Banff, AB, Canada, June 22–25, 2015
Proceedings, Part IV

 Springer

Editors

Osvaldo Gervasi
University of Perugia
Perugia
Italy

Beniamino Murgante
University of Basilicata
Potenza
Italy

Sanjay Misra
Covenant University
Canaanland
Nigeria

Marina L. Gavrilova
University of Calgary
Calgary, AB
Canada

Ana Maria Alves Coutinho Rocha
University of Minho
Braga
Portugal

Carmelo Torre
Polytechnic University
Bari
Italy

David Taniar
Monash University
Clayton, VIC
Australia

Bernady O. Apduhan
Kyushu Sangyo University
Fukuoka
Japan

ISSN 0302-9743   ISSN 1611-3349   (electronic)
Lecture Notes in Computer Science
ISBN 978-3-319-21409-2   ISBN 978-3-319-21410-8   (eBook)
DOI 10.1007/978-3-319-21410-8

Library of Congress Control Number: 2015943360

LNCS Sublibrary: SL1 – Theoretical Computer Science and General Issues

Springer Cham Heidelberg New York Dordrecht London

Printed on acid-free paper

Springer International Publishing AG Switzerland is part of Springer Science+Business Media
(www.springer.com)

Preface

The year 2015 is a memorable year for the International Conference on Computational Science and Its Applications. In 2003, the First International Conference on Computational Science and Its Applications (chaired by C.J.K. Tan and M. Gavrilova) took place in Montreal, Canada (2003), and the following year it was hosted by A. Laganà and O. Gervasi in Assisi, Italy (2004). It then moved to Singapore (2005), Glasgow, UK (2006), Kuala-Lumpur, Malaysia (2007), Perugia, Italy (2008), Seoul, Korea (2009), Fukuoka, Japan (2010), Santander, Spain (2011), Salvador de Bahia, Brazil (2012), Ho Chi Minh City, Vietnam (2013), and Guimarães, Portugal (2014). The current installment of ICCSA 2015 took place in majestic Banff National Park, Banff, Alberta, Canada, during June 22–25, 2015.

The event received approximately 780 submissions from over 45 countries, evaluated by over 600 reviewers worldwide.

Its main track acceptance rate was approximately 29.7 % for full papers. In addition to full papers, published by Springer, the event accepted short papers, poster papers, and PhD student showcase works that are published in the IEEE CPS proceedings.

It also runs a number of parallel workshops, some for over 10 years, with new ones appearing for the first time this year. The success of ICCSA is largely contributed to the continuous support of the computational sciences community as well as researchers working in the applied relevant fields, such as graphics, image processing, biometrics, optimization, computer modeling, information systems, geographical sciences, physics, biology, astronomy, biometrics, virtual reality, and robotics, to name a few.

Over the past decade, the vibrant and promising area focusing on performance-driven computing and big data has became one of the key points of research enhancing the performance of information systems and supported processes. In addition to high-quality research at the frontier of these fields, consistently presented at ICCSA, a number of special journal issues are being planned following ICCSA 2015, including TCS Springer (*Transactions on Computational Sciences,* LNCS).

The contribution of the International Steering Committee and the International Program Committee are invaluable in the conference success. The dedication of members of these committees, the majority of whom have fulfilled this difficult role for the last 10 years, is astounding. Our warm appreciation also goes to the invited speakers, all event sponsors, supporting organizations, and volunteers. Finally, we thank all the authors for their submissions making the ICCSA conference series a well recognized and a highly successful event year after year.

June 2015

Marina L. Gavrilova
Osvaldo Gervasi
Bernady O. Apduhan

Organization

ICCSA 2015 was organized by the University of Calgary (Canada), the University of Perugia (Italy), the University of Basilicata (Italy), Monash University (Australia), Kyushu Sangyo University (Japan), and the University of Minho, (Portugal)

Honorary General Chairs

Antonio Laganà	University of Perugia, Italy
Norio Shiratori	Tohoku University, Japan
Kenneth C.J. Tan	Sardina Systems, Estonia

General Chairs

Marina L. Gavrilova	University of Calgary, Canada
Osvaldo Gervasi	University of Perugia, Italy
Bernady O. Apduhan	Kyushu Sangyo University, Japan

Program Committee Chairs

Beniamino Murgante	University of Basilicata, Italy
Ana Maria A.C. Rocha	University of Minho, Portugal
David Taniar	Monash University, Australia

International Advisory Committee

Jemal Abawajy	Deakin University, Australia
Dharma P. Agrawal	University of Cincinnati, USA
Claudia Bauzer Medeiros	University of Campinas, Brazil
Manfred M. Fisher	Vienna University of Economics and Business, Austria
Yee Leung	Chinese University of Hong Kong, SAR China

International Liaison Chairs

Ana Carla P. Bitencourt	Universidade Federal do Reconcavo da Bahia, Brazil
Alfredo Cuzzocrea	ICAR-CNR and University of Calabria, Italy
Maria Irene Falcão	University of Minho, Portugal
Marina L. Gavrilova	University of Calgary, Canada
Robert C.H. Hsu	Chung Hua University, Taiwan
Andrés Iglesias	University of Cantabria, Spain
Tai-Hoon Kim	Hannam University, Korea
Sanjay Misra	University of Minna, Nigeria
Takashi Naka	Kyushu Sangyo University, Japan

Rafael D.C. Santos   Brazilian National Institute for Space Research, Brazil
Maribel Yasmina Santos   University of Minho, Portugal

Workshop and Session Organizing Chairs

Beniamino Murgante   University of Basilicata, Italy
Jorge Gustavo Rocha   University of Minho, Portugal

Local Arrangement Chairs

Marina Gavrilova   University of Calgary, Canada (Chair)
Madeena Sultana   University of Calgary, Canada
Padma Polash Paul   University of Calgary, Canada
Faisal Ahmed   University of Calgary, Canada
Hossein Talebi   University of Calgary, Canada
Camille Sinanan   University of Calgary, Canada

Venue

ICCSA 2015 took place in the Banff Park Lodge Conference Center, Alberta (Canada).

Workshop Organizers

Agricultural and Environment Information and Decision Support Systems (AEIDSS 2015)

Sandro Bimonte   IRSTEA, France
André Miralles   IRSTEA, France
Frederic Hubert   University of Laval, Canada
François Pinet   IRSTEA, France

Approaches or Methods of Security Engineering (AMSE 2015)

TaiHoon Kim   Sungshin W. University, Korea

Advances in Information Systems and Technologies for Emergency Preparedness and Risk Assessment (ASTER 2015)

Maurizio Pollino   ENEA, Italy
Marco Vona   University of Basilicata, Italy
Beniamino Murgante   University of Basilicata, Italy

Advances in Web-Based Learning (AWBL 2015)

Mustafa Murat Inceoglu   Ege University, Turkey

Bio-inspired Computing and Applications (BIOCA 2015)

Nadia Nedjah   State University of Rio de Janeiro, Brazil
Luiza de Macedo   State University of Rio de Janeiro, Brazil
  Mourell

Computer-Aided Modeling, Simulation, and Analysis (CAMSA 2015)

Jie Shen   University of Michigan, USA, and Jilin University, China
Hao Chen   Shanghai University of Engineering Science, China
Xiaoqiang Liun   Donghua University, China
Weichun Shi   Shanghai Maritime University, China

Computational and Applied Statistics (CAS 2015)

Ana Cristina Braga   University of Minho, Portugal
Ana Paula Costa   University of Minho, Portugal
  Conceicao Amorim

Computational Geometry and Security Applications (CGSA 2015)

Marina L. Gavrilova   University of Calgary, Canada

Computational Algorithms and Sustainable Assessment (CLASS 2015)

Antonino Marvuglia   Public Research Centre Henri Tudor, Luxembourg
Beniamino Murgante   University of Basilicata, Italy

Chemistry and Materials Sciences and Technologies (CMST 2015)

Antonio Laganà   University of Perugia, Italy
Alessandro Costantini   INFN, Italy
Noelia Faginas Lago   University of Perugia, Italy
Leonardo Pacifici   University of Perugia, Italy

Computational Optimization and Applications (COA 2015)

Ana Maria Rocha   University of Minho, Portugal
Humberto Rocha   University of Coimbra, Portugal

Cities, Technologies and Planning (CTP 2015)

Giuseppe Borruso   University of Trieste, Italy
Beniamino Murgante   University of Basilicata, Italy

Econometrics and Multidimensional Evaluation in the Urban Environment (EMEUE 2015)

Carmelo M. Torre   Polytechnic of Bari, Italy
Maria Cerreta   University of Naples Federico II, Italy
Paola Perchinunno   University of Bari, Italy

Simona Panaro   University of Naples Federico II, Italy
Raffaele Attardi   University of Naples Federico II, Italy
Claudia Ceppi   Polytechnic of Bari, Italy

Future Computing Systems, Technologies, and Applications (FISTA 2015)

Bernady O. Apduhan   Kyushu Sangyo University, Japan
Rafael Santos   Brazilian National Institute for Space Research, Brazil
Jianhua Ma   Hosei University, Japan
Qun Jin   Waseda University, Japan

Geographical Analysis, Urban Modeling, Spatial Statistics (GEOGAN-MOD 2015)

Giuseppe Borruso   University of Trieste, Italy
Beniamino Murgante   University of Basilicata, Italy
Hartmut Asche   University of Potsdam, Germany

Land Use Monitoring for Soil Consumption Reduction (LUMS 2015)

Carmelo M. Torre   Polytechnic of Bari, Italy
Alessandro Bonifazi   Polytechnic of Bari, Italy
Valentina Sannicandro   University Federico II of Naples, Italy
Massimiliano   University of Salerno, Italy
  Bencardino
Gianluca di Cugno   Polytechnic of Bari, Italy
Beniamino Murgante   University of Basilicata, Italy

Mobile Communications (MC 2015)

Hyunseung Choo   Sungkyunkwan University, Korea

Mobile Computing, Sensing, and Actuation for Cyber Physical Systems (MSA4CPS 2015)

Saad Qaisar   NUST School of Electrical Engineering and Computer
                           Science, Pakistan
Moonseong Kim   Korean Intellectual Property Office, Korea

Quantum Mechanics: Computational Strategies and Applications (QMCSA 2015)

Mirco Ragni   Universidad Federal de Bahia, Brazil
Ana Carla Peixoto   Universidade Estadual de Feira de Santana, Brazil
  Bitencourt
Roger Anderson   University of California, USA
Vincenzo Aquilanti   University of Perugia, Italy
Frederico Vasconcellos   Universidad Federal de Bahia, Brazil
  Prudente

Remote Sensing Data Analysis, Modeling, Interpretation and Applications: From a Global View to a Local Analysis (RS2015)

Rosa Lasaponara   Institute of Methodologies for Environmental Analysis,
                               National Research Council, Italy

Scientific Computing Infrastructure (SCI 2015)

Alexander Bodganov   St. Petersburg State University, Russia
Elena Stankova   St. Petersburg State University, Russia

Software Engineering Processes and Applications (SEPA 2015)

Sanjay Misra   Covenant University, Nigeria

Software Quality (SQ 2015)

Sanjay Misra   Covenant University, Nigeria

Advances in Spatio-Temporal Analytics (ST-Analytics 2015)

Joao Moura Pires   New University of Lisbon, Portugal
Maribel Yasmina Santos   New University of Lisbon, Portugal

Tools and Techniques in Software Development Processes (TTSDP 2015)

Sanjay Misra   Covenant University, Nigeria

Virtual Reality and Its Applications (VRA 2015)

Osvaldo Gervasi   University of Perugia, Italy
Lucio Depaolis   University of Salento, Italy

Program Committee

Jemal Abawajy   Deakin University, Australia
Kenny Adamson   University of Ulster, UK
Filipe Alvelos   University of Minho, Portugal
Paula Amaral   Universidade Nova de Lisboa, Portugal
Hartmut Asche   University of Potsdam, Germany
Md. Abul Kalam Azad   University of Minho, Portugal
Michela Bertolotto   University College Dublin, Ireland
Sandro Bimonte   CEMAGREF, TSCF, France
Rod Blais   University of Calgary, Canada
Ivan Blecic   University of Sassari, Italy
Giuseppe Borruso   University of Trieste, Italy
Yves Caniou   Lyon University, France
José A. Cardoso e   Universidade Nova de Lisboa, Portugal
  Cunha
Leocadio G. Casado   University of Almeria, Spain

Ivana Kolingerova	University of West Bohemia, Czech Republic
Dieter Kranzlmueller	LMU and LRZ Munich, Germany
Antonio Laganà	University of Perugia, Italy
Rosa Lasaponara	National Research Council, Italy
Maurizio Lazzari	National Research Council, Italy
Cheng Siong Lee	Monash University, Australia
Sangyoun Lee	Yonsei University, Korea
Jongchan Lee	Kunsan National University, Korea
Clement Leung	Hong Kong Baptist University, Hong Kong, SAR China
Chendong Li	University of Connecticut, USA
Gang Li	Deakin University, Australia
Ming Li	East China Normal University, China
Fang Liu	AMES Laboratories, USA
Xin Liu	University of Calgary, Canada
Savino Longo	University of Bari, Italy
Tinghuai Ma	NanJing University of Information Science and Technology, China
Sergio Maffioletti	University of Zurich, Switzerland
Ernesto Marcheggiani	Katholieke Universiteit Leuven, Belgium
Antonino Marvuglia	Research Centre Henri Tudor, Luxembourg
Nicola Masini	National Research Council, Italy
Nirvana Meratnia	University of Twente, The Netherlands
Alfredo Milani	University of Perugia, Italy
Sanjay Misra	Federal University of Technology Minna, Nigeria
Giuseppe Modica	University of Reggio Calabria, Italy
José Luis Montaña	University of Cantabria, Spain
Beniamino Murgante	University of Basilicata, Italy
Jiri Nedoma	Academy of Sciences of the Czech Republic, Czech Republic
Laszlo Neumann	University of Girona, Spain
Kok-Leong Ong	Deakin University, Australia
Belen Palop	Universidad de Valladolid, Spain
Marcin Paprzycki	Polish Academy of Sciences, Poland
Eric Pardede	La Trobe University, Australia
Kwangjin Park	Wonkwang University, Korea
Ana Isabel Pereira	Polytechnic Institute of Braganca, Portugal
Maurizio Pollino	Italian National Agency for New Technologies, Energy and Sustainable Economic Development, Italy
Alenka Poplin	University of Hamburg, Germany
Vidyasagar Potdar	Curtin University of Technology, Australia
David C. Prosperi	Florida Atlantic University, USA
Wenny Rahayu	La Trobe University, Australia
Jerzy Respondek	Silesian University of Technology Poland
Ana Maria A.C. Rocha	University of Minho, Portugal

Humberto Rocha	INESC-Coimbra, Portugal
Alexey Rodionov	Institute of Computational Mathematics and Mathematical Geophysics, Russia
Cristina S. Rodrigues	University of Minho, Portugal
Octavio Roncero	CSIC, Spain
Maytham Safar	Kuwait University, Kuwait
Chiara Saracino	A.O. Ospedale Niguarda Ca' Granda - Milano, Italy
Haiduke Sarafian	The Pennsylvania State University, USA
Jie Shen	University of Michigan, USA
Qi Shi	Liverpool John Moores University, UK
Dale Shires	U.S. Army Research Laboratory, USA
Takuo Suganuma	Tohoku University, Japan
Sergio Tasso	University of Perugia, Italy
Ana Paula Teixeira	University of Tras-os-Montes and Alto Douro, Portugal
Senhorinha Teixeira	University of Minho, Portugal
Parimala Thulasiraman	University of Manitoba, Canada
Carmelo Torre	Polytechnic of Bari, Italy
Javier Martinez Torres	Centro Universitario de la Defensa Zaragoza, Spain
Giuseppe A. Trunfio	University of Sassari, Italy
Unal Ufuktepe	Izmir University of Economics, Turkey
Toshihiro Uchibayashi	Kyushu Sangyo University, Japan
Mario Valle	Swiss National Supercomputing Centre, Switzerland
Pablo Vanegas	University of Cuenca, Equador
Piero Giorgio Verdini	INFN Pisa and CERN, Italy
Marco Vizzari	University of Perugia, Italy
Koichi Wada	University of Tsukuba, Japan
Krzysztof Walkowiak	Wroclaw University of Technology, Poland
Robert Weibel	University of Zurich, Switzerland
Roland Wismüller	Universität Siegen, Germany
Mudasser Wyne	SOET National University, USA
Chung-Huang Yang	National Kaohsiung Normal University, Taiwan
Xin-She Yang	National Physical Laboratory, UK
Salim Zabir	France Telecom Japan Co., Japan
Haifeng Zhao	University of California, Davis, USA
Kewen Zhao	University of Qiongzhou, China
Albert Y. Zomaya	University of Sydney, Australia

Reviewers

Abawajy Jemal	Deakin University, Australia
Abdi Samane	University College Cork, Ireland
Aceto Lidia	University of Pisa, Italy
Acharjee Shukla	Dibrugarh University, India
Adriano Elias	Universidade Nova de Lisboa, Portugal
Afreixo Vera	University of Aveiro, Portugal
Aguiar Ademar	Universidade do Porto, Portugal

Aguilar Antonio	University of Barcelona, Spain
Aguilar José Alfonso	Universidad Autónoma de Sinaloa, Mexico
Ahmed Faisal	University of Calgary, Canada
Aktas Mehmet	Yildiz Technical University, Turkey
Al-Juboori AliAlwan	International Islamic University Malaysia, Malaysia
Alarcon Vladimir	Universidad Diego Portales, Chile
Alberti Margarita	University of Barcelona, Spain
Ali Salman	NUST, Pakistan
Alkazemi Basem Qassim	University, Saudi Arabia
Alvanides Seraphim	Northumbria University, UK
Alvelos Filipe	University of Minho, Portugal
Alves Cláudio	University of Minho, Portugal
Alves José Luis	University of Minho, Portugal
Alves Maria Joo	Universidade de Coimbra, Portugal
Amin Benatia Mohamed	Groupe Cesi, France
Amorim Ana Paula	University of Minho, Portugal
Amorim Paulo	Federal University of Rio de Janeiro, Brazil
Andrade Wilkerson	Federal University of Campina Grande, Brazil
Andrianov Serge	Yandex, Russia
Aniche Mauricio	University of São Paulo, Brazil
Andrienko Gennady	Fraunhofer Institute for Intelligent Analysis and Informations Systems, Germany
Apduhan Bernady	Kyushu Sangyo University, Japan
Aquilanti Vincenzo	University of Perugia, Italy
Aquino Gibeon	UFRN, Brazil
Argiolas Michele	University of Cagliari, Italy
Asche Hartmut	Potsdam University, Germany
Athayde Maria Emilia Feijão Queiroz	University of Minho, Portugal
Attardi Raffaele	University of Napoli Federico II, Italy
Azad Md. Abdul	Indian Institute of Technology Kanpur, India
Azad Md. Abul Kalam	University of Minho, Portugal
Bao Fernando	Universidade Nova de Lisboa, Portugal
Badard Thierry	Laval University, Canada
Bae Ihn-Han	Catholic University of Daegu, South Korea
Baioletti Marco	University of Perugia, Italy
Balena Pasquale	Polytechnic of Bari, Italy
Banerjee Mahua	Xavier Institute of Social Sciences, India
Barroca Filho Itamir	UFRN, Brazil
Bartoli Daniele	University of Perugia, Italy
Bastanfard Azam	Islamic Azad University, Iran
Belanzoni Paola	University of Perugia, Italy
Bencardino Massimiliano	University of Salerno, Italy
Benigni Gladys	University of Oriente, Venezuela

Bertolotto Michela	University College Dublin, Ireland
Bilancia Massimo	Università di Bari, Italy
Blanquer Ignacio	Universitat Politècnica de València, Spain
Bodini Olivier	Université Pierre et Marie Curie Paris and CNRS, France
Bogdanov Alexander	Saint-Petersburg State University, Russia
Bollini Letizia	University of Milano, Italy
Bonifazi Alessandro	Polytechnic of Bari, Italy
Borruso Giuseppe	University of Trieste, Italy
Bostenaru Maria	"Ion Mincu" University of Architecture and Urbanism, Romania
Boucelma Omar	University of Marseille, France
Braga Ana Cristina	University of Minho, Portugal
Branquinho Amilcar	University of Coimbra, Portugal
Brás Carmo	Universidade Nova de Lisboa, Portugal
Cacao Isabel	University of Aveiro, Portugal
Cadarso-Suárez Carmen	University of Santiago de Compostela, Spain
Caiaffa Emanuela	ENEA, Italy
Calamita Giuseppe	National Research Council, Italy
Campagna Michele	University of Cagliari, Italy
Campobasso Francesco	University of Bari, Italy
Campos José	University of Minho, Portugal
Caniato Renhe Marcelo	Universidade Federal de Juiz de Fora, Brazil
Cannatella Daniele	University of Napoli Federico II, Italy
Canora Filomena	University of Basilicata, Italy
Cannatella Daniele	University of Napoli Federico II, Italy
Canora Filomena	University of Basilicata, Italy
Carbonara Sebastiano	University of Chieti, Italy
Carlini Maurizio	University of Tuscia, Italy
Carneiro Claudio	École Polytechnique Fédérale de Lausanne, Switzerland
Ceppi Claudia	Polytechnic of Bari, Italy
Cerreta Maria	University Federico II of Naples, Italy
Chen Hao	Shanghai University of Engineering Science, China
Choi Joonsoo	Kookmin University, South Korea
Choo Hyunseung	Sungkyunkwan University, South Korea
Chung Min Young	Sungkyunkwan University, South Korea
Chung Myoungbeom	Sungkyunkwan University, South Korea
Chung Tai-Myoung	Sungkyunkwan University, South Korea
Cirrincione Maurizio	Université de Technologie Belfort-Montbeliard, France
Clementini Eliseo	University of L'Aquila, Italy
Coelho Leandro dos Santos	PUC-PR, Brazil
Coletti Cecilia	University of Chieti, Italy
Conceicao Ana	Universidade do Algarve, Portugal
Correia Elisete	University of Trás-Os-Montes e Alto Douro, Portugal
Correia Filipe	FEUP, Portugal

Correia Florbela Maria da Cruz Domingues	Instituto Politécnico de Viana do Castelo, Portugal
Corso Pereira Gilberto	UFPA, Brazil
Cortés Ana	Universitat Autònoma de Barcelona, Spain
Cosido Oscar	Ayuntamiento de Santander, Spain
Costa Carlos	Faculdade Engenharia U. Porto, Portugal
Costa Fernanda	University of Minho, Portugal
Costantini Alessandro	INFN, Italy
Crasso Marco	National Scientific and Technical Research Council, Argentina
Crawford Broderick	Universidad Catolica de Valparaiso, Chile
Crestaz Ezio	GiScience, Italia
Cristia Maximiliano	CIFASIS and UNR, Argentina
Cunha Gaspar	University of Minho, Portugal
Cutini Valerio	University of Pisa, Italy
Danese Maria	IBAM, CNR, Italy
Daneshpajouh Shervin	University of Western Ontario, Canada
De Almeida Regina	University of Trás-os-Montes e Alto Douro, Portugal
de Doncker Elise	University of Michgan, USA
De Fino Mariella	Polytechnic of Bari, Italy
De Paolis Lucio Tommaso	University of Salento, Italy
de Rezende Pedro J.	Universidade Estadual de Campinas, Brazil
De Rosa Fortuna	University of Napoli Federico II, Italy
De Toro Pasquale	University of Napoli Federico II, Italy
Decker Hendrik	Instituto Tecnológico de Informática, Spain
Degtyarev Alexander	Saint-Petersburg State University, Russia
Deiana Andrea	Geoinfolab, Italia
Deniz Berkhan	Aselsan Electronics Inc., Turkey
Desjardin Eric	University of Reims, France
Devai Frank	London South Bank University, UK
Dwivedi Sanjay Kumar	Babasaheb Bhimrao Ambedkar University, India
Dhawale Chitra	PR Pote College, Amravati, India
Di Cugno Gianluca	Polytechnic of Bari, Italy
Di Gangi Massimo	University of Messina, Italy
Di Leo Margherita	JRC, European Commission, Belgium
Dias Joana	University of Coimbra, Portugal
Dias d'Almeida Filomena	University of Porto, Portugal
Diez Teresa	Universidad de Alcalá, Spain
Dilo Arta	University of Twente, The Netherlands
Dixit Veersain	Delhi University, India
Doan Anh Vu	Université Libre de Bruxelles, Belgium
Durrieu Sylvie	Maison de la Teledetection Montpellier, France
Dutra Inês	University of Porto, Portugal
Dyskin Arcady	The University of Western Australia, Australia

Eichelberger Hanno	University of Tübingen, Germany
El-Zawawy Mohamed A.	Cairo University, Egypt
Escalona Maria-Jose	University of Seville, Spain
Falcão M. Irene	University of Minho, Portugal
Farantos Stavros	University of Crete and FORTH, Greece
Faria Susana	University of Minho, Portugal
Fernandes Edite	University of Minho, Portugal
Fernandes Rosário	University of Minho, Portugal
Fernandez Joao P.	Universidade da Beira Interior, Portugal
Ferrão Maria	University of Beira Interior and CEMAPRE, Portugal
Ferreira Fátima	University of Trás-Os-Montes e Alto Douro, Portugal
Figueiredo Manuel Carlos	University of Minho, Portugal
Filipe Ana	University of Minho, Portugal
Flouvat Frederic	University New Caledonia, New Caledonia
Forjaz Maria Antónia	University of Minho, Portugal
Formosa Saviour	University of Malta, Malta
Fort Marta	University of Girona, Spain
Franciosa Alfredo	University of Napoli Federico II, Italy
Freitas Adelaide de Fátima Baptista Valente	University of Aveiro, Portugal
Frydman Claudia	Laboratoire des Sciences de l'Information et des Systèmes, France
Fusco Giovanni	CNRS - UMR ESPACE, France
Gabrani Goldie	University of Delhi, India Galleguillos Cristian, Pontificia Universidad Catlica de Valparaso, Chile
Gao Shang	Zhongnan University of Economics and Law, China
Garau Chiara	University of Cagliari, Italy
Garcia Ernesto	University of the Basque Country, Spain
Garca Omar Vicente	Universidad Autònoma de Sinaloa, Mexico
Garcia Tobio Javier	Centro de Supercomputación de Galicia, CESGA, Spain
Gavrilova Marina	University of Calgary, Canada
Gazzea Nicoletta	ISPRA, Italy
Gensel Jerome	IMAG, France
Geraldi Edoardo	National Research Council, Italy
Gervasi Osvaldo	University of Perugia, Italy
Giaoutzi Maria	National Technical University Athens, Greece
Gil Artur	University of the Azores, Portugal
Gizzi Fabrizio	National Research Council, Italy
Gomes Abel	Universidad de Beira Interior, Portugal
Gomes Maria Cecilia	Universidade Nova de Lisboa, Portugal
Gomes dos Anjos Eudisley	Federal University of Paraba, Brazil
Gonçalves Alexandre	Instituto Superior Tecnico Lisboa, Portugal

Gonçalves Arminda Manuela	University of Minho, Portugal
Gonzaga de Oliveira Sanderson Lincohn	Universidade Do Estado De Santa Catarina, Brazil
Gonzalez-Aguilera Diego	Universidad de Salamanca, Spain
Gorbachev Yuriy	Geolink Technologies, Russia
Govani Kishan	Darshan Institute of Engineering Technology, India
Grandison Tyrone	Proficiency Labs International, USA
Gravagnuolo Antonia	University of Napoli Federico II, Italy
Grilli Luca	University of Perugia, Italy
Guerra Eduardo	National Institute for Space Research, Brazil
Guo Hua	Carleton University, Canada
Hanazumi Simone	University of São Paulo, Brazil
Hanif Mohammad Abu	Chonbuk National University, South Korea
Hansen Henning Sten	Aalborg University, Denmark
Hanzl Malgorzata	University of Lodz, Poland
Hegedus Peter	University of Szeged, Hungary
Heijungs Reinout	VU University Amsterdam, The Netherlands
Hendrix Eligius M.T.	University of Malaga/Wageningen University, Spain/The Netherlands
Henriques Carla	Escola Superior de Tecnologia e Gestão, Portugal
Herawan Tutut	University of Malaya, Malaysia
Hiyoshi Hisamoto	Gunma University, Japan
Hodorog Madalina	Austria Academy of Science, Austria
Hong Choong Seon	Kyung Hee University, South Korea
Hsu Ching-Hsien	Chung Hua University, Taiwan
Hsu Hui-Huang	Tamkang University, Taiwan
Hu Hong	The Honk Kong Polytechnic University, China
Huang Jen-Fa	National Cheng Kung University, Taiwan
Hubert Frederic	Université Laval, Canada
Iglesias Andres	University of Cantabria, Spain
Jamal Amna	National University of Singapore, Singapore
Jank Gerhard	Aachen University, Germany
Jeong Jongpil	Sungkyunkwan University, South Korea
Jiang Bin	University of Gävle, Sweden
Johnson Franklin	Universidad de Playa Ancha, Chile
Kalogirou Stamatis	Harokopio University of Athens, Greece
Kamoun Farouk	Université de la Manouba, Tunisia
Kanchi Saroja	Kettering University, USA
Kanevski Mikhail	University of Lausanne, Switzerland
Kang Myoung-Ah	ISIMA Blaise Pascal University, France
Karandikar Varsha	Devi Ahilya University, Indore, India
Karimipour Farid	Vienna University of Technology, Austria
Kavouras Marinos	University of Lausanne, Switzerland
Kazar Baris	Oracle Corp., USA

Keramat Alireza   Jundi-Shapur Univ. of Technology, Iran
Khan Murtaza   NUST, Pakistan
Khattak Asad Masood   Kyung Hee University, Korea
Khazaei Hamzeh   Ryerson University, Canada
Khurshid Khawar   NUST, Pakistan
Kim Dongsoo   Indiana University-Purdue University Indianapolis, USA
Kim Mihui   Hankyong National University, South Korea
Koo Bonhyun   Samsung, South Korea
Korkhov Vladimir   St. Petersburg State University, Russia
Kotzinos Dimitrios   Université de Cergy-Pontoise, France
Kumar Dileep   SR Engineering College, India
Kurdia Anastasia   Buknell University, USA
Lachance-Bernard   École Polytechnique Fédérale de Lausanne, Switzerland
  Nicolas
Laganà Antonio   University of Perugia, Italy
Lai Sabrina   University of Cagliari, Italy
Lanorte Antonio   CNR-IMAA, Italy
Lanza Viviana   Lombardy Regional Institute for Research, Italy
Lasaponara Rosa   National Research Council, Italy
Lassoued Yassine   University College Cork, Ireland
Lazzari Maurizio   CNR IBAM, Italy
Le Duc Tai   Sungkyunkwan University, South Korea
Le Duc Thang   Sungkyunkwan University, South Korea
Le-Thi Kim-Tuyen   Sungkyunkwan University, South Korea
Ledoux Hugo   Delft University of Technology, The Netherlands
Lee Dong-Wook   INHA University, South Korea
Lee Hongseok   Sungkyunkwan University, South Korea
Lee Ickjai   James Cook University, Australia
Lee Junghoon   Jeju National University, South Korea
Lee KangWoo   Sungkyunkwan University, South Korea
Legatiuk Dmitrii   Bauhaus University Weimar, Germany
Lendvay Gyorgy   Hungarian Academy of Science, Hungary
Leonard Kathryn   California State University, USA
Li Ming   East China Normal University, China
Libourel Thrse   LIRMM, France
Lin Calvin   University of Texas at Austin, USA
Liu Xin   University of Calgary, Canada
Loconte Pierangela   Technical University of Bari, Italy
Lombardi Andrea   University of Perugia, Italy
Longo Savino   University of Bari, Italy
Lopes Cristina   University of California Irvine, USA
Lopez Cabido Ignacio   Centro de Supercomputación de Galicia, CESGA
Lourenço Vanda Marisa   University Nova de Lisboa, Portugal
Luaces Miguel   University of A Coruña, Spain
Lucertini Giulia   IUAV, Italy
Luna Esteban Robles   Universidad Nacional de la Plata, Argentina

M.M.H. Gregori Rodrigo	Universidade Tecnológica Federal do Paraná, Brazil
Machado Gaspar	University of Minho, Portugal
Machado Jose	University of Minho, Portugal
Mahinderjit Singh Manmeet	University Sains Malaysia, Malaysia
Malonek Helmuth	University of Aveiro, Portugal
Manfreda Salvatore	University of Basilicata, Italy
Manns Mary Lynn	University of North Carolina Asheville, USA
Manso Callejo Miguel Angel	Universidad Politécnica de Madrid, Spain
Marechal Bernard	Universidade Federal de Rio de Janeiro, Brazil
Marechal Franois	École Polytechnique Fédérale de Lausanne, Switzerland
Margalef Tomas	Universitat Autònoma de Barcelona, Spain
Marghany Maged	Universiti Teknologi Malaysia, Malaysia
Marsal-Llacuna Maria-Llusa	Universitat de Girona, Spain
Marsh Steven	University of Ontario, Canada
Martins Ana Mafalda	Universidade de Aveiro, Portugal
Martins Pedro	Universidade do Minho, Portugal
Marvuglia Antonino	Public Research Centre Henri Tudor, Luxembourg
Mateos Cristian	Universidad Nacional del Centro, Argentina
Matos Inés	Universidade de Aveiro, Portugal
Matos Jose	Instituto Politecnico do Porto, Portugal
Matos João	ISEP, Portugal
Mauro Giovanni	University of Trieste, Italy
Mauw Sjouke	University of Luxembourg, Luxembourg
Medeiros Pedro	Universidade Nova de Lisboa, Portugal
Melle Franco Manuel	University of Minho, Portugal
Melo Ana	Universidade de São Paulo, Brazil
Michikawa Takashi	University of Tokio, Japan
Milani Alfredo	University of Perugia, Italy
Millo Giovanni	Generali Assicurazioni, Italy
Min-Woo Park	SungKyunKwan University, South Korea
Miranda Fernando	University of Minho, Portugal
Misra Sanjay	Covenant University, Nigeria
Mo Otilia	Universidad Autonoma de Madrid, Spain
Modica Giuseppe	Università Mediterranea di Reggio Calabria, Italy
Mohd Nawi Nazri	Universiti Tun Hussein Onn Malaysia, Malaysia
Morais João	University of Aveiro, Portugal
Moreira Adriano	University of Minho, Portugal
Moerig Marc	University of Magdeburg, Germany
Morzy Mikolaj	University of Poznan, Poland
Mota Alexandre	Universidade Federal de Pernambuco, Brazil
Moura Pires João	Universidade Nova de Lisboa - FCT, Portugal
Mourão Maria	Polytechnic Institute of Viana do Castelo, Portugal

Mourelle Luiza de Macedo	UERJ, Brazil
Mukhopadhyay Asish	University of Windsor, Canada
Mulay Preeti	Bharti Vidyapeeth University, India
Murgante Beniamino	University of Basilicata, Italy
Naghizadeh Majid Reza	Qazvin Islamic Azad University, Iran
Nagy Csaba	University of Szeged, Hungary
Nandy Subhas	Indian Statistical Institute, India
Nash Andrew	Vienna Transport Strategies, Austria
Natário Isabel Cristina Maciel	University Nova de Lisboa, Portugal
Navarrete Gutierrez Tomas	Luxembourg Institute of Science and Technology, Luxembourg
Nedjah Nadia	State University of Rio de Janeiro, Brazil
Nguyen Hong-Quang	Ho Chi Minh City University, Vietnam
Nguyen Tien Dzung	Sungkyunkwan University, South Korea
Nickerson Bradford	University of New Brunswick, Canada
Nielsen Frank	Université Paris Saclay CNRS, France
NM Tuan	Ho Chi Minh City University of Technology, Vietnam
Nogueira Fernando	University of Coimbra, Portugal
Nole Gabriele	IRMAA National Research Council, Italy
Nourollah Ali	Amirkabir University of Technology, Iran
Olivares Rodrigo	UCV, Chile
Oliveira Irene	University of Trás-Os-Montes e Alto Douro, Portugal
Oliveira José A.	University of Minho, Portugal
Oliveira e Silva Luis	University of Lisboa, Portugal
Osaragi Toshihiro	Tokyo Institute of Technology, Japan
Ottomanelli Michele	Polytechnic of Bari, Italy
Ozturk Savas	TUBITAK, Turkey
Pagliara Francesca	University of Naples, Italy
Painho Marco	New University of Lisbon, Portugal
Pantazis Dimos	Technological Educational Institute of Athens, Greece
Paolotti Luisa	University of Perugia, Italy
Papa Enrica	University of Amsterdam, The Netherlands
Papathanasiou Jason	University of Macedonia, Greece
Pardede Eric	La Trobe University, Australia
Parissis Ioannis	Grenoble INP - LCIS, France
Park Gyung-Leen	Jeju National University, South Korea
Park Sooyeon	Korea Polytechnic University, South Korea
Pascale Stefania	University of Basilicata, Italy
Parker Gregory	University of Oklahoma, USA
Parvin Hamid	Iran University of Science and Technology, Iran
Passaro Pierluigi	University of Bari Aldo Moro, Italy
Pathan Al-Sakib Khan	International Islamic University Malaysia, Malaysia
Paul Padma Polash	University of Calgary, Canada

Peixoto Bitencourt Ana Carla	Universidade Estadual de Feira de Santana, Brazil
Peraza Juan Francisco	Autonomous University of Sinaloa, Mexico
Perchinunno Paola	University of Bari, Italy
Pereira Ana	Polytechnic Institute of Bragança, Portugal
Pereira Francisco	Instituto Superior de Engenharia, Portugal
Pereira Paulo	University of Minho, Portugal
Pereira Javier	Diego Portales University, Chile
Pereira Oscar	Universidade de Aveiro, Portugal
Pereira Ricardo	Portugal Telecom Inovacao, Portugal
Perez Gregorio	Universidad de Murcia, Spain
Pesantes Mery	CIMAT, Mexico
Pham Quoc Trung	HCMC University of Technology, Vietnam
Pietrantuono Roberto	University of Napoli "Federico II", Italy
Pimentel Carina	University of Aveiro, Portugal
Pina Antonio	University of Minho, Portugal
Piñar Miguel	Universidad de Granada, Spain
Pinciu Val	Southern Connecticut State University, USA
Pinet Francois	IRSTEA, France
Piscitelli Claudia	Polytechnic University of Bari, Italy
Pollino Maurizio	ENEA, Italy
Poplin Alenka	University of Hamburg, Germany
Porschen Stefan	University of Köln, Germany
Potena Pasqualina	University of Bergamo, Italy
Prata Paula	University of Beira Interior, Portugal
Previtali Mattia	Polytechnic of Milan, Italy
Prosperi David	Florida Atlantic University, USA
Protheroe Dave	London South Bank University, UK
Pusatli Tolga	Cankaya University, Turkey
Qaisar Saad	NURST, Pakistan
Qi Yu	Mesh Capital LLC, USA
Quan Tho	Ho Chi Minh City University of Technology, Vietnam
Raffaeta Alessandra	University of Venice, Italy
Ragni Mirco	Universidade Estadual de Feira de Santana, Brazil
Rahayu Wenny	La Trobe University, Australia
Rautenberg Carlos	University of Graz, Austria
Ravat Franck	IRIT, France
Raza Syed Muhammad	Sungkyunkwan University, South Korea
Rinaldi Antonio	DIETI - UNINA, Italy
Rinzivillo Salvatore	University of Pisa, Italy
Rios Gordon	University College Dublin, Ireland
Riva Sanseverino Eleonora	University of Palermo, Italy
Roanes-Lozano Eugenio	Universidad Complutense de Madrid, Spain
Rocca Lorena	University of Padova, Italy
Roccatello Eduard	3DGIS, Italy

Rocha Ana Maria	University of Minho, Portugal
Rocha Humberto	University of Coimbra, Portugal
Rocha Jorge	University of Minho, Portugal
Rocha Maria Clara	ESTES Coimbra, Portugal
Rocha Miguel	University of Minho, Portugal
Rodrigues Armanda	Universidade Nova de Lisboa, Portugal
Rodrigues Cristina	DPS, University of Minho, Portugal
Rodrigues Joel	University of Minho, Portugal
Rodriguez Daniel	University of Alcala, Spain
Rodrguez Gonzlez Alejandro	Universidad Carlos III Madrid, Spain
Roh Yongwan	Korean IP, South Korea
Romano Bernardino	University of l'Aquila, Italy
Roncaratti Luiz	Instituto de Física, University of Brasilia, Brazil
Roshannejad Ali	University of Calgary, Canada
Rosi Marzio	University of Perugia, Italy
Rossi Gianfranco	University of Parma, Italy
Rotondo Francesco	Polytechnic of Bari, Italy
Roussey Catherine	IRSTEA, France
Ruj Sushmita	Indian Statistical Institute, India
S. Esteves Jorge	University of Aveiro, Portugal
Saeed Husnain	NUST, Pakistan
Sahore Mani	Lovely Professional University, India
Saini Jatinder Singh	Baba Banda Singh Bahadur Engineering College, India
Salzer Reiner	Technical University Dresden, Germany
Sameh Ahmed	The American University in Cairo, Egypt
Sampaio Alcinia Zita	Instituto Superior Tecnico Lisboa, Portugal
Sannicandro Valentina	Polytechnic of Bari, Italy
Santiago Jnior Valdivino	Instituto Nacional de Pesquisas Espaciais, Brazil
Santos Josué	UFABC, Brazil
Santos Rafael	INPE, Brazil
Santos Viviane	Universidade de São Paulo, Brazil
Santucci Valentino	University of Perugia, Italy
Saracino Gloria	University of Milano-Bicocca, Italy
Sarafian Haiduke	Pennsylvania State University, USA
Saraiva João	University of Minho, Portugal
Sarrazin Renaud	Université Libre de Bruxelles, Belgium
Schirone Dario Antonio	University of Bari, Italy
Schneider Michel	ISIMA, France
Schoier Gabriella	University of Trieste, Italy
Schuhmacher Marta	Universitat Rovira i Virgili, Spain
Scorza Francesco	University of Basilicata, Italy
Seara Carlos	Universitat Politècnica de Catalunya, Spain
Sellares J. Antoni	Universitat de Girona, Spain
Selmaoui Nazha	University of New Caledonia, New Caledonia
Severino Ricardo Jose	University of Minho, Portugal

Shaik Mahaboob Hussain	JNTUK Vizianagaram, A.P., India
Sheikho Kamel	KACST, Saudi Arabia
Shen Jie	University of Michigan, USA
Shi Xuefei	University of Science Technology Beijing, China
Shin Dong Hee	Sungkyunkwan University, South Korea
Shojaeipour Shahed	Universiti Kebangsaan Malaysia, Malaysia
Shon Minhan	Sungkyunkwan University, South Korea
Shukla Ruchi	University of Johannesburg, South Africa
Silva Carlos	University of Minho, Portugal
Silva J.C.	IPCA, Portugal
Silva de Souza Laudson	Federal University of Rio Grande do Norte, Brazil
Silva-Fortes Carina	ESTeSL-IPL, Portugal
Simão Adenilso	Universidade de São Paulo, Brazil
Singh R.K.	Delhi University, India
Singh V.B.	University of Delhi, India
Singhal Shweta	GGSIPU, India
Sipos Gergely	European Grid Infrastructure, The Netherlands
Smolik Michal	University of West Bohemia, Czech Republic
Soares Inês	INESC Porto, Portugal
Soares Michel	Federal University of Sergipe, Brazil
Sobral Joao	University of Minho, Portugal
Son Changhwan	Sungkyunkwan University, South Korea
Song Kexing	Henan University of Science and Technology, China
Sosnin Petr	Ulyanovsk State Technical University, Russia
Souza Eric	Universidade Nova de Lisboa, Portugal
Sproessig Wolfgang	Technical University Bergakademie Freiberg, Germany
Sreenan Cormac	University College Cork, Ireland
Stankova Elena	Saint-Petersburg State University, Russia
Starczewski Janusz	Institute of Computational Intelligence, Poland
Stehn Fabian	University of Bayreuth, Germany
Sultana Madeena	University of Calgary, Canada
Swarup Das	Ananda Kalinga Institute of Industrial Technology, India
Tahar Sofiène	Concordia University, Canada
Takato Setsuo	Toho University, Japan
Talebi Hossein	University of Calgary, Canada
Tanaka Kazuaki	Kyushu Institute of Technology, Japan
Taniar David	Monash University, Australia
Taramelli Andrea	Columbia University, USA
Tarantino Eufemia	Polytechnic of Bari, Italy
Tariq Haroon	Connekt Lab, Pakistan
Tasso Sergio	University of Perugia, Italy
Teixeira Ana Paula	University of Trás-Os-Montes e Alto Douro, Portugal
Tesseire Maguelonne	IRSTEA, France
Thi Thanh Huyen Phan	Japan Advanced Institute of Science and Technology, Japan

Thorat Pankaj	Sungkyunkwan University, South Korea
Tilio Lucia	University of Basilicata, Italy
Tiwari Rupa	University of Minnesota, USA
Toma Cristian	Polytechnic University of Bucarest, Romania
Tomaz Graça	Polytechnic Institute of Guarda, Portugal
Tortosa Leandro	University of Alicante, Spain
Tran Nguyen	Kyung Hee University, South Korea
Tripp Barba, Carolina	Universidad Autnoma de Sinaloa, Mexico
Trunfio Giuseppe A.	University of Sassari, Italy
Uchibayashi Toshihiro	Kyushu Sangyo University, Japan
Ugalde Jesus	Universidad del Pais Vasco, Spain
Urbano Joana	LIACC University of Porto, Portugal
Van de Weghe Nico	Ghent University, Belgium
Varella Evangelia	Aristotle University of Thessaloniki, Greece
Vasconcelos Paulo	University of Porto, Portugal
Vella Flavio	University of Rome La Sapienza, Italy
Velloso Pedro	Universidade Federal Fluminense, Brazil
Viana Ana	INESC Porto, Portugal
Vidacs Laszlo	MTA-SZTE, Hungary
Vieira Ramadas Gisela	Polytechnic of Porto, Portugal
Vijay NLankalapalli	National Institute for Space Research, Brazil
Vijaykumar Nandamudi	INPE, Brazil
Viqueira José R.R.	University of Santiago de Compostela, Spain
Vitellio Ilaria	University of Naples, Italy
Vizzari Marco	University of Perugia, Italy
Wachowicz Monica	University of New Brunswick, Canada
Walentynski Ryszard	Silesian University of Technology, Poland
Walkowiak Krzysztof	Wroclaw University of Technology, Poland
Wallace Richard J.	University College Cork, Ireland
Waluyo Agustinus Borgy	Monash University, Australia
Wanderley Fernando	FCT/UNL, Portugal
Wang Chao	University of Science and Technology of China, China
Wang Yanghui	Beijing Jiaotong University, China
Wei Hoo Chong	Motorola, USA
Won Dongho	Sungkyunkwan University, South Korea
Wu Jian-Da	National Changhua University of Education, Taiwan
Xin Liu	École Polytechnique Fédérale de Lausanne, Switzerland
Yadav Nikita	Delhi Universty, India
Yamauchi Toshihiro	Okayama University, Japan
Yao Fenghui	Tennessee State University, USA
Yatskevich Mikalai	Assioma, Italy
Yeoum Sanggil	Sungkyunkwan University, South Korea
Yoder Joseph	Refactory Inc., USA
Zalyubovskiy Vyacheslav	Russian Academy of Sciences, Russia

Zeile Peter	Technische Universität Kaiserslautern, Germany
Zemek Michael	University of West Bohemia, Czech Republic
Zemlika Michal	Charles University, Czech Republic
Zolotarev Valeriy	Saint-Petersburg State University, Russia
Zunino Alejandro	Universidad Nacional del Centro, Argentina
Zurita Cruz Carlos Eduardo	Autonomous University of Sinaloa, Mexico

Sponsoring Organizations

ICCSA 2015 would not have been possible without the tremendous support of many organizations and institutions, for which all organizers and participants of ICCSA 2015 express their sincere gratitude:

University of Calgary, Canada (http://www.ucalgary.ca)

University of Perugia, Italy (http://www.unipg.it)

University of Basilicata, Italy (http://www.unibas.it)

Monash University, Australia (http://monash.edu)

Kyushu Sangyo University, Japan (www.kyusan-u.ac.jp)

Universidade do Minho, Portugal (http://www.uminho.pt)

Contents – Part IV

**Workshop on Quantum Mechanics: Computational Strategies
and Applications (QMCSA 2015)**

**Workshop on Remote Sensing Data Analysis, Modeling, Interpretation
and Applications: From a Global View to a Local Analysis (RS 2015)**

Workshop on Scientific Computing Infrastructure (SCI 2015)

Workshop on Software Engineering Processes and Applications (SEPA 2015)

Workshop on Land Use Monitoring for Soil Consumption Reduction (LUMS 2015)

Demographic Changes and Urban Sprawl in Two Middle-Sized Cities of Campania Region (Italy)

A Measurement in the Cities of Benevento and Avellino

Massimiliano Bencardino[✉]

Department of Political, Social and Communication Sciences, University of Salerno,
Via Giovanni Paolo II, 132 – 84084, Fisciano (SA), Italy
mbencardino@unisa.it

Abstract. This paper aims to show a measure of the spatial expansion of the buildings in inland areas of Campania and, through this, an analysis of the most complex phenomenon of urban sprawl. This work is a pilot study aiming to test a research methodology. Thus, at this stage, the area of investigation was restricted to two medium-sized cities: Benevento and Avellino. So, the Author proposes to investigate whether there is a sprawl in this particular context, in line with the European trend, and proposes a physical and anthropic correlation index between the changes of the built areas, seen as a measure of the taken land, and the demographic changes, to analyze the phenomenon of urban sprawl in relation to housing demand. Therefore, for examined urban areas, the Author analyzes the correlation between the change in population density between the years 2001 and 2011, extracted from the Census of the population at the fractional scale, and the change in the building coverage ratio extracted from the RTC (Regional Technical Cartography) in 1998 and in 2005.

Keywords: Urbanization · Urban sprawl · Demographic changes · Coverage ratio · Medium-sized cities · Shrinking cities · Inland areas

1 Introduction

The issue of urban sprawl, which refers, as it is known, to a model of expansion of the cities characterized by low-density housing and high urban fragmentation and dispersion, is regaining its centrality within the broader debate on land consumption. In fact, one of the main drivers of that are certainly the intense processes of urbanization that have redesigned Western cities over the last two centuries [6].
Although Europe has historically been characterized by dense and compact forms of urbanization, the American model of sprawl was imposed in most cities of the Old continent, both in Northern and in Mediterranean Europe.

Furthermore, many «European regions are already facing population decline and a quasi surplus of urban land» [28]. In fact, as stated in the EUKN 2010 conference [15], the implications of an aging population and migration to the suburbs are felt in many cities of the continent, both in large attracting cities and in *shrinking cities*

© Springer International Publishing Switzerland 2015
O. Gervasi et al. (Eds.): ICCSA 2015, Part IV, LNCS 9158, pp. 3–18, 2015.
DOI: 10.1007/978-3-319-21410-8_1

(cities in decline), generating many problems concerning the cost of public services, the increase of traffic and pollution.

So, if in some regions there is a growth that occurs within the limited borders of the city, leading to an urban densification, in others, there is a suburbanisation with a consequent urban sprawl, which it is necessary to measure.

In Italy, recent studies [11], [26] have shown that, for each new inhabitant, the phenomenon of land take in villages and small villages is significantly higher than in cities or middle towns, and real estate overvaluation has now generated a *decoupled land take,* that is not proportional to the real housing or productive demand.

Therefore, this paper aims to show the demographic dynamics and a measure of the spatial expansion of the built-up area in two middle-sized towns of the inland areas of Campania and, through this, to conduct an analysis of the most complex phenomenon of urban sprawl. We propose to investigate whether there is a sprawl in this particular context, in line with the European trend, and to understand if this territory represents an attractor of the polycentric pressures of the metropolitan area of Naples.

So, in this paper, a physical and anthropic correlation index between the changes of the built-up areas, seen as a measure of the taken land, and the demographic changes is proposed to analyze the phenomenon of urban sprawl in relation to housing demand. For this purpose, the correlation between the change in population density and the change in the building coverage ratio is analyzed [10].

2   A Measurement of Urban Sprawl

The question of the measurement of urban sprawl can be traced to two key issues. The first is related to the complexity of measuring land consumption, ie the research for a unique method of quantifying the transformation of natural and agricultural surfaces into artificial surfaces, through the construction of buildings, infrastructures and other facilities. The second is related to the lack of an unambiguous and universally accepted definition of sprawl.

In reference to the complexity of the measurement, there are several projects at the European and national level aimed at defining solid methodologies and comparable data for the measurement of land use; among the others, the European Corine Land Cover (CLC) project, the monitoring conducted by ISPRA in collaboration with the National System for the protection of the environment and the LEAC (Land and Ecosystem Accounts) methodology, set up by the European Agency for the Environment, are worth mentioning. However, completely satisfactory results do not match a very wide range of monitoring methods. In fact, they are often dissimilar and produce inhomogeneous, hard-to-compare measurements of the consumed land [9], [22].

As mentioned above, the second problem is related to the very definition of sprawl, and then to the search for variables that can measure the land take, that is not proportional to the real housing or productive demand. Despite the difficulty in finding a unanimous definition of the sprawl term, there is consensus about identifying the reducing population density as one of its main features. Therefore, population density is the first indicator for which we can have a measurement, and it is certainly the most used [17], [23], [25]. Other factors are related to the continuity of the built-up area, the concentration or the fragmentation of the urban centers, the complexity of urban form, and the centrality or diffusion of urban functions [19], [23].

In the search for a correlation between the built-up areas and the socio-economic variables in order to quantify in an analytical and scientific way the diachronic evolution of urbanized space, several authors [1], [20] have made use of Technical Regional Cartography. Even some regional administrations, such as the Piemonte Region, have started monitoring land take through the use of regional technical maps [27]. This has resulted in the definition of a set of indices, able to provide information both on the characteristics of urban patterns (quantification and measurement of dispersion, fragmentation of the urbanized area, etc.) and on the socio-economic characteristics of the population (residents, families, employees, enterprises) to correlate with the former.

Therefore, for the examined urban areas, the correlation between the change in population density between the years 2001 and 2011 (1), extracted from the Census of the population at the fractional scale, and the change in the building Coverage ratio, extracted from the RTC (Regional Technical Cartography) in 1998 and in 2005 (2), is analyzed:

$$\Delta\delta = \left(\frac{Pop.\,2011}{Census\,area}\right)_{2011} - \left(\frac{Pop.\,2001}{Census\,area}\right)_{2001} \tag{1}$$

$$\Delta C.r. = \left(\frac{Built\,2005 - Built\,1998}{Census\,area}\right)_{2011} \tag{2}$$

According to this approach, the change in population density presents complexities of calculation, due to differences between the territorial basis of 2001 and of 2011 on which the people were surveyed[1]. To standardize the size of the analysis units for the two periods, the territory was divided into comparable clusters of a similar size (with pixels of 5m x 5m resolution), through the rasterization of shape files.

By contrast, the Coverage ratio is given by the relation between the built-up surfaces in the RTC and the surfaces of the corresponding Census geographic units of the 2011. The spatial correlation between these two indices may be a way to measure the urban sprawl at the scale of the Census unit.

Moreover, the time gap between Census data and data derived from technical maps is only an apparent problem for analysis. In fact, the updates of the territorial bases of the Census of 2001 and 2011 were developed through the projects Census 2000[2] and

---

[1]   In fact, the Census geographic units of the 2011 are not necessarily the same as those of 2001; indeed their variation is almost always associated with the identification by ISTAT (National Statistics Institute) of new inhabited localities.

[2]   The Census 2000 project, which became operational in 2000 and was concluded in early 2001, was achieved through an agreement between ISTAT, AIMA (Company for interventions in the agricultural market) and the Ministry of Agriculture and Forestry Policies. Multiple information sources were used, including digital aerial orthophotos taken by AIMA between 1996 and 1998, the Regional and Municipal Technical Cartography and graphs with drawings of roads, railways and hydrography [12]. The territorial bases thus produced were returned to the Municipalities in 2005.

Census 2010[3] respectively which were realized through photointerpretation made on orthophotographic bases of 1996/'98 and of 2006/'08. In effect, the acquisitions that gave rise to the RTC and those which gave birth to the Census territorial basis are almost contemporary or differ by a few years (Figure 1).

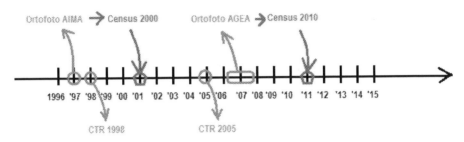

Fig. 1. Time scale of the sources

Therefore, it is possible to assert that the changes in population density and the changes in the building Coverage ratio can be put in mutual correlation with a minimum margin of error. Thus, the two-dimensional analysis of the two variables draws the geographical space. The urban area is classified in homogeneous zones, similar to other studies [16], [21], in relation to the processes of urbanization taking in place. So, seven main classes are identified, to which a sub-class, functional to a more detailed analysis, is added (Table 1).

Then, where the variables (change in population density $\Delta\delta$ and variation of coverage ratio ΔCr) are both positive[4], we have new urban development areas, called "areas of residential expansion"; when the only $\Delta\delta$ grows, we have "areas of residential densification"; instead, if $\Delta\delta$ is negative, the areas can be defined as "abandoned" or "redevelopment" ones; finally, when we detect a flurry of building and a simultaneous decline or stasis of population density, the areas can be defined "of sprawl" or "of depopulation and sprawl".

---

[3] For the Census 2010 project, started in March 2009 and ended two months later, they used the colored orthophotos, owned AGEA and produced between 2006 and 2008, as a basis for interpretation. [18]. An update of the territorial bases has been proposed to Municipalities, together with the documents, the data, the cartography and the software to perform any necessary review or validation of the same ones. It is important to note that the project specified that the Municipalities would have to try to keep the design and the coding of the localities and the Census geographic units as untouched/unaltered as possible. Changes could make only in the presence of obvious changes in the settlements compared to 2001, eg. the expansion of inhabited localities or the formation of new inhabited localities (ISTAT's Protocol no. 2679 of 21 April 2009). The territorial bases thus produced were published in December 2013.

[4] The variation of the population density is considered greater than zero when it is greater than 2 inh./ha, and negative when it is less than 2 inh./ha. The variation of the Coverage ratio is considered different from zero when it is greater than ± 0.4%. These thresholds are the result of the empirical analysis and may be different according to the context.

Table 1. Table of the physical and anthropic correlations in urban evolution (Source: Our elaboration)

	Change in population density $\Delta\delta$ (2011- 2001)		
Change in the building Coverage ratio ΔCr (2004- 1998)	$\Delta\,\delta\leq0$; $\Delta\,Cr<0$ Redevelopment areas		$\Delta\,\delta>0$; $\Delta\,Cr\leq0$ Areas of residential densification
	$\Delta\,\delta<0$; $\Delta\,Cr=0$ Abandoned areas	$\Delta\,\delta=0$; $\Delta\,Cr=0$ Invariants Areas	
	$\Delta\,\delta<0$; $\Delta\,Cr>0$ Areas of depopulation and sprawl	$\Delta\,\delta=0$; $\Delta\,Cr>0$ Areas of sprawl	$\Delta\,\delta>0$; $\Delta\,Cr>0$ Areas of residential expansion
	$\Delta\,\delta\leq0$; $\Delta\,Cr>0$; δ inh.≈0 Areas of industrial or tertiary expansion		

These are the areas of greatest interest for the decoupled land take, not proportional to the real demand, which could be residential or of other nature, such as that associated with an "industrial or tertiary expansion". Therefore, a subclass of the previous two is identified when, for a $\Delta\delta\leq0$ and a $\Delta Cr>0$, is also associated a population density close to zero, δ inh.≈0[5]. In this way, we create a real zoning of urban area, connected both to the nature of the functional construction and the demographic evolution of the resident population [10].

Finally, the study was extended to the dynamics of the core and the crown of examined urban systems, trying to contextualize the growth of the construction to the phases of the urban life cycle, as described in the Curb model set out by van den Berg [29].

In this test phase of the methodology, other potentially useful parameters were not taken into account (such as number of households, the number of employees or companies rather than functions of the dispersion and fragmentation of the urban area). In fact, the purpose of this study is related to the verification of the usability of the information derived from RTC for setting objectives and for the verification of the compatibility of the same with the Census data.

3   Case Studies

As anticipated, the proposed methodology was applied to two cities of the Campania Region: Benevento and Avellino. They are two medium-sized cities, provincial capitals and city of services. These cities are of particular interest both because they are at the center of development strategies of the European Union, as medium-sized cities[6], and because are included in the particular context of the inland areas of the Region.

---

[5] The population density is considered to be close to zero if it is not greater than 1 inh./ha. In this case, in sparsely populated areas, the growth of built areas could be clearly associated with industrial, commercial or of public utility facilities.

[6] These cities are the target of specific funding of the 2014-2020 Community program.

In the last two decades, they have lived intense processes of urban expansion, of suburbanisation, in a substantial demographic stasis. Furthermore, in these areas the effects of the recent economic downturn have been very obvious.

Therefore, it's interesting to measure urban sprawl in a context where there are no special requirements related to housing need or particular conditions of economic development.

3.1   The City of Benevento and Its Urban System

The first city taken for analysis is Benevento. Despite having had a project of urban planning (the Piccinato Plan) since 1933, over the last century it has been transformed in a spontaneous way. The Piccinato plan never became operational and the built-up areas have expanded, especially in the 50s and 60s, to the outside of each organic planned development. Thanks to the availability of funds bestowed as war damage, in those years the city had a period of strong expansion outside the walls of the old town and the great neighborhoods were born: Libertà, Ferrovia e Mellusi [2].

Although the current analysis of the development of the city is confined to a rather limited period, ie the time between the two regional technical cartography (1998-2005), in the schematically summarized results in the following table (Tab. 2) some changes of the urban structure can be distinctly read.

Table 2. Zoning of the municipality of Benevento according to the present model

Areas	Built 2005 (ha)	New Built (ha)	%	Pop 2011	ΔPop ('11-'01)	Urban area (ha)
Invariants a.	144	11	+7	11.387	-521	10.603
Redevelopment a.	9	-1	-10	2.158	-207	62
Abandoned a.	78	0	0	21.233	-6.217	328
a. of depopulation and sprawl	19	2	+11	3.736	-1.218	102
a. of sprawl	75	17	+22	3.130	+134	1.474
a. of industrial or tertiary expansion [7]	*39*	*11*	+28	*14*	-288	289
a. of residential densification	38	-1	-2	14.401	+5.157	344
a. of residential expansion	27	6	+23	5.444	+2.570	172
Tot.	390	34	+10	61.489	-302	13.083

According to the used model, the municipality of Benevento is characterized by a prevalence of "invariants areas" where there isn't significant growth both of built area and of the population. Nevertheless, there are clear signs of a significant growth of the urban perimeter and of a displacement of the resident community.

In fact, although slightly decreasing population, the city continues to expand its urban area to occupy lands once used for agriculture. In the period examined, a total of

---

[7]   The area of industrial or tertiary expansion is a subclass of the previous two.

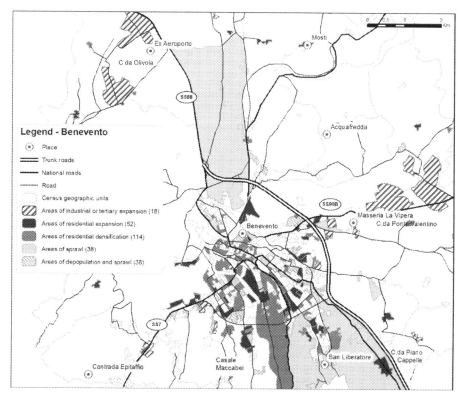

Fig. 2. Areas of expansion, densification or sprawl, in the municipality of Benevento[8]

36 hectares of newly built area is detected; of these, 6 intended for residential expansion and 19 to decoupled expansion in "areas of sprawl", ie a significant growth of built-up areas with no corresponding proportional population growth (Fig. 2).

In addition, although 11 of the 19 surveyed hectares are attributable to an expansion in the industrial or service sector, a significant economic development for the city has not been demonstrated during the period of this study[9]. In fact, the new buildings in the industrial areas (A.S.I.) of Ponte Valentino or in new i. areas of C.da Olivola (Ex Aeroporto) are often abandoned or not fully used. Morover, the closure of many commercial activities in the historic center of the city corresponds to the opening of two shopping malls, and the closure of many cinemas in the city

---

[8] The redevelopment and abandoned areas are not highlighted in the map.

[9] Similar to 20 other cities, situated for the most part in the South (Avellino, Bari, Benevento, Cagliari, Caserta, Catanzaro, etc.), Benevento is among the "swallows", ie the cities that attempt an alignment with the Italian average, present a dynamic economy, but showed a lack of a fundamental strength, according to the classification made by Rur-Censis at the "Second Convention of the Italian cities".

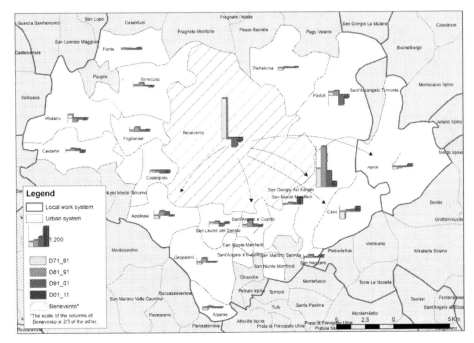

Fig. 3. Population changes in the urban system of Benevento

center corresponds to the opening of two multiplex cinemas. All this feeds an expansion of the urban area and a more intensive use of the automobile as a means of transportation.

By an analysis at the scale of the urban system, we can note a process of replacement of housing from the areas where there is an ongoing abandonment or depopulation[10] to areas where there is residential densification or expansion taking place[11]. In particular we can see a "shrinkage sprawl" in the crowning of the oldest town or along the main arteries of what has been called "*sistema urbano a direttrici e costellazione*" (an urban constellation along main axes) [3,4,5].

In fact, these migrations are not only contained within the municipal borders of Benevento, but affect the entire ring of the urban system (Fig. 3), which is still in a phase of demographic growth, unlike what occurs in the core (Table 3).

In particular, from the 80s, the depopulation of the city has accompanied growth of peripheral municipalities, mainly San Giorgio del Sannio, but also Sant'Angelo a Cupolo, Apollosa, Ceppaloni, Foglianise Paduli and, in the last decade, also San Nicola Manfredi, Calvi and Apice (Fig.3).

---

[10] It concerns generally many neighborhoods of the city, such as the historical center and the quarters Libertà, Mellusi, Ferrovia e Pacevecchia.

[11] It concerns Lungosabato road, Meomartini road, Santa Colomba road and the contrade Montecalvo, Madonna della Salute, Gran Potenza, San Liberatore, Piano Cappelle.

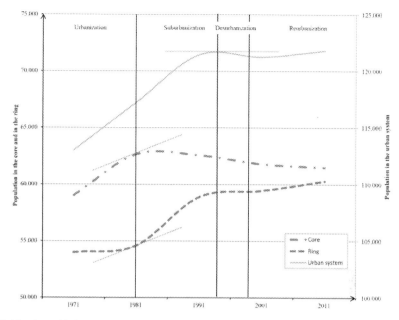

Fig. 4. Number of inhabitants in the core, in the ring and the entire urban system of Benevento, in accordance with the van den Berg model

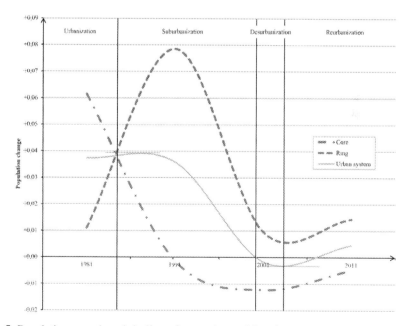

Fig. 5. Population growth and decline of core, ring and in urban system of Benevento, in accordance with the van den Berg model

Table 3. Population growth in the core, in the ring and in the urban system of Benevento

	1971	1981	1991	2001	2011
Urban system	113.000	117.223	121.431	121.228	121.780
Core	59.009	62.636	62.561	61.791	61.489
Ring	53.991	54.587	58.870	59.437	60.291

According to the model of van den Berg [29], the urban system of Benevento had a phase of urbanization until the mid-80s, when a process of suburbanization developed that has lasted at least until the second half of the 90s. Since the 2000s, a slight disurbanization process has developed, which lasted a few years and it wasn't exactly corresponding to the theoretical model. In fact, we have not recorded a decline of the ring and a subsequent recovery of the core, but a slight decline in the growth of the ring followed by a new phase of suburbanization (Fig. 4 and Fig. 5).

Finally, during the period in which it was measured, the building growth coincides with a phase of desurbanization, albeit slight, of Benevento. Therefore, the 10% increase of the built area in seven years does not have corresponding drivers in the housing needs. So, the 34 new hectares of built-up area are certainly not justified by the decrease of the population in the city and, at the urban level, the value of 1600 square meters for each new resident is absolutely disproportionate.

3.2   The City of Avellino and Its Urban System

The development of the city in the last century was heavily influenced by natural and anthropogenic factors. In fact, the geomorphology of the area, surrounded by the two Walloons (dei Lupi e Finestrelle) upstream and downstream of the city, along with destructive natural events (such as earthquakes of 1930 and 1980) have been defining the boundaries of expansion for many years. This has meant that urban development is concentrated in this area, and facilities and infrastructure were often rebuilt on the existing urban plan. In addition, administrative difficulties in the approval of the Plans[12] and a lack of adequate planning instruments were added. They have fueled expansion processes limited to individual interventions, often linked to the spontaneity and management of the needs of the moment [13].

Like in Benevento, during the short period between the two regional technical maps, the analysis clearly shows the changes of the urban structure and the axes of the development of the city (Tab. 4). Conversely, Avellino is a city that has experienced a turnaround in the demographic dynamics having grown by about 1,500 inh. in the last 10 years, after two decades of population decline.

According to the model used, half of the territory is subject to change. In the municipal borders of Avellino, totalling only 3'036 hectares, 30 hectares of newly built

---

[12] Planning tools of Avellino date back to the second half of the nineteenth century; the first building regulation (the Denti Plan) is dated 1877; then the Rossi Plan (1883), Cucciniello and Ferrara Plans (1913), the Valle Plan (1933) have succeeded to arrive, since the 60s, the most modern Petrignani Plans (that of '71 and that of '91) until the policy directions of the "garden city" in the '90s.

area were measured (during the period of analysis), of which 8 are associated with a residential expansion (in some cases also affected by the PIU Europe projects) and 19 of sprawl, e an expansion decoupled from the increase in population.

Table 4. Zoning of the municipality of Avellino according to the present model

Areas	Built 2005 (ha)	New Built (ha)	%	Pop 2011	ΔPop ('11-'01)	Urban area (ha)
Invariants a.	38	3	+9	5.489	-334	1.660
Redevelopment a.	-	-	-	-	-	0
Abandoned a.	31	0	0	11.535	-1.865	145
a. of depopulation and sprawl	33	3	+8	11.404	-1.487	153
a. of sprawl	75	16	+21	5.909	+357	696
a. of industrial or tertiary expansion	*22*	*8*	*+39*	*69*	*-155*	*146*
a. of residential densification	19	0	+1	7.432	+1.419	80
a. of residential expansion	47	8	+16	12.453	+3.429	302
Tot.	243	30	+12	54.222	+1.519	3.036

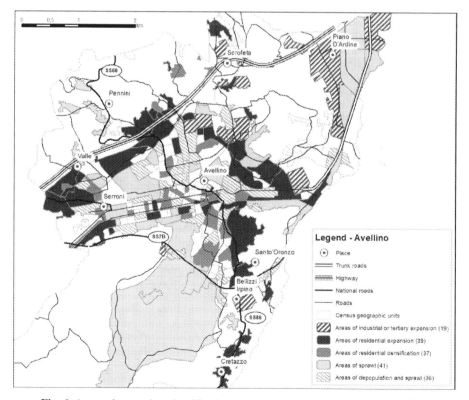

Fig. 6. Areas of expansion, densification or sprawl, in the municipality of Avellino

So, the expansion in the suburbs, from Pennini to Serroni and from Bellizzi until Cretazzo (Fig. 5), is strong and not always commensurate with the increase in population.

Also, we found 8 hectares of industrial or tertiary expansion localized mainly in the ASI area of Piano D'Ardine, in the commercial district of Scrofeta and in that of sports and leisure, Contrada Santa Caterina, as well as in the new hospital area of the A.O. Moscati, in the cemetery and in the prison of Bellizzi Irpino.

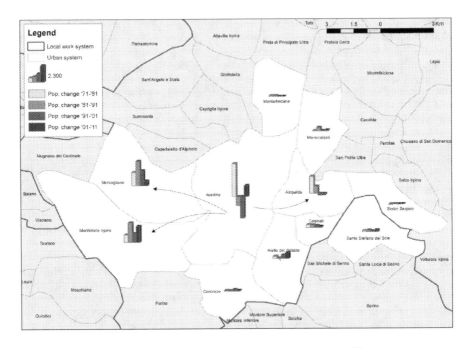

Fig. 7. Population changes in the urban system of Avellino

According to the model of van den Berg [29], since the 70s, in the urban system of Avellino a process of suburbanization has taken place, which has never stopped. The city of Avellino has begun to expand into a wider area, which affected at first the small-size town of Atripalda, then those of Mercogliano and Monteforte Irpino, and finally the small-size town of Aiello del Sabato (Fig. 6). Here too, the urban expansion takes the form of "shrinkage sprawl".

Between the 80s and 90s, these municipalities have absorbed a large portion of the population of the city (Tab. 5).

Table 5. Population growth in the core, in the ring and in the urban system of Avellino

	1971	1981	1991	2001	2011
Urban system	81.441	92.239	99.376	100.564	106.202
Core	52.382	56.892	55.662	52.703	54.222
Ring	29.059	35.347	43.714	47.861	51.980

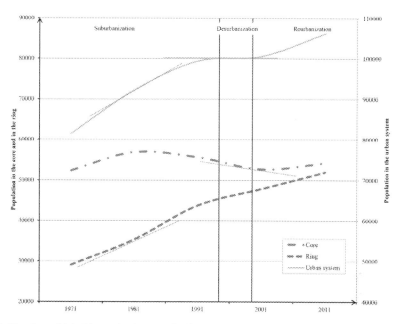

Fig. 8. Number of inhabitants in the core, in the ring and the entire urban system of Avellino, in accordance with the van den Berg model

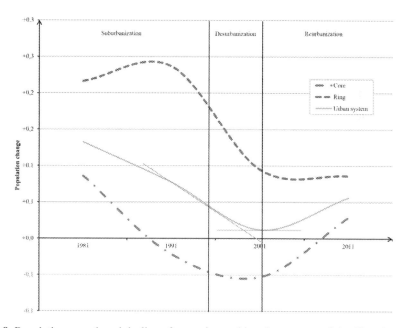

Fig. 9. Population growth and decline of core, ring and in urban system of Avellino, in accordance with the van den Berg model

Never entering into a real disurbanization phase, the urban system of Avellino, starting in the 2000s, is going through a phase of anomalous contemporary re-urbanization and suburbanization (Fig. 7 and Fig. 8). In fact, the population of the entire urban system has never decreased, but only had a horizontal inflection in growth.

The conclusions drawn for Benevento are also applicable to Avellino. The period in which a building growth was measured in the city coincides with a phase of demographic stagnation. Therefore, the 12% increase of the built area in seven years can not be justified by the light population growth that the core of the urban system has achieved in the last decade, after two decades of population decline and housing market crisis [14].

In both case, the urban sprawl appears strongly linked to a demand for more services or the desire of a growing part of the urban population to live in the surrounding rural places where it is guaranteed a greater sense of well-being [6].

4   Open Issues

Trying to quantify the phenomenon of sprawl in the considered areas, it's possible to say that the method of analysis has presented satisfying results, when compared to the complexity of the phenomenon that was represented. Therefore, the attempt to arrive at a numerical quantification allowed some preliminary evaluations on these areas and opens up various issues to collective reflection, both technical, procedural and theoretical-analytical.

The main problems are given by the time gap between the Census data and the cartographic surveys at the regional scale (as discussed in previous chapters), the usability of the data derived from the RTC, some errors in the geo-coding of the ISTAT data at the fractional scale and, finally, by the choice of ideal partition level of the space in which to analyze the considered indices. The quality of the encoding of RTC data appeared sometimes insufficient. In fact, coherence and consistency in the allocation of drawn surfaces to the various analytical categories (generic, industrial or agricultural building, shed, warehouse, shack) has not always been observed.

For the Census data at the fractional scale, errors may exist on survey, as happened. Pressed on the issue, ISTAT declares that «geo-coding at the fractional scale of the addresses of surveyed households was carried out by the municipalities which, in some cases, have not worked according to the required standard and have proceeded to award a Census geographic units adjacent to correct ones», and that «the Census data at the fractional level are considered provisional». ISTAT ensures the validity of the final data to higher territorial aggregations.

Despite this, the choice of the level of disaggregation of the indexes appears good. In fact, the main directions of the urban expansion and of the urban sprawl can thus be highlighted. Probably, a greater reflection on the threshold of membership in each class should be made.

Likewise, the choice of the urban system as scale of analysis has appeared useful. In fact, to contextualize the measurement of urban evolution to a specific phase of the urban life cycle of van den Berg model, appeared very effective to show the disconnection between the growth of the built and the real housing needs.

Finally, this study confirms that the decoupled land take in the inland areas and in the peripheral urban systems is stronger than the one measured in the big cities by other authors [6], [11], [26], and points out the urgency of a new law on ground rent.

References

1. Ballocca A., Foietta P.: Land use and sprawl – The experience of Province of Torino. In: Badiani B., Tira (eds.) Urban containment. The italian approach in the european perspective. Maggioli, Santarcangelo di Romagna, pp. 42–47 (2009)
2. Bencardino F.: Benevento. Funzioni urbane e trasformazioni territoriali tra XI e XX secolo, ESI, Napoli (1991)
3. Bencardino M.: Il sistema urbano di Benevento. In: D'Aponte T. (eds.) Il Cavallo di Troia. Disagio sociale, politiche carenti, marginalità diffusa, nello sviluppo territoriale della Campania. Aracne Editrice, Roma, pp. 285–296 (2009)
4. Bencardino M.: Un approccio metodologico quantitativo per la definizione dei sistemi urbani di Benevento e Salerno. In: Las Casas G., Pontrandolfi P., Murgante B. (eds.) Informatica e Pianificazione Urbana e Territoriale. Atti della Sesta Conferenza Nazionale in Informatica e Pianificazione Urbana e Territoriale INPUT 2010. Libria, Melfi, vol. II, pp. 265–277 (2010)
5. Bencardino, M.: A quantitative methodological approach for the definition of the urban systems of Benevento and Salerno. In: Borruso, G., Bertazzon, S., Favretto, A., Murgante, B., Torre, C.M. (eds.) Geographic Information Analysis for Sustainable Development and Economic Planning: New Technologies, pp. 20–30. IGI Global, Hershey (2012)
6. Bencardino, M.: Consumo di suolo e sprawl urbano: drivers ed azioni politiche di contrasto. Bollettino della S.G.I. **VII**(2), 201–222 (2015)
7. Bencardino M., Greco I.: Processes of adaptation and creation of a Territorial Governance. The experience of the cities of Benevento and Salerno (Campania region, Italy). Journal of Sociology Study, David Publishing Company, USA, vol. 11(2), pp. 819-833 (2012)
8. Bencardino M., Greco I., Valanzano L.: Processi di sviluppo urbano e tecnologie informatiche di analisi dei dati territoriali: l'evoluzione urbanistica della città di Battipaglia e l'efficienza dei programmi di sviluppo urbano (PRG). Bollettino dell'A.I.C., 144-145-146, pp. 5–75 (2012)
9. Bencardino M., Iovino G.: Analysing and managing urban sprawl and land take. Discussion Papers, CELPE RePEc, n. 131, pp. 1–44 (2014)
10. Bencardino M., Valanzano L.: Una misura dello sprawl urbano nelle aree interne della Campania: i casi di Benevento, Avellino e Battipaglia. In: Munafò M., Marchetti M. (eds.) Recuperiamo Terreno. Analisi e prospettive per la gestione so-stenibile della risorsa suolo, Franco Angeli, Milano, pp. 73–88 (2015)
11. Bonora P.: Atlante del consumo di suolo per un progetto di città metropolitana, Il caso Bologna. Bologna, Baskerville (2013)
12. Crescenzi F.: L'evoluzione della base territoriale dell'Istat da CENSUS a CENSUS 2000. MondoGIS, n. 22, Creative Commons (2004). http://www.geoforus.it/
13. Cresta A.: Le fonti cartografiche per una lettura delle trasformazioni urbanistiche della città di Avellino. Bollettino dell'A.I.C., 144-145-146, pp.77–95 (2012)
14. De Mare, G., Manganelli, B., Nesticò, A.: Dynamic analysis of the property market in the city of avellino (Italy). In: Murgante, B., Misra, S., Carlini, M., Torre, C.M., Nguyen, H.-Q., Taniar, D., Apduhan, B.O., Gervasi, O. (eds.) ICCSA 2013, Part III. LNCS, vol. 7973, pp. 509–523. Springer, Heidelberg (2013)

15. Europe's Urban Knowledge Network. http://www.eukn.org/
16. European Enviroment Agency: Analysing and managing urban growth. http://www.eea.
    europa.eu/articles/analysing-and-managing-urban-growth
17. Fulton, W.: The Regional City. Island Press (2001)
18. Di Pede F., Lipizzi F.: Nuovi strumenti territoriali per i censimenti del 2010-2011 (2009),
    http://www.geoforus.it/
19. Lelli, C., Pezzi, G.: Urban sprawl, come valutare l'urbanizzazione. Ecoscienza, Arpa Emi-
    lia-Romagna **5**, 80–83 (2012)
20. Massimo D., Barbalace A.: Urban sprawl e crescita economica territoriale. La sfida della
    scala in una stima a livello sub-regionale. In: Federalismo, integrazione europea e crescita
    regionale. Atti della XXX Conferenza Italiana di Scienze Regionali, 09-11.09.2009,
    AISRe, Firenze (ITA) (2009)
21. Mazzeo G.: Dall'area metropolitana allo sprawl urbano: la disarticolazione del territorio.
    TeMA. Journal of Land Use Mobility and Environment, University of Naples "Federico
    II" Print, 7-20 (2009)
22. Munafò M., Ferrara A.: Consumo di suolo: proposte di tassonomia e misura. In:
    Istituzioni, Reti Territoriali e Sistema Paese: La governance delle relazioni locali-
    nazionali. XXXIII Conferenza Scientifica annuale AISRe, 13-15.09.2012, AISRe, Roma
    (2012)
23. Molinari, M.: La città che cambia: la diffusione urbana. Mobilità residenziale e stili di vita
    emergenti nel Comune di Argelato (Bologna). Dissertation thesis, Alma Mater Studiorum
    Università di Bologna (2012). http://amsdottorato.unibo.it/4404/
24. Nesticò, A., De Mare, G.: Government tools for urban regeneration: the cities plan in italy.
    a critical analysis of the results and the proposed alternative. In: Murgante, B., Misra, S.,
    Rocha, A.M.A., Torre, C., Rocha, J.G., Falcão, M.I., Taniar, D., Apduhan, B.O., Gervasi,
    O. (eds.) ICCSA 2014, Part II. LNCS, vol. 8580, pp. 547–562. Springer, Heidelberg
    (2014)
25. Pendall, R.: Do land-use controls cause sprawl? Environment and Planning B: Planning
    and Design **26**(4), 555–571 (1999)
26. Pileri P.: La frammentazione amministrativa consuma suolo. In: Convegno ISPRA, CRA e
    Università La Sapienza, Il Consumo di suolo, lo stato, le cause e gli impatti, 5.02.2013,
    ISPRA,, Roma (2013)
27. Regione Piemonte: Monitoraggio del consumo di suolo in Piemonte. http://www.regione.
    piemonte.it/.../consumoSuolo.pdf
28. Siedentop S., Fina S.: Urban Sprawl beyond Growth: the Effect of Demographic Change
    on Infrastructure Costs. Flux, 79-80(1/2010), pp. 90–100 (2010). www.cairn.info/revue-
    flux-2010-1-page-90.htm
29. van den Berg, L., et al.: Urban Europe, a Study of Growth and Decline. Elsevier Ltd.,
    London (1982)

Coastal Monitoring: A Methodological Proposal for New Generation Coastal Planning in Apulia

Marco Lucafò[1], Giovanna Mangialardi[2(✉)], Nicola Martinelli[1], and Silvana Milella[1]

[1] Department DICAR, Technical University of Bari, Bari, Italy
marcolucafo@gmail.com,
{nicola.martinelli,silvana.milella}@poliba.it
[2] Department DII, University of Salento, Lecce, Italy
giovanna.mangialardi@unisalento.it

Abstract. This paper aims to provide a methodological dynamic management framework for ecological processes affecting the coastal strip, purveying tools and guidelines for the redaction of Municipal Coastal Plans in Apulia Region, and more specifically of Barletta-Andria-Trani Province. In detail, the objective is to develop added value products and services for the analysis of soil consumption and all the environmental and landscape parameters tailored on the coastal landscapes. This will be done on the basis of structure, management, control and monitoring of the coastal area to guarantee the right to access and enjoy the use of the public patrimony within and without the state-owned area. The model here proposed would include the implementation of a semi-automatic multi-scale, multi-time and multi-source tool, built on a set of indicators, which will be useful for a new generation coastal planning based on the key principles of sustainability, territorial vocation and local specificities.

Keywords: Soil consumption · Coast · Planning · Monitoring · Satellite · Indicators

1 Introduction

The tendency to intensive coast occupation and building over the last decades is nowadays an important issue that needs damming, given the absence of planning which caused huge damage in terms of loss of identity, landscape homologation, weakening of the ecosystem functions of the land-sea interface areas and, at times, absence of primary urban development infrastructures.

The coasts and seaside areas of Apulia are traditionally regarded as a primary source of tourist traffic; in contrast to the wasteful uses of the past, the growing awareness is fostering the outbreak of a more sustainable conception of tourism economy, attentive to the coastal landscape enhancement, envisaging and actualizing actions towards an integrated management of the littoral zones, able to propose strategies and projects designed to protect and improve the environment quality and to recover the existing patrimony. A valorization able to sustainably integrate tourism with the other coastal practises, such as fishing, commerce, shipbuilding and, in line

© Springer International Publishing Switzerland 2015
O. Gervasi et al. (Eds.): ICCSA 2015, Part IV, LNCS 9158, pp. 19–34, 2015.
DOI: 10.1007/978-3-319-21410-8_2

with ICZM[1] Strategy [1], also takes into account the environmental, economic, and social aspects, the coastal dynamics, the intervention policies, the policy-makers and the institutions.

This paper aims to propose a methodology for the development of added value products and services for the analysis and monitoring of soil consumption and all the environmental and landscape parameters affecting the coastal landscapes, through the implementation of a semi-automatic tool, built on a set of indicators, which could be useful for a new generation coastal planning based on some key principles [2] such as: (a) systemic approach; (b) understanding of context specificities; (c) identification of the accord with natural processes; (d) interventions characterized by a precautionary approach; (e) multi-participant planning; (f) multi-level governance; (g) wide range of intervention tools.

The application field is, in detail, the coast of Apulia region and the new procedures of Coastal Municipal Plan redaction within the more complex Coastal Planning in Apulia[2]. The experimentation of innovative process optimization approaches will be tested and applied by the Civil Engineering and Architecture Department DICAR from Bari Institute of Technology in two important reference contexts: the collaboration Protocol with Apulia Region, Barletta-Andria-Trani Province and the Municipalities of Barletta, Bisceglie, Margherita di Savoia and Trani, aimed to the territorial requalification and regeneration in the coastal area of the province[3]; the financed project "Apulia Space"[4], in which the Institute of Technology participates as a partner.

The methodological approach, exposed in the following paragraphs, is the first result of the theoretical and operational research. It corresponds to the studies addressed in the first phase of the research project Apulia Space, started formally on July 1, 2013, whose end is scheduled for 31 December 2016. The first priority of the project will be the identification of the coastal strip soil consumption and to provide useful tools to carry out an integrated investigation of "objectives" and "means to achieve them" that will take into account physical, environmental and landscape parameters of the whole coastal habitat. This will be possible thanks to the analysis of very high

---

[1] Verso una Strategia Europea per la Gestione Integrata delle Zone Costiere (GIZC) – Principi Generali e Opzioni Politiche (1999) European Commission. Demonstrative Program of the Integrated Coastal Zone Management, EU 1997-1999.

[2] Apulia Region approved the Regional Coastal Plan and the guideline document "Linee guida per la individuazione di interventi tesi a mitigare le situazioni di maggiore criticità delle coste basse pugliesi" ("Guidelines for the identification of interventions aimed to mitigate the most critical situations in the low coasts of Apulia"), respectively with the Regional Council Decisions DGR n. 2273/2011 and DGR n. 410/2011.

[3] Elaboration, presentation and negotiation of the Territorial Strategic Project "Il sistema costiero" ("The Coastal System") PST 3 in Province Coordination Territorial Plan PTCP BAT.

[4] "Apulia Space" is included in the National Operational Programme "Ricerca e Competitività 2007-2013" ("Research and Competitiveness 2007-2013") Convergence Regions, ASSE I "Support to Structural Changes", Operational Objective: "Networks for the reinforcement of the scientific-technological potential of the Convergence Regions", and it undoubtedly represents an opportunity for the project thanks to very high resolution multi-spectral satellite data that can be used as a support in the area planning and monitoring.

resolution ortho-photos and of other spatial data, both already available and still to be obtained, contextually evaluating the restrictions and information deriving from the higher-level planning In a second step this project will facilitate the systematization of those information and investigations in order to provide *ad hoc* guidelines that would help standardizing the coastal planning process, according to a dynamic, continuous and iterative integrated management. The process will cover the whole cycle of: (a) data gathering, (b) programming, (c) decision-making, (d) management, (e) actuation supervision.

These phases will be managed thanks to data organization and elaboration, implementation of indicators and elaboration of indexes for each planning level beyond the Coastal Municipal Plan boundaries, which otherwise would only include the state-owned coastal strips.

2   Background

2.1   The Coast of Apulia. Strengths and Criticalities

A crucial ecological dominance of Apulia is undoubtedly represented by the coastal strip, with dominance being intended as a system that is dominant for its extension and connective function in a given regional environment. The coastal strip of Apulia, including both the Adriatic and the Ionic coasts, covers almost one tenth of the national coast perimeter with its 950 kilometers and includes landscapes that are remarkably diversified in terms of geomorphologic features. The morphological types observed in the area are: flat graded rocky coasts (the most widespread); convex graded rocky coasts; rocky cliffed coasts (contexts of great landscape relevance); sandy beaches (in correspondence to alluvial riverbeds and river outlets); limited dune formations, relic of what used to be a well more extended regional system. The morphological richness of the coast in terms of physical and geochemical components goes along with a significant biological richness. In fact, the coastal strips of Apulia have a considerable importance due to their thriving habitats and benthic biocenoses.

In the Italian distribution of morphological coast types [3], central Apulia falls within the domain of Terraces, a middle segment of the regional coast characterized by an intense fragmentation of the Murge cretaceous plateau, slightly sloping, generally low and rocky (1.5 – 2.5 m above sea level) marked by small incisions due to the numerous valley outlets, locally defined as *Lame* (knives). The Bari Basin is situated amid this structure, and its setting changes following a latitudinal gradient; south of Bari the plateau becomes more distanced and forms a step, while the coast – formed by Pliocene sediments – becomes lower and has a rich subterranean hydrographical network pouring out at the coastal level. The only exceptions to this morphological uniformity are the cliffs of Trani and Bisceglie.

Moreover, in the considered area there are segments of remarkable ecological interest, for instance the Ofanto river outlet and some basins partly linked to the sea (Ariscianne, Canale Camaggio, Boccadoro, Pantano Ripalta) that are ecotones of crucial environmental permeability in various interactions: land-sea, river-land-sea, pond-land-sea.

These ecotone habitats are places of interaction between adjacent environments where the index of naturalness is particularly high, as well as very high is the danger level intrinsic to the abrupt changes produced by human settlement [4]. Thus they have become a "scarce resource" and, as a consequence, a starting point for an innovative coastal planning in Apulia.

In addition, the global issue of the coastal squeeze – already described in the Atlante delle Spiagge Italiane (Atlas of Italian Beaches) by the National Research Centre CNR [5] in 1985 and then analyzed in detail for Apulia by the Regional Coastal Plan (PRC) – requires a series of strategic decisions that would need to look simultaneously at the complex set of concurrent causes of the phenomenon: the rarefaction of the Ofanto river sediment supply – progressively diminished because of the damming, and the erosion of anthropogenic origin. The latter aspect of the erosion caused by human settlement has been pointed out by the vast area descriptions (Researches and Plans) displaying all along the central Apulia coastal strips an unbearable soil consumption that goes at the same pace with the "thickening" of the coastal settlement.

The problems generated by such processes were not solved by an adequate local coastal planning system; on the one hand there was just municipal planning – never particularly well-known for its attention to the coastal ecosystem balance – that included urban development right on the coastline and too close to the sea, on the other hand there have been significant delays in the regional landscape planning, which was completed after a great part of the damage had already been done. From a simple overlay mapping between Corine Land Cover, municipal technical cartographies (combined with historical maps from the Military Geographic Institute IGM) and on-field observations, it is clear how there is a tendency to seamless urban development that creates a coastline urban continuum north of Bari (the towns of Giovinazzo, Molfetta, Bisceglie, Trani and Barletta); moreover, most part of the scattered coastal and sub-coastal settlement in the area consists of vacation houses, summer residences that in some areas (e.g. Santo Spirito, Palese) privatizes whole coastline segments with enclosures isolating residences and tourist resorts from the rest of the territory. This scattered landscape reveals a complete loss of public areas, open and shared structures, centrality [6]. These "urban materials" are in fierce competition with the traditional uses of the coastline open spaces, such as irrigation system-based vegetable gardens or sub-coastal cultivations, which at the present day are to be looked at as relics of the past. The schematic and reductive approach of the old planning in Apulia showed little or no awareness of the need for conservation of the coastal space and its delicate ecotone function. In fact, the uniqueness of the coastal area has attracted an aggressive kind of urban development that has been typical of the Twentieth century and has expanded at an unprecedented and unparalleled speed rate. The consequence is an intermittence landscape [4] alternating full development and voids – the latter being the ruins of the old broad coastal agriculture spaces (e.g. irrigation system based vegetable gardens, grazing lands, coastal meadows) – split between the environmentally protective restrictions and the growing pressure exerted by housing market and tourism industry.

2.2   Coastal Planning in Apulia

The new era of Coastal Planning in Apulia is linked to a new conception of the coastal landscape[5] introduced in 2005 and to the regional government decision to adopt policies more attentive to regional landscape conservation and protection, and led to the approval of a new Regional Territorial Landscape Plan (PPTR) in February 2015. This plan was preceded by the Regional Coastal Plan (PRC, redacted according to the Regional Law LR 17/2006, Regulation of Coastal Use and Protection). Nevertheless, as an analysis of the first ICZM attempts will show, the 2006 Plan was negatively affected by the limited range of means provided for the municipalities according to section 3 of the Regional Coastal Plan PRC.

Some of the issues raised by the IMCA research in 2010 have proved to be well-grounded:

"Amid a general confusion between intentions and outcomes, [our] Atlas of Coastal Landscape helps problematizing the most considerable paradoxes of the modernity project, with the flourishing contradictions of a urban project that shows to be "inattentive" to the values of the contexts; its outcomes are not completely evident yet, but it is already possible to read the counterintuitive effects of the most protected and planned territories and, at the same time, the most violated, less controlled and most urbanized at a speed rate and with an intensity that are unparalleled and unprecedented." [7]

With the intent of correctly setting the reflection, it is necessary to understand the exact scope of these plans, therefore finding a correct definition a coastal area: an accurate review of the literature on the subject [8] will outline a wide range of possible physical borders, but also reveal how all the definitions converge on the idea of coastal area as the crucial point where sea and land interact, in a relationship that have to be preserved, and possibly fostered, according with the principles of sustainability. On an international level, the Coastal Zone Management (CZM) [9] was born in the seventies and, with the purpose of planning and managing correctly the coastal areas, claimed a contextual integration of: (a) ecosystem entirety, economic efficiency, social justice; (b) economic uses and/or sectors; (c) spatial and/or environmental settings: surface area, marine environment, biosphere; (d) time settings: long-term, mid-term and short-term programs; (e) juridical and decision-making settings; (f) responsibilities to be assigned to the numerous local institutions and/or communities and the different governance levels; (g) research fields: natural sciences, engineering, economics, law.

---

[5] The research outcomes are being tested on the field with technology transfer experiences and support to the local institutions carrying out Municipal Coastal Plans (PCC), which is part of their tasks for what concerns the use and management of both state-owned coastal strips and local coastal areas for tourism and leisure purposes. This authority was delegated to the Regions by the State according to section 105 clause 2, letter 1 of the Legislative Decree DLgs 112/98, and were then delegated by the Regions to the Municipalities, since the 1st of January 2001, according to the sections 40, 41 and 42 of the DLgs 30/03/99 n. 96.

The Municipal Coastal Plan experience launched by the authors, working in the DICAR Department of Bari Institute of Technology in collaboration with four local coastal municipalities, led to face the hardships linked to the realization of the Regional Law LR 17/2006 and to an actual debate among those subjects involved in the coastal territory planning and managing procedures (such as the already mentioned PPTR and PRC), proving that the co-planning was still rarely realized in practice among bodies and sectors..

This issue appears even more evident considering that the Municipal Coastal Plans only concern the state-owned coastal strips, a rather limited part of the whole territory. As a matter of fact, an efficient coastal planning approach cannot be limited in its analyses and actions to the exiguous state-owned portion of the coast, as it needs to take into consideration a broader and more complex system centered on a variable depth territorial segment, able to represent the whole coastal landscape setting to which the state-owned portion belongs. In fact, the new Regional Territorial Landscape Plan (PPTR) has often broadened its scope beyond the 300m coastal strip provided for in the Ministerial Order DM 431/1985.

The coastal landscape already contains all the features, strengths and potential that can make it autonomous. In fact, the coast has to be looked at not only as a seaside tourism destination, but as a much wider assortment of territory uses, as well as other aspects of the touristic offer (cultural, natural, agritourist, sport-related etc.) need to be taken into account, according to one of the five territorial projects for the regional landscape included in the PPTR regarding the integrated enhancement of the coastal landscapes [10].

2.3   Tools, Plans, Projects and Techniques for the Coastal Landscape Monitoring

Modern trends in planning call for continuous territory monitoring processes, integrating all the institutional and administrative levels involved. The project is thus focused on the attempt to systematically integrate monitoring analysis of soil consumption and other parameters of environmental protection and enhancement of the coast in the Municipal Coastal Planning tool. First of all, an in-depth investigation will be focused on the coastal anthropization, recognized as one of the main human-generated actions that affect coastal erosion [11]. The European project "Eurosion" contains a description of the European coastal situation, and one of the main causes of the coastal erosion is said to be the human intervention, both widespread and local, carried out on the coast and on the hydrographical basins [12]. Eurosion and Copernicus maps show respectively high exposure to erosion in Apulia and significant coastal soil sealing value north of Bari.

The concept of soil consumption is then a fundamental parameter to be analyzed in this research. The literature on the subject generally defines it an alteration from a non-artificial soil covering (non-consumed soil) to an artificial soil covering (consumed soil) [13] that leads to the loss of a fundamental environment resource, due to the occupation of an originally agricultural, natural or semi-natural surface. A rich source of information and data about soil consumption is available thanks to a monitoring network created by the Institute for Environmental Protection and Research (ISPRA) with the collaboration of the Regions and Autonomous Provinces through their Agencies for Environment Protection. Nowadays, the indicator evaluation is based on analyses deriving from a network availing itself of national, regional and municipal data integrated with other cartographies coming from the European initiative Copernicus [14]. As a result, the latest cartographies are extremely more detailed compared to the Corine Land Cover, and this guarantees their validation and assures a more coherent data spatialization. Although such data represent an important reference for a specific analysis of the coastal soil consumption, they are not sufficient for this purpose.

A significant attempt of integrated coastal strip management was carried out by the MOGEFAICO project [15], presented in 2001 within the Regional Operating Program 2000-2006 and tested on the south western coastal strip of Sicily. This project was focused on searching for tools able to evaluate the pressure exerted on the coastal ecosystems through some of the most relevant studies on the subject: the 1983 Couper model [16] with eight indicators, the 1990 Sorensen and McCreary model [17] with eleven indicators, the coastal use plan created in 1992 by Vallega with sixteen indicators, the 1992 Pido and Chua model [18] with ten indicators. The mentioned models showed, in many cases, a particular emphasis on the land perspective or the sea perspective, depending on the authors [19].

The Pilot Project for Action 4.4, carried out by the Working Group ARPA Puglia was called "SHAPE" (Shaping an Holistic Approach to Protect the Adriatic Environment between coast and sea) and represents another good example of integrated management of the coastal strip [20]. In detail, the objective of this pilot project was to evaluate the environmental features and criticalities in the marine coastal segment between Torre Guaceto and the urban and industrial area of Brindisi, with the purpose of realizing an example of good practices supporting a sustainable management of the coastal area between land and sea. The project was divided in two main phases: the first consisted in the gathering and elaboration of all the available information finalized to the definition of the environmental framework; the second aimed at the application of a model based on DPSIR indicators (Determinants, Pressures, State, Impact, Responses) to the analyzed area.

A first set up of semi-computerized tools for land use analysis was activated by the DICAR Department and other partners within the MaTRis project (Very High Resolution Thematic Mapping from Aerospace) financed by Apulia Region POR funds [21]. The project defined and standardized techniques for the satellite data interpretation and rendering; in particular, a first phase served to provide a soil sealing layer (automatic identification of sealed areas); subsequently, the extraction of a semi-computerized soil consumption classification defined a layer of land use with 43 class legend keys, in line with the CORINE Land Cover Standard (up to the fourth level for classes concerning the urban fabric). Thanks to the photo-interpretation and the integration with the "historical" numerical cartography it was possible to render further information layers concerning the buildings and the transport network of the analyzed area in a geodatabase that allows immediate integration with other data typologies and is available for all the phases of a planning process.

The GIS (Geographic Information System) tools like ESRI ArcGIS and its extensions Spatial Analyst and 3D Analyst represent a valid technological support to carry out the spatial analyses identified in the above mentioned projects and for the analysis proposed in this research. In fact, the GIS environment allows the gathering, storage, analysis and visualization of the information concerning a geographical area, such as the coastal area, in a dynamic and iterative way. This facilitates a systematic approach to the gathering and management of environmental information, and fosters the comparison and compatibility of data groups, improving the accessibility and spread of the information, as well as guaranteeing an easier spatial analysis of environmental impacts.

3   Research Design

The research here presented is the result of the experience of DICAR Department and of the authors in two apparently detached fields: coastal planning and the use of satellite images for land use mapping. As stated before, two different projects triggered further advancements in both fields: on one hand, the Apulia Space research aimed to integrate remote-sensing and land use data into planning processes; on the other hand, the protocol for the regeneration of coastal areas in the Barletta-Andria-Trani Province demanded the construction of a data framework that could guide integrated and multi-scale planning/monitoring in a highly complex and critical environment. Moreover, the constantly worsening issue caused by the human consumption of the coastal areas and the necessity to develop innovative and dynamic systems to deepen the knowledge of, and possibly appease, this phenomenon represented the starting point for the integration of these two experiences into the development of new ICZM tools and methods.

In order to pursue these goals, the research takes into account specific issues of coast management, such as (a) high dynamism of coastal processes, both natural and human-driven; (b) lack of coherent, updated, precise data; (c) subjectivity and ineffectiveness of current planning policies. Previous ICZM approaches in similar areas [15][20] show that indicators can be useful evaluation tools, but they seem to lack the capability to be easily updated and to produce as a result possible courses of action, therefore failing to some extent to address these issues. The authors' research focuses on remote sensing as the key to an efficient coastal monitoring, thanks to high data availability and standardization.

The main objective of the project as a whole is to set up semi-automatic analysis techniques, decision support systems (DSS) and monitoring procedures to be used during the coastal landscape planning, able to innovatively systematize and integrate the myriad of data coming from different sources and sensor typologies (multi-source), with different degrees of precision (multi-scale) and distributed along time (multi-time), in particular by using high resolution satellite images. While the Barletta-Andria-Trani coast will be the "test field" in the operational and validation phase of the research, this paper presents the theoretical definition of the processes involved:

the result is a methodological framework that is standardized and replicable in different contexts and guarantees its integration into planning tools and with landscape and coastal protection.

4   Methodological Results

The result of the research up to this point is a methodological and technological framework that consists of four modules, as shown in Fig. 1. Each module is designed to address specific issues such as data acquisition, spatial and time analysis and planning integration, in order to achieve the proposed objectives for coastal management and monitoring.

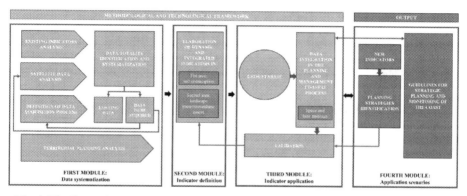

Fig. 1. The figure shows the logical flow of the methodological framework hereby proposed

4.1   Data Systematization

The first module included in the hypothesized framework consists in the systematization of useful data by scanning the existing sources and planning tools and subsequently the acquisition of satellite data. Remote sensing represents the added value of the whole process because of its precision, information quantity, fast and easy access.

The high resolution images used in the project come from the WorldView-3 satellite that, thanks to its multispectral sensors and the 0.3m precision of its panchromatic images (rendered as adjusted ortho-photos), represents the commercial remote sensing excellence [22]. The Matris [21] project experience allows to obtain from these images a series of thematic layers concerning soil sealing and soil consumption in a precise, rapid and standardized manner.

Moreover, Planetek Italia has later developed the very same satellite data interpretation method for the creation of thematic and ad hoc indicators for the coastal areas [23], such as the identification and classification (natural, artificial, fake, but also high/low through cross-referencing with DTM models) of coastline - and backshore line for low sandy/pebble typologies - but also of ports and coast defense works. All this information turned out to be extremely useful in defining the reference frameworks for the municipal coastal planning as provided for in the current regulation.

Moreover, a specific indicator for the erosion of sandy beaches has already been developed, and it is able to evaluate entity and seriousness of the erosion phenomenon also in relation to the beach width, as it also takes into consideration historical data and can integrate more remote sensing sessions.

It is evident that the possibility to acquire this and other information through satellite images can be more efficiently capitalized within a system that would include and take into account all the available and relevant data concerning the analyzed territory. In particular, for what appertains to the coastal planning in Apulia (even if the methodology here explained can be replicable to other regions with few variations) the background has to be the Regional Coastal Plan, which provides crucial regional data such as: (a) coastline, (b) state-owned area borders, (c) morphological coast type classification, (d) erosion criticality and environmental sensitivity, (e) technical map of the state-owned area and high resolution ortho-photos (2010).

A further data set comes from the higher-level protection tools, that is to say the landscape plan (for Apulia, the recently approved PPTR, which identifies the landscape heritage and other contexts needing protection according to the Cultural and Landscape Heritage Code) and the Hydro geomorphological Structure Plan redacted by the Basin Authority (for what concerns the definition of flood and landslide hazard areas). These will be integrated inside a geodatabase where each protection category will be given a different "weight" according to the current regulation and to the coastal planning related regulations.

In the overall framework definition, these input types (even if updatable and available for integration if necessary) will have a "static" role, being outlined both as (a) "historical" reference data and as support for the subsequent elaboration phase (considering for example the coastline data or the Regional Technical Map CTR) and (b) landscape "unchangeable" information (for example the environmentally important or hydrogeomorphologically endangered areas). In contrast, the main benefit of using very high resolution satellite images consists in a dynamic and multi-time approach to the analysis of the physical, environmental and landscape coastal features, being these highly variable. Thus the project aims to structure the remote sensing data acquisition in various sessions, which will already be integrated in the planning and, later on, in the monitoring processes (for example, with a summer/winter biannual frequency in the first phase that would become annual during the second phase), with the purpose of defining a priori a set of dynamic indicators to be constantly updated with each subsequent elaboration, and will also render an up-to-date image of the area physical features.

4.2 Indicator Definition

Once both the available and remote sensing produced data will have been systematized, the following framework module will be the implementation of indicators that take the gathered information as reference. These indicators are designed to be standardized and reproducible in different places and times, able to effectively represent the territory features and to guide the decision-makers to pay specific attention to the soil consumption issue (that in this context can also be seen as public coast consumption) and to environmental rescue actions. The analysis of the available data, along with existing indicators, suggests a possible and yet not complete list of indicators.

- *"Valuable" coast consistency.* Identifies the percentage, both in linear and area terms (in case of beaches with identifiable backshore line), of those areas that can be used for touristic or other purposes as provided for in the Regional Law LR 17/2006; this is achieved by cross-referencing among data concerning coastal typology (natural/artificial, high/low), depth and possible presence of protected areas;
- *Coastal soil consumption index.* Quantifies the areas affected by soil sealing and human interventions, with reference to the beach areas (where beaches are present), to the state-owned areas (supporting the removal of permanent structures as provided for in the current regulation) or to broader contexts that would offer a more organic vision of the coastal landscape dynamics. The constant updating of this index, together with the data on soil consumption, can show in practice the results of the adopted policies and of the naturalization/denaturalization processes in the coastal habitat;
- *Backshore soil consumption index.* Quantifies the anthropized areas in the backshore of beaches and state-owned coastal areas, to signal possible critical situations in terms of accessibility and usability of the public areas;
- *Coastal strip occupation index.* Quantifies through photo interpretation those areas that are involved in various kinds of licensed uses (e.g. beach resorts) within the "valuable" coast area, according to a summer/winter seasonal criterion. This way it would be possible to evaluate both in qualitative and quantitative terms (a) the respect of the current regulation concerning the free access beaches (b) possible violations of the current license regulation;
- *Erosion criticality and environmental sensitivity classification.* It is an update of the criticality and sensitivity indicators provided by the Regional Coastal Plan (that represent the starting point): for what concerns the sandy/pebble beach erosion phenomena, the constant update of the coastline related data can draw attention to current coastal squeezing/expanding phenomena that can confirm or modify the existing classification. The environmental sensitivity value can be integrated with a landscape features database, but also with a land use screening able to dynamically stress the presence of environmental value elements (e.g. woodland covering, dune bars etc.);
- *Criticality identification.* Accurate screening and quantification of human-originated elements that interfere with protected zones or other possible areas of environmental enhancement; for the flood hazard zones, the main concern will be to identify the areas affected by soil sealing, while for other areas a broader range of possible interventions will be highlighted according to the situation (e.g. densely populated areas in environmentally protection restricted areas, single interventions for landslide hazard areas etc.) with the final aim to define a criticality map corresponding to the physical and juridical territory features.

4.3   Indicator Application

In the third module of the proposed methodology the hypothesized indicators would be modeled within a geodatabase including the already available data, and organized so that it can be updated with new data later on. For modeling it is here intended the definition and standardizing of geo-processing and database management rules and procedures (with tools that are equivalent to ModelBuilder in an ESRI ArcGIS environment) to be applied to the information layers identified during the first framework building module. This would allow obtaining values univocally determined on an adequate scale (that might be normalized) for the indicator taken into consideration and immediately comparable with each other.

The indicator definition model, coherently with the enounced multi-scale and multi-time principle, has to be applicable to different spatial and time units according to the situation. Following this assumption, the best option for the coastal environment peculiarities seems to be the definition of an investigation spatial matrix with not necessarily isotropic features. In fact, while a first phase indicator definition can be actualized on a homogeneous spatial network of predetermined size "cells" (e.g. 2x2m) that is useful to obtain an accurate and detailed representation, the identification of different analysis settings from a "longitudinal" and "cross" perspective can offer more relevant values concerning the considered phenomena and their interpretation. Imagining applying Cartesian axes to a coastal segment of an indefinite size, it is possible to identify values linked to "spatial abscissas and ordinates" not mathematically determined, but based on the physical and juridical features of the considered coastal area (Fig.2).

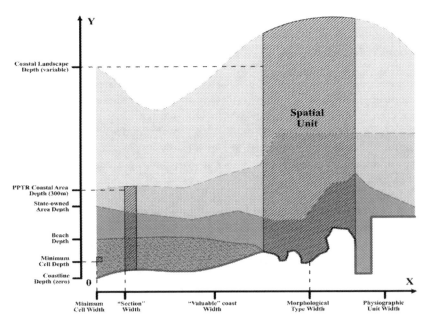

Fig. 2. A possible Cartesian representation of a generic coast area and of different spatial units

Here reported, some examples of elements that can be identified through longitudinal scan:

- "Minimal" cell width;
- Width of a relevant coastal segment or "section" (e.g. five meters, so that it may include relevant physical features such as fixed or transportable structures, mobility etc.);
- Width of seaside available or "usable" coastal segments, excluding areas such as anthropized segments, cliffs etc.;
- Width of the coastal segments that belong to the same morphological coastal type (sandy, rocky etc.);
- Width of the coastal segments characterized by the same physiographic unit or sub-unit (as identified by the Regional Coastal Plan);

Similarly, from a cross point of view, it is possible to identify various levels of "depth" of the coastal area:

- Depth "zero", corresponding to the coastline itself (useful to quantify phenomena linked to the coastline data, such as erosion or occupied beachfront);
- Minimal cell depth, which matches the longitudinal one;
- Beach depth, where beaches are present, using the backshore as reference;
- State-owned area depth (useful to quantify significant phenomena from the juridical and administrative point of view in the considered area);
- The depth of the coastal territory strip (300 metres from the coastline, as defined by the PPTR for the considered context);
- Other strips that can be considered with constant depth (e.g. 300 or 500 metres) starting from the coastline, to evaluate territorial elements from a cross perspective (e.g. settlement or environmental system etc.);
- (Variable) depth of the "coastal landscape" in a broader sense, based on the territorial features.

Once this spatial coordinate system has been defined, also in terms of its relationship with the GIS data, it will be possible to calculate the indicators for each spatial unit, identified by a "x/y" couple of values, guaranteeing an output that allows analyzing the territory on both a micro and macro scale, according to those scans that will turn out to be more relevant to physical/juridical features.

However, it is evident that such a great amount of data would result overabundant and dispersive in respect to the necessity of providing an interpretation tool designed to be immediately understandable and usable during the planning phase; thus, it is fundamental in this project module to associate to each indicator the most suitable spatial reference units/sub-units (for example to analyze the variation "hot spots" of the considered phenomena compared to an average value), as well as the related time units. Choices will have to be made taking into account the intrinsic needs of the Municipal Coastal Plans, so that the indicators and their spatial/time contrast may provide actual interpreting and/or operating frameworks; if necessary, through a validation process of the obtained results it is possible to modify and perfect the hypothesized modeling by intervening on the system variables, that is to say the indicators (multi-sensing), the spatial x/y units (multi-scale) and the time units (multi-time). (Fig. 3)

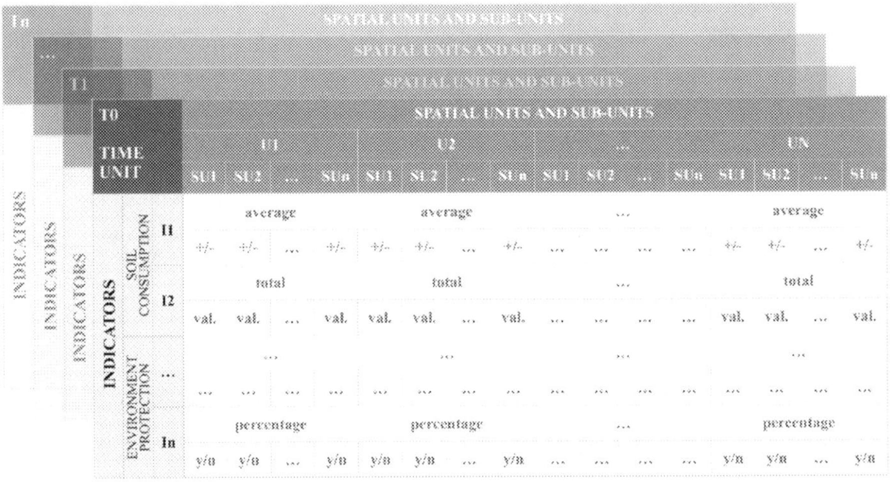

Fig. 3. The figure shows the possible output of the system in a three-dimensional matrix

4.4   Application Scenarios

This system will be able to provide interpretation data concerning specific aspects of the coastal areas (in particular soil consumption and landscape protection/enhancement) that are relevant both to the Coastal Areas Integrated Management and to the current planning in Apulia.

The experimentation within the collaboration Protocol with Barletta-Andria-Trani province will allow to evaluate the results of the proposed methodology in an actual application scenario that is also particularly pertinent, given its coastal soil consumption issues; the application of the proposed system to the vast area taken into consideration will be actualized in the last phase by linking the indicators and their possible spatial/time variations to strategies and lines of action finalized to the achievement of priority objectives that can be subsequently applied to a local context with a strategic co-planning attitude.

The final objective is to provide to all those operators and institutions working in the field of coastal landscape management, planning and monitoring an array of accurate, dynamic and updatable decision support tools, such as constantly updated indicators, ready-to-use descriptive and interpreting cartographies (that can also be used as means to foster dissemination and participation) and strategic intervention guidelines responding to the real needs of the territory.

5   Conclusion and Further Developments

A methodology combining the use of (a) very high resolution satellite images that are also updatable in time, (b) semi-automatic and standardized analysis, interpretation and rendering systems and (c) spatial and time investigation matrixes tailored on the specificities of a territory, might suggest a new approach to strategic integrated planning, launching new forms of dynamic update within the territory governance tools. A system such as the one here proposed gives the possibility to integrate *a priori* the monitoring activity of a series of relevant data or indicators into every phase of territorial planning; this regular update allows envisaging the end of crystallized plans, and the birth of plans that are automatically able to modify actions and impacts on the territory, on the basis of specific phenomena or of the responses received by the tool. This would be done according to established and validated rules and trying to exclude arbitrariness to the biggest extent possible.

The methodological proposal, here presented as a first result of the research, will be ready for application during the Barletta-Andria-Trani Province Coastal Planning experiences, with the purpose of validating the procedure with actual data and results, and bringing the scientific-technological investigation forward within the Apulia Space Project. Moreover, the research focus and scope could be easily shifted, for example from the "land" perspective of the coast to the marine perspective, and subsequently to the interaction between the two. The marine area, here overlooked given its little relevance to the soil consumption issue, could be integrated in a second phase as well as other ICZM aspects such as economic or social dynamics, in order to improve the expected results

Acknowledgments. The authors would like to acknowledge Planetek Italia s.r.l. Their contribution for supporting data acquisition and future elaboration is much appreciated.

Author Contributions. This paper is the result of a shared reflection of the authors. All authors contributed equally to design, literature review, and document analysis. To make writing more efficient, Section 1 was drafted by S.M., Sections 2.1 and 2.2 were jointly written by N.M. and S.M.; Sections 2.3, 3 and 4.1 were drafted by G.M and Sections 4.2, 4.3 and 4.4 were drafted by M.L; Section 4 was jointly written by G.M and M.L.; while Section 5 was jointly written by all authors. All authors have read and approved the final manuscript.

References

1. http://ec.europa.eu/environment/iczm/prop_iczm.htm
2. Commissione Europea: Insegnamenti specifici del programma di Dimostrativo sull'ICZM (1999)
3. CNEN: Classificazione Biotipologica delle coste italiane, Roma (1980)

4. Mininni, M.V.: Paesaggi Costieri. Un atlante provvisorio. In: La costa obliqua Un atlante per la Puglia. Donzelli, Roma (2010)
5. CNR: Atlante delle spiagge italiane, Roma (1985)
6. Lamacchia, M.R., Martinelli, N., Rignanese, L.: Un territorio della Puglia Centrale. In: Seminario di studi 50 × 50 km, Politecnico di Bari (2001) e Viganò, P.: La dispersione nel territorio del Nord-Barese tra resistenze e mutazioni. In: Mostra New territories. Situations, projects, scenarios for the European city and territory, Convento delle Terese, Venezia, Dottorato in Urbanistica IUAV coord. B. Secchi (2002)
7. Unità di Ricerca DICAR, Politecnico di Bari: Imca (Integrated Monitoring Coastal Areas). Ricerca D.M.593 08.08.2000 Ricerche precompetitive (2009)
8. Lamacchia, M.R., Mairota, P., Martinelli, N., Mininni, M., Sallustro, D.: A multi-discipline and multi-scale approach to coastal zone and river basin planning in a suburban Mediterranean region. In: Options méditerrenéennes n. 53 (2002)
9. Knecht, R.W.: Archer: Integration. In: The US coastal zone management program. Ocean and Coastal Management n. 21 (1993)
10. Regione Puglia: i cinque progetti territoriali per il paesaggio della regione. In: Piano Paesaggistico Territoriale Regionale (2015)
11. Petrillo, A. F.: Aree costiere: attuali e future criticità. In: Geologi e Territorio, Periodico dell'Ordine dei Geologi della Puglia n. 3-4/2007, pp. 117–130 (2007)
12. National Institute for Coastal and Marine Management of the Netherlands (RIKZ) et al.: Living with coastal erosion in Europe: Sediment and Space for Sustainability PART II – Maps and statistics (2004)
13. http://www.isprambiente.gov.it/it/temi/suolo-e-territorio/il-consumo-di-suolo
14. http://www.eea.europa.eu/about-us/what/seis-initiatives/copernicus
15. Pernice, G., Patti, I.: Mogeifaco: un GIS esperto per la gestione integrata della fascia costiera, Reports IAMC-CNR – Mazara del Vallo (2006)
16. Couper, A.D.: Atlas of the Oceans. Times Books, London (1983)
17. Sorensen, J.C., McCreaty, S.T.: Institutional Arrangements for Managing Coastal. In: Resources and Environments. National Park Service, US Department of Interior, US Agency for International Development, Washington DC (1990)
18. Pido, M.D., Chua, T.E.: A Framework for Rapid Appraisal of Coastal Environments. In: Chua, T.E., Fallon Scura, L. (eds.) Integrative Framework and Methods for Coastal Area Management. International Center for Living Resources Management, Manila, pp. 144–147 (1992)
19. Vallega, A.: Sea Management. A Theoretical Approach. Elsevier Applied Science, London (1992)
20. http://www.shape-ipaproject.eu/
21. La Mantia, C., Iasillo, D., Milella, S., Martinelli, N., Bux, M.: Il progetto MaTRis (Mappe Tematiche da Aerospazio ad Altissima Risoluzione) come supporto alla pianificazione urbanistica per il Comune di Apricena. In: Atti 14° Conferenza Nazionale ASITA, Brescia, pp. 1123–1128 (2010)
22. https://www.digitalglobe.com/sites/default/files/DG_WorldView3_DS_forWeb_0.pdf
23. http://www.planetek.it/prodotti/tutti_i_prodotti/preciso_coast

Ecosystem Services Assessment Using InVEST as a Tool to Support Decision Making Process: Critical Issues and Opportunities

Andrea Arcidiacono, Silvia Ronchi[(✉)], and Stefano Salata

Department of Architecture and Urban Studies, Politecnico of Milano, Milan, Italy
{andrea.arcidiacono,silvia.ronchi,stefano.salata}@polimi.it

Abstract. The awareness of Ecosystem services concept has gained prominence in the decision making process. The inclusion of this issue strictly depends on the way in which may be incorporated in the development strategies of a region by the policy makers. The paper want to test one of the most used model to quantify the Ecosystem services, with a spatial distribution output, in order to recognize the critical issues and the opportunities to use it as a tool to support decision making process. The InVEST model was experimented for the Habitat Quality and Carbon Sequestration functions. The survey area is the Municipality of Lodi in the south part of Lombardy Region (north of Italy) due to the high accessibility to the database information and also to attempt the adaptability of the software to product reliable output at micro-scale.

Keywords: Ecosystem services · Invest model · Land use · Habitat quality · Carbon sequestration · Decision making

1 Introduction

In recent years defining, classifying, detecting, mapping, and evaluating Ecosystem Services (ES) has been the goal of several enlightening publications.

Ecosystem services studies and related application were investigated by major international initiatives, as the Millennium Ecosystem Assessment (MA)[1], The Economics of Ecosystems and Biodiversity (TEEB)[2] and the Intergovernmental Platform on Biodiversity and Ecosystem Services (IPBES)[3]. Despite increasing political attention on ES recently boosted because the socio-economic relevance on such issue is directly connected with the general objective of "sustainable development" which seems to be, at all, one of the pillar that steer contemporary decision making processes.

The awareness of the importance of ES generated different approaches to classification and evaluation of the Ecosystem functions with a doubled challenge

---

[1] www.millenniumassessment.org/

[2] www.teebweb.org/

[3] http://www.ipbes.net/

© Springer International Publishing Switzerland 2015
O. Gervasi et al. (Eds.): ICCSA 2015, Part IV, LNCS 9158, pp. 35–49, 2015.
DOI: 10.1007/978-3-319-21410-8_3

regarding both conceptual and technical aspects. ES are commonly defined as the benefits that humans obtain from ecosystem functions [1], [2], or as direct and indirect contributions from Ecosystems to human well-being [3] [4]. With the rise of the ES concept and the increase of proposals for a more sustainable management of natural resources integrated with a sustainable development goal [2], it remains a double challenge that seems to capture the attention of research on such field: once ES has been defined by few pioneering studies, the literature is divided by who try to classify ES [5], and who try to evaluate ES [6]. In fact, the ES concept gained prominence in the ecological and economic literature for the attempt to classify and assess the services in compliance with several disciplines research methodologies, their methods of inquiry and their technical procedures.

At least, the aim of the ecological and economical disciplines focused on ES intend to standardize ES as a requirement to measure and to assess them in order to support the policy makers in the decision making process.

Recently, an important discussion concerning the definition of a common international classification of ES (CICES) has emerged [7]. ES classification is known as the list of benefits in terms of environmental goods or services; it helps the potential assessment of specific land use transformation during the time.

International agencies mainly focus on a common classification of ES as the standard baseline to share a common knowledge of disciplines around the issue. Nowadays the great deal of research for many studies is how to recognize, assess and map ES. In particular, the most important key of discussion for both ecological and economical disciplines, is the values of ES: what kind of value is correct to attribute?

Many existing tools and approaches for measuring, mapping and evaluating ES are still subject to deep scientific testing, nevertheless too often such analytical framework remains at the theoretical stage because it is composed by suppositions and proposals without an active perspective that could support the theory. This research paper try to put bridges between actual gap that separate theoretical stages and practical experiences on case of study.

Despite the most common application of ES mapping is done at macro-scale using national inventories of Land Use rather than European ones (Corine Land Cover), the tentative of the research is to apply it at micro-scale. According to the subsidiarity principle, which ask for a better detailed information to support the local policies, ES mapping in this paper provides an output highly precise.

By the way a tentative of integration of sustainable planning procedures will be presented using ES maps as proxy for the overall value of soils. Hereafter it will be showed how InVEST could be used as software for support the construction of an analytical framework for local town planning management.

2   Ecosystem Services, Land Use and Decision Making Process

The changes in land use affect ES values which increase or decrease on the base of the land use variation between different years. On the basis of knowledge on urban land use changes, the subsequent step is to evaluate their impact on natural ecosystems [8].

ES is the conditions, process, and components of the natural environment that provide both tangible and intangible benefits for sustaining and fulfilling human life [4]. Its measurement is codified: "The value of the world's ecosystem services and natural capital" [3] present an economic evaluation of the goods and services that human population derive, directly or indirectly, from Ecosystem functions.

The ES approach can explore the influence of land use and practices on natural capital stocks, on the processes that build and degrade these stocks, and on the flow of ES from the use of these stocks [9].

Connected to the land use changes and the observation of the land take by new urbanization, the evaluation of the ES help to enforce the decision making mechanism. In fact, land use change leading from urbanization often have a significant negative impact on the affected ecosystems and the goods and services that they provide [10]. Different land uses also influence the shaping of land cover and the amount of impervious surfaces; soil sealing is closely related to land take or land degradation. The management of soil sealing includes ecological, economic and social dimensions which need to be considered in line with sustainable urban management, This is why planning and policy have do identify a balance between these three dimensions [11]. In this sense, the integration of the ES approach into planning is crucial, and therefore should not be considered as an option, but an essential element; the lack of this element has to be tackled and compensated by planning discipline, even if it get involved the entire current planning approach.

Taking into consideration that soil performs many environmental, economic, social and cultural functions, the policy makers that act for steer land use planning process have to include contributions from different disciplines and theoretical background for more huge knowledge about ES and a better use of it .

Required integration should include, among others, the ecological systems that provide the services, the economic systems that benefit from them, and the institutions needed to develop effective codes for a sustainable use [12].

The design of environmental management policies frequently involves weighing up the consequences of proposed actions. It is necessary to consider impacts upon ecosystems as well as the social and economic systems to which they are linked.

Regarding to this issue, Costanza classified the global land use into sixteen primary categories and grouped ES into seventeen goods and services (gas regulation, climate regulation, disturbance regulation, water regulation, water supply, erosion control and sediment retention, soil formation, nutrient cycling, waste treatment, pollination, biological control, refugia, food production, raw materials, genetic resources, recreation and cultural), using this approach a lot of recent international bibliography has been dedicated to extract an equivalent weight factor per hectare in different areas [13]. The total ES for each land use category can be obtained through multiplying the area of each land category by the value coefficient:

$$ESV = \sum (A_i \bullet VC_i) \tag{1}$$

Where ESV is the estimated ecosystem service value, A_i is the area (ha) and VC_i is the value coefficient for land use category "i" (Helian et al. 2011). ESV is associated to a land use transition matrix, notable changes on ESV can be observed and the economic loss of specific transitions (in particular the diminishing of cropland or other natural covers in favor of new urbanized land) can be noted and explained.

New indicators (as the percent decrease of the total ESV) can enforce the evidence of economical long term effect of land use change and urbanization. Even simplified and theoretical, such method helps to improve the knowledge of qualitative effect of land use change, thus increasing the attention of cause-effect mechanism due by planning options.

Mostly ES analysis is useful to analyze the percent rate of increment or decrement of values rather than the total amount of ESV which can be substantially influenced by the methodology adopted (using Costanza's method the accuracy on estimating the coefficient values of the major land cover is crucial).

Up to now, few analyses are focused on environmental effect of land take to ES provided by natural soils [10], especially the ones which ask for integrative analysis across different disciplines [14]. It is quite recent, the research dedicated to estimate the environmental effects of land take process, especially using ES as a proxy [11] [15] [16].

From systematic studies on surface and covers, a huge amount of research on assessment of urban transformation in hydrologic system is focused on "what happened on topsoil and under it, when a process of urbanization occurs" [17].

In general, despite ES approach emerge as the main paradigm to estimate quantitative and qualitative land transformation [4] [3], there is a lack of technical assessment to introduce indicators that hold different multidimensional features of soil transformation (i.e. the alteration of productive capacity – land capability, waterproofing, biodiversity decrease, landscape and cultural values). Composite indicators on land take are far away from being rooted in scientific literature (even if they are well defined) [18], despite a broad rhetoric claiming for an interdisciplinary approach on land management, no systematic results seem to be achieved. The demand for profound soil knowledge is high [19] [20] and a major interaction of scientists from other disciplines is requested in order to achieve a broad holistic role in society, and the context of "fusion" between different background needs to be enforced [21] [8] [22] [23].

3   The InVEST Model

Starting from the assumption that the concept of ES can change the way ecosystems are considered in policy and planning by the promotion of regulative options that will reduce environmental degradation and biodiversity loss while enhancing human well-being. There are several obstacles that prevent the transition from theory to action.

One major obstacle is the lack of a more systemic and holistic agreement on a common considerations that land uses is co-determined by natural and socio-economic factors and their interaction. Such interaction request a high integration of knowledge between ecological, social, and economic theories and studies.

ES can contribute to enforce the above mentioned holistic approach, because its systematic assessment over the analytical framework for town planning can be pursued using 5 important steps: 1) framing of key policy issue related to ES preservation or restoration; 2) identify ES and users (e.g. the definition); 3) mapping and assessing status; 4) valuation; 5) assess policy options including distributional impacts.

As mentioned, between (1/2) framing/identifying and (4/5) evaluating/assessing policy, there is an in between phase (3), which is crucial for introduce a progressive shift from description to prescription and local regulation of ES: mapping and assessing the status of ES on a context based situation.

The importance of the mapping and assessment of ES with an integrated framework was reflected on the proliferation of different mechanisms, methods and procedures to ensure that the value of ES is visible in decision-making.

One of the possible way of mapping and assessing ES is the use of InVEST system (Integrated Valuation of Ecosystem Services and Tradeoffs) which is a free software developed during the Natural Capital Project, with the aim to align economic forces with conservation, by developing tools that make incorporation of natural capital into decisions, demonstrating also the power of these tools and by engaging leaders globally.

InVEST is a tool for geographic, economic and ecological accounting on ES, according with specific types of land uses/covers. It is especially designed for territorial and town planning evaluation. In particular the ones focused on environmental protection at local scale, and all the decisions aimed to restore or defend the natural capacity of soil to provide non market goods as biodiversity, carbon poll, etc.

The InVEST model may be useful for informing resource management strategies and quantitative ranking of scenarios that can aid decision making, also because is a powerful tool to explore possible results of scenario between different land use alternatives (it is especially useful to compare degrade or ecological upgrade of specific soil functions, even in economic terms) [24].

The ideation of InVEST depends strictly to the concept that ES must be explicitly and systematically integrated into decision making by individuals, corporations, and governments [25]. The aim of this tool is to inform managers and policy makers about the impacts of alternative resource management choices on the economy, human well-being, and the environment, in an integrated way.

InVEST has 17 models that valuate ES, both biophysical processes and processes with monetary/economic value. The results of this model is a map of the geographic area of focus, the model requires spatially explicit and works on a GIS platform, as well as data describing the biophysical properties of land use/land cover (LULC) types [26].

The software works with a standalone modality and provides specific output, asking for different data input. As well as the evaluation request a high account of precision, the software request a significant number and high quality of raw data. The software can also evaluate, for specific ES, the trend of upgrade or degrade for different LULC map (baseline, current, future).

Each model requires inputs relevant to the ES of interest and LULC data. Most model outputs are a series of maps that represent relative values for the aggregate data over the area of interest. The research presented in the paper use the last release available (in 2015) of the InVEST model (version 3.1.0). The InVEST functions selected for this preliminary research were Habitat Quality and Carbon Sequestration.

4   An InVEST Possible Application

The challenge in the application of the InVEST model is to make the ES framework credible, replicable, scalable and suitable.

The area chosen to experiment this software is the Municipality of Lodi, an Italian town of 44,000 inhabitants with a territorial extension of 41kmq located in the south part of the Lombardy region.

The modeling approach is time intensive and requires knowledge of local ecology as well as technical skill with geospatial software. The choice of Lodi as a tester area is due by the presence of different database useful to create the requested dataset for "running" the model. In fact, the creation of the input dataset is the most important aspect for the quality of the outputs. The data inputs required for each model vary depending on the service, with data formats in GIS raster grids, GIS shape files or database tables.

Moreover, it is important to specify that the application of InVEST strictly depends on the context and on the degree of detail of single data. In this sense, for example, in a low dense residential area the detection of the simple land use/cover change is highly affected by territorial morphology, by settlement typology, and by infrastructural distribution. By the way, territorial conditions weigh heavily the productions of maps of the model and the organization of input dataset (e.g the weights assigned to each single data) is crucial too.

In this case, it was used the Topographic Database (DBtop) elaborated for the Province of Lodi. The DBtop is the more detailed LULC framework used as a cartographic base for town planning instruments. The survey dates back to 2008 with a map scale up to 1:500 until 1:1.000. More than that, The Province of Lodi, and for instance the municipality of Lodi, is also a territorial context with high degree of additional geospatial information (Land Capability, constrains, protected areas, slopes, water protection layers and other GIS) freely downloadable from Geo web site of Italian Government[4].

A high degree of precision was required since the testing phase because, as mentioned, efficiency of the program is highly dependent from the reliability of maps. Raster output, for each services tested, was setted in high resolution (raster cell size 5*5 meters). The micro-scale range is a novelty aspect in the use of InVEST. This is extremely helpful if considering the planning phase central for sustainable policy making, and require a high degree of information on possible land use allocation.

4.1   Habitat Quality Index

This experimental application start with the "Habitat Quality model" that is the ability of the ecosystem to provide conditions appropriate for individual and population

---

[4] http://www.dati.gov.it/

persistence. Habitat with high quality is relatively intact and it's depends on a habitat proximity to human land uses and the intensity of these land uses.

Effects of land-use change range from habitat destruction and pollution to extensive modifications of global biogeochemical cycles. In a global perspective, land use change affect the matter cycles and the global climate and hydrology. As a result, this decline produce impacts in biodiversity and accordingly habitat loss, modification and simplification with reflection on the local system [28].

The model identify the habitat of a specific species or in a more generically way in order to estimate how common threats affect wide range of viable habitat in the selected area. In summary the InVEST indicator is related to the biodiversity module in order to assess terrestrial habitat quality combining information on land use-land cover (LULC) and threats to biodiversity (anthropogenic pressures).

Below are listed the input data required for InVEST model [27]:

- Current LULC map
- Threat data
- Accessibility to sources of degradation, that is the legal/institutional/social/physical barriers provide against threats.

Firstly was clipped the DBtop with the boundaries of the municipality of Lodi and after there was a class dissolve for each LULC element. LULC classes are:

- Class 1 (means urbanized areas: urban fabric plots, without streets and green private/public spaces);
- Class 1.2.2 (means street, parks, railways and technological spaces dedicated);
- Class 1.4 (means green urban areas, similar to class 1.4 of CLC, even with high degree of detail, for example, in this classes green private gardens on open urban fabric are recognized);
- Class 2 (means agricultural areas, as for CLC legend);
- Class 3 (means natural or seminatural areas, as for CLC legend);
- Class 4 and Class 5 (means water, as for classes 4 and 5 of CLC).

The input file were elaborated in a GIS platform using ArcGis 10.1 release. All single shape files were unified with union function, than a rasterization process for LULC map creation can be conducted. Rasterization was applied with a 5*5 meters cell size, with LULC code as pixel unit, using the maximum area of pixel as proxy to attribute the value.

After creating the LULC maps, it is necessary to define the "threat data". For the case study, the threats identified are the settlements (SET), streets (STR), urban green (UGR) and, finally, agriculture land (AGR).

Threats are articulated in:

- maximum distance over which each threat affects habitat quality (in kilometers);
- weight that is the impact of each threat on habitat quality relative to other threats, expressed with 1 at the highest to 0 at the lowest;
- decay distinguished in linear or potential depending on the function expressed;
- maps of threats.

Table 1. Scores assigned for each category

THREAT	MAX_DIST	WEIGHT	DECAY
SET	0,3	0,7	Linear
STR	0,5	1,0	Linear
UGR	0,1	0,2	Linear
AGR	0,2	0,4	Linear

The Accessibility of habitat to threat (social, political, geographical restrictions) was evaluated from 0 to 1 point in which 1 is fully accessible without any restrictions to the threats while 0 correspond to the area less likely to be access by threats. The input requires is a .csv file with the level of access and a shape file with the spatial distribution of the restriction.

Table 2. Accessibility to sources of degradation and Habitat type and sensitivity

LULC	NAME	HABITAT	L_set	L_str	L_ugr	L_agr
1	residential	0	0	0.4	0	0
122	street	0	0	0	0	0
14	urban green	0.5	0.5	0.4	0	0.1
2	agricultural	0.6	0.8	1	0	0
3	natural	1	1	1	0.2	0.5
45	water	1	1	1	0.3	0.5

The single inputs were included in the InVEST model. The outputs are two maps:

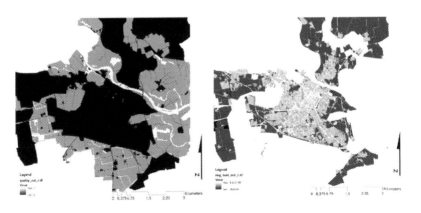

Fig. 1. Habitat quality map (left) and Habitat degradation (right)

This function can be used to evaluate how different scenarios of changes in land cover or habitat threats might affect the availability of quality habitat, and consequently biodiversity. A sort of parallel investigation related to the first one that illustrate the habitat degradation in a grey scale of color. The darken one are the landscape with high degradation while the light one the landscape that has managed to preserve a certain quality. Obviously, the two maps are complementary since the two elements closely dependent.

4.2   Carbon Sequestration Function

The second ecosystem function investigated is the Carbon Storage and Sequestration. estimated by investigating the carbon stock in present land use. Specifically, the carbon stock is valued on the size of 4 primary carbon "pools" defined by the IPCC [29]:

1. Above-ground biomass. All living biomass above the soil including stem, stump, branches, bark, seeds and foliage.
2. Below-ground biomass. All living biomass of live roots. Fine roots of less than (suggested) 2mm diameter are sometimes excluded because these often cannot be distinguished empirically from soil organic matter or litter.
3. Soil organic matter. It includes organic matter in mineral and organic soils (including peat) to a specified depth chosen by the country and applied consistently through the time series.
4. Dead organic matter. This category combines in one section Dead organic matter includes litter as well as

For each of these pools, was estimated the total carbon storage. Considering that is not available a specific database at local level we aggregated different sources.

As required by the InVEST model, a LULC map composed by the single categories of land use cover defined in the DBtop has been created. A selection of LULC categories in the area of the case study was completed also with the table of associated values. In this case study, the impermeable area (buildings, infrastructures, industrial platform), the water system (rivers, lakes, streams) and the desolate and unfertile areas are not considered. Below are listed the input data for InVEST program.

- Current land use/land cover (LULC) map that is the same dataset charged for Habitat Quality function previously presented. A table of LULC classes, containing data on carbon stored in each of the four fundamental pools for each LULC class.

As for the Habitat Quality index, the input file were elaborated in a GIS platform using ArcGis 10.1 release with a high detailed resolution of the raster file (5*5 meters cell size) with LULC code as pixel unit, using the maximum area of pixel as proxy to attribute the value. For the four carbon pool requested by the model the data were collected using different sources. Particularly, some data were provided by Silvia Solaro and Stefano Brenna - ERSAF[5] and the Italian National Inventory of Forests and Forest Carbon Sinks[6] in the second annual report of 2005.

---

[5] http://www.aip-suoli.it/editoria/bollettino/n1-3a05/n1-3a05_07.htm
[6] http://www.sian.it/inventarioforestale/jsp/dati_introa.jsp?menu=3

Table 3. Input .csv file

lucode	LULC_name	C_above	C_below	C_soil	C_dead
1	residential	0	0	0	0
122	street	0	0	0	0
14	urban green	10	2	12.4	0.4
2	agricultural	30	6	37	1
3	natural	50	10	65	5
45	water	2	1	3	0

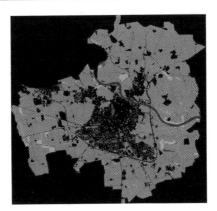

Fig. 2. Carbon sequestration

The output maps consisted of the total amount of carbon currently stored in milligrams as a sum of all carbon pools per grid cell of 5*5 meter. Due to the reliance on LULC data, carbon storage estimates strictly depends on the detailed LULC classification utilized.

4.3   Outputs

Even at preliminary stage, InVEST produces outputs that seem to fits for evaluation of ES at local scale. The two elaborations where enough detailed to support local strategies for planning options management. This means that two of the most important soil function are mapped and evaluated using a specific software. The procedures facilitate a successive multi-criteria analysis [30] should help to produce a Soil Quality Indicator (SQI) [31] [32] [33] necessary to understand if:

- urban transformation occurs on good or bad locations (referring to soil quality);
- how land use change impact with the environmental condition of soil;
- where mitigation, compensation or restoration occur.

Even simplified, the model helps to clearly define an environmental zoning where definition of the green corridors and infrastructures necessary to maintain or restore the quality of open spaces is crucial, and thus planning procedures could be empowered by this tools which give a technical robustness to specific sustainable analysis for planning are requested [34] [35].

This approach can also serve to evaluate the potential recovery of ES after land transformation. The model can support decision making for more sustainable development and is useful for making decisions for selecting lower impact sites [30]. This can significantly contributes to bridge the gap between theories and practices of sustainable town planning using ES indicators as proxies of Soil Quality.

In this preliminary analysis, only two functions were considered in order to test the InVEST model and the database available so it is quite difficult to understand at all how changing parameters of each single variables of input dataset can have significant effect on model's output. Probably a continuous work within the dataset input could enforce such objective. Further, more the program is tested at local scale, more the detail of information augment.

Anyway, it is possible to state that the technical support of the software is crucial to fill the gap between analysis and project of land transformation, its use in the screening phase of local planning, or even in advanced practices of spatial transformation, could produce significant political awareness of soil related function.

This simple consideration could enforce the aim of having significant technical tool that directly influence practices, process and project of land transformation even at the local stage.

5   Conclusion

The InVEST model propose a geo-informatization of different Ecosystem functions in order to determine the baseline services and, subsequently, the potential changes caused primary by land use changes. Each service is modeled separately, so that stacking of services takes place with the combination of model results.

The applications of InVEST can be useful to a wide variety of users, including conservation organizations, government agencies and research centers. Unfortunately, the Author's Guide of the program is delivered and oriented not for who intend to use it as a tool for local strategies of land use planning. Of course, the program can support such use, but all the sources needs to be re-defined, re-selected, and scaled to an expected output with a high degree of precision.

As mentioned, the output strictly depends on the detail of the LULC data and information, a more disaggregation of the different land uses influence the spatialization of the output in the final maps. More than that, the data required are very specific and detailed so their assumptions are often simplified with a margin of uncertainty. This is the principal limit to the InVEST model and, generally, for all the models that work with a large amount of data.

The organization of the input is one of most complex aspect of the model and would be very time consuming and challenging to obtain. This was experimented especially in the Carbon Sequestration function.

Moreover, usually the data needs are very specific and may often require to create new data with different investigations to complete a specified analysis.

By the way, some limitations are also connected with bibliography, especially on practical application, which is quite general or absent even if the time spend to collect the information and data, as well as the expertise beyond what may be gathered from the documentation and bibliography [36].

Even more the environmental databases of sources (climatic, hydrologic, pedologic…) are often collected and restituted at macroscale rather than at microscale. But, as it is broadly confirmed by literature, is the recognition and definition at local scale of ES that can really support policies against land take, preserving natural functions. Especially in context with high administrative fragmentation the local plan can give significant contribute to ES preservation when it take care soil quality on microzones of land. Also some research point out that it is on micro transformation the place where impact of land take on ES is higher. This is why InVEST is just a tool to starting an analysis which need to be refined, articulate, handled with adjustment, even simplified, with adding information, or with a synthesis of results made with multilayered analysis.

References

1. de Groot, R., Wilson, M., Boumans, R.: A typology for the classification, description and valuation of Ecosystem functions, goods and services. Ecological Economics **141**, 393–408 (2002)
2. MA - Millennium Ecosystem Assessment: Ecosystem and human well-being, Synthesis. Island press, Washington DC (2005)
3. Costanza, R., d'Arge, R., de Groot, R., Farber, S., Grasso, M., Hannon, B.: The value of the world's ecosystem services and natural capital. Nature **1387**, 253–260 (1997)
4. Daily, G.: Introduction: what are ecosystem services? In: Daily, G. (ed.) Nature's services: Societal dependence on natural ecosystems, pp. 1–10. Island Press, Washington D.C. (1997)
5. Hayha, T., Franzese, P.P.: Ecosystem services assessment: A review under an ecological-economic and system perspective. Ecological Modelling **1289**, 124–132 (2014)
6. Laurans, Y., Rankovic, A., Billè, R., Pirard, R., Mermet, L.: Use of ecosystem services economic valuation for decision making: Questioning a literature blindspot. Journal of Environmetal Management **1119**, 208–219 (2013)
7. Haines-Young, R., Potschin, M.: Common International Classification of Ecosystem Services (CICES). European Environment Agency, Nottingham (2011)
8. Shuying, Z., Changshan, W., Hang, L., Xiadong, N.: Impact of urbanization on natural ecosystem service values: a comparative study. Environmental monitoring and assessment **1179**, 575–588 (2011)
9. Dominati, E., Patterson, M., Mackay, A.: A framework for classifying and quantifying the natural capital and ecosystem services of soils. Ecological Economics **69**, 1858–1868 (2010)
10. Helian, L., Shilong, W., Hang, L., Xiaodong, N.: Changes in land use and ecosystem service values in Jinan, China. Energy Procedia **15**, 1109–1115 (2011)
11. Artmann, M.: Institutional efficency of urban soil sealing management - From raising awareness to better implementation of sustainable development in Germany. Landscape and urban Planning **1131**, 83–95 (2014)
12. Braat, L.C., de Groot, R.: The ecosystem services agenda: bridging the worlds of natural science and economics, conservation and development, and public and private policy. Ecosystem Services **1**, 4–15 (2012)

13. Clerici, N., Paracchini, M.: L. Maes, J.: Land-Cover change dynamics and insights into ecosystem services in European stream riparian zones. Ecohydrology & Hydrobiology **14**, 107–120 (2014)
14. Breure, A.M., De Deyn, G.B., Dominati, E., Eglin, T., Hedlund, K., Van Orshoven, J., Posthuma, L.: Ecosystem services: a useful concept for soil policy making! Current Opinion in Environmental Sustainability **14**, 578–585 (2012)
15. Jansson, A.: Reaching for a sustainable, resilient urban future using the lens of ecosystem services. Ecological Economics **186**, 285–291 (2013)
16. Li, F., Wang, R., Hu, D., Ye, Y., Wenrui, Y., Hongxiao, L.: Measurement methods and applications for beneficial and detrimental effects of ecological services. Ecologica Indicators **147**, 102–111 (2014)
17. Gardi, C., Panagos, P., Van Liedekerke, M.: Land take and food security: assessment of land take on the agricultural production in Europe. Journal of Environmental Planning and Management **58**(15), 898–912 (2014)
18. Giovannini, E., Nardo, M., Saisana, M., Saltelli, A., Tarantola, A., Hoffman, A.: Handbook on constructing composite indicators: methodology and user guide. Organization for Economic Cooperation and Development (OECD) (2008)
19. Havlin, J., Balster, N., Chapman, S., Ferris, D., Thompson, T., Smith, T.: Trends in soil science education and employment. Soil Science Society of, America **15**(74), 1429–1432 (2010)
20. Hopmans, J.: A plea to reform soil science education. Soil Science Society of America **171**, 639–640 (2007)
21. McBratney, A., Field, D.J., Koch, A.: The dimension of soil security. Geoderma **1213**, 203–313 (2014)
22. Dominati, E., Mackay, A., Greenb, S., Pattersonc, M.: A soil change-based methodology for the quantification and valuation of ecosystem services from agro-ecosystems: A case study of pastoral agriculture in New Zealand. Ecological Economics **1100**, 119–129 (2014)
23. Rutgersa, M., van Wijnen, H.J., Schouten, A.J., Mulder, C., Kuiten, A.M.P., Brussaard, L., Breure, A.M.: A method to assess ecosystem services developed from soil attributes with stakeholders and data of four arable farms. A method to assess ecosystem services developed from soil attributes with stakeholders and data of four arable farm **415**, 39–48 (2012)
24. Montanarella, L.: Special session Soil: sealing and consumption. In : 7° EUREGEO 2012. 04 luglio 2012. http://ambiente.regione.emilia-romagna.it/geologia-en/video/suoli/video-sessions-on-soil-euregeo-2012/luca-montanarella-special-session-soil-sealing-and-consumption-7b0euregeo-2012
25. NRC - National Research Council: Valuing ecosystem services: toward better environmental decision making. National Academies Press, Washington D.C. (2005)
26. Crawford-Gallagher, J., Martin, M., Nal, A., Ratliff, N.: Informing Conservation Decisions Based on Ecosystem Services. University of Washington (2014)
27. Sharp, R., Tallis, H.T., Ricketts, T., Guerry, A.D., Wood, S.A., Chaplin-Kramer, R., Nelson, E., Ennaanay, D., Wolny, S., Olwero, N., Vigerstol, K., Pennington, D., Mendoza, G., Aukema, J., Foster, J., Forrest, J., Cameron, D., Arkema, K., Lonsdorf, E., Kennedy, C., Verutes, G., Kim, C.K., Guannel, G., Papenfus, M., Toft, J., Marsik, M., Bernhardt, J., Griffin, R., Glowinski, K., Chaumont, N., Perelman, A., Lacayo, M. Mandle, L., Hamel, P., Vogl, A. In: Sharp, R., Chaplin-Kramer, R., Wood, S., Guerry, A., Tallis, H., Ricketts, T. (eds.) InVEST User's Guide. The Natural Capital Project, Stanford (2014)
28. Vitousek, P.M., Mooney, H.A., Lubchenco, J., Melillo, J.M.: Human domination of Earth's ecosystems. Science **15325**(277), 494–499 (1997)
29. IPCC: Guidelines for national greenhouse gas inventories, IGES, Japan (2006)

30. Keller, A.A., Fournier, E., Fox, J.: Minimizing impacts of land use change on ecosystem services using multi-criteria heurisitc analysis. Journal of Environmental Management **1156**, 23–30 (2015)
31. Vrscaj, B., Poggio, L.: Ajmone Marsan, F.: A method for soil environmental quality evaluation fon management and planning in urban areas. Landscape and Urban Planning **188**, 81–94 (2008)
32. Culshaw, M., Nathanail, C., Leeks, G., Alker, S., Bridge, D., Duffy, T., Fowler, D., Packman, J., Swetnam, R., Wadsworth, R., Wyatt, B.: The role of web-based environmetal information in urban planning-the environmetal information system for planners. Science of the Total Environment **1360**, 233–245 (2006)
33. Peccol, E., Movia, A.: Evaluating land consumption and soil functions to inform spatial planning. In: 3rd International Conference on Degrowth for Ecological Sustainability and Social Equity, Venezia (2012). http://www.venezia2012.it/wpcontent/uploads/2012/03/WS_3_FP_PECCOL.pdf
34. Fugazza, B., Ronchi, S., Salata, S.: La ricomposizione degli assetti ecosistemici a partire dalla valutazione delle funzioni dei suoli: una proposta di green infrastructure per il territorio lodigiano. Reticula **17**, 103–109 (2014)
35. European Environment Agency: Spatial analysis of green infrastructure in Europe. Euroepan Environment Agency, Luxembourg (2014)
36. Finn, S., Keiffer, S., Koroncai, B., Koroncai, B.: Assessment of InVEST 2.1 Beta: Ecosystem service valuation software. http://webspace.ship.edu/cajant/documents/white_papers/finnetal_invest_2011.pdf
37. Sharp, R., Tallis, H., Perelman, A., Lacayo, M., Mandle, L., Griffin, R., Hamel, P. In: Sharp, R., Chaplin-Kramer, R., Wood, S., Guerry, A., Tallis, H., Ricketts, T. InVEST User's Guide. The Natural Capital Project, Stanford (2014)
38. Nordhaus, W.: Critical Assumption in the Stern Review on Climate Change. Science **1317**, 201–202 (2007)
39. Stern, N.: The Economics of Climate Change: The Stern Review. Cambridge University Press, Cambridge (2007)
40. Hope, C.: The social cost of carbon: what does it actually depend on? Climate Policy **16**, 565–572 (2006)
41. Murphy, J., Sexton, D.M., Barnett, D.N., Jones, G.S., Webb, M.J., Collins, M., Stainforth, D.A.: Quantification of modelling uncertainties in a large ensemble of climate change simulation. Nature **1430**, 768–772 (2004)
42. Stainforth, D., Aina, T., Christensen, C., Collins, M., Faull, N., Frame, D.J., Kettlebourg, J., Knight, S., Martin, A., Murphy, J., Piani, C., Sexton, D., Smith, L., Spicer, R., Thorpe, A., Allen, M.: Uncertainty in predictions of the climate response to rising levels of greenhouse gases. Nature **1433**, 403–406 (2005)
43. Crossman, N., Connor, J., Bryan, B., Summers, D., Ginnivan, J.: Reconfiguring an irrigation landscape to improve provision of ecosytem services. Ecological Economy **169**, 1031–1042 (2010)
44. Collins, M., Steiner, F., Rushman, M.: Land-use suitability analysis in the United States: historical development and promising technological achievements. Environmental Management **128**(5), 611–621 (2001)
45. Malczewski, J.: GIS-based land-use suitability analysis: a critical overview. Progress in Planning **11**(62), 3–65 (2004)
46. Toman, M.: Why not calculate the value of the world's ecosystem services and natural capital. Ecological Economics **125**, 57–60 (1998)
47. Pimm, S.: The value of everything. Nature **1387**, 231–232 (1997)

48. Laurans, Y., Rankovic, A., Billè, R., Pirard, R., Mermet, L.: Use of ecosystem services economic evaluation for decision making: questioning a literature blindspot. Journal of Environmental Management **1119**, 208–219 (2013)
49. Baral, H., Keenan, R.J., Sharma, S.K., Stork, N.E., Kasel, S.: Economic evaluation of ecosystem goods and services under different landscape management scenarios. Land Use Policy **139**, 54–64 (2014)
50. Bateman, I.J., Harwood, A.R., Mace, G.M., Watson, R.T., Abson, D.J., Andrews, B., Binner, A., Crowe, A., Day, B.H., Dugdale, S., Fezzi, C., Foden, J., Hadley, D., Haines-Young, R., Hulme, M., Kontoleon, A., Lovett, A.A., Munday, P., Pascual, U., Paterson, J., Perino, G., Sen, A., Siriwardena, G., van Soest, D., Termansen, M.: Bringing Ecosystem Services into Economic Decision-Making: Land Use in the United Kingdom. Science **341**, 45–50 (2013)
51. Costanza, R., d'Arge, R., de Groot, R., Faber, S., Grasso, M., Hannon, B., Limburg, K., Naeem, S., O'neill, R., Paruelo, J., Raskin, R., Sutton, P., Van den Belt, M.: The value of the world's ecosystem services and natural capital. Nature **1387**, 253–260 (1997)
52. Pedroli, B., Van Doorn, A., De Blust, G., Paracchini, M., Wascher, D.: Europe's Living Landscapes. Essays Exploring Our Identity in the Countryside. Landscape Europe, Wageningen/KNNV Publishing, Zeist (2007)
53. Daily, G.C., Polasky, S., Goldstein, J., Kareiva, P.M., Mooney, H., Pejchar, L., Ricketts, T., Salzman, J., Shallenberger, R.: Ecosystem services in decision making: time to deliver. Ecological environment **7**, 21–28 (2009)
54. Reyers, B., Roux, D., O'Farrell, P.: Can ecosystem services lead ecology on a transdisciplinary. Environmental conservation **37**, 501–511 (2010)
55. Abson, D., von Wehrden, H., Baumgärtner, S., Fischer, J., Hanspach, J., Härdtle, W., Heinrichs, H., Klein, A.M., Lang, D., Martens, P., Walmsley, D.: Ecosystem services as a boundary object for sustainability. Ecological Economics **13**, 29–37 (2014)
56. S.S.B.S. Associazione Italiana Pedologi. http://www.aip-suoli.it/editoria/bollettino/n1-3a05/n1-3a05_07.htm
57. Nahlik, A.M., Kentula, M.E., Fennessy, M., Landers, D.: Where is the consensus? A proposed foundation for moving ecosystem service concepts into practice. Ecological Economics **77**, 27–35 (2012)
58. de Groot, R., Wilson, M., Boumans, R.: A typology for the classification, description and valuation of ecosystem functions, goods and services. Ecological Economics **41**, 393–408 (2002)
59. TEEB - The Economics of Ecosystems and Biodiversity: The Economics of Ecosystems and Biodiversity. Ecological and Economic Foundations. Earthscan, London (2010)
60. Solaro, S., Brenna, S.: Associazione Italiana Pedologi. http://www.aip-suoli.it/editoria/bollettino/n1-3a05/n1-3a05_07.htm

Climate Change and Transformability Scenario Evaluation for Venice (Italy) Port-City Through ANP Method

Maria Cerreta[✉], Daniele Cannatella, Giuliano Poli, and Sabrina Sposito

Department of Architecture (DiARC), University of Naples Federico II,
via Toledo 402 80134, Naples, Italy
{cerreta,daniele.cannatella,giuliano.poli,
sabrina.sposito}@unina.it

Abstract. The paper explores the consequences of resilience loss in some so-cial-ecological components of Venice port-city in Italy and suggests integrated sustainable strategies on increasing their stress response capability. The urban model of Venice depends on two influential factors: the natural balance be-tween land and water, and the social-economic dependence between mainland and islands. The authors adopt a broad framework of urban spaces and water environment, considering Venice as a complex regional unit, highly dynamic and sensitive. The adaptive balance historically reached by population and na-ture in Venice has been altered, because of a non-sustainable economic expan-sion, and tourism policy. The paper adopts a multi-dimensional approach, integrating the cognitive and evaluative dimensions with the technical and eco-nomic ones, in order to define possible strategies of action through Multi-Criteria Analysis and ANP method, able to play a strategic role in enhancing resilience at various scale.

Keywords: Resilience · Integrated evaluation · Sustainable scenario · ANP method

1   Introduction

Nowadays cities are completely involved in the fight against Climate Change. Even if the European Union is trying to promote both adaptive and mitigation strategies, it is difficult to define uniform actions because of the pronounced diversity that Climate Change effects arise across regions. Indeed, many areas are in danger: from Arctic to South Europe; from Alps to the Mediterranean Basin; and more, densely populated floodplains and urbanized coastal areas [1]. Climate Change leads a sensible increas-ing of global mean surface air temperatures over land and oceans. This phenomenon has strong impact on the rise in mean sea level, coastal erosion, and intensification of natural disaster due to meteorological drivers. Predicting Climate Change scenarios is extremely hard, however it seems to be clear how all these impacts involve environ-mental and anthropogenic systems. For example, abundance or lack of water can influence different economic sectors, i.e. agriculture, industrial production, tourism

© Springer International Publishing Switzerland 2015
O. Gervasi et al. (Eds.): ICCSA 2015, Part IV, LNCS 9158, pp. 50–63, 2015.
DOI: 10.1007/978-3-319-21410-8_4

and energy. For these reasons, the global resilience improvement set by European Community to protect environment is also an opportunity for a new green-based economy.

Urban areas hosting large and growing population become more vulnerable to climate change, for different causes as social inequality conditions and dependency on infrastructures [2]. In Italy, major consequences are related to the sea level rise. It has been calculated that, considering a rise rate between 18 and 30 cm by the end of the century, about 4500 square kilometres of the nation would be subject to flooding risk, especially coastal areas situated near the Ionian Sea (more than 60%) and the Northern Adriatic Sea (about 25%) [3]. In this framework, transformations due to the socio-economic development influence in a massive way the environmental system, including coastal systems. While the Low Elevation Coastal Zone constitutes the 2% of the world's land area, it contains 600 million people, the 10% of world's population, and 13 trillion US$ worth of assets [4, 5]. Therefore, the anthropogenic pressure, rather than climate-related drivers, becomes the largest driver of change in coastal systems such as beaches, sand dunes, coral reefs, estuaries and lagoons.

In response to this set of issues, Venice is an interesting example of lagoon/urban system seriously at risk. Eustatism, subsidence, touristic pressure, oversize ships are the main impacts that affect the Venetian environment at three different levels: city region, Lagoon ecosystem and city core. Therefore, the current research observes Venice as a complex system made up of islands, lagoon environment and mainland, within which it is possible to balance the effects of tourists moving toward the historical center. The port of *Marittima* plays a decisive role as system's center of gravity becoming a potential hub for improving the management of touristic and trade flows. This action is a first step towards reduction of touristic pressure and reconnection of the port area to the city.

Considering the above issues, it is possible to make explicit some relevant critical aspects, analyse emerging conflicts and evaluate impacts on the territory of Venice, through the elaboration of a Decision Support System (DSS) that considers: processing of a sustainable strategy and related actions; identification of a stakeholders map; selection of suitable indicators set. Through the implementation of a synergistic relation between social, environmental, economic and political components, it is possible to verify how a new sustainable strategy could modify the existing context.

In section 2, the paper introduces a cross-scale strategy for Venice city-region together with the description of the case study. In section 3, we examine the methodological proposal of a Decision Support System (DSS) in relation to its application to the case study. Finally, in section 4 we discuss the results, drawing some methodological conclusions and possible future developments of the research.

2   The Context: Venice and Its Touristic Port Area

2.1   Cross-Scale Strategy for Venice City-Region

The impacts on the ecosystem services need a decision-making strategy able to support economic and social development while preserving sustainability [6]. However, the interdependence across territories goes beyond the constraint of metropolitan governments, and unfolds within the city region This level combines adaptable deci-sion-making processes that can rearrange and enhance local units by involving com-munities in new sustainable regional and metropolitan scenarios [7, 8]. Moreover, the integrated valorisation of environmental components can inspire new strategies, able to give a resilient response to the Climate Change effects [9].

The metropolitan city of Venice highlights these implications. It was established by tracing out the Venetian territorial boundaries, keeping out the strategic location of Venice at the confluence of complex long-range networks [8, 10] (Figure 1).

Fig. 1. The study area

Venice intersects the trans-European Corridor V (Lisbon-Kiev) that passes through Padua and Treviso. Moreover, the city is characterized by a strategic localization for the different transport systems (air, sea and land). Therefore, OECD recommendations and regional plans proposed to strengthen and coordinate the urban environment along the Corridor V. OECD's main purpose is setting a metropolitan governance within a wider polycentric strategy, in order to manage urban, social-economic, envi-ronmental and climate features. The requirement of a competitive entity is evident in the Venice city-region configuration, which puts together the provinces of Venice, Padua and Treviso [10]. In this context, the identity of Venice – based on historical heritage, lagoon, maritime facilities, architecture and constructive techniques – be-comes reliable indicators of city-region performance and quality, encouraging sus-tainable regional and local policies. Within the city-region, Venice can be a driving force of economic and cultural growth, as international tourist site. Meanwhile, so-cial-ecological alteration and instability pushed by natural processes and human ac-tions can tamper with the metropolitan system balance. However, any action should consider the greater area of influence, which can affect by means of direct/indirect, long/short-term impacts [10].

Based on these considerations, the paper introduces a multidimensional methodol-ogy for an integrated assessment to enhance the metropolitan role of Venice while reversing the general urban decline, ecosystem damage and resources depletion, in order to improve the good governance process of phenomena and impacts.

The identification of three levels of investigation (level A - Venice city region; level B - Venice lagoon ecosystem; level C - Venice city core), which fit three territo-rial "clusters" involving different stakeholders, defines the geographical framework for the indicators selection and facilitates to examining trade-offs between the area of interest and its surrounding within a drop-down frame. The conceptual framework points out that sustainable projects, covered by level C, can increase the whole system's resilience if taken by sustainable policies at level A and B. It allows testing consistency of their actions across different scales. Considering three levels of investigation, some OECD indicators have been selected useful for understanding the different following components in order to identify a possible strategy of resilience (Figure 2):

1. Level A: Venice city region (Padua-Venice-Treviso): regional and metropolitan level of suburbanization, low population growth and dispersed form of production. It reaches a population of 2.6 million. Many projects will improve the city-region infrastructural system, especially the railway lines. This level requires a metropolitan resilience strategy, able to strengthen urban nodes, reduce land consumption and prevent risks/costs of extreme events.
2. Level B: Venice lagoon ecosystem (lagoon and mainland): hydrological system level, made up of inland and lagoon water, vulnerable to flooding. The level B is more subject to the water instability, determined by soil morphology and effects of climate change. The human pressures sharpen these criticalities, altering the environmental balance of the wetland. This level requires an ecosystem resilience strategy to face the global change of climate, sea and soil levels and to adapt at changes according to a sustainable vision.
3. Level C: Venice city core (historical centre and port): local level where lives around 2% of the city-region's population. It is subject to tourist pressure, which frequently causes the carrying capacity exceeded. The level C requires an urban resilience strategy based on regeneration and recycle.

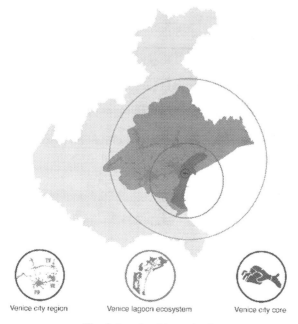

Venice city region   Venice lagoon ecosystem   Venice city core

Fig. 2. Levels of investigation

2.2   The Case Study: Venice Port-City

The study area is the port of *Marittima*, localized in the North-West side of Venice island. It is the touristic hub of the city port system, constituted also by the commercial terminal on the mainland, in Marghera. The *Marittima* contains the main flows deriving from mainland and sea. The area was built in the second half of the XIX century, during the industrial development of the city.

Venice evolved around a set of cores on indefinite islands emerging from the lagoon, developing through a thickening process of built tissues. This approach outlines how Venetians made possible prospering for centuries in a hostile environment. During the industrial age this approach changed because of environmental transformations imposed by new technical navigation requirements and new port activities [11]. For this reason, areas close to the historical centre, as *Marittima* and Tronchetto, appear to be out of any proportion and context.

3   Toward a Decision Support System (DSS)

3.1   Cross-Scale Indicators for Transformability Scenarios Evaluation

Due to analyse a complex reality as Venice, and specifically the area of interest, a Decision Support System (DSS), based on the identification and selection of heterogeneous data, has been structured [12, 13, 14]. Data are organized into a set of indicators that describe the social, economic and environmental features of the study area and its surrounding context. The methodological framework of the DSS classifies indicators according the DPSIR model [15] and indicates: reference year or trend, source, unit of measure, value and territorial coverage. According to a cross-scale approach, the territorial coverage refers not only to the local environment but also to an extended landscape that includes the Lagoon and the mainland.

The criteria of the DSS subdivide themselves into three levels, according to a tree-structure. The first level represents three macro-category including social, economic and environmental dimensions. The second level concerns the thematic areas derived from first level criteria and it contains issues related to: residents, local involvement and cultural consciousness included in Society meta-criterion; cruise tourism, tourism and occupation, included in Economy meta-criterion; infrastructure, urban amenities, biodiversity, climate, water, waste, air and soil, included in Environment meta-criterion. Finally, the third level concerns the indicators that are parameters used to understand and analyse those phenomena, which characterize the territory. The three level indicators show the state of the spatial system and its trend. Nevertheless, the DSS adopted a selection procedure for some categories of indicators pertinent to the study area, marked in grey in Table 1.

The cross-scale strategy becomes essential regarding to the planning and decision-making process of port areas since a localized action can influence and have effect on a larger scale. The displacement of cruise route or touristic flows toward proper directions and sites, for example, can enhance the efficiency of the ecological and social systems involved.

Table 1. Core-set of indicators

1st Criteria	2nd Criteria	Indicators	DPSIR	Year	Source	U.M.	Value	Level
Society	Residents	Inhabitants	P	2013	Municipality	num.	57.999	C
		Demographic trend	D	2001-2010	ISTAT	num.	59.621	C
		Foreigner residents	P	2007	Municipality	num.	12.700	C
		Growth population scenario	D	2020	Municipality	num.	54.799	C
		Metropolitan inh.	P	2011	OECD	num.	2.600.000	A
	Local involvement	Planning and Environmental Partecipation Index	R	2011	Legambiente	index 0-100	75	C
		Eco-management Index	R	2011	Legambiente	Index 0-100	54	C
	Cultural consciousness	UNESCO area	R	2014	UNESCO	ha	70.230,46	C
		Cultural events	R	2007	OECD	num.	1.800	C
Economy	Cruise tourism	Passenger traffic	P	2013	VTP	num.	2.073.953	C
		Harbours with over 25 m berths	S	2013	Authors comp.	num.	32	B
		Ship arrivals	P	2012	Municipality	num/yr	1.280	C
		Cruise tourists	P	2013	VTP	num.	1.815.823	C
		Berths over 40.000 GT	S	2014	Port Authority	num.	4	C
		Cruise arrivals	S	2012	Municipality	num/yr	661	C
		Negative externalities cost	I	2013	"Ca Foscari" University	€*resident/yr	6.000	C
		Mega-yacht arrivals	P	2013	Port Authority	megayacht/yr	200	C
	Tourism	Touristic arrivals	P	2012	Municipality	people/yr	2.485.136	C
		Touristic turn-out	P	2012	Municipality	people/yr	6.221.821	C
		Touristic carrying capacity	P	1990	"Bicocca" University	people/yr	22.000	C
		Metropolitan touristic arrivals	P	2007	OECD	people/yr	39.500.000	A
		Hotels	R	2012	Municipality	num.	418	C
		Beds in hotels	R	2012	Municipality	num.	28.442	C
	Occupation	Employment rate (15-64 age)	P	2013	ISTAT	%	59,7	A
		Unemployment rate (15-74 age)	P	2013	ISTAT	%	8,6	A
		Structural dependency index	P	2013	Tuttitalia.it	num.	56,1	A
		Recharge of working age residents index	P	2013	Tuttitalia.it	num.	151,4	A
		Structure of working age residents index	P	2013	Tuttitalia.it	num.	142,1	A
		Operative enterprises	R	2011	ISTAT	num.	22.947	C
		Employees in operative enterprises	R	2012	ISTAT	num.	101.858	C
Environment	Infrastructure	Railway passengers in S. Lucia	P	2012	Gianni Pellicani foundation	num.	34.990	C
		Railway passengers in Mestre	P	2012	Gianni Pellicani foundation	num.	27.189	C
		People mover maximun load	R	2014	AVM	passenger/h	3.000	C
		Bus fleet	R	2014	ACTV	num.	600	B
		Boat fleet	R	2014	ACTV	num.	152	B
		Parking spaces in Venice	R	2014	Authors comp.	num.	7.109	C
		Parking in Mestre and Marghera	R	2014	Authors comp.	num.	850	C
		Park and ride	R	2014	AVM	num.	12	C

Table 1. (*Continued*)

Environment		Car rate	P	2013	Legambiente	vehicles/ 100 inh.	42	C
		Motorcycle rate	P	2013	Legambiente	motor-cycle/ 100 inh.	7	C
		Num. of vehicles	P	2012	ACI	num.	143.444	C
	Infrastruc-ture	Public car park transit	R	2011	Municipality	vehi-cles/yr	235.503	C
		Pass for Limited Traffic Area	R	2011	Municipality	num./yr	80.767	C
		Public Transport offer	R	2013	Legambiente	km-vehicles/ inh.per yr	64	C
		Public Transport Passengers	P	2013	Legambiente	passen-ger/inh per yr	592	C
		Sustainable Mobility index	R	2011	Legambiente	index 0-100	73,3	C
		Bike path	R	2013	Legambiente	meters/ 100 inh.	12,47	C
	Urban amenities	Usable urban green area	R	2013	Legambiente	mq/ inh.	37,4	C
		Green areas	S	2013	Legambiente	%	65	C
	Biodiversity	Barene erosion 1930-2000	I	2012	Magistrato alle Acque	kmq	35	B
		New artificial barene	R	2000	Consorzio Venezia Nuova	ha	500	B
		ZPS, SIC, IBA	R	2014	Natura 2000	ha	115.326,34	B
	Climate	Average of precipita-tion	S	2013	Authors comp. on Eurometeo	mm/yr	66,6	C
		Relative humidity rate	S	2013	Authors comp. on Eurometeo	%	75,8	C
		Max temperature	S	2013	Authors comp. on Eurometeo	°C	17	C
		Min temperature	S	2013	Authors comp. on Eurometeo	°C	8,5	C
	Waste	Blackwater	P	2011	Authors comp.	l per day	145.658	C
		Greywater	P	2011	Authors comp.	l per day	1.844.996	C
		Bilge water	P	2011	Authors comp.	l per day	48.553	C
		Solid waste	P	2011	Authors comp.	kg per day	24.276	C
	Air	Average of NO2	S	2013	Legambiente	mg/mc	36,5	C
		Average of PM10	S	2013	Legambiente	mg/mc	32,8	C
		Average of O3 up to 120 mg/mc	S	2013	Legambiente	mg/mc	34,5	C
	Soil	Urban waste produc-tion	P	2013	Legambiente	kg/ inh. per yr	618,7	C
		Soil consumption per inh.	P	2007	SINAnet	mq per inh.	199,54	C
		Land use	D	2012	EEA	class	26	B

Level A indicators show that the local loss of population in Venice is actually a re-distribution phenomenon involving the city-region. The main reason is the high concentration of historical and cultural values in the city core that attracts almost 2.5 millions of tourists every year. Therefore, recent studies point out a possible cause-effect relationship between population decrease and tourism growth. The cruise tourism has mostly influence on tourist incoming, since Venice Terminal Passenger (VTP) computed more than 1.8 million cruise passengers in 2013. The cruise traffic makes a significant contribution to the tourist carrying capacity exceeding. The threshold of 22.000 tourists a day in the city center, indeed, has been overcome, reaching peaks of 100.000 arrivals [16]. Likewise, the lack of accommodation facilities in the mainland and lagoon archipelago encourages the tourist presence in the city core, where numerous buildings were turned into hotels. The tourist pressure induces a series of dangerous impacts as: environmental damage, pollution, sea bottom erosion, high-frequency currents, deterioration of buildings and identity weakening. The negative externalities linked to pollution of air, sea and climate change amount to 278 million euro per year, compared to 290 of cruise tourism revenue, for an average of 6.000 euro per inhabitant a year estimated for the historic city of Venice [17].

In 1987, Venice and the Lagoon entered in UNESCO's World Heritage List. It demanded to protect the historical architectures, techniques and traditions as well as the wet habitat over time. The UNESCO criteria underlined the urban and environmental values of Venice, and recognized the primary role of *barene* (clay soil formations that are periodically submerged by water) for the hydrologic balance maintenance.

Decision-making processes must deal with this sensitive matter, recommending new visions and strategies to reverse pressure conditions while allowing touristic purposes. The shift toward more sustainable forms of tourism can launch conciliation procedures on crucial issues, raising the Eco-management Index [18, 19].

3.2   Resilient Urban Design as Response to Pressures

Considering potentials and critical aspects underlined by the main indicators, a proposal of a new sustainable strategy has been elaborated. The project aims at shifting the *Marittima* from terminal-cruise to maxi-yacht marina, outsourcing logistics functions and cruise lines in *Marghera*, in order to reduce the environmental impacts, including the risks of people and artifacts generated by transit and berthing of oversize ships (Figure 3).

According to a systemic vision based on multi-scale approach, *Marittima* has been considered a strategic infrastructural hub and a gateway to the historic city by the mainland and the sea.

The objectives try to make a significant contribution in terms of increasing social, economic and environmental resilience of Venice. These objectives are:

— Sustainable management of daily touristic flow;
— Integration of the *Marittima* with its surrounding context;
— Improvement of the environmental performances;
— Enhancement of the identity-related values of port.

The Master Plan defines three different functional fields: touristic port, behind-port area and port-city connection area, and integrates itself with a set of logistic interventions, i.e. Ro-Ro terminal in *Fusina*, offshore platform and high-speed railway system. It aims at optimizing mobility networks by reducing road transport and creating intermodal parking. Moreover, the design of a new green park within the port area enhances urban and environmental amenities, by locating various facilities that can host mixed and temporary uses.

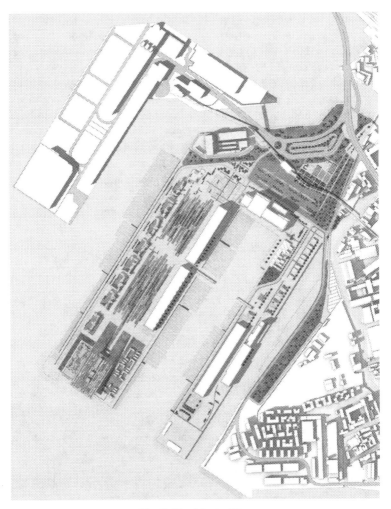

Fig. 3. The Master Plan

According to this vision, the conceptual framework is based on several elements: the preservation of the existing urban grid that combines old and new elements; the use of containers following low-impact approaches; the reuse and restyle of cruise terminals. Specifically, touristic port have a variety of functions due to ensure its continued vitality: residences, micro-exhibition centers, shopping galleries, recreational and cultural facilities and a Yacht Club. The area of connection port-city becomes an arts, science and technology center, with new housing for students, artist's studios and exhibition spaces. The green park hosts some prototypes of eco-sustainable houses while the cruise terminal becomes the new center of the lagoon biology research.

3.3   Scenario Evaluation Through Analytic Network Process (ANP)

The evaluation process, elaborated with the ANP application software provided by Creative Decision Foundation [20], aims at showing the preference of one alternative scenario compared with the others. According to the analysis of the social, economic and environmental contest (see 3.1), it is possible to define three scenarios for the study area:

(a) the current scenario, called *Scenario 0*;
(b) the cruise tourism scenario, called *Scenario 1*;
(c) the Master Plan scenario, called *Scenario 2*.

The *Scenario 0* represents the current state of *Marittima* port area composed by five cruise terminals, parking areas for tourists and warehouses; in this scenario the port area is physically and morphologically separated by the city. The *Scenario 1* improves the cruise tourism through the restyle of the terminals and the edification of a new building composed of a multi-storey car park and roof garden; this scenario partially tries to connect the port with the city. The *Scenario 2* is the Master Plan that suggests to shift the cruise tourism to *Marghera*, making the *Marittima* a maxi-yacht port; it considers the port as a strategic part and it achieves the connection with the city through green filtering and functional variety (Figure 4).

Fig. 4. Scenarios. From left to right: current scenario (Source: Google maps), cruise tourism scenario (Source: www.port.venice.it) and Master Plan (Source: authors' elaboration)

The ANP multi-criteria method is an implementation of the Analytic Hierarchy Process (AHP), powered by Saaty [21,22,23], which includes the interrelationship between elements within the hierarchy system of criteria. The ANP is structured in four main phases. First, it is necessary to define a goal for the analysis; subsequently, the method allows sorting the decision problem into two fundamental elements: the nodes, compounded by main categories, and the clusters that constitute sub-categories of the nodes. This framework is similar to the human thought processing and allows examining the complexity of a problem through the pairwise comparison technique. The third and fourth phases consist, respectively, of the super-matrix, that combines the interrelationships between clusters and nodes, and finally the weights and the priority vectors related to each main category [24, 25].

The ANP method, combined with the Benefit-Opportunity-Costs-Risks (BOCR) analysis, provides problem simplification by structuring and classifying the selected issues according to these categories, through the built of an ANP-BOCR model [21]. Four main clusters of BOCR model compose the decision framework and each cluster contains some nodes, in sub-network level, that are the indicators previously divided and categorized according to DPSIR model. The goal of the evaluation process is to identify the best scenario for the study area.

After the model structuring, it is possible to make pairwise comparisons among the nodes related to each cluster. In the ANP method, there are two levels of pairwise comparisons: the one relating to clusters, which is general, and the other relating to nodes, which is more specific as it involves the indicators level [22]. The rating assigned in the pairwise follows the Saaty's fundamental scale of 1-9, where the 1's represents the same value in the judgment, while the 9's defines extremely importance [23]. Moreover, the sensitivity analysis evaluates the stability of the final output changing the weight of assigned judgments. According to ANP-BOCR model, the decision problem has been divided in categories, and the judgments – assigned to clusters and nodes during the evaluation process – provide four final graphs describing scenarios in terms of benefits, opportunities, costs and risks (Figure 5).

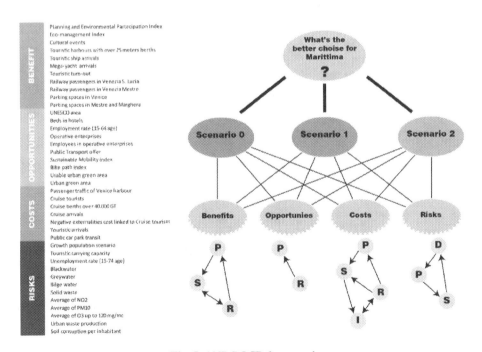

Fig. 5. ANP-BOCR framework

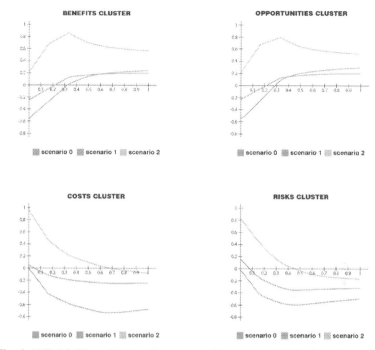

Fig. 6. ANP-BOCR model: benefits, opportunities, costs and risks for each scenario

Regarding to Benefits and Opportunities networks, the bends of the three scenarios grow with the increase of assigned weights (Figure 6). Therefore, they have positive features in terms of Benefits and Opportunities. However, the Master Plan scenario is the alternative with the highest score compared with other two, while there is an irrelevant difference between scenario 0 and scenario 1.

Concerning Costs and Risks networks, the bends of scenarios decrease with the increase of the assigned weights. In reference to both criteria, Master Plan scenario is preferable because it represents the alternative that throws down costs and risks more than others. Finally, there is a significant difference between scenario 0 and scenario 1, as the first wins in Costs graph while it loses in Risks graph compared to scenario 1.

Fig. 7. Scenario ranking

The Master Plan scenario becomes the least preferable solution since it proposes an urban regeneration action that tries to lower environmental risks and costs while enhancing benefits and opportunities. The distributive method of the AHP has been used in order to obtain the ratings. Normalized results show that the preferable scenario for all networks is the Master Plan with a score of 0.54 compared to 0.36 of scenario 1 and 0.10 of scenario 0 (Figure 7).

4   Conclusion

The assessment of tangible and intangible values makes it possible to understand social, economic and environmental dimensions through a multi-dimensional approach [26, 27]. Moreover, the interpretation of critical and potential elements generates strategic actions that can have consistent or conflicting impacts on different scales. In the climate and demographic change-related framework, multi-dimensional and cross-scale approaches can play an essential role to conceive new forms of economies able to activate circular processes, enhancing resilience of anthropic and natural systems.

Indeed, the SSD is useful to understand the impacts related to climate and demographic changes, analyse the possible forward scenarios and check the coherence of a multidimensional-approach. At the same time, the selection of specific knowledge-based indicators and the implementation of an ANP model allows identifying the most relevant issues, and helps organizing the alternatives and choosing the preferable one.

The proposed SSD becomes particularly useful to understand the complex reality of Venice, in which all conflicting dimensions show themselves in multiple shapes and different scales.

References

1. Commission of European Community: White Paper on adapting to climate change: Towards a European framework for action (2009). http://ec.europa.eu/
2. Melillo, J.M., Richmond, T.C., Yohe, G.W.: Climate Change Impacts in the United States: The Third National Climate Assessment. U.S. Government Printing Office, U.S. (2014)
3. ENEA (Agenzia nazionale per le nuove tecnologie, l'energia e lo sviluppo economico sostenibile). Terza Comunicazione Nazionale dell'Italia alle Nazioni Unite (UNFCCC). Ministero dell'Ambiente e del Territorio, Roma (2003)
4. IPCC (Intergovernmental Panel on Climate Change): Climate Change 2014: Impacts, Adaptation, and Vulnerability (2014). http://www.ipcc.ch/report/ar5/wg2/
5. MA (Millennium Ecosystem Assessment): Ecosystems and Human Well-being: A Framework for Assessment. Island Press, Washington DC (2003)
6. Costanza, R., d'Arge, R., de Groot, R., Farberk, S., Grasso, M., Hannon, B., Limburg, K., Naeem, S., O'Neill, R.V., Paruelo, J., Raskin, R.G., Suttonkk, P., van den Belt, M.: The value of the world's ecosystem services and natural capital. Nature **387**, 253–260 (1997); Vogel, R.K., Savitch, H.V., Xuc, J., Yehd, A.G.O., Wue, W., Sanctonf, A.: Governing global city regions in China and the West. Progress in Planning **73**, 1–75 (2010)
7. Messina, P.: Metropolitan city or city region? The case of the Veneto in the European context. Economia e Società regionale **XXXI**, 46–63 (2013)

8. Gasparrini, C.: Una nuova città: diffusione e densificazione. In: Talia, M., Sargolini, M. (eds.) Ri-conoscere e ri-progettare la città contemporanea. Studi Urbani e Regionali, Franco Angeli, Milano (2012)
9. Fondazione Venezia 2000: OECS | OECD: Territorial review: il caso di Venezia. Materiali e contributi (2010). http://www.fondazionevenezia2000.org
10. Sposito, S., Cannatella, D.: Building the lagoon city: from venice to the lagoon, from the lagoon to venice. In: Proceedings of the Meeting Port Cities as Hotspots of Creative and Sustainable Local Development, BDC – Bollettino del Dipartimento di Conservazione dei Beni Architettonici ed Ambientali 12 (2012)
11. Cerreta, M.: Thinking through complex values. In: Cerreta, M., Concilio, G., Monno, V. (eds.) Making Strategies in Spatial Planning. Knowledge and Values 3. Springer, Dordrecht (2010)
12. Cerreta, M., Poli, G.: A Complex Values Map of Marginal Urban Landscapes: An Experiment in Naples (Italy). International Journal of Agricultural and Environmental Information Systems **4**, 41–62 (2013)
13. Fusco Girard, L., Cerreta, M., De Toro, P.: Integrated Assessment for Sustainable Choices, Scienze Regionali. Italian Journal of Regional Science, 111–142 (2014)
14. EEA (Environmental European Agency). http://www.eea.europa.eu
15. Costa, P.: La carrying capacity. Il caso di Venezia. In: Costa, P., Manente, M., Furlan, M.C. (eds.) Politica economica del turismo, pp. 135–150. Touring Club Italiano, Milano (2001)
16. Tattara, G.: É solo la punta dell'iceberg! Costi e ricavi del crocierismo a Venezia. Note di Lavoro 2 (2013)
17. Legambiente: Ecosistema Urbano. XIX Rapporto sulla qualità ambientale dei comuni capoluogo di provincia (2012). http://www.legambiente.it
18. Legambiente: Ecosistema Urbano. XXI Rapporto sulla qualità ambientale dei comuni capoluogo di provincia (2014). http://www.legambiente.it
19. http://www.creativedecision.org
20. Saaty, T.L., Ozdemir, M.S.: The Encyclicon. A Dictionary of Complex Decisions Using the Analytic Network Process. RWS Publications, Pittsburgh (2008)
21. Saaty, T.L.: The Analytic Hierarchy Process. McGraw Hill, New York (1980)
22. Saaty, T.L., Vargas, L.G.: Decision Making with the Analytic Network Process. Springer, US (2006)
23. Bottero, M., Ferretti, V.: Assessing urban requalification scenarios by combining environmental indicators with the Analytic Network Process. Journal of Applied Operational Research **3**, 75–90 (2011)
24. Lombardi, P., Lami, I.M., Bottero, M., Grasso, C.: Application of the analytic network process and the multi-modal framework to an urban upgrading case study. In: Horner, M., Hardcastle, C., Price, A., Bebbington, J. (eds.) International Conference on Whole Life Urban Sustainability and its Assessment, Glasgow (2007)
25. Attardi, R., Canta, A., Torre C.M.: Urban design, institutional context and decision-making process. In: Complex Evaluations for Hybrid Landscapes, BDC – Bollettino del Dipartimento di Conservazione dei Beni Architettonici ed Ambientali 14, pp. 129–143 (2014)
26. Camarda, D., Romandini, M., Torre C.M.: Wetlands, coastline, historical heritage vs. urban spread: a complex integrate planning experience in Taranto, Italy. In: Camarda, D., Grassini, L. (eds.) Coastal zone management in the Mediterranean region. Bari: CIHEAM, pp. 123–135 (Options Méditerranéennes: Série A. Séminaires Méditerranéens; n. 53) (2002)

The Evaluation of Landscape Services: A New Paradigm for Sustainable Development and City Planning

Roberta Mele[1] and Giuliano Poli[2(✉)]

[1] Department of Architecture, University of Rome La Sapienza, Rome, Italy
roberta.mele@uniroma1.it
[2] Department of Architecture, University of Naples Federico II, Naples, Italy
giuliano.poli@unina.it

Abstract. The paper provides a methodological framework in order to investigate, map and evaluate the landscape, understood as multi-dimensional complex system. A possible landscape knowledge-related approach seeks to recognize the *landscape services* spatial distribution on the territory. The management of multi-dimensional geospatial data, indeed, allows to face contemporary environmental matters regarding continuous biodiversity loss, ecological fragmentation and acceleration of climate change. The GIS provides helpful tools for systematization, management and analysis of spatial indicators, which describe and quantify the type of *ecosystem* and *anthropic service*s. The paper proposes the development of a Spatial Decision-Making Support System (SDSS), in sync with GIS data-set and multi-criteria method AHP. The SDSS is useful to develop a *landscape services* thickness map and to define possible territorial transformation scenarios. The recognition of the *landscape services* can support the decision-making process referred to a sustainable management and planning, which are still an open investigation field.

Keywords: Landscape services · GIS · SDSS · AHP · Scenario evaluation

1 Introduction

The management of multi-dimensional geospatial data allows to face structured environmental issues that mainly concern: the continuous biodiversity loss at a global scale; the discontinuity of ecological connections/network due to fragmentation [1]; the acceleration of climate change generated by anthropic impact on the environment [2]; the different *autopoiesis* periods of ecosystems compared to human transformation actions; the alteration of regeneration time of soils which are therefore considered non-renewable resources [3].

In response to these issues, a possible interpretation of landscape concept, understood as multi-dimensional complex system, implies a selection of spatial indicators that represent the type of ecosystem and human services as well as their resilience to environmental pressures. The recognition of the *landscape services* spatial distribution on the territory supports the decision-making process referred to a sustainable management and planning, which are still an open investigation fields [4].

© Springer International Publishing Switzerland 2015
O. Gervasi et al. (Eds.): ICCSA 2015, Part IV, LNCS 9158, pp. 64–76, 2015.
DOI: 10.1007/978-3-319-21410-8_5

In literature, the *ecosystem services* concept has many and vague definitions because of its availability in adapting to investigation field of the different thematic areas [5]. The researchers in environmental fields provide some definitions of *ecosystem services* that involve their areas of interest. Indeed, it is possible to define three general meaning of services similar to the issues proposed in this paper. Specifically, these meanings are included in the definition founded by de Groot, considering «the capacity of natural processes and components to provide goods and services that satisfy human needs, directly or indirectly» [6,7]; the definition of MA, regarding «the benefits people obtain from ecosystems» [8]; the economic-environmental definition provided by TEEB (The Economics of Ecosystems and Biodiversity) considers the *ecosystem services* as «the direct and indirect contributions of ecosystems to human wellbeing» [9].

Lastly, according to Termorshuizen and Opdam, it is possible to include the *ecosystem services* into wider classification criteria regarding more categories and values such as socio-economic, environmental, cultural and morphological ones. The purpose of this theoretical approach is the brainchild of a common knowledge-based platform to multiple disciplines. In this way of thinking, the *ecosystem services* become a category of the *landscape services* [10].

According to this methodological framework, it is possible understand how the landscape functions are linked to the feature of the ecosystem processes and to their capacity for producing services. These services are the benefits that the ecosystems give to human well-being. Nowadays, decisions makers are increasingly giving more attention to *landscape services* evaluation in planning policy, although it is complicated to recognize an economic value for these services, due their lack of two basic parameters for the estimation of private goods such as market and production. These services are fundamental because they constitute essential benefits to human survival that can be perceived in terms of use values and independent use values [11].

In regard to these issues, the interdisciplinary methodological framework allows to relate different values involved in the process of landscapes knowledge, which can be defined as a multi-functional entities due to their complex nature [12].

The first part of the paper (section 2) defines the methodological approach of the SDSS; the second one (section 3) shows study case; the third (section 4) analyzes the results of the research while the fourth (section 5) explains the conclusions about the usefulness in city policy and planning.

2   Methodological Framework

The methodological approach, combining the *landscape services* concept with Multi-Criteria Analysis, builds a Spatial Decision-Making Support System (SDSS) that is able to understand, examine, map and evaluate the landscape as a dynamic complex system [13,14,15]. Next, we present a short literature review about the systemic landscape concept and the Multi-Criteria Decision-Making Analysis (MCDA) and GIS tools, while in the final section we introduce KME-SDSS.

2.1   Landscape as System and Landscape Ecology

The systemic concept considers the environmental, social, cultural, morphological components not only as the results of continuous anthropic and natural transformation processes of the landscape, but also as the result of different time actions which are strictly interrelated each other [16]. Indeed, the systemic way of thinking give special attention to the link among components or features apparently different, considering that «the whole is more than the sum of its parts» [17].

Therefore, it becomes essential to use a multi-disciplinary approach that is able to analyze and understand, considering different points of view, the phenomena that generate the dynamism of landscape continuous changes, in order to have more varied and complete vision of the object investigated.

The Landscape Ecology, in its more general formulation, provides necessary tools to understand the landscape transformation flows not only in strictly ecological terms but also including pressure, impact and response factors generated or absorbed by human-being.

The approach of Termorshuizen and Opdam, as previously seen, considers the *ecosystem services* as a sub-category of *landscape services*, and it takes into account that they have a specific value as a function of the relationships established with other components of the landscape system. The analysis of the landscape multi-scale pattern can be very useful to define how the human choices and actions influence the processes, the functionality and the structure of the services [18,19].

On the other hand, the European Convention of Landscape highlights the relevant role of people as part of the landscape and it opens the way to a new vision for which the landscape perception cannot be separated by the knowledge of continuous changes that people produce on the land for its benefit [20,21].

The method applied consists of a multi-scale approach to the landscape. The patches mosaic, in which the landscape can be conceptually assimilated [22], is structured in a hierarchical multi-level framework, according to the environmental structures, characterized by sets of systems and subsystems and incorporated, each other, in larger systems [23]. These hierarchies and their interdependence are preliminary compared with the multi-scale approach and they allow to understand the nature of the network among human and environmental components within the landscape complex system, in order to act by weighting choices without interfere with natural and social ecosystems. Finally the multi-criteria methods aim to analyze the patches in detail, for the purpose of quantifying the information derived from the relevant environmental indicators and assigning a weight to each indicator considered in the evaluation phase.

2.2   Multi-Criteria Decision-Making Analysis (MCDA) and GIS

As part of the integrated approaches and strategic evaluations, the Multi-Criteria Decision Making Analysis (MCDA), integrated with GIS, has now become vital to address problems related to planning, land management and evaluation of different alternatives [24,25,26]. In these fields, characterized by complexity and uncertainty, multi-criteria analysis becomes a tool to explain the contributions of different criteria according to the alternatives which are compared with the final goals of decision makers [27].

At the same time, the fundamental role of GIS concerns its capability to explain the spatial component of the data, which is why it can be considered as one of the most advanced tools to manage the decision-making process of sustainable planning [28,29]. Although these two instruments can be used in an independent way to deal with simple problems, their interaction is essential in multi-disciplinary and multi-dimensional problems, where the purpose is to achieve a balance between economic development, environmental protection and social balance in order to reach the optimal solution [30]. The decision support systems GIS-MCDA are becoming capable tools for solving spatial issues related to problems where conflicting criteria are involved [31] and for merging the spatial information according to multiple criteria in a single evaluation index [32, 33]. The structured SDSS (GIS-MCDA) provides a methodological framework to contemplate the different points of view and the components of a problem and let to organize them in hierarchical structures that take individual parts into account as much as the mutual interrelationships. They define new approaches to provide techniques and tools that combine multidimensional information about geographic data, identify priorities and explain the preferences of actors involved in a decision making processes [26].

2.3   The KME-SDSS Method

Within the framework of Spatial Decision Support System, we propose the KME-SDSS model which is divided into three main phases [27]: (i) Knowledge, which consists in the identification of *landscape services* related to investigation area; (ii) Mapping, which serves the purpose of mapping the spatial selected indicators; (iii) Evaluation, which assigns different weights to the spatial indicators through a suitable evaluation method.

The KME-SDSS consists of various steps carried out in GIS environment, which are preparatory to the definition of the final output in the evaluation process. In this way it becomes possible to standardize the operational process, according to Post-normal science approach that considers the procedure as a methodological framework useful for different study fields [34]. For this purpose, the operations performed in a GIS environment are synthetically shown as follows:

- study area selection;
- coordinate system and data georeferencing;
- shapefile editing;
- mapping of main shapefiles related to *landscape services*;
- classifying and weighting indicator;
- final output concerning transformability evaluation maps.

The development of this model will be described in detail in the next section (section 3).

3   Case Study

3.1   Case Study and Research Purpose

The proposed methodology has been applied to identify *landscape services* rest in the metropolitan area of Naples, which covers 560 sqkm, including the city itself, the fourteen municipalities directly neighboring and other municipalities that are connected to the city in terms of economic, social and environmental flows. This area is characterized by the presence of high biodiversity zones alternated by extensive fields of intensive agriculture, significant portions of natural and semi-natural areas, numerous urban zones of variable density population, which are full in potential landscape and environmental quality.

The city of Naples has been chosen as a case study for application of the methodology due to its barycentric position in the local context, and the reasons for proposing a KME-SDSS on this area has let the opportunities: to assess the multidimensionality of landscapes service, which is a complex problem that must be deal with specific tools, to investigate limits and potentials of this innovative approach and to comprehend in which way the analyzed municipalities are influencing the city of Naples and vice versa.

3.2   Processing and Selection of Spatial Indicators

The KME-SDSS, as stated previously, is divided into three main phases strictly linked theirself. These phases are [35]: (i) Knowledge, (ii) Mapping, (iii) Evaluation and they have been shown in a workflow diagram (Figure1).

The first phase (i) consists of cognitive framework construction through the selection of spatial indicators and their organization into multiple levels categories. At a first level of knowledge, the identified indicators have been classified into two main categories that represent the *ecosystem* and *anthropic* services. Subsequently, at a second level, for both previous macro-categories different services typologies have been identified. Specifically, for *ecosystem services,* two sub-categories have been identified, which are *provision* and *regulation services*, while for *anthropic* services, four sub-categories have been selected, which are *infrastructure, soil, habitation* and *recreation services*.

Provisions. include all environmental resources supply within the ecological system in support of anthropic system, while *regulations.* are related to the ability of landscape to regulate ecological processes in order to promote natural cycles. *Infrastructures* are linked to the capacity of landscape to provide a suitable substrate for transportation; *habitations* are connected to the capacity of landscape to provide suitable places for human living; *soils* are related to the capacity of landscape to provide a suitable substrate to different services and *soil* use and *recreations* contain all services referring to cultural sites and leisure activities [4], [6, 7], [12], [36].

Each *landscape service* has been explicated, mapped and classified into a spatial indicators set. The hierarchical structure of indicators set shows multiple levels of analysis and describes: three classification criteria, source, reference year, unit of measure and value (Table 1).

Table 1. The spatial indicators set

Criteria 1th level	Criteria 2nd level	Criteria 3rd level	Year	Source	Analysis distance (m)	Value	Unit of Measure
Ecosystem services	Provision services	Waterways	2012	Open Source (VdsTech)	300	33.027	m
		Mineral extraction sites	2009	Urban Atlas, EEA	100	1.968.530	m²
		Agricolture and wetlands areas	2009	Urban Atlas, EEA	1000	222.905.504	m²
	Regulation services	Sic-Zps areas	2012	Natura 2000	2500	89.135.728	m²
		Water bodies	2009	Urban Atlas, EEA	300	5.566.472	m²
		Forests	2009	Urban Atlas, EEA	2500	22.584.628	m²
		Land without current use	2009	Urban Atlas, EEA	100	6.007.729	m²
		Green urban areas	2009	Urban Atlas, EEA	2000	6.484.058	m²
	Infrastructure services	Railways	2012	Open Source (VdsTech)	100	82.831	m
		Roads	2012	Open Source (VdsTech)	100	4.369.847	m
		Airport	2009	Urban Atlas, EEA	1000	2.369.115	m²
		Port areas	2009	Urban Atlas, EEA	1000	2.405.899	m²
	Recreation services	Cultural sites	2014	Elaborazioni Gis	2000	5.278.486	m²
		Sport and leisure	2009	Urban Atlas, EEA	500	6.277.319	m²
		Park areas	2009	Urban Atlas, EEA	1000	6.484.058	m²
Anthropic services	Habitation services	Habitation density	2009	Urban Atlas, EEA	100	157.187.020	m²
	Soil services	Isolated structures	2009	Urban Atlas, EEA	100	1.412.309	m²
		Construction sites	2009	Urban Atlas, EEA	100	2.390.998	m²
		Waste disposal	2012	Open Source (VdsTech)	100	1.356.218	m²
		Industrial, commercial, and other use	2009	Urban Atlas, EEA	100	84.527.681	m²

The second phase (ii) consists of the processing, through GIS editing tools, of the spatial indicators related to the different categories of considered services. The selected shapefiles have been georeferenced according to a coordinate system in order to proceed to overlay multiple levels of information that will be evaluated in the last stage of the model.

In this phase we chose WGS84 as common geographic coordinates system, operating spatial adjustment for the data having a different projective coordinates.

The Mapping phase is essential since it allows to represent the different indicators homogeneously through the attribution of common parameters that can facilitate a better understanding.

At this level, in order to include spatial areal indicators homogeneous to punctual ones, we consider the Istat census areas on which points insist as a reference surfaces, thus by choosing the surfaces as a unambiguous unit of measurement that makes possible a subsequent evaluation phase.

In the last stage (iii), the *landscape services* have been compared with each other by the AHP evaluation method in order to set priorities and get the two final maps that represent the output of the KME-SDSS.

The indicators represented by GIS maps have been compared in pairs according to their level of classification and to environmental issues they describe. Weights have been assigned to each indicator by the fundamental scale of Saaty [37,38,39], considering their level of importance in performing anthropic and ecosystem functions. The final output of the evaluation process consists of two maps that indicate for each macro-category the different scale of services density (low, low-medium, medium, medium-high or high density). These maps can be considered a decision-making tool to guide territorial changing choices toward planning management [40,41] (Figure 2).

3.3   Weighting Phase and Software MASCOT

More specifically, in the final stage of the process (iii), after the decisional-tree has been established and the maps of spatial indicators for each sub-category has been constituted, we proceed to overlay maps to make pairwise comparisons between elements, respecting the hierarchy of the decision-making structure, in order to establish the thickness of considered *landscape services*. Among the multi-criteria analysis for decision-making purposes (MCDA) we chose the Analytical Hierarchy Process method (AHP), because it allows relating quantitative and qualitative analysis, which are difficult to compare, and combining heterogeneous measures in a single scale value [37]. This method, which is particularly useful in problems with a large number of indicators, consists of three phases related to construction of a suitable hierarchy, establishment of priorities between elements in the hierarchy through pairwise comparisons and check of the logical consistency of pairwise comparisons [37], [39].

For the evaluation phase, we use Multi-criteria Analytical Scoring Tool (MASCOT) in integration with GIS as a tool based on spatial analysis [42]. In the weighting process we compare each sub-criteria either by direct method (direct input) or the AHP method. MASCOT software allows defining a distance analysis on which the assessment has been based. The distance analysis is a fundamental parameter to evaluate the spatial impact that each criterion has on the context, indeed if the scoring shape file is a vector, features of the base layer that overlap scored features are associated the maximum value weight, which decreases linearly up to the limit of distance analysis, where it will be equal to zero. The distance analysis aims to contain the indicator effects in neighboring zones, making a simplification and defining core-areas that produce distance-decay in which the indicator loses gradually its intensity. The core area, indeed, «represent the area characterized by the absence of edge effects extending from surrounding areas» [43].

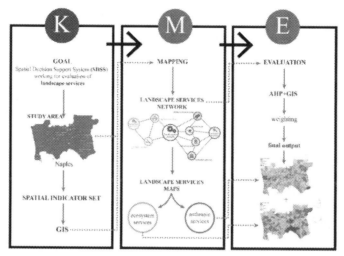

Fig. 1. KME-SDSS workflow

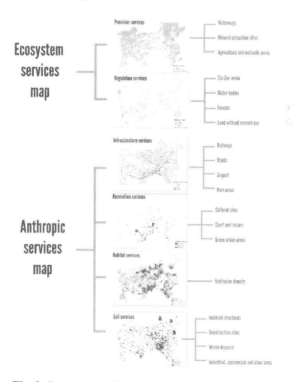

Fig. 2. *Ecosystem and anthropic services* dendrogram

4   Results

After having processed maps for each indicator and weights and priorities have been determined, all the information must be combined in order to obtain final maps of territorial transformation, related to ecosystem and anthropic categories, which consider the overall impact of the *landscape services* on the study area to guide the decision-makers choices towards a vision of integrated sustainability (Figures 3, 4).

The gradual scale of colors, from red to green, allows to identify areas with a different intensity of services, considering the red as low-density and green as high one, while the others mean intermediate judgments. Analysis of the final maps shows that low concentration of *ecosystem services* in the city of Naples is balanced by a high presence of anthropic ones. These services also include fragmented green urban spaces that, although not providing the same level of *regulation* services of green areas surrounding the larger city walls, can also play an important role in the emissions drain due to city congestion and pollution. *Ecosystem services* map also shows that Naples might be considered as a watershed between two relevant systems for provisioning and *regulation* services. However, the enhancement of green urban areas provides an opportunity for linking these systems through the creation of new green infrastructures for their connection or through the consolidation of existing ones. The ecological fragmentation and land consumption are two of the major problems in urban issues, which are accentuated by sprawl and new construction in already overpopulated context. Another interesting result which can be deduced from the observation of the *ecosystem* s. map, and partially from the *anthropic* one, is connected to the conformation of intermediate areas that delimit the urban belt. These filter zones play a key role in the planning processes because they are the most important transformable areas giving benefits to other ones. Thus, if sustainable types of transformations will be done, these areas will be able to influence indirectly the enhancement of ecosystem and *anthropic* services in the city.

Fig. 3. *Ecosystem services* map

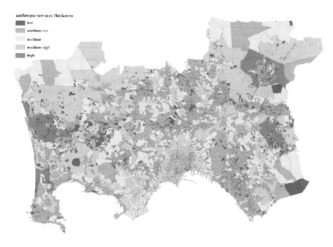

Fig. 4. *Anthropic services* map

5   Conclusions

In summary, the present study aims at the development of a KME-SDSS to assess the landscape through its services, considering it as a multidimensional and dynamic system made up of multiple components whereby an integrated approach between GIS and AHP was essential to spatially explicit the selected indicators on the one hand and prioritize them on the other hand.

Thus, it was possible to improve awareness of the complexity and to provide a framework and to express the potential of this approach in spatial planning field.

The proposed SDSS shows that, through the spatial representation of indicators, it is possible determine distribution, quantification and interrelationships of phenomena. The GIS-AHP synergy let to make a multi-level analysis in order to better display and map the complex values of the landscape. Thus, the data can be organized in a specific set of spatial indicators able to describe the territory through its *anthropic* and *ecosystem services* and simplify the evaluation process. Finally, the dynamic change of the landscape elements requires flexible tools in order to update the information continuously and the decision-makers can have a clear knowledge of their territory in order to identify and implement responsible actions.

The results allow to consider the used methodology as an initial basis to implement opportunities offered by analysis and evaluation of *landscape services* within a wider decision-making process that is capable of integrating different skills in order to improve the data entirety and the overall evaluation process coherence.

Acknowledgment. The authors would like to thank Professor Maria Cerreta, Department of Architecture, University of Naples Federico II, for her precious and collaborative support in structuring and checking coherence of this research.

References

1. Battisti, C., Romano, B.: Frammentazione e connettività. Dall'analisi ecologica alle strategie di pianificazione. Città Studi, Torino (2007)
2. IPCC (Intergovernmental Panel on Climate Change), Climate Change 2014: Impacts, Adaptation, and Vulnerability. http://www.ipcc.ch/report/ar5/wg2/
3. European Commission: Life and soil protection, Publications Office of the European Union, Luxembourg (2014)
4. Hermann, A., Kuttnera, M., et al.: Assessment framework for landscape services in European cultural landscapes: An Austrian Hungarian case study. Ecological Indicators **37**, 229–240 (2014)
5. Costanza, R., et al.: The value of the world's ecosystem services and natural capital. Nature **387**, 253–260 (1997)
6. De Groot, R.S.: Functions of Nature: Evaluation of Nature in Environmental Planning, Management and Decision Making. Wolters-Noordhoff, Groningen (1992)
7. De Groot, R., Wilson, M.A., Boumans, R.M.J.: A typology for the classification, description and valuation of ecosystem functions, goods and services. Ecological Economics **41**, 393–408 (2002)
8. MA (Millennium Ecosystem Assessment): Ecosystems and Human Well-being: The Assessment Series (Four Volumes and Summary). Island Press, Washington, DC (2005)
9. TEEB (The Economics of Ecosystems and Biodiversity): Mainstreaming the Economics of Nature: A synthesis of the approach, conclusions and recommendations of TEEB. Progress Press, Malta (2010)
10. Termorshuizen, J.W., Opdam, P.: Landscape services as a bridge between landscape ecology and sustainable development. Landscape Ecol. **24**, 1037–1052 (2009)
11. Fusco Girard, L.: Risorse architettoniche e culturali: valutazioni e strategie di conservazione. Angeli, Milano (1990)
12. De Groot, R.S.: Function-analysis and valuation as a tool to assess land use conflicts in planning for sustainable, multi-functional landscapes. Landscape Urban Plan. **75**, 175–186 (2006)
13. Bartel, A.: Analysis of landscape pattern: towards a 'top down' indicator for evaluation of landuse. Ecological Modelling **130**, 87–94 (2000)
14. Cerreta, M., Mele, R.: A landscape complex values map: integration among *Soft* values and *Hard* values in a spatial decision support system. In: Murgante, B., Gervasi, O., Misra, S., Nedjah, N., Rocha, A.M.A., Taniar, D., Apduhan, B.O. (eds.) ICCSA 2012, Part II. LNCS, vol. 7334, pp. 653–669. Springer, Heidelberg (2012)
15. Cerreta, M., Poli, G.: A Complex Values Map of Marginal Urban Landscapes: An Experiment in Naples (Italy). International Journal of Agricultural and Environmental Information Systems **4**, 41–62 (2013)
16. Turner, M.G., Gardner, R.H., O'Neill, R.V.: Landscape ecology in theory and practice. Springer-Verlag, New York (2001)
17. Capra, F.: La rete della vita. Bur Rizzoli, Pordenone (2010)
18. Jones, K.B., et al.: Landscape approaches to assess environmental security: summary, conclusions, and recommendations. In: Petrosillo, I., et al. (eds.) Use of Landscape Sciences for the Assessment of Environmental Security, pp. 475–486. Springer, Dordrecht (2008)
19. Van Eetvelde, V., Antrop, M.: A stepwise multi-scaled landscape typology and characterization for trans-regional integration, applied on the federal state of Belgium. Landscape and Urban Planning **91**, 160–170 (2009)
20. Council of Europe: European landscape convention. Florence, Italy (2000)

21. Antrop, M.: The language of landscape ecologists and planners: a comparative content analysis of concepts used in landscape ecology. Landscape Urban Plan. **55**, 163–173 (2001)
22. Forman, R., Godron, M.: Landscape Ecology. Wiley & Sons, New York (1986)
23. Ferrari, C., Pezzi G.: L'ecologia del paesaggio. Il Mulino, Bologna (2012)
24. Malczewski, J., Ogryczak, W.: The multiple criteria location problem: 1. A generalized network model and the set of efficient solutions, Environment and Planning A **27**, 1931–1960 (1995)
25. Malczewski, J., Ogryczak, W.: The multiple criteria location problem: 2. Preference-based techniques and interactive decision support. Environment and Planning A **28**, 69–98 (1996)
26. Malczewski, J.: GIS-based multicriteria decision analysis: a survey of the literature. International Journal of Geographical Information Science **20**, 703–726 (2006)
27. Feizizadeh, B., Jankowski, P.: A GIS based spatially-explicit sensitivity and uncertainty analysis approach for multi-criteria decision analysis. Computers & Geosciences **64**, 81–95 (2014)
28. Campagna, M.: GIS for Sustainable Development. Taylor & Francis Group, LLC, USA (2006)
29. Higgs, G.: GIS for Environmental Decision-Making. Andrew Lovett and Katy Appleton, USA (2008)
30. Li, Y., Shen, Q., Li, H.: Design of spatial decision support systems for property professionals using MapObjects and Excel. Autom. Constr. **13**, 565–573 (2004)
31. Kordi, M., Brandt, S.A.: Effects of increasing fuzziness on analytic hierarchy process for spatial multicriteria decision analysis. Computers Environment and Urban Systems **36**, 43–53 (2012)
32. Chen, Y., Yu, J., Khan, S.: Spatial sensitivity analysis of multi-criteria weights in GIS-based land suitability evaluation. Environmental Modelling & Software **25**, 1582–1591 (2010)
33. Fusco Girard, L., Cerreta, M., De Toro, P.: Analytic Hierarchy Process (AHP) and Geographical Information Systems (GIS): an Integrated Spatial Assessment for Planning Strategic Choices. International Journal of the Analytic Hierarchy Process **4**, 4–26 (2010)
34. Funtowicz, S.O., Ravetz, J.: A New Scientific Methodology for Global Environmental Issues. Ecological Economics. The Science and Management of Sustainability, pp. 137–152. Columbia University Press, New York (1991)
35. Attardi, R., Cerreta, M., Franciosa, A., Gravagnuolo, A.: Valuing cultural landscape services: a multidimensional and multi-group SDSS for scenario simulations. In: Murgante, B., Misra, S., Rocha, A.M.A., Torre, C., Rocha, J.G., Falcão, M.I., Taniar, D., Apduhan, B.O., Gervasi, O. (eds.) ICCSA 2014, Part III. LNCS, vol. 8581, pp. 398–413. Springer, Heidelberg (2014)
36. Ferretti, V., Pomarico, S.: Ecological land suitability analysis through spatial indicators: An application of the Analytic Network Process technique and Ordered Weighted Average approach. Ecological Indicators **34**, 507–519 (2013)
37. Saaty, T.L.: The Analytical Hierarchy Process. McGraw Hill, New York (1980)
38. Saaty, T.L.: Decision making for leaders: The analytic hierarchy process for decisions in a complex world. RWS Publications, Pittsburgh (1999)
39. Saaty, T.L., Vargas, L.G.: Models, methods, concepts and applications of the Analytic Hierarchy Process. Kluwer Academic Publisher, Dordrecht (2001)
40. Keeney, R.L.: Value-focused Thinking: A Path to Creative Decision Making. Harvard University Press, Cambridge (1992)

41. Girard, L.F., Torre, C.M.: The Use of Ahp in a Multiactor Evaluation for Urban Development Programs: A Case Study. In: Murgante, B., Gervasi, O., Misra, S., Nedjah, N., Rocha, A.M.A., Taniar, D., Apduhan, B.O. (eds.) ICCSA 2012, Part II. LNCS, vol. 7334, pp. 157–167. Springer, Heidelberg (2012)
42. Lacroix, P., Santiago, H., Ray, N.: MASCOT: Multi-Criteria Analytical SCOring Tool for ArcGIS Desktop. International Journal of Information Technology & Decision Making **13**, 1135–1159 (2014)
43. Geneletti, D.: Using spatial indicators and value functions to assess ecosystem fragmentation caused by linear infrastructures. International Journal of Applied Earth Observation and Geoinformation **5**, 1–15 (2004)

Workshop on Mobile Communications (MC 2015)

A Quorum-Based Adaptive Hybrid Location Service for VCNs Urban Environments

Ihn-Han Bae[1(✉)] and Jeong-Ah Kim[2]

[1] School of IT Engineering, Catholic University of Daegu, 13-13 Hayang, Gyeongsan,
Gyeongbuk 712-702, Republic of Korea
ihbae@cu.ac.kr
[2] Department of Computer Education, Catholic KwanDong Umiversity, 597,
Beomil, Gangneung, Gangwon 210-701, Republic of Korea
clara@cku.ac.kr

Abstract. Location information services or location management systems in VCN (Vehicular Communication Networks) are used to provide location information about vehicles such as current location, speed, direction and report this information to other vehicles or network entities that require this information. This paper designs firstly an EDQS (Extended Dynamic Quorum System) which is a logical structure for location management of VCNs. Secondly, we propose a QAHLS (Quorum-based Adaptive Hybrid Location Service) on the basis of the EDQS which uses direct location scheme mixed with indirect location scheme according to the location preference and the location mobility for a vehicle. The performance of QAHLS is evaluated by an analytical model and compared with that of existing GQS (Grid Quorum System) based and DQS (Diamond Quorum System) based location services.

Keywords: Direct location scheme · Indirect location scheme · Location services · Quorum systems · VCN

1 Introduction

Vehicular Communication Networks (VCNs) are used to supply a communication platform for Intelligent Transportation Systems (ITSs) services also for value added services in different road systems. To build up a VCN, various entities are used among others: vehicles, roads infrastructure such as intelligent traffic lights, traffic signs, RSUs (Road Side Units) and wired or wireless backhaul networks that are applied to provide communication between these entities and to enable interconnection with the Internet. VCNs are responsible for the communication between moving vehicles in a certain environment. A vehicle can communicate with another vehicle directly which is called Vehicle to Vehicle (V2V) communication, or a vehicle can communicate to an infrastructure such as a RSU, known as Vehicle-to-Infrastructure (V2I) [1, 2].

During the past few years, ITS have become a significant topic in research and development as their ambit is to enhance safety and efficiently in transportation systems via the use of advanced technologies [3]. ITSs are expected to offer fundamental

© Springer International Publishing Switzerland 2015
O. Gervasi et al. (Eds.): ICCSA 2015, Part IV, LNCS 9158, pp. 79–89, 2015.
DOI: 10.1007/978-3-319-21410-8_6

services, which can be distinguished into ITS services and value added services. Some examples of ITS services are pre/post-crash warning and intelligent traffic control. Internet service provisioning, intelligent fleet management or detailed real-time traffic flow information and advanced navigation services are some examples for value added services [4].

In traditional positioning systems, location information has typically been taken by a device and with the help of a satellite system, that is a global positioning system (GPS) receiver, so each vehicle has only its own knowledge. With the help of location information systems (LISs) this information can be distributed and shared among other vehicles. Vehicle's location is an important matter in this data service world: it allows us to understand completely new service concept (e.g., tracking applications). Moreover, it has the capability to make many messaging and vehicle Internet services more relevant to clients as information adjusted to context (e.g., weather information adjusted to the region one in it). In addition location information can considerably improve service usability [1].

Vehicle is a node moving at a high speed and to design a reliable transmission should be an urgent and hard goal. A robust routing protocol can achieve this goal. In VCNs, routing protocols usually depend on accurate location information of mobile node. Hence, location information provision is a basic and important service. Accordingly, we design firstly an EDQS which is a logical structure for location management of VCNs. Secondly, we propose a QAHLS on the basis of the EDQS which uses direct location scheme mixed with indirect location scheme according to the location preference and the location mobility for a vehicle. The performance of QAHLS is evaluated by an analytical model, and compared with that of existing Grid Quorum System (GQS) based and Diamond Quorum System (DQS) based location services.

The rest of this paper is organized as follows. Section 2 gives a brief description of related works for location services in VCNs. Section 3 describes an EDQS-based QAHLS for VCNs urban environments. Section 4 presents the performance of QAHLS that is evaluated through an analytical model. Finally, Section 5 concludes this paper and describes our future works.

2   Related Works

As we mentioned before LIS should provide location information about vehicles. For this purpose, LIS need to gather, store, analyze and distribute these location information to vehicles or network entities that require this information. Some positioning systems such as GPS are used for gathering position information of vehicles, when a vehicle enters to cover range of some RSUs, sends its position to one of the RSUs and consequently that RSU sends this information to location servers to maintain and store location information of vehicles. Now we have a large database with position information of vehicles and also a timestamp for each of this information that has been generated. So we have opportunity to analyze them and predict next location of vehicle for some applications such as vehicle tracking, improve handoff latency and mobility aware forwarding in advance [1].

The location-based services can be classified according to Fig. 1 into two classes: flooding-based and rendezvous based [5]. The first class is composed of reactive and proactive services. In the proactive flooding-based location service, very node floods its geographic information through the whole network periodically. Thus, all nodes are able to update their location tables. In the second class (rendezvous-based location service), all the nodes agree on a unique mapping of a node to other specific nodes. The geographic information is disseminated through the elected nodes called the location servers. Thus, the location-based services consist of two components:

- Location Update: A node has to recruit location servers and, then, needs to update its location through theses servers.
- Location Request: A node, seeking the location information of another node, broadcasts a location request. The location server will replay as soon as it receives this request.

There are two major approaches for the rendezvous-based location services. The first approach, the quorum-based approach, the location update is sent to a group of nodes (update quorum). The location request is sent to a different group (request quorum). These two groups are not necessarily disjoint. Quorum systems have been used to implement a wide variety of distributed objects and services. Typical quorum systems include GQS [6, 7], DQS [7, 8] and etc. The second approach is the hierarchical approach. In this approach, the network is divided into several levels. At each level, a node recruits location servers. The location request is forwarded up and down in the hierarchy. This limits the forwarded packets and avoids flooding.

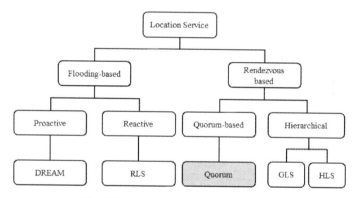

Fig. 1. Location-based services taxonomy

3   QAHLS Design

In section 3, we design not only the EDQS that is the logical structure of the VCN for QAHLS but also the QAHLS protocol for VCN urban environments.

3.1   EDQS

Quorum is grouping the nodes or sites. There are two types of quorum that are the read and write quorums. Replicating data at different nodes can enhance data availability and fault tolerance. When data are replicated, consistency and high data availability is required whereas at the same time the network communication cost is reduced.

Diamond protocol [7, 8] had proposed for managing replicated data. In this protocol, the nodes in the network are logically organized into a two-dimensional diamond structure. The protocol can be viewed as a specialized version of the grid protocol because the logical structure can be seen as a grid with holes. There are two main properties of the diamond protocol that make it a good choice for replicated data management:

- Compared with the majority quorum, tree quorum and grid quorum protocols, the diamond protocol results in the greatest number of disjoint read quorums which shall lead to a better throughput and response time.
- The protocol achieves the smallest optimum read quorum size and the second smallest write quorum size among the above protocols.

The location information of hot vehicles which location information is requested frequently by other vehicles should be managed in many nodes, while the location information of cold vehicles should be managed in small nodes. Thus, the write quorums of various sizes are needed to location information management for VCNs. Accordingly, we propose an EDQS that improves the second property of diamond protocol.

The proposed QAHLS is based on the EDQS which is logical structure of RUSs. To design EDQS, we firstly divide the VCN into grids and organize the grids into two hierarchical levels in the same manner as region-based Hierarchical Location Service with Road-adapted Grids (HLSRG) [9]. Fig. 2 shows the hierarchical framework of regions in VCN, where four level 1 regions form a region of level 2, and any level 1 region can belong to only one region of level 2. In each region of level 1, an intersection which is nearest to the center of the region is selected to be the center of region of level 1. Region centers have to collect location update packets in their regions and transmit the packets to their corresponding upper level. RSUs are established to maintain the location information for region of level 2. RSUs in each region of level 2 are wired connect to other RUSs where are in other regions of level 2.

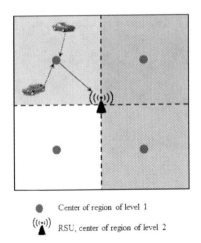

●   Center of region of level 1

((•))
▲   RSU, center of region of level 2

Fig. 2. Hierarchical framework of regions

The EDQS are constructed by using the RSUs, centers of regions of level 2. In EDQS, all nodes are logically organized into two-dimensional diamond structure. The EDQS provides 4 update quorums: Geographic Quorum (GQ), Intersection Quorum (IQ), ROw-based Quorum (ROQ) and Column-based Quorum (CQ) such as Fig. 3.

- GQ: The GQ is composed of all location nodes in sub-dimensional diamond which covers a local location node.
- IQ: The IQ is composed of the location nodes which intersect among sub-dimensional diamonds.
- ROQ: The ROQ construction starts the line horizontally along the row first. When there are no more location nodes to join on the right, the line is turned 90 degree to the bottom.
- CQ: The CQ construction starts the line vertically along the column first. When there are no more location nodes to join on the bottom, the line is turned 90 degree to the right.

Also, we can choose an arbitrary location node of each row to form a Request Quorum (REQ).

If the number of location nodes on a side of EDQS is m, the total number of location nodes or RSUs is $L=m^2$, and the number of GQs on a side is $k = \lfloor \sqrt{m} \rfloor$. The size of each quorum is as follows:

- The size of GQ, $q_g = \left(\left\lceil \frac{m}{2} \right\rceil \right)^2$.
- The size of IQ, $q_i = m$.
- The size of ROQ, $q_{ro} = m$
- The size of CQ, $q_c = m$.
- The size of REQ, $q_{re} = m$.

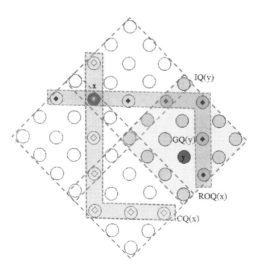

Fig. 3. Structure of EDQS

3.2   QAHLS

We design QAHLS that is an adaptive hybrid location service on the base of EDQS. The QAHLS takes advantages of the mobility and the preference of vehicles. Accordingly, we assume that each vehicle knows its location preference and its location mobility rates through the number times its location is requested and the number times its location is updated per hour. According to the mobility and the preference for a vehicle, QAHLS uses not only other location update schemes: direct and indirect but also other write group quorums.

QAHLS places location information for a vehicle V in a set of regions of level 2. We call these regions as the responsible regions (RR) of V. To update its location centers according to the direct location scheme, a vehicle computes its responsible regions. Position updates scheme is called direct because a location center of RR directly knows the position of the vehicle. The network load can be reduced with indirect location scheme where the only local location center on the region of level 2 where a vehicle is located knows the position of the vehicle, while the rest location centers in its responsible regions only know the region of level 2 where a vehicle is located. More precise location information is not necessary on the region of level 2. The pointers which represent the location information do no longer point to the last known position. They point to the responsible region on the region of level 2.

Three update group quorums: UGQ_{hh}, UGQ_{hc} and UGQ_m from the quorums of EDQS are formed. Location update of QALHS uses three $UGQs$ and the ROQ from EDQS. While location request of QALHS uses only the REQ from EDQS. The construction of three WGQs is as follows:

- $UGQ_{hh} = IQ \cup ROQ \cup CQ$
- $UGQ_{hc} = GQ \cup ROQ \cup CQ$
- $UGQ_m = ROQ \cup CQ$

All nodes for location information management in VCN perform the location update and the location request procedures of QALHS such as Fig. 4 and 5, where λ_x and μ_x represent the location preference and the location mobility rates of vehicle x, δ_{λ_h} and δ_{λ_l} represent the threshold values of high preference and low preference rates for vehicles, and δ_{μ_h} and δ_{μ_l} represent the threshold values of high mobility and low mobility rates for vehicles, respectively. Additionally, $UGQ(x)$, $ROQ(x)$ and $REQ(x)$ represent the quorums for the region of level 2 which cover vehicle x.

```
LocUpdate(L, x)
/* Update a location information L from vehicle x */
```

Event: change the location of vehicle x.
If $(\lambda_x \geq \delta_{\lambda_h})$ **then**

    If $(\mu_x \leq \delta_{\mu_l})$ **then** /* direct location scheme */
        Send LocUpdate(L, x) packet to all location centers of $UGQ_{hc}(x)$
          for the region center of level 2;
    Else if $(\mu_x \geq \delta_{\mu_h})$ **then**
        Send LocUpdate(L, x) packet to all location centers of $UGQ_{hh}(x)$
          for the region center of level 2;
    Else
        Send LocUpdate(L, x) packet to all location centers of $UGQ_m(x)$
          for the region center of level 2;
    End if
Else
    If (the location region of level 2 for the vehicle is changed) **then**
        If $(\lambda_x \leq \delta_{\lambda_l})$ **then** /* indirect location scheme */
          Send LocUpdate(the region center of level 2 for L, x) packet
            to all location centers of $ROQ(x)$ for the region center of
            level 2;
        Else /* indirect location scheme */
          Send LocUpdate(the region center of level 2 for L, x) packet
            to all location centers of $UGQ_m(x)$ for the region center of
            level 2;
        End if
    End if
End if

Event: receive a LocUpdate(L, x) packet or a LocUpdate(the region center of level 2 for L, x) packet.
Store the (L, x) or the (the region center of level 2 for L, x) in the received node;

Fig. 4. Location update procedure of QAHLS

```
LocRequest(x, y)
/* Request location information of vehicle y from vehicle x */
```

Event: request the location of vehicle y
If (the location information of y, Loc(y) is found in the local region
    center of level 2 of x) **then**
  Return the Loc(y) to vehicle x;
Else if (the local center of the region center of level 2 has the pointer for Loc(y))
  Send LocRequest(x, y) packet to the pointer for Loc(y);
Else
  Send LocRequest(x, y) packet to all location centers in REQ(x) for
    the region center of level 2;
End if

Event: receive a LocRequest(x, y) packet
If (the Loc(y) is found in the received center) **then**
  Return the Loc(y) to vehicle x;
Else if (the pointer for Loc(y) is found in the received center) **then**
  Send LocRequest(x, y) packet to the pointer for Loc(y);
End if

Fig. 5. Location request procedure of QAHLS

4   Performance Evaluation

The performance of QAHLS is evaluated by an analytical model, where the performance of QAHLS is evaluated by average location management cost that adds the location update and location request costs. Average location management cost, LCM_{ave} is computed as follows:

$$LCM_{ave} = \left(P_{s\_hh}C_{u\_hh} + P_{s\_hl}C_{u\_hl} + P_{s\_hm}C_{u\_hm} + P_{s\_l}C_{u\_l} + P_{s\_m}C_{u\_m}\right)$$
$$\frac{\lambda}{\mu}\left(P_{t\_hh}C_{r\_hh} + P_{t\_hl}C_{r\_hl} + P_{t\_hm}C_{r\_hm} + P_{t\_l}C_{r\_l} + P_{t\_m}C_{r\_m}\right) \qquad (1)$$

Where $\frac{\lambda}{\mu}$ represent call-to-mobility ratio, and P_{s_hh}, P_{s_hl} and P_{s_hm} represent the probabilities that location preference of the source vehicle which is updating its location is high and its location mobility is high or low or medium, and $P_{s\_h} = P_{s\_hh} + P_{s\_hl} + P_{s\_hm}$. Accordingly, P_{s_h}, P_{s_l} and P_{s_m} represent the probabilities that the location preference for a source vehicle is high or low or medium, $P_{s\_h} + P_{s\_l} + P_{s\_m} = 1$. P_{t_hh}, P_{t_hl} and P_{t_hm} represent the probabilities that location preference of the target vehicle which is requested its location is high and its location mobility is high or low or medium, and $P_{t\_h} = P_{t\_hh} + P_{t\_hl} + P_{t\_hm}$. Accordingly, P_{t_h}, P_{t_l} and P_{t_m} represent the probabilities that the location preference for a target vehicle is high or low or medium, $P_{s\_h} + P_{s\_l} + P_{s\_m} = 1$. Also, C_{u_hh}, C_{u_hl} and C_{u_hm} represent the location update costs that the location preference of the source vehicle is high and its location mobility is high or low or medium, C_{u_l} and C_{u_m} represent the location update cost that the location preference for a source vehicle is low or medium. And C_{r_hh} C_{r_hl} and C_{r_hm} represent the location request costs that the location preference of the target vehicle is high and its location mobility is high or low or medium, C_{r_l} and C_{r_m} represent the location request cost that the location preference for a target vehicle is low or medium.

These costs can be computed as follows:

$$C_{u\_hh} = C_{local\_u\_cost} + \left(q_{ughh} - 1\right)C_{far\_u\_cost}$$
$$C_{u\_hc} = C_{local\_u\_cost} + \left(q_{ughc} - 1\right)C_{far\_u\_cost}$$
$$C_{u\_hm} = C_{local\_u\_cost} + \left(q_{ughm} - 1\right)C_{far\_u\_cost}$$
$$C_{u\_c} = P_{upadte\_r1}C_{local\_u\_cost} + P_{update\_r2}\left(q_{ro} - 1\right)C_{far\_u\_cost}$$
$$C_{u\_m} = P_{update\_r1}C_{local\_u\_cost} + P_{update\_r2}\left(q_{ugm} - 1\right)C_{far\_u\_cost}$$
$$C_{r\_hh} = \frac{q_{ughh}}{L}C_{local\_r\_cost} + \left(1 - \frac{q_{ughh}}{L}\right)q_{re}C_{far\_r\_cost}$$
$$C_{r\_hc} = \frac{q_{ughc}}{L}C_{local\_r\_cost} + \left(1 - \frac{q_{ughc}}{L}\right)q_{re}C_{far\_r\_cost}$$
$$C_{r\_hm} = \frac{q_{ughm}}{L}C_{local\_r\_cost} + \left(1 - \frac{q_{ughm}}{L}\right)q_{re}C_{far\_r\_cost}$$
$$C_{r\_c} = \frac{q_{ro}}{L}\left(C_{local\_r\_cost} + C_{far\_r\_cost}\right) + \left(1 - \frac{q_{ro}}{L}\right)q_{re}2C_{far\_r\_cost}$$
$$C_{r\_c} = \frac{q_{ugm}}{L}\left(C_{local\_r\_cost} + C_{far\_r\_cost}\right) + \left(1 - \frac{q_{ugm}}{L}\right)q_{re}2C_{far\_r\_cost}$$

In above cost equations, $q_{ug_{hh}}, q_{ug_{hc}}, q_{ug_{hm}}$ represent the sizes of update group quorums: UGQ_{hh}, UGQ_{hc} and UGQ_{hm}.

The parameters and values for the performance evaluation of QAHLS are shown in Table 1, where P_{update_r2} and P_{update_r2} represent the probability of location update to the region of level 1 and the probability of location update to the region of level 2, respectively.

Table 1. Parameters and values for performance evaluation

Parameter	Value
m, k	7, 2
L	49
The number of regions in a region of level 2, g	4
The total number of regions of level 1, N	196
$P_{update\_r1}, P_{update\_r2}$	0.5, 0.5
q_g	16
q_i, q_{ro}, q_c, q_{re}	7
$q_{ug_{hh}}, q_{ug_{hc}}$	23
q_{ug_m}	13
$C_{local\_u\_cost}, C_{local\_r\_cost}$	1
$C_{far\_u\_cost}, C_{far\_r\_cost}$	2

In Table 1, we use stochastic vehicular mobility [10] that is the urban section model as a vehicular mobility model. The stochastic vehicular mobility constrains vehicles movement on a grid-shaped road topology, in which all edges are considered to be bidirectional, single-lane road. Vehicles randomly select one of the intersections of the grid as their destination and move towards it at constant speed, with one horizontal and one vertical movement. In case stochastic vehicular mobility, the value of g is 4, P_{update_r2} and P_{update_r2} are 0.5, respectively. If $500m \times 500m$ grids are region level 1, the QAHLS can manage vehicle location information in an urban of $49 Km^2$ size because total number of regions of level 1 is 196.

Fig. 6 shows average location management cost over call-to-mobility ratio (CMR). As shown in the Fig. 6, the average location management cost of QAHLS is less than that of the existing quorum-based location services regardless of call-to-mobility ratio. In the legend QAHLS(x_1, x_2, y_1, y_2) of Fig. 6, parameters x_1, x_2, y_1 and y_2 represent $P_{s\_h}, P_{s\_c}. P_{t\_h}$ and P_{t_c} respectively, where the P_{s_h} and the P_{t_h} of QAHLS(0.3, 0.5, 0.8, 0.05) have the following detailed probabilities: $P_{s\_hh} = 0.15$, $P_{s\_hl} = 0.1$, $P_{t\_hh} = 0.6$ and $P_{t\_hl} = 0.05$, the P_{s_h} and the P_{t_h} of QAHLS(0.25, 0.6, 0.7, 0.1) have the following detailed probabilities: $P_{s\_hh} = 0.05$, $P_{s\_hl} = 0.15$, $P_{t\_hh} = 0.5$ and $P_{t\_hl} = 0.05$. Also, the performance of QAHLS(0.25, 0.6, 0.7, 0.1) is little better than that of QAHLS(0.3, 0.5, 0.8, 0.05). From the results, we confirm that as the QAHLS has less P_{s_h} and P_{t_h} and more P_{s_c} and P_{t_c}, the better performance of QALHS is derived.

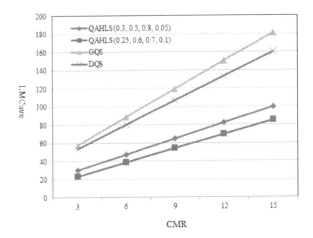

Fig. 6. Average location management cost for vehicular location services

5   Conclusion

We designed the extended dynamic quorum system EDQS for efficiently supporting vehicular location services, also proposed the quorum-based adaptive hybrid location service QAHLS. The proposed location service used direct location scheme mixed with indirect location scheme according to the location preference and the location mobility for a vehicle. The performance of our location service was evaluated by an analytical model. As the results of performance evaluation, the average location management cost of proposed QAHLS was less than that of the existing quorum-based location services regardless of call-to-mobility ratio. And as QAHLS has less P_{s_h} and P_{t_h} and more P_{s_c} and P_{t_c}, the better performance of QALHS was derived. Our future work includes design on a hierarchical adaptive location service considering vehicle characteristics.

References

1. Heidari, E., Gladisch, A., Moshiri, B., Yavangarian, D.: Survey on location information services for Vehicular Communication Networks. Wireless Networks **20**, 1085–1105 (2014)
2. Rehman, S., Khan, M.A., Zia, T.A., Zheng, L.: Vehicular Ad-Hoc Networks (VANETs) - An Overview and Challenges. Journal of Wireless Networking and Communications **3**, 29–38 (2013)
3. Anagnostopoulos, C., Anagnostopoulos, I., Kayafas, E., Lumos, V.: A License Plate-Recognition Algorithm for Intelligent Transportation System Applications. IEEE Transactions on Intelligent Transportation Systems **7**, 377–392 (2006)
4. Sohoch, E., Kargl, F., Weber, M., Leinmuller, T.: Communication patterns in VANETs. IEEE Communication Magazine **46**, 119–125 (2008)

5. Ayaida, M., Fouchal, H., Afilal, L., Ghamri-Doudane, Y.: A comparison of reactive, grid and hierarchical location-based services for VANETs. In: IEEE Vehicular Technology Conference, pp. 1–5. IEEE Press, Piscataway (2012)
6. Mabni, Z., Latip, R.: A comparative study on quorum-based replica control protocols for grid environment. In: Abd Manaf, A., Sahibuddin, S., Ahmad, R., Mohd Daud, S., El-Qawasmeh, E. (eds.) ICIEIS 2011, Part III. CCIS, vol. 253, pp. 364–377. Springer, Heidelberg (2011)
7. Pandey, A.P., Tripathi, B.M.: A novel quorum protocol for improved performance. Proceedings of The 2014 World Congress in Computer Science. Computer Engineering, and Applied Computing, pp. 1–7. UCMSS, San Diego (2014)
8. Fu, A.W.-C., Wong, Y.S., Wong, M.H.: Diamond Quorum Consensus for High Capacity and Efficiency in a Replicated Database System. Distributed and Parallel Databases **8**, 471–492 (2000)
9. Chang, G.-Y., Chen, Y.-Y.: A region-based hierarchical service with road-adapted grids for vehicular networks. In: International Conference on Parallel Processing Workshops, pp. 554–561. IEEE Press, Piscataway (2010)
10. Olariu, S., Weigle, M.C.: Vehicular Networks: From Theory to Practice. Chapman and Hall/CRC, Boca Raton (2009)

Advanced Persistent Threat Mitigation
Using Multi Level Security – Access Control Framework

Zakiah Zulkefli, Manmeet Mahinderjit Singh$^{(\boxtimes)}$,
and Nurul Hashimah Ahamed Hassain Malim

School of Computer Sciences, Universiti Sains Malaysia, Penang, Malaysia
zz14_com061@student.usm.my, {manmeet,nurulhashimah}@usm.my

Abstract. Bring Your Own Device (BYOD) concept has become popular amongst organization. However, due to its portability and information available through social network, BYOD has become susceptible to information stealing attacks such as Advanced Persistent Threat (APT) attack. APT attack uses tricky methods in getting access into the target's machine and mostly motives and stand as a threat to politics, corporate, academic and even military. Various mitigation techniques have been proposed in tackling this attack but, most of them are relying on available information of the attacks and does not provide data protection. Hence, it is challenging in providing protection against APT attack. In this paper, we will investigate on the available mitigation techniques and its problems in tackling APT attack by looking on the root cause of the attack inside BYOD environment. Lastly, based on the information obtained we will propose a new framework in reducing APT attack.

Keywords: Bring Your Own Device (BYOD) model · Advanced Persistent Threat (APT) · MLS · Access control · Framework

1   Introduction

Mobile computing is a new network computing paradigm which allows users to access information and collaborates with others on the move, through combination of wireless communication infrastructure and portable computing devices [1]. Traditionally it focuses on portability of the computing devices, and now it is possible for the convergence of ubiquitous and pervasive computing to bring out the full potential of mobile computing. In this paper, smartphones and mobile will be emphasized as the perfect example of mobile computing.

Back in the early days, the sole purpose of the mobile is to connect people through phone calls, and later through messaging services like Short Messaging Service (SMS) or Multimedia Messaging Service (MMS). In other words, phones in the past were big, bulky, and were a luxury to have, but had only one function: calling. Nowadays, mobiles which are more commonly deemed as smartphones are equipped with interactive services such as file transfer, internet browsing and mailing services, interactive web-based applications and other ubiquitous functions like camera, has marked a huge leap in the course of mobile phone history. Through these remarkable and

© Springer International Publishing Switzerland 2015
O. Gervasi et al. (Eds.): ICCSA 2015, Part IV, LNCS 9158, pp. 90–105, 2015.
DOI: 10.1007/978-3-319-21410-8_7

additional features, it has garnered newfound glory and popularity among users from all walks of life.

The security issue of the mobile computing has become important because people are concerned about the data stored in their mobile phone. Smartphone has become a need in the society and the user stored a lot of personal data in it. The security to protect the data of the mobile devices has highlighted the importance of the security measurement for mobile computing. Beside the confidentiality of the data is at risk, the availability is also one of the worries for the users. Some devices cannot operate without constant connection from wireless network and the stability of the network is questionable.

According to CBS News, out of the 5 billion phones now, 1 billion of them are smartphones [2]. As a consequence, this plethora of appealing features has led to a widespread of diffusion of smartphones, which makes them the ideal targets for attackers and privacy intruder [3]. It is further inferred that the security attacks of smartphones will evolve as in the same trend of that as computers. As a proof that attackers are starting to target mobile platforms, there has been a sharp ascend in the number of reported new mobile Operating System (OS) vulnerabilities: from 115 in 2009 up to 163 in 2010, which makes up 42% more vulnerabilities [4]. There is also a significant increase in attention to security issues from security researchers.

Although the amount of vulnerabilities reported is alarming, companies are still allowing their employees to use their own device which is known Bring Your Own Device (BYOD) policies due to several reasons. As a result, the company is vulnerable to information theft attacks either through the malware or network. One of such attacks is known as Advance Persistent Threat (APT) attack. The main aim of this paper is to propose Advanced Persistent Threat (APT) mitigation using multi level security (MLS) and access control framework. The objectives are; 1) to investigate current security mobile architecture in BYOD and its problem in countering APT attack; 2) briefly discuss about APT attacks and how it can occur in BYOD environment and 3) to propose a new framework to tackle APT attack in BYOD using multi level security and access control.

Differ from traditional attack, APT has wider attack methods, usually comes with the motive of espionage, manipulating user's trust to access into the target computer and hard to detect. Therefore, it is important for us to look into various security architectures in BYOD so that we can propose an effective method to mitigate APT attack. Thus, the paper is organized as follows: - Section 2 describes the literature review in relation to BYOD model and the Mobile Security Reference Architecture (MRSA). In Section 3 we will present the taxonomy of Advanced Persistent Threat and we will present our proposed solution in Section 4. Lastly, Section 5 is set as the conclusion of the paper.

2   Background and Study

Adopting BYOD can be profitable to an organization as it reduces the cost in providing devices to employees. However, it creates challenges in protecting the data. In this section we will present a review of BYOD model and the security mechanisms that have been outlined by Mobile Security Reference Architecture (MSRA).

2.1   Bring Your Own Device (BYOD) Model

Starting from the year of 2011, the trend of bringing own devices to work has begun to gain recognition among organizations, especially in countries with high-growth economies such as Brazil, Russia and India [18]. Apart from reducing the operating cost, this phenomenon happened because of gain more flexible working hour and working environment besides increasing employee's productivity [19]. However, this trend has its own downside in term of preserving the integrity and confidentiality of sensitive data due to the portability of the devices and pervasive network which allows the data to be accessed anywhere. Therefore, Mobile Security Reference Architecture (MSRA) [5] has provided a review on how to protect our device from these kinds of problems. Next, we will present the things that have been outlined by MSRA to protect our device.

2.2   Overview of Mobile Security Reference Architecture (MRSA)

The MSRA presents the architectural components which are necessary to provide the data confidentiality, integrity, and availability that are critical to Government mission success. Among the architecture of security components for mobile are Virtual Private Networks (VPN), Mobile Device Management (MDM), Mobile Application Management (MAM), Mobile Application Store, Mobile Application Gateway (MAG), Data Loss Prevention (DLP), Intrusion detection system, Gateway and Security Stack (GSS) [5]. The explanation for each component is as the following:-

- Virtual Private Networks (VPN) - VPN technologies aim to create trusted secure connection between mobile devices and any intended recipient server. VPN connection could also be established and allow ad-hoc connection for external users.
- Mobile Device Management (MDM) - manages and optimizes the functionality of applications, data and configuration setting on mobile devices centrally. In terms of security, MDM enforced D/A policies and procedures.
- Mobile Application Management (MAM) - provides in depth distribution,   configuration, data control, and life-cycle management for specific applications installed on a mobile device. MAM also provides authentication   mechanism, command and control tool and diagnostic features, such as remote log-ins, reporting, and troubleshooting.
- Mobile Application Store - is a repository of mobile applications. Public application stores (i.e.: External Application Stores) offer mobile applications for sale (or for free) to the public.
- Mobile Application Gateway (MAG) - is a piece of software that provides application-specific network security for mobile application infrastructures. The purpose of a MAG is to act as a network proxy, accepting connections on behalf of the application's

network infrastructure, filtering the traffic, and relaying the traffic to mobile application servers.

- Data Loss Prevention (DLP) - Mobile infrastructure data loss prevention focuses on preventing restricted information from being transmitted to mobile devices, or from mobile devices to unauthorized locations outside the organization.
- Intrusion Detection System (IDS) - uses a set of heuristics to match known attack signatures against incoming network traffic and raises alerts when suspicious traffic is seen.
- Gateway and Security Stack (GSS) - Access to the enterprise must be restricted through one or more known network routes (i.e., Gateways) and inspected by standard network defenses such as stateful packet inspection, intrusion detection, and application and protocol. This prevents damage to the enterprise from a compromised mobile device.

Therefore, any security measures in mobile device should comply with the things that have been outlined by MSRA [5]. However, we believed that apart from the security measure that has been outlined, additional mitigation techniques such as combining control access with multi level security should be taken due to the sophistication of APT attack. Hence, in the next section (Section 3), we will look at the taxonomy APT attack to support our argument.

3   Advanced Persistent Threat (APT) in BYOD Environment

In this section, we will provide a detailed review regarding APT attack and analyze the motives and the causes that trigger the attack to support our proposed solution. Our main research question is "How does multi-level security and sentient computing can be used to tackle the APT security attacks on BYOD smartphones?" Under this assumption, we argue that to have a secure and privacy preserving information containers for mission critical applications, a new prototype should be proposed by adopting and enhancing of trusted model theory.

3.1   Definitions of Advanced Persistent Threat (APT)

The attack gets its named as Advanced Persistent Threat because of its ability to change its method of attacking the victims based on the sophistication of the targeted system (Advanced) and their stand in attacking a strong or powerful target with patience (Persistent) [8]. According to Tankard [7], it is a new kind of intelligent and sophisticated attack which is hard to combat since it is able to hide itself from being detected by always changing itself and using encryption so that it is unrecognizable. Meanwhile, Mustafa [9], define it as a "Purposeful Evasion Technique" or in other word, Evasion Attack which results in Malicious Data Leak (MDL) or unauthorized access to confidential data through non-human actors (i.e. malware) by manipulating human's trust. Its ability of gaining access to the victim's network and stay for a long time until it manages to perform an attack to the system with the motive of espionage, sabotage or to "drop" its target by revealing sensitive information [8] has remained a concern to organizations especially government sectors. Thus, we will further discuss about various motives, tools and methods that are used by the attacker to get into the victim's computer in the next section (Section 3.2).

3.2   How Does Advanced Persistent Threat (APT) Occurs Within BYOD Environment?

The case of APT attack is not a new case in computer science area. It has been reported starting from the year of 1998 with attack known as Moonlight Maze. However, during the year of 2006, it began to gain recognition as it has been coined as an Advance Persistent Threat by US Air Force circa due to its objective of stealing national's confidential data or causing damages to other nation-states without being detected [20]. After that, several cases have been reported in broad target areas such as government, military, educational and private sectors usually with the intend of espionage, sabotage important organization's operation and to leak or steal data consistently as shown in Table 1 below [10]:

Table 1. Analysis of past APT attacks [10]

APT attack	Motive	Action that triggers the attack	Effect of the attack
Stuxnet	Sabotaging the Iranian Nuclear Program (Natanz uranium enrichment plant)	Unknown but, by theory, it is caused by USB device	Slowing down the program by four years
Duqu	Espionage on SCADA system [12]	MS word files thatcontained zero day True Type font passing vulnerability or spear phishing	Information stealing malware is downloaded into the machine [12]
Flame	Stealing information by targeting on Middle East users who used Window system	Unknown but, by theory, it is caused by USB device	Gain information (i.e. screenshots, email content and sound)
Red October or Rocra	Information gathering that is targeted on diplomatic, governmental and scientific agencies	Spear phishing with MS Word and Excel document attached	Gain information (i.e. personal information , system configuration) either on mobiles or workstation [13]
Miniduke	Espionage on government, military and etc. [14]	Spear phishing	Gain information on the target machine by creating backdoor [14]

The main method on how APT threat occurs is due to social engineering and malware [7]. By using social engineering, the attacker will trick the targeted user to open his email with a file attached. Usually, the file attached is in the form of which we are familiar and seems trustable such as .doc, .pdf and .img with malicious code in it [6]. Once the target opens the file, the malicious code will try to download Remote Administration Toolkit. This will allow the attacker to take control on the target machine remotely [8].

According to Sood et. al. [8], there are various ways in gaining the victim's attention or manipulating the victim so that the threat can infiltrate into the victim's machine and get the information that they want. To facilitate our understanding, we have summarized the methods of gaining access into the victim's system in Table 2 [8].

Table 2. Methods used by the attacker to get into the victim's computer [8]

Methods	Attacks	Motives	Technique
Drive-by-download and spear phishing	Drive-by-downloads	Making the victim to download the malware directly into the system by redirecting to a malicious domain.	• Compromised the domain with malicious Iframe • Spear phishing
	Spear phishing or whaling	Making the victim to download the attachment provided without realizing that it is a malware	• Attaching files embedded with small but virulent malicious code
Exploiting web infrastructure	SQLI injection	To come out with more attack by extracting database details using SQLI vulnerabilities.	• Visiting vulnerable website
	Hybrid of SQLI and XSS (SQLXSS)	Making the victim directing to a malicious domain by inserting malicious iframe into the database of vulnerable website.	
Exploiting communication medium	Compromise SMTP server	Broad range spear phishing	• Malware • DNS cache poisoning
	Insecure FTP & HTTP server	Turn into the place to host the malware	
	Exploitation on DNS Protocol	Directing user into the malicious site	

Table 2. (*Continued*)

Online social network exploitation	Spear phishing	Making victim click on the link provided.	• Social engineering
Exploiting co-location services	Virtual hosting	Providing the malware(s) a place to "stay" and take control on the hosting server.	• C-99 • Malicious iframe injection
	Cloud provider		
	Rogue wi-fi and open or weak wireless network	Information gathering or hosting of malware for drive-by download.	• Soft AP/virtual Wi-fi function in window 7 • Hidden behind a single IP address as as several networks are allowed to do so.
	Bluetooth	Information gathering or act as a place to host malware.	
	Instant messaging and online chatting	Gain victim trust in clicking the malware	• Spear phishing
Physical attack	Portable devices	Allows malware to copy itself into another system when the USB stick plug on the machine	• Worm
	Teensy device	Capture keystroke and execute payload	• Pineapple Wi-Fi • Pwn Plug
	Hardware with Backdoor	Direct malware installation as it has gained trust by all Internet security because it is considered as something that is needed by the hardware	

As shown in Table 2, there are lots of methods used by the attacker to get access into the victim's system. These methods are done with the help of malware and protocol exploitation as shown in Table 3 below. Table 3 will summarize about the method utilized by the attacker to launch the attack on the targeted victim [8].

Table 3. Methods utilized to launch BYOD attacks [8]

Methods	Attack	Motive	Technique
Malware infection framework	Internet Relay Chat protocol	Allows communication to C&C server to happen thus, they can take control on the machine	• Botnet (i.e. SpyEye and Zeus)
	P2P Protocol		
	Hypertext Transfer Protocol (HTTP)		
Browser Exploit Pack and Glype proxies	Browser Exploit Pack	Medium to provide information regarding the target machine	• Install on commonly visit site • Malicious iframe • Spear phishing • Malware (i.e. Blackhole/ Phoenix)
	Glype proxies	Medium to search details about the target on the World Wide Web anonymously	
Remote Access Control (RAT) and rootkit	Remote Access Control (RAT)	Allows remote management	• PoisonIvy • GhostRAT
	Rootkit	Hide the infection and taking control of the system and downloading malware into the system	• Zero Access • TDL
Morphing and obfuscation toolkits		Prevent the malware from being detected	• Toolkits (i.e. Packers, Crypters, Code protectors, packagers)
Underground market		Pay for service	• Mule

As what we have discussed in this section, there are various methods, tools and motives that can be done by the attacker to get into the victim's computer. The most popular method to perform APT attack is through spear phishing. In the next section (Section 3.3), we will discuss on why BYOD devices are vulnerable to APT attacks.

3.3   Methods Used to Gain Access into Victim's Device

As seen in the previous section, past APT attacks start with the mistake of the user in trusting certain information (i.e. spear phishing) or using shared device. Actions such as either purposely or accidently downloading malware into the organization's network, visiting untrusted website, using unsecure network and untrusted portable device can create the entrance point for APT into the system. Unfortunately, the chance of getting APT attack in mobile device will be greater due to:-

- The power of social networks
  Social networks are the first source of information to the attacker. Some of the social network used the function of GPS to track the user's location. User ignorance in turning off the GPS functionality or willingly allows the website to track his or her location can cause a lot of troubles. This allows the attacker to monitor their actions. Besides that, by following the status of the targeted user and career information available through social networking sites such as LinkIn, the attacker will have a hint in providing a successful spear phishing or data gathering.
- Automatic download
  Some phones applies automatic download to certain file type such as .doc and .pdf. The name of the file usually comes out in hash like values instead of real name. This will lead the user to execute the file just to see its content. For example, in the case of spear phishing, if the attacker is using the social network and become our "close friend" and send a link that ask us to click to get a picture of our vacation, then our tendency to click it is higher compare to general phishing. Once clicked, the user has less control on the items downloaded since some phone applies the auto-download property files such as .doc and .pdf . Furthermore, the files are usually small thus, presenting challenge in stopping the download. This will become easy access into the phone.
- All applications mix up together
  Due to its portability, user tends to install software that can give them entertainment, connecting with others and works. All of them are mixed up together thus, increasing the chance of being attacked. For example, some users may install games without realizing that it is a malware and store it in the same drive that he used to store his work. Once the user downloaded his work file into his phone and stores it in the drive, the malware will be able to capture his work. Things will get worse if the malware gets root access into the phone which allows it to hide itself and have greater control on the user's phone [15].
- Using insecure network
  As we know, WEP is less secure compared to WPA or WPA2 due to its low key value in encrypting the data. However, it is still being used [16]. Accessing internet using the public network may increase the risk of having data being gathered through packet sniffing and rogue wi-fi, man-in-the middle attack such as TCP hijacking and unprotected data deliver. In this case, the attacker usually sniffs the user activity using available software. Then, after further evaluation, he makes changes in the TCP/IP packet by modifying the IP value which leads the user to be directed to the malicious domain instead of the correct domain so that the users

download the malware containing APT. The chance of having IP spoofing is bigger if the user is using unencrypted endpoint network or does not protected using SSL protocol.

- Lack of awareness from authorized users
  According to Morrow [15], although organization knows that their data is at risk to information leak, but most of them are lack of action in combating them. Meanwhile, according to the research made by ISACA [17], although most of respondents know that mobile devices can be a breeding ground for APT attack but, more than 70% of them do not use mobile control. It is also stated that, although lately there are increasing amount of awareness among organization in APT threat, but lack of information in combating them will become an obstacle.

Thus, it is hard to detect it as a threat since the file type used is usually something that we are familiar with (i.e. pdf and MS Word), some of them are enjoyable (i.e. games and applications) and usually come out as something or someone that we trust since it involves with social engineering. Furthermore, Morrow [15] also stated that information leak usually happened because of carelessness of the users and not by malicious users. Thus, action needs to be taken to prevent any unauthorized access to data thus, preserving its confidentiality and integrity.

Thus, the best way to prevent this from happening or reducing the possibility of being attack is through access control and multi level security to give protection on the data by limiting the access to data. Such framework has been proposed by Mustafa [10] and known as 3-D Correlation. The framework used the concept of access control and MLS to protect the data by focusing on the security clearance of the stakeholders and classification with confidentiality level of the data. However, we would like to argue that the user behavior should be included in BYOD environment because the users are able to access the network or data in different environment. Therefore, we will discuss our proposed solution extensively in the next section (Section 4).

4   Advanced Persistent Threat (APT) Mitigation Using Multi Level Security (MLS) – Access Control (AC) Framework

Based on our evaluation in previous sections, we can conclude that user's access to data needs to be restricted. The user's behaviors that can cause APT attack are: - 1) downloading file or software and store it in the same drive as their work device, 2) accessing social network or malicious application while at work, 3) accessing company's information using public network and 4) no restriction in creating or deleting high risk files such as .pdf and .doc.

The best way to prevent this from happening or reducing the possibility of being attack is to create a security policy that gives protection to data by limiting the access to data based on certain criteria. Our proposed solution is to create a trusted security system using multi layer security based on context aware Multi-Level Security (MLS) and Access Control. Employing the security mobile architecture alone as shown in Section 2.2 is not enough due to the following reasons :-

- The ability of the attack to evolve from time to time has created a problem in recognizing a new attack as less information is available [11].
- We cannot fully rely on VPN alone to ensure data security as the users are exposed to various security threats when they are connected through the public network or insecure network [18].
- MDM is hard to be implemented and managed as there are various operating systems and platform in mobile devices [18].
- MAM does not have the ability to protect data and cannot ensure fine-grained control [18].

Therefore, they have proposed several solutions in solving these issues. An expert from SafeLogic [18] mentions that most researchers are agreeing that multi layer security can reduce information leak and the layers should include the following properties: - device management, containerization, security policies, access control and encryption. This will create a secure environment in the device, where only whitelisted applications are available. According to Parmar [11], application whitelisting should be made as it ensures that the data is safe from being accessed by unauthorized party such as malware. These statements proved that multi layered security is essential in securing the data's confidentiality, integrity and availability from being leak by APT attacks. Thus, this information has been used to help us in designing the new framework on catering the APT attack inside the BYOD environment as shown in Fig. 1 below.

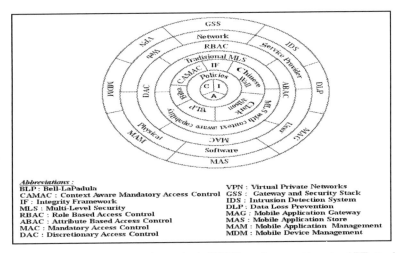

Fig. 1. Framework for enhanced security in BYOD environment to cater APT attack

As what have been discussed in the previous section (Section 3), APT attack starts with passive attack or data gathering to build up the active attack. Examples of active attacks are spear phishing and malware. By having multi layer security, each attack needs to pass each layer before gaining access into the target machine. In general, our proposed framework (Fig. 1) can be divided into three categories. Layer 1 (outside layer) and Layer 2 are the mobile component category. Meanwhile, Layer 3 to Layer 5 will act as authentication, authorization and accounting. Lastly, Layer 6 will act as security policies. The description of each layer of the framework will be presented from the outside layer (Layer 1) to an inside layer (Layer 6) as shown below:-

- Layer 1 – MSRA architecture
  As discussed previously in Section 2.2, to ensure mobile security, a mobile system needs to have these properties. This layer will allow the admin to monitor, manage and audit the activities that can be done by the employees such as managing authorized application. Thus, we would be able to reduce the threats occur in the device. MDM, MAM and DLP are great candidates as they can give a greater control on the device, application and data in case of a breach in security occur.
- Layer 2 – Players
  Each stakeholder has its own functions and will be trying to gather information with good or bad intent. The information gathered will be limited as authorized activities have been filtered by layer 1. Any unauthorized activity will be audited and block immediately by admin. Thus, only trustable application and activities can be done. Since, we are not able to ensure the player are handling the data "wisely", each player will not have the same right on the data as it will be filtered based on their roles, context and clearance level in Layer 3 to 6 and it will be audited.
- Layer 3 – Access Control
  Each stakeholder needs to pass this layer to protect the privacy of the data. Thus, only authorized one will be allowed. Users can consider to use either MAC, DAC, RBAC and ABAC according to the importance of their working environment such as roles, security and files. For example, MAC works well with MLS and provides security while RBAC is the most popular choice to work around a context aware environment or in an organization will multiple roles [22]. Meanwhile, ABAC can be used to define fine-grained access [21]. On the other hand, DAC is not suitable to be used as it is weak in security, although the user will have greater flexibility [23].
- Layer 4 – Multi Level Security
  Once the user has passed layer 3, the system will determine the things that can be done on the data (i.e. read or write) using MLS. Here, the MLS can be divided into two: - traditional MLS and MLS with context capability. User needs to choose one of them according to the suitability of the organization for example, requires the user to be monitored. According to Jafarian et. al. [22], traditional or single MLS is not capable to give both integrity and confidentiality of the data, not flexible and does not support context aware ability. Meanwhile, MLS with context aware capability will further evaluate the accessibility of data based on the data obtained through sensor(s). Thus, it is suitable for organization which has many sensitive data such as government sector. Based on our early evaluation, there are three important contexts to prevent APT attack: - time (i.e. working hour), location (i.e. public network or company's network) and day (i.e. weekend or weekdays).

- Layer 5 – Security Models
  In this layer, users will have the capability in selecting the MLS that he wants to use. The MLS can be divided into two: - single and hybrid. Examples of single MLS are Bell-LaPadula, Biba, Clark Wilson and Chinese Wall model. Bell-Lapadula provides data confidentiality, while Biba and Clark Wilson model provides data integrity. Meanwhile, Chinese Wall model provides protection based on conflict of interest [24]. On the other hand, hybrid MLS are CAMAC [22] and context-aware Integrity Framework [25] .CAMAC combines BLP and Biba with context aware to give protection in term of confidentiality and availability of the data which is suitable for military environment. Meanwhile, Anderson et al [25], are using the hybrid between Biba and Clark-Wilson model to come out with context-aware Integrity Framework which focus on military environment.
- Layer 6 – Policies
  Thus, through a combination of Layer 1 to Layer 5 a security policy that governs the integrity, confidentiality and availability is created.

In this section, we have discussed briefly about our proposed framework. Next, we will evaluate our proposed framework based on the multi layer security requirement to prevent information leakage that has been outlined by an expert from SafeLogic [18] and its advantages.

4.1   Evaluation on Proposed Framework and Its Advantages

Previously, in Section 4, we have stated that an expert from SafeLogic [18] has outlined that multi layer security that is used to prevent information leak should include the following properties: - device management, containerization, security policies, access control and encryption. Thus, to determine the usefulness of our proposed framework, we have made evaluation as shown in the table below (Table 4):-

Table 4. Evaluation of our proposed framework

Properties	Achieved	Description
Device Management	✓	Achieved through Layer 1 and Layer 2
Encryption	✓	
Security Policies	✓	Achieved through Layer 3 to Layer 6.
Access Control	✓	
Containerization	✓	Achieved through Layer 1 to Layer 6.

As shown Table 4, we are able to achieve the device management and encryption properties through Layer 1 and Layer 2. MSRA architecture has the functionality of Mobile Application Management (MAM) and Mobile Device Management (MDM) which allows the organization to take control on the application that can be accessed by employees and monitor their device. Thus, this will allow the organization to create application whitelisting. Meanwhile, encryption can be achieved by using the Data Loss Prevention (DLP). To solve the problem of managing the MDM, the organization can create standardization in term of operating system that is allowed to be used. Since MAM cannot provide protection on the data [18], Layer 3 to Layer 6 will take part in solving this problem.

Layer 3 to 6 will provide protection on the data by allowing only authorized user to access it using access control. MLS will further tighten the security control on the data by allowing access to the data based on security clearance to classified data. Thus, the combination on of both MLS and access control will provide security policies on the data that govern confidentiality, integrity and availability.

Lastly, through the combination of Layer 1 to Layer 6, we will able to create containerization based on the organization's requirement (i.e. user's role) thus, further tighten the security policy. For example, organization specifies that during working hour (i.e. 7 a.m. to 5p.m.) a user cannot access any personal application and the activities on the device should be comply with the user's role and security clearance. Thus, this will allow separation between the working and personal environment.

The advantages of this framework are organization has greater control of the employee's device. For example, if any malicious behaviour occur, the admin can take control on the device immediately using the function of MDM. The actions include data wipe and device switch off. Besides that, it can reduce the effect of data gathering when the user is trying to access the information outside the organization's network as the amount of information that the user can receive is limited. After that, the risk of having spear phishing also can be reduced as the system will prevent access to unauthorized social network site based on the company's policy. Lastly, the risk of having APT is also can be reduced as read and write access to into the device will be restricted. Next, we will discuss about the difference between our proposed framework with the 3-D Correlation proposed by Mustafa [10].

4.2   Comparative Evaluation of Multi Level Security (MLS) – Access Control (AC) with 3-D Correlation

In general, the table (Table 5) below shows the differentiation of Mustafa's [10] works with our proposed framework.

Table 5. Differences between each framework in preventing APT attacks

Framework	Applicable with MRSA	Environ- ment	Context Awareness	MLS Concept	Attack type
3-D Corre- lation [10]	No	Not specified	Nil	BLP	Non- human
MLS - AC	Yes	Mobile device	Yes	Multi computer security models	Human based centric

Based on the table above (Table 5), we can see that the main difference of our work is our proposed framework focused on human based centric, since the results of our research points to the user's behaviour as the main cause of APT attack. Therefore, due to this reason, we are looking on mobile security in BYOD environment. Lastly, we will discuss about the conclusion that can be made based on our research.

5   Conclusion

In this paper, we have shown that the user's behavior can contribute to APT attack. An APT attack starts with a passive attack through data gathering with the reason of getting information regarding the victim. Next, with the data obtained, they come out with an active attack so that they can enter into the victim's computer by manipulating the victim's trust. Besides that, we have shown that APT attack at on BYOD can occur in various positions. Thus, the best ways to prevent it are to educate and create awareness among users about actions that can present a threat to the company. Apart from that, it is also important to design a good security policy that complies with confidentiality, integrity and availability to protect the sensitive information.

Lastly, our future work will be the implementation and design of the proposed framework by focusing on context aware access control and MLS in BYOD device. The hurdles will be integrating the policy with MDM and finding the best MLS that suited with the access control choose.

References

1. Fischer, N., Smolnik, S.: The impact of mobile computing on individuals, organizations, and society - synthesis of existing literature and directions for future research. In: IEEE 2013 46th Hawaii International Conference System Sciences (HICSS), pp. 1082–1091 (2013)
2. Dover, S.: CBS News, Study: Number of smartphone users tops 1 billion. http://www.cbsnews.com/8301-205_162-57534583/
3. La Polla, M., Martinelli, F., Sgandurra, D.: A Survey on Security for Mobile Devices. IEEE Communications Surveys & Tutorials 15(1), 446–471 (2013)
4. Fossi, M., Egan, G.Y., Haley, K., Johnson, E., Mack, T., Adams, T., Blackbird, J., Low, M.K., Mazurek, D., McKinney, D., Wood, P.: Symantec Internet Security Threat Report – Trends for 2010, Technical Report Volume 16, Symantec (2011)
5. Mobile Security Reference Architecture, Federal CIO Council and Department of Homeland Security National Protection and Program Directorate Office of Cybersecurity and Communications Federal Network Resilience (2013)
6. Schmidt, A., Schmidt, H., Batyuk, L., Clausen, J.H., Camtepe, S.A., Albayrak, A.: Smartphone malware evolution revisited: android next target?. In: IEEE 2009 4th International Conference on Malicious and Unwanted Software (MALWARE), pp. 1–7 (2009)
7. Tankard, C.: Advanced Persistent Threats and How to Monitor and Deter Them. Network Security 2011(8), 16–19 (2011)
8. Sood, A.K., Enbody, R.J.: Targeted Cyberattacks: A Superset of Advanced Persistent Threats. IEEE Security & Privacy 11(1), 54–61 (2013)
9. Mustafa, T.: Malicious data leak prevention and purposeful evasion attacks: an approach to advanced persistent threat (APT) management. In: 2013 Saudi International Electronics, Communications and Photonics Conference, SIECPC, pp. 1–5 (2013)

10. Virvilis, N., Gritzalis, D.: Trusted computing vs. advanced persistent threats: can a defender win this game? In: 2013 IEEE 10th International Conference on Autonomic and Trusted Computing (UIC/ATC) Ubiquitous Intelligence and Computing, pp. 396–403 (2013)

11. Parmar, B.: Protecting against spear-phishing. Computer Fraud & Security **2012**(1), 8–11 (2012)

12. Hipolito, J.M.: DUQU Uses STUXNET-Like Techniques to Conduct Information Theft. TrendMicro. http://www.trendmicro.com/vinfo/us/threat-encyclopedia/web-attack/90/duqu-uses-stuxnetlike-techniques-to-conduct-information-theft

13. Storm., D.: Red October 5-year cyber espionage attack: Malware resurrects itself. Computerworld (2013). http://www.computerworld.com/article/2474163/cybercrime-hacking/red-october-5-year-cyber-espionage-attack—malware-resurrects-itself.html

14. GReAT Miniduke is back: Nemesis Gemina and the Botgen Studio. KASPERSKY lab (2013). http://securelist.com/blog/incidents/64107/miniduke-is-back-nemesis-gemina-and-the-botgen-studio/

15. Morrow, B.: BYOD security challenges: control and protect your most sensitive data. Network Security **2012**(12), 5–8 (2012)

16. Noor, M.M., Hassan, W.H.: Wireless Networks: Developments, Threats and Countermeasures. International Journal of Digital Information and Wireless Communication (IJDIWC) **3**(1), 119–134 (2013)

17. Meadows, R.: ISACA Global Study: Organizations Not Prepared for Advanced Cyberthreats; Big Gaps in Education and Mobile Security Remain (2014). http://www.isaca.org/About-ISACA/Press-room/News-Releases/2014/Pages/ISACA-Global-APT-Survey.aspx

18. Leavitt, N.: Today's Mobile Security Requires a New Approach. Computer **46**(11), 16–19 (2013)

19. Osterman Research by Dell : The Need for IT to Get in Front of the BYOD Problem. White paper, Osterman Research Inc. (2012)

20. Websense: Advanced Persistent Threats and Other Advanced Attacks: Threat Analysis and Defense Strategies for SMB, Mid-Size, and Enterprise Organizations. White paper, Websense Inc. (2011)

21. Zhauniarovich, Y., Russello, G., Conti, M., Crispo, B., Fernandes, E.: MOSES: Supporting and Enforcing Security Profiles on Smartphones. IEEE Transactions Dependable and Secure Computing **11**(3), 211–223 (2014)

22. Jafarian, J.H., Amini, M., Jalili, R.: A context-aware mandatory access control model for multilevel security environments. In: Harrison, M.D., Sujan, M.-A. (eds.) SAFECOMP 2008. LNCS, vol. 5219, pp. 401–414. Springer, Heidelberg (2008)

23. Suhendra, V.: A survey on access control deployment. In: Kim, T.-h., Adeli, H., Fang, W.-c., Villalba, J.G., Arnett, K.P., Khan, M.K. (eds.) SecTech 2011. CCIS, vol. 259, pp. 11–20. Springer, Heidelberg (2011)

24. Stallings, W., Brown, L.: Computer Security: Principles and Practice. Prentice Hill, 2008.

25. Anderson, M., Montague, P., Long, B.: A context-based integrity framework. In: 2012 19th Asia-Pacific Software Engineering Conference (APSEC), vol.1, pp. 1–9. IEEE (2012)

26. SafeLogic. SafeLogic Inc. (2015). http://www.safelogic.com/

Design of Disaster Collection and Analysis System Using Crowd Sensing and Beacon Based on Hadoop Framework

Eun-Su Mo[1], Jae-Pil Lee[1], Jae-Gwang Lee[1], Jun-Hyeon Lee[1],
Young-Hyuk Kim[2], and Jae-Kwang Lee[1(✉)]

[1] Department of Computer Engineering, Hannam University, Daejeon, South Korea
{esmo,jplee,jglee,jhlee,jklee}@netwk.hannam.ac.kr
[2] Public Procurement Service, Information Management Division, Daejeon, Korea
kimyh86@korea.kr

Abstract. Currently, disaster data is collected by using site-based, limited regional collection. In this study, a system that collects location information of users that have a mobile device is proposed. The proposed system collects real-time disaster data by using crowd sensing, a user-involved sensing technology. In order to quickly and accurately determine a large amount of unstructured data, among big data frameworks, the Hadoop framework is applied as it efficiently sorts a large amount of data. Also, to enable fast local evacuation alert for users, a beacon-based ad-hoc routing interface was designed As an integrated interface of the proposed systems, a hybrid app based on HTML5, which uses JSON syntax.

Keywords: Crowd sensing · Big data · Beacon · Disaster · Calamity · Warning system

1 Introduction

In general, disaster refers to a naturally-caused accident, while calamity means artificially-caused one. The former includes typhoon, flood, storm, tsunami, heavy snow, and yellow dust and the latter fire, breakdown, explosion, traffic accident, malfunction of national infrastructure, and infectious disease [1].

Many accidents from disaster and calamity are taking place all around the world. In general, disaster and calamity result in a wide range of damage and loss and spreads quickly. Therefore, it is crucial to secure the "golden time" to reduce danger.

However, the current system is focused on reporting rather than problem-solving, and, therefore, unable to effectively secure the golden time [2].

In Japan, which has the best system related to meteorology and earthquake technology, March 11, 2011, earthquake of 9.0 magnitude hit the Pacific Ocean near Tohoku, Japan, resulting in over-10m tsunami and some of the worst damages in history, including deaths, property damage, and radiation leakage The first early warning for earthquake and tsunami was sent three minutes after strong earthquake [3].

O. Gervasi et al. (Eds.): ICCSA 2015, Part IV, LNCS 9158, pp. 106–116, 2015.
DOI: 10.1007/978-3-319-21410-8_8

South Korea has total 11,542km of coastline, which is relatively long for the land, in comparison to other countries. The coastline length of South Korea compared to the area, in percentage, is 117%, which is much higher than Japan 87%, UK 57%, and New Zealand 56%. For that reason, South Korea suffers more serious damage from disaster and calamity than other countries. However, the country currently has many problems regarding disaster and calamity detection and subsequent evacuation of residents [2] [4].

Information related to all accidents is referred to as safety information or safety service information. Research has been conducted to take measure against increasing safety-related accidents. Most notably, in IoT (Internet of Things), importance, value, and utility of big data are increasing. Recent research is focused on how to integrate and converge big data into a system.

Most cases of integrating big data into safety service are concentrated in crime analysis, such as crime analysis and domain awareness system of New York and crime prediction service of San Francisco and Los Angeles.

New York has built an anti-terrorism detection system named DAS (Domain Awareness System). Over 4,000 surveillance cameras installed in Manhattan are used to analyze information on suspicious people, objects, or cars, which is immediately provided to local police, fire station, or other relevant institutions [5].

San Francisco analyzes past crime data to predict areas with highest risks of new crimes. Among 10 districts that were warned of possible crimes, seven actually saw criminal cases [5].

In South Korea, geographical profiling service of National Police Agency is one of the safety services that integrated big data, which also include SOS Resident Safety Service of Ministry of Security and Public Administration, and iNavi Safe [6].

Safety-related information and service in most of the current services based on big data are only focused on crimes, and irrelevant to natural disaster and calamity.

For that reason, in this study, the following system was designed for the purpose of collecting and analyzing disaster and calamity data and the system is divided into three steps: collection, analysis, and provision.

The main component of this three-step system is crowd sensing, a user-involved sensing technology, for collecting real-time disaster and calamity information.

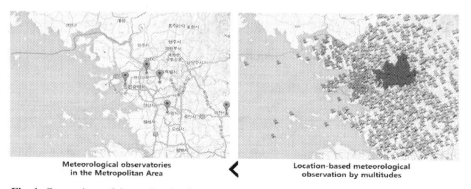

Fig. 1. Comparison of data collection between conventional data import and crowd sensing [7]

Crowd sensing is a compound word of crowd and sensing and collects data as users who have sensors and computing devices participate, creating new knowledge by extracting sensor data from a wide area. [Figure 1] shows the core of crowd sensing. It is 'the multitudes in different regions' and not 'a small number of experts' who can find out about rapidly-changing information of an area in the most accurate and fastest way. Their collaboration brings about accurate and fast data.

In Chapter 2, previous research on crowd sensing, big data, and beacon, which are core technologies in the three steps, will be reviewed. And then, in Chapter 3, the front-end, back-end system will be designed to provide the target service. Finally, the authors will discuss advantages and problems of the design.

2   Related Studies

2.1   Analysis of Conventional Disaster and Calamity System

2.1.1   Forest Fire Prediction Service Based on Big Data

The research in Korea on forest fire prediction service based on big data analyzed forest vulnerability according to climate change, and aimed to improve the national forest fire warning system that was used since 2003 [8]. This study has the following characteristics: First, location information of people who are hiking or climbing the mountain is collected from weather information provided by Korea Meteorological Administration (KMA), forest soil digital mapping, mountain slope, and direction, and data without personal information provided by telecommunication providers. Second, forest disaster is predicted and the information is sent to users based on the collected data. In the forest soil mapping, mountain slope, and direction, trails can change according to the weather and population in motion. These data must be re-established based on different periods and require a lot of time and workforce. In this study, crowd sensing and smartphone sensors will be used to receive transformed data.

2.1.2   Heavy Rain Prediction System of Rio de Janeiro

Brazil's Rio de Janeiro heavy rain forecast system is a large-scale study initiated due to a natural disaster forecast system suffered casualties due to landslides caused by

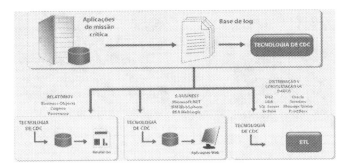

Fig. 2. Structure of heavy rain prediction system of Rio de Janeiro [9]

heavy rainfall concentrated in 2011. [Figure 2] shows the system structure. Data and processes of more than 30 institutions in Rio de Janeiro were integrated into an intelligent system, in order to manage and supervise natural disaster, traffic, and power supply. The intelligent operation system enables prediction of heavy rain 48 hours in advance based on urban management and high-resolution weather forecast system and state-of-art modeling system. Unlike the initial design, the system expanded in its scope to all areas, crimes, and accidents that can take place in the city [10].

2.2   Crowd Sensing

2.2.1   Semantic Map Research

In semantic map research using crowd sensing, data including operation types, customer visit patterns, and sale characteristics of stores near universities was collected by using crowd sensing and implemented in a map for visualization

Stores near Yonsei and Ehwa Womans' Universities show clear characteristics as shown in [Figure 3]. Near Ehwa Woman's University, most of the businesses are stores for shopping. By contrast, near Yonsei University, businesses are focused on night life and entertainment. Behind the universities are crowded by accommodations, and workplaces are distributed close to the universities. Verifying and researching this type of data by an individual or small group requires a substantial amount of time and effort, and this study demonstrated that crowd sensing helps collect valuable and accurate data quickly and even visualize it [11].

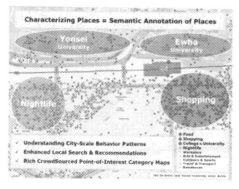

Fig. 3. Semantic map visualization using crowd sensing [11]

2.2.2   Service for Finding Missing Children

The research designed to find missing children based on crowd sensing is focused on situations in which children who have a smartphone or smart tag are gone missing. When a child is missing from a theme park, department store, or in other large spaces, adjacent users are asked for help and sent the basic information about the child's appearance and message. Information about the route of the child is provided by using sensing data of the users. The server, then, compares data collected from the child's smartphone and data provided by users, and provides matching data so that the child can find the parents. This study helps find missing children by using crowd sensing [11].

2.3   Big Data

2.3.1   Security Log Analysis System Using Hadoop

In the research on security log analysis system using Hadoop, the security log of security equipment is supplemented based on SIEM (security information & event management) by using the Hadoop framework rather than the conventional ESM (Enterprise Security Management). So far, logs from security equipment were analyzed by individual equipment or analyzed as a whole based on ESM. However, it was not possible to analyze a large amount of data that was accumulated over a long period, due to limited data capacity and data processing of ESM, or search data quickly. These limitations were supplemented by using Hadoop framework, which enables fast collection, storage, analysis, and visualization of data. [Figure 4] shows composition of Big Data Platform Software [12].

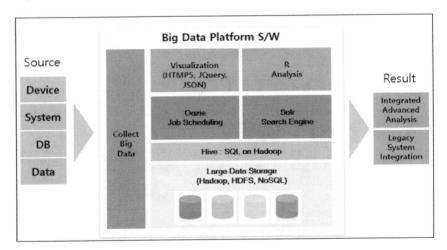

Fig. 4. Big Data Platform S/W [12]

2.3.2   Reduction of Network Use by Hadoop Cluster

The research on block relocation algorithm was designed to reduce network consumption by reducing data locality, which is weakness of Hadoop cluster.

The job scheduler of map/reduce uses FIFO (First In First Out) to sequentially assign tasks. This method has low processing efficiency when the internal/external data of the cluster is called when there is a large amount of tasks. Because the scheduler operates in a delayed state, that is, when the initially located blocks are not relocated, there is difference in load received by different nodes. The relocation algorithm is used to improve efficiency by redistributing node blocks according the cluster work pattern [13].

2.4   Beacon Structure and Service Trend

Beacon is a type of wireless sensor that applies real-time location awareness system.

It is based on the Bluetooth communication protocol and uses BLE (Bluetooth Low Energy). Location is measured by exchanging data with smartphones via RSSI (Received Signal Strength Indicator). Although smartphone LBS (location-based service) using NFS and GPS and real-time data communication technology has been introduced, it has the advantages of location accuracy and scope in comparison to indoor location determination technology. NFC can be used with 1:1 contact within 10cm, Beacon has a long available distance of maximum 50 to 70m. While GPS signals cannot locate a smartphone user, Beacon is capable of accurate location, by 5cm, and can be used both indoors and outdoors. Positioning service can be divided into three methods: checkpoint, zone, and track [14] [15].

Beacon service has been adopted at the main store and Icheon store of Lotte Premium Outlet. The smartphone app is automatically recognized by the sensor, and provides welcome message, store map, and shopping information to the visitors [15].

3   Design and Implementation

3.1   System Model

[Figure 5] is a block diagram of the system for the proposed service.

Section A collects data from users and receives data provided by the control center.

Section B is a control center that manages I/O of the entire data. All data must pass the control center and, as it is distributed and arranged, the level of load balancing is important.

Section C is a backbone where data is stored, analyzed, and processed. To quickly and accurately process a large amount of data, distributed processing based on Apach Hadoop Framework is performed. Also, Hadoop has high fault-tolerance and, thereby, system stability.

The proposed system is executed as follows:

In the first step, collection, crowd sensing is used for real-time collection of disaster and calamity data. One of the prerequisites for this is that the user has a mobile device. For a wider range of users, an HTML5-based hybrid app is provided to expand applicable mobile devices. Second, the latest BLE is used to limit system resources used for user participation. Third, BLE-based direct communication network is proposed to create a user network within a site, even if there is a problem with connection with the control center.

In the second step, analysis, big data framework is used to quickly and accurately detect and determine a large amount of unstructured data collected from users. The Hadoop framework is applied to the framework applied in this study as it is effective for large data arrangement. Hadoop arranges data in couple based on the key value, showing high speed in crowd sensing and Beacons data arrangement and processing. Particularly, this advantage is useful for larger amounts of data, when there are more users and, thereby, collected data.

In the last step, Beacon is used for sending prompt evacuation warning to local users based on the analyzed data.

BLE-based Beacon uses less battery than the conventional Bluetooth version. The relevant service is provided as it automatically locates the users without users' action. Also, location is determined by using RSSI. In this study, information is provided so that users can prepare for and evacuate from unspecified situations, by using the user network (hereafter, BLE Ad-hoc routing). JSON is used for communication between these user systems. Unlike XML, JSON does not use tags but text-based exchange with key values. It has the advantage of not relying on language.

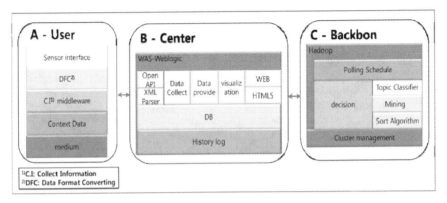

Fig. 5. Service Model

3.2   Data Collection

[Figure 6] shows the process that collects and stores data by crowd sensing. All sensing data collected from users' mobile devices are converted into JSON by the DFC module. The converted data combines users' unique IDs, and, thereby, verifies data destination. Finally, the finished data is sent to the control center through either BLE or network. Also, users can exchange data and user information within the same site by using BLE Ad-Hoc on Beacon.

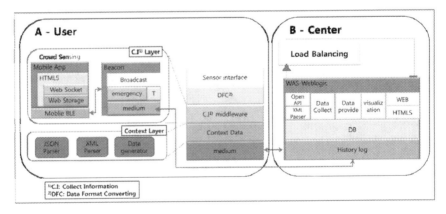

Fig. 6. Data collection process

3.3   Data Analysis

[Figure 7] shows the flow by which all data is sent through the control center to the backbone, for distributed storage. The control center manages and distributes all data. It is responsible for receiving and sending all data. The control center stores the received data in DB, and this is included in record management in the history log.

In the backbone, Section C, the polling scheduler performs real-time monitoring of DB of the control center. It takes newly stored data for distributed storage in multiple clustered nodes via Map/Reduce.

For storage, the topic classifier performs data mining, and, as main categories, the topics are classified as climate, earthquake, fire, typhoon, and tsunami. After the first classification, specific topics are classified by a top-down method.

Semantic resolution of the data divided into the minimum unit is performed by using data mining. For fast semantic resolution, the pattern algorithm that defined particular patterns is used for filtering.

Finally, data of distributed nodes is stored, and, cluster managing algorithm, which arranges data by grouping them based on nodes, is used for distributed storage.

As a distributed storage method, Hadoop Framework, which is an Apache Open Project, is used. The system was composed of the master server, where the polling scheduler is operated, and slave nodes according to Hadoop standard. Disaster and calamity data are distributed-stored by Map/Reduce, and the structural characteristics allow the system to operate without a problem even if there is a small amount of data node error.

Especially, the Map/Reduce method uses API or its own query language that is different from the conventional SQL query method, because it is processed in a key-value form.

Lastly, nation-level public data is additionally collected by using OpenAPI (OPEN Application Programming Interface) of public data portals relating to disaster and calamity. This supports reliability of the proposed system in that it can freely use highly reliable, large-scale data in real time. Also, Open API is provided as a method that facilitates implementation. Notably, SBIS (System Build & Integration Service) Weather API, a climate/weather distribution service of KMA that was used as reference in this study, has the following characteristics: First, the weather data is made into XML according to the relevant website. Second, to minimize users' burden, the data is taken in real-time XML instead of being stored in the server.

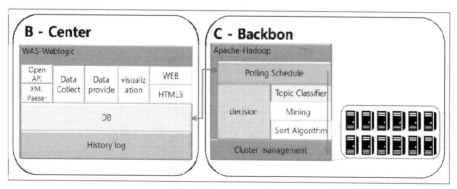

Fig. 7. Data processing

3.4   Service Scenario

The disaster and calamity warning scenario of the proposed system assumes that the backbone, Section C, determines disaster or calamity, and can be divided into three types as shown in [Figure 8].

In the first scenario, the decision server that exists in Section C of [Figure 5] detects disaster or calamity, and sends the data via the control center in Section B to Beacon and, finally, to users' mobile devices. Beacon devices have their unique IDs and the Beacon of the relevant site sends them via BLE Ad-Hoc to Beacon and users' mobile devices.

In the second scenario, the decision server detects disaster and calamity and directly warns the users through the control center. Users' mobile devices have unique IDs and the warning is sent based on the user mobile device ID and collected location data. This method implies that data can be sent even if the local Beacon does not function normally. Users' mobile devices recognize the special situation in which data was directly sent without Beacon, and continue to the following, third scenario.

The final scenario is activated by the second scenario. Without the decision/control server, users warn disaster and calamity on their own, when the infrastructure does not function normally during disaster and calamity. In this situation, the data is sent directly to the local Beacon and users from the time disaster or calamity was notification was received.

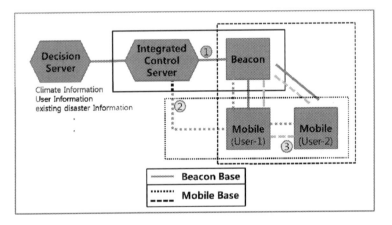

Fig. 8. Provision of service

4   Conclusion

This study proposed a system that subdivides disaster and calamity by town units, and collects data quickly and accurately. The entire process was defined by dividing it into three steps of collection, analysis, and provision. Golden time is especially important in disaster and calamity. For this, crowd sensing was used to improve reliability by integrating user-based location data and local data using OpenAPI. All data uses the control center, the central server, which links users and large-scale backbone clusters. Therefore, all data passes through the control center for distributed storage by Hadoop. In Hadoop framework, the polling scheduler, which is a master that monitors new data of the control server, is placed above, and, below it, slave nodes process big data.

JSON structure is used for data exchange between systems, and allows users to use their mobile devices. Especially, as the service is provided through a hybrid app. It can be provided to various devices and OS. The system was designed to deliver data from the control center to Beacons that consist of local infrastructure, and Beacons send the data to users via broadcasting. This is part of load balancing of the control center, and a backup plan was made so that, when there is an error, users can form a network and exchange data. As a result, this can radically address problems with conventional disaster and calamity system, and also detects disaster and calamity by small area basis.

However, to improve the service, the data processing algorithm of Hadoop must be researched so as to accommodate it to particular situations of disaster and calamity. Also, the three step scenarios proposed in this study have the disadvantage that they cannot determine users with ill intention. A policy to resolve this problem will be needed.

Acknowledgment. This research was supported by Basic Science Research Program through the National Research Foundation of Korea(NRF) funded by the Ministry of Science, ICT & Future Planning(No. 2014R1A1A2055522).

References

1. Framework Act on the Management of Disasters and Safety
2. Hyeon-Cheol, S.: Improvement of Disaster Management System in Korea: The Institute for the Future of State (May 2014)
3. Korea Meteorological Administration: Mid-to-Long-Term Policy Development for Improved Meteorological Service (October 2011)
4. National Emergency Management: Plan for Mid-to-Long-Term Research and Development Against Earthquakes and Tsunamis (April 2013)
5. Jang, K.: Choi: Big Data-based Convergence Service Industry Creation: Science & Technology Policy Institute. Policy Research **2013–20**, 210–214 (2013)
6. Jang, Kim: Main policy and suggestions for Big Data-based Convergence Service Industry Creation; Science & Technology Policy **192**, 4–13 (2013)
7. Wearherplanet. http://www.weatherplanet.co.kr/about/
8. Sang-Woo, B.: Implementation of integrated systems for forest disaster management. The Korea Contents Association: The Journal of the Korea Contents Association **12**(2), 73–77 (2014)
9. MD2. http://www.md2net.com.br/ibm_InfoSphere_CDC.asp
10. BIGDATA Strategy Forum: Data analysis for a better future, National Information Society Agency (2013)
11. Ho-Jung, C.: Mobile CrowdSensing: Korea Internet Conference, TRACK J1-2 (2014)

12. Han, K.-H., Jeong, H.-J., Lee, D.-S., Chae, M.-H., Yoon, C.-H., Noh, K.-S.: A Study on implementation model for security log analysis system using Big Data platform: Korean Society of Computer Information. Journal of Digital Convergence **12**(8), 351–359 (2014)

13. Kim, J.-S., Kim, C.-H., Lee, W.-J., Jeon, C.-H.: A Block Relocation Algorithm for Reducing Network Consumption in Hadoop Cluster: The Society of Digital Policy and Management. Journal of the Korea Society of Computer and Information **19**(11), 9–15 (2014)

14. ITworld. http://www.itworld.co.kr/slideshow/85994

15. KB Financial Group Management Institute, KB Vitamin knowledge, vol. 94 (2014)

Network Traffic Prediction Model
Based on Training Data

Jinwoo Park[1], Syed M. Raza[1], Pankaj Thorat[1], Dongsoo S. Kim[2],
and Hyunseung Choo[1(✉)]

[1] College of Information and Communication Engineering,
Sungkyunkwan University, Seoul, Republic of Korea
{streejp,s.moh.raza,pankaj,choo}@skku.edu
[2] School of Engineering and Technology,
Indiana University-Purdue University Indianapolis, Indianapolis, USA
dskim@iupui.edu

Abstract. Real-time audio and video services have gained much popularity in last decade, and now occupying a large portion of the total network traffic in the Internet. As the real-time services are becoming mainstream the demand for Quality of Service (QoS) is greater than ever before. To satisfy the increasing demand for QoS, it is necessary to use the network resources to the fullest. In this regards, the available bandwidth based routing is a promising solution. Unfortunately the instantaneous available bandwidth of a network is not enough as it may change the next moment in highly dynamic networks. To solve this issue, we present a prediction model for network traffic, on the basis of which network available bandwidth can be estimated. This paper utilizes the efforts done in regard to road traffic prediction to formulate a prediction model for network traffic.

Keywords: Network traffic prediction · Modeling · K-Nearest neighbors

1   Introduction

Availability of high transmission speed on wired and wireless links have spawned the concepts like Cloud and Always Connected, which in turn generate high volumes of data on the networks. With the IoT and M2M gaining momentum in future this data volume will grow substantially as many are predicting it to grow to 10 fold by 2020 [1][2]. In last couple of years, network research in academia and industry has gotten much attention, as a result of the increasingly obvious fact that the current network systems, architecture and protocols will not be able to cope with the high demands of network traffic in the near future [3].

It has been reported that an overwhelming portion of today's network traffic consist of video content [4]. With the emergence of more and more real-time services, the growth in the real-time audio/video network traffic will be greater than ever before. To provide the user a reasonable real-time experience, a certain level of QoS is required, which today's best effort networks will not be able to provide. In the past

© Springer International Publishing Switzerland 2015
O. Gervasi et al. (Eds.): ICCSA 2015, Part IV, LNCS 9158, pp. 117–127, 2015.
DOI: 10.1007/978-3-319-21410-8_9

there have been multiple efforts to introduce QoS into the networks [5-8], but none of them made it to the mainstream because they require network elements to perform additional tasks, which make the commodity network elements expensive.

To provide end to end QoS to the user, the main problem lies in the networks where routing algorithms like OSPF [9] and RIP [10] are based on the shortest path. Clearly, routing based on the shortest path does not guarantee the QoS as the shortest path can also be the most congested path. To ensure QoS, the routing algorithms are required to be based on the state of the network at any given time. In this regard, the routing based on the available bandwidth of the links can continually ensure that the best path in terms of quality of services has been selected.

In todays distributed networks, the available bandwidth based routing has two main hurdles:

1. All the network elements must know the available bandwidth of all links at any given time.
2. Due to dynamic nature of the networks it is almost impossible to calculate the available bandwidth of the link, as it can be changed instantaneously.

In the distributed networks, keeping all the networks elements aware of the latest available bandwidth of all the links will cause huge control traffic overhead. Software defined networking (SDN) is a concept in which the control plane of all the network elements is centralized at a controller [11]. Therefore, in SDN networks only the controller is required to know the available bandwidth of each link, resolving the issue of the huge control traffic overhead for available bandwidth based routing in distributed networks. To solve the second hurdle, where the instantaneously calculated value of the link's available bandwidth can be extremely short lived and unusable, this paper proposes a network traffic prediction model. If the network traffic on a link for a particular period can be predicted then the estimated available bandwidth of a link for that period can also be calculated.

Prediction modeling is a well-researched in the field of road networks, economics and meteorological studies. Proposals for use of prediction models in the field of networks are scarce, hence in order to predict the traffic in the network, this paper utilizes the K- Nearest Neighbors (K-NN) algorithm [12] based prediction model proposed for road traffic prediction. The data used for evaluation in this paper is based on network traffic statistics (collected for 35 days), and the results show that the proposed prediction model can reasonably predict the network traffic for next certain period.

The rest of this paper is managed in the following way. In Section 2, we will discuss the background and motivation for the proposed network traffic prediction model along with the related works. A detailed discussion of the proposed prediction model is presented in Section 3. A performance evaluation and its analysis are presented in Section 4, and in Section 5 we discuss our conclusion and future works.

2   Background and Related Work

2.1   Motivation

Networks are based on the distributed architecture, which has benefits like scalability; the biggest downfall is that the routing algorithms have to be distributed in nature. Even for a centralized algorithm first information is gathered over the whole network. Any change in the network will cause the information to be distributed among all the network elements and this dissemination of the information has significant amount of latency. Therefore current routing algorithms are based on the network information which doesn't change very often e.g. hop count to the destination. Unfortunately hop count is not significant factor for QoS. The available bandwidth of the links is a significant factor for QoS and it is desirable that the routing is performed on the available bandwidth of the links as this make sure that best QoS is provided to the user and network resources are maximally used. In highly dynamic networks, the information regarding available bandwidth of the links expires quickly, and requires recalculation. To mitigate this problem an estimated available bandwidth can be calculated based on the network traffic prediction model.

2.2   Prediction Models

In the field of economics, as well as other fields, Autoregressive-Moving Average (ARMA) models are used for predicting the future values. The model consists of two parts, an autoregressive (AR) part and a moving average (MA) part. The model is usually then referred to as the $ARMA(p, q)$ model where p is the order of the autoregressive part and q is the order of the moving average part. In the field of machine learning, Support Vector Machines (SVM) are used for supervised learning with associated learning algorithms that analyze data and recognize patterns, it is also used for classification and regression analysis. In a set of training data, where each datum is marked as belonging to one of two categories, an SVM training algorithm builds a model that assigns new examples into one category or the other, making it a non-probabilistic binary linear classifier. An SVM model is a representation of the examples as points in space, mapped so that the examples of the separate categories are divided by a clear gap that is as wide as possible. New examples are then mapped into that same space and predicted to belong to a category based on which side of the gap they fall on. K-Nearest Neighbors (K-NN) is another algorithm used for classification and regression. K-NN is one of the simplest algorithms in machine learning. In K-NN, the input consists of the closest training examples in the feature space and the output depends on whether K-NN is used for classification or regression. In case of both classification and regression it is useful to assign the weight values to the neighbors as it defines their contribution. This way the nearer neighbors contribute more to the average than the more distant ones. A common weighting scheme consists of giving each neighbor a weight of $1/d$, where d is the distance to the neighbor. The neighbors are taken from a set of objects for which the class or the object property value is known. This can considered as the training set for the K-NN algorithm, however no explicit training step is required in K-NN algorithm.

2.3   Related Work

Self-loading periodic streams (SLoPS) is scheme presented by M. Jain et al to calculate the end-to-end available bandwidth [13]. The basic principle of SLoPS is that an increasing trend in the one-way delay of the periodic stream can be observed when the stream rate is higher than the available bandwidth. SLoPS operate on the end systems and therefore cannot be used in the network elements to estimate the available bandwidth of the links in real-time for the available bandwidth based routing.

In the field of road traffic, S. Ishak et al investigated the factors that have a significant impact on the forecasting accuracy of travel times using a nonlinear time series traffic prediction model [14]. Parameters with a statistically significant effect on the model's performance were determined through statistical analysis. S. Ishak et al then optimized their short-term traffic prediction model using multiple artificial neural network topologies under different network and traffic condition settings [15]. In order to enable the networks to learn from historical information, a long-term memory component is utilized for the input patterns to allow the networks to build an internal representation of the recurrent conditions, in addition to the short-term memory that is encoded in the most recent information. Their statistical analysis with a naive and heuristic approach, shows that the optimized neural network approach resulted in a better prediction performance. DynaMIT [16], presented by M.B Akiva et al., is another system which provides prediction-based guidance with respect to departure time, pre-trip path and mode choice decisions and en-route path choice decisions. DynaMIT is organized around two main functions: state estimation, and prediction-based guidance generation. It utilizes both off-line and real-time information. The most important off-line information, in addition to the detailed description of the network, is a database containing historical network conditions. This is the system's memory. The real-time information is provided by the surveillance system and the control system. The quality of the prediction depends on the quality of the current state description, and on the horizon. Therefore, the state of the network is regularly estimated so that all available information is incorporated in a timely fashion, and a new prediction is computed.

S. Clark presents an intuitive method for road traffic prediction, using a pattern matching technique [17]. The adopted technique is a multivariate extension of the nonparametric regression that exploits the three-dimensional nature of the traffic state. Their model is based on pattern matching, where recent observations are matched with those contained in a database of historical observations. From all the matches, either the k nearest matches or all the matches below a given distance threshold are located. The successive observations from these "best" matches are then averaged to obtain the forecasts. The only parameters in their model are the number of observations to match it with, the number of best matches to retain, and the distance threshold. Once a sequence of recent observations has been matched and forecasts have been made, they move the recent observations to the historical matching database for use in subsequent matching operations.

3   Proposed Network Traffic Prediction Model

To fully utilize the network resources and to provide the required bandwidth to the user, the available bandwidth based routing scheme is a viable solution. The available bandwidth of a link, at any given time, depends on the capacity of the link and amount of the network traffic on the link at that time. The capacity of the link is a constant value, however, the network traffic on the link changes with time. Due to the increase in the usage of mobile devices, the nature of network traffic is getting more dynamic, hence the fact that it causes the available bandwidth of the link to change instantaneously. Prediction techniques can be employed to forecast the network traffic, which provides the opportunity to estimate the available bandwidth of the link for a certain period of time.

Prediction modeling is well-researched in other domains (e.g., economics, road networks and meteorology). When prediction models are evaluated, the correctness, operational speed, scalability, robustness, and the interoperability of the models are considered. Our proposed model uses the K-NN because it nicely provides a balance between operational cost and correctness, with consideration to the wired data network environment and evaluation standards. Also, authors in [17] uses the nonparametric regression method as a K-NN regression for predicting the road traffic. Road networks are very similar to data networks, as the vehicles and roads in the road networks can be considered to be like data packets and links in the data networks. Even though prediction modeling is frequently used in many related research fields, its applications in data networking are not well explored. This paper proposes a network traffic prediction model for measuring the estimated available bandwidth by using a K-NN regression algorithm.

The amount of traffic on a link at any given time is the number of bytes on a link, and can be defined as a utilization of this link. The available bandwidth of a link at any given time depends on the capacity and utilization of the link at that time. When provided with the information about the capacity and utilization of the link in a time interval, the available bandwidth (A) can be calculated by using the equation (1) [13].

$$A(t) = c_i\{1 - u_i(t, t + \tau)\} \tag{1}$$

Eq. (1) represent the available bandwidth in the time interval $(t, t + \tau)$. c_i is the capacity of link i, and $u_i(t, t + \tau)$ is the utilization of link i during the time interval. A prediction model is required to calculate the utilization of the link in a particular time interval. Predictions for the network traffic can be made based on a variable number of factors. When a uni-variant approach is used for prediction, there can be many matched points in the matching data which causes confusing predictions. If more parameters such as number of packets, flows, etc. are considered in the prediction model as a multivariate approach, we can find the nearest observation in the matching data set and the predicted value is closer to the recent observation.

Our proposed prediction model, as presented in this paper, uses a multivariate approach for the K-NN algorithm and considers the number of packets, bytes and flows to be the parameters. The training data set, consisting of raw data packets, is processed using the lag period and converted to the matching data set. Lag is defined as the period in which the values for the number of packets, bytes and flows are calculated from the training data set and added to matching data set. In terms of the training data set (consisting of 24 hours): if a lag value of 10 minutes is used, then the data points for the number of packets, bytes and flows in the matching data set will be 144. Fig. 1 shows, as the example, the matching data set for the number of packets, obtained from the training data set of 24hrs, with a lag of 10min. The training data is distributed based on the day of the week because the daily traffic patterns within a week would be similar, and the corresponding matching data set is calculated accordingly [18]. A recent observation value for the number of packets, bytes and flows are also measured during a lag. For precise prediction results, more than one recent observation is considered, and the pattern of the recent observation is matched in the matching data set. For example, we considered the values 9528, 10349 and 10421, calculated over the three lag periods, as a recent observation for the parameter number of packets. This pattern of values (recent observation) is matched in the matching data set for the number of packets, which is shown in the Fig. 1. When performing the match between the patterns, using the proposed prediction model, the goal was find the matching point that has the minimum difference between recent observations and data in matching data set.

$$tss_m = \sum_{i=m}^{L+m-1} \left[\frac{(p_{ri}-p_{mi})}{w_p}\right]^2 + \sum_{i=m}^{L+m-1} \left[\frac{(b_{ri}-b_{mi})}{w_b}\right]^2 + \sum_{i=m}^{L+m-1} \left[\frac{(f_{ri}-f_{mi})}{w_f}\right]^2 \tag{2}$$

$$for\ 1 \leq m \leq T - L + 1$$

Eq. (2), shown above, is used in the nearest matched point with the recent observation data, using a K-NN algorithm. In Eq. (2), tss is the total sum of squares, as the statistic data in the proposed model. p_{ri}, b_{ri}, and f_{ri} are the recent observation values for the number of packets, bytes and flows respectively, at lag i, where p_{mi}, b_{mi}, and f_{mi} are data values for the number of packets, bytes, and flows respectively in the matching data set, at lag i. L is the number of recent observations and data points in the matching data set used for the matching. T represents the total number of lags (data points) in the matching data set. w_p, w_b, and w_f are specific weights, and each weight can have different magnitudes (such as mean, spread, etc.) in the training data set. It is also used as a specific percentage, but the sum of the percentage of the weights could be 1. By using Eq. (2), for the recent observation values and matching data set values in the Fig. 1, we can find out the candidate matched points which are shown as the thick lines in the Fig. 1. These thick lines have tss_m values of 0.454802, 0.55233 and 0.667907 respectively, and represent the number of packet data values in the matching data set, which most closely match the recent observation values for the number of packets. These tss_m values are measured using the term $\sum_{i=m}^{L+m-1} \left[\frac{(p_{ri}-p_{mi})}{w_p}\right]^2$ from Eq. (2), which refers to the weighted average squared error (ASE) between the recent observations and the matching observations.

$$Nearest\ Matched\ Point = Min(tss_m) \tag{3}$$

The minimum value of tss_m is the nearest matched point between recent observations and the data point in the matching data set; from the above mentioned tss_m

values 0.454802 is the minimum, thus, the first thick line in the Fig. 1 is the matched point in the matching data set. The data in the next lag of the matched point in the matching data set is considered to be the predicted value.

The estimated available bandwidth of a link can be calculated by applying the predicted number of bytes as utilization in Eq. (1). By finding the estimated available bandwidth of all links, the optimal routing path in terms of bandwidth can be calculated from a source to a destination.

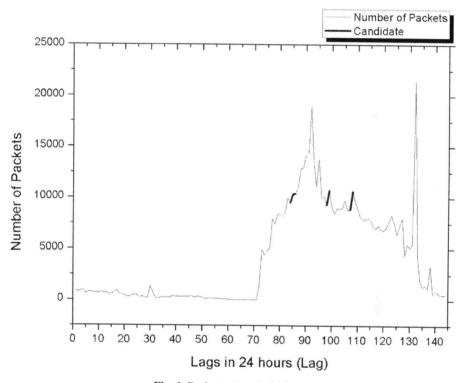

Fig. 1. Packet pattern in 24 hours

Even though the K-NN algorithm has higher accuracy as a prediction model, the cost for calculating the predicted value is higher than other prediction model. The proposed prediction model is as limited as the matching data; the precision of the proposed prediction model increases alongside the data in the training data set and the matching data set.

4   Performance Evaluation and Analysis

4.1   Simulation Environment

In this paper, we tested our proposed data network prediction model for the wired network environment. Our proposed network traffic prediction model is implemented in C++ for applicability. The simulation uses the offline PCAP files [19], instead of the real time traffic information on the switch ports. The PCAP files are generated by the Lincoln Laboratory (at M.I.T in the U.S.), and consists of data collected from Monday to Friday for seven weeks. We extracted the numbers of packets, bytes, and flows from the files to make a matching data set and sample data set (recent observations).

In the simulation, the training data set was constructed from the data for six weeks in the files, and the sample data set consists of the data for a week. Information about the three attributes (packets, bytes, and flows) is stored as a lag in the matching and sample data sets. For training purposes, in matching data set, we took the average of all the data in each lag. In the data mining area, there are many methods for training the data. The number of lags depends on the duration of the lag. We considered lag durations of 5minutes and 10minues, and the data from each day was split using several lag values. In the simulation, we only used some of each sample set because the traffic patterns from140 to 288 (in the lag duration of 5 minutes) and from 72 to 144 (in the lag duration of 10 minutes), show dynamic activity. Based on the training and sample sets, simulations were performed with different values of L (3, 4, 5, 6, 7, 8, 9 and 10), and K is considered to have a fixed value of 1.

4.2   Simulation Results

The data values in the sample data set used in the simulations are matched with all data in the matching data set. During the matching, the model finds the best matched lag point. In Fig. 2, graph (a) shows the lag point, and graph (b) shows the average distance on the separate lag values. When the data values from the sample data are matched with the data values in the matching data set, the dot is placed in the base line. As a dot arises from the line, the error (distance) increases, whereas the error decreases when a dot is placed within the proximity of the line.

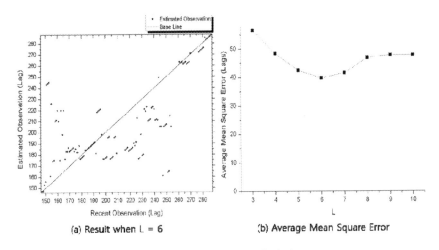

(a) Result when L = 6   (b) Average Mean Square Error

Fig. 2. Results (Lag = 5 min.)

As shown in Fig. 2, when L is 3, the average of the distance between the sample data and the training data is 56.6 lags. When L is 4, 5, 6, 7, 8, 9 and 10, the averages of the distances are 48.3, 42.4, 39.7, 41.5, 46.9, 47.8 and 47.8 respectively. The smallest average distance is when L is 6. As shown in Fig. 3, when L is 6, it has the smallest average distance (which is 9.4). When L is 3, 4, 5, 7, 8, 9 and 10, the averages are 12.2, 9.6, 9.9, 17.5, 14.4, 12.4 and 10.2 respectively. Through these results, we can conclude that more perceive matches and prediction values are obtained when the L value is 6.

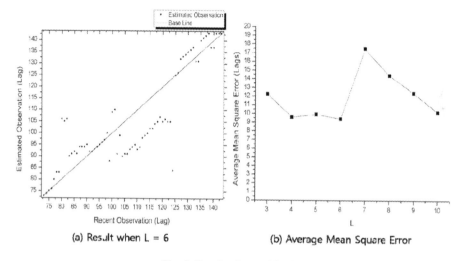

(a) Result when L = 6   (b) Average Mean Square Error

Fig. 3. Results (Lag = 10 min.)

According to the data, if L increases or decrease, then the candidates also increase or decrease. Every training data set has a different optimal values for lag, L, and K. In the training data set used in this simulation, the optimal value of L is 6 based on the above simulation results. When provided with more data, the precision of the proposed model can be tested.

5   Conclusion and Future Work

This paper proposes a network traffic prediction model. Our proposed model estimates the point in the historical data set that is similar to the current observation and then calculates the estimated available bandwidth based on a K-NN algorithm. The simulation results of the proposed model show sub-optimal performance and that is mainly because of the lack of the historical data necessary to perform extensive simulations.

In the future, we will rebuild the proposed model to achieve a higher amount of precision in the results, by gathering enough data to evaluate the performance. Further on, we will implement the proposed prediction model on the controller in the SDN network, and will consider real-time data from the network to be the recent observation. Comparison between the proposed model and other prediction models in the data networks, will also be covered in the future.

Acknowledgement. This research was supported in part by IITP (B0101-15-1366), PRCP (NRF-2010-0020210), and NGICDP (2010-0020727), Korea.

References

1. Data set to grow 10-fold by 2020 as internet of things takes off. http://www.computerweekly.com/news/2240217788/Data-set-to-grow-10-fold-by-2020-as-internet-of-things-takes-off
2. Saleh, A.A.M., Simmons, J.M.: Technology and Architecture to Enable the Explosive Growth of the Internet. IEEE Communications Magazine **49**(1), 126–132 (2011)
3. Siekkinen, M., Goebel, V., Plagemann, T., Skevik, K.-A., Banfield, M., Brusic, I.: Beyond the future internet–requirements of autonomic networking architectures to address long term future networking challenges. In: 11th IEEE International Workshop on Future Trends of Distributed Computing Systems, FTDCS 2007, pp. 89–98 (2007)
4. Gantz, J., Reinsel, D.: The Digital Universe In 2020: Big Data, Bigger Digital Shadows, and Biggest Growth in Far East. http://www.emc.com/collateral/analyst-reports/idc-the-digital-universe-in-2020.pdf
5. Rajvaidya, P., Almeroth, K.C., Claffy, K.C.: A scalable architecture for monitoring and visualizing multicast statistics. In: Ambler, A.P., Calo, S.B., Kar, G. (eds.) DSOM 2000. LNCS, vol. 1960, pp. 83–94. Springer, Heidelberg (2000)
6. Braden, R., Zhang, L., Berson, S., Herzog, S., Jamin, S.: Resource ReSerVation Protocol (RSVP) Version 1 Functional Specification. In: IETF RFC 2205 (September 1997)
7. Rajan, R., Verma, D., Kamat, S., Felstaine, E., Herzog, S.: A Policy Framework for Integrated and Differentiated Services in the Internet. IEEE Network, 36–41, September/October 1999
8. Heinanen, J.: Use of the IPv4 TOS Octet to Support Differential Services, IETF Internet Draft. draft-heinanen-diff-tos-octet-01.txt (November 1997)
9. Moy, J.: "OSPF Version 2" RFC 2328. IETF (1998)
10. Malkin, G.: "RIP Version 2", RFC 2453. IETF (1998)
11. Open Networking Foundation White Paper.: Software-Defined Networking: The New Norm for Networks (2012)
12. Peterson, L.E.: K-nearest Neighbor. Scholarpedia (2009)
13. Jain, M., Dovrolis, C.: End-to-End Available Bandwidth: Measurement Methodology, Dynamics, and Relation With TCP Throughput. IEEE/ACM Transactions on Networking **11**(4), August 2003
14. Ishak, S., Al-Deek, H.: Performance Evaluation of Short-Term Time-Series Traffic Prediction Model. Journal of Transportation Engineering, November/December 2002
15. Ishak, S., Alecsandru, C.: Optimizing Traffic Prediction Performance of Neural Networks under Various Topological, Input, and Traffic Condition Settings. Journal of Transportation Engineering (ASCE), July/August 2004

16. Ben-Akiva, M., Bierlaire, M., Koutsopoulos, H., Mishalani, R.: DynaMIT: a simulation-based system for traffic prediction. In: DACCORD Short Term Forecasting Workshop (February 1998)
17. Clark, S.: Traffic Prediction Using Multivariate Nonparametric Regression. Journal of Transportation Engineering (ASCE), March/April 2003
18. Kumpulainen, P., Hatonen, K.: Characterizing mobile network daily traffic patterns by 1-dimesional som and clustering. In: Jayne, C., Yue, S., Iliadis, L. (eds.) EANN 2012. CCIS, vol. 311, pp. 325–333. Springer, Heidelberg (2012)
19. Lincoln Laboratory. http://www.ll.mit.edu/index.html

Clustering Wireless Sensor Networks
Based on Bird Flocking Behavior

Soon-Gyo Jung[1], Sanggil Yeom[1], Min Han Shon[1],
Dongsoo Stephen Kim[2], and Hyunseung Choo[1(✉)]

[1] College of Information and Communication Engineering,
Sungkyunkwan University, Seoul, Korea
{soongyo,sanggil12,minari95,choo}@skku.edu
[2] School of Engineering and Technology,
Indiana University-Purdue University Indianapolis, Indianapolis, USA
dskim@iupui.edu

Abstract. One of the most important issues in Wireless Sensor Networks
(WSNs) is the efficient use of limited energy resources. A popular approach for
efficient energy consumption is clustering. In this paper, we propose an energy
efficient clustering algorithm, called Bird Flocking Behavior Clustering
(BFBC). By adopting the bird flocking behavior, our clustering algorithm forms
clusters with simple local interactions. With an improvement on the existing
bio-inspired clustering algorithm, that forms a cluster using several messages,
BFBC forms a cluster with only one message. Simulation results show that
BFBC significantly decreases the number of messages for cluster head election,
and also reduces the energy consumption for communication between cluster
members and their dedicated cluster head.

Keywords: Wireless Sensor Networks · Clustering · Collective behavior · Bird
flocking behavior · Bio-inspired · Swarm Intelligence · RSSI

1 Introduction

Sensor nodes are typically low energy devices with small memory and low processing
power; therefore, the efficient use of limited energy resources is the one of the key challenges of WSNs [1-2]. Clustering the sensor nodes is the well-known method in making
energy efficient protocols for WSNs [3-5]: clustering can enhance the energy efficiency
and increase the network lifetime. Swarm Intelligence (SI) is inspired by the observations of the collective behavior involved in biological activities, such as foraging of the
ant, the division of labor of the bee, and migration of the bird, and so forth [6]. SI has
the desirable properties of being adaptive, scalable, distributed and robust; these properties can be key in designing energy efficient clustering protocols for WSNs. Couzin's
model describes a self-group formation model in three-dimensional space, and investigates the spatial dynamics of grouped animals, such as fish schools and bird flocks [7].
Several clustering techniques for WSNs have been developed by adopting SI [8-11].

© Springer International Publishing Switzerland 2015
O. Gervasi et al. (Eds.): ICCSA 2015, Part IV, LNCS 9158, pp. 128–137, 2015.
DOI: 10.1007/978-3-319-21410-8_10

The bio-inspired Low-Complexity Clustering (B-LCC) algorithm is based on the flocking behavior of birds [9]. The authors employed a minimal transmission power, based coarse grained localization approach, and the sensor nodes determined their distance from its cluster head by increasing the transmission power in a stepwise manner. This algorithm achieves well-distributed cluster heads, has the lowest processing time complexity $O(1)$ per cluster, and generates low overhead for cluster head election. However, all cluster heads should broadcast their advertisement messages three times, and not all nodes use their ability to adjust their transmission power efficiently; these drawbacks would increase energy consumption.

This paper proposes an energy efficient clustering algorithm for WSNs; our algorithm employs the concept of the collective behavior of bird flocks. BFBC follows the definition of self-organization as the emergence of a system-wide adaptive structure and functionality, from simple local interactions between individual entities [6]. Our proposed algorithm utilizes the received signal strength to determine the distance of a node from its cluster head. Subsequently, the ordinary nodes determine their role, and the cluster member adjusts its transmission power when communicating with its cluster head. Simulation results show that BFBC reduces the total energy consumption for messages by about 52% for cluster head election, and the total required energy for intra-cluster communication by about 51%.

The remainder of this paper is organized as follows. The preliminaries for the proposed scheme are first introduced in Section 2. After that, the proposed scheme BFBC will be described in Section 3. Section 4 shows the results of the simulations. The conclusions are presented in Section 5.

2   Preliminaries

2.1   Network Model

We are making the following assumptions for our network model [9]. All nodes have equal abilities in terms of battery capacity without recharging, processing and communication capability. The sensor nodes have no mobility, and no external positioning system; moreover, they are left unattended after deployment. Links are symmetrical between sensor nodes. Each node has an adjustable transmission power. Additionally, the complete topology of neighbors is not known to each node.

2.2   Related Work

Flocking Behavior of Birds: Our approach refers to Couzin's model, which describes a self-group formation model, and investigates the spatial dynamics of grouped animals such as fish schools and bird flocks [7]. In the model, there are three behavioral rules for an individual: Rule 1. Individuals always try to maintain a minimum distance between themselves and others. Rule 2. Individuals tend to be attracted towards other individuals to avoid being isolated. Rule 3. Individuals tend to align themselves with their neighbors. Based on three behavioral rules of an individual, the three-dimensional space can be corresponded to the three behavioral zones, as shown in Fig. 1 (a): the zone of repulsion, the zone of attraction and the zone of orientation.

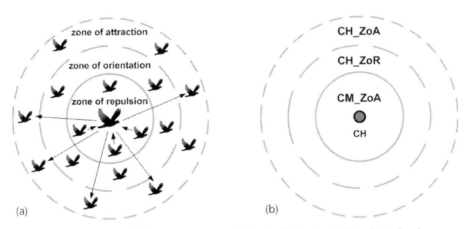

Fig. 1. The three different zones around (a) an individual and (b) a cluster head

Bio-Inspired Low-Complexity Clustering Algorithm: Bio-inspired Low-Complexity Clustering (B-LCC) algorithm does not require any knowledge of the network, such as location of node, the number of neighbors, and residual energy of node [8]. This algorithm performs in the distributed way and accomplishes a well-distributed cluster heads. Referring to Couzin's model, this algorithm defines the three types of zones around a cluster head, as shown in Fig. 1 (b): the cluster member attractive zone (CM_ZoA), the cluster head repulsion zone (CH_ZoR), and the cluster head attractive zone (CH_ZoA). Employing the minimal transmission power, based coarse grained localization approach, each cluster head broadcasts three ad messages at three fixed transmission power levels, which corresponds to the three zones. By receiving the messages, the nodes in the proximity of the cluster head can know which zone they are located in and can determine their role. In this B-LCC algorithm, there are two drawbacks. First is that the cluster head always transmits the messages three times. If the number of elected cluster heads is 40, the number of transmitted messages is 120. Second, the cluster members in the same zone transmit at the same power level, depending on how close they are to their cluster head. Hence, it causes energy waste and interference.

Received Signal Strength Indicator (RSSI): RSSI is a well-known measurement, used to measure the distance of the receiver from the source, using the strength of a received wireless signal [12]. It is obvious that the received signal strength decreases in a manner that is inversely proportional to the distance. RSSI is an appropriate method for measuring the distance of the receiver from the source. RSSI is typically an 8 or 10 bit number, and is obtained from the physical layer; the number of bits is hardware dependent [13]. The lower RSSI value means the sender is far from its receiver, and the higher RSSI value means the sender is close to its receiver. In clustering-based power-controlled routing (PCR), RSSI is used to determine the distance of the node from its cluster head, in the clustering and inter-cluster communication phase [14]. All nodes have an ability to adjust their transmission power; this means that each node can adjust their transmission power according to the RSSI. The authors mentioned that variable transmission power could not affect the performance of their proposed protocols, because the transmission power is not adjusted during the clustering and inter-clustering communication phase.

3   The Proposed Scheme

Energy efficiency is one of the key challenges of WSNs for which clustering has been recognized as an effective solution. The clustering should be fast, have low complexity, good energy efficiency, and should use variable transmission power. We propose the BFBC algorithm, inspired by the collective behavior of bird flocks, for WSNs. The BFBC algorithm utilizes the RSSI to determine the distance of a node from its cluster head. In addition, for energy efficiency in intra-cluster communication, all cluster members would adjust their transmission power according to the RSSI.

We use the concept of the three zones around a cluster head from the B-LCC. Instead of the stepwise increase of transmission power, the RSSI can help the sensor nodes around a cluster head to know their distance from the cluster head. As the sensor nodes are stationary, the zone of orientation is not adopted. A brief description of each of the zones (as depicted in Fig. 2) is listed below:

- Member Attraction (MA) Zone: The sensor nodes within this zone will become the cluster members of the cluster head.
- Head Repulsion (HR) Zone: The sensor nodes within this zone will become the cluster members of the other cluster head.
- Head Attraction (HA) Zone: The sensor nodes within this zone will become tentative cluster heads. Some of sensor nodes have the possibility to be a new cluster head.

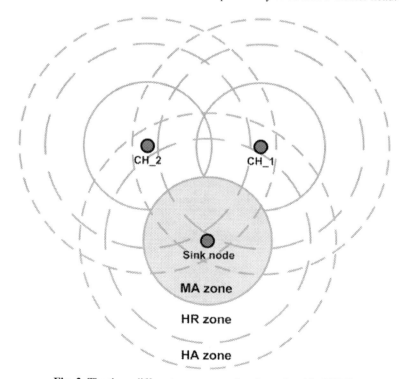

Fig. 2. The three different zones around a cluster head in BFBC

3.1   Cluster Head Election Phase

The sink node elects itself as a new cluster head by default; the new cluster head broadcasts the ad message within its transmission range. When the message is broadcasted, the ID of the cluster head and the hop distance to the sink node are inserted into the message. Upon the RSSI of the received message, the sensor nodes can recognize which zone they are in. If the sensor nodes are within the MA zone, the sensor nodes join the cluster of the cluster head. If the sensor nodes are within the HR zone, they wait for the other ad messages. The sensor nodes within the HA zone become tentative cluster heads as soon as they receive the message, and set a random back-off timer. In setting the random back-off timer, the hop distance (which is included in the message) is employed to calculate the random back-off timer. If the random back-off timer expires first, the tentative cluster head becomes a new cluster head, and broadcasts the ad message. Upon receiving the new message, the other tentative cluster heads will cancel their random back-off timers. As mentioned, if the new message is broadcasted again, the other sensor nodes determine their zone and their role. This cluster head election phase is finished after the expiration of an election phase timer.

When a tentative cluster head receives the new message before the back-off timer expires, the tentative cluster head compares its hop distance with the new hop distance, which is included in the new message. If the new hop distance is lower than previous one, the tentative cluster head will keep running its random back-off timer. If the new hop distance is higher than previous one, the tentative cluster head will reset its random back-off timer; the reason for this is that the cluster head having smaller hop distance is closer to the sink node and can reach the sink node in an energy efficient way.

The left sensor nodes, which do not decide their role after finishing the cluster head election phase, will be cluster members or tentative cluster heads, according to the number of the received ad messages. If the number of the received messages is one, the sensor node will be a tentative cluster head because the sensor node has the possibility that it is at the edge of the network topology. When some of the tentative cluster heads become new cluster heads, the new cluster heads can connect to the other left sensor nodes around them. If the number of received messages is more than two, the left sensor nodes will be cluster members of the closest cluster head. By adopting these additional steps, all sensor nodes can decide their roles and also can connect to their cluster head or the next hop cluster heads.

3.2   Communication Phase

Intra-cluster Communication Phase: The Intra-cluster Communication phase is used to send the sensed data from the cluster member to its cluster head. Each cluster member has the information of the cluster heads, which are around itself, such as ID, hop distance and RSSI. Each cluster member adjusts its transmission power, according to the recorded RSSI value, which comes from the cluster member having lowest hop distance.

We show an example of the intra-cluster communication in Fig. 3. In the example, we show a small clustered topology, which is formed by the cluster head election phase. In the cluster head election phase, each node records the RSSI value from its cluster head. For instance, cluster member A has a low RSSI value, because it is in the MA zone of CH_2, and very close to the cluster head. On the other hand, cluster member E has a high RSSI value because the member is located in the HR zone of CH_1. Cluster member E has joined the closest CH_1, after the finishing election phase, because it receives more than two ad messages. The CH_1 becomes the cluster head of cluster member C because the hop distance from CH_1 to C is smaller than the hop distance of CH_2 to C. Hence, the sensed data of the cluster member C can be transmitted to the sink node more energy-efficiently.

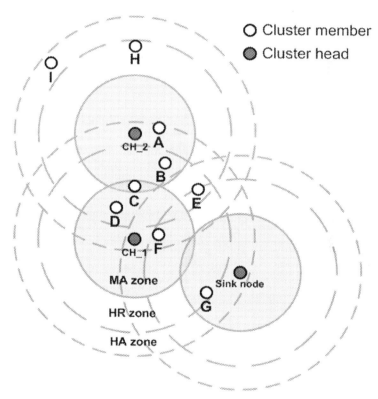

Fig. 3. An example of intra-cluster communication

Inter-cluster Communication Phase: In the Inter-cluster Communication phase, each cluster head transmits its aggregated data to the other cluster heads or sink node. All cluster heads have the information of the other cluster heads, such as hop distance and ID; the hop distance can be used to discover the topology and find the shortest route. This cluster head, which has lowest hop distance among its neighbor cluster heads, consumes more energy; thus, it is required to rotate the cluster head toward the other cluster heads. Using data packet overhearing, the residual energy of the cluster head would be shared by a data packet; the residual energy will be used as a weight to select a cluster head.

4   Performance Evaluation

4.1   Simulation Settings

The proposed BFBC algorithm was evaluated using a simulation, and compared to B-LCC. In this simulation, 1000 nodes were uniformly deployed in a field with the dimensions of $500m \times 500m$. The sink node is located in the center of the field; the transmission range of the sensor node is $100m$. The ranges of three zones were $50m$, $80m$ and $100m$. The network topology is shown in Figure 4. We can see that the cluster heads are well-distributed; we evaluated the number of transmitted ad messages, the number of received ad messages, the number of hops from the cluster heads to the sink node and the total transmission range from the cluster members to their cluster heads. For energy consumption evaluation, we employed the same energy model and network parameters as [15-16].

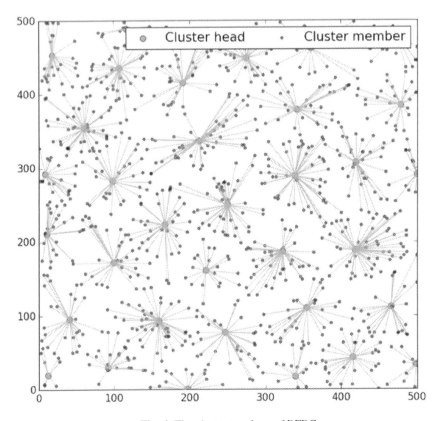

Fig. 4. The cluster topology of BFBC

4.2   Simulation Results

Compared to the B-LCC algorithm, our proposed scheme uses a smaller number of the ad messages. In the B-LCC algorithm, the cluster head broadcasts the ad message three times per cluster; however the cluster head in the BFBC algorithm broadcasts the ad message just once per cluster. Hence, this algorithm decreases the number of transmitted ad messages by about 65% and the energy consumption by about 52%.

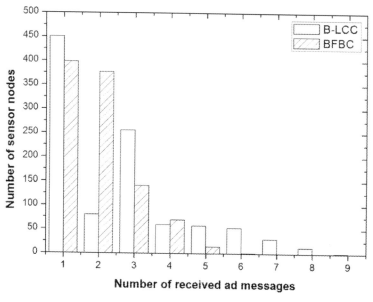

Fig. 5. The comparison of the number of received ad messages

The sensor nodes receive several ad messages until they determine their role. Figure 5 illustrates a comparison of the number of received ad messages between BFBC and B-LCC. Very few nodes are unable to determine their role after the basic clustering process; however, there are two exceptions. Some of the nodes are within the HR zones of some cluster heads. The reset of the nodes are located around the border of the sensing field. After the expiration of the election phase timer, the left sensor nodes can choose the nearest cluster head or can be tentative cluster heads if they have received only one ad message. From these cases, the evaluation of the received ad messages can be explained. In BFBC, each node receives about 1.9 ad messages, on average. On the other hand, the average of the received ad messages for B-LCC is 2.5. A cluster head in BFBC broadcasts an ad message one time. Therefore, each sensor node can determine its distance from the cluster head by receiving only one ad message. In B-LCC, a cluster head broadcasts the ad message three times by increasing its transmission power in a step-wise manner. Consequently, the sensor nodes around the CM_ZoA of the cluster head can determine only one ad message; however, the almost sensor nodes around the CH_ZoR of the cluster head determine their role after receiving three ad messages. These results clearly demonstrate that our proposed algorithm significantly decreases the number of ad messages, with a little overhead.

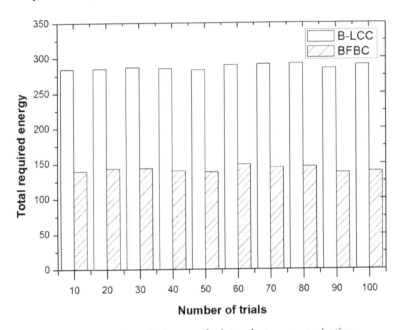

Fig. 6. The total required energy for intra-cluster communication

Fig. 6 represents the total required energy for transmission from the cluster members to their cluster heads. As explained in Section 3.2, all clusters adjust their transmission power according to the RSSI value. Compared to B-LCC, our algorithm is more fine-grained and is not limited to three power levels. In the simulation, our proposed algorithm, BFBC, reduces the total required energy for intra-clustering communication by about 51% compared to the B-LCC. From this result, we can say that the network lifetime can be prolonged.

5   Conclusion

In this paper, we proposed a BFBC algorithm, inspired by the collective behavior of bird flocks, for WSNs. Our proposed algorithm adopts the RSS, to determine the distance from its cluster head to a node. Thereafter, the sensor nodes would learn their role using the determined distance, and the cluster member would adjust its transmission power when communicating with its cluster head. The simulation showed that this algorithm not only decreases the total energy consumption for ad messages by about 52%, but also reduces the total required energy for the intra-clustering communication by about 51%.

Acknowledgments. This study was supported in part by IITP (B0101-15-1366), ICT R&D program (10041244), NGICDP (2010-0020727) and PRCP (NRF-2010-0020210), Korea.

References

1. Abbasi, A.A., Younis, M.: A survey on clustering algorithms for wireless sensor networks. Computer Communications **30**(14), 2826–2841 (2007)
2. Villas, L.A., Boukerche, A., Ramos, H.S., de Oliveira, H.A., de Araujo, R.B., Loureiro, A.A.F.: DRINA: A lightweight and reliable routing approach for in-network aggregation in wireless sensor networks. IEEE Transactions on Computers **62**(4), 676–689 (2013)
3. Younis, O., Fahmy, S.: HEED: a hybrid, energy-efficient, distributed clustering approach for ad hoc sensor networks. IEEE Transactions on Mobile Computing **3**(4), 366–379 (2004)
4. Jin, Y., Wang, L., Kim, Y., Yang, X.: EEMC: An energy-efficient multi-level clustering algorithm for large-scale wireless sensor networks. Computer Networks **52**(3), 542–562 (2008)
5. Afsar, M.M., Tayarani-N, M.H.: Clustering in sensor networks: A literature survey. Journal of Network and Computer Applications **46**, 198–226 (2014)
6. Bonabeau, E., Dorigo, M., Theraulaz, G.: Swarm intelligence: from natural to artificial systems. Oxford University Press (1999)
7. Couzin, I.D., Krause, J., James, R., Ruxton, G.D., Franks, N.R.: Collective memory and spatial sorting in animal groups. Journal of Theoretical Biology **218**(1), 1–11 (2002)
8. Selvakennedy, S., Sinnappan, S., Shang, Y.: A biologically-inspired clustering protocol for wireless sensor networks. Computer Communications **30**(14), 2786–2801 (2007)
9. Zhang, Q., Jacobsen, R.H., Toftegaard, T.S.: Bio-inspired low-complexity clustering in large-scale dense wireless sensor networks. In: IEEE GLOBECOM, pp. 658–663 (2012)
10. AbdelSalam, H.S., Olariu, S.: Bees: Bioinspired backbone selection in wireless sensor networks. IEEE Transactions on Parallel and Distributed Systems **23**(1), 44–51 (2012)
11. Karaboga, D., Okdem, S., Ozturk, C.: Cluster based wireless sensor network routing using artificial bee colony algorithm. Wireless Networks **18**(7), 847–860 (2012)
12. Wang, Y., Guardiola, I.G., Wu, X.: RSSI and LQI Data Clustering Techniques to Determine the Number of Nodes in Wireless Sensor Networks. International Journal of Distributed Sensor Networks **2014** (2014)
13. Saxena, M., Gupta, P., Jain, B.N.: Experimental analysis of RSSI-based location estimation in wireless sensor networks. In: COMSWARE 2008, pp. 503–510 (2008)
14. Madani, S.A., Hayat, K., Khan, S.U.: Clustering-based power-controlled routing for mobile wireless sensor networks. International Journal of Communication Systems **25**(4), 529–542 (2012)
15. Le, H.N., Zalyubovskiy, V., Choo, H.: Delay-minimized energy-efficient data aggregation in wireless sensor networks. In: The Proceedings of International Conference on Cyber-Enabled Distributed Computing and Knowledge Discovery, pp. 401–407 (2012)
16. Cheng, C., Tse, C.K., Lau, M.: A Delay-Aware Data Collection Network Structure for Wireless Sensor Networks. IEEE Sensors Journal **11**(3), 699–710 (2011)

Design and Implementation of an Easy-Setup Framework for Personalized Cloud Device

Bonhyun Koo[✉], Taewon Ahn, Simon Kong, and Hyejung Cho

IoT Vertical Solution Lab, DMC R&D Center, Samsung Electronics, P.O. BOX 105, 416, Maetan-3dong, Dong Suwon, Yeongtong-gu, Suswon-si, Gyeonggi-do 442-600, Korea
{bonhyun.koo,taewon.ahn,simon.kong,cho1115}@samsung.com

Abstract. With the popularization of Public Cloud services, requirements for Personal Cloud services with more enhanced security and independent personalized storage are also increasing. In order to provide Personal Cloud services, a variety of Home Cloud devices available in the home have been introduced. Based on the analysis of these devices and previous researches, we extracted critical restrictions regarding the initial configuration such as the complex procedures, limitation of installation environment, and the problem that users wasted a lot of time for its setup. Thus, we propose a novel wireless network-based Easy-Setup framework using smartphones and Android-based personal cloud devices-HomeSync in order to improve these restrictions. In addition, we constructed an experiment test-bed to validate the effectiveness of our proposed framework. Finally, we present the results of the comparative experiments compared with previous researches in terms of consumption time and initial setup procedures.

Keywords: Personal cloud · Cloud computing · Contents sharing · Cloud privacy · Personal security · Easy-Setup · Initial configuration

1 Introduction

With the advent of various converged Cloud services, requirements for more independent and secure enhanced personal storages are increasing [1][2]. According to these requirements, home cloud devices which can provide additional services and media gateway functions to network router such as Network Attached Storage in our home are introducing [3-6]. However, these devices generally require a large and complex process during the initial setup of the device, the registration of user account information, network settings, and the management server registration [7]. In this paper, we propose the framework architecture in which can reduce these complex procedures and user input interaction by using smart phones in order to improve these complex initial setup process and user conveniences [8]. To validate our proposed framework, we also introduce a personal cloud device which can be shared with group members and up/download contents both inside and outside.

The rest of the paper is organized as follows: Section 2 talks about related works for conventional initial setup and configuration technologies for personal cloud devices.

O. Gervasi et al. (Eds.): ICCSA 2015, Part IV, LNCS 9158, pp. 138–147, 2015.
DOI: 10.1007/978-3-319-21410-8_11

Section 3 describes the framework and architecture of proposed Easy-Setup Framework. Section 4 presents experimental analysis and implementation results. Finally, section 5 concludes this paper and provides a future outlook.

Table 1. Comparison of initial setup procedure

	Pogoplug Series4	Google Nexus Q	LG NT3 N1T1	Synology DS-112
App. Type	WEB UI (+Mobile App.)	Mobile App	PC App. (+Mobile App.)	PC App
Ethernet Support	Default	Support	Default	Default
WiFi Support	N/A	Support	N/A	N/A
User Input (for Device)	3	2	2	2
User Input (for Client)	8	3	7	N/A
Setup time (Ethernet)	40sec	45sec	150sec	18min
Setup time (WiFi)	N/A	1min 30sec	N/A	N/A

2   Motivation and Considerations

In this section, we investigate the previous research problems in the domain of personal cloud devices and describe the necessary factors to solve or mitigate the problems. The idea of applications and products of various personal devices within home cloud services has been conducted by a number of researches. Based on the market share and in terms of published time of research result, representative cloud devices are Pogoplug Series9, LG NT3-N1T3, Google Nexus Q, Synology DS-112, and so on[9][10]. In this section, we have analyzed and compared overall input procedures and required time for during the initial setup of these personal cloud devices. Based on the result of these analyses, we have extracted and summarized for additional requirements should be improved on these setup procedures. The following Table1 is shown in the result of our comparisons and experiments regarding application S/W and the network type, the number of user inputs, and required time for initial setup. The network environment of this experiment consists of ADSL (10~50Mbps), Ethernet 100M, and WiFi 802.11n.

(A) Cable connection   (B) Device registration (Web)   (C) Exclusive App. (Mobile)

Fig. 1. Pogoplug initial device setup (cable connection) and related applications

First, we compared client application type which is used for initial network setup and registration of cloud devices. In case of NT3 N1T3 and DS-112 except for Nexus Q, PC client should be used. Pogoplug case, it supports mobile applications for contents sharing, but it also should use Web browser on PC during initial setup for cloud device as shown in Figure 1 (B) and (C).

Second is related with supporting of wired/wireless network interfaces in the initial setup process for these devices. Table shows that most of target devices except Nexus Q, Ethernet cable should be used for the devices setup. (Not supporting for WiFi based setup) [12]. Unlike most of home appliances recently are executed on the wireless network, it is able to know that is accompanied by limitations in terms of operation and space associated with WiFi not supporting[13].

Third, it is the result of a comparison of the required time and the number of user input during the initial setup. The number of inputs on the cloud device side includes connection procedure such as power, LAN cable, etc. In case of client input, we compared the number of times that should be input through the mouse and screen touch from the mobile device or PC. Based on this, DS-112 and N1T3 are required two times for the number of device input. In case of Pogoplug, it is required three times including connection procedure for additional external HDD and LAN connection as shown in Figure 1(A). These connection procedures of cable on initial setup, it can be a part to reduce significantly the convenience of products and to waste user time. In case of Nexus Q, it is required one more step for NFC tagging procedure to download exclusive application from App market (as illustrate in Figure 2(A) and (B)).

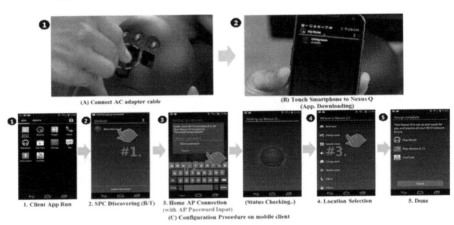

Fig. 2. Pogoplug initial device setup (cable connection) and related applications

The number of input on client, Pogoplug was needed eight times input as most by Web Browser excluding serial number input procedure, N1T3 were required seven times user input through the PC. DS-112 is a type which embedded OS is installed in the device via PC based dedicated applications. After installation, complex and multiple user input processes regarding DDNS and Port forwarding configuration of Home AP for connection from outside were required depends on your needs. Finally, in case of Nexus Q, we need to install a dedicated application for initial setup on smartphone. It also required three times of inputs for device identification, input and submission of Home AP password, final location selection as shown in Figure 2(C). However, reduction of the number of additional steps, which is caused by input procedures for AP password receiving and user's input errors, is should be improved in order to support the convenience of wireless network environment.

Finally, we compared the total consumption time for initial procedures, including account creation and process of registration to service server to use the cloud services of each device. As experiments result, Pogoplug showed the fastest time about 40 seconds, DS-112 was needed a long time of 18 minutes or more required in accordance with the installation of the OS. In case Nexus Q, although it supports the setting function of the device based on the wireless network, but it was confirmed that a long time compared with wired network is consumed 1 minute and 30 seconds. Even if considering of wireless network environment, the consumption time of 1 minute or over causes the reduce problem in terms of convenience of the configuration process of the devices to the users.

We analyzed and extracted major problems through the comparative analysis of the experiment: decrease of convenience and increase of setup time in response to complex user input procedures from the client and cloud devices, and restrictions that must use PC and Ethernet connection in the process of initial setup. In this paper, we introduce an Easy-Setup framework utilizing the mobile device in a wireless network environment in order to overcome these constraints.

3   Framework Architecture

The purpose of this paper is to propose and validate a designed framework which can configure initial setup effectively and improve problem of previous researches and technologies as introduced in Section 2. In this section, we present the proposed Easy-Setup framework architecture which can be used to setup cloud devices and describe overall initial setup and provisioning flow based on our framework.

3.1   Framework Design

The proposed Easy-Setup framework architecture is composed of HomeSync, HomeSync client, and management server as shown in Figure 3. The basic role of each of these components is:

— **HomeSync:** personal cloud service device
— **HomeSync client:** taking care of initial setup and media contents handling on the smartphones
— **Management server:** processing of the management of required information.

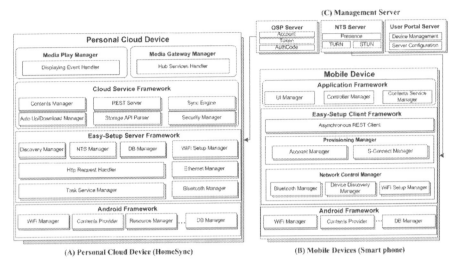

Fig. 3. Overall Easy-Setup Framework Architecture

We now describe in detail the functions of each component in Figure 3. HomeSync is a personal cloud device that can be installed on your house as comparative analyzed devices in section 2. S/W architecture structure has been designed based on Android Framework which is used generally on the smartphones. In case of HomeSync, it is composed of Easy-Setup Server Framework for initial setting (ESSF), Cloud Service Framework (CSF) required content management, and Service Manager needed for application services (Media Play Manager and Media Gateway Manager) as illustrated in Figure 3(A). Ethernet Manager and WiFi Setup Manager of ESSF are responsible for connection to Home AP and receiving of network information from Client. Bluetooth Manager (BM) takes care of setting information exchange via a Bluetooth interface that is used primarily during the initial setup. Discovery Manager is in charge of processing and response for connection and discovery request from client on initial network setup. NTS Manager takes charge of management of connections to server for requests of mobile devices to access HomeSync from inside and outside of home. DB Manager handles information storing and command processing of required various client and network status information in the setup process. Task Service Manager is responsible for ensuring the reliability and prioritization of responses through the creation and scheduling of the task threads to process request messages of clients. Http Request Handler is in charge of REST Query processing such as change requests for network status and account information from clients. Cloud Service Framework (CSF) is composed of a sub modules that are responsible for media contents management mainly. Contents Manager, Sync Engine, and Auto Up/Download Manager take care of management for the download, upload and synchronization of contents from remote clients. It also performs the processing of the main API through standardized API and parser modules. In addition, Security Manager performs the processing of the confidentiality and integrity of the request and response and the encryption and decryption of contents data. Finally, Media Play/Gateway Manager reproduces to monitor or TV regarding the internal stored media contents by using the output ports such as HDMI of HomeSync. These managers also provide contents handling services to manage the services and Download additional content.

HomeSync Client as a delegator device of HomeSync is in charge of processing the server registration and connection configuration of device information. Figure 3(B) illustrates Android based framework architecture of HomeSync client. Easy-Setup Client Framework (ESCF) consists of Network Control Manager (NCM), Provisioning Manager (PM), and Asynchronous REST Client (ARC) in order to process of initial setup on client side. NCM is composed of sub modules such as Bluetooth Manger, WiFi Setup Manager, and Device Discovery Manager. This is responsible for processing and configuration of the network connection and discovery of HomeSync on initial setup by using a mobile device. PM processes the tasks of management to register HomeSync information to server by using Account and S-Connect Manager based on accounts logged in the client. It also handles request/response information with the device management server and HomeSync with REST client. Application Framework of Client has three sub modules, Contents Service Manager (CSM), UI Manager (UM), and Controller Manager (CM). CSM is taking care of the management of contents sharing and playing with HomeSync. UM is responsible for processing the user input and displaying to the screen the contents information via application UI. CM provides the controlling service by using smart phones as controllers.

In order to manage required information between HomeSync and clients, there are three kinds of server: OSP, User Portal and NTS server. OSP server is taking care of processing and authentication of user account server that is operated on the basis of Samsung SSO (Single Sign On) policy. In case of User Portal server, it is responsible for not only the ID issuance required for NTS server, but also services coordination with HomeSync and client. NTS (Network Traversal Service) Server supports routing functions to access from outside through IP identification of HomeSync based on the account information and Device ID.

3.2   Initial Network Setup Procedure

In order to improve the constraints conditions in which should use always legacy PC and Ethernet, to reduce the long setup time caused by complex user inputs of the client and cloud device as discussed in Section 2, we have proposed Easy-Setup Framework. In the following, we introduce how to setup automatic wireless network by using a smart phone based proposed Framework.

In order to access to HomeSync of in-home from outside, overall device configuration and procedure of wireless network setup are as shown in Figure 4. To proceed with the configuration of HomeSync, the dedicated client application have to be installed to a smart phone.

When we tag the mobile device to NFC (Passive) Tag, which is equipped with HomeSync, it can download and install the HomeSync Client S/W from the App Store automatically (1). NFC Tag includes two types of information: URL information (for market) and MAC address of HomeSync.

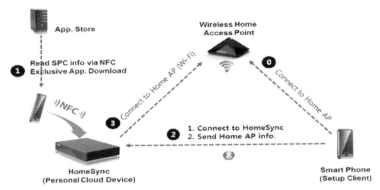

Fig. 4. Overall wireless AP setup flow

4   Implementation and Experiment Result

In this section, we present a result of our experiments and implementation based on Easy-Setup framework in order to validate and prove our proposed framework compared with previous researches. To proof an efficient implement and validation for smart phone based Easy-Setup solution, we have built our experimental test-bed. Multimedia content of HomeSync is able to play to TV and the Audio system via the HDMI and S/PDIF output in the home. Mobile device can play the contents stored in HomeSync from the outside, and it is also possible to backup new contents to HomeSync in real-time by the 'Auto Upload' function. Based on this infrastructure, experiment devices of our test-bed are composed of Samsung HomeSync (GT-B9150) as personal cloud device, smart phones (SGH-I337) as control devices and Smart TV (UN46C8000) in our framework. We have implemented a dedicated Client Application that provides Cloud services and auto-configuration feature of personal cloud device as a research result of this paper. H/W basic specifications of HomeSync consists 1.7GHz dual-core Gaia chipset, 1GB RAM, and 1TB HDD. In addition, it also supports multiple interface types such as WiFi, Bluetooth, USB 3.0, Optical Audio, and HDMI out. S/W Framework has been implemented based on Android JBP (Jelly Bean Plus). Figure 5 shows HomeSync and a smartphone Galaxy S4 used in HomeSync Client, and it is a screen that the smartphone tag to HomeSync by using NFC in order to process the configuration of initial Wireless Network Setup and Provisioning. We have developed our system so that all procedures can be completed at one time by NFC Tagging without user inputs as shown in Figure 5. We also have to consider the problem of backwards compatibility since the upgrade cycle of H/W and S/W Specifications becomes shorter than before. For these reasons, we have applied flexible user interfaces design in order to support smart phones which are not support JBP Platform and NFC environment required by proposed our framework.

Fig. 5. HomeSync Easy-Setup with a smart phone

In case of using the smart phone supporting NFC and JBP platform more higher version, it is possible to be complete the all configurations by the first NFC tagging as shown in Figure 6 (A). However, if NFC is not supported (or if Disable status), user can perform 'Pairing' procedure with HomeSync directly regarding the manual Bluetooth discovery process as shown in phase 3 and 4 of Figure 6 (B). Input procedure of Bluetooth password in step 4 is needed to protect against session hijacking attacks from malicious users. It is possible to move to the next step by entering the random four digit values that is output via the HDMI output, or by pushing the setting button of HomeSync directly. In order to support Android versions that does not support WiFi password auto-configuration feature proposed in this paper, we also have implemented another processing logic which is able to handle by the user's input as illustrated Figure 6 (B). We can overcome the problem of backward compatibility through this approach.

From experimental results, we can see that the whole time of HomeSync configuration is required average 30 seconds (average 16 seconds for Ethernet base). It means that we have reduced the initial setup time to 60% as compared with Pogoplug, and 30% compared to Nexus Q. Regarding the number of input, we have shortened with one time processing (NFC Tagging) for the device and account registration procedures which seven or eight times are required. Based on our implemented result, it is possible to overcome the restrictions of previous researches, and it will be able to apply as the more easy and convenient 'One-Step' Setup Solution.

Fig. 6. Easy-Setup User interfaces on a smart phone

5   Conclusion and Future Works

In this paper, we extracted the hurdles to reduce user inconvenience and long setup time through comparative analysis of the initial setup procedures regarding representative personal cloud devices. In order to improve these restrictions, we proposed a novel Easy-Setup framework utilizing smartphone based on wireless network. Moreover, we built a test-bed, which was composed of TV, HomeSync, and smart phones, to proof the efficiency of our proposed framework. Through the test-bed, this paper presents the implementation and experimental result that was shortened the whole initial setup time without complicated user inputs. Specifically, it also verified the effectiveness of NFC based 'One-Step' setup.

Future work will focus on applying the extension of HomeSync services integrated with home control services. Through this approach, HomeSync can be applied as the central device of Smart Home Service which will be able to control ZigBee and WiFi based home appliances[16][17]. Therefore, this will be able to take advantage of the Smart Home Point Solution which can configure and control easily for the all devices in the home [18] [19].

References

1. Kazi, A., et al.: Supporting the personal cloud. In: Proc. of 2012 2nd IEEE Asia Pacific Cloud Computing Congress (APCloudCC), pp. 25–30 (2012)
2. Dillon, T., et al.: Cloud computing: issues and challenges. In: Proc. of 2010 24th IEEE International Conference on Advanced Information Networking and Applications (AINA), pp. 27–33 (2010)
3. Google Nexus Q Official Web site. https://play.google.com/store/devices/details/Nexus_Q?id=nexus_q
4. Pogoplug Product Web site. http://pogoplug.com/devices#features
5. NT1 Offical Web site. http://www.lge.co.kr/lgekr/product/detail/LgekrProductDetailCmd.laf?prdid=EPRD.24399
6. DS-11 Specification. http://www.synology.com/products/spec.php?product_name=DS112
7. Abolfazli, S., et al.: Mobile cloud computing: a review on smartphone augmentation approaches. In: Proc. 1st Int'l Conf. Computing, Information Systems, and Communications, pp. 199–204 (2012)
8. Guan, L., et al.: A survey of research on mobile cloud computing. In: Proc. 2011 10th IEEE/ACIS International Conference on Computer and Information Science, pp. 387–392 (2011)
9. Tian, Y., et al.: Towards the development of personal cloud computing for mobile thin-clients. In: Proc. International Conference on Information Science and Applications (ICISA), pp. 1–5 (2011)
10. Gopalan, K.S., et al.: A cloud based service architecture for personalized media recommendations. In: Proc. 2011 Fifth International Conference on Next Generation Mobile Applications and Services, pp. 19–24 (2011)
11. Nexus Q Setup Guide book. http://commondatastorage.googleapis.com/support-kms-prod/SNP_2672134_en_v0

12. Pogoplug Series 4 review. http://www.theverge.com/2012/4/5/2910802/pogoplug-series-4-review
13. Yoon, H., Kim, J.W.: Collaborative Streaming-based Media Content Sharing in WiFi-enabled Home Networks. IEEE Transactions on Consumer Electronics, 2193–2200 (2010)
14. Synology NAS DMZ setup guide. http://www.synology.com/support/tutorials_show.php?q_id=456
15. Naito, K., et al.: Proposal of seamless ip mobility schemes: network traversal with mobility (NTMobile). In: Proc. 2012 IEEE Global Communications Conference (GLOBECOM), pp. 2572–2577 (2012)
16. Koo, B.H., et al.: R-URC: RF4CE-based universal remote control framework using smartphone. In: Proc. International Conference on Computational Science and Its Applications (ICCSA 2010), pp. 311–314 (2010)
17. ZigBee RF4CE Standard. http://www.zigbee.org/Markets/ZigBeeRF4CE/tabid/413/Default.aspx
18. Gu, H., et al:. The design of smart home platform based on cloud computing. In: Proc. 2011 International Conference on Electronic & Mechanical Engineering and Information Technology, pp. 3919–3922 (2011)
19. Yang, Y., et al.: A cloud architecture based on smart home. In: Proc. 2010 Second International Workshop on Education Technology and Computer Science, pp. 440–443 (2010)

Security Improvement of Portable Key Management Using a Mobile Phone

Jiye Kim[1], Donghoon Lee[1], Younsung Choi[1], Youngsook Lee[2], and Dongho Won[1(✉)]

[1] College of Information and Communication Engineering,
Sungkyunkwan University, Seoul, South Korea
jykim.isg@gmail.com, {dhlee,yschoi,dhwon}@security.re.kr
[2] Department of Cyber Investigation Police,
Howon University, Gunsan-si, South Korea
ysooklee@howon.ac.kr

Abstract. Users often store sensitive information on their laptops, but it can be easily exposed to others if a laptop is lost or stolen. File encryption is a common solution to prevent the leakage of data from lost or stolen devices. For the management of strategies like this, key management is very important to protect the decryption key from attacks. Huang et al. proposed a portable key management scheme, whereby a laptop shares secret values with a mobile phone. Their scheme is convenient as well as practical because it is not reliant on a special device or password input. However, we found that it is still vulnerable to an attack if a laptop is stolen. In this paper, we analyse the security of Huang et al.'s scheme and propose a solution to the outstanding vulnerability. Our proposed scheme exploits two types of keys including a one-time symmetric key to protect the file decryption key. Additionally, the security improvement does not compromise the convenience of the portable key management scheme.

Keywords: Lost/stolen portable devices · Data protection · Key management · Secret sharing · Symmetric/asymmetric ciphers · Session keys

1 Introduction

Recently, the use of portable devices such as laptops has increased significantly because of their convenience to users, who can use them in any location. A user often stores sensitive data on their laptop, from private financial, health, and security information to the business secrets of their company [1][2]; all of which can be easily exposed to others if the laptop is lost or stolen [1][3][4][5]. To prevent the leakage of data from lost or stolen devices, several technological solutions have been proposed and the encryption of sensitive files that can only be decrypted by a legitimated user is a commonly used security measure [1][6][2][7]. With data encryption, key management is very important to protect decryption keys from potential attackers [8][1][9]. Several key management

© Springer International Publishing Switzerland 2015
O. Gervasi et al. (Eds.): ICCSA 2015, Part IV, LNCS 9158, pp. 148–159, 2015.
DOI: 10.1007/978-3-319-21410-8_12

schemes have been proposed thus far [1][6][2][7]: Corner and Noble proposed that key encryption key(KEK), which is stored in a small token worn by the user instead of being stored on the laptop, is used to encrypt the file decryption key [1]; MacKenzie and Reiter proposed that a laptop should interact with a remote server to retrieve a cryptographic key [6]; and Studer and Perrig proposed that a cryptographic key should automatically be recovered, but only when a user is located in specific locations like their home or office [2]; Chang et al. proposed a key management scheme whereby a laptop shares secret values with a universal serial bus(USB) device using Shamir's secret sharing method [10][11]. However, usability was inadequately considered during the design of such schemes, as users are typically required to manage a password or a special device, such as a wearable token.

Huang et al. proposed a portable key management scheme, whereby a laptop shares secret values with a mobile phone [7]. In their scheme, a mobile phone sends several secret values to a laptop for the decryption key reconstruction [7]. In addition, the key reconstruction process can be performed automatically because it does not require a user's password or biometric data, while the wireless functionality of both the laptop and mobile phone facilitate communication between the two devices [7]. Practicality is another advantage, as the mobile phone is the only additional device that is required for the scheme [7]. Huang et al. insisted that their scheme is secure against attacks in the case of a stolen laptop because an attacker cannot derive the decryption key without the corresponding mobile phone [7]. However, we found that their scheme is still vulnerable to attacks in the case of a stolen laptop. In this paper, we analyse the security of Huang et al.'s scheme and propose a way to improve its security. Our proposed scheme exploits two types of keys including a one-time symmetric key and double encrypts messages using both of them. Moreover, the improvement of the security of the scheme does not compromise its convenience and practicality in any way.

The remainder of the paper is organized as follows: section 2 presents a review of Huang et al.'s scheme; section 3 is an analysis of the security of their scheme; section 4 proposes the improved scheme; section 5 analyzes the security of the proposed scheme against possible attacks; section 6 briefly introduces related works about key management for a portable device; and section 7 concludes this paper.

2   Review of Huang et al.'s Scheme

In Huang et al.'s scheme [7], sensitive files are stored on a laptop after they are encrypted with a secret key. The key is generated using two secret values that are respectively stored on a laptop and a mobile phone. The laptop must receive the other secret value from the mobile phone to reconstruct the key. Table 1 shows the notations used in the remainder of the paper. The sensitive files are encrypted using the secret key k, where $k = a \oplus b$. In the initialization phase, a and b are stored on the laptop L and the mobile phone M, respectively. Also,

Table 1. Notations

Symbol	Description
L	User's laptop
M	User's mobile phone
TTP	Trusted third party
ID_l	Identity of L
sk	Symmetric key shared between L and M (or TTP)
k	Secret key to encrypt or decrypt sensitive files ($k = a \oplus b$)
k_{pr}, k_{pb}	Private or public key used for asymmetric cryptography
$E_K(\cdot), D_K(\cdot)$	Symmetric encryption or decryption using the key K
$\hat{E}_K(\cdot), \hat{D}_K(\cdot)$	Asymmetric encryption or decryption using the key K
$h(\cdot)$	One-way hash function
$\oplus$	XOR operation
$\|$	Concatenation operation
$\leq, =?$	Verification operation
RN_l, RN_m	Random nonce of L or M
t_l, t_m	Current timestamp of L or M
Δt	Expected time interval for transmission delay
K_REQ	key request

L and M share a symmetric key sk to encrypt or decrypt messages transmitted between them.

Fig. 1 shows the overall process of Huang et al.'s scheme. L sends the key request with ID_l to M when the file access is required. When M receives the message from L, it computes $I = ID_l \parallel t_m$ and $H\_I = h(I)$, where t_m is the current timestamp of the mobile phone system and returns the response message $\{I, H\_I\}$ to L. Then, L verifies the integrity of the response message by checking $h(I) =? H\_I$. If the verification is not passed successful, then the process is aborted; otherwise, the next step proceeds. L generates a random nonce RN_l and encrypts $RN_l \parallel ID_l \parallel t_m$ using the key sk. L sends the encryption result X and the hash value of X to M. When M receives the message, it verifies the message integrity after it decrypts X using the symmetric key sk ((1) to (5)).

$$RN_l \| ID_l^* \| t_m^* = D_{sk}(X) \tag{1}$$

$$ID_l^* =? ID_l \tag{2}$$

$$t_m^* =? t_m \tag{3}$$

$$H\_X^* = h(X) \tag{4}$$

$$H\_X^* =? H\_X \tag{5}$$

If the verification is not successful, then the process is aborted; otherwise, the next step proceeds. M computes $Tmp = RN_l \oplus b \oplus RN_m$ and generates a random nonce RN_m. Then, it encrypts $RN_m \parallel Tmp \parallel t_m$ using the key sk and sends the encryption result d, as well as the hash value of d to L. When L receives the message from M, it verifies its integrity and decrypts d using sk.

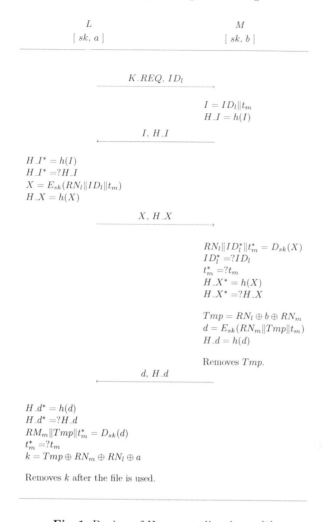

Fig. 1. Review of Huang et al's scheme [7]

L derives the file decryption key k by computing $k = Tmp \oplus RN_m \oplus RN_l \oplus a = a \oplus b$ because $Tmp = RN_l \oplus b \oplus RN_m$. Both the laptop and mobile phone use wireless technology like Bluetooth to communicate with each other. The laptop periodically sends "ALIVE" messages to the mobile phone and also receives responses to them. If the laptop does not receive a response message from the mobile phone within a specific time interval, it considers the mobile phone to be outside the wireless communication range and re-encrypts its decrypted files. L deletes the key k after the file is used.

3   Security Analysis of Huang et al.'s Scheme

Huang et al. insisted that their scheme prevents an attacker from knowing the secret key k even if they have the laptop in their possession, as they are without the mobile phone [7]. However, we found that a sophisticated attacker can still obtain the key k using only the stolen laptop. Assuming that an owner is using their laptop in a public place and the attacker is located nearby, the attacker can eavesdrop on messages transmitted between the laptop and the mobile phone. This assumption is reasonable enough since both devices communicate with each other via a wireless connection, while the communication ranges of several wireless technologies including Bluetooth, Wi-Fi, and Zigbee are between a few meters and several tens of meters [12]. If an attacker is located within the wireless communication range of a laptop or mobile phone, he/she can eavesdrop on and record any messages $\{K\_REQ, ID_l\}$, $\{I, H\_I\}$, $\{X, H\_X\}$ or $\{d, H\_d\}$ transmitted between two devices during the key reconstruction process. Then, if he/she steals the laptop and extracts the message decryption key sk from it, he/she can decrypt the messages X and d, using the key sk ((6) and (7)).

$$RN_l \| ID_l^* \| t_m^* = D_{sk}(X) \tag{6}$$

$$RM_m \| Tmp \| t_m^* = D_{sk}(d) \tag{7}$$

The attacker can simply compute $k = Tmp \oplus RN_m \oplus RN_l \oplus a = a \oplus b$ by using the decryption result Tmp, RN_m, RN_l, and the secret value a extracted from the laptop. The attacker can therefore access the owner's sensitive laptop files using the file decryption key k.

Also, Huang et al.'s scheme does not consider the possibility of the attacker stealing both the laptop and the mobile phone at the same time. Although this situation does not occur frequently compared with the loss of only one of the devices, it is still likely to occur because of the portability and ubiquity of both. If an attacker steals both, he/she can extract two secret values a and b from L and M, respectively, and will be able to compute the file decryption key k.

4   The Proposed Scheme

In this section, we present the assumptions, goals, and design outline of our proposed scheme. Then, we describe the proposed scheme in detail.

4.1   Goals and Assumptions

The goal of the design in our proposed scheme is to achieve user convenience and practicality, as in Huang et al.'s scheme [7]. Therefore, the proposed scheme should not require a user's password or biometric data for the file decryption key reconstruction. Also, it should not require the use of any special device.

Moreover, all parties have to communicate using wireless technology which is usually already installed, such as Bluetooth [7].

The most important goal in designing the proposed scheme is to achieve enhanced security with user convenience and practicality, as mentioned above, compared with Huang et al.'s scheme. As described in section 2, their scheme is a key management in preventing the sensitive files stored in user's portable device from being exposed to an attacker. The proposed scheme adopts the main ideas of Huang et al.'s scheme [7] to be secure even when an attacker captures the portable device; the file decryption key k is divided with two secret values a and b, and they are stored them on the laptop and the mobile phone respectively. However, as described in section 3, Huang et al.'s scheme is vulnerable to sophisticated attacks which uses both stolen laptop and eavesdropping of messages during the key reconstruction process. Additionally, Huang et al. assumed that an attacker cannot obtain both the laptop and the mobile phone at the same time, but this is likely to occur because both devices are portable and can be used anywhere. Therefore, we have to consider the possibility of a more powerful attack, where an attacker obtains both devices at the same time. A security improved key management scheme has to be designed to overcome the vulnerability in Huang et al.'s scheme.

We assumes the following, in order to further clarify the security requirements for the proposed scheme. An attacker can eavesdrop on or forge messages transmitted between communication parties; an attacker can obtain the laptop or the mobile phone and use it for other attacks such as message forgery attacks; the laptop or the mobile phone is not tamper-resistant, so an attacker can easily extract secret values from it, such as cryptographic keys; an attacker can obtain both the laptop and the mobile phone at the same time in our proposed schemes. Malware defense is also an important issue, but it is beyond the scope of this paper; so, we assume that there is no malware inside a communication party's system before it is captured by an attacker, as in [2] and [7].

4.2   The Main Ideas

Our proposed scheme needs to be resistant against possible attacks while also achieving user convenience and practicality. In terms of security, it has to be secure against message forgery attacks, parallel attacks, replay attacks, and device capture attacks. Moreover, it must overcome the vulnerabilities presented in Section 3. The main ideas to achieve these goals are as follows:

– **One-time message encryption key**: As presented in 3, Huang et al.'s scheme is vulnerable to laptop capture attacks, which uses messages eavesdropped during the key reconstruction process, because all of the messages are encrypted using the same key, sk, in all sessions. In our proposed scheme, the message encryption key is replaced with a different one in the last step of every key reconstruction process. This means that a key distribution process between the communication parties has to be added to the scheme.

– **The mobile phone system's notification**: The file decryption key is automatically reconstructed through the interaction between the laptop and the mobile phone in Huang et al.'s scheme. Although it is an effective solution for user convenience, it is insufficient for robust security. If an attacker has just stolen a laptop, he/she may send the key request message $\{K\_REQ, ID_l\}$ to the mobile phone without the user's knowledge. To avoid this possibility, the mobile phone system notifies the user with an option to confirm or reject the processing of the key request. User convenience is not compromised here, because the confirmation process is not effort-intensive like the entry of a user password.

– **TTP and double encryption**: Other communication parties can usually authenticate a user through their mobile phone, because he/she is often carrying the device which is equipped with mobile communication technology. However, it can be lost or stolen since it is a portable device. In the proposed scheme, the laptop shares the message encryption key with a trusted third party(TTP) instead of with the mobile phone. Also, another secret value b is stored in the TTP, so that when TTP sends the secret value b to the laptop as the response to its key request, it double encrypts the message in asymmetric and symmetric cryptography [13][14]. Therefore, an attacker is unable to compute b and the file decryption key k without having both of them.

– **Additional security policy**: If a mobile phone is lost or stolen, the user can request TTP not to respond to any key requests, by using a secure channel. The key request process is aborted and an attacker cannot receive any information about the secret value b, even if he/she obtains both the laptop and the mobile phone.

4.3   Details of the Proposed Scheme

There are three communication parties in the proposed scheme: the laptop, the mobile phone, and the TTP. The user owns the laptop and the mobile phone, both of which can communicate with each other because of wireless technologies. After encryption, the sensitive files are stored in the laptop using the k secret key, where $k = a \oplus b$. In the initialization phase, a and b are respectively stored in the laptop, L, and the TTP, TTP, as secret values. Moreover, L and TTP share the symmetric cryptographic key sk. The mobile phone, M, and the TTP, TTP, have a pair of asymmetric keys, whereby the private key k_{pr} is stored in M and the public key k_{pb} is stored in TTP.

Fig. 2 illustrates our proposed scheme in detail. When file access is required, L sends a key request with the identity of L ID_l to M. When M receives the key request message, the mobile phone system notifies the user who confirms or rejects the key request process. If the notification is rejected, the process is aborted; otherwise, M generates a random nonce, RN_m, encrypts the key request with RN_m using the private key k_{pr}, and then sends the resulting C_m to TTP. When TTP receives the message C_m from M, it decrypts C_m using the public key k_{pb}. If C_m was generated using a different key from k_{pr}, TTP

obtains a noise value instead of the original plaintext, $\{K\_REQ \parallel ID_l \parallel RN_m\}$, as the decryption result. Because k_{pr} is known to only M, TTP can verify that the message C_m was generated by M if the decryption result is a normal and meaningful value. Next, TTP encrypts $b \parallel RN_m$ using the key sk and re-encrypts the result X using k_{pb} ((8) and (9)). Then, TTP returns the final result C_t to M. TTP computes $sk' = h(\ sk \parallel RN_m\ )$ and replaces sk with sk'.

$$X = E_{sk}(b \| RN_m) \tag{8}$$

$$C_t = \widehat{E}_{k_{pb}}(X) \tag{9}$$

When M receives the response C_t from TTP, it decrypts C_t using k_{pr} and sends the result $X = \widehat{D}_{k_{pr}}(\ C_t\ )$. Then, L decrypts X using the key sk and obtains b and RN_m. L computes $sk' = h(\ sk \parallel RN_m\ )$ and replaces sk with sk'. L reconstructs the file decryption key k by computing $k = a \oplus b$. The laptop periodically sends "ALIVE" messages to the mobile phone and receives responses to them as in [7]. If the laptop does not receive responses from the mobile phone in an expected time interval, it re-encrypts the decrypted files [7]. L deletes the computation result, such as k, sk, RN_m and b, after the file is used.

5   Security Analysis of the Proposed Scheme

In the proposed scheme, the file decryption key k is secure against possible attacks such as message forgery attacks, parallel attacks, and replay attacks. In addition, it resists attacks that use stolen devices, and it is secure even if an attacker steals both the laptop and the mobile phone. Also, it resists sophisticated attacks that use stolen laptop and eavesdropping of messages during the key reconstruction process, as described in 3. Consequently, the proposed scheme can ensure the confidentiality of sensitive files stored in the laptop.

– **Message forgery attacks**: All messages transmitted between communication parties are encrypted by using secret keys. Therefore, an attacker cannot forge these messages without knowing the keys. For example, an attacker cannot successfully forge the message C_m or C_t, transmitted between M and TTP, without knowing the asymmetric key k_{pr} or k_{pb}. Also, an attacker cannot forge the message X, transmitted from M to L, without knowing the symmetric key sk.
– **Parallel attacks**: Parallel attacks are done by computing meaningful values using messages transmitted in more than one session. However, all of C_m, C_t, and X are ciphertexts and an attacker cannot know of their decryption keys. Therefore, an attacker is unable to compute or derive any meaningful values from the parallel session messages.
– **Replay attacks**: The proposed scheme is secure against replay attacks. TTP generates the message X, transmitted from M to L, using the symmetric key sk. sk is replaced with different key sk' in each session. Also, the message C_m or C_t, transmitted between M and TTP, is changed in each session because

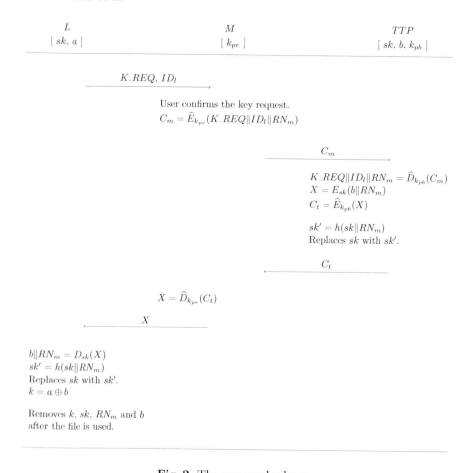

Fig. 2. The proposed scheme

its plaintext includes a random nonce, RN_m. An attacker cannot successfully perform replay attacks using X, C_m, or C_t because the messages generated in the previous session are no longer meaningful in the next session. An attacker cannot use even a key request message $\{K\_REQ, ID_l\}$ for replay attacks. In our scheme, the mobile phone system notifies the user of an initiation of a key request, and the user can reject it because it is unknown to the user.

- **Device capture attacks (attacks using stolen devices):** If an attacker steals only the laptop and extracts the key sk, the proposed scheme is still secure. As described above, the key request process requires the user's confirmation. An attacker cannot receive the information about b and derive the file decryption key k. Moreover, an attacker cannot decrypt the recorded messages using sk because the symmetric key sk is replaced with a different key, sk', in each session. Also, the proposed scheme is secure if an attacker

Table 2. Security comparison of the proposed scheme

Security attacks	Huang et al.'s scheme [7]	The proposed scheme
Message forgery attacks	Yes	Yes
Parallel session attacks	Yes	Yes
Replay attacks	Yes	Yes
Attacks using lost/stolen devices	Partially	Yes
- Lost of laptop	Partially	Yes
- Lost of mobile phone	Yes	Yes
- Lost of both	No	Yes

Yes: The scheme resists the attack; No: The scheme does not resist the attack.

steals only the mobile phone and extracts the key k_{pr} from it [1]. The laptop shares the key sk and the secret value b with TTP instead of the portable device M. TTP double encrypts the message in asymmetric and symmetric cryptography (respectively using k_{pb} and sk). Therefore, an attacker needs both k_{pb} and sk to compute b and the file decryption key k. In addition, even when an attacker steals both the laptop and the mobile phone, the proposed scheme is secure. An attacker cannot receive any information about the key k from TTP if the user requests TTP to completely ignore all key requests, by using a secure channel, when the mobile phone is lost or stolen.

The laptop transmits the key reconstruction request message $\{K\_REQ, ID_l\}$ to the mobile phone in the form of plaintext. However, the message does not include any information that can be used to derive the file decryption key k. Therefore, an attacker cannot attempt other attacks by using the message.

Table 2 compares the proposed scheme with Huang et al.'s scheme, in terms of resistance to the attacks as mentioned above. Huang et al.'s scheme is vulnerable to laptop capture attacks if an attacker eavesdrops on the messages transmitted between the laptop and the mobile phone in the previous session, as described in section 3. In addition, it does not resist attacks that uses both the laptop and the mobile phone. Table 2 shows that we remove the vulnerabilities by adopting two cryptographic techniques and two security policies described in section 4.2.

6   Related Works

In 2003, MacKenzie and Reiter proposed a cryptographic key retrieval scheme [6]. In their scheme, a portable device such as a laptop interacts with a remote

---

[1] The final goal of the proposed scheme is to protect the encrypted files stored in the laptop. Thus, in practice, it is meaningless to consider the case where only the user's mobile phone is stolen. However, the case was classified as a separate item for detailed security analysis.

server to retrieve a cryptographic key that is used for signatures or decryptions [6]. Also, the key should be derived from and activated with a password [6]. Consequently, their scheme is secure against dictionary attacks [15][16][17] as well as device capture attacks [6]. Moreover, its security is not reduced even if the remote server is untrusted [6]. However, a user needs to input a password whenever a key retrieval is required. In 2014, Chang et al. proposed a key management scheme whereby a laptop shares secret values with a universal serial bus(USB) device using Shamir's secret sharing method [10][11]. In their scheme, a laptop can retrieve a cryptographic key offline while unconnected to any network. However, the key retrieval requires user password. In addition, it is vulnerable to password guessing attacks [15][16][17] by computing two hash functions and one decryption repeatedly if an attacker steals the USB and extracts values for user authentication from it.

Several schemes have sought to achieve convenience along with security, as password-based approaches have been perceived as effort-intensive [1][2]. In 2002, to solve the data exposure problem caused by stolen laptops, Corner and Noble proposed a user authentication scheme that they named "zero-interaction authentication (ZIA)" [1]. In ZIA, the file decryption key is encrypted using the KEK, which is stored in a small token worn by the user instead of on the laptop [1]. The token is used to authenticate the laptop and negotiate the session key before the KEK is sent [1]. The user needs to adopt a wearable device, which have not been popular until now, in which the token is stored [2]. Studer and Perrig proposed that a cryptographic key should automatically be recovered, but only when a user accesses sensitive data in specific locations, like their home or office, and named their scheme "mobile user location-specific encryption(MULE)" [2]. They assumed that users tend to only access their sensitive data in specific and trusted locations [2]. In MULE, the file decryption key is transmitted from a device set up within the trusted location to the laptop via a constrained channel such as infrared [2]. The MULE scheme does not require any user effort including password entry, biometric entry, or keeping possession of cryptographic tokens [2]. However, the trusted locations are too limited though a portable device can be used anytime and anywhere.

7   Conclusion

Following an analysis of the security of Huang et al.'s scheme, we propose a security improvement for portable key management that uses a mobile phone. Our proposed scheme exploits two types of keys including a one-time symmetric key and double encrypts messages using both of them. The proposed scheme is resistant to message forgery attacks, parallel attacks and replay attacks, and is secure against possible attacks that uses stolen devices. Our scheme does not require any special device or user effort such as password input, as in [7]. Instead, the mobile phone system only requires confirmation from a user. In conclusion, the security improvement does not compromise the convenience or practicality of the portable key management scheme.

Acknowledgments. This research was supported by the Basic Science Research Programthrough theNational Research Foundation of Korea (NRF) funded by the Ministry of Science, ICT, and Future Planning (2014R1A1A2002775).

References

1. Corner, M.D., Noble, B.D.: Zero-interaction authentication. In: Proceedings of the 8th Annual International Conference on Mobile Computing and Networking, pp. 1–11. ACM (2002)
2. Studer, A., Perrig, A.: Mobile user location-specific encryption (mule): using your office as your password. In: Proceedings of the Third ACM Conference on Wireless Network Security, pp. 151–162. ACM (2010)
3. Foster, A.L.: Increase in stolen laptops endangers data security. The Chronicle of Higher Education (2008)
4. Wyld, D.C.: Help! someone stole my laptop!: how rfid technology can be used to counter the growing threat of lost laptops. Journal of Applied Security Research **4**(3), 363–373 (2009)
5. Wyld, D.C.: Preventing the worst scenario: combating the lost laptop epidemic with rfid technology. In: Novel Algorithms and Techniques in Telecommunications and Networking, pp. 29–33. Springer (2010)
6. MacKenzie, P., Reiter, M.K.: Networked cryptographic devices resilient to capture. International Journal of Information Security **2**(1), 1–20 (2003)
7. Huang, J., Miao, F., Lv, J., Xiong, Y.: Mobile phone based portable key management. Chinese Journal of Electronics **22**(1) (2013)
8. Choi, D.-H., Choi, S., Won, D.: Improvement of probabilistic public key cryptosystems using discrete logarithm. In: Kim, K. (ed.) ICISC 2001. LNCS, vol. 2288, pp. 72–80. Springer, Heidelberg (2002)
9. Nam, J., Choo, K.K.R., Park, M., Paik, J., Won, D.: On the security of a simple three-party key exchange protocol without servers public keys. The Scientific World Journal **2014** (2014)
10. Shamir, A.: How to share a secret. Communications of the ACM **22**(11), 612–613 (1979)
11. Chang, C.C., Chou, Y.C., Sun, C.Y.: Novel and practical scheme based on secret sharing for laptop data protection. IET Information Security (2014)
12. Lee, J.S., Su, Y.W., Shen, C.C.: A comparative study of wireless protocols: bluetooth, uwb, zigbee, and wi-fi. In: 33rd Annual Conference of the IEEE Industrial Electronics Society, IECON 2007, pp. 46–51. IEEE (2007)
13. Park, S., Park, S., Kim, K., Won, D.: Two efficient rsa multisignature schemes. Information and Communications Security, 217–222 (1997)
14. Lee, Y., Ahn, J., Kim, S., Won, D.: A PKI system for detecting the exposure of a user's secret key. In: Atzeni, A.S., Lioy, A. (eds.) EuroPKI 2006. LNCS, vol. 4043, pp. 248–250. Springer, Heidelberg (2006)
15. Kwon, T., Song, J.: Security and efficiency in authentication protocols resistant to password guessing attacks. In: Proceedings of the 22nd Annual Conference on Local Computer Networks, pp. 245–252. IEEE (1997)
16. Pinkas, B., Sander, T.: Securing passwords against dictionary attacks. In: Proceedings of the 9th ACM Conference on Computer and Communications Security, pp. 161–170. ACM (2002)
17. Narayanan, A., Shmatikov, V.: Fast dictionary attacks on passwords using time-space tradeoff. In: Proceedings of the 12th ACM Conference on Computer and Communications Security, pp. 364–372. ACM (2005)

Workshop on Quantum Mechanics: Computational Strategies and Applications (QMCSA 2015)

States of a Non-relativistic Quantum Particle in a Spherical Hollow Box

Haiduke Sarafian[✉]

The Pennsylvania State University, University College, York, PA 17403, USA
has2@psu.edu

Abstract. In this article we derive the wave functions of a non-relativistic quantum particle confined in a spherical hollow box. Utilizing these states and deploying a Computer Algebra System (CAS) such as *Mathematica* [1] we display the three dimensional radial wave functions. We compute the energy levels and the position expectation values.

1 Motivation and Goals

Traditionally, analysis of the states of a non-relativistic quantum particle influenced by various potentials entails solving 1D static Schrödinger equations. For a bound state particle one begins with a 1D hollow rigid wall box "the potential well" and progresses the analysis considering various potentials [2],[3]. Naturally, one anticipates a seamless transition to a 2D space addressing issues of the same sort; a literature search proves otherwise e.g. [4],[5],[6], meaning, less attention is paid to the 2D quantum problems than 1D. One departs from unanswered issues of the 2D space and makes a quantum leap into the 3D. In 3D space, instead of tackling the hollow box problem, "the 3D potential well", the trend is to analyze the Coulombian singular potential of a binary system, i.e. hydrogen atom [2],[7],[8]. Analysis of the latter con-jolts physics with the famous mathematical functions, so that the physics problem becomes a mathematical-physics problem. We defer addressing the issues of the 2D quantum physics to our future work. Here we focus on the 3D spherical hollow box. A literature search shows that there is one such study [3]. However, its objectives are limited; it is incomplete and has errors. Our work is complete and is a mixed blend of analytic and numeric calculations. Its analytic side encounters the famous mathematical functions; its numeric side flourishes from utilizing a Computer Algebra System (CAS). Its completeness entails evaluation of the normalization factors and the position expectation values. As a corollary product it embodies the energy levels. We include also plots of 3D radial wave functions as well as an image of a 3D solid print of a prototype wave function of a specific state. This work is composed of four sections. In addition to Motivation and Goals, in Section 2, Physics of the Problem and its Solution is presented. In Section 3, we present the numeric results accompanied with corresponding graphs. We conclude our work with a few closing remarks.

© Springer International Publishing Switzerland 2015
O. Gervasi et al. (Eds.): ICCSA 2015, Part IV, LNCS 9158, pp. 163–173, 2015.
DOI: 10.1007/978-3-319-21410-8_13

2   Physics of the Problem and Its Solution

We begin with a 3D static Schrödinger equation, $\hat{H}\psi(r) = E\psi(r)$, [2]. For a potential free particle of mass m the Hamiltonian is $H = \frac{p^2}{2m}$. Its quantization yields,

$$(\nabla^2 + k^2)\,\psi(r) = 0 \tag{1}$$

where the squared wave number is $k^2 = \frac{2m}{\hbar^2}E$.

Since the objective is to seek for the states of the particle in a spherical hollow box, we write (1) in the spherical coordinates. Applying the separation variable method we write,

$$\nabla^2 = \frac{1}{r^2}\left(\nabla_r^2 + \nabla_{\vartheta,\varphi}^2\right) \tag{2}$$

$$\psi(r) = R(r)f(\vartheta, \varphi) \tag{3}$$

where, ∇_r^2 and $\nabla_{\vartheta,\varphi}^2$ are, respectively,

$$\partial_r\left(r^2\partial_r\right) \tag{4}$$

$$1/\sin(\vartheta)\partial_\vartheta\left[\sin(\vartheta)\partial_\vartheta\right] + 1/\sin(\vartheta)^2\,\partial_\varphi^2 \tag{5}$$

Substituting (3) in (1) applying (2),(4) and (5) yields,

$$\left[\frac{d^2}{dr^2} + \frac{2}{r}\frac{d}{dr} + \left(k^2 - \frac{n(n+1)}{r^2}\right)\right]R(r) = 0 \tag{6}$$

$$\left[\nabla_{\vartheta,\varphi}^2 + n(n+1)\right]f(\vartheta, \varphi) = 0 \tag{7}$$

Solution of (7) is the standard spherical harmonics, i.e., $f(\vartheta, \varphi) \equiv Y_{nm}(\vartheta, \varphi)$, with $n = 0, 1, 2, \dots$ and $|m| = 0, 1, 2, \dots n$, [2]. Substituting $\xi = kr$ in (6) gives,

$$\left[\frac{d^2}{d\xi^2} + \frac{2}{\xi}\frac{d}{d\xi} + \left(1 - \frac{n(n+1)}{\xi^2}\right)\right]R(\xi) = 0 \tag{8}$$

This is the Bessel equation [9],[10]. Its solution is,

$$R_n(\xi) = Aj_n(\xi) + B\eta_n(\xi) \tag{9}$$

Here A and B are constants and $j_n(\xi)$ and $\eta_n(\xi)$ are the spherical Bessel functions of the first and second kind, respectively. These functions are related to the half integer regular functions, [9],[10],

$$j_n(\xi) = \sqrt{\frac{\pi}{2\xi}}J_{n+\frac{1}{2}}(\xi) \tag{10}$$

$$\eta_n(\xi) = (-1)^{n+1}\sqrt{\frac{\pi}{2\xi}}J_{-(n+\frac{1}{2})}(\xi) \tag{11}$$

Putting these pieces together yields the wave function subject to (1) namely,

$$\psi_{nm}(\boldsymbol{r}) = R_n(r)Y_{nm}(\vartheta, \varphi) \tag{12}$$

Since the Bessel function of the second kind is singular at the origin we set $B = 0$. On the other hand, because the spherical box of radius a is an impenetrable shell, it implies, $j_n(ka) = 0$. Utilizing (10) yields, $ka = \lambda_\ell^{n+\frac{1}{2}}$; here $\lambda_\ell^{n+\frac{1}{2}}$ for a chosen n is the ℓth root of the regular Bessel function of the first kind. It is worthwhile mentioning that the regular and the spherical Bessel functions of the first kind are oscillatory functions, and such, a chosen n embodies multitudes of numerable ℓ roots. With this remark in mind the radial term of the wave function becomes, $R_n(r) = j_n(\xi_{n\ell})$ yielding,

$$\psi_{n\ell m}(\boldsymbol{r}) = A_{n\ell}\sqrt{\frac{\pi}{2\left(\frac{1}{a}\lambda_\ell^{n+\frac{1}{2}}\right)r}}J_{n+\frac{1}{2}}\left(\frac{1}{a}\lambda_\ell^{n+\frac{1}{2}}r\right)Y_{nm}(\vartheta, \varphi) \tag{13}$$

where A is the normalization factor. Knowing the spherical harmonics are normalized, i.e. $\int |Y_{nm}(\vartheta, \varphi)|^2 d\Omega = 1$, the coefficient A in (9) becomes a $\{n, \ell\}$ dependent, this requires,

$$|A_{n\ell}|^2 \frac{\pi a}{\left(2\lambda_\ell^{n+\frac{1}{2}}\right)} \int_0^a \left[J_{n+\frac{1}{2}}\left(\frac{1}{a}\lambda_\ell^{n+\frac{1}{2}}r\right)\right]^2 rdr = 1 \tag{14}$$

Applying the known property of the Bessel function [11] namely,

$$\int_0^a \left[J_n\left(\frac{1}{a}\lambda_\ell^{n+\frac{1}{2}}r\right)\right]^2 rdr = \frac{a^2}{2}\left[J_n'\left(\lambda_\ell^{n+\frac{1}{2}}\right)\right]^2 \tag{15}$$

with J_n' being the derivative with respect to the argument, (14) yields,

$$A_{n\ell} = \sqrt{\frac{4\lambda_\ell^{n+\frac{1}{2}}}{\pi a^3}}\frac{1}{J_{n+\frac{1}{2}}'\left(\lambda_\ell^{n+\frac{1}{2}}\right)} \tag{16}$$

This is missing in [3]. It is worthwhile pointing out that certain kinematic quantities such as energies may be calculated without utilizing (16). On the other hand there are quantities such as position expectation values that keenly require (16). We pursue the computation in the next paragraph.

3   Analysis and Results

Energies: A general observation. The corollary information deduced from applied boundary conditions are conducive to the energies. Namely, since we have already established the fact that, $ka = \lambda_\ell^{n+\frac{1}{2}}$ and because $k^2 = \frac{2m}{\hbar^2}E$, manipulating these two equations we arrive at, $E \equiv (E_{n\ell})_{\text{sphere}} = \frac{\hbar^2}{2ma^2}\left[\lambda_\ell^{n+\frac{1}{2}}\right]^2$. This can also be related to the quantized energies of a particle confined in a 1D rectangular impenetrable hollow box, namely, $(E_\ell)_{\text{rectangle}} = \left(\frac{\pi^2 \hbar^2}{2ma^2}\right)\ell^2$, $\ell = 1, 2, 3...$, [2],[7],[8]. Manipulating the latter two energy expressions yields, $(E_{n\ell})_{\text{sphere}} = (E_\ell)_{\text{rectangle}} \left[\frac{1}{\pi\ell}\lambda_\ell^{n+\frac{1}{2}}\right]^2$, $n = 0, 1, 2, ...$ Meaning, a particle in a 3D space because of its additional moving degrees of freedom sustains a broader energy spectrum. A similar observation applies to the wave functions as well. I.e. for a 1D rectangular box of width a, the normalized wave functions are, $\psi_\ell(x) = \sqrt{\frac{2}{a}}\sin(\frac{\ell\pi}{a}x)$, $\ell = 1, 2, 3, ...$ [2],[7],[8] while the radial term of (9) depends on two indices, $\{n, \ell\}$, and therefore sustains a wider range of possibilities. Table 1 embodies the energies of the 3D case.

Table 1. Quantized energies of a particle of mass m in a spherical hollow box of radius $a = 1$. For the sake of simplicity the energies are scaled to $\frac{\hbar^2}{2ma^2}$.

0	9.86	39.47	88.82	157.91	246.74
1	20.19	59.67	118.90	197.85	296.55
2	33.21	82.71	151.85	240.70	349.28
3	48.83	108.51	187.63	286.40	404.88
4	66.95	137.00	226.19	334.93	463.344

The first row of Table 1 lists the values of $(E_{0\ell})_{\text{sphere}}$ for $\ell = 1, 2, 3, 4, 5$. As we noted in the text, since Bessel functions are oscillatory functions for a chosen n, in this case $n = 0$, contains multitude of roots. These roots are multiples of π, i.e. $\lambda_\ell^{\frac{1}{2}} = \ell\pi$. We utilize these values to evaluate the corresponding energies of the first row. One notes the energy levels of a spherical box for $n = 0$ matches identically the energy levels of a 1D case, namely $(E_{0\ell})_{\text{sphere}} = (E_\ell)_{\text{rectangle}}$. Moreover, a particle in a sphere is allowed to obtain a multitude of additional energies with no counter-piece in 1D. These are shown in the rows beneath the first. Figure 1 displays the content of the Table 1.

In Figure 1 the energy axis is scaled to $\frac{\hbar^2}{2ma^2}$ and the coordinates are $\{n, \ell\}$. For instance the first left most dot has the energy of 9.86 and its coordinates are $\{0, 1\}$. Similarly the top dot of the first vertical line corresponds to energy of 246.7 and its coordinates are $\{0, 5\}$. The dots on the first vertical line are also the energies of a 1D case. Figure 1 shows the energies associated with the larger n are higher than the smaller n. It is noted that [3] contains an inaccurate energy graph; its corresponding Table has also a wrong heading as well.

4   Normalization Factors

To further the computations for $n > 0$ and attempting to put the derived wave functions to practice, one encounters challenges. For instance the wave functions (12) are composed of two terms. On the other hand, their angular terms,

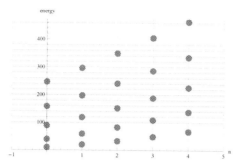

Fig. 1. Energy levels of a particle in a 3D spherical box. Coordinates of each dot is $\{n, \ell\}$. The coordinates of the left most bottom dot and the right most top dot are $\{0, 1\}$ and $\{4, 5\}$, respectively.

$Y_{nm}(\vartheta, \varphi)$ are analytic functions with well known and easy to use properties [12]. On the other hand their radial terms are not so trivial; they are composed of half integer Bessel functions with arguments associated to their zeros. The normalization factors (16) suffer the same symptoms. Computer Algebra System (CAS) specially *Mathematica* makes the computation of the desired quantities possible.

For the sake of simplicity we set the radius of the spherical box to unity, e.g. $a = 1$. First we calculate the values of the normalization factors (16) for a wide range of possibilities. Table 2 contains a variety of cases.

Table 2. Values of the normalization factors $A_{n\ell}$ for $n = 0, 1, 2, 3, 4$ and $\ell = 1, 2, 3, 4, 5$

0	4.44288	8.88577	13.3286	17.7715	22.2144
1	6.5101	11.0163	15.4855	19.9428	24.3949
2	8.54265	13.1018	17.6019	22.079	26.5445
3	10.5685	15.1625	19.6917	24.1896	28.6705
4	12.5983	17.2092	21.7633	26.2813	30.778

The first row of Table 2 lists the values of $A_{0\ell}$ for $\ell = 1, 2, 3, 4, 5$. As we explained in the text, since Bessel functions are oscillatory functions, a chosen n, in this case $n = 0$ contains multitude roots. We utilize the values of these individual roots to evaluate the corresponding values in the first row. The meaning of the rest of the numbers in the subsequent rows are the same as the aforementioned explanation. The CPU time for computing the entire Table 2 is a fraction of a second! One may easily augment to the table include cases of interest.

Next, utilizing the procedure conducive to the numeric values of Table 2 we check the accuracy of the normalization of the wave functions. In other words we check the normalization of the wave functions given in (16). Theoretically the wave functions (13) are normalized, i.e. $\int |\psi_{n\ell m}(\mathbf{r})|^2 r^2 dr d\Omega = 1$. Here utilizing our numeric approach we validate the claim. Table 3 contains triple integrations of density functions, namely, $\int |\psi_{n\ell\,m}(r)|^2 r^2 \mathrm{d}r\mathrm{d}\Omega$.

Table 3. Values of $\int |\psi_{n\ell\,m}(r)|^2 r^2 \mathrm{d}r\mathrm{d}\Omega$ for $n = 0, 1, 2...6$ and $\ell = 1, 2, 3, ...8$

0	1.	1.	1.	1.	1.	1.	1.	1.
1	1.	1.	1.	1.	1.	1.	1.	1.
2	1.	1.	1.	1.	1.	1.	1.	1.
3	1.	1.	1.	1.	1.	1.	1.	1.
4	1.	1.	1.	1.	1.	1.	1.	1.
5	1.	1.	1.	1.	1.	1.	1.	1.
6	1.	1.	1.	1.	1.	1.	1.	1.

The first row of Table 3 lists the values of $\int |\psi_{o\ell\,m}(r)|^2 r^2 \mathrm{d}r\mathrm{d}\Omega$ for $\ell = 1, 2, 3, ...$. These quantities as expected evaluate to unity. The rows following the first have the same interpretations. It is assuring that irrespective of the chosen indices (states) the corresponding wave functions are justifiably normalized. The CPU time to compute the entire Table 3 is about a couple of seconds; note that these are numeric triple integrations.

Plots of Radial Wave Functions:

Fig. 2. Display of a typical 3D wave function confined in a 3D spherical box

The lip of the 3D wave function shown in Figure 2 touches the equator as required by the boundary condition.

Next, utilizing the radial term of the wave function, i.e. $R_n(r) = j_n(\xi_{n\ell})$ for two different values of $n = 0$ and 1 in Figure 3 and 4 we display a few 2D and their corresponding 3D graphs, respectively.

The author made an extensive atlas similar to the ones shown in Figures 3 and 4 for $n = 2, 3, 4...6$. However, because of the space limitation they are not included.

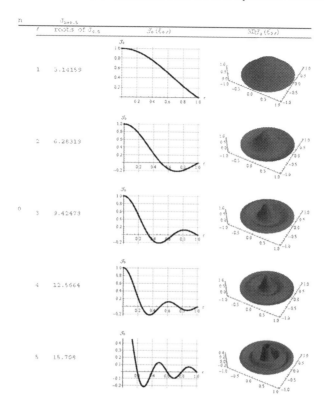

Fig. 3. Display of 2D (the 4th column) and their corresponding (the 5th column) 3D radial wave functions of $j_0 (\xi_{0\ell})$ for $\ell = 1, 2, 3, 4, 5$. Roots of $J_{0.5}$ determining the values of $\xi_{0\ell}$ are the numbers underneath the "roots of $J_{0.5}$" heading.

Table 4. Values of $< r >_{n\ell}$ for $n = 0, 1, 2, ...6$ and the associated values of ℓ

0 0.5	0.5	0.5	0.5	0.5
1 0.591	0.539	0.522	0.514	0.510
2 0.647	0.573	0.545	0.531	0.523
3 0.686	0.602	0.567	0.548	0.536
4 0.715	0.627	0.586	0.564	0.549
5 0.737	0.647	0.604	0.578	0.561
6 0.756	0.665	0.620	0.592	0.573

Expectation Values of the Position:

In addition to analyzing the aforementioned topics we may now evaluate additional quantities of interest. For instance in a coordinate space, we compute the expectation values of the position, namely $< r >_{n\ell}$. Table 4 contains these values for a wide range of cases.

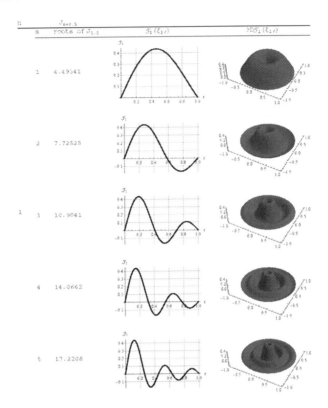

Fig. 4. Display of 2D (the 4th column) and their corresponding (the 5th column) 3D radial wave functions of j_1 ($\xi_{1\ell}$) for $\ell = 1, 2, 3, 4, 5$. Roots of $J_{0.5}$ determining the values of $\xi_{1\ell}$ are the numbers underneath the "roots of $J_{1.5}$" heading.

The first row corresponds to $n = 0$. Irrespective to the chosen values of ℓ the expectation values are equal to 0.5. This stems from the fact that for this special case the probability densities in a 1D equals the one in a 3D. To make the point clear, in Figure 5 we display the associated probability densities, namely, P(x) and P(r), respectively. The former is $\sin^2(\ell\pi x)$ while the latter is $\left[j_0 \left(\lambda_\ell^{\frac{1}{2}} r \right) r \right]^2$. As shown, these two functions perfectly overlap, hence their position expectation values are the same, i.e. the values in the first row of Table 4. In other words, the values in the first row correspond to the expectation values of a 1D case namely, $< x >_\ell = \int_0^a \sqrt{\frac{2}{a}} \mathrm{Sin}[\frac{\ell\pi}{a}x] \times \sqrt{\frac{2}{a}} \mathrm{Sin}[\frac{\ell\pi}{a}x] \, dx = \frac{a}{2}$. Setting $a = 1$ yields 0.5. The rest of the values in Table 4 have self explanatory meanings.

Figure 5 shows the probability densities are the same and hence their corresponding expectation values are equal. Solid lines are the P(r) and the thick gray curves are the corresponding P(x). The plot shows the P(r) are stretched out toward the rim of the shell evaluating larger expectation values than 0.5.

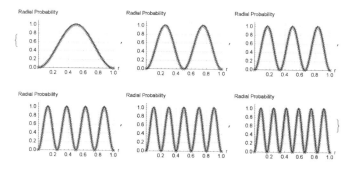

Fig. 5. Plots of probability densities, $P(x) = \sin^2[\ell\pi\,x]$ vs. $P(r) = \left[j_0\left(\lambda_\ell^{\frac{1}{2}}r\right)r\right]^2$, respectively for $\ell = 1, 2, 3, ...6$. The gray thick curves and the solid black thin curves correspond to $P(x)$ and $P(r)$, respectively. For the sake of clarity the $P(r)$ are scaled up by a factor of $10\ell^2$.

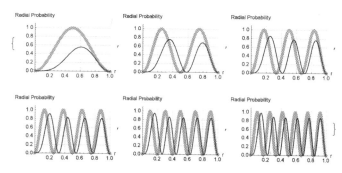

Fig. 6. Plots of probability densities, $P(x) = \sin^2[\ell\pi\,x]$ vs. $P(r) = \left[j_1\left(\lambda_\ell^{\frac{1}{2}}r\right)r\right]^2$, respectively for $\ell = 1, 2, 3, ...6$. The gray thick curves and the solid black thin curves correspond to $P(x)$ and $P(r)$, respectively. For the sake of clarity the $P(r)$ are scaled up by a factor of $10\ell^2$.

Figure 6 shows the contrast between the $P(x)$ and $P(r)$ for $n > 0$. These miss alignments are the visual differences between the expectation values in Table 4. The interested reader may compute higher position dependent moments such as $< x^k >_{n\ell}$.

Figure 7 is a display of the first three rows of Table 4. Polar plot of the expectation values of $< r >_{n\ell}$ for $n = 0, 1, 2$) and $\ell = 1, 2, 3...8$ are shown. For instance the eight values of the first row are shown with the black dots about the center of the plot at a constant distance of 0.5. Note that because of space limitation only the first four values are included in each row. By the same token the eight values of the second row are displayed in gray and etc. Quick scanning the dots in a counter clockwise direction shows their distances from the center are elongated, meaning the particle is less localized.

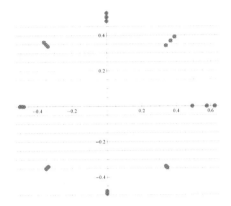

Fig. 7. A polar display of the expectation values of $< r >_{n\ell}$ for $n = 0, 1, 2$ and their corresponding $\ell = 1, 2, 3...8$

5   Conclusions

One of the objectives of crafting this article was to analyze the states of a 3D non-relativistic quantum particle in a hollow spherical box. The motivation of tackling this project was to fill in the gap transiting from a 1D to a 3D space. The formulation of the problem is completed analytically utilizing the famous mathematical functions. In addition to accomplishing our set goal we also deployed a Computer Algebra System (CAS) such as *Mathematica* computing 1) the normalization factors of the wave functions 2) position expectation values 3) energy levels and 4) displaying 2D and 3D radial wave functions. Whenever appropriate we correlated the calculation of a 3D to the corresponding 1D case.

Acknowledgments. The author acknowledges kind efforts of professor Ikusaburo Kurimoto-san from Kisaazu National College of Technology, Japan for printing the 3D replica of the 3D wave function displayed in Figure 2.

References

1. Wolfram, S.: Mathematica, V10.01, a computational software program to do scientific computation. Wolfram Research (2014)
2. Davydov, A.S.: Quantum Mechanics. NEO Press (1966)
3. Flugge, S.: Practical Quantum Mechanics. Springer-Verlag (1974)
4. Gol'dman, I.I., Krivchenkov, V.D.: Problems in Quantum Mechanics. Dover Publications, Inc., New York (1993)
5. Kogan, V.I., Galitskiy, V.M.: Problems in Quantum Mechanics. Dover Publications, Inc. (2011)
6. Thaller, B.: Visual Quantum Mechanics, V1. Springer-Verlag New York, Inc. (2000)
7. Shiff, L.: Quantum Mechanics, 3rd edn. McGraw-Hill, New York (1968)

8. Powell, J., Crasmann, B.: Quantum Mechanics. Addison-Wesley Publishing Company, Inc., Reading (1965)
9. Abramowitz, M., Stegun, I.: Handbook of Mathematical Functions, 9th edn. Dover Publications, Inc., New York (1970)
10. Wang, Z.X., Guo, D.R.: Special Functions. World Scientific, Singapore (1989)
11. Tenenbaum, M., Pollard, H.: Ordinary Differential Equations, p. 620. Dover Publications, Inc., New York (1963)
12. Brink, D.M., Satchler, G.R.: Angular Momentum, 2nd edn. Oxford University Press (1967)

Workshop on Remote Sensing Data Analysis, Modeling, Interpretation and Applications: From a Global View to a Local Analysis (RS 2015)

Adaption of a Self-Learning Algorithm for Dynamic Classification of Water Bodies to SPOT VEGETATION Data

Bernd Fichtelmann[1(✉)], Kurt P. Guenther[2], and Erik Borg[1]

[1] German Aerospace Center, German Remote Sensing Data Center,
Kalkhorstweg 53 17235, Neustrelitz, Germany
{Bernd.Fichtelmann,Erik.Borg}@dlr.de
[2] German Aerospace Center, German Remote Sensing Data Center, Oberpfaffenhofen,
Postfach 1116 82230, Wessling, Germany
Kurt.Guenther@dlr.de

Abstract. Within the ESA CCI "Fire Disturbance" project a dynamic self-learning water masking approach originally developed for AATSR data was modified for MERIS-FR(S) and MERIS-RR data and now for SPOT VEGETATION (VGT) data. The primary goal of the development was to apply for all sensors the same generic principles by combining static water masks on a global scale with a self-learning algorithm. Our approach results in the generation of a dynamic water mask which helps to distinguish dark burned area objects from other different types of dark areas (e.g. cloud or topographic shadows, coniferous forests). The use of static land-water masks includes the disadvantage that land-water masks represent only a temporal snapshot of the water bodies. Regional results demonstrate the quality of the dynamic water mask. In addition the advantages to conventional water masking algorithms are shown. Furthermore, the dynamic water masks of AATSR, MERIS and VGT for the same region are presented and discussed together with the use of more detailed static water masks.

Keywords: Self-learning algorithm · Land-water mask · Interpretation · Remote sensing · VGT data · Cloud cover

1 Introduction

It is undisputed that satellite earth observation is a major data source for analysing different aspects of global environment. The full-coverage of Earth's surface can be achieved within some days by sensors with swath-widths of more than 1000km (e.g. MODIS or MERIS). In case of SPOT VEGETATION a nearly complete global coverage is available for each day. But it is connected with the general drawback of optical sensors that clouds obstruct an undisturbed observation. That means at least when using MODIS data as basis for a global land-water mask (MOD44) with 250m resolution that the resulting mask can be fragmentary in the required scale as shown e.g. in [11], [14]. Further progress can be expected when time series of remote sensing data (e.g. daily near infrared (NIR) reflectance from Terra (MODGQ09) and

© Springer International Publishing Switzerland 2015
O. Gervasi et al. (Eds.): ICCSA 2015, Part IV, LNCS 9158, pp. 177–192, 2015.
DOI: 10.1007/978-3-319-21410-8_14

AQUA (MYD09GQ) with a spatial resolution of 250m and the daily 500m snow cover product from Terra (MOD10A1) and AQUA (MYD10A1)) are used to derive a yearly water mask for 2013 showing the number of days of water coverage [10].

During the development of our self-learning algorithm for water classification the best available static land-water mask was the SRTM Water Body Detection (SWBD) with a spatial accuracy better than 30m for water bodies in the geographical region between 54° South and 60° North. Caused by the limited temporal duration of the SRTM mission (only 11 days in February 2000), the full global coverage includes data gaps. According to personal information Carroll et al. [2] describes that the SWBD team has tried to fill "these gaps with help of Landsat Geocover data". However, if the Geocover data were too cloudy, then the gaps could not be filled.

The most actual global land-water mask with 30m spatial resolution is based on Landsat 7 Geocover data covering the time interval from 2000 to 2012 developed together with High-Resolution Global Maps of Forest Cover Change GFC_2013 [9]. This mask (called Hansen_GFC) corrects some missed water bodies of the SWBD mask, which were not detected during the short SRTM mission due to unfavourable weather conditions, empty dam lakes or e.g. the Okavango basin which was relative dry in February 2000 due to local summer drought. Fig. 1 shows a comparison between SWBD and Hansen_GFC masks for a small region in North-East of Germany around Neustrelitz. The main differences are obvious. The white marked water bodies are only available in the Hansen_GFC mask. The dark blue marked water pixels are only classified in the SWBD mask as water. Ground observations showed that they are mostly swimming reed islands.

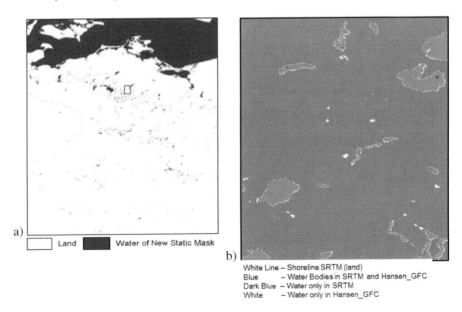

a)

| | Land | | Water of New Static Mask |

b)

White Line – Shoreline SRTM (land)
Blue   – Water Bodies in SRTM and Hansen_GFC
Dark Blue – Water only in SRTM
White   – Water only in Hansen_GFC

Fig. 1. Comparison of SWBD and Hansen_GFC mask for a region in North-East of Germany with the marked region around Neustrelitz (a), differences of the two water masks (b)

As already mentioned in [5], [6], the precise global detection of water bodies is a complex task because of changing atmospheric conditions as haze, aerosol and cloud but also due to the surface roughness of water bodies. Additionally, different components as chlorophyll, yellow substance or suspended sediments influence the spectral characteristics of water bodies. Furthermore the use of static land-water masks of different quality in scale and region (compare e.g. with [12]) has the disadvantage that the masks reflect only the actual distribution of water bodies during production period ([1], [2]). This is not only a problem of the best available land-water mask based on SRTM data [14] used for these studies. This problem is also connected with actual and future masks.

Therefore, a self-learning process was developed for AATSR data [5], for global automatic detection of water bodies combining an optimized use of static masks and results of spectral pre-classification. This procedure was adapted within the same ESA-CCI project to the MERIS full resolution and reduced resolution (FR(S), RR) data and at least to VGT data. First results of the method applied to VGT data are shown and discussed including the different intermediate masks.

2   Material and Methods

2.1   Remote Sensing Data

The self-learning algorithm was developed first for multispectral L1b data of the AATSR sensor on board of the ENVISAT satellite. This paper shows the adaption of the self-learning algorithm to classify dynamically water bodies using daily, global L3 maps of SPOT VEGETATION data (VGT) for the period from 1999 to 2006. The ground resolution of VGT data at nadir is 1km. The VGT sensor measures the reflected radiation in 4 spectral bands (Table 1). For our investigations the global mapped level 3 reflectances ρ of all bands were used including latitude and longitude for each pixel. For the development of the algorithm datasets with different size were selected for different test regions. The resulting land-water masks were tested for 10 test sites defined by the ESA-CCI project "Fire Disturbance" [3], and additional for regions in the Alps with terrain shadow, Scandinavia with inaccuracies of geometry in static mask and the region around Caspian Sea with partly strong changes (desiccation) of water area. Especially the Scandinavian region was used to study the strength of the self-learning procedure when using water masks of lower accuracy for regions north of 60°N.

Table 1. Specification of the 4 VGT bands according [13]

VGT Channel Identifier	Centre Wavelength (nm)	Bandwidth (nm)	Spectral domain
B0	450	40	BLUE
B2	645	70	RED
B3	835	110	NIR
B4	1665	170	SWIR

2.2   Available Static Land-Water Masks

As in [5], [6] the actual best available land-water mask, the SRTM Water Body Data (SWBD) acquired during the Shuttle Radar Topographic Mission was used for the geographical region between 60°S and 60°N. The spatial accuracy of SWBD is better than 30m [14]. The SWBD represents the instantaneous water extent in February 2000 and is delivered in 1 * 1 degree cells. Cells including only water or land are not available in the SWBD data base. Therefore, for the 2051 missing land cells corresponding dummies were constructed.

For the region 60°N to 90°N the latest version of the Global Self-consistent Hierarchical High-resolution Shorelines (GSHHS, released in 2011) is used as additional static land-water mask. The spatial resolution is better than 100m. The data are available in segments of 10 * 10 degrees. The data were delivered by the National Geophysical Data Center and described by [15]. All static water masks were used to derive the fraction of water area for each VGT pixel in percentage. The generation of a consistent static land-water mask is described in [4] in more detail.

The GSHHS land-water mask was substituted for some studies by the UMD Global 250 meter Land Water Mask (MOD44) [2].

3   Dynamically Self-Learning Evaluation Method (DySLEM) – An Adoption to VGT Data

The Dynamically Self-Learning Evaluation Method (DySLEM) and the corresponding DySLEM processor were developed at first for top-of-atmosphere (TOA)-AATSR reflectance data. But it was already shown [6] that the generic principle can be transferred to other sensor data as MERIS-FR(S) or -RR data respectively. The basic principle of DySLEM described in [5] consists first in the creation of a static and two dynamic land-water masks where water pixels classified by the dynamic masks are labelled as "candidates" for water. The dynamic water masks are applied for subsets of the scene or path in order to find "candidates" which are typical for small regions. Based on the regional "candidates" of a 512 * 512 pixel subsets a mean regional water spectrum is calculated which is used in the final decision (learning) process. This will result in the final land-water mask based on the static mask, the regional "candidates" and the mean, regional spectrum. If a frame is outside the image then it has to be postponed into the borders of the image. That means that the last frame of a line or column produces an overlap with its neighbour. A post-processor includes the result of the subset into the final output product.

3.1   Structure of the Processor

The structure of processor is generic, already presented in [5] and a modification to MERIS data is described in [6]. Apart from the changing frame size only the sensor specific classifications for BOA-VGT reflectance data have to be used. Beside of pre- and post-processing the processor consists of three basic steps.

3.2   Generation of Regional Static Land-Water Mask (WS1)

The generation of a regional static land-water mask (lwmss: land-water mask, static, section) is generic. The first work step (WS1) is already described in detail in [5]. In contrast to AATSR the pre-processing of VGT data selects subsets of 512 * 512 pixels from the mapped data. Constructing only for test a first map of size x * y of same size. Based on this map a second high resolution map (9x * 9y) with land-water distribution is generated. On basis of this map and the 9 * 9 sub pixels the fraction of water of each VGT pixel will be determined. For example, Fig. 2 shows the land-water masks of the Valdecanas Dam Lake of river Tejo, Spain, for two spatial resolutions.

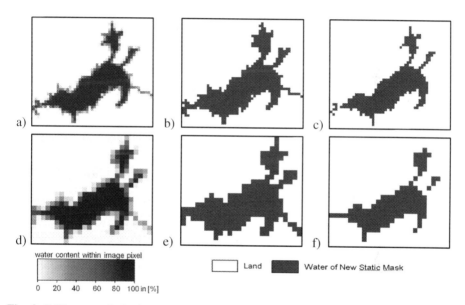

Fig. 2. Different static land-water masks for Valdecanas Dam Lake of river Tejo based on SWBD with resolution of MERIS-FR(S) (compare [6]) and of VGT (second row) data. Fraction of water for a VGT pixel in percentage (d), static masks with ≥10% (e) and ≥60% fraction of water (f).

The masks in the upper row of Fig. 2 show the results for a spatial resolution of 300m x 300m (MERIS-FR(S) in sensor projection), already presented in [6]. The second line presents the same masks for 1000m * 1000m (VGT data in map coordinates). Nevertheless, the results are nearly identical. In Fig. 2d the static mask based on SWBD data shows the fraction of water for each VGT pixel. Because of the higher resolution of MERIS-FR(S) data (300m) the contour of the lake is represented much sharper and includes more pixel with 100 percentage water of pixel area. In the centre of the lake all pixels have a fraction of water of 100%. At the shoreline the fraction of water is variable between 0% and below 100%. During the processing step **WS1** two additional masks will be created based on the static mask by selecting pixels with ≥10% (e) and ≥60% fraction of water (f). All water pixels of Fig. 2f and 2c are "static candidates" for "stable water". When not accepted as "stable water" all water pixels of Fig. 2e and 2b will be examined for "accepted static water".

3.3   Classification Algorithms of VEGETATION Data Within DySLEM

There are many reasons for the different spectral behaviour of water bodies including physical (concentration of suspended particles) and biological factors (chlorophyll concentration) as well as atmospheric phenomena (haze and clouds), surface roughness (glint) or regions with turbid and shallow water. Additionally, the pixel area can consists of an unknown composition of land and water.

Therefore, any global classification of water objects needs special regional attention otherwise it will be incomplete or will include errors. Thus, work step 2 (**WS2**) is used as regional pre-classification resulting in "dynamic candidates" for the decision process performed in following work step 3 (**WS3**).

According to the studies with AATSR and MERIS data two different algorithms for dynamic land-water mask were applied to VGT data to identify all candidates of water within the section (lwm**ds**).

PWS2-1-Generation of lwmd1s: As for the other sensors before the first algorithm for the dynamic land-water mask uses the spectral decrease of the reflectances ρ towards the SWIR band for water pixels. The L3-VGT data are atmospherically corrected and all bands represent BOA reflectances. Improper aerosol correction might result in negative reflectances for the blue band (B0) which are corrected to zero. In the global L3-VGT maps large water bodies (e.g. oceans) are masked with zero in each band. Therefore, an additional condition is introduced with a threshold for the reflectance at 450nm (B0) (eq. 1). The result is the first **d**ynamic land-water mask (lwmd1-VGT).

$$(\rho_{450} > \rho_{645}) \wedge (\rho_{645} > \rho_{835}) \wedge (\rho_{835} > \rho_{1665}) \vee (\rho_{450} = 0) \cdot ?\cdot 1 : 0 \ . \tag{1}$$

PWS2-2-Generation of lwmd2s: For VGT data [8] describes a multi-temporal algorithm for small water bodies for Africa. This algorithm is based on SPOT VGT-S10 (10day synthesis) products and uses "anomalies" to the mean value of a pre-defined region. However, an accurate classification of water is not necessary within PWS2, rather just the identification of candidates for water, namely dark regions. Therefore, we defined rules which are based on an RGB$_{new}$ image with bands according to eq. 2:

$$R_{new} : \cdots (\rho_{645} > 0) \cdot ?\cdot \log(\rho_{645}) : 0$$
$$G_{new} : \cdots (\rho_{835} > 0) \cdot ?\cdot \log(\rho_{835}) : 0 \qquad . \tag{2}$$
$$B_{new} : \cdots (\rho_{450} > 0) \cdot ?\cdot \log(\rho_{450}) : 0.8$$

The final rule for lwmd2s-VGT is:

$$(R_{new} \le -1.0) \wedge (G_{new} \le -0.8) \wedge$$
$$(B_{new} \le -1.0) \wedge (\rho_{1665} < 0.18) \vee (\rho_{450} = 0) \cdot ? \cdot 1 : 0 \quad . \tag{3}$$

3.4   A Self-learning Algorithm to Identify Temporal Dynamic of Water Bodies

The third work step (**WS3**) includes the self-learning algorithm based on the results of lwmss, lwmd1s and lwmd2s masks of the previous work steps. **WS3** consists of three sub-classification processors and a fusion processor.

The first sub-processing step (PWS3-1) is a quality check of the static mask. If a pixel is found with $\ge 60\%$ water fraction in lwmss and additionally identified as water in at least one of the two dynamic land-water masks (lwmd1s or lwmd2s) then such a pixel is defined as "stable water" pixel. This rule can be written by the following Boolean equation:

$$(lwmd1s = 1 \vee lwmd2s = 1) \wedge lwmss \ge 60? \cdot 1 : 0 \quad . \tag{4}$$

All "static water" pixels are encoded by 1, stored by the fusion processor in the final matrix lwmd3s and excluded from further processing. These pixels will be named as "stable pixels", because available in static mask and detectable in one of dynamic masks. In order to distinguish and visualize the different results of this land-water masking processing the different results will be encoded by different numbers and colours. In all figures shown below these "stable water" pixels are presented by a dark blue colour. "Stable water" bodies are shown in Fig. 3c, 4f, and 5e. The probability of the existence of water is very high in this case. On the other hand the probability is low that the "static water" pixel will change since creation time of static mask to vegetation (forest) and then by fire to another dark region (e.g. burned area).

It may be possible that some "static water" pixels are covered by clouds, haze or ice. In case of haze and thin clouds an inclusion of these water pixels in the resulting mask is useful to prevent identification as burned area. Furthermore it is possible that water pixels of the static land-water mask have changed their fraction of water including a complete desiccation since the generation of the static mask. The task of PWS3-2 consists in decision which non-stable "static water" pixels have to be included in the resulting mask lwmd3s, working at least as second quality control of static mask.

Principally, the fusion processor calculates a regional mean spectrum for the actual selected segment of image (all spectral bands i) in preparation of PWS3-2, based on all n "stable water" pixels with lwmd3s = 1 according eq. 5.

$$\overline{\rho}_i = \left(\sum_1^n \rho_i (lwmd3s = 1) \right) / n \quad . \tag{5}$$

A next step is the labelling of all pixels of the "static mask" with $\ge 10\%$ water fraction which are not included in lwmd3s in the step before. After that the fusion processor initiates the second sub-processing step (PWS3-2) in which the spectrum of each selected "static water" pixel is compared with the mean water spectrum of eq. 5. Our analysis and tests have shown that the use of only one mean spectral band (reflectance) is sufficient instead of using the full spectral information. This finding also reduces the computing time. The mean reflectance of NIR band (ρ_{835}) is used as reference according to the rule in eq. 6. For improving the results an offset of 0.08 is applied including an upper threshold for ρ_{835}.

$$lwmd3s = 0 \wedge lwmss \geq 10 \wedge \rho_{835} \leq \overline{\rho}_{835}(lwmd3s = 1) + 0.08 \wedge \qquad (6)$$
$$\rho_{835} \neq 0 \wedge (\rho_{835} < 0.21)?\cdot 2 : 0$$

All pixels fulfilling the condition described above are encoded by 2 in mask lwmd3s and coloured green in the following figures. During the development it was obvious that a higher threshold for the ratio is helpful in consensus with the processing of AATSR and MERIS data to identify additional water pixels but with a little bit lower confidence.

$$lwmd3s = 0 \wedge lwmss \geq 10 \wedge \rho_{835} \leq \overline{\rho}_{835}(lwmd3s = 1) + 0.16 \wedge \qquad (7)$$
$$\rho_{835} \neq 0 \wedge (\rho_{835} < 0.24)?\cdot 3 : 0$$

All pixels fulfilling the condition above are marked by 3 in the mask lwmd3s and coloured light blue in the following figures. Additional relations can be defined on basis of eq. 6 and 7 creating further sub-classes for additional studies with changed 'softer' thresholds.

After this processing step all pixels of the static water mask with a fraction of water $\geq 10\%$ have been investigated in detail. Fig. 3 shows as example the result of such a quality test of the static mask for a region of Caspian Sea. This includes the identification of static water below dust and water with high salt concentration of Kara-Bogaz Gol (shallow water region with a depth of 7 to 10 meters and 34% salt concentration). All pixels with $\geq 10\%$ fraction of water which not satisfy the criteria in any form are excluded by the fusion processor from further studies.

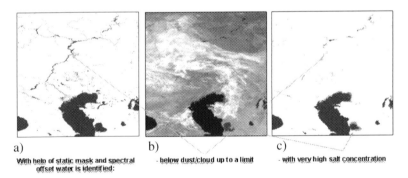

a)   b)   c)
With help of static mask and spectral   - below dust/cloud up to a limit   - with very high salt concentration
offset water is identified:

Fig. 3. Static mask (a), RGB image (b) and dynamic land-water mask lwmd3 (c) for a VGT scene (18th July 2005) for a region of Caspian Sea. lwmd3 is shown after eq. 7. Blue: stable water pixels; green: water pixel fulfilling eq. 6; light-blue: water pixel fulfilling eq. 7.

In the third sub-processing step, termed PWS3-3, all "candidates" for water of the dynamic masks lwmd1s and lwmd2s are evaluated when the fraction of water is below 10% and when not already stored in the resulting mask lwmd3s. The fusion processor selects the following pixels according to eq. 8 as already described for MERIS data.

$$lwmss < 10 \wedge (lwmd1s = 1 \vee lwmd2s = 1) \cdot ? \cdot 1 : 0 \ . \tag{8}$$

These pixels are marked generic as "dynamic candidates" and stored in an intermediate layer lwmdis. This new layer is input for the sub-processor PWS3-3. The task here is the identification of "new" dynamic water pixels which are not marked as water in the static land-water mask. Examples for such pixels are already given in [6]. It is an additional task of this step to exclude shadow or burned area pixels from the final result. The decision tree for the identification of water in the intermediate layer "lwmdis" is given in the following equations:

$$\begin{aligned} & lwmdis = 1 \wedge \\ & (\rho_{450} < 0.03) \wedge (\rho_{645} < 0.03) \wedge \\ & (\rho_{835} < 0.06 \vee \rho_{835}/\rho_{1665} < 1,50) \wedge \\ & (\rho_{1665} < 0.0581) \wedge (\rho_{835}/\rho_{1665} > 0.008)? \, 6 : 0 \end{aligned} \tag{9}$$

All pixels fulfilling these relations are coded with 6 as "accepted dynamic water" pixel in the mask lwmd3s and coloured with red in the resulting images. According to [5] work step **WS3** will be finished with the DySLEM-output lwmd3s. When all subsets are processed the n subsets will be integrated (generic) into a complete final mask lwmd3 by equation 10.

$$. \, lwmd3 = \sum_{1}^{n} lwmd3s \ . \tag{10}$$

The other masks (lwmss, lwmd1s, lwmd2s) can be combined for further quality tests in the same way. In the following figures all results of the new dynamic land-water masking algorithms are shown in order to visualize the accuracy of the approach.

4   Results

In the following some examples will demonstrate the efficiency and the stability of the proposed procedure for VGT data compared with the results from AATSR and MERIS. In Fig. 4 the land-water mask derived from VGT data is presented for water below haze and clouds (Alps). In Fig. 5 the identification of water bodies for regions above 60°N (Scandinavia) is shown where GSHHS is used as static water mask. GSHHS is not as accurate and detailed as the SWBD water mask.

Fig. 4 shows the comparison of results for MERIS-FRS (upper row) and VGT data for a western region of the Alps (see Fig. 4g for the geographic position) e.g. with Lake Geneva (lower left) and Lake Lucerne in the upper right of Fig. 4d or Lake Constance on the right border and river Rhine in the middle of Fig. 4a. The red marked rectangle in Fig. 4a corresponds to the region of the VGT data, approximately. The RGB-image (Fig. 4e) based on VGT data uses the bands B3, B2 and B0. To increase the separability of the different dark objects new RGB bands are defined according eq. 2 (as already shown e.g. in Fig. 3b). The Alps were selected as test case to demonstrate that water pixels can be differentiated from terrain shadow. Fig. 4a and 4d show the static water mask with a fraction of water ≥60%. The visual comparison of the RGB-images (Fig. 4b and e) and the final water masks (Fig. 4c and 4f) show a good agreement of identified water pixels. The cloud cover which is different for both sensors determines how many water pixels can be identified. Most water pixels are identified as "stable water" (dark blue) based on the static and dynamic masks within PWS3-1. The VGT water mask shows in general more stable water pixels than the MERIS water mask due to more dust in the MERIS scene. But for the VGT scene clouds and haze obscure the identification of water pixels in the region of Lake Geneva. The green and light blue signed water pixels were identified in step PWS3-2 as "accepted static water", the result of MERIS data on basis of the same step in [6]. The cloud cover suppresses partially "accepted static water" pixels with the result that a gap between cloud and water masks developed independently of each other can be prevented in the transition region from water to haze or cloud. Red marked pixels ("accepted dynamic pixel") cannot be found in the resulting masks.

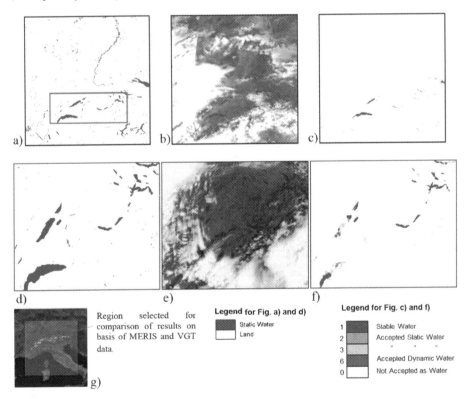

Fig. 4. Comparison of RGB images (b and e) and results for a subset of a MERIS-FRS scene (a-c) and for a subset of the daily VGT map (d-f) of the same date (18ᵗʰ July 2005) covering the western region of the Alps (g). The static masks lwms based on SWBD data with ≥60% fraction of water is shown in (a and d) and the dynamic land-water masks lwmd3 are shown in (c and f).

The identification of such "new" water bodies with respect to the static land-water mask is shown for a Scandinavian region (Fig. 5) as for AATSR and MERIS data before ([5], [6]) where no SRTM data are available (above 60°N). For these areas the GSHHS static land-water mask is used. Fig. 5a shows the RGB_{new} image (according eq. 2) of the region of interest. Fig. 5b and 5c show the static water masks for pixels with ≥10% (b) or ≥60% fraction of water (c). The water bodies of b) are a little bit larger than in c) because more pixels of the lake shore zones are included (compare with Fig. 2). The ratio of pixel number with 100 percentage water content to shoreline pixels with lower water content increases when the geometric resolution decreases. The comparison of static mask with the RGB image shows that a part of the small water bodies is not visible with 100 percentage security in the RGB_{new} data. The three large lakes in the centre of the image are Lake Storuman, Lake Vojmsjön and Lake Malgomaj (from above, [7]). Without additional knowledge the identification as water bodies is difficult in this region.

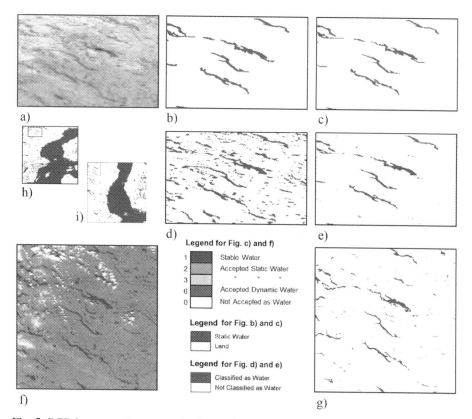

Fig. 5. RGB images and water masks for a subset in Scandinavia (red rectangle) of VGT (h) and MERIS-FR(S) (i) data from 4th July 2005: - VGT-RGB_{new}-image (a), VGT-lwms based on GSHHS data with ≥10% (b) and ≥60% fraction of water (c), VGT-2nd dynamic mask based on the proposed algorithm according eq. 3 (d), VGT-lwmd3 using DySLEM (e), - MERIS-RGB image (bands 15, 6 and 2) (f), MERIS-FR(S)-lwmd3 using DySLEM (g)

The lwmd1 mask does not include any "candidates" for dynamic water and was not included in Fig. 5. On other hand, the second dynamic mask (Fig. 5d) shows a lot of possible "candidates" for water, supporting the static information according eq. 6. Most "candidates" are included as "stable water" in the final result (Fig. 5e). Some static pixels of the shoreline of these long and narrow lakes are included in the final mask as light blue and green marked pixels as "accepted static water". Furthermore, this partial result is in a good agreement with the result for MERIS-FR(S) data of the comparable segment of the data path of the same date, already shown in [6] and presented in Fig. 5g. The resulting mask reflects the identification of water bodies/pixels not available in the static mask. A view to the RGB_{new} data demonstrates that a visual detection of water bodies is difficult in this region. The available information of MERIS-FR(S) data and the masks of better accuracy [6] support the correctness of water body identification. Additionally, the comparison of MERIS RGB data (f) and static masks (b and c) shows that a lot of water bodies are not available in the static water mask but available as "dynamic candidates" for water in the masks of pre-classification (Fig. 5d). All water pixels not included in the resulting mask before will be examined a second time with stronger relations and limits within PWS3-3. A small part of these candidates can be identified as "accepted dynamic water" pixels and can be seen as red marked pixels in the resulting mask (Fig. 5e). The results on basis of the higher resolved MERIS-FR data (g) show that more water objects can be identified.

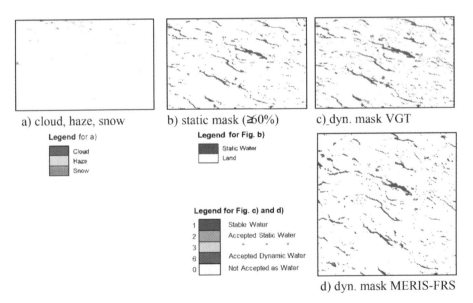

Fig. 6. Comparison of the final water masks for VGT (c) and MERIS-FRS (d) data shown in Fig. 5 when using MOD44 as static water mask (b)

Using the more detailed MOD44 water mask (Fig. 6b) instead of GSHHS as static land-water mask the final land-water mask is shown in Fig. 6c. The sub-classes of the static and final masks are shown in Tab. 2 where only cloud-, haze- and snow-free pixels are selected for comparison. For our fuzzy based cloud, haze and snow mask (Fig. 6a) the following thresholds were used: cloud >0.7, haze ≥0.5, snow >0.8. For example, the static GSHHS mask used in Fig. 5b includes 2445 VGT pixels with ≥10% fraction of water. In this number 1814 VGT pixels with ≥60% fraction of water (Fig. 5c) are included. 1422 of the 1814 pixels were identified as "stable water". 365 pixels with ≥60% fraction of water were identified in PWS3-2 as "accepted static water". That means only 27 VGT pixels with ≥60% fraction of water were not in-cluded in the final mask. At least 965 VGT pixels with ≥10% fraction of water (it includes the already mentioned 365 pixels) are identified as "accepted static water". Using the GSHHS mask 84 "new" water pixels in the VGT scene are identified. This is a relative small number compared to the 2848 water pixel in the MERIS data which is due to the reduced geometric resolution of VGT data. Furthermore the comparison of the RGB data shows that it is difficult to decide whether a dynamic candidate is accepted or not. Only 2 pixels as part of the shoreline of existing water bodies were identified when using the MOD44 mask (Fig. 5c). One pixel was already identified as "accepted dynamic water" when using the GSHHS mask. The second water pixel of this type can only identified in the mask on basis of MOD44. The same pixel is identi-fied as "stable water" pixel because in the GSHHS mask this pixel is included as a pixel with ≥60% fraction of water. Also it can be seen that in case of VGT data the largest proportion of static pixels about 97 per cent is included in the final masks as well as on basis of GSHHS or MOD44 mask (Fig. 5e and 6c) in contrast to 41.3 or 35.3 per cent in case of MERIS data (Fig. 5g and 6d).

The dynamic land-water mask based on MOD44 mask as static mask shows (Fig. 6c) that in contrast to Fig. 5e most part of the static water bodies is classified as "accepted static water". The dynamic mask lwmd2 of VGT data includes 5781 "dynamic candi-dates" for water. If using the GSHHS mask 68.4 per cent and in case of MOD44 mask only 36.9 per cent are not accepted and therefore not included in the resulting mask.

Table 2. Comparison of results of learning algorithm based on different static masks

sub-classes of water	MERIS_FRS		VGT	
	GSHHS	MOD44	GSHHS	MOD44
static water ≥10%	12713	27387	2445	6565
static water ≥60%	11415	19384	1814	2837
stable water	6993	15041	1422	2177
accepted static water ≥10%	468	2670	965	4175
accepted static water ≥60%	122	302	365	627
accepted dynamic water	2848	43	84	2
not accepted static water ≥10%	5252	9676	58	213
not accepted static water ≥60%	4300	4041	27	33
identified water (total)	10309	18342	2471	6354

An additional result is obvious in the final water mask of Fig. 3. It reflects accurately the decrease of Lake Aral (on the right border) due to desiccation since 2000 (generation of SWBD mask).

As observed before using AATSR and MERIS data the static GSHHS mask is shifted with respect to VGT data. The shift of the lakeshore into south-west direction for Lake Pyhäjärvi in Finland was also obvious in the VGT data. But the shift is corrected by the self-learning algorithm. Therefore, some water pixels of the static mask cannot be identified on the south-west lakefront and will not be included in the final mask. Some other pixels of the lake (north-eastern lakefront), outside the static land-water information, will be identified as "accepted dynamic water" pixels. For the MOD44 water mask no shift is seen.

Finally, it should be mentioned that for the different test sites cloud or terrain shadow was never identified as water.

5   Conclusions

The self-learning algorithm for dynamic classification of water bodies was modified and adapted for L3-VGT data. The final VGT land-water masks show that this new algorithms can be applied to L3-VGT data, to L2-AATSR data [5] and also to L2-MERIS-FR(S) and -RR data [6]. For our development we used 10 test sites of the ESA CCI project (partially complete Western Europe), and additionally regions where other dark pixels (e.g. terrain shadowed or burned area pixels) made the identification of water pixels difficult [6]. As for AATSR data (1.2 km spatial resolution) our self-learning algorithm detects also "new" water objects when working with SPOT-VGT data (1 km resolution) and a coarse static land-water mask as GSHHS where many water bodies are not included.

Beside the good quality of the final water mask the algorithm for the first dynamic mask (eq. 1) should be modified when working with L3-VGT maps. The reason for this modification is given by the fact that L3-VGT reflectances are atmospherically corrected while AATSR and MERIS data are top of atmosphere reflectances. Atmospheric correction can result in a strong decrease of B0 reflectance when aerosol correction is properly applied.

Good agreement of the final VGT water mask with MERIS and AATSR results was observed for the test regions. For regions above 60°N the water mask can be improved using more detailed static land-water masks as e.g. the MOD44 or Hansen_GFC [9] mask.

Acknowledgements. This study was supported by the ESA CCI ECV Fire Disturbance project (fire_cci), N°4000101779/10/I-LG. Many thanks to VITO, Belgium, and the VEGETATION Programme for the SPOT-VGT data. The daily VEGETATION Images were provided by the VEGETATION Programme.

The authors would like to acknowledge the use of the following free datasets:

- USGS for providing the SRTM Water Body Data (SWBD), version 2.1
- NOAA National Geophysical Data Center (NGDC) for providing the GSHHS data (A Global Self-consistent, Hierarchical, High-resolution Geography Database).
- GLCF (Global Land Cover Facility) for providing the UMD Global 250 meter Land Water Mask.
- University of Maryland, Department of Geographical Sciences for providing Hansen_GFC mask (Land Water Mask as part of Global Forest Change 2000-2013 Data)

References

1. Borg, E., Fichtelmann, B.: Determination of the usability of remote sensing data. EP 1591961 B1 (2005)
2. Carroll, M.L., Townshend, J.R., DiMiceli, C.M., Noojipady, P., Sohlberg, R.A.: A new global raster water mask at 250m resolution. Int. J. of Digital Earth **2**, 291–308 (2009)
3. ESA CCI ECV Fire Disturbance (fire_cci), N° 4000101779/10/I-LG
4. Fichtelmann, B., Borg, E., Kriegel, M.: Verfahren zur operationellen Bereitstellung von Zusatzdaten für die automatische Fernerkundungsdatenverarbeitung. In: Angewandte Geoinformatik 2011 (Strobl, Blaschke, Griesebner), 23. AGIT Symposium, Salzburg, pp. 12–20 (2011)
5. Fichtelmann, B., Borg, E.: A New Self-Learning Algorithm for Dynamic Classification of Water Bodies. In: Murgante, B., Gervasi, O., Misra, S., Nedjah, N., Rocha, A.M.A., Taniar, D., Apduhan, B.O. (eds.) ICCSA 2012, Part III. LNCS, vol. 7335, pp. 457–470. Springer, Heidelberg (2012)
6. Fichtelmann, B., Borg, E., Guenther, K.P.: Adaption of a Self-Learning Algorithm for Dynamic Classification of Water Bodies to MERIS Data. In: Murgante, B., Misra, S., Rocha, A.M.A., Torre, C., Rocha, J.G., Falcão, M.I., Taniar, D., Apduhan, B.O., Gervasi, O. (eds.) ICCSA 2014, Part I. LNCS, vol. 8579, pp. 376–392. Springer, Heidelberg (2014)
7. Google Earth. http://www.google.de/intl/de/earth/ (last access: January 21, 2014)
8. Haas, E.M., Bartholomé, E., Combal, B.: Time series analysis of optical remote sensing data for the mapping of temporary surface water bodies in sub-Saharan western Africa. J. Hydrology **370**, 52–63 (2009)
9. Hansen, M.C., Potapov, P.V., Moore, R., Hancher, M., Turubanova, S.A., Tyukavina, A., Thau, D., Stehman, S.V., Goetz, S.J., Loveland, T.R., Kommareddy, A., Egorov, A., Chini, L., Justice, C.O., Townshend, J.R.G.: High-Resolution Global Maps of 21st-Century Forest Cover Change. Science 342, 850–53 (2013). http://earthenginepartners.appspot.com/science-2013-global-forest (last access: April 23, 2014)
10. Klein, I., Dietz, A., Gessner, U., Dech, S., Kuenzer, C.: Results of the Global WaterPack: a novel product to assess inland water body dynamics on a daily basis. Remote Sensing Lett. (6), 78–87 (2015). http://www.dlr.de/eoc/desktopdefault.aspx/tabid-5258/15811_read-41169/ (last access: January 22, 2015)
11. Justice, C., Giglio, L., Korontzi, S., Owens, J., Morisette, J., Roy, D., Descloitres, J., Alleaume, S., Petitcolin, F., Kaufman, Y.: The MODIS fire products. Remote Sensing of Environment **83**(1&2), 244–262 (2002)
12. Lehner, B., Doll, P.: Development and validation of a global database of lakes, reservoirs, and wetlands. Journal of Hydrology **296**, 1–22 (2004)
13. VITO, Terms of use (last update: July 8, 2014). http://www.spot-vegetation.com/userguide/book_1/1/13/133/e133.htm (last access: January 22, 2015)

14. USGS (U.S. Geological Survey): Documentation for the Shuttle Radar Topography Mission (SRTM) Water Body Data Files. http://dds.cr.usgs.gov/srtm/version2_1/SWBD/SWBD_Documentation/Readme_SRTM_Water_Body_Data.pdf (last access: January 21, 2014)
15. Wessel, P., Smith, W.H.F.: A global, self-consistent, hierarchical, high-resolution shoreline database. J. Geophys. Res. **101**(B4), 8741–8743 (1996)

Assessment of MODIS-Based NDVI-Derived Index for Fire Susceptibility Estimation in Northern China

Xiaolian Li[1], Antonio Lanorte[2], Luciano Telesca[2],
Weiguo Song[1], and Rosa Lasaponara[2(✉)]

[1] State Key Laboratory of Fire Science,
University of Science and Technology of China, Hefei, China
wgsong@ustc.edu.cn
[2] Institute of Methodologies for Environmental Analysis,
National Research Council of Italy, Potenza, Italy
rosa.lasaponara@imaa.cnr.it

Abstract. Some satellite-based indices are useful for fire susceptibility estimation in some regions. However, the obtained results are region-dependent to some extent. The aim of this study is to assess the effectiveness of two NDVI-derived indices: the relative greenness index (RGI) and the vegetation danger index (VDI) applied to the northern China. Thus, the Moderate Resolution Imaging Spectroradiometer sensor (MODIS) data MYD13Q1, which is the 16-days composite product, were used. The results indicated that the RGI values were higher than 70% during the before-fire period from spring up to autumn, whereas it have decreased sharply when fire happened and after a period time and also it was below normal level during a period of after-fire. The VDI values were negative when fire happened and after a short period and that of the before fire and after fire were positive. Thus, it can be concluded that the two MODIS-based NDVI-derived indices have a higher possibility for fire susceptibility estimation even though it needs to combine other fire-related parameters.

Keywords: MODIS · NDVI-derived · Relative greenness · Vegetation danger index · Fire susceptibility

1 Introduction

Forest fire is one of the most critical natural hazards in the word and occurs more and more frequently in recent years[1]. It causes some damages such as the loss of biodiversity, land degradation and increase in greenhouse effect, etc.[2-5]. Thus, prevention measures, together with the fire prediction can help to make a decision for allocating the fire fighting resources reasonably and reducing the loss caused by fires.

Satellite remote sensing techniques have recently been considered as an effective approach to make a prediction of fire risk in wildland regions because a number of factors involved in fire susceptibility estimation (fuel moisture content, vegetation cover) can be derived from the abundant satellite data such as NOAA/AVHRR, EOS/MODIS

© Springer International Publishing Switzerland 2015
O. Gervasi et al. (Eds.): ICCSA 2015, Part IV, LNCS 9158, pp. 193–203, 2015.
DOI: 10.1007/978-3-319-21410-8_15

and Landsat/TM, etc[6-8]. Some previous researches have been developed for fire susceptibility estimation based on satellite images, but most of them were focus on NOAA/AVHRR data[9-11]. The Moderate Resolution Imaging Spectroradiometer (MODIS) onboard Aqua and Terra satellites is designed for fire detection mission and has higher spatial resolution at 250 m, 500 m and 1km than that of AVHRR[12]. It provides more abundant channel information for fire research activities.

The Normalized Vegetation Index (NDVI) is the most widely used tool for assessing the fire activities in before-fire, co-fire and after-fire phases [13-15]. It is calculated from the visible (RED) and the near infrared (NIR) channels by using the following formula NDVI=(RNIR-RRED)/(RNIR+RRED). However, it is too insufficient to estimate the fire susceptibility only based on the NDVI. Thus, some researchers have proposed to use the NDVI-derived indices[9,11,16], also together with the surface temperature[17] for fire susceptibility estimation.

It can obtain different results when applied the same index to different biomes or geographic regions. In other words, the usefulness of the index for fire susceptibility estimation is region-dependent. Thus, it has created some confusion about the effectiveness of the index. Meanwhile, the main problem is that the NDVI-derived indices are hardly adopted for fire susceptibility estimation in northern China. Therefore, the aim of this study is to assess the usefulness of two MODIS NDVI-derived indices (Relative greenness Index and Vegetation Danger Index) applied to the northern China. For this reason, the MODIS images of which the fire happened in Daxing'anling region were acquired to analyze the change of the indices from before-fire to after-fire.

2   Study Area and Dataset

This study was carried out in part of the Daxing'anling region in northern China. This region is located in the north of Heilongjiang Province and Inner Mongolia (50.5°-52.25°N, 122°-125.5°E, Fig.1). This region is selected because forest fires happen very usually and also vast areas are usually affected. The total area of Daxing'anling region is about 250, 000 km2. The forest coverage of Daxing'anling region is about 62%. There is about 81 300 km2 of forest located in Inner Mongolia. The prevailing land covers are: coniferous forest (Larix, Pinus Sylvestris, Spruce, Brich), broad-leaved deciduous forest (Oak, black birch), meadow vegetation, mire vegetation, grassland, agricultural areas as well as urban areas (Fig.2).

Daxing'anling region experiences cool continental monsoon climate, which is very cold in winter and warm in summer. During the spring and autumn, the weather is dry and the rainfall is low. Thus, fires frequently occur during this period as well as in summer.

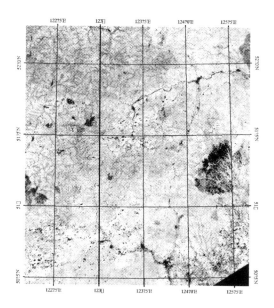

Fig. 1. The study area

Landcover of Study Area

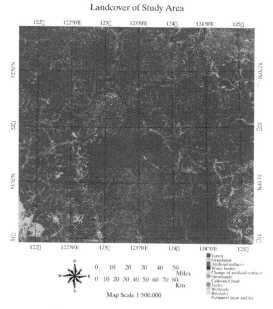

Fig. 2. The landcover of study area

In this study the data of MODIS sensor (MYD13Q1, 250m spatial resolution) on-board Aqua acquired from Land Processes Distributed Active Archive Center (LPDAAC) (http://e4ftl01.cr.usgs. gov/MOLA/) were used. MYD13Q1 is a 16 days composite vegetation index product, which is the maximum value of vegetation indic-es during the 16 days. A time series from 2008 to 2011 was used to analyze in this study. The pre-processing steps were as follows.

1) Projection Conversion. MYD13Q1 data are calibrated and georeferenced, thus, the original projection was converted to UTM (Universal Transverse Mercator) WGS84.

2) Extraction study area. We extracted the study area from MYD13Q1 of every image.

3) Data layer stacking. In this study, we used the data that NDVI gradually increase to the maximum of one year then decrease to some extent from spring (1st, May) to autumn (22nd, Sept). Thus, the data from spring to autumn in one year (2008-2011) were used to make layer stacking and calculate the vegetation danger index for making comparison among before-fire, co-fire and after-fire.

3   Fire Susceptibility Estimation Method Based on NDVI-Derived

Relative Greenness Index (RGI)

Relative greenness index (RGI) can be used as an indicator for monitoring the vegeta-tion coverage. Several previous studies [18,19] have demonstrated that the relative greenness has the feasibility to estimate fire susceptibility by comparing with fuel moisture content. Burgan et al.(1997) used relative greenness index as an input para-meter for estimating live fuel load. High relative greenness values indicate that the vegetation is healthy and vigorous whereas low greenness values indicate that the vegetation is under stress, dry (possibly from drought), is behind in annual develop-ment, or dead. Relative greenness index is calculated with respect to a historical database for that particular pixel. In this study relative greenness index have been calculated by using a long time series NDVI maps from 2002 to 2014 based on the following formula:

$$RGI = \frac{NDVI_{cur} - NDVI_{min}}{NDVI_{max} - NDVI_{min}} \times 100\% \tag{1}$$

where RGI is the value of relative greenness, NDVIcur is the observed NDVI for a given pixel, NDVImin and NDVImax are the minimum and maximum NDVI values for a given pixel during the study period.

The value trends of relative greenness index (historical series data for the same time of year) have been compared to confirm the below normal level during the fire occurrence.

Vegetation Danger Index (VDI)

Illera et al. (1996)[11] have proposed that the NDVI temporal evolution has the possibility and potentiality of estimating forest fire susceptibility. It was performed by using the NDVI time series data which is the maximum value composite computed in every 10 days beginning from the spring and lasting up to the period in which fire occurred. In order to perform estimation in a higher temporal resolution, Lasaponara et al. (2005)[17] adopted AVHRR daily NDVI instead of the maximum value composite images to estimate and compute the fire susceptibility from spring up to the date of fire occurrence by revising the method proposed by Illera et al. (1996)[11].

In our study, we used the method proposed by Illrea et al. (1996)[11] and Lasaponara et al. (2005)[17] based on the MODIS NDVI 16-days composite images from beginning from spring up to the date of fire occurrence for computing the fire susceptibility map, and also applied to the dates until autumn for analyzing the change of trends of VDI values in before-fire, co-fire and after-fire.

The vegetation danger index is shown in the following formula:

$$VDI = \sum_{i=1}^{n} \frac{NDVI(d_i) - NDVI(d_{i-1})}{d_i - d_{i-1}} \qquad (2)$$

where VDI is the Vegetation danger index, i is the image number in the considered series (from May to Sept), NDVI is the 16-days maximum composite data, and di is the date of the given image.

4   Results and Discussion

The investigation was performed in the Daxing'anling region located in the northern China by using a set of MODIS images from spring up to the date of fire occurrence in 2010 (The fire happened on 28th June, 2010), and also the dates until autumn, which the NDVI values decrease during this period. In order to show the change of the NDVI-derived indices among the before-fire, fire occurrence and after-fire, the data during the same period in 2008, 2009 and 2011 were also used in this study.

The relative greenness index of the date of fire occurrence in 2010 and that of the same dates in 2008, 2009 and 2010 in before-fire and after-fire were calculated by using the formula described in the previous part. The maximum and minimum NDVI were calculated by using the time series data from 2002 to2014. The results are shown in Fig.3. It can be seen from the zoom area in the lower right corner of the image that the RGI in this period of before fire is higher than 70% (Fig.3 (a), Fig.3 (b)). Whereas, the RGI of most of the fire-affected pixels are lower than 30% during the fire occurrence (Fig.3 (c)). And the RGI of fire-affected pixels in after-fire are almost between 20-50%. Thus, the relative greenness index didn't recover after one year. From this point of view, the RGI is related to the fire occurrence, but it seems that it is not RGI-dependent for fire susceptibility estimation during the after-fire due to the recovering of the vegetation.

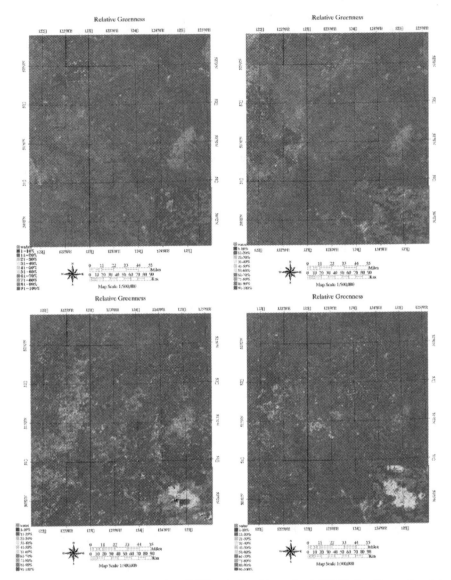

Fig. 3. The Relative Greenness Index (RGI) calculated by using MODIS the time series data MYD13Q1 from 2002 to 2014; (a) the RGI in 4th July, 2008 (before-fire); (b) the RGI in 4th July, 2009 (before-fire); (c) the RGI in 4th July, 2010 (fire occurrence); (d) the RGI in 4th July, 2011 (after-fire). Note: the red rectangle region in the lower right corner of very image is the zoom of the rectangle region in the central of image.

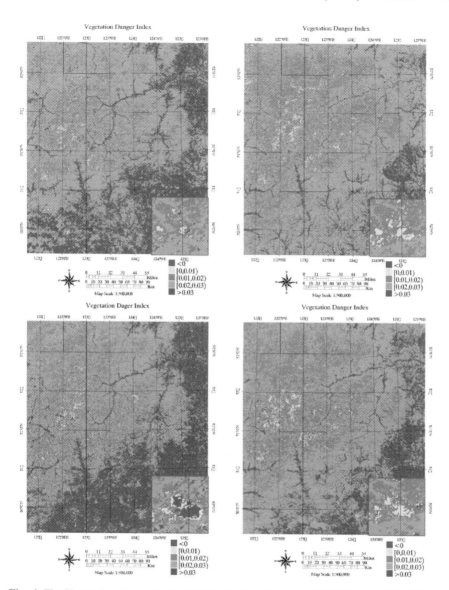

Fig. 4. The Vegetation Danger Index (VDI) calculated by using MODIS data MYD13Q1 from spring up to the date of fire occurrence; (a) the VDI in 4th July, 2008 (before-fire); (b) the VDI in 4th July, 2009 (before-fire); (c) the VDI in 4th July, 2010 (fire occurrence); (d) the VDI in 4th July, 2011 (after-fire). Note: the red rectangle region in the lower right corner of very image is the zoom of the rectangle region in the central of image.

The Vegetation Danger Index (VDI) which is calculated from spring up to the date of fire occurrence. The VDI of the same date in 2008 (before-fire), 2009 (before-fire), 2010 (fire occurrence) and 2011 (after-fire) were shown in Fig.4. It was divided into five classes as shown in Fig.4 and Table. 1. It can be seen that the VDI of fire-affected areas during before-fire were in moderate and low danger levels even though only a small part lied in high danger level (Fig.4 (a), Fig.4 (b)). However, the VDI of fire-affected areas during the fire occurrence almost lied in very danger level (Fig.4 (c)). And the VDI of fire-affected area during the after-fire lied from high to low danger level. From this point of view, the VDI lied in very danger level has a high possibility for fire susceptibility. But the high level needs to be confirmed together with other parameters such as the relative greenness index.

Table 1. Five classes of Vegetation Danger Index

VDI values	<0	[0,0.01)	[0.01,0.02)	[0.02,0.03)	>0.03
Level	Very high	high	moderate	low	No danger

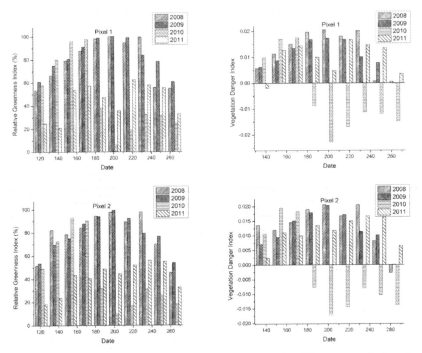

Fig. 5. The trends of RGI and VDI from spring up to autumn in different years (before-fire, fire occurrence and after-fire). (a) The RGI trends of pixel 1 in different years; (b) the change of VDI of pixel 1 in different years; (c) the RGI trends of pixel 2 in different years; (d) the change of VDI of pixel 2 in different years.

In order to show the change of RGI and VDI of fire-affected pixels during the be-fore-fire, fire occurrence and after-fire phases, the values of two fire-affected pixels in different years from spring up to autumn were extracted. The results were shown in Fig.5. It can be seen that the RGI values of two pixels in 2008 and 2009 (before-fire) are almost higher than 50% and reach to the maximum (close to 100%) during the summer. But it has a sharply decrease which is below 40% in 2010 during the fire occurrence. And also the RGI values of 2011 (after fire) didn't recover to the level of 2008 and 2009 (Fig.5 (a), Fig.5 (c)). Fig.5 (b) and Fig.5 (d) show that the VDI values are negative during the fire occurrence and after a period in 2010. And the VDI values during 2008, 2009, 2011 and a short period in 2010 before fire are positive even though there are two negative values in 2009 and 2011. It can be excluded the danger level together with RGI (50%) in 2009 during the before-fire. But the danger level needs to be confirmed during the after-fire period by using RGI and VDI together with other parameters such as surface temperature in the future.

From previous analysis, these two indices can be used as input parameters for fire susceptibility estimation.

5   Conclusion

Two NDVI-derived indices: the relative greenness index (RGI) and vegetation danger index (VGI) were applied to the Daxing'anling region located in the northern China based on the MODIS time series data MYD13Q1. It can be seen that the forest area has high RGI values from spring up to autumn, and reaches to the maximum in summer during the before-fire period. However, the RGI values have decreased sharply when fire happened. And also the RGI values need a long time to recover to the normal level during the after-fire period. It is shown that the VDI values were negative when fires happened and after a period. Whereas, the VDI values of before fire and after fire are almost positive even though a small part of values are negative. But it can be confirmed if the pixel is fire susceptible by using the VDI together with RGI and other parameters in the future. Thus, we can conclude that there two NDVI-derived have a high effectiveness for fire susceptibility estimation in Daxing'anling region. Results from our investigation confirm the reliability of satellite data for fire monitoring as already highlighted by [20,21,22]

Acknowledgement. The authors wish to thank the China Scholarship Council and the ARGON laboratory of the National Research Council of Italy for   providing the opportunity for Xiao-lian Li to study in Italy.

References

1. Telesca, L., Kanevski, M., Tonini, M., Pezzatti, G.B., Conedera, M.: Temporal patterns of fire sequences observed in Canton of Ticino (southern Switzerland). Natural Hazards and Earth System Sciences **10**, 723–728 (2010)
2. Crutzen, P.J., Andreae, M.O.: Biomass Burning in the Tropics - Impact on Atmospheric Chemistry and Biogeochemical Cycles. Science **250**, 1669–1678 (1990)

3. Kaskaoutis, D.G., Kharol, S.K., Sifakis, N., Nastos, P.T., Sharma, A.R., Badarinath, K.V.S., et al.: Satellite monitoring of the biomass-burning aerosols during the wildfires of August 2007 in Greece: Climate implications. Atmospheric Environment **45**, 716–726 (2011)

4. Li, Z.Q.: Influence of absorbing aerosols on the inference of solar surface radiation budget and cloud absorption. Journal of Climate **11**, 5–17 (1998)

5. Li, L.M., Song, W.G., Ma, J., Satoh, K.: Artificial neural network approach for modeling the impact of population density and weather parameters on forest fire risk. International Journal of Wildland Fire **18**, 640–647 (2009)

6. Stow, D., Niphadkar, M., Kaiser, J.: MODIS-derived visible atmospherically resistant index for monitoring chaparral moisture content. International Journal of Remote Sensing **26**, 3867–3873 (2005)

7. Chowdhury, E.H., Hassan, Q.K.: Use of remote sensing-derived variables in developing a forest fire danger forecasting system. Natural Hazards **67**, 321–334 (2013)

8. Chuvieco, E., Cocero, D., Riano, D., Martin, P., Martinez-Vega, J., de la Riva, J., et al.: Combining NDVI and surface temperature for the estimation of live fuel moisture content in forest fire danger rating. Remote Sensing of Environment **92**, 322–331 (2004)

9. Lopez, S., Gonzalez, F., Llop, R., Cuevas, J.M.: An Evaluation of the Utility of Noaa Avhrr Images for Monitoring Forest-Fire Risk in Spain. International Journal of Remote Sensing **12**, 1841–1851 (1991)

10. GonzalezAlonso, F., Cuevas, J.M., Casanova, J.L., Calle, A., Illera, P.: A forest fire risk assessment using NOAA AVHRR images in the Valencia area, eastern Spain. International Journal of Remote Sensing **18**, 2201–2207 (1997)

11. Illera, P., Fernandez, A., Delgado, J.A.: Temporal evolution of the NDVI as an indicator of forest fire danger. International Journal of Remote Sensing **17**, 1093–1105 (1996)

12. Kaufman, Y.J., Ichoku, C., Giglio, L., Korontzi, S., Chu, D.A., Hao, W.M., et al.: Fire and smoke observed from the Earth Observing System MODIS instrument - products, validation, and operational use. International Journal of Remote Sensing **24**, 1765–1781 (2003)

13. Telesca, L., Lasaponara, R.: Vegetational patterns in burned and unburned areas investigated by using the detrended fluctuation analysis. Physica a-Statistical Mechanics and its Applications **368**, 531–535 (2006)

14. Veraverbeke, S., Gitas, I., Katagis, T., Polychronaki, A., Somers, B., Goossens, R.: Assessing post-fire vegetation recovery using red-near infrared vegetation indices: Accounting for background and vegetation variability. Isprs Journal of Photogrammetry and Remote Sensing **68**, 28–39 (2012)

15. Leon, J.R.R., van Leeuwen, W.J.D., Casady, G.M.: Using MODIS-NDVI for the Modeling of Post-Wildfire Vegetation Response as a Function of Environmental Conditions and Pre-Fire Restoration Treatments. Remote Sensing **4**, 598–621 (2012)

16. Prosper-Laget, V., Douguedroit, A., Guinot, J.P.: A satellite index of risk of forest fire occurrence in summer in the Mediterranean area. International Journal of Wildland Fire **8**, 173–182 (1998)

17. Lasaponara, R.: Inter-comparison of AVHRR-based fire susceptibility indicators for the Mediterranean ecosystems of southern Italy. International Journal of Remote Sensing **26**, 853–870 (2005)

18. Kogan, F.N.: Remote-Sensing of Weather Impacts on Vegetation in Nonhomogeneous Areas. International Journal of Remote Sensing **11**, 1405–1419 (1990)

19. Burgan, R.E., Andrews, P.L., Bradshaw, S.L., Chase, C.H., Hartford, R.A., Latham, D.J.: Current status of the wildland fire assessment system (WFAS). Fire Management Notes **57**, 14–17 (1997)
20. Lanorte, A., Danese, M., Lasaponara, R., Murgante, B.: Multiscale mapping of burn area and severity using multisensor satellite data and spatial autocorrelation analysis. International Journal of Applied Earth Observation and Geoinformation **20**, 42–51 (2013)
21. Telesca, L., Lasaponara, R.: Pre and post fire behavioral trends revealed in satellite NDVI time series. Geophysical Research Letters **33**(14) (2006)
22. Tuia, D., Ratle, F., Lasaponara, R., Telesca, L., Kanevski, M.: Scan statistics analysis of forest fire clusters, pp. 1689–1694 (2008)

On the Use of the Principal Component Analysis (PCA) for Evaluating Vegetation Anomalies from LANDSAT-TM NDVI Temporal Series in the Basilicata Region (Italy)

Antonio Lanorte, Teresa Manzi, Gabriele Nolè, and Rosa Lasaponara[✉]

Institute of Metodologies for Enviromental Analysis, National Research Council of Italy,
Tito Scalo (PZ), Italy
rosa.lasaponara@imaa.cnr.it

Abstract. In this paper, we present and discuss the investigations we conducted in the context of the MITRA project focused on the use of low cost technologies (data and software) for pre-operational monitoring of land degradation in the Basilicata Region. The characterization of land surface conditions and land surface variations can be efficiently approached by using satellite remotely sensed data mainly because they provide a wide spatial coverage and internal consistency of data sets. In particular, Normalized Difference Vegetation Index (NDVI) is regarded as a reliable indicator for land cover conditions and variations and over the years it has been widely used for vegetation monitoring. For the aim of our project, in order to detect and map vegetation anomalies ongoing in study test areas (selected in the Basilicata Region) we used the Principal Component Analysis applied to Landsat Thematic Mapper (TM) time series spanning a period of 25 years (1985-2011).

Keywords: Satellite based analysis · Land degradation · Principal component analysis · GIS · Spatial variation · Vegetation index · Basilicata

1   Introduction

In relation to the land cover changes that can occur over large areas, the use of remote sensing data is an essential tool for change monitoring and mapping [3], [13].

Furthermore, remote sensing has been recognized worldwide as an effective, accurate and economical method to monitor changes in land cover from global down to a local scale [9], [14].

Many studies use remote sensing to monitor and evaluate the impact of urban growth on agricultural land [10], [15] as well as, to a lesser extent, the phenomena of agricultural neglect [1].

The Basilicata Region is characterized by a significant hydrogeological risk level and ongoing land degradation processes which have been particularly evident in the 30 last years. The so-called "erosion" processes affect large areas and are frequently strongly related to land degradation processes.

This paper is focused on the results we obtained from investigation conducted using the Principal Component Analysis applied to Landsat Thematic Mapper (TM)

O. Gervasi et al. (Eds.): ICCSA 2015, Part IV, LNCS 9158, pp. 204–216, 2015.
DOI: 10.1007/978-3-319-21410-8_16

time series spanning over a 25 year period (1985-2011) to detect and map vegetation anomalies. The Normalized Difference Vegetation Index (NDVI) was used as input to a selective Principal Component Analysis (PCA) procedure. The PCA was used as a first step of data transform to enhance regions of localized change in multi-temporal data sets [8], [5]. Results from PCA were further processed using Geospatial analysis to identify and map land degradation phenomenon.

NDVI is very effective for the identification of vegetation health being based on the normalized difference (see formula 1) between the infrared red band reflectance:

$$NDVI = (\rho_{RED} - \rho_{NIR})/(\rho_{RED} + \rho_{NIR})\tag{1}$$

NDVI provides information about the spatial and temporal distribution of vegetation cover in term of types, amount (biomass) and conditions. It provides a reliable estimation of the amount and vigor of vegetation because it is strongly related to the photosynthetic activity (Fig. 1).

The success of the NDVI is due to its reliability in detecting vegetation as well as in its simplicity in terms of computation and interpretation. That is the main reason of its wider use and larger popularity compared to other satellite-based spectral vegetation index.

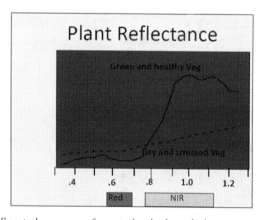

Fig. 1. Spectral response of vegetation in the red-nir spectrum regions

$$\text{cov}\, k1, k2 = 1/nm \sum_{i=1}^{n} \sum_{J=1}^{m} \left(MVC_{i,j,k1} - \mu_{k2}\right)\left(MVC_{i,j,k2} - \mu_{k2}\right)\tag{2}$$

Where k1, k2 are two input time series dates, MVC_{ij} the annual maximum NDVI value in row i and column j , n the number of row, m the number of columns and μ is the mean of all pixel MVC values in the subscripted input dates.

The percent of total dataset variance explained by each component is obtained by formula 3:

$$\% i = 100 * \lambda_i / \sum_{i=1}^{k} \lambda_i\tag{3}$$

where λ_i are eigenvalues of S.

Finally, a series of new image layers (called eigenchannels or components) are computed (using formula 4) by multiplying, for each pixel, the eigenvector of S for the original value of a given pixel in the input bands

$$P_i = \sum_{i=1}^{n} P_k \times u_{k,i} \tag{4}$$

where P_i indicates a spectral channel in component i, u_{ki} eigenvector element for component i in input band k, P_k spectral value for channel k, number of input band.

A loading, or correlation R, of each component i with each input band k can be calculated by using formula 5.

$$R_{k,i} = u_{k,i} \times (\lambda_i)^{1/2} \times (\mathrm{var}_k)^{1/2} \tag{5}$$

where var_k is the variance of input date k (obtained by reading the kth diagonal of the covariance matrix).

PCA transformation produces new principal components (PC1, PC2, PC 3, …etc.), which are uncorrelated and ordered in terms of the amount of variance they represent with respect to the set of the original bands. These bands are often more interpretable than the source data.

PCA is widely used in detecting changes in time series data and has become one of the most popular techniques because of its simplicity and capability of enhancing even subtle changes.

In our investigations, PCA was used to emphasize the areas that present significant changes in multi-temporal data sets. This is a direct result of the (i) high correlation that exists among pixels related to regions that do not change significantly and the (ii) relatively low correlation associated with regions that change substantially.

The major portion of the variance in a multi-temporal image data set is associated with constant cover types and is represented in PC1, while the regions with localized change will be enhanced in later components. In particular, each successive component contains less of the total data set variance. In other words, the first component contains the major portion of the variance, whereas, later components contain a very low proportion of the total data set variance.

PC1 explains most of the variation in the NDVI integrals and represents the average spatial integrated NDVI pattern. PC1 shows the typical NDVI over the entire series.

PC2 explains the maximum remaining variation not explained by PC1 and subsequent components follow the same rationale.

Therefore the components subsequent to PC1 (especially PC2) tend to reflect specific events such as fires, drought periods and, in general, land degradation phenomena related to the vegetation, rather than to depict the general development during the period.

PCA is then used here as a tool for mapping areas that show a significant i.e measurable degree of inter-annual variability, able to discriminate unidirectional changes.

2   Dataset and Study Area

Data Set

The investigations were performed by using NDVI data derived from the Landsat TM images (Tab. 1) selected on the base of the data quality and low percentage of cloud cover in the months between June and September (period 1985-2011).

Table 1. Analytical categories

Thematic Mapper (TM)		
Landsat bands	Wavelength (micrometers)	Resolution (meters)
Band 1	0.45-0.52	30
Band 2	0.52-0.60	30
Band 3	0.63-0.69	30
Band 4	0.76-0.90	30
Band 5	1.55-1.75	30
Band 6	10.40-12.50	120
Band 7	2.08-2.35	30

The Landsat TM data were acquired free of charge from the United States Geological Survey (USGS) web site (Tab. 2).

Table 2. Landsat-TM Dataset

Year	Month	Day
1985	August	10
1986	August	13
1986	September	14
1987	June	13
1987	September	17
1993	July	15
2002	June	22
2003	June	25
2003	July	11
2003	July	27
2009	July	27
2010	September	16
2011	August	02
2011	August	18

Study Area

The analysis was performed in the Basilicata Region (see Fig. 2) that is characterized by typical Mediterranean climate with a pronounced bi-seasonality regime having hot/dry summers and cold/wet winters. Due to a combined effect of natural hazards (drought, wind and rain erosion, floods) and human activity (industry, fires, over tilling, land abandonment), this area recently increased its vulnerability [2].

The environmental equilibrium of the Basilicata is fragile and highly vulnerable to perturbations, as in other Mediterranean regions, and, therefore, it is expected that natural ecosystems, such as forest, shrubland and herbaceous cover, should be more sensitive to the changes that are presently affecting the whole Mediterranean basin.

Fig. 2. Study area

3   Methodology

Figure 3 shows a flow chart of the methodology that we adopted.

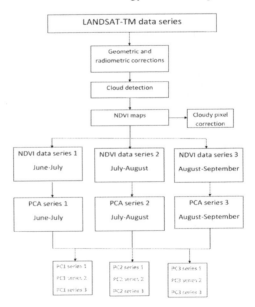

Fig. 3. Methodology flow chart

The Landsat-TM data series (summer images) were acquired on the basis of their free-of-charge availability, images quality and low percentage of cloud cover.

On each image we applied a geometric, radiometric and atmospheric correction, then a cloud detection if need.

Then we created three NDVI time series corresponding to June-July, July-August and August-September periods: this in order to avoid to apply PCA in the temporal windows too different on the phenological point of view.

For each NDVI series the PCA was applied and, finally, the first, the second and the third components of each series was combined and analyzed separately.

4   Results and Discussion

As mentioned before the most significant results in the evaluation of vegetation anomalies can derive from the second component of the PCA. The results shown in figure 4 concern the analysis of the second component of the PCA in some significant test sites selected in the Basilicata region.

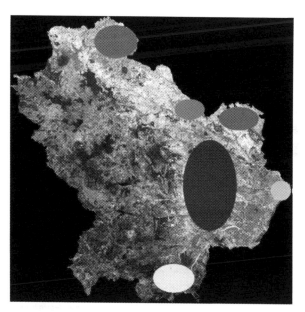

Fig. 4. Study areas. (purple=Matera; orange=Metaponto; yellow= Pollino; green=Irsina; red=North-East; Blue=Lucanian Badlands).

In figure 5 there is a comparison of the second components obtained for each NDVI data series for the Matera site. The graphics show the loadings of the second component of each series. The loadings representing the correlation of each component (in this case the second component) with each input date (NDVI). In this case the loadings show in general a decreasing trend in all three series.

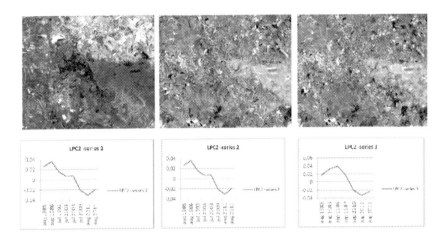

Fig. 5. Matera site. PC2 and loadings (3 series).

This trends justify the presence of negative anomalies (Fig. 6). This results concern only the anomalies detected in natural or semi-natural areas excluding, then, the agricultural areas in order to avoid to confuse the changes of agricultural land use with vegetation anomalies.

We distinguished highly negative anomalies (in red) and moderately negative anomalies (in orange). We defined the areas with highly negative anomalies as "critical areas" and the areas with moderately negative anomalies as "fragile areas".

The critical areas correspond to areas where negative anomalies are high in all three considered series, while the fragile areas correspond to areas where the negative anomalies are high only in two series. High negative values are the values below a set threshold.

Fig. 6. Matera site. Negative anomalies (left: critical areas; right: fragile areas).

In the case of the Metaponto site, PC2 loadings trends (Fig. 7) are very different compared to the first case. Here, the loadings of series 1 show a decreasing trend until the 2003 and an increasing trend after; the loadings of series 2 tend to increase or (after 2003) they are stable; the loadings of series 3 show a decreasing trend more pronounced but in the last years they are stable.

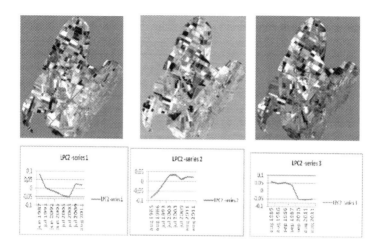

Fig. 7. Metaponto site. PC2 and loadings (3 series).

The map of anomalies reflects these trends because we did not detect any highly negative anomaly but only fragile areas (Fig. 8).

Fig. 8. Metaponto site. Negative anomalies (fragile areas).

Figure 9 shows the results obtained in the Pollino site. In this case PC2 loadings show an increasing trend in series 1 and 2 but a decreasing trend in series 3.

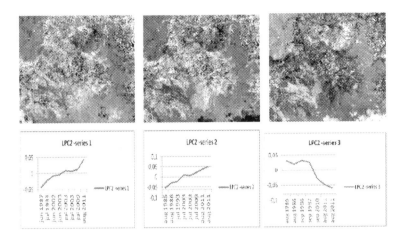

Fig. 9. Pollino site. PC2 and loadings (3 series).

These trends are confirmed in the map of anomalies where we observe some fragile areas (Fig. 10).

Fig. 10. Pollino site. Negative anomalies (fragile areas).

The same observations are suitable for the Irsina site (Figs. 11 and 12).

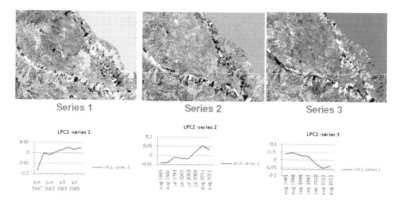

Fig. 11. Irsina site. PC2 and loadings (3 series).

Fig. 12. Irsina site. Negative anomalies (fragile areas).

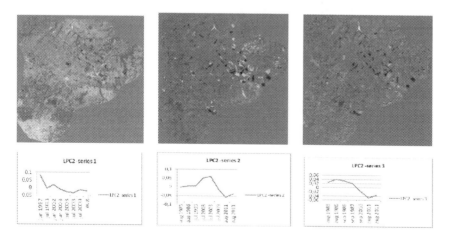

Fig. 13. North-East Basilicata site. PC2 and loadings (3 series).

In agreement with this analysis we found several highly and moderately negative anomalies (Fig. 14).

Fig. 14. North-East Basilicata site. Negative anomalies (left: critical areas; right: fragile areas).

The last site correspond to the Lucanian badlands area. In this case all loadings clearly show a decreasing trend (Fig. 15).

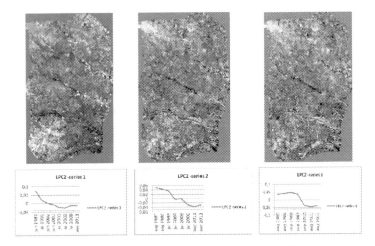

Fig. 15. Lucanian Badlands site. PC2 and loadings (3 series).

In fact we found many negative anomalies (critical and fragile areas) (Fig. 16).

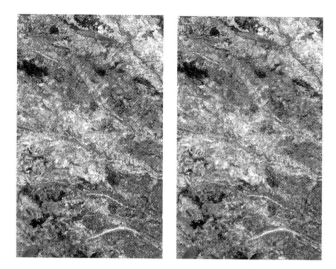

Fig. 16. Lucanian Badlands site. Negative anomalies (left: critical areas; right: fragile areas).

Finally, the results shown are consistent with the analysis performed with SPOT-VGT NDVI time series for land degradation monitoring in the Basilicata Region [6]. Moreover we found significant spatial correlations with the map of areas susceptible to desertification in Basilicata [4]. Thus confirming the reliability of spatial data and robust statistical analysis for operational risk monitoring as, for example, in [7,11,12, 16, 17,18].

5   Final Remarks

The technique and the methodology we adopted show a promising ability to identify and monitor vegetation anomalies also considering the relatively small number of Landsat images. The use of the PCA allowed us to monitor degradation phenomena in heterogeneous landscapes such those investigated for the Basilicata region at the medium-high scale provided by Landsat-TM data. The use of PCA can be an effective and low cost tool for extracting valuable information from NDVI temporal series regarding vegetation inter-annual variations. This study exemplifies the potential use of Landsat time series for environmental analyses performed on local scale and provides basic information applied in change detection analyses and in monitoring land degradation processes. The results accuracy could be increased by using other Landsat datasets series (ETM, Landsat 8), by improving data pre-processing and by integrating the results with ground surveys.

Acknowledgement. This research has be made in the context of Bilateral Italy-Argentina MAE project PGR00189 titled "Remote sensing technologies applied to the management of natural and cultural heritage in sites located in Italy and Argentina. Risk monitoring and mitigation strategies".

References

1. Alcantara, P.C., Radeloff, V.C., Prishchepov, A., Kuemmerle, T.: Mapping abandoned agriculture with multi-temporal MODIS satellite data. Remote Sensing of Environment **124**, 334–347 (2012)
2. APAT report 2006. La vulnerabilità alla desertificazione in Italia. Manuali e linee guida 40/2006
3. Dewan, A.M., Yamaguchi, Y.: Using remote sensing and GIS to detect and monitor land use and land cover change in Dhaka Metropolitan of Bangladesh during 1960-2005. Environ. Monit. Assess. **150**, 237–249 (2009). doi:10.1007/s10661-008-0226-5
4. Ferrara, A., Bellotti, A., Faretta, S., Mancino, G., Baffari, P., D'Ottavio, A.: Carta delle aree sensibili alla desertificazione della Regione Basilicata. Forest@-Journal of Silviculture and Forest Ecology **2**(1), 66 (2005)
5. Howarth, P., Piwowar, J., Millward, A.: Time-Series Analysis of Medium-Resolution, Multisensor Satellite Data for Identifying Landscape Change. Photogrammetric Engineering and Remote Sensing **72**(6), 653–663 (2006)
6. Lanorte, A., Aromando, A., De Santis F., Lasaponara, R.: Investigating satellite SPOT VEGETATION multitemporal NDVI maps for land degradation monitoring in the Basilicata Region: Preliminary Results from the MITRA project. In: Proceedings of the 33rd EARSel Symposium, Matera (Italy) (2013)
7. Lanorte, A., Danese, M., Lasaponara, R., Murgante, B.: Multiscale mapping of burn area and severity using multisensor satellite data and spatial autocorrelation analysis. International Journal of Applied Earth Observation and Geoinformation **20**, 42–51 (2013)
8. Lasaponara, R.: On the use of principal component analysis (PCA) for evaluating interannual vegetation anomalies from SPOT/VEGETATION NDVI temporal series. Ecol. Model. **194**, 429–434 (2006)
9. Lunetta, R.S., Johnson, D.M., Lyon, J.G., Crotwell, J.: Impacts of imagery temporal frequency on land-cover change detection monitoring. Remote Sensing of Environment **89**, 444–454 (2004)
10. Stefanov, W.L., Ramsey, M.S., Christensen, P.R.: Monitoring urban land cover change: An expert system approach to land cover classification of semiarid to arid urban centers. Remote Sensing of Environment, 173–185 (2001)
11. Telesca L., Lasaponara R.: Pre and post fire behavioral trends revealed in satellite NDVI time series. Geophysical Research Letters 33(14) (2006)
12. Tuia, D., Ratle, F., Lasaponara, R., Telesca, L., Kanevski, M.: Scan statistics analysis of forest fire clusters, pp. 1689–1694 (2008)
13. Tziztiki, J.G.M., Jean, F.M., Everett, A.H.: Land cover mapping applications with MODIS: a literature review. International Journal of Digital Earth **5**(1), 63–87 (2012)
14. Wilson, E.H., Sader, S.: Detection of forest harvest type using multiple dates of Landsat-TM imagery. Remote Sensing of Environment **80**, 385–396 (2002)
15. Yeh, A.G., Li, X.: An integrated remotes sensing and GIS approach in the monitoring and evaluation of rapid urban growth for sustainable development in the Pear River Delta, China. International Planning Studies **2**(2), 193–210 (1997)
16. Lanorte, A., Danese, M., Lasaponara, R., Murgante, B.: Multiscale mapping of burn area and severity using multisensor satellite data and spatial autocorrelation analysis. International Journal of Applied Earth Observation and Geoinformation **20**, 42–51 (2013)
17. Telesca, L., Lasaponara, R.: Pre and post fire behavioral trends revealed in satellite NDVI time series. Geophysical Research Letters 33(14) (2006)
18. Tuia, D., Ratle, F., Lasaponara, R., Telesca, L., Kanevski, M.: Scan statistics analysis of forest fire clusters, 1689–1694 (2008)

A New Approach of Geo-Rectification for Time Series Satellite Data Based on a Graph-Network

Bernd Fichtelmann[✉] and Erik Borg

German Aerospace Center, German Remote Sensing Data Center,
Kalkhorstweg 53, 17235, Neustrelitz, Germany
{Bernd.Fichtelmann,Erik.Borg}@dlr.de

Abstract. Earth observation is indisputably one of the most important data sources for current synoptic geo-data in diverse environmental applications. Nevertheless, obtained thematic information by means of Earth observation is only as good as the quality of the pre-processed data. Important pre-processing steps are e.g. data usability assessment, geo-referencing and atmospheric correction. While data usability assessment or atmospheric correction expects radiometric corrected data for a thematic interpretation, the accuracy of geo-location of interpreted data is ensured by geo-referencing. This applies especially to multi-temporal analyses of environmental processes. In this case, precise spatial allocation of data represents a prerequisite for a correct interpretation of process dynamics. The paper is dealed with a new geo-referencing algorithm. Corresponding graph-networks in reference data as well as in remote sensing data which are based on virtual object points (e.g. centroids) will be used for geo-rectification.

Keywords: Remote sensing · Geo-correction · Image series · LANDSAT

1 Introduction

In geo- and environmental sciences, in context of supporting of decision-making of authorities and politicians or in diverse other applications, remote sensing is well accepted as actual information source covering large regions. However, the quality of remote sensing based environmental information cannot be better than the quality of pre-processing before thematic processing.

The development of sensors must be adapted to increasing requirements of respective applications. Therefore, geometric resolution of satellite sensors was recently improved in order to be able to detect urban changes such as small-scale changes. However, there is a discrepancy between data collection and data validation. This aspect is very complex and expensive, if it is manually or interactively. In this case, automated processors for evaluation of remote sensing data can rectify this. Beside cost reduction, automation of data evaluation has still further advantages such as faster processing, by which results can be delivered in near real time or elimination of subjective influences by e.g. interpreters. The automation of pre-processing is thus

© Springer International Publishing Switzerland 2015
O. Gervasi et al. (Eds.): ICCSA 2015, Part IV, LNCS 9158, pp. 217–232, 2015.
DOI: 10.1007/978-3-319-21410-8_17

urgent to be solved. Beside atmospheric correction, geo-referencing of remote sensing data is an essential precondition to prepare data for thematic processing and to derive quantitative parameters and/or indicators from these data.

Geo-correction is an important pre-processing operation, aimed to eliminate geometric distortions in image data caused by e.g. rough terrain, various aberrations that can be occur by sensors, central perspective recording and/or incorrect orientation of imaging system. With reference to geo-correction, two extreme cases can be defined:

- Case 1: The imaging process is completely well known and can be described exactly by a mathematical image model.
- Case 2: The imaging process and/or imaging parameters is/are totally unknown. The local reference of image data must be derived from detection and correlation of similarities of image structures and reference structures. These reference data are a geo-corrected pre-requisite called reference image.

Between both cases, combinations of available information are possible. A number of different algorithms is used in practice.

However, one problem in determination of coordinates for each image pixel is resulting from a limited accuracy of available orbit parameters. Especially if non-correct orbit parameters are used for processing the results are insufficient for multi-temporal analysis. The overlaid shorelines of classified water bodies for a selected region in north east of Germany derived from five data sets demonstrate these problem (Fig 1).

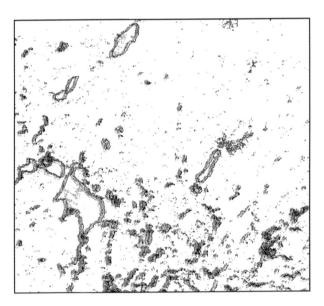

Fig. 1. Overlaid shorelines of water bodies derived from 5 LANDSAT 7 data sets (year 2000) on basis of available geographic coordinates. Compare with Fig. 10 for more details.

With respect to product quality, geo-correction is a critical pre-processing step since even relatively small position errors can produce large errors in thematic products. Thus there is a demand on automatic geo-correction to achieve a position accuracy as high as possible. In best case, the position accuracy should be in sub-pixel range. A survey of geo-correction principles used for remote sensing and/or other applications is given in literature (e.g. [1], [2], [3], [4], [5], [6]).

This paper deals with development of a new geo-correction algorithm based on a graph-network between object centroids, at least virtual points, which is foreseen for an automated processing chain.

2   Material and Methods

2.1   Test Site DEMMIN in Mecklenburg-Western Pomerania

The selected test region for applying the geo-referencing processor is located in the federal state Mecklenburg-Western Pomerania in North-East Germany (see Fig. 2). The test region includes the investigation facility DEMMIN (Durable Environmental Multidisciplinary Monitoring Information Network) for calibrating and validating e.g. remote sensing processing algorithms, and remote sensing data as well as corresponding value-added products [7].

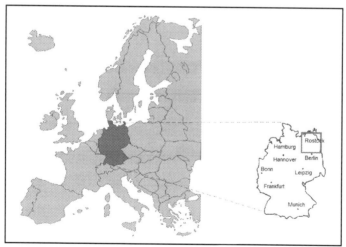

Fig. 2. Geographical location of the federal state Mecklenburg-Western Pomerania

The landscape is part of German lowlands and relative flat. The climate is embossed humid. The test region was formed during the young Pleistocene[1]. The drainage network comprises numerous lakes. These landscape objects are relative stable. With respect to geo-rectification the high number of lakes, the coastline of the Baltic Sea, and some large islands are optimal. Additional, the flat landscape minimizes the requirement of an elevation model.

---

[1]The Pleistocene (geological epoch) began at 2,588,000 and continues to 11,700 years ago.

2.2   Reference Data Basis

An important pre-requisite for geo-referencing are accurate control data. Thus, the accuracy of geo-referenced satellite data cannot be better than the position accuracy of the reference basis. In this case, some fundamental problems of geo-referencing are transferred into the preparation of the reference basis. These can be minimized by extensive manual manufacturing and/or by using a GPS-based reference.

Besides the high position accuracy, the greatest possible similarity between pre-processed and reference data is required. However, generally a perfect similarity cannot be expected. Therefore, at least relatively stable landscape objects with relatively well known reflectance characteristics in space and time should be available. In experiment 1 the new algorithm was applied to classified LANDSAT quick-look data including significant water and land (islands) objects to demonstrate the principle functionality. The procedure starts with an automatically produced reference map based on the 1993 CIA World map (Land-Water-Classification) data [8]. The map includes large lakes and islands. In experiment 2 a segment of a classified LANDSAT 7 scene in UTM projection (in ETM_FAST_L7A format [9]) was used as reference data. For demonstrating the algorithm was based on a smaller segment of a second classified cloud free scene available in same format and projection. In experiment 3 the CIA world map was used as reference basis to rectify 5 selected full resolution LANDSAT 7 scenes ETM_FAST_L7A format relative to each other.

2.3   Test Data

2.3.1   LANDSAT 7/ ETM⁺ Quick-Look Data

The investigations presented here are based on quick-look[2] and metadata[3]. The LANDSAT 7 quick-look data have a size of 1000 x 1000 pixels available by ESA [10].

The LANDSAT-data are radiometric corrected by a predefined linear stretch. The geometric pre-processing includes, e.g. a data reduction 30 m to 180 m (6 x 6 pixel to 1 pixel) of the bands 1-5 and 7. Additional to the geometric data reduction, the original data are stored in JPEG format compressed with a ratio of 10:1 [11]. Therefore, quality, information losses and artefacts are possible. A detailed procedure description can be found in [12], [13]. Although the level of compression depends on the image content, this represents a JPEG quality metric Q-factor of 35 [14]. The quality loss is of significance for thematic post-processing.

2.3.2   LANDSAT 7 /ETM⁺ Data

Beside the quick-look data 10 LANDSAT 7 full resolution data (ETM_FAST_L7A format [9]) for Mecklenburg-Vorpommern (tracks 193 and 194, floating type of frames 22 and 23) of the year 2000 are used. For classifying the data we used an algorithm proposed by [15].

---

[2] Quick-look data are preview images derived from original remote sensing data.

[3] Metadata describes remote sensing data (e.g. satellite mission, orbit, track, frame).

3 Automated Geo-referencing Processor Based on a Graph-Network

3.1 Aspects of Automation

The users of geo-data come from various sectors of economy. Therefore, the requirements on geo-information are increasing steadily. In order to fulfil the requirements, of automated interpretation of remote sensing data beside available remote sensing data suitable processors for data processing are required. These processors have to deliver different status information during processing: i.) control of start and end, ii.) status at any time iii.) response to any process failures. While information of start and/or end of processing can be easily delivered, information about undefined processor failures is more complicated because not all types of errors can be anticipated.

Especially these control algorithms are required for serving product quality.

3.2 Processor Structure and Mathematical Principles

Remote sensing sensors collect data of Earth's surface from a distance. However, in many cases, an increase of distance is related to a decrease in spatial resolution and image sharpness. A further difference from ground-based remote sensing consists in disturbing influences of atmospheric effects. This influence extends from the change in surface radiation going up from Earth surface to complete coverage of Earth by clouds. In general, these characteristics of remote sensing must be considered when geo-referencing.

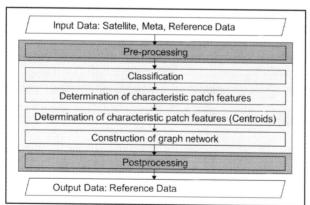

Fig. 3. Flow chart of the geo-referencing processor based on a graph-network for remote sensing data (adopted and modified [16])

Humans can recognize the same object very well in various images under a variety of conditions (e.g. lighting, perspective). This is valid even still if the object shape is changed, since yes still the local relations with other objects are given in addition. This is true if the shape of objects has changed, when the local relationships to other objects still exist. Therefore, for a visual analysis of remote sensing data no geo-correction is required.

The procedure presented here is a sub-module of a more complex geo-correction processor for single-channel or multi-channel remote sensing data [16]. However, with the sub-processor remote sensing data can be referenced as well as ground control points (GPS) can be delivered to supply these to a highly precise geo-correction. In Fig. 3 the procedure is schematically represented according to processing steps. The appropriate geo-correction data can be made available for a high precise geo-correction method as described in [16].

In order to operate the processor, we want to presuppose several basic requirements:

1. The higher the degree of similarity between reference and remote sensing data to be geo-referenced, the higher is the degree of matching precision.
2. Reference and remote sensing data cover the same landscape objects, which can be detected, identified and linked in both data sets.
3. The landscape objects and relations between these objects are distorted in satellite data but not additional by e.g. disturbances in the mapping process.
4. The smaller a landscape object is, the sooner it reaches the size adequate to the ground resolution of the sensor the lesser is its uniqueness for recognition. That means, identified objects cannot be distinguished by characteristic features.
5. Additional, identified landscape objects should be invariant to annual and seasonal variations in detectable properties.
6. It is obvious, that the usability of objects for geo-rectification decreases with an increasing of cloud coverage within the satellite images.
7. The larger a landscape object, the higher is the probability that an object is obscured by clouds.

With respect to these preconditions, the following consequences can be expected:

1. A precise reference basis is given by an ortho-rectificated data set based on relatively recent satellite image data.
2. The thematic and geometric properties of landscape objects covered by remote sensing data are important for their identification and comparison with objects covered by reference data. The object characterization can be done by i) classification according to their land cover or land, and ii) structure analysis according to their geometry for feature matching.
3. The size of identified landscape objects should fulfil the sampling theorem in minimum, in order to identify them as well in reference as in image data.
4. It is advantageous, that if the size of a large object is relative stable than the centroid usually varies in a very small spatial range depending e.g. from seasonal situation.

The processing aims to classify stable landscape objects in reference as well as in satellite data. Corresponding classes are e.g. water or forest. In addition, it is assumed that at a lower hierarchy class level, the classification error can be reduced. The classification includes a segmentation, thematic differentiation, and object initialization to identify and to label objects in remote sensing (in following defined as image or image data) and reference data (in following defined as reference). A prerequisite of our geo-rectification and use of following equations for shift and rotation consist in preparation of a map in suitable projection (Lambert projection with preliminary centre coordinates of image as projection centre) and size to be determined iterative. The length r of an image line and the distance c from first to last line must be the same after projection into the map in pixel length as shown in Fig. 4. In the following the map can filled in e.g. with a land-water mask or with reference data[4]. This is not necessary if reference and image are available in same projection and size, e.g. as shown in a second experiment. In a next preparation step, all cloud obscured or to small objects are eliminated from the reference as well as the image data. All other objects are labelled by unique object-IDs. For these, different robust patch measures (e.g. area, perimeter, centroids) will be calculated, which make a linking of objects in both data sets possible. The object centroids are used as virtual object points (nodes) to calculate a graph network in both data sets.

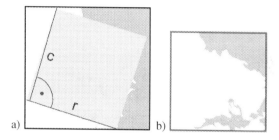

a)   b)

Fig. 4. Schematic projection (a) of image data (b) on basis of preliminary corner coordinates

The computation of centroids for pixel based image objects is described by [17]. Since all N pixels $P_i=(x_i,y_i)$ of an object m are marked by the same object ID, the X_m and Y_m coordinates of the centroid can be computed by eq. 1:

$$X_m = \frac{1}{n}\sum_{i=1}^{n} x_i \quad \text{and} \quad Y_m = \frac{1}{n}\sum_{i=1}^{n} y_i \tag{1}$$

The Euclidean Distance between two points is described by eq. 2:

$$d = \sqrt{(x_2 - x_1)^2 + (y_2 - y_1)^2} \tag{2}$$

If an object is accepted, then it is stored in an allocation matrix for further processing. In case of several candidates being identified, the candidate with highest correlation to reference object is selected, if the correlation is in a defined limit.

The shift vector (dx,dy) between centroids, preferably computed from image coordinates (x_0^I,y_0^I) of the starting object[5] in the image data and appropriate image coordinates (x_0^R,y_0^R) of reference, is given by eq. 3.

$$dx = x_0^R - x_0^I \quad \text{and} \quad dy = y_0^R - y_0^I \tag{3}$$

---

[4] Reference data have to re-projected if available in other map projection and/or resolution.
[5] Starting object: first acceptable object for searching additional objects usable for construction a graph network.

The rotation angle α of image can be computed according eq. 4 using the vectors created by centroids of starting object 0 and of another identified object 1 in reference ($\vec{a}$) and image ($\vec{b}$).

$$\cos\alpha = \frac{\vec{a}\vec{b}}{|\vec{a}||\vec{b}|} \tag{4}$$

After determination of angle α the image coordinate (x_i^I, y_i^I) of object i can be transferred into coordinate $(x_i^{R''}, y_i^{R''})$ of reference using centroid coordinates (x_0^I, y_0^I) as new coordination centre $(0,0)$ for rotation and the transformation of eq. 5 including the corresponding coefficient matrix (eq. 6).

$$\begin{pmatrix} x_i^{R''} \\ y_i^{R''} \end{pmatrix} = \begin{pmatrix} x_0^I \\ y_0^I \end{pmatrix} + D \begin{pmatrix} x_i^I - x_0^I \\ y_i^I - y_0^I \end{pmatrix} + \begin{pmatrix} dx \\ dy \end{pmatrix} \tag{5}$$

$$D = \begin{pmatrix} \cos\alpha & -\sin\alpha \\ \sin\alpha & \cos\alpha \end{pmatrix} \tag{6}$$

In this way, a quality control for high correlated linked objects is possible by delivering registration parameters. More accurate and stable results will be achieved by computing a mean rotation angle. Additionally, the distance between starting object and other objects as well as object size can be used as weighting factors before using eq. 5 for other image pixels.

For quality check of the method two different measures were computed. The first is the mean of distances between centroids $(x_i^{R''}, y_i^{R''})$ and (x_i^R, y_i^R) of accepted N objects. As a second error measure we determine the Root Mean Square Error $(RMSE_{xy})$ according eq. 7.

$$RMSE_{xy} = \sqrt{\frac{1}{N}\sum_i \left(x_i^{R''} - x_i^R\right)^2 + \frac{1}{N}\sum_i \left(y_i^{R''} - y_i^R\right)^2} \tag{7}$$

4   Results and Discussion

The development and demonstration of sub-processor was based on LANDSAT 7/ETM+-quick look data. These data were classified (Fig. 5b). Additional a reference map with a land-water mask (CIA World map database [8] included in IDL software was prepared (Fig. 5a) using available preliminary corner coordinates of image (marked by yellow lines in the map). It is assumed that each pixel of map has the same size as the image pixel. In order to geo-reference remote sensing data, a relation must be delivered between remote sensing and reference data. For this, structure information in both data sets must be identified and characterized to link remote sensing with reference data.

In Fig. 5a and 5b different islands e.g. Ruegen (largest island, centre), Bornholm (Island of Denmark, upper right) are shown as potentially corresponding object in image (right) and reference data (left).

The processor i.) selects islands and water bodies, ii.) eliminates small objects or having contact to image border (e.g. Bornholm) or clouds, iii.) sorts remaining objects based on size, iv.) labels them with an identification number (ID), v.) determines shoreline of selected objects and vi.) computes centroids of objects. Additionally a graph network will be constructed using centroids as nodes of directly neighbouring objects and marked by blue lines in Fig. 5c and 5d. The complete data set of graphs is stored in a matrix. The relations of graph network (graph characterized by magnitude and direction) is a characteristic structure which permit the match of constructed graph network of image and reference data.

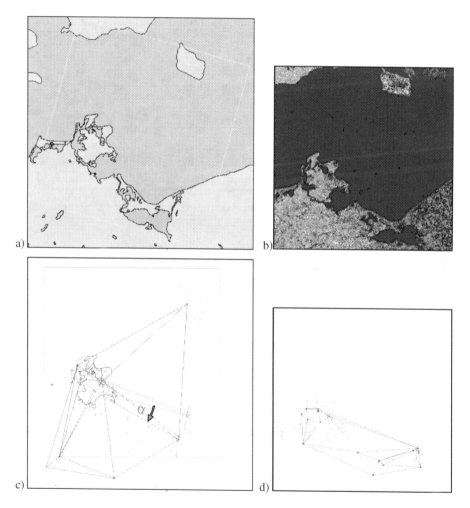

Fig. 5. Geo-referencing of classified LANDSAT 7 quick-look (b) based on CIA land-water mask (a). The contours of usable objects, centroids and graph network (blue) are shown in c) and d). In c) is added the graph network of identified objects (red) and additionally objects of d) are included after shift with dx and dy. The determination of rotation angle α is shown.

The geo-referencing begins with identification of a starting object in image and reference data for linking both data sets. The more exactly allocation succeeds, the higher is the accuracy of a constructed graph network and further processing based on the network. For identification of objects in image data and property matching of objects in reference data different object properties are used. The demonstration presented here based on only object class and size with a minimum size of 50 pixels and an allowed size deviation of 10%. In case of positive match of object properties in image and reference data, the corresponding object is identified as starting object. In our example island Ruegen was identified as starting object. If several objects are identified as candidates for starting object, then these can be considered in searching and decision making process. In case, that the relations (e.g. distances) of identified objects to other objects in their neighbourhood will be checked.

Analogous to identification of a starting object the search procedure for additional objects is processed. Identified objects in image data are compared with related objects in reference data according to their thematic and geometric properties.

If it is possible to link additional objects in image data with their corresponding objects in reference data, these objects are included and stored in an allocation matrix for further processing. In case of several identified candidates, the candidate of highest correlation to reference object is selected. Especially the comparison of centroid distance of candidate and starting object in reference and image data can be used as decision criterion. In other case, an object is excluded from processing. It should be mentioned here, that, if a second object is already identified the rotation angle α can be calculated according eq. 4 and used as second secure decision criterion.

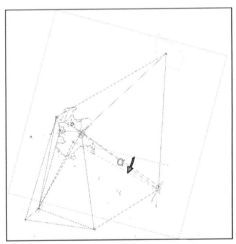

Fig. 6. Result of geo-referencing according Fig. 5 after shift and rotation

In Fig. 5c the identified objects are connected by red marked graph network or better, this is the identified part of the virtual reference graph network The shift vector (dx,dy), preferably computed from image coordinates (x_0^I, y_0^I) of starting object in the scene and appropriate image coordinates of reference (x_0^R, y_0^R) according eq. 3. The application of shift vector to intermediate result can be seen in Fig. 5c as green marked objects and graphs. The rotation angle α can be well determined based on largest possible distance between starting object 0 and a second object 1. In this example coordinate (x_1^I, y_1^I) of this centroid according to its shift into reference system $(x_1^I + dx, y_1^I + dy)$ and corresponding centroid coordinate (x_1^R, y_1^R) are used in eq. 4. After considering the rotation, the coordinate (x_1^I, y_1^I) is transferred into coordinate $(x_1^{R''}, y_1^{R''})$ in this special case equal to (x_1^R, y_1^R). Fig. 6 demonstrates the correct rotation of used part of graph network in the image overlaying the corresponding red marked graph in reference image, now marked with green colour. The rotated image border is also included.

When using satellite data as segment of a path or using for example classified data based on it as shown in Fig. 5b, than a shift from image line to image line has to be calculated. This preparation is caused by the rotation of earth during overflight by satellite. This assumes that the latitude of corner coordinates of the scene are known within a limit of some pixels. All preliminary pixel coordinates can if not available, calculated on basis of the available corner coordinates, the equations for distances, angles on surface of a sphere and transformations of the spherical coordinate system can be used [18].

The procedure was also successful when using the land-water mask of Germany as reference.

In a second experiment a 1400 x 1400 pixel segment of a classified LANDSAT 7 scene from the 14th August 2000 was used as reference (Fig. 7a) and a smaller 1000 x 1000 segment of a classification of a scene from 17th May 2000 was used as image (Fig. 7b). The LANDSAT scenes are available in ETM_FAST_L7A format as result of ESA processing in 2002.

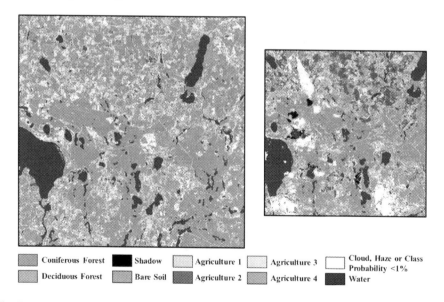

Coniferous Forest	Shadow	Agriculture 1
Deciduous Forest	Bare Soil	Agriculture 2

Agriculture 3   Cloud, Haze or Class Probability <1%
Agriculture 4   Water

Fig. 7. Water segments of automated classified LANDSAT 7 data are used as reference (a: 14th August 2000) and are used as test image for geo-referencing (b: 17th May 2000)

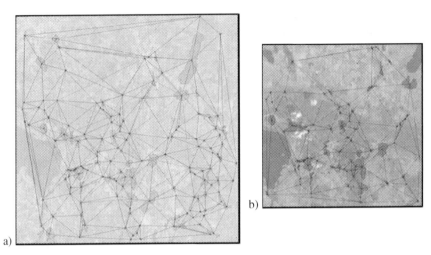

Fig. 8. Constructed graph network based on the centroids of identified landscape objects (e.g. water objects - nodes) in reference (left) and image data to be geo-referenced (right) without crossing of graphs. The contours of usable objects are green marked. The graph network of identified objects with a distance <*1* pixel is red marked in the reference (a). The parameters of the scenes are given in Fig. 7.

The segmentation and classification of image objects in remote sensing data are often an essential precondition for the processing and thus one of the most required routine tasks in interpretation of remote sensing data. If this processing step is interactive, the procedures require a considerable amount of time depending on the desired classification accuracy. In some cases, a faster classification procedure with lower requirements in classification hierarchy can have priority in relation to a time-consuming high-detailed classification. Therefore, the class aggregation was useful in order to prepare an appropriate land-water mask as reference. However, the classification demonstrates that also other references can be used. As shown in the example before, the graph network based on centroids of usable objects is shown in reference (Fig. 8a) and in image data of band 4 (Fig. 8b). Additionally, all usable objects are linked with an ID number. For example, the search for reference objects was limited to 80 corresponding image objects. The smallest image object determines the minimum size of usable reference objects. Image objects characterized by a threshold of distance to reference object *dxy_ref>1* will be eliminated. The centroid coordinates[6] of 29 accepted objects are listed in Table 1 with their ID number in the first row. It includes for all centroids of N linked objects i the distance between (x_i^R, y_i^R) of reference object and corresponding coordinates of image object in reference $(x_i^{R''}, y_i^{R''})$ according eq. 5. The geographic coordinates of reference centroids are also listed. *m_alpha* indicates the determined mean rotation angle for transformation based on all distances from starting object to all other 28 objects. Additionally the angle is listed when using the distance between start object to other objects and their object size as weighting factors. *dx0* and

---

[6] In image coordinates (x,y).

dy0 are elements of the shift vector from allocation of starting object. For 29 remaining objects the transformation parameters were iterative varied, until the sum of all deviations (distances) has reached a minimum. This is delivered as *total(dxy_ref)* of all remaining distances. At least, this is the result of matching of identified part of virtual graph network. For example, from the number of remaining objects a middle distance deviation *0.34 pixel units* ($RMSE_{XY} = 0.39$ *pixel units*) can be determined as possible quality measure for allocation. During iteration, the distance between centroids of linked object with *ID=1* changes from *0.0* to *0.26629*. In this way, the procedure permits a quality control for high correlated linked objects by delivering registration parameters. The visual control is given in Fig. 9 with an overlay of image after its transformation based on determined shift vector and rotation angle.

Table 1. Example of possible completion parameters of approximate rectification delivers for post processes precision rectification

	Image Row	Image Column	Reference Row	Reference Column	dxy_ref	Latitude	Longitude
1	245.46973	926.54047	491.94821	1047.70337	0.26629	53.44844	13.14670
2	757.80664	624.22815	998.03162	734.05774	0.34864	53.33530	12.95548
3	646.83075	751.25885	889.35516	863.61536	0.23657	53.35575	13.02373
4	889.50775	621.24969	1128.73083	728.09546	0.56170	53.30138	12.93913
6	503.54922	497.52338	740.17419	612.44495	0.71282	53.41050	12.92924
8	464.65396	55.02124	691.86456	171.18092	0.32315	53.45037	12.74036
10	821.92108	301.27399	1054.89783	409.48148	0.31551	53.34044	12.80720
12	666.47223	402.21744	902.09595	514.36072	0.66190	53.37407	12.86918
15	759.70422	478.56580	996.21631	588.15045	0.26084	53.34482	12.89170
16	521.96844	427.69473	757.10065	542.31317	0.62698	53.41040	12.89666
17	814.73859	567.31982	1053.80505	675.86072	0.44852	53.32428	12.92411
18	386.38068	480.83383	623.38165	598.86572	0.18928	53.44197	12.93557
21	721.73987	554.31757	960.75311	665.31378	0.77153	53.34934	12.92926
29	591.90717	552.96204	830.25989	666.50659	0.10547	53.38350	12.94350
30	660.49573	484.44446	896.98138	596.24652	0.38357	53.37035	12.90565
31	916.87067	300.74570	1149.38123	407.12872	0.24007	53.31579	12.79632
35	722.32385	231.15909	953.51135	342.11365	0.30844	53.37120	12.78821
36	291.78183	137.25455	520.78064	257.63870	0.41536	53.48997	12.79625
37	574.72668	534.02484	812.74512	647.87579	0.07870	53.38926	12.93716
38	397.09149	83.64706	625.34882	201.61240	0.20066	53.46597	12.76068
41	871.68243	533.66217	1109.76990	640.75397	0.31147	53.31177	12.90286
44	23.39850	687.87219	264.55789	813.47369	0.39455	53.52274	13.06792
54	767.01904	510.84763	1004.41302	619.97827	0.48041	53.34070	12.90479
58	352.80612	57.56123	580.33722	176.31395	0.05243	53.47933	12.75424
60	709.00000	267.63528	941.16882	378.70129	0.15628	53.37218	12.80556
70	862.79102	235.08955	1094.22949	342.44263	0.25225	53.33424	12.77370
71	754.31818	582.01514	993.21429	691.85712	0.06597	53.33918	12.93749
76	723.01752	253.21053	954.85187	363.94446	0.12483	53.36950	12.79765
78	691.94342	984.01886	939.29785	1095.25537	0.43681	53.32817	13.11996
m_alpha:		1.2559539			dx0,dy0:	120.943, 153.371	
+weight:		1.2545627			TOTAL(dxy_ref):	9.73102	

The aim of the post-processor is to deliver a reference table including coordinates *(row,column)* of image data and geographical coordinates *(lat,lon)* of objects in reference.

Especially Fig. 10 shows the registration quality of a stack of multi-temporal data. For this demonstration the classification results of five selected full resolution LANDSAT 7 scenes in ETM_FAST_L7A format were used. In Fig. 10b object contours (lake shore lines and islands) are drawn in the prepared map according first experiment based on provided coordinates (source: metadata). The included cloud masks (grey coloured) in Fig. 10a show, that a lot of usable objects can be masked/destroyed. In Fig. 10c the contours of the 5 scenes are also included, but now on basis of results of presented method of geo-rectification. In the lower part of figure the same detail is shown. While the superimposed contours in Fig. 10e give a blurred impression Fig. 10f shows a good correlation between the drawn object contours. This result can be attributed to high quality of derived transformation parameters for each remote sensing data.

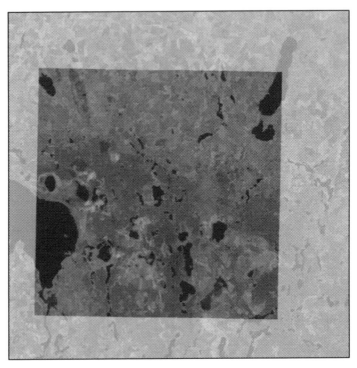

Fig. 9. Transformation of a scene in reference according graph-network matching in Fig. 8

5   Summary

The present work deals with an automatic geo-referencing algorithm based on matching a graph network using the centroids of landscape objects (virtual points) identified in image and reference data. A distortion-free representation of objects is presupposed. Preparation of image and reference are basic assumptions (when not available in the same format or projection) for transforming image coordinates. Preliminary coordinates were used to restrict the search area in the reference. The resulting table of corresponding objects lists the centroid coordinates of objects in image and corresponding geographic coordinates in reference data. This table can be delivered to an additional geo-referencing procedure matching the real points to eliminate possible distortions of image. It could be shown that results on basis of LANDSAT-7 data in FAST-format have a good quality relative to each other. For a good absolute accuracy of the method, reference data of higher accuracy than the used CIA World map is preconditioned. A stable procedure for object detection has to be included in the processing.

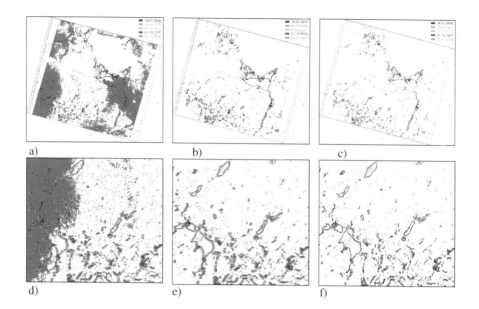

a)   b)   c)

d)   e)   f)

Fig. 10. Multi-temporal contours of detected object (5 LANDSAT scenes: acquisition period 2000). b) comparison based on delivered corner coordinates, a) same as b) with included cloud masks, c) comparison based on transformation parameters derived by proposed geo-referencing method, d-f) same details of a-c)

Identification and exact linkage of objects in image and reference data are based on the identified start object. Precondition for exact linkage is the determination of additional object properties because the probability of cloud obscuring of objects grows with the size of objects. The identified start object is the basis for using appropriate relations to other objects and if necessary, a decision process has to identify the correct start object. In our demonstration, the search for starting object was not supported by the preliminary geographic coordinate of image object that is being investigated as a candidate for starting object. If the accuracy of geographic coordinates is within a limit of a few pixels the search in the reference can be reduced to the immediate pixel environment.

Presupposing a distortion-free representation of objects is given, the construction of graph network can be considered as presentation of similarity independent on scale and rotation. The identification and exact linking of objects in image data to those of appropriate reference data are based on identified start object. This linking supplies the first approximation of scale and distortion, and supplies beside object characteristics the basis for using appropriate relations to other objects.

Acknowledgements. The authors wish to thank our colleagues V. Beruti, G. Pitella and R. Biasutti for providing the LANDSAT-quick-look data. Additional to this, the authors would also like to thank H. Asche for his suggestions and advices.

References

1. Bannari, A., Morin, D., Bénié, G.B., Bonn, F.J.: A Theoretical review of different mathematical models of geometric corrections applied to remote sensing images. Remote Sensing Reviews **13**, 27–47 (1995)
2. Bartoli, G.: Image Registration Techniques: A Comprehensive Survey. Universita degli Studi di Siena, Visual Information Processing and Protection Group, S. 57 (2007)
3. Brown, L.G.: A survey of image registration techniques. ACM Computing Surveys **24**(4), 325–376 (1992)
4. Harrison, B.A.: Remote Sensing: Image Rectification and Registration, vol. 4. CSIRO Publishing (1995)
5. Fonseca, L.M.G., Manjunath, B.S.: Registration techniques for multisensor remotely sensed imagery. Photogrammetric Engineering and Remote Sensing **62**(9), 1049–1056 (1996)
6. Toutin, T.: Geometric processing of remote sensing images: Models, algorithms and methods. Int. J. Remote Sensing **25**(10), 1893–1924 (2004)
7. Borg, E., Lippert, K., Zabel, E., Löpmeier, F.J., Fichtelmann, B., Jahncke, D., Maass, H.: DEMMIN – Teststandort zur Kalibrierung und Validierung von Fernerkundungsmissionen.- In: Rebenstorf, R.W. (Hrsg.)15 Jahre Studiengang Vermessungswesen – Geodätisches Fachforum und Festakt, Neubrandenburg, January 16–17, 2009, Eigenverlag, pp. 401–419 (2009)
8. Pape, D.: CIA World DataBank II, http://www.evl.uic.edu/pape/data/WDB/ (last modification 2004)
9. Earth Science Data and Information System (ESDIS) Level 1 Product Output Files Data Format Control Book, vol. 5, Book 2 Rev. 2 (1998)
10. Borg, E., Fichtelmann, B., Asche, H.: Cloud Classification in JPEG-compressed Remote Sensing Data (LANDSAT 7/ETM+). In: Murgante, B., Gervasi, O., Misra, S., Nedjah, N., Rocha, A.M.A., Taniar, D., Apduhan, B.O. (eds.) ICCSA 2012, Part II. LNCS, vol. 7334, pp. 347–357. Springer, Heidelberg (2012)
11. Spaventa, V. D.: Personal communication (2004)
12. Lammi, J., Sarjakoski, T.: Image Compression by the JPEG algorithm. Photogrammetric Engineering and Remote Sensing **61**, 1261–1266 (1995)
13. Lane, T.: JPEG image compression FAQ, part 1 and part 2 (1999). http://www.faqs.org/faqs/jpeg-faq/ (last access: February 20, 2012)
14. Lau, W.-L., Li, Z.-L., Lam, K.W.-K.: Effects of JPEG compression on image classification. Int. J. Remote Sensing **24**(7), 1535–1544 (2003)
15. Borg, E., Fichtelmann, B.: Verfahren zur automatischen Detektion und Identifikation von Objekten in multispektralen Fernerkundungsdaten.- DE 199 39 732 C2 (2003)
16. Guenther, A., Borg, E., Fichtelmann, B.: Process and Device for the Automatic Rectification of Single-Channel or Multi-Channel Image - UP 7, 356, 201 B2 (2008)
17. Pavlidis, T.: Algorithmen zur Grafik und Bildverarbeitung. Heinz Heise Verlag, Hannover - 5. Aufl., 508 (1994)
18. Bronstein, I.N., Semendjajew, K.A., Musiol, G., Muelig, H.: Taschenbuch der Mathematik, Verlag Harry Deutsch, Thun u. Frankfurt a.M. (2000, 2001)

Semi-supervised Local Aggregation Methodology

Marzieh Azimifar[1], Ali Heidarzadegan[2(✉)], Yasser Nemati[2],
Sajad Manteghi[1], and Hamid Parvin[1]

[1] Department of Computer Engineering, Mamasani Branch,
Islamic Azad University, Mamasani, Iran
s.manteghi@mail.sbu.ac.ir, parvin@iust.ac.ir
[2] Department of Computer Engineering, Beyza Branch,
Islamic Azad University, Beyza, Iran
heidarzadegan@beyzaiau.ac.ir

Abstract. In this paper we propose a novel approach for automatic mine detection in SONAR data. The proposed framework relies on possibilistic based fusion method to classify SONAR instances as mine or mine-like object. The proposed semi-supervised algorithm minimizes some objective function which combines context identification, multi-algorithm fusion criteria and a semi-supervised learning term. The optimization aims to learn contexts as compact clusters in subspaces of the high-dimensional feature space via possibilistic semi-supervised learning and feature discrimination. The semi-supervised clustering component assigns degree of typicality to each data sample in order to identify and reduce the influence of noise points and outliers. Then, the approach yields optimal fusion parameters for each context. The experiments on synthetic datasets and standard SONAR dataset show that our semi-supervised local fusion outperforms individual classifiers and unsupervised local fusion.

Keywords: Supervised learning · Ensemble learning · Classifier fusion

1 Introduction

Several wars and armed conflicts over the last century have yielded around one hundred million unexploded land mines in approximately 70 different countries [1]. Moreover, around five million new mines are yearly buried in the ground [3]. Most of these mines are anti-personal, and claim the lives of around 70 civilians on daily basis, in regions like Afghanistan, Angola, Bosnia, and Cambodia [1]. The tactical and psychological effectiveness of these land mines, along with their simple and low cost fabrication are the main reasons behind their proliferation. In other words, they emerged as an interesting alternative for country and armed organizations which cannot acquire sophisticated defense systems [4]. Thus, despite mine-clearing efforts around the world, million landmines are still deployed. The United Nations claim that conventional mine neutralization methods would require over 1000 years to clear out the mine fields around the world [3]. Hand probes and metal detectors which represent the classical detection methods of buried land mines cannot be used for large

© Springer International Publishing Switzerland 2015
O. Gervasi et al. (Eds.): ICCSA 2015, Part IV, LNCS 9158, pp. 233–245, 2015.
DOI: 10.1007/978-3-319-21410-8_18

operations. Moreover, recent anti-personal mines are small sized and plastic made with small metal portion which makes their detection challenge even more acute [3]. Other techniques such as dogs trained to sniff out explosives, and ground-penetrating radar proved to be slow and dangerous because they should operate very close to mines to be able to detect them. During the last decade, infrared camera based detection of buried mines proved to be effective when the field conditions are appropriate. Also, novel mine neutralization technologies include energetic photon detection, thermal neutron activation, infrared emission, and ground-penetrating radar [3]. The latest researches attempt to aggregate these technologies with conventional metal detectors in order to enhance their detection performance [6].

For underwater naval mines, the U. S. navy has used marine mammals such as Atlantic and Pacific bottlenose dolphins, white whales, and sea lions to recognize buried mines [5]. However, shallow water of the surf zone affects the performance of these mammals drastically. Recently, the SOund NAvigation and Ranging (SONAR) which is a sound propagation technology, has been exploited for underwater mine detection. It has yielded an impressive growth of researches related to the detection and classification of SONAR signals [2]. Side-scan sonar imagery conveys high resolution shots of the sea floor scene. Despite the contribution of these images in promoting applications like Mine Counter Measure (MCM) where speed is a key factor, they have not facilitated that much the detection of objects of interest in these scenes. In fact, for underwater scenes, object detection and classification is even more challenging due to the acoustical medium and the wide variability and heterogeneity of this environment. Also, the objects located on the sea floor or buried under the sand, are difficult to detect because their appearance may vary considerably based on the neighboring scene. Despite these challenges, the ability of SONAR imaging to operate in poor visibility conditions without additional light sources requirements triggered tremendous efforts to develop under water mine detection systems based on this image modality. Additionally, the low fabrication cost along with the low power consumption represents other main advantages that motivated researchers to include SONAR module in Autonomous Underwater Vehicles (AUVs) for under water mine detection. Processing this high-dimensional data on board has become an urgent need for this solution. Also, embedding unsupervised decision-making features and reducing the expert involvement in the recognition process emerged as new active research field. Novel clearing operations of underwater mines would rely on AUVs equipped with SONAR capabilities, and adopting Computer Aided detection (CAD) solutions. The recognition task consists in classifying signatures of region of interest as mine or not. The subsequent classification algorithms are intended to recognize the false alarms as possible while detecting real mines. Choosing an appropriate supervised classification model for sonar data pattern recognition is a critical issue for objects of interest detection under the sea [7].

In this paper, we propose a possibilistic based local approach that adapts multi-classifier fusion to different regions of the feature space. The proposed approach starts by categorizing the training samples into different clusters based on the subset of features used by the single classifiers, and their confidences. This phase is a complex optimization problem which is prone to local minima. To alleviate this problem we include a semi-supervised learning term in the proposed objective function. This categorization process associates a possibilistic membership, representing the degree of typicality, with each data sample in order to identify and reduce the influence of noise points and outliers [8]. Also, an expert is appointed for each obtained cluster. These experts represent the best classifiers for the corresponding cluster/context. Then, aggregation weights are estimated by the fusion component for each classifier. These weights reflect the performance of the classifiers within all contexts. Finally, for a given test sample, the fusion of the individual confidence values is obtained using the aggregation weights associated with the closest context.

2   Proposed Method

Classifier fusion has yielded in promising finding, and has outperformed single classifier systems both theoretically and experimentally. The classifiers complementarity is the main characteristic which allows an ensemble of classifiers to outperform individual learners by inheriting their strengths and limiting their weaknesses. Nonetheless, two critical conditions for fusion algorithms to outperform individual classifiers are diversity and accuracy. Classifier fusion relying on effective aggregation of classifiers outputs considers all learners equally trained and competitive over the feature space. For testing, single experts classifications are performed simultaneously, and the resulting outputs are aggregated to yield a final fusion decision. Typical fusion approaches include Borda count, average, majority vote, Bayesian, probabilistic, and weighted average.

Approaches for combining diverse learners can be categorized into two main classes: local approaches and global approaches. Unlike global approaches which assign an average relevance degree for each learner with respect to the whole training set, local approaches consider a relevance degree to the different training set subspaces. This relies on the assumption that higher classification accuracy can be reached using appropriate data-dependent weights. Clustering of the input data samples into homogeneous categories during the training phase is required for local fusion approach. This clustering may be achieved on the space of individual learner classification, based on which classifiers behave similarly, or using attributes of the input space. Next, in each space region, the most accurate classifier is appointed as expert. For classification, unknown instances get assigned to regions, and the corresponding expert learner of this region generates the final decision. A dynamic data partitioning and classification during the testing phase is proposed in. The authors estimate the classifier accuracies using sample vicinity in the local regions of the feature space, and the most accurate one is used to predict the class of the test sample. The local fusion approach called Context-Dependent Fusion (CDF) in starts by clustering the training instances into homogeneous categories of contexts. This clustering phase and the selection of a local expert learner are two sequentially independent stages of CDF. The researchers in described a generic framework for context-dependent fusion which simultaneously clusters the feature space, and aggregates the outputs of the expert learners. This fusion approach uses a simple linear aggregation to generate fusion weights for individual learners. However, these weights may be inefficient to capture the classifiers mutual interaction.

The authors in outlined an approach that assesses the performance of each expert in local regions of the feature space. For each local region, the most accurate classifier is exploited to predict the final decision. However, the performance evaluation for test instances is timely complex, and affects the practicality of the approach with large data. The clustering and selection phase determines the statically most accurate classifier. First, the clustering of the training instances discovers the decision regions. Then, the most accurate learner on this local region is chosen. The main drawback of this solution is that it does not handle more than one classifier per region. The researchers extended the clustering and selection algorithm so it divides the training dataset into correctly and incorrectly categorized instances. The feature space is then grouped by clustering the training instances. For testing, the most effective classifier in the vicinity of the test instance is selected in order to generate the final decision. In other words, each learner maintains its corresponding cluster. This approach reduces the computational effectiveness of the solution. Recently, in [11], the authors outlined a local fusion approach that categorizes the feature space into homogeneous clusters based on their features, and takes into consideration the obtained clusters when aggregation individual learners outputs. The fusion stage consists in assigning an aggregation weight to each individual learner with respect to each context based on its relative performance within it. Notice that the used fuzzy approach increases the sensitivity of the fusion component to outliers which decrease the overall classification performance. Some authors proposed a possibilistic based local approach that adapts the fusion method to different regions of the feature space. It categorizes the training samples into different clusters based on the subset of features used by the single classifiers, and their confidences. This clustering phase generates possibilistic memberships representing the degree of typicality of each data sample in order to identify and discard noise points. Also, an expert classifier is associated with each obtained cluster. More specifically, aggregation weights are simultaneously learned for each classifier. Finally, for a given test sample, the fusion of the individual confidence values is obtained by using the aggregation weights associated with the closest context/cluster. Although this approach yielded promising results, the adopted optimization approach is prone to local minima.

3   Local Fusion Based on Possibilistic Context Extraction

Let $T = \{t_j | j = 1, ... N\}$ be the desired output of N training observations. These outputs were obtained using K classifiers. Each classifier k uses its own feature set $X_k = \{x_j^k | j = 1 ... N\}$ and generates the confidence values $Y_k = \{y_j^k | j = 1 ... N\}$. The K feature sets are then concatenated to generate one global descriptor, $\chi = \bigcup_{k=1}^K \chi^k = \{x_j = [x_j^1, ... , x_j^K | j = 1, ... , N]\}$. The possibilistic-based context extraction for local fusion algorithm in [8] has been formulated as partitioning the data into C clusters minimizing one objective function. However, this clustering approach requires the estimation of various parameters using complex optimization and is prone to several local minima. To overcome this potential drawback, we propose a semi-supervised version of the algorithm in [8]. The supervision information relies on two sets of pairwise constraints. The first one is *Should-link* constraints which specify that two data instances are expected to belong to the same cluster. The second set of constraint is the *ShouldNot-link* which specifies that two data instances are expected to belong to different clusters.

Let *SL* be the set of *Should-link* pairs of instances. If the pair (X_i, X_j) belongs to *SL*, then X_i and X_j are expected to be assigned to the same cluster. Similarly, let *NL* be the set of *ShouldNot-link* pairs. If the pair (X_i, X_j) belongs to *NL*, then X_i and X_j are expected to be assigned to different clusters. In this work, we reformulate the problem of identifying the C components/clusters as a constrained optimization problem. More specifically, we modify the objective function in [8] as follows

$$J = \sum_{j=1}^{N} \sum_{i=1}^{C} u_{ij}^m \sum_{k=1}^{K} v_{ik}^q d_{ijk}^2$$

$$+ \sum_{j=1}^{N} \sum_{i=1}^{C} \beta_i u_{ij}^m \left(\sum_{k=1}^{K} \omega_{ik} y_{kj} - t_3 \right)^2 + \sum_{i=1}^{C} \eta_i \sum_{j=1}^{N} (1 - u_{ij})^m$$

$$+ \mu \left[\sum_{(X_t, X_k \in NL)} \sum_{i=1}^{C} u_{ij}^m u_{kj}^m + \sum_{(X_t, X_k \in SL)} \sum_{i=1}^{C} \sum_{p=1, p \neq i}^{C} u_{ij}^m u_{kj}^m \right] \quad (1)$$

subject to $\sum_{i=1}^{C} u_{ij} = 1 \forall j, u_{ij} \in [0,1] \forall i, j$,
$\sum_{k=1}^{K} v_{ik} = 1 \forall i, v_{ik} \in [0,1] \forall i, k,$ and $\sum_{k=1}^{K} \omega_{ik} = 1 \forall i$.
In (1), u_{ji} represents the possibilistic membership of X_j in cluster i [8]. The $C \times N$ matrix, $U = [u_{ij}]$ is called a possibilistic partition if it satisfies:

$$\begin{cases} u_{ij} \in [0,1], \forall j \\ 0 < \sum_{i=1}^{C} u_{ij} < N, \forall i, j \end{cases} \quad (2)$$

On the other hand the $C \times d$ matrix of feature subset weight, $V = [v_{ik}]$ satisfies

$$\begin{cases} v_{ik} \in [0,1], \forall i, k \\ \sum_{k=1}^{K} v_{ij} = 1, \forall i \end{cases} \quad (3)$$

The first term in (1) corresponds to the objective function of the SCAD algorithm [8]. It aims to categorize the N points into C clusters centered in c_i such that each sample x_j is assigned to all clusters with fuzzy membership degrees. Also, it is intended to simultaneously optimize the feature relevance weights with respect to each cluster. SCAD term is minimized when a partition of C compact clusters with minimum sum of intra-cluster distances is discovered. The second term in (1) intends to learn cluster-dependent aggregation weights of the K algorithm outputs. ω_{ik} is the aggregation weight assigned to classifier k within cluster i. This term is minimized when the aggregated partial output values match the desired output. The third term in (1) yields the generation of the possibilistic memberships u_{ji} which represent the degree of typicality of each data point within every cluster, and reduce the effect of outliers on the learning process. In (1), $m \in [1, \infty)$ is called the fuzzier, and η_i are positive constants that controls the importance of the third term with respect to the first/second one. This term is minimized when u_{ij} are close to 1, thus, avoiding the trivial solution of the first term (where $u_{ij} = 0$). Note that $\sum_{i=1}^{C} u_{ij}$ is not constrained to sum to 1. In fact, points that are not representative of any cluster will have $\sum_{i=1}^{C} u_{ij}$ close to zero and will be considered as noise. This constraint relaxation overcomes the disadvantage of the constrained fuzzy membership approach which is the high sensitivity to noise and outliers. The parameter η_i is related to the resolution parameter in the potential function and the deterministic annealing approaches. It is also related to the idea of "scale" in robust statistics. In any case, the value of 0.7 determines the distance at which the membership becomes 0.5. The value of η_i determines the "zone of influence" of a point. A point X_j will have little influence on the estimates of the model parameters of a cluster if $\sum_{k=1}^{K} v_{ik}^2 \left(d_{ijk}^s\right)^2$ is large when compared with η_i. On the other hand, the "fuzzier" m determines the rate of decay of the membership value. When $m = 1$, the memberships are crisp. When $m \to \infty$, the membership function does not decay to zero at all. In this possibilistic approach, increasing values of m represent increased possibility of all points in the data set completely belonging to a given cluster. The last term in (1) represents the cost of violating the pairwise Should-link, and *ShouldNot-link* constraints. These penalty terms are weighted by the membership values of the instances that violate the constraints. In other words, typical instances of the cluster which have high memberships yield larger penalty term. The value of μ controls the importance of the supervision information compared to the other terms.

The performance of this algorithm relies on the value of β. Over estimating it results in the domination of the multi-algorithm fusion criteria which yields non-compact clusters. Also, sub-estimating β decreases the impact of the multi-algorithm fusion criteria and increases the effect on the distances in the feature space. When appropriate β is chosen, the algorithm yields compact and homogeneous clusters and optimal aggregation weights for each algorithm within each cluster.

Minimizing J with respect to U is equivalent to minimizing the following individual objective functions with respect to each column of U:

$$
\begin{aligned}
J^{(i)}(U_i) = &-\sum_{j=1}^{N} u_{ij}^m \sum_{k=1}^{K} v_{ik}^q d_{ijk}^2 \\
&+ \sum_{j=1}^{N} \beta_i u_{ij}^m \left(\sum_{k=1}^{K} \omega_{ik} y_{kj} - t_j\right)^2 \\
&+ \eta_i \sum_{j=1}^{N} \left(1 - u_{ij}\right)^m \\
&+ \mu \left[\sum_{(X_t, X_k \in NL)} u_{ij}^m u_{kj}^m + \sum_{(X_t, X_k \in SL)} \sum_{p=1, p \neq i}^{C} u_{ij}^m u_{kj}^m\right]
\end{aligned}
\tag{4}
$$

For $i = 1, \dots, C$. By setting the gradient of $J^{(i)}$ with respect to the possibilistic memberships u_{ij} to zero, we obtain

$$\frac{\partial J^{(i)}(U_i)}{\partial u_{ij}} = m(u_{ij})^{m-1} \left(\sum_{k=1}^{K} v_{ik}^q d_{ijk}^2 + \sum_{j=1}^{N} \beta_i \left(\sum_{k=1}^{K} \omega_{ik} y_{kj} - t_j \right)^2 \right.$$

$$\left. + \mu \left[\sum_{(X_i, X_k \in NL)} u_{kj}^m + \sum_{(X_i, X_k \in SL)} \sum_{p=1, p \neq i}^{C} u_{kp}^m \right] \right) \tag{5}$$

$$- m \eta_i (1 - u_{ij})^{m-1} = 0 \tag{6}$$

This yields the following necessary condition to update u_{ij}:

$$u_{ij} = \left[1 - \left(\frac{D_{ij}^2}{\eta_i} \right)^{\frac{1}{m-1}} \right]^{-1} \tag{7}$$

where

$$D_{ij} = \sum_{k=1}^{K} v_{ik}^q d_{ijk}^2 + \beta_i \sum_{k=1}^{K} v_{ik}^q \left(\sum_{l=1}^{K} \omega_{il} y_{lj} - t_j \right)^2$$

$$+ \mu \left[\sum_{(X_t, X_k \in NL)} u_{kj}^m + \sum_{(X_t, X_k \in SL)} \sum_{p=1, p \neq i}^{C} u_{kp}^m \right]$$

D_{ij} represents the total cost when considering point X_j in cluster i. As it can be seen, this cost depends on the distance between point X_j and the cluster's centroid c_i, the cost of violating the pairwise *Should-link*, and *ShouldNot-link* constraints (weighted by μ), and the deviation of the combined algorithms' decision from the desired output (weighted by β). More specifically, points to be assigned to the same cluster: (*i*) are close to each other in the feature space, and (*ii*) their confidence values could be combined linearly with the same coefficients to match the desired output.

Minimizing J with respect to the feature weights

$$v_{ik} = \sum_{t=1}^{K} \left[(D_{ik}^2 / D_{il})^{\frac{1}{q-1}} \right] \tag{8}$$

where $D_{il} = \sum_{j=1}^{N} u_{ij}^m d_{ijl}^2$.

Minimization of J with respect to the prototype parameters, and the aggregation weights yields

$$c_{ik} = \frac{\sum_{j=1}^{N} u_{ij}^m X_{ik}}{\sum_{j=1}^{N} u_{ij}^m} \tag{9}$$

and

$$w_{ik} = \frac{\sum_{j=1}^{N} u_{ij}^m y_{kj} \left(t_j - \sum_{\substack{l=1 \\ l \neq k}}^{K} \omega_{il} y_{lj} \right) - \zeta_i}{\sum_{j=1}^{N} u_{ij}^m y_{kj}^2} \tag{10}$$

where ζ_i is a Lagrange multiplier that assures that the constraint in (2) is satisfied, and is defined as

Algorithm 1. The proposed semi-supervised possibilistic clustering, feature weighting and classifier aggregation.

> **Inputs**: X: The data instances.
> > Y : The confidences obtained using the different classifiers.
> > > NL: The set of ShouldNot-Link constraints.
> > > SL: The set of Should-Link constraints.
> > > T: The labels of the data instances.
> > > C: The number of clusters.
> > > m: The fuzzyfier.
> > > q: The exponent of the feature weights.
> > > T: The labels of the data instances.
> > > > β: The weight assigned to the second term of the objective function (1).
> > > > η: The weight assigned to the third term of the objective function
> (1).
> **Outputs**: U: The possibilistic membership matrix of the data instances.
> > c_i: The Clusters centers.
> > V : The feature weights.
> > W: The aggregation weights.

Begin
Initialize the centers;
Initialize the possibilistic partition matrix U;
Initialize the relevance weights;
Repeat
Compute d^2 ijk, for $1 \leq i \leq C$ and $1 \leq j \leq N$ and $1 \leq k \leq K$;
Update the relevance weights v_{ik} using equation (8);
Compute D^2_{ij}
Update the partition matrix U using equation (7);
Update the aggregation weights matrix W and the feature weights matrix V using equations (10) and (8), respectively;
Update the centers using equation (9);
Until (centers stabilize)
End

$$\zeta_i = \frac{\sum_{k=1}^{K}\left(\sum_{j=1}^{N} u_{ij}^m y_{lj}\left(t_j - \sum_{k=1}^{K} \omega_{ik} y_{kj}\right)\right)\left(\sum_{j=1}^{N} u_{ij}^m y_{lj}^2\right)^{-1}}{\sum_{l=1}^{K}\left(\sum_{j=1}^{N} u_{ij}^m y_{lj}^2\right)^{-1}} \tag{11}$$

The obtained iterative algorithm starts with an initial partition and alternates between the update equations of u_{ij} ; v_{ik}; w_{ik} and c_{ik} as shown in Algorithm 1.

The time complexity of one iteration of this first component is $O(N \times d \times K \times C)$. Where N is the number of data points, C is the number of clusters, d is the dimensionality of the feature space, and K is the number of feature subsets. The computational complexity of one iteration of other typical clustering algorithms (e.g. FCM, PCM) is $O(N \times d \times C)$. Since we use small number of feature subsets ($K = 3$), one iteration of our algorithm has a comparable time complexity to other similar algorithms. However, we should note that since we optimize for more parameters, it may require a larger number of iterations to converge.

After training the algorithm described above, the proposed local fusion approach adopts the steps below in order to generate the final decision for test samples:

- Run the different classifiers on the test sample within the corresponding feature subset space, and obtain the decision values, $Y^j = \{y_{kj} | k = 1, \ldots k\}$.
- The unlabeled test sample inherits the class label of the nearest training sample.
- Assign the membership degrees u_{ij} to the test sample j in each cluster i using eq. (7).
- Aggregate the output of the different classifiers within each cluster using $\hat{y}_{ij} \sum_{k=1}^{K} w_{ik} y_{kj}$.
- The final decision confidence is estimated using $\hat{y} = \sum_{i=1}^{C} u_{ij} \hat{y}_{ij}$.

4   Experiments

We illustrate the performance of the proposed semi-supervised local fusion algorithm using synthetic data sets. For these data sets, we compare our approach to individual classifiers, and the method in [8] in Table 1.

In this experiment, we illustrate the need for semi-supervised possibilistic local fusion. We use our semi-supervised local fusion approach to classify the synthetic 2-dimensional dataset. Let each sample be processed by two single algorithms (K-Nearest Neighbors (K-NN) with $K = 3$). Each algorithm, k, considers one feature X_k; and assigns one output value y_k. Samples from Class 0 are represented using blue dots and samples from Class 1 are displayed in red. Black samples represent noise samples. The dataset consists of four clusters. Each one of them is a set of instances from the two classes [9].

Table 1. Learned weights for each classifier with respect to the different clusters obtained using the method in [8] and the proposed semi-supervised method

	Cluster #	1	2	3	4
Method in [45]	Classifier 1	0.3112	**0.7301**	0.0266	0.0229
	Classifier 2	**0.6888**	0.2699	**0.9734**	**0.9771**
Proposed semi-supervised method	Cluster #	1	2	3	4
	Classifier 1	0.2946	**0.7366**	0.0290	0.0247
	Classifier 2	**0.7054**	0.2634	**0.9710**	**0.9753**

In order to construct the set pairwise constraints, we randomly select samples that are at the boundary of each cluster. We consider 7% of the total number of instances as *Should-link* and *ShouldNot-link* sets. Pairs of instances belonging to the same cluster (based the ground truth) form the Should-link set. Similarly, pairs that belong to different clusters form the *ShouldNot-link* set.

We use the accuracy as performance measure to evaluate the performance of our semi-supervised method. The overall accuracy of the partition is computed as the average of the individual class rates weighted by the class cardinality. To take into account the sensitivity of the algorithm to the initial parameters, we run the algorithm 10 times using different random initializations. Then, we compute the average accuracy values for each supervision rate. Based on experimentation, the accuracy increased at a much lower rate with supervision rate larger than 7%. Thus, for the rest of the experiments we set the supervision rate used to guide our clustering algorithm to 7%.

Table 2. Performance comparison of the individual learners, the method in [8], and the proposed method for SOANR data set [10]

	Accuracy	Precision	Recall	F-measure
K-NN 1	0.8269	0.7680	0.8803	0.8203
K-NN 2	0.8416	0.7972	0.8842	0.8384
K-NN 3 [10]	0.8511	0.8142	0.8808	0.8461
Method in	0.8604	0.8636	0.8431	0.8532
Proposed method	0.9024	0.9090	0.8947	0.9017

In this section, we use our approach to classify standard dataset frequently used by researchers from the machine learning community. Namely, we consider the SONAR dataset [10] which consists of 208 instances and 60 attributes. 97 instances were obtained by bouncing sonar signals off a metal cylinder under various conditions and at various angles. A variety of different aspect angles, spanning 90 degree for the cylinder and 180 degrees for the rock were considered to contain the dataset signal. Each attribute represents the energy within a particular frequency band, integrated over a given period of time. SOANR dataset is summarized in [10].

In our experiments, for individual learners and local fusion approaches we adopt a 5-fold cross-validation in which each fold is treated as a test set with the rest of the folds used for training.

We divide the SONAR features into three subsets, and we dedicate one learner for each one of them. We run simple K-NN learner to generate confidence values for each instance. We categorize the training samples using 3 K-NN classifiers (K=3) within their corresponding feature subspaces. Then, the proposed semi-supervised local fusion is used to categorize the training instances into 3 homogeneous clusters, and learn the optimal aggregation weights. Then, test instances are classified using the three individual learners, and assigned to the closest cluster. Finally, the fusion decision is generated by combining the partial confidences with the aggregation weights of the closest cluster. Notice that Should-link and *ShouldNot-link* constraints are generated using a clustering algorithm. More specifically, we cluster the training dataset using the possibilistic-based algorithm in [12], and we include pairs of typical instances (with high possibilistic membership), belonging to the same cluster, in the *Should-link* set. On the other hand, pairs of typical instances (with high possibilistic membership), belonging to different clusters, are included in the *ShouldNot-link* set. We limit the number of pairwise constraints to 7% of the total number of instances.

We report the mine detection accuracies, precision, recall and F-measure obtained using K-NN classifier with different values of the parameter K. As it can be concluded experimentally, $K = 5$ yields the best overall performance measures. Thus, for the rest of the experiments, we set this K to 5.

We compare the obtained average accuracy, precision, recall, and F-measure values obtained using individual K-NN learners, the method in [8], and the proposed method with the SONAR dataset in Table 2. Our semi-supervised approach outperforms the other classifiers on this dataset based on the four performance measures. This proves that the association of supervision information with local fusion technique yields better clustering results and let individual learners cooperate more efficiently to generate more accurate final decision. This confirms the results obtained with synthetic datasets in the previous experiment.

Table 3(a) shows the learners aggregation weights with respect to the different clusters generated by our algorithm. These weights reflect the impact of each individual learner within each cluster. For instance, the second individual K-NN is perceived by our approach as the most accurate classifier for instances from cluster 1. Similarly, the highest aggregation weight is assigned to the first individual K-NN within cluster 3.

Table 3. (a-above) Learned weights for each classifier in each cluster obtained using the proposed semi-supervised local fusion with SONAR data set [10]. (b-below) Per-cluster accuracy of the three K-NN classifers with SONAR data [10].

Cluster #	Cluster 1	Cluster 2	Cluster 3
K-NN 1	0.2080	0.2758	**0.8003**
K-NN 2	**0.6142**	**0.5309**	0.1038
K-NN 3	0.1778	0.1933	0.0959

Cluster #	Cluster 1	Cluster 2	Cluster 3
K-NN 1	0.6947	0.7523	**0.8846**
K-NN 2	**0.8713**	**0.8600**	0.5992
K-NN 3	0.6401	0.6765	0.5889

Table 4. (a-above) Learned weights for each classifier in each cluster with SONAR data [10]. (b-below) Per-cluster accuracy obtained using SVM, K-NN and Naive bayes classifiers on SONAR data [10].

Cluster #	1	2	3
SVM	**0.3775**	0.3311	0.4589
K-NN	0.3488	0.3298	**0.4702**
NBayes	0.2737	**0.3391**	0.0709

Cluster #	Cluster 1	Cluster 2	Cluster 3
SVM	0.8989	0.8430	0.8799
K-NN	0.8577	0.8366	**0.8831**
NBayes	0.6389	0.8466	0.5984

To demonstrate that the semi-supervised local fusion exploits the strengths of the individual learners within local regions of the features space, we report the accuracy of the three individual learners (K-NN) within the 3 clusters. These performance measures shown in Table 3(b) are calculated based on the classification of test samples belonging to each cluster separately (given the membership degrees generated by the proposed semi-supervised clustering algorithm). As one can notice, the local performances of the individual K-NN depends on the cluster. K-NN classifier 2 performs better than the other learners for samples from cluster 1. Consequently, K-NN classifier 2 is the most relevant classifier with respect to cluster 1. Thus, the highest aggregation weight is assigned to this classifier as reported in Table 3(a). Similarly, in cluster 3, the most accurate individual classifier is K-NN classifier 2.

Table 5. Performance measures of the individual learners, the method in [8], and the proposed method with SONAR dataset

	Accuracy	Precision	Recall	F-Measure
K-NN	0.8259	0.7589	0.8788	0.8144
SVM	0.8581	0.7995	0.8865	0.8407
Naive Bayes	0.8337	0.7987	0.8623	0.8292
Method in [1]	0.8659	0.8641	0.8503	0.8571
Proposed Method	0.9087	0.9126	0.9010	0.9067

In the following experiment we use the same feature subsets defined in the previous experiment. However, we use them with different classifiers. Namely, we classify SONAR instances in each features subset using K-NN, Naive Bayes and SVM classifiers. Then, the our semi-supervised local fusion algorithm clusters the training data, generates 3 categories, and learns optimal aggregation weights. This experiment is intended to show that our approach does not require specific classifiers, and can deal with various supervised learning algorithms.

In Table 4(a), we report the aggregation weights learned by our semi-supervised local fusion approach for each classifier with respect to the different clusters. The achievements of the different supervised learning techniques vary drastically depending on the context/cluster. More specifically, SVM is the most important learner with respect to cluster 1. This can be explained by the highest weight assigned for SVM classifier within cluster 1. Similarly, K-NN is the most relevant classifier for cluster 3.

In Table 4(b) shows the per-cluster accuracy values obtained within the different clusters generated by our semi-supervised algorithm. As one can notice, the reported values are consistent with the relevance weights in Table 4(a). For instance, SVM which obtained the highest aggregation weight with respect to cluster 1, yields the highest accuracy with respect to this cluster. Similarly, NBayes and K-NN are the most accurate classifiers in cluster 2 and cluster 3, respectively.

Table 5 displays four performance measures obtained by the different individual learners, the method in [8], and our semi-supervised local fusion approach. Namely, accuracy, precision, recall and F-measure are reported for SONAR data [10]. Our approach outperforms the other methods with respect to all the performance measures.

5   Conclusion

In this paper we have proposed a novel approach of automatic mine detection in SONAR dataset. This approach consists in a semi-supervised local fusion algorithm which categorizes the feature space into homogeneous clusters, learns optimal aggregation weights for individual classifiers and optimal fusion parameters for each context in a semi supervised manner. The experiments have shown that the semi-supervised fusion approach yields more accurate classification than the unsupervised version and the individual classifiers on synthetic and real datasets.

Although the proposed approach yields promising results, there is still room for improvement. Future work may consist in extending the proposed approach so it handles multiple class (more than two classes) categorization problems. Finally, in order to overcome the need to specify the number of clusters/contexts apriori, we can investigate the ability of the possibilistic logic to generate duplicated clusters in order to find the optimal number of clusters.

References

1. Peyvandi, H., Farrokhrooz, Roufarshbaf, M., Park, S.-J.: SONAR systems and underwater signal processing: classic and modern approaches. In: Kolev, N.Z., (ed.) Sonar Systems, pp. 173–206. InTech, Hampshire (2011)
2. Walsh, N E., Walsh, W S.: Rehabilitation of landmine victims: the ultimate challenge
3. Zamora, G.: Detecting Land Mine (November 2014).
http://www.nmt.edu/mainpage/news/landmine.html
4. Miles, D.: Confronting the Land Mine Threat (November 2014).
http://www.dtic.mil/afps/news/9806192.html
5. Dye, D.: High Frequency Sonar Components of Normal and Hearing Impaired Dolphins, Master's thesis, Naval postgraduate school, Monterey, CA (2000)
6. Miles, D.: DOD Advances Countermine Technology (November 2014).
http://www.dtic.mil/afps/news/9806193.html
7. Lv, C., Wang, S., Tan, M., Chen, L.: UA-MAC: an underwater acoustic channel access method for dense mobile underwater sensor networks. International Journal of Distributed Sensor Networks 2014, Article ID 374028, 10 pages (2014)
8. Ben Ismail, M.M., Bchir, O.: Insult Detection in Social Network Comments Using Possibilistic Based Fusion Approach. In: Lee, R. (ed.) Computer and Information Science. SCI, vol. 566, pp. 15–25. Springer, Heidelberg (2015)
9. Ben Ismail, M.M., Bchir, O., Emam, A.Z.: Endoscopy Video Summarization based on Unsupervised Learning and Feature Discrimination. IEEE Visual Communications and Image Processing, VCIP 2013, Malaysia (2013)
10. http://archive.ics.uci.edu/ml/datasets/Connectionist+Bench+(ines+vs.+Rocks)
11. Ben Abdallah, A.C., Frigui, H., Gader, P.D.: Adaptive Local Fusion With Fuzzy Integrals. IEEE T. Fuzzy Systems $20(5)$, 849–864 (2012)
12. Maher Ben Ismail, M., Frigui, H.: Unsupervised Clustering and Feature Weighting based on Generalized Dirichlet Mixture Modeling. Information Sciences 274, 35–54 (2014). doi:10.1016/j.ins.2014.02.146

Workshop on Scientific Computing Infrastructure (SCI 2015)

Development of the Configuration Management System of the Computer Center and Evaluation of Its Impact on the Provision of IT Services

Nikolai Iuzhanin$^{(\boxtimes)}$, Tatiana Ezhakova, Valery Zolotarev,
and Vladimir Gaiduchok

Saint Petersburg State University, Saint Petersburg 199034, Russia
{yuzhanin,trezhakova,viz,gajduchok}@cc.spbu.ru

Abstract. This paper discusses the problem of development of the Configuration Management system and the Configuration Management Process in computer center which provides services based on the virtualization technologies. In view of strong integration of the IT components in infrastructure with virtualized part, control of such infrastructure becomes difficult. Following this IT service support becomes more complicated, which entails higher costs and lower quality of services provided. As a workaround of the problem of configuration management, a system that combines multiple information systems around the service desk is offered. Considerable attention is paid to the resolution of specific requirements to the system caused by the specifics of the data center: collective use of supercomputer resources and wide use of the virtualization. In addition, the article focuses on the analysis of the impact of the configuration management system on the computer center business processes related to the provision and support of its services. However, the main attention is focused on the development prospects of the system itself as well as a part of the ITSM complex.

Keywords: ITSM · Configuration management · Computer center · Infrastructure

1 Introduction

The main goals of any IT department, regardless of the specifics of the work is to provide IT services and support the pre-specified level of quality of the service. In the case where the amount of its services is small, as well as the requirements for quality, service management of such a unit - is a simple problem and it can be solved even without the use of complex information systems with small number of employees.

However, in the case where the IT department is a computer center and its work is not in the interests of one group or scientific laboratory, and the computer center serves a large organization (in our case - the University), the spectrum of users of computational problems solved is very wide. Accordingly the range of

© Springer International Publishing Switzerland 2015
O. Gervasi et al. (Eds.): ICCSA 2015, Part IV, LNCS 9158, pp. 249–258, 2015.
DOI: 10.1007/978-3-319-21410-8_19

services provided by the computer center was expanded to the fullest possible coverage of the needs of users. Another aspect of the work at the service of the University is a large number of users of IT services, which, combined with limited resources, especially human, does not allow an individual approach to each user. An exception is made only for the top research groups with highest computational activity.

Due to the very wide range of users' tasks those require the IT services, the amount of hardware and software in computer center is also very large. In addition, hardware, technologies and programs are very diverse and universal approach to the management of such infrastructure is quite difficult to find. In addition to the foregoing, the university computer center also has infrastructure for learning technologies related to information systems, virtualization, HPC [1].

Considering all activities of the computer center of the University, it can be concluded that the number of IT components in its infrastructure is very high. All this makes the control of infrastructure very difficult, the maintenance of up to date information about each component for use in the processes of ITSM - a non-trivial task perfectly. The difficulty lies in the fact that the use of powerful industrial infrastructure virtualization solutions, IT departments are forced to use a special system of virtualization management, which also control the server park, networks and storage systems [2,3]. In general, all this forms a tightly integrated hardware and software system in which each component affects many others, and in many cases, implicitly. It greatly complicates the analysis of the impact of problems and changes in the maintenance of the services support.

With the growth of the computer center of the University from the unit providing only high-performance computer services to a limited number of research groups, to the center for collective use, the processes of incident management and problem management were gradually developed (though even without the use of the terminology of ITIL). However, these processes have been isolated and had a great number of difficulties in obtaining the data about the components of the infrastructure on which the problem occurred, and had almost no opportunity to quickly analyze the incident and find the "guilty" IT component.

The data on IT components, their relationships and settings stored partly in multiple databases, partly in the form of web-site ETI, partly in the form of text files in the shared directories and a part of the components were just not documented. Lack of information on many IT components caused a number of difficulties even in the process of providing services and in the process of services support:

1. *Difficulty of Control of the Changes in Infrastructure.* Due to the lack of authorized information about the infrastructure (i.e. obviously true and actual) it was impossible to be sure that the state of the component meets the requirements for it, and has not been affected by an unauthorized changes. Also, there were no data on who and when was made changes in the configuration of the IT component.

2. *Impact Accounting Difficulty.* Without information on the relationships between components of the infrastructure could not adequately assess which

components, and, ultimately, the services affected by the implementation or modification of a service.

3. *Known Errors Accounting Difficulty.* The lack of documented relationships of problems and components that caused them, meant no trace which components cause the known errors and which components and services affected by these errors. This is very serious hurt to the planning of changes to remove known bugs, in particular the prioritization of changes was very difficult.

4. *Inflated Time Costs of Incident Solving.* Most incidents are somehow connected with the violation of the reference configuration of the components used for the provision of services. In the absence of a unified database, which stores descriptions of authorized reference configurations of IT components, looking for information on what reference configuration should be, took a considerable amount of time. There were times when all the information has not been found, which turned a simple incident in issue, taking a long time to investigate the causes.

5. *Difficulties in Collecting Information for the Formation of Reports.* In the absence of a unified repository of information on all infrastructure components creating of summary reports became difficult and time-consuming task due to the the need to collect information from multiple sources. In this case, the validity and relevance of the information could be guaranteed not always. Some components may be undocumented, or information about them has been lost, which required, among other things, the time to restore or re-creation of descriptions.

6. *The Complexity of Collecting Information for Planning of Service Delivery.* Gathering the information about what components are used by whom, how many and to what extent it is possible to rely on their reliability (whether the incidents caused by them, and how much) when planning capacity, in the case where the data is stored in multiple databases, not related between them becomes quite a challenge.

7. *Hidden Costs.* Each of the employees took notes on the part of the infrastructure, which is responsible for, respectively; the data collected by different people could have redundancy or contradiction. Resolving the contradictions needed to spend time and extra time spent on the creation of copies of redundant data.

2 Statement of the Problem

The objective of this research may be indicated as follows: development of the Configuration Management System of the computer center and its integration with the rest of the information systems, providing IT service management. It is necessary to consider the following specific properties of the computer center of the University:

1. Use of virtualization for providing business services, and also for providing operating services.

2. A wide range and diversity of the services provided by the computer center: high performance computing, education.
3. Various hardware used to provide services.
4. Collective use: the number of users is close to a half of thousand.

Obvious need to build a system in accordance with the recommendations of ITIL 2011 for logical integration with systems that provide the other ITSM processes in the computer center. It is particularly important to ensure maximum interaction between configuration management systems and change management because the changes have a direct impact on the composition of the data in the CMDB, which in turn are used to carry out the changes.

Another important requirement is the need for a system to put in it an opportunity to develop together with the entire set of ITSM. To meet this requirement the configuration management system must be equipped with an interface for interaction with other information systems. This interface must be sufficiently reliable, have the flexibility to reconfigure quickly as possible to change the composition and communication system in the process of growth and development. As this interface is more convenient to use a web service which, besides flexibility, are able to maintain the reliability under certain conditions.

In addition, only the configuration management information system is not enough to solve the problem of control of IT infrastructure, and a configuration management process that includes the drafting of regulations on the process also required to build. This drafting should describe all activities as part of the process, their sequence, communications, documents created and used in the process, as well as the individuals involved and their roles in the process (RACIS).

3   Choice of Tools

The technical implementation of the configuration management system for the St. Petersburg State University's computer center has been based on a service desk system OTRS :: ITSM which is available for free. There were several reasons for this choice.

The main reason is that this software is one of the most functional service desk systems available for free. The basis of the software is OTRS Helpdesk, which provides the Incident Management and Problem Management's running, as well as the processing of service requests. The service desk modules are deployed around the kernel, making up ITSM bundle. OTRS::ITSM contains tools for the operation of Configuration Management, Change Management, Incident Management and Problem Management, which form the basis of processes for IT services support. Incident Management and Problem Management systems based on the considered software solution had been already introduced and used in the St. Petersburg State University's computer center by the time of designing the configuration management system. Moreover, there was built the complex for tracking computational tasks [4]. Accordingly, all the necessary infrastructure for the operation of the configuration management system based on OTRS, had already existed and had been verified by a long period of operation.

In addition to all, OTRS supports knowledge base as an extra module, which can store data about known bugs, instructions for dealing with them, workaround instructions for incidents and instructions for service requests.

Another reason for this choice is that the OTRS::ITSM system is certified in accordance with the ITIL Library, which guarantees the feasibility of the main recommendations of ITIL, but at the same time it gives a freedom to do as you wish (with the specifics of the computer center). ITIL certification also means that the system architecture in general is designed for development without violations of the ITIL recommendations. For the computer center it means the opportunity to develop ITSM-system gradually without worrying about a version of the service desk.

One more argument for the OTRS::ITSM is the existence of the universal interface of web services for integration with other information systems in case of neessity. This interface supports a variety of architectures of web services (SOAP and REST). The range of information systems, that could be intergrated with other, is expanded by the ability to use the services of different architectures. Furthermore, an increase in reliability of data transmission using the interface could be done by means of specific mechanisms [5]. Embedded mechanisms for creating configuration items using web services are especially useful for creating configuration management system based on the OTRS. Also similar mechanisms are quite easily configured using the GUI.

Another important advantage of the OTRS::ITSM superstructure is the presence of the configuration item's import/export module, which allows the creation of a mass configuration items, changing their descriptions, as well as uploading information on a highly customizable criteria.

At the same time, there are some flaws in the OTRS::ITSM system. Some of them have a significant influence on the processes of submitting and supporting computer center's services. One of the main disadvantages of OTRS is low productivity and poor scalability. OTRS particularly sensitive to the number of entries in the database that represent active elements in OTRS (i.e. open applications configuration items which are in service). There are about 10,000 records in the CMDB database needed for a complete description of the St. Petersburg State University's computer center's infrastructure, which significantly affect the system performance.

Besides, the considered system is used for registration of the report about computational tasks in queues systems of supercomputer center. Accordingly, in certain periods of time, the system processes a sufficiently large number of requests per unit time, and the processing of these requests can not be accelerated using the system scaling, which together with the contribution of the database has a negative effect on the system performance in general.

Insufficiently functional CMDB search module is another weakness of the OTRS in terms of building OTRS system configuration management. In particular, it is impossible to search configuration items taking into account the parameters that do not depend on the class of configuration item.

The software platform IBM FileNet P8 has become the second important component. It has been planned to place all documents which regulate the process of configuration management into FileNet. For such a simple task ECM-system is too powerful, but it was selected with the prospect of long term construction of the computer center integrated service management solution with the management of center's business processes.

4   Technical Implementation

The creation of the computer center's service catalog had become the first step to a new configuration management system. Here one should pay attention to the fact that we had to describe not only the services available to external users (business services), but also the services provided within the center which are essential to its successful operation (operation services). For example, the providing resources service and support resource service are two different services, the first of which relates to the business catalog, and the other relates to the operation catalog. All staff of the computer center (depending on the scope of activities) took part in creating the catalogs. All services, regardless of the type are linked with the computer center system. There have been done some schemes to illustrate these links for the convenience of the subsequent design of the CMDB and in order to create a class list of configuration items. It was important to strike a balance between detailed description of the infrastructure and the load on the staff for making and maintaining the relevance of records, that is why the level of detailing had been defined previously. Next, one defined the coverage of the CMDB, i.e. what classes would be in the CMDB.

The next step was the creation of metadata sets for each of the classes. One have to start this work with the naming agreement of configuration items, then generate metadata for each class, as well as links between them. At this stage, the preparatory work can be considered as over. The immediate implementation of the configuration items's classes in the OTRS could be done next.

One of the main problems in the implementation of the new configuration management system is employees psychological resistance. Difficulties with the usual work often baffled even the most experienced professionals, that leads to delays in the execution of requests, incident and problem resolution. All these factors are critical for computer center serving a large organization. In connection with this, the first worth thinking about before the introduction of the new system is ease of working with the system. It have to be considered at each step of the method of transition to the new system described above, especially for determining the list of the metadata.

Implementation of the Configuration Management system is not enough to begin the use of Configuration Management in computer center. There must be the second important part - documentation on the Configuration Management process as a part of ITSM of computer center. Scheme of the process representing the main activities within the process, inbound and outbound documents of activities and relationships between Configuration Management and other ITSM processes applied to SPBU computer center displayed on a figure below.

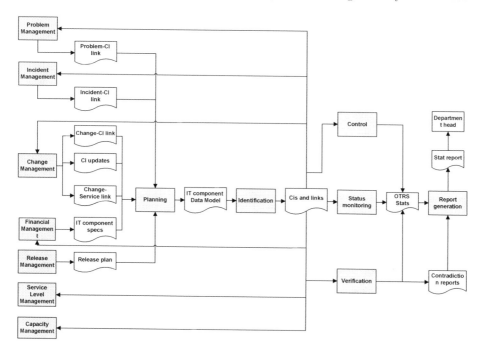

Fig. 1. Scheme of Configuration Management Process

Good help in staff adaption to a new way of working is understandable and accessible compiled instructions and regulations that must be prepared in advance in order to eliminate outages. Each employee must clearly understand the scope of their competence and area of responsibility. The role of each employee should be spelled out in the employee's job description. The matrix of responsibility RACI could be used to describe the roles of people involved in current business process.

As mentioned in the introduction, it is important to the center to restore the strict control of all processes, with reports providing. That is important not only for keeping all the variety of equipment in order, but also to supervise a work of each employee. Performed work reflects in the number of closed tickets and written notes. Thus, proving the efficiency (for example by writing a report on the changes), employee supports information completeness of the current configuration item, and the connection of this configuration item with others allows to monitor and promptly remove the source of the problem. Of course, the employees of the center are not always guilty of equipment failure, but in this case they point in the configuration item of broken equipment that it has the repair status. This, in conjunction with well-timed information for users who use this resource contributes to the smooth operation, which is particularly important for projects which computations require a long period of time.

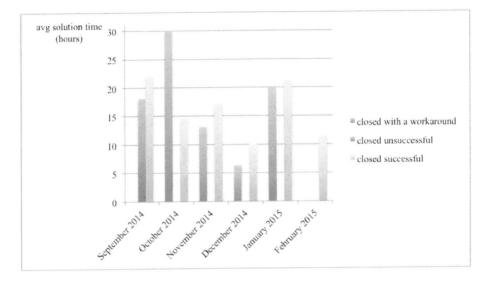

avg solution time (hours)

closed with a workaround
closed unsuccessful
closed successful

Fig. 2. Statistics on incidents in six months

Access Group is the configuration items class in the CMDB, which combine performers belonging to the same project. It is used in order to link users and resources allocated to them under the ongoing project. The very same Project configuration item is linked with head of current project group only, excluding from the base description all other users of the project, who, in turn, linked to the project through the Project-Access Group link. This is the example of an exception redundancy and creating customer support facilities. The abundance of links between configuration items can be confusing at first, but in these connections are urgently needed and served, in particular, to accelerate the work.

Doubts about the relevance of the CMDB could be negated all the positive aspects of the introduction of the new configuration management system metioned above. Supporting the relevance is a major task for data center, which managed to create a comfortable, working and well-described configuration management system.

5   Conclusions

Let us illustrate the efficiency of the new configuration management system by showing the statistics on incidents in six months, in the period from September 2014 to February 2015. The new system was introduced at the end of January. Statistics show that the number of applications, closed with a workaround or closed unsuccessful and not in time has been reduced to a minimum, which is an indicator of the staff productivity's increase, as well as improvment the quality of customer support.

At the moment there are two large completely built parts of the configuration management system: configuration management database (CMDB) and also the set of configuration management process definition and instructions for working with the data about IT components' configurations. These parts allow getting the latest information about the configuration of any component of IT-infrastructure. Techniques for maintaining the relevance of the CMDB data are reflected in the instructions for the configuration management system work configuration and in the standing order of the configuration management process. This feature provides a sufficient level of control of the infrastructure from the responsible persons. Also, all the configuration items in the CMDB linked in accordance with the relationship described by their IT-infrastructure components in the computer center. In case of necessity of determining the effect of any change one could pass on the links and determine the multiplicity of IT-items, which are affected by the change. Similarly, IT-components with known errors can be found using the links between the configuration items. In addition, it is easy to estimate the complexity of the problem, evaluate the change cost, and on the basis of these data to create change in their solvings and to conduct proper prioritization of these changes.

General database storing the information about each IT-component is one of the main positive results of the system implementation. This is especially important for applications requiring a summary of the computer center infrastructure. The main consumers of this information is capacity, availability and continuity processes. In addition, there is no need for each employee to do a separate work in order to keep information about the system he is responsible for up to date. Because according to the standing order of the configuration management process updating information about the IT-component deals with the employee who directly holds change with it. This allowed to get rid of the implicit labor sources.

Having a single point of gathering information about the infrastructure is very convenient for generating summary reports. Besides, configuration management system, based on OTRS, allows the creation of several types of reports automatically (manual start or schedule), as well as the automatic sending to the head of center, or using web services, displaying information on the website.

We can say that in this direction have already done a lot, but some problems at the moment are still not resolved. A system based on the OTRS software has, as described above, a number of drawbacks. Low performance of such solution is especially harmful for productivity of work with the configuration management system. Inconvenience of OTRS CMDB search module is not as critical, but periodically affects the time costs.

To address these issues the gradual shift of functionality of the configuration management system towards FileNet platform and jump to the case management technology, implementing it with IBM FileNet Case Manager are planned. Using a balanced mix of the case management and process management technology in the future we plan to bring the rest of ITSM processes and functions of data center ITSM to the IBM FileNet + Case Manager.

Integrated ITSM-solution of the computer center is undoubtedly require the organization of interaction with external information systems, as well as these systems are not under control by the computer center staff, we have to take care of the maintenance of very high level of reliability of data exchange. For this reason it is planed to use RESTful web services with support of long-life transactions.

This solution promises several advantages: firstly reliability, both from the information system and from the interface of its interaction with other information systems. Second, the use of applets, web services and third-party databases, makes FileNet-based configuration management system unique and flexible solution and allows to change the system as it evolves without termination of its services or reducing the services' quality.

In addition to the transition of the configuration management system on the IBM FileNet there are also several steps of work on the system, related to the solution of specific problems and enhancing of functionality of the system.

The first step is a creating of the opportunity to generate the configuration items automatically while creating users of virtual machine, as well as creating users and resources Access Groups in LDAP, which in general can be described as automatic registration activities related to the delivery of resources to users.

Second step is an integration with a system of tracking and reflecting the changes in infrastructure, recorded in the CMDB for updating data on the IT-components.

In the future, it could be a transition to independent creation of virtual machines and private virtual clusters by users using a special configurator (e.g. FishDirector or the same product) [6], which would require compulsory automatic registration to create and modify objects in the CMDB.

Acknowledgments. The research was carried out using computational resources of Resource Center Computational Center of Saint-Petersburg State University within frameworks of grants of Russian Foundation for Basic Research (project No. 13-07-00747) and Saint Petersburg State University (projects No. 9.38.674.2013 and 0.37.155.2014).

References

1. Resource Center Computer Center of St.Petersburg State University. http://cc.spbu.ru/en
2. Bogdanov, A., et al.: Virtual workspace as a basis of supercomputer center. In: Proceedings of the 5th International Conference on Distributed Computing and Grid-Technologies in Science and Education, Dubna, Russia, pp. 60–66 (2012)
3. Gankevich, I., Korkhov, V., Balyan, S., Gaiduchok, V., Gushchanskiy, D., Tipikin, Y., Degtyarev, A., Bogdanov, A.: Constructing virtual private supercomputer using virtualization and cloud technologies. In: Murgante, B., et al. (eds.) ICCSA 2014, Part VI. LNCS, vol. 8584, pp. 341–354. Springer, Heidelberg (2014)
4. Bogdanov, A., et al.: The use of service desk system to keep track of computational tasks on supercomputers. LNCS Transactions on Computational Science (to appear)
5. Tipikin, Yu.: Transactional mechanism for RESTful web services. In: The XLV Annual International Conference on Control Processes and Stability (CPS 2014). Abstracts., 78 (2014)
6. Sardina Systems. http://www.sardinasystems.com

Novel Approaches for Distributing Workload on Commodity Computer Systems

Ivan Gankevich[✉], Yuri Tipikin, Alexander Degtyarev, and Vladimir Korkhov

Saint Petersburg State University, Universitetskii 35,
Petergof, 198504 Saint Petersburg, Russia
igankevich@yandex.com, yuriitipikin@gmail.com

Abstract. Efficient management of a distributed system is a common problem for university's and commercial computer centres, and handling node failures is a major aspect of it. Failures which are rare in a small commodity cluster, at large scale become common, and there should be a way to overcome them without restarting all parallel processes of an application. The efficiency of existing methods can be improved by forming a hierarchy of distributed processes. That way only lower levels of the hierarchy need to be restarted in case of a leaf node failure, and only root node needs special treatment. Process hierarchy changes in real time and the workload is dynamically rebalanced across online nodes. This approach makes it possible to implement efficient partial restart of a parallel application, and transactional behaviour for computer centre service tasks.

Keywords: Long-lived transactions · Distributed pipeline · Node discovery · Software engineering · Distributed computing · Cluster computing

Introduction

There are two main tasks for a computer centre: to run users' parallel applications, and to service users. Often these tasks are delegated to some distributed systems, so that users can service themselves, and often these systems are heterogeneous and there are many of them in a computer centre. Yet another system is introduced to make them consistent, but this system is usually not distributed. Moreover, the system usually does not allow rolling back a distributed transaction spanning its subordinate systems. So, the use of reliable distributed systems under control of an unreliable one makes the whole system poorly orchestrated, and absence of automatic error-recovery mechanism increases amount of manual work to bring the system up after a failure.

To orchestrate computations and perform service tasks in distributed environment the system should be decomposed into two levels. The top level of this system is occupied by transaction manager which executes long-lived transactions (a transaction consisting of nested subordinate ones which spans a long

© Springer International Publishing Switzerland 2015
O. Gervasi et al. (Eds.): ICCSA 2015, Part IV, LNCS 9158, pp. 259–271, 2015.
DOI: 10.1007/978-3-319-21410-8_20

period of time and provides relaxed consistency guarantees). Transactions are distributed across cluster nodes and are retried or rolled back in case of a system failure. On the second level a distributed pipeline is formed from cluster nodes to process compute-intensive workloads with automatic restart of failed processes. The goal of this decomposition is to separate service tasks which need transactional behaviour from parallel applications which need only restart of failed processes (without a roll back). On both levels of the system a hierarchy is used to achieve these goals.

The first level of the system is capable of reliably executing ancillary and configuration tasks which are typical to university's computer centres. Long-lived transactions allow for a task to run days, months and years, until the transaction is complete. For example, a typical task of registering a user in a computer centre for a fixed period of time (the time span of his/her research grant) starts after the user submits registration form and ends when the research is complete. Additional tasks (e.g. allocation of computational resources, changing quotas and custom configuration) are executed as subordinate of the main one. Upon completion of the work tasks are executed in reverse, reconfiguring the system to its initial state and erasing or archiving old data.

The rest of the paper describes the structure of two levels of the system and investigates their performance in a number of tests. Long-lived transactions are discussed in Section 1 and distributed pipeline (lower level) in Section 2.

1 Long-Lived Transactions

While performing computing in HPC environment various errors may occur. Hardware errors have the greatest impact among others. To fix this type of errors several approaches exist today which often consist of restarting the job completely. In distributed systems computational nodes have even more risk to be lost because of additional factors such as unreliable network. Thus, a complete job restart every time one or more nodes fail is ineffective.

While searching solution of this problem let us refer to the traditional transaction mechanism. Designed for operations on data it uses logging and locking to prevent data corruption and loss. ACID properties — Atomicity, Consistency, Isolation, Durability — of the transaction apply to relational databases, but for distributed system they are implemented with several restrictions. Now let us take a look at high-performance distributed environment. Here operations are performed on the tasks and the objective is to get valid computation results. At first glance, to apply transactions for computing one needs to segment initial task code and take a subtask as an atomic operation, but this is not sufficient.

Unlike database operations, computations can take much more time to execute, and long-lived transactions, which theoretically can take as much time as needed, is the solution. The main aspect of this technology is a correct logging and further journal processing. There is no unified definition of "long-lived transaction", so we define this term here.

Long-lived transaction is a transaction operations of which are executed for long time periods and there are long gaps between completion of operations.

So, it is not safe to assume fast execution time of such transactions. For them only atomicity and reliability properties are guaranteed.

At the modeling stage web service can be a perfect container for computational subtask. Using REST [2] — representational state transfer — as a specific implementation of web services, we design and implement job scheduling on nodes as a call of a web service with target URL. Main accent here is on restore process of failed operation. The aim of this paper is to offer time-efficient algorithm of such restoration. Next, we will compare properties of REST realisation to "reliable" set of properties ACID and will draw a conclusion about meaningful changes in our model, which guarantee satisfiability of the initial task.

During testing REST web services inside transaction container, the fact of inapplicability of ACID "as is" was revealed. These properties conflict with both REST basics and a definition of a web service. Lets consider these properties step by step.

1.1   ACID in REST

Atomicity. Atomicity guarantees that transactions will not be partially saved and in terms of data this works best. However, web services are operations which create, modify and delete data while running, so transaction involving one or more web services must be moved to a higher level of abstraction. In fact, REST web service transactional system has two levels of atomicity: database level and level where web services are called directly. First of them is provided by a particular database implementation by default, second one is on the logical level, which programmer must implement in terms of a particular algorithm. It is important to understand what web service implementation must guarantee logical atomicity of its internal operations by providing only two available states of termination: absolutely correct result of entire web service and error state result. Thus, a collection of web services on logical level can be considered as a single web service recursively applying such requirements.

So, there is a need for a rollback operation for a web service, and in [6,11] the authors also came to this conclusion. However, REST architecture has no mandatory requirements to system functionality implementation, and this generally leads to impossibility of automatic operation rollback in those systems. Even if rollback operations were implemented by web service developer, there would be no guarantee of valid result after calling these methods, because logical side of an algorithm can be non-trivial.

Durability. Durability is an interesting property. It prevents losing state information (even from hardware failure) by logging all actions in special journal. There are some challenges to implement an efficient distributed journal, but for now there are a few related articles where logging REST web services was partially described [6,11]. In our approach to achieve efficient failed-task restoration distributed logging system plays an important role, and implementation of this property meets REST requirements.

Isolation. In REST this property is difficult to achieve without making an additional proxy server objects. Those objects serve as queues filtering "transactional" web services and executing them sequentially. This method is described in more detail in [6]. But if web service is designed to use parallel operations, a proxy server can be a performance bottleneck. In this case, isolation must be implemented on web service level, not on abstract level of a collection of web services. Transaction isolation through web service isolation imposes additional requirements to web service developer, but efficiency of transactional system may suffer without it. Isolation of a collection of transactions is applied by transactional job scheduler by default, because it executes transactions sequentially from a queue.

Consistency. Much like isolation, consistency is difficult to guarantee on web service level. It is a property of database rather than a web service.

1.2   Implementation

Main feature of our transactional manager is a special initial task code structure for long-lived transactions. Implementation consist of a server which executes transactions sequentially by placing them to a queue, logging and collecting responses, and a rewritten client code, which uses special functions (we called it *act* and *react*) to divide a task into a set of subtasks. On server's input we have structured code of a subtask, transferred in JSON format, for example. Each subtask is an autonomous part of code. Initial task after rewriting by this algorithm is represented by a tree, which itself is a transaction and leaves are operations. Thus, sequence of operations are saved: sequential operations have parent-child relation, and parallel have child-child relation.

This approach is largely different from those described in [6,11], it is not focused only on web service calls. In real world applications, reliable summary result is what user wants, without middleware web services calls information. But those middleware operations must be processed in any case. Only simplistic algorithms use exclusively web service calls, often a call is a result of subtask processing from another web service. Dividing task to a group of subtasks and after that formatting a transaction allows processing not only invokes of web services. In fact, any part of initial task can be secure.

In REST web services there is no universal way to achieve general atomicity. Rollback is implemented by moving the tree backward from a leave with failed state. Rollback performs for each subtask separately, not affecting other subtask on that particular computational node. In case of a node hardware failure, first rollback will wait the node to come online for some time, to not to move a task to another node's queue. In case of impossibility of that scenario, rollback function goes one level up and tries the same approach. Code segmentation can improve restoration time significantly, but also prevent legacy application to run in such environment.

The ineffectiveness of running rollback straight away on higher levels (entire subtree) shown in [2]. As previously mentioned, rollback function is empty by default and must be written by a programmer of a web service himself. This step is a necessary and logical, because correct rollback for each function must be written by a person who exactly knows in which state abstract transaction will be after failed operation and what that operation was doing before stop. Transactional system described in this paper is a tool, not an universal solution for all types of operations because each computational algorithm is unique. Use of this tool requires rethinking of original algorithm in terms of partitioning to autonomous segments.

1.3   Building the Transaction and Early Results

There is a step-by-step algorithm of transaction manager.

1. Programmer select self-sufficient parts of original algorithm – marking it as transaction operations.
2. Programmer writes a rollback function for all of such parts.
3. Structured code is sent to transaction server.
4. Server executes transaction by placing a root node (which includes a whole algorithm) to queue and then starts moving down across the code tree.
5. If processing is done without failures, iteration reaches leaves and starts going back by transferring successful results from children to their parents.
6. If processing is done with failures, transaction will run a rollback function on failed nodes and try to prevent massive rollback by slowly moving up the tree by one level at time.

The system was tested on 3 level algorithm tree which produces 25 lines in journal until it is completed. Measured time indicates how long transaction manager works to complete the task if a rollback function was called at specific log line (Fig. 1). Computation time without any rollback calls shown as dotted line. Time peaks in Fig. 1 belong to lines in journal that designate "execution" step of one or more subtasks. Total overhead oscillate from 5% to 15% because of fast prototype implementation on Node.js technology. The task itself consists of simple integration. Web services as task was tested as well, to investigate properties of long-lived transactions that are described above.

2   Distributed Pipeline

The main idea of distributed pipeline is to create virtual topology of processes running on different computer cluster nodes and update it in real-time as infrastructure changes. The changes include nodes going offline and online, joining or leaving the cluster, replacement of any hardware component or an entire network node (including switches and routers), and other changes affecting system performance. Each change results in virtual topology being updated for a new infrastructure configuration.

The main purpose of distributed pipeline is to optimise performance of distributed low-level service tasks running on a computer cluster. Typically these

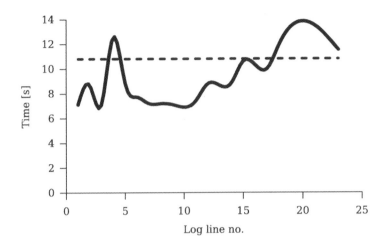

Fig. 1. Rollback time of a subtask at different log lines

tasks involve querying each cluster node for some information or updating files on each node, and often single master node sends and collects messages from all slave nodes. To reduce network congestion intermediate nodes should be used to communicate master and slave nodes, that is to collect information and send it to master or the other way round. In case of large cluster a distributed pipeline with virtual topology of a tree is formed.

The other purpose of distributed pipeline is to improve efficiency of high-performance applications. These applications typically launch many parallel processes on a subset of cluster nodes, which communicate messages with each other. To make these communications efficient virtual topology should resemble a real one as much as possible and take into account nodes' performance (relative to each other) and communication link speed (see Section 2.3). That is, if two nodes are adjacent to each other in virtual topology, then the node which is closest to the tree root should have higher performance than the other one, and the link between them should be of the highest throughput.

Another aspect of high-performance applications is their fault tolerance and recovery time from node failure. Often these applications consist of long-running processes which are checkpointed with specified time period. If a node fails, then a new one is reserved and the application is recovered from the last checkpoint on each node. To reduce recovery time checkpointing can be made incremental and selective, that is to recover only those parts of a program that have failed and checkpoint the data that have actually changed. In distributed pipeline it is accomplished by logging messages sent to other nodes and resending them in case of a node failure (see Section 2.1).

To summarise, distributed pipeline creates virtual topology of a computer cluster in a form of a tree with high-performance nodes being closest to the root, and every virtual link having the highest-possible throughput. The purpose of this pipeline is to optimise performance of various cluster management tasks, but it can be beneficial for high-performance applications as well.

2.1 Implementation

From technical point of view distributed pipeline is a collection of reliable network connections between principal and subordinate nodes, which are made unique and shared by multiple applications (or service tasks) in a controlled way. The absence of duplicate connections between any two nodes conserves operating system resources and makes it possible to build distributed pipeline crossing multiple NAT environments: If the node hidden behind NAT can reach a node with public Internet address, they can communicate in both directions. Additionally, it allows creating persistent connections which are closed only when a change in topology occurs or in an event of system failure.

Connection between nodes can be shared by multiple applications in a controlled way. Each message is tagged with an application identifier (a number), and each application sends/receives messages from either standard input/output or a separate file descriptor (a pipe) which can be polled and operated asynchronously. The data is automatically converted to either portable binary (with network byte order) or text format. So, if high performance of asynchronous communication and small size of binary messages are not required, any programming language which can read and write data to standard streams can be used to develop an application for distributed pipeline.

To simplify writing high-performance applications for distributed pipeline the notion of a message is removed from the framework, instead the communication is done by sending one object to another and passing it as an argument to a defined method call. Each object can create subordinates to process data in parallel, perform parallel computations, or run tasks concurrently, and then report results to their principals. The absence of messages simplifies the API: an object always processes either principal's local data or results collected from subordinate objects.

Various aspects of reliable asynchronous communication have to be considered to make distributed pipeline fault-tolerant. If the communication between objects is not needed, the object is sent to a free node to perform some computation, and sent back to its principal when it is done. Objects which are sent to remote computer nodes are saved in a buffer and are removed from it when they return to their principals. That way even if the remote node fails after receiving the objects, they are sent back to their principals from the buffer with an error. After that it is principal that decides to rollback and resend objects to another node, or to fail and report to its own principal. In case the failure occurs before sending an object (e.g. node goes offline), then the object is sent to some other node bypassing its principal. To summarise, subordinate objects *always* return to their principals either with a success or a failure which further simplifies writing high-performance applications.

Principal/subordinate objects easily map to tree topology. If not specified explicitly, a principal sends a subordinate object to a local execution queue.

In case of high load, it is extracted from the queue and sent to a remote node. If the remote node is also under high load, the object is sent further. So, the hierarchy of principals and subordinates is compactly mapped to tree topology: lower-level nodes are used on-demand when a higher-level node can not handle the flow of objects. In other words, when a higher-level node "overflows", the excessive objects are "spilled" to lower-level nodes.

So, the main goal of the implementation is to simplify application programming by minimising dependencies for small service programmes, making asynchronous messaging reliable, and by using automatic on-demand load balancing.

2.2   Peer Discovery

The core of distributed pipeline is an algorithm which builds and updates virtual topology, and there are several states in which a node can be. In *disconnected* state a node is not a part of distributed pipeline, and to become the part it discovers its peers and connects to one of them. Prior to connecting the link speed and relative performance of a node are measured. After connecting the node enters *connected* state. In this state it receives updates from subordinate nodes (if any) and sends them to its principal. So, updates propagate from leaves to the root of a tree.

In initial state a node uses discovery algorithm to find its peers. The node queries operating system for a list of networks it is part of and filters out global and Internet host loopback addresses (/32 and 127.0.0.0/8 blocks in IPv4 standard). Then it sends a subordinate object to each address in the list and measures response time. In case of success the response time is saved in the table and in case of failure it is deleted. After receiving all subordinate objects principal repeats the process until minimal number of performance samples is collected for all peers. Then the rating (see Section 2.3) of each peer is calculated and the table is sorted by it. The principal declares the first peer in the table a leader and sends a subordinate object to it which increases peer's level by one. Then the whole algorithm repeats for the next level of a tree.

Sometimes two nodes in disconnected state can be chosen to be principals of each other, which creates a cycle in tree topology of distributed pipeline. This can happen if two nodes have the same rating. The cycle is eliminated by rejecting offer from the node with higher IP address, so that a node with lower IP address becomes the principal. To make rating conflicts rare IP address is used as the second field when sorting the table of peers.

On initial installation the total number of subordinate objects sent by each node amounts to mn^2 where n is the total number of nodes and m is the minimal number of samples, however, for subsequent restarts of the whole cluster this number can be significantly reduced with help of peer caches (Section 2.7). Upon entering connected state or receiving updates from peers each node stores current peer table in a file. When the node restarts it runs discovery algorithm only for peers in the file. If no peers are found to be online, then the algorithm repeats with an empty cache.

2.3   Node's Rating

Determining performance of a computer is a complex task on its own right. The same computers may show varying performance for different applications, and the same application may show varying performance for different computers. For distributed pipeline performance of a node equals the number of computed objects per unit of time which depends on a type of a workload. So, performance of a computer is rather a function of a workload, not a variable that can be measured for a node.

In contrast to performance, concurrency (the ability to handle many workloads of the same type or a large workload) is a variable of a node often amounting to the number of processor cores. Using concurrency instead of processor speed for measuring performance is equivalent to assuming that all processors in a cluster have the same clock rate so that a number of cores determines performance. This assumption gives rough estimate of real performance, however, it allows determining performance just by counting the total number of cores in a computer node and relieves one from running resource-consuming performance tests on each node.

In [1,3,9] the authors suggest generalisation of Amdahl's law formula for computer clusters from which node's rating can be devised. The formula

$$S_N = \frac{N}{1 - \alpha + \alpha N + \beta\gamma N^3}$$

shows speedup of a parallel programme on a cluster taking into account communication overhead. Here N is the number of nodes, α is the sequential portion of a program, β is the diameter of a system (the maximal number of intermediate nodes a message passes through when any two nodes communicate), and γ is the ratio of node performance to network link performance. Speedup reaches its maximal value at $N = \sqrt[3]{(1 - \alpha)/(2\beta\gamma)}$, so the higher the link performance is the higher speedup is achieved. In distributed pipeline performance is measured as the number of objects processed by a node or transmitted by network link during a fixed time period. The ratio of these two numbers is used as a rating.

So, when nodes are in disconnected state the rating is estimated as the ratio of a node concurrency to the response time. The rating of a remote node is re-estimated to be the ratio of number of transmitted objects to the number of processed objects per unit of time when nodes enter connected state and start processing objects.

2.4   Rating and Level Updates

A node in connected state may update its level if a new subordinate node with the same or higher level chooses it as a leader. The level of each node equals to the maximal level of subordinates plus one. So, if some high-level node connects to a principal, then its level is recalculated and this change propagates towards root of a tree. That way the root node level equals the maximal level of all nodes in the cluster plus one.

The rating of principal can become smaller than the rating of one of its subordinates when the workload type is changed from compute-intensive to data-intensive or the other way round. This also may occur due to a delayed level update, a change in node's configuration, or as a result of higher level node being offline. When it happens, the principal and the subordinate swap their positions in virtual topology, that is the higher level subordinate node becomes principal. Thus high-performance node can make its way to the root of a tree, if there are no network bottlenecks in the path.

So, rating and level updates propagate from leaves to the root of a tree in virtual topology automatically adapting distributed pipeline for a new type of workload or new cluster configuration.

2.5   Node Failures and Network Partitions

There are three types of node failures in distributed pipeline: a failure of a leaf node, a principal node, and the root node. When a leaf node fails, objects in the corresponding sent buffer of its principal node are returned to their principals, and objects in unsent buffer are sent to some other subordinate node. If error processing is not done by principal object, then returning subordinate objects are also re-sent to other subordinate nodes. In case of principal or root node failure the recovery mechanism is the same, but subordinate nodes that lost their principal enter disconnected state, and a new principal is chosen from these subordinate nodes.

In case of root node failure all objects which were sent to subordinate nodes are lost and retransmitted one more time. Sometimes this results in restart of the whole application which discards previously computed objects. The obvious solution to this problem is to buffer subordinate objects returning to their principals so that in an event of a failure retransmit them to a new principal node. However, in case of a root node failure there is no way to recover objects residing in this particular node other than replicating them to some other node prior to failure. This makes the solution unnecessary complicated, so for now no simple solution to this problem has been found.

In case of network partition the recovery mechanism is also the same, and it also possess disadvantages of a root node failure: The results computed by subordinate objects are discarded and potentially large part of the application has to be restarted.

So, the main approach for dealing with failures consists of resending lost objects to healthy nodes which is equivalent of recomputing a part of the problem being solved. In case of root node failure or network partition a potentially large number of objects are recomputed, but no simple solution to this problem has been found.

2.6   Evaluation

Test platform consisted of a multi-processor node, and Linux network namespaces were used to consolidate virtual cluster of varying number of nodes on a physical node [4,5,7]. Distributed pipeline needs one daemon process on each node to operate correctly, so one virtual node was used for each daemon. Tests were repeated multiple times to reduce influences of processes running in background. Each subsequent test run was separated from previous one with a delay to give operating system time to release resources, cleanup and flush buffers.

Discovery test was designed to measure effect of cache on the time an initial node discovery takes. For the first run all cache files were removed from file system so that a node went from disconnected to connected state. For the second run all caches for each node were generated and stored in file system so that each node started from connected state.

Buffer test shows how many objects sent buffer holds under various load. Objects were sent between two nodes in a "ping-pong" manner. Every update of buffer size was captured and maximum value for each run was calculated. A delay between sending objects simulated the load on the system: the higher the load is the higher the delay between subsequent transfers is.

2.7 Test Results

Discovery test showed that caching node peers can speed up transition from disconnected to connected state by up to 25 times for 32 nodes (Fig. 2), and this number increases with the number of nodes. This is expected behaviour since the whole discovery step is omitted and cached values are used to reconstruct peers' level and performance data. The absolute time of this test is expected to increase when executed on real network, and cache effect might become more dramatic.

Buffer test showed that sent buffer contains many objects only under low load, i.e. when the ratio of computations to data transfers is low. Under high load computations and data transfer overlap, and the number of objects in sent buffer lowers (Fig. 3).

Related Work

In [8] the authors discuss the use of message logging to implement efficient recovery from a failure. They observed that some messages written to log are commutative (can be reordered without changing the output of a program) and used this assumption to optimise recovery process. In our system objects in sent buffer are used instead of messages in a log, they are commutative within bounds of the buffer but do not represent history of all messages sent to another node. They are deleted upon receiving a reply, and deleted object can be safely ignored when recovering from a hard fault. So, when an object depends on parallel execution of its subordinates, results can be collected in any order, and it is often desirable to log only execution of principal to reduce recovery time.

In [10] the author introduces general distributed wave algorithms for leader election. In contrast to distributed pipeline, these algorithms assume that there can be only one leader and as a result do not take into account link speed

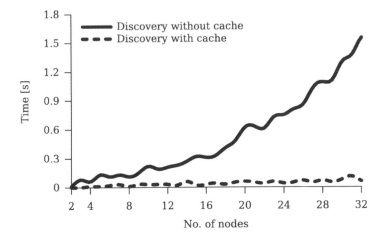

Fig. 2. Time taken for various number of nodes to discover each other with and without help of cache

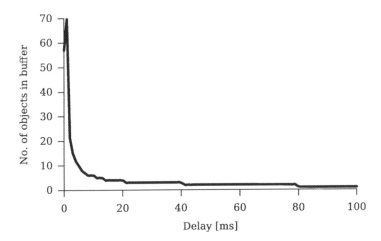

Fig. 3. Number of objects in sent buffer for various delays between objects transfers

between nodes. The goal of discovery algorithm is to create a framework of leaders to distribute workload on a cluster, and each leader is found by recursively repeating election process for the new level of a hierarchy. Using hierarchy of leaders simplifies election algorithm, and taking into account links between nodes allows efficient mapping of resulting hierarchy to physical topology of a network.

Conclusions and Future Work

Distributed pipeline is a general method for distributing workload on a commodity cluster which relies on dynamically reconfigurable virtual topology and reliable data transfer. It primarily focuses on service tasks but can be used for high-performance computing as well. The future work is to find a simple way to make distributed pipeline resilient to the root node failures, and test applicability of this approach to high-performance computing applications.

Acknowledgments. The research was carried out using computational resources of Resource Centre "Computational Centre of Saint Petersburg State University" (T-EDGE96 HPC-0011828-001) within frameworks of grants of Russian Foundation for Basic Research (project no. 13-07-00747) and Saint Petersburg State University (projects no. 9.38.674.2013 and 0.37.155.2014).

References

1. Andrianov, S., Degtyarev, A.: Parallel and distributed computations. Saint Petersburg State University (2007). (in Russian)
2. Armstrong, J.: Making reliable distributed systems in the presence of software errors. PhD thesis, The Royal Institute of Technology Stockholm, Sweden (2003)
3. Degtyarev, A.: High performance computer technologies in shipbuilding. In: Birk, L., Harries, S. (eds.) OPTIMISTIC – optimization in marine design. Mensch & Buch Verlag, Berlin
4. Handigol, N., Heller, B., Jeyakumar, V., Lantz, B., McKeown, N.: Reproducible network experiments using container-based emulation. In: Proceedings of the 8th International Conference on Emerging Networking Experiments and Technologies, pp. 253–264. ACM (2012)
5. Heller, B.: Reproducible Network Research with High-fidelity Emulation. PhD thesis, Stanford University (2013)
6. Kochman, S., Wojciechowski, P.T., Kmieciak, M.: Batched transactions for RESTful web services. In: Harth, A., Koch, N. (eds.) ICWE 2011. LNCS, vol. 7059, pp. 86–98. Springer, Heidelberg (2012)
7. Lantz, B., Heller, B., McKeown, N.: A network in a laptop: rapid prototyping for software-defined networks. In: Proceedings of the 9th ACM SIGCOMM Workshop on Hot Topics in Networks, p. 19. ACM (2010)
8. Lifflander, J., Meneses, E., Menon, H., Miller, P., Krishnamoorthy, S., Kalé, L.V.: Scalable replay with partial-order dependencies for message-logging fault tolerance. In: IEEE International Conference on Cluster Computing (CLUSTER), pp. 19–28. IEEE (2014)
9. Soshmina, I., Bogdanov, A.: Using GRID technologies for computations. Saint Petersburg State University Bulletin (Physics and Chemistry) **3**, 130–137 (2007). (in Russian)
10. Tel, G.: Introduction to distributed algorithms. Cambridge University Press (2000)
11. Wilde, E., Pautasso, C.: REST: from research to practice. Springer Science & Business Media (2011)

Managing Dynamical Distributed Applications with GridMD Library

Ilya A. Valuev[1(✉)] and Igor V. Morozov[1,2]

[1] Joint Institute for High Temperatures of Russian Academy of Sciences, Moscow, Russia
{valuev,morozov}@ihed.ras.ru
[2] Higher School of Economics, Moscow, Russia

Abstract. The open source C++ class library GridMD for distributed computing is reviewed including its architecture, functionality and use cases. The library is intended to facilitate development of distributed applications that can be run at contemporary supercomputing clusters and standalone servers managed by Grid or cluster task scheduling middleware. The GridMD library used to be targeted at molecular dynamics and Monte-Carlo simulations but at present it can serve as a universal tool for developing distributed computing applications as well as for creating task management codes. In both cases the distributed application is represented by a single client-side executable built from a compact C++ code. In the first place the library is targeted at developing complex applications that contain many computation stages with possible data dependencies between them which can be run efficiently in the distributed environment.

Keywords: Distributed computing · Grid computing · Execution graph · Workflow · Multiscale simulations

1 Introduction

A large numerical experiment can usually be divided into weakly dependent tasks that may be executed in a distributed computing environment. The simplest case is a parameter sweep needed for statistical averaging, optimal parameter search or obtaining parameter dependence. The main idea of the GridMD library is to automate separation of individual tasks, identification of data links between them, generation and processing of the workflow scenario by using additional instructions (library function calls) that the programmer should insert into the application source code [1], [2]. Therefore there are two main purposes of the GridMD library: to generate a workflow and to execute it in a distributed environment.

The workflow execution systems [3], [4] can be divided into two classes depending on whether they require or not an additional middleware installed and persistently running on the target computing infrastructure. The systems of the first class can be called as 'infrastructural'. In this case the middleware controls the user authorization and submission of their jobs to appropriate job schedulers. Examples of such middleware are Grid managers (e.g. Unicore [4], Triana [5], gUSE [6]) and distributed

© Springer International Publishing Switzerland 2015
O. Gervasi et al. (Eds.): ICCSA 2015, Part IV, LNCS 9158, pp. 272–289, 2015.
DOI: 10.1007/978-3-319-21410-8_21

database agents (e.g. Taverna project [7]). The systems of the second class may be called 'universal' as they do not require persistent agents on the target infrastructure and interact directly to a wide range of job schedulers. Many universal systems such as Kepler [8]and Pegasus[9] use the Condor scheduler to execute jobs on different types of computing infrastructure.

Another classification is possible according to who is the system user: whether it is an end user (researcher) or it is an application developer (programmer). The end user oriented systems usually offer a graphical interface to define the execution workflow. Using this interface the end user has to define and customize the workflow manually going into details of the workflow definition process or rely on third-party workflow developers   (see e.g. the gUSE [6] project) who can be unqualified in the subject field. The workflow systems oriented to the application developers (e.g Pegasus [9]) allow for more flexibility with the choice of the application design and user interface.

As opposed to most of the existing workflow management systems, the GridMD library is a compact universal tool for application developers. The library is not meant as a replacement of very useful and well-known infrastructural middleware, but rather as a lightweight workflow-supporting interface to various common computational resources that an "average" application user has access to. Traditionally, for the researchers working with compute-intensive programs (for example, quantum chemistry, molecular dynamics or hydrodynamics simulators) these resources are high-performance workstations, university or company clusters, sometimes distributed Grids. Currently many types of computational resources can also be easily configured in a cloud and acquired from cloud providers.

A GridMD-based application can be executed fully on the client side with no direct need for installing and maintaining a middleware such as Grid systems and portals. At the same time the GridMD application can be adopted to run jobs via the services provided by this middleware on behalf of the user. Moreover, the library can be used for developing computational platforms oriented at specific research field that require low-level control of the workflow execution and hiding this workflow from the end user. An example of such application may be found in [10], where a multi-physics software platform for the organic light-emitting diodes (OLEDs) simulation is described. There GridMD library is used in this project to manage workflows composed of several simulation codes each representing an important physical model. The software is designed as a desktop application allowing building composite models from large blocks, the actual GridMD workflow jobs are more fine-grained and do not show up on the user interface level.

The basic concepts of GridMD and some usage examples were described in our previous works ([1], [11], [12]). These works were concentrated on workflows encapsulated into a single C++ code, fully created 'from scratch' and deployed for execution under control of GridMD. This implicit workflow model proved to be useful for Molecular Dynamics simulations where the source code of the model is available and it is possible to add a few GridMD instructions and skeletons there. In the recent time, following feedback from the OLED project [10], the library was substantially extended with the new functionality, such as dynamical workflow management, comprehensive explicit workflow model, local nodes with callback functions or job timings measurement. These changes simplified the usage of GridMD as a standalone workflow manager for operating external applications, available as binaries installed on remote servers. The aim of this paper is to systematically review the current state of all most important GridMD capabilities, giving deeper insight into the newly added functions. The paper is organized as follows: in Section 2 we revisit the architecture of GridMD; in Section 3 we give an example of a compact but functional dynamical parameter sweep code measuring job timings; in Section 4 we describe a lower-level job management interface; we give conclusions and future work remarks in Section 5.

2   GridMD Architecture

In this section we review the basic architecture of GridMD library, the reader is referred to GridMD documentation [2] for the details of the functionality and interface. The main purpose of the library is to provide a C++ programming interface to manage a distributed workflow, i.e. a sequence of computational tasks, related to each other by input and output data and possibly executed on different computers. The workflow management is performed solely from the program executing GridMD library calls. We will further call this program a *client application*. The client application makes requests to run commands and jobs on (possibly remote) *compute systems* using the credentials that the user commits to the client application. No permanently running demons or other special intermediates from GridMD on a compute system are needed for the execution of GridMD workflow. The only exception from this is when the remote computational tasks are formulated in the form of callable C++ code. In this case the compute server application implementing this code should be deployed to the remote compute system.

GridMD has two main components: a workflow manager and a job manager. Workflow manager is a high-level component, controlling the computational scenario as a whole, including creation of compute tasks and their dependencies. The job manager is a lower level component which processes requests on actual job execution and provides a universal interface to different job execution environments, including job scheduling systems, Grid systems or clouds.

Workflow manager is implemented by a class gmManager, there can be multiple instances of it, each representing an independent workflow. The workflow is described by a directed acyclic graph (DAG) consisting of nodes and edges. The nodes and edges of the DAG are created and manipulated via interface function calls of gmManager. For example, to create a DAG, containing 3 nodes the following code may be used:

```
gmManager dag("test");
dag.node("A");
dag.node("B","A", gmStatusLink());
dag.node("C","A", gmFileLink("a.out","file_a.in"));
dag.link("A","C", gmFileLink("a.out","file_b.in"));
```

Fig. 1. GridMD code example and the corresponding DAG

At each time moment, the workflow graph stored in the client application by the instance of `gmManager` has its particular composition and state. This state may be exported in ".dot" format and visualized with a help of GraphViz [10] program. By default workflow manager is configured to update the related ".dot" file at each change of DAG. This is useful for debugging purpose to ensure the correctness of DAG composition and to understand the graph processing sequence. The graph state is also written in XML format to a graph state file, which simplifies checkpointing and restarting the GridMD application.

As usual for the workflow systems [3], the nodes of the workflow graph represent certain actions to be executed and its edges represent data dependencies. Generally, the actions associated with the nodes compose the compute jobs that are passed to job manager component for execution and monitoring. In GridMD, the nodes may optionally be configured as 'local', in this case their execution is performed directly by the client application:

```
dag.local_node("B"); // create a node as local
dag.set_node_property(&gmNodeProp::local,true,"A"); // reconfigure
as local
```

The local nodes usually represent some lightweight management actions on the client side to be included in the workflow graph, their 'locality' guarantees that their actions are not scheduled by the job manager but executed 'in place' when the DAG dependencies are satisfied. Note that for time consuming operations on the client side, rather than defining a local node, it is more practical to assign to the node a 'local shell' compute resource, which is operated by job manager in a standard way.

The actions associated with the node are setup via the workflow manager. A node may be assigned either of three types of actions: it may be formulated as a script, as a function or as a code fragment. The first type of node action simply defines a script (it may be a single command) to be executed on a target compute system when the node is run.

```
dag.set_node_action("date > a.out","A"); // script action:
put the date in 'a.out'
```

In some cases, for example when running a pre-configured external application on a target system, the call script requires system-specific tuning. This is accomplished by configuring the application section of the compute resource configuration (see Section 4).

For the pre-configured applications the action setup requires the corresponding configuration name:

```
dag.set_node_action("VASP MPI","A");// refers to 'VASP MPI'
comp. resource section
dag.set_node_property(&gmNodeProp::args,"INPUT","A"); //
specify arguments
dag.set_node_property(&gmNodeProp::np,10,"A"); //the number of
MPI processes
```

In the case of the local node, the script action is performed via a synchronous `system` call from the client application.

The second type of node action is a function, which may be programmed in the client application and defined by inheriting from `gmNodeAction` class and overriding its `OnExecute()` function:

```
class myAction: public gmNodeAction{
  GM_DECLARE_DYNAMIC_CLASS(myAction) // needed for remote execution only
public:
  virtual gmRESULT OnExecute(){...} // called when the node is run
};
...
dag.set_node_action(MyAction(),"B"); // function as node action
```

For the local nodes the node functions are executed in place by the same process that executes the client application. The functions may also be executed remotely; in this case a program, implementing all inherited node classes must be installed on a target compute system. We will call this program a *function server application*.

By default the function server application should have the same name as the client application and initialize GridMD on its startup (call `gmInitialize(argc,argv)` function). The initialization guarantees that nodal actions are constructed by their class names and called when certain node action is requested. The request on node execution is passed by command line arguments to the function server. In most cases *a copy of the client application* run in *worker mode* may serve as a function server because it already implements all necessary classes. The worker mode is then implicitly switched on by a first call to `gmInitialize(argc,argv)` which analyzes the program arguments, and in case they contain *worker mode flags* (`-w<node_spec>`) executes the required nodes and terminates the application. Note the compilation and installation of the function server on a remote system may be configured as a workflow node, in the future releases of GridMD a special support for deployment nodes is planned.

The third type of node action is the code fragment placed directly in the client application following the node creation and analyzing its return code. We will call this type of node creation an *implicit workflow specification*:

```
int main(int argc, char *argv){
  gmInitialize(argc,argv);
  return gridmd_main();
}

int gridmd_main(){ //combined client/worker                          A
  gmManager dag("test");                                             |
  if(dag.node("A")){ // construct a node                             |
    do_something(); // perform node action                          |file_from_A
    write_file("file_from_A"); // write output                       ▼
  }                                                                  B
  if(dag.local_node("B")){ // construct a node
    read_file("file_from_A"); // read input
    do_something_else(); // perform node action
  }
  dag.link("A","B", gmFileLink("file_from_A"));
  dag.execute(); // node actions not specified: use gridmd_main
  return 0;
}
```

Fig. 2. Implicit workflow specification and automatically generated graph

In the implicit workflow case the client application serves also as a function server, as described above for function calls, but in the worker mode it executes until a corresponding node() creation or process_node() function is called. When it is called, it returns true (only for the requested node and in worker mode) initiating the node actions. For implicit workflow all DAG node actions should be placed in a special gridmd_main() function, otherwise they will not be called. The nodal actions may as well be separated from the node creation, but still should be placed in gridmd_main() function. In this case the return codes of process_node() functions rather than node() functions should be utilized

```
  int main(int argc, char *argv){
    gmInitialize(argc,argv); // in worker mode calls gridmd_main() and
terminates
    // client part
    gmManager dag("test");
    dag.node("A"); // construct a node
    dag.local_node("B"); // construct a node
    dag.link("A","B", gmFileLink("file_from_A"));
    dag.execute(); // node actions not specified: use gridmd_main
  }

  int gridmd_main(){ // worker part
    gmManager dag("test"); // DAG  names should match with client
    if(dag.process_node("A")){ // true for all nodes with name "A"
      do_something(); // perform node action
      write_file("file_from_A"); // write output
    }
    if(dag.process_node("B")){ // true for all nodes with name "B"
      read_file("file_from_A"); // read input
```

```
    do_something_else(); // perform node action
  }
  return 0;
}
```

The implicit workflow mechanism is useful for creating small workflow programs or for modifying the existing computational code without much programming effort. For example, the workflow described above, may be considered as reprogramming the existing code:

```
do_something(); // hard computations
do_something_else(); // processing
```

for optionally delegating the first portion of work to remote compute server (node A) and using its result ("file_from_A") on a client (local node B). The implicit workflow creation was described in detail in our earlier works [1] and GridMD documentation.

Any type of node can be assigned different properties, affecting its execution. For example, the target compute system (compute resource) for a node may be manually specified (rather than left for automatic scheduling):

```
dag.set_node_property(&gmNodeProp::resource,"cluster1",id); //
assign resource
```

A special type of property is an arbitrary vector of character strings which the user is free to compose and assign to a node at any moment of graph execution. This array is stored and retrieved when the graph state file is written to a file, so it is some cases it is useful for storing the results of node execution.

```
std::vector<std::string> vec;
dag.get_node_string_data(vec, id); // get data
dag.set_node_string_data(vec, id); // set data
```

As seen from the above examples, the edges of the graph represent some data dependence between the node. There are four types of graph edges that are currently available in GridMD. A *hard link* requires that the nodes are executed by the same process (optionally by different threads), thus connects the nodes into a single supernode and runs them within the same compute job. A *file link* represents a requirement to move a file from the input node to the output node. The constructor of a file link has an option of manipulating the files (reformatting, renaming and using wildcards) when transferring them between nodes. A *status link* represents an order of execution only, not requiring to transfer any data. A data link is a special form of file link, which is formalized to transfer data of certain type. This type is used as a template parameter for constructing data links, and for composite types file read and write routines should be specialized.

```
dag.link(gmHardLink(),"A","B"); // hard link
dag.link(gmFileLink("output.dat","input.dat"),"A","B"); // file link
dag.link(gmFileLink("out*","./localdir/"),"A","B"); // file link
with wildcards
dag.link(gmStatusLink("?"),"A","B"); // status link
dag.link(gmDataLink<double>(),"A","B"); // data link for double
```

The workflow manager has internal scheduler that analyzes the graph and dispatches the node jobs between configured resources. The remote compute resources are communicated to gmManager via special commands or by reading configurations in the format supported by job manager (see section 3):

```
gmResource cluster(gmRES_PBS,gmSHELL_UNIX); // configure manually
cluster.session.host = "10.1.1.0";
cluster.session.login = ...;
dag.add_resource(cluster,"my_local_cluster");
dag.load_resources("resources.xml"); // read configuration from file
```

GridMD has a mechanism of tuning the scheduling algorithm managing different compute resources simultaneously. By default, a simple round-robin approach between the resources passing job requirements (installed software, number of nodes, etc.) is utilized.

At every time moment, the state of the workflow graph is defined by its composition and the states of individual nodes. These states represent the living cycle of node jobs. The execution of the jobs of the workflow graph is initiated by the execute() function of the workflow manager. Execution of jobs is usually an iterative process, implying communications with remote hosts. By default the execute() function returns when all the nodes have reached some final state ('processed' or 'failed').

Since the workflow manager functions are designed to be thread safe, the execute function may be run in a special thread to proceed with job execution. The workflow manager may then be operated from another thread to monitor or modify the workflow graph. Note that there is no restriction for *dynamical* modification of the workflow graph, by adding new nodes and edges. Another option of managing the execution is quantizing the scheduling operations into execute iterations. Each iteration consists of some closely connected portion of the scheduling work, for example a check of some job status. For the quantized execution the workflow operation in a single thread may be used as follows:

```
while(dag.execute_iteration()!=gmDONE){
  if(dag.job_state_changed()){
    // update job view
  }
  if(dag.has_errors()){
```

```
    // process errors
  }
  sleep(some_time); // optionally sleep between iterations
  ...                // or do something useful
};
```

The client application running graph execute iterations or waiting for execute function to complete may be interrupted at any moment and restarted using the graph state file:

```
// compose a graph
dag.node("a");
dag.node("b");
...
// load saved state
dag.load_state("mydag.xml");
```

Note that only the graph *state* (including all running jobs), but not its *composition* may be retrieved at the restart. This is explained by the complexity of possible modifications that may occur during the graph configuration, so the responsibility for restoring the graph composition at restart is left to the client application programmer.

3   Parameter Sweep Example

In this Section, we discuss a sample C++ code that uses the GridMD library to execute a bunch of external executables and measure their scheduling, running and other times. The code we present here is fully functional; advanced version of this code is available with the GridMD distribution as sweep_farm example, which may be used as a base for performance management experiments. The code works as follows: the user specifies an arbitrary unix command which is queued and performed as a job at a remote compute resource. When the job is finished, its timings (total run time, time in the queue, GridMD management time) are recorded in a summary table under unique name (workflow branch name). Each workflow branch is composed of three graph nodes: 'read', 'calculate' and 'collect'. The first one is responsible for parameter initialization, the second one for computation and the third one for storing the results. The table state, which is displayed after each user command, is maintained through different runs of the code: new entries (branches) may be added, some may be removed. The jobs are executed asynchronously, i.e. several jobs may run in parallel. The code may be terminated (command 'quit') and restarted again, in this case all existing from previous runs jobs and their results are recovered and displayed in the summary table.

The program is listed in Figures 3 to 6. We use explicit workflow creation here by assigning node actions as unix commands (user input) and callback functions

(ReadAction, CollectAction classes). We also demonstrate the dynamic workflow management: in this code we use the local 'read' node as a creator of the branch corresponding to one command test. Let us go through the code, commenting it contents.

The main program code, which is very brief, is listed in Figure 3. We construct the gmManager object specifying its name and the flag that triggers reloading the object state from a restart file. This will be needed to recover the previous jobs when the code is restarted. We also load the resource description from the file "resources.xml". This file contains all the information about where to run the remote jobs, in our simple example the jobs will be assigned the compute resource on a round-robin basis from all resources marked as active in the file "resources.xml". Then 3 local nodes are created, the last one ('end') is not connected to any node, 'read' is connected to 'start'. Local nodes are expected to run as part of the client process, we assign a ReadAction to the 'read' node, other nodes remain with no actions assigned (they will serve as connecting hubs for workflow graph edges). Then we execute the graph by the call to execute() function. If there were no previous runs, the graph state before this call corresponds to the one depicted in the inset of Figure 3.

When the 'read' node is executed, its callback function, ReadAction::OnExecute(), which is displayed in Figure 4, is called. In this node, inside the while loop, we read and interpret the user's commands. For this example we left only four possible user actions which are quitting the code, adding a new test, removing a test from the summary table and refreshing the table by pressing Enter when 'Command:' is prompted. The call to results_summary() (Figure 5) prints a table containing current states for all branches of the calculation.

```
#include <gridmd.h>

int main(){
    gmManager dag("sweep",true); //true = load restart info
    dag.load_resources("resources.xml");//where jobs are run
    //add initial nodes
    dag.local_node("start");
    dag.local_node("end",gmNODE_NONE);
    dag.local_node("read","start");
    dag.set_node_action(ReadAction());// attach read function
    dag.execute();
    return 0;
}
```

Fig. 3. Main code of the parameter sweep example and the initial workflow graph state generated automatically in course of code execution. On the graphs remote nodes are represented by rounded shapes and local nodes by squared shapes; green color means the nodes in 'processed' state, yellow – in 'executing' state, grey – 'blocked' state.

Although GridMD itself allows for duplicated node names (only node identifiers are unique), in our code the branch names specified by the user should be unique, this is checked by trying to find the branch with the same name after the user adds a new branch. If the check is passed, we find the current empty read node and add the branch name and the node command as data entries to the nodal string vector, thus activating it creation of the calculation and collect nodes of the branch (see the ReadAction code). To prevent the 'read' node from entering the 'PROCESSED' state,

```
class ReadAction: public gmNodeAction {
public:
  virtual gmRESULT OnExecute() {
    gmManager *dag = GetManager();
    int this_nodeid = GetParent()->GetID();
    std::string command, branch, execcmd;
    std::vector<std::string> data = dag->get_node_property_by_id(&gmNodeProp::string_data,
      this_nodeid);
    if(data.size()==3){ // this is recorded "add" command with 2 parameters
      command = data[0]; // "add"
      branch  = data[1];
      execcmd = data[2];
    }
    else{ // reading from user's input
      results_summary(*dag,"end");
      data.clear();
      do{
        cout << "Command (add/remove/quit): ";
        std::getline(cin,command);
        data.push_back(command); // record the command
        if(command == "add" || command=="remove"){
          cout << "Branch name: "; std::getline(cin,branch);
          if(branch == "")continue;
          data.push_back(branch);
          if(command=="add"){
            if(dag->find_node(std::string("calc ") + branch)
               != gmNODE_NONE){
              cout << "Duplicate name '" << branch << "' for branch\n"; continue;
            }
            cout<<"Execute: "; std::getline(cin,execcmd);
            if(execcmd == "")continue;
            data.push_back(execcmd);
          }
          else{ // remove
            // stoping calculation in case it is running
            dag->stop_node(std::string("calc ")+branch);
            if(dag->set_node_state(gmNS_WAIT,
                std::string("{read|calc|collect} ")+branch)!=3)
              cout<<"Branch '"<<branch<<"' not found\n";
            else
              dag->set_node_name("inactive",
                std::string("calc ")+branch);
          }
          break;
        }
        else if(command=="quit"){
          dag->stop_scheduling(); break;
        }
        else if(command =="")
          break;
      }while(true);
      dag->set_node_property(&gmNodeProp::string_data,data,this_nodeid);
    }
    if(command=="add"){
      // changing the name for this node
      dag->set_node_name(std::string("read ")+branch,this_nodeid);
      // adding calculation and collect nodes
      dag->node(std::string("calc ")+branch,this_nodeid,gmStatusLink());
      dag->set_node_action(std::string("/usr/bin/time -f\"%e\" -o timing ") + execcmd);
      dag->local_node(std::string("collect ")+branch,gmNODE_PREV,gmFileLink("timing",branch +
".time")); // tmp name
      dag->set_node_action(CollectAction());
      dag->link(gmStatusLink(),gmNODE_CUR,"end");
      // adding new read node for next parameter
      dag->local_node("read","start",gmHardLink());
      dag->set_node_action(ReadAction());
      return gmDONE;
    }
    return gmREDO;
  }
  virtual gmNodeAction* Clone() const { return new ReadAction(); }
};
```

Fig. 4. The code of 'read' node callback function and the intermediate workflow graph state generated automatically in course of code execution. On the graphs remote nodes are represented by rounded shapes and local nodes by squared shapes; green color means the nodes in 'processed' state, yellow – in 'executing' state, grey – 'blocked' state.

the return value of the callback function may be set as gmREDO. This will toggle restart of the node on the next scheduling iteration and the node will be able to read more user's commands. When the callback function returns gmDONE, the node is marked as processed and is not rerun until its state is not explicitly changed.

In addition to reading the user's input, the 'read' nodes serve for reloading the graph from the corresponding state file. The associated data is stored in the node string_data property which is recovered when the node is restored. When the restart file is specified to the gmManager object, it tries to find there the stored data for each of the newly created nodes. So, if the restart file is not empty, the first 'read' node will obtain its data at the time of creation (in the main function). Then, when its callback function is called, it triggers a creation of other read nodes.

When the 'add' command issued by user or read form the restart file is encountered, the 'read' node appends a branch containing 'calculation' and 'collect' nodes to itself. The branch name and the Unix command are the parameters that are used to compose the branch, the command is prepended by 'time' to measure the timing of the command directly. This direct timing is the main result of our sample calculation: it will then be compared with timing produced by GridMD. The measurement result has to be recorded into a file and then parsed by a function specified in Figure 6. The transfer of the file to the client side and its availability to the 'collect' node is provided by gmFileLink functionality. When the user prints 'quit', the execution of the graph is interrupted by a call to gmManager::stop_scheduling() function and the control returns to the main function.

```
inline gmRESULT parse_output_file(std::vector<std::string> &data,
                                   const std::string& filename){
   ifstream timestr(filename.c_str());
   if(!timestr.good())
     return gmFAIL;
   std::string value;
   timestr >> value;
   data.assign(1,value);
   timestr.close();
   return gmDONE;
}

class CollectAction: public gmNodeAction {
public:
   virtual gmRESULT OnExecute() {
     gmEdgeID edgeid = GetManager()->find_edge(GetParent()->GetID()-1,GetParent()->GetID());
     gmEdge *edge = GetManager()->get_edge_ptr(edgeid);
     if(!edge)
       return gmFAIL;
     std::string filename = edge->GetDestName();
     return parse_output_file(GetParent()->GetStringData(),filename);
   }
   virtual gmNodeAction* Clone() const { return new CollectAction(); }
};
```

Fig. 6. Auxiliary functions for the parameter sweep example: node action classes with callback function for 'collect' and 'read' nodes; tunable parse_output function transferring the results from file to the node string data vector.

```
#----------------------------------------------------------------------
#         Name         State     Unix_time   Gridmd_time   Queue_time   Aux_time
#----------------------------------------------------------------------
          test1        PROC        20.09        20.222        6.158         0.01
          test2        EXE            -1        -0.001        6.167        0.009
#----------------------------------------------------------------------

Command (add/remove/quit):
#----------------------------------------------------------------------
#         Name         State     Unix_time   Gridmd_time   Queue_time   Aux_time
#----------------------------------------------------------------------
          test1        PROC        20.09        20.222        6.158         0.01
          test2        PROC        25.09        25.221        6.167        0.009
#----------------------------------------------------------------------

Command (add/remove/quit):
```

Fig. 5. Output of the parameter sweep code (summary table). The first table shows one executing branch and corresponds to the graph state depicted in the inset of of Figure 4. The second table is displayed when the 'test2' branch is already processed, the timings are updated and added to the table.

An intermediate graph state after adding two branches (test1 and test2) is depicted in the inset of Figure 4, while the consecutive result tables for this graph are shown in Figure 5. In the tables the first column stands for the branch name, the second column is the state of the 'calculation' node of the branch (PROC means processed, EXE stands for executing), the third column corresponds to the direct execution time measurement (in seconds), and the rest columns contain GridMD-measured times of job execution. The procedures of GridMD time measurement are described in Section 4. The execution times are generally in good agreement (here with a difference of approximately 0.11 s which may be attributed to additional time for starting GridMD job scripts on the server side). The procedures for extracting time are given in the listing of Figure 6.

```
void results_summary(const gmManager &dag, const std::string &endnode){
  printf("#----------------------------------------------------------------------\n");
  printf("#         Name         State     Unix_time   Gridmd_time   Queue_time   Aux_time\n");
  printf("#----------------------------------------------------------------------\n");
  std::vector<gmEdgeID> edgeids;
  dag.select_edges(edgeids,".*",endnode);
  for(size_t i=0;i<edgeids.size();i++){
    gmNodeID collect_node = dag.get_source(edgeids[i]), read_node = collect_node-2, calc_node
= collect_node-1;
    std::vector<std::string> data = dag.get_node_property_by_id(&gmNodeProp::string_data,
read_node);
    if(data.size()!=3) // should have command, branch name and unix_command as data
      continue;
    std::vector<std::string> result = dag.get_node_property_by_id(&gmNodeProp::string_data,
collect_node);
    if(result.size()<1)
      result.push_back("-1");
    gmNODE_STATES state = dag.get_node_state_by_id(calc_node);
    if(state==gmNS_WAIT) // removed branch
      continue;
    gmJobTiming gridmd_times = dag.get_node_property_by_id(&gmNodeProp::timing, calc_node);
    printf("  %12s %12s %12s %12g %12g %12g\n",data[1].c_str(),gmGetStateName(state).c_str(),
          result[0].c_str(),(double)gridmd_times.running/1000.,
          (double)gridmd_times.queued/1000, (double)gridmd_times.submitting/1000.);
  }
  printf("#----------------------------------------------------------------------\n ");
}
```

Fig. 6. Function printing summary table based on the incoming edges of the node (endnode). GridMD job timings are requested via 'timing' property of the calculation node, while the direct unix time measurement is read from the 'collect' node data.

4   Job Manager and Low-Level Job Control

Job Manager is the GridMD library component responsible for running workflow elements on local or remote resources. It implies uploading of input data, submitting jobs to an external job management system, monitoring job statuses are retrieving results. When the application is run under control of the workflow manager the Job Manager is used implicitly. At the same time the Job Manager functions can be called directly from user code. It enables to use this GridMD component separately from the whole library as a low-level job management system.

At present the GridMD Job Manager support various types of external job queues and Grid middleware such as Portable Batch System (PBS), Simple Linux Utility for Resource Management (SLURM), Globus, etc. Simple jobs can be run on local or remote system just as a background process which reduces overhead caused by a job queue.

On initialization of a particular Job Manager instance the user should define a communication protocol to be used for remote command execution and data transfer. For remote systems the SSH protocol is implemented via the LibSSH library [11]. Alternatively the GridMD can use external applications such as plink and pscp from the PuTTY package for Windows [12] or build-in SSH client for Linux. The classes that implement communication protocols provide even lower level interface for file transfer, text file conversion, remote directory management and command execution which may be needed as example for configuring the remote system prior to submitting jobs.

The properties and status of each job is stored in the corresponding instance of the gmJob class which contain information about input and output files, commands to be executed, required CPU cores and other resources. An initialized job object can be submitted to any available Job Manager. After submission the job becomes linked to this particular manager and then the gmJob interface can be used to control the job, get its current status, upload additional files to the job temporary directory on the remote system or retrieve intermediate results, wait for the job completion.

A sample part of code that uses main Job Manager functions is given in the listing below:

```
1    // Job initialization
2    gmJob job;
3    job.command = gmdString("./user_prog input.dat resdir/res.dat");
4    job.AddInFile("user_prog");
5    job.AddInFile("input.dat", "", gmJob::TEXT);
6    job.AddOutFile("output.dat", "resdir/res.dat", gmJob::TEXT);
7    job.AddOutFile("stdout.txt", "STDOUT");
8    job.AddOutFile("stderr.txt", "STDERR");
9
10   // Job manager initialization
11   gmShellLibssh shell("username", "my_remote.system.org");
12   shell.SetParam("privkey", ".ssh/id_dsa");
13   gmPBSManager mngr(shell);
```

```
14
15      // Job submission and output of the GridMD and remote job
identifiers
16      state = job.Submit(mngr, "test_job", true);
17      if(state != JOB_SUBMITTED) return -1;
18      printf(" Job id = %s, subm_id = %s\n",
19              job.GetID().c_str(), job._subm_id().c_str());
20
21      // Inquiring job status and retrieving intermediate results
22      printf("Job state is %s\n", job.GetStateStr());
23      job.StageOut("output1.dat", "resdir/res.dat", gmShell::TEXT);
24
25      // Waiting for job completion and fetching the final results
26      puts("Waiting for the job to complete...");
27      state = job.Wait();
28      if(state == JOB_COMPLETED) state = job.FetchResult();
29
30      // Clearing gmJob structure and removing the remote temporary
directory
31      job.Clear();
```

The local job initialization part (lines 1-8) includes creation of the gmJob object, definition of a command to be executed on a working node of the remote cluster (line 3),
definition of input and output files (lines 4-8) including files that will receive all output
to stdout and stderr upon job completion. The communication protocol (SSH) and the
job manager (PBS-managed system at my_remote.system.org) are defined in lines 10-
13. Both objects have a few named parameters that can be set by SetParam() function.
When the job is submitted (line 16) a unique internal id is assigned to it which can be
used then to restore the job if the application is restarted. The internal id and the remote
system id assigned by the remote job management system are displyed in lines 18-19.
In course of job execution the main GridMD application can check the job status and
retrieve intermediate results such as log files as illustrated in lines 21-23. In lines 25-28
the main application is blocked until the job is completed. The the job.FetchResult()
function initiates downloading of all output files defined in the job object.

A typical sequence of events for this example at both the host system where the
GridMD application is running and the target system where the job actually executes
is presented in Figure . Initialization of the job object is made entirely on the host
system. The submission process involves execution of several command on the target
system and data transfer via the communication protocol Execute and StageIn calls.
First the temporary directory is created, then the input files are uploaded and finally
the job is submitted to the target system queue. As the job is submitted it can be managed from the host system. In our case it includes several job status inquires caused
low level Execute calls and downloading an intermediate result using the StageIn
call. When the job status is changed to 'Completed' the results are fetched and the job
object can be cleared which means also removal of the temporary directory at the
target system.

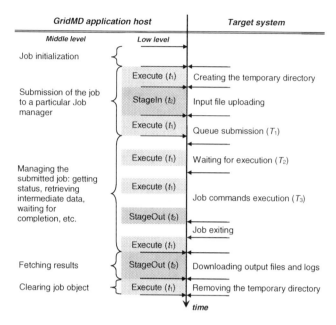

Fig. 7. Example of the job life cycle. Left side: operations with the job object (middle level) and communication protocol (low level) performed on the host system where the GridMD application is executed. Right side: operations on the target system.

In order to facilitate profiling of GridMD applications, timings of the main operations on the host and target systems can be obtained. The counters $t_1 - t_2$ (see Figure) accumulate delays caused by data transfers and execution of service commands on the target system. The times $T_1 - T_2$ are related to the performance of the target job management system and the queue load. Finally the time T_3 allows to benchmark execution of the user job commands.

An example of the library service timings measured for a test job are given in Table 1. The GridMD application was executed on two host systems working under Windows and Linux OS. The ping times between the host and target systems were (minimal/average/maximal): 5/5/6 ms for the Windows host and 4.8/5.3/6.3 ms for the Linux host. The input and output data sizes were 10 and 22 bytes respectively so the delays were caused mainly by service data transfers. The delay times were averaged over 15 job runs. These particular results confirm that the communication protocol implemented via the LibSSH library and supporting persistent SSH sessions has smaller overheads than the one based on usage of the external SSH/SCP client calls for each operation.

Table 1. Timings of the test job run which show typical overhead caused by GridMD service procedures

Communication protocol class/host operating system	Time of remote command execution (t_1), s	Time of data transfers (t_2), s	Overall service operations time $(t_1 + t_2)$, s	Total job execution time, s
Plink/Windows	4.25 ± 0.85	4.71 ± 0.53	8.97 ± 1.38	13.58
LibSSH/Windows	0.69 ± 0.18	0.73 ± 0.02	1.41 ± 0.20	5.40
SSH/Unix	1.63 ± 0.39	1.31 ± 0.19	2.93 ± 0.58	6.92
LibSSH/Unix	0.69 ± 0.13	0.72 ± 0.06	1.42 ± 0.18	5.40

5   Summary and Outlook

GridMD is a programming tool aimed at the developers of distributed software that utilizes local or remote compute capabilities to perform loosely coupled computational tasks. Unlike other workflow systems and platforms, GridMD is not integrated with heavy infrastructure such as Grid systems, web portals, user and resource management systems and databases. It is a very lightweight tool accessing and operating on a remote site by delegated user credentials.

For starting compute jobs the library supports Globus Grid environment; a set of cluster queuing managers such as PBS(Torque) or SLURM and Unix/Windows command shells. All job starting mechanisms may either be used locally or remotely via the integrated SSH protocol. Working with different queues, starting of parallel (MPI) jobs and changing job parameters is generically supported by the API. The jobs are started and monitored in a "passive" way, not requiring any special task management agents to be running or even installed on the remote system. The workflow execution is monitored by an application (task manager performing GridMD API calls) running on a client machine. Data transfer between different compute resources and from the client machine and a compute resource is performed by the exchange of files (gridftp or ssh channels). Task manager is able to checkpoint and restart the workflow and to recover from different types of errors without recalculating the whole workflow. Task manager itself can easily be terminated/restarted on the client machine or transferred to another client without breaking the workflow execution. Apart from the separated tasks such as command series or application launches, GridMD workflow may also manage integrated tasks that are described by the code compiled as part of task manager. Moreover, the integrated tasks may change the workflow dynamically by adding additional jobs or dependencies to the existing workflow graph. The dynamical management of the workflow graph is an essential feature of GridMD, which adds large flexibility for the programmer of the distributed scenarios.

Acknowledgements. The work is supported by Russian Foundation for Basic Research (grant No. 15-02-08493) the Programs of Fundamental Research of the Presidium of RAS No. 6, 13 and Program of Science Strategic Sectors No. 4. The computations with GridMD workflow management were performed in part on resources of the Joint Supercomputer Center of Russian Academy of Science.

References

1. Morozov, I.V., Valuev, I.A.: Automatic Distributed Workflow Generation with GridMD Library. Computer Physics Communications **182**, 2052–2058 (2011)
2. GridMD project. http://gridmd.sourceforge.net
3. van der Aalst, W.: The application of Petri nets to workflow management. The Journal of Circuits, Systems and Computers **8**(1), 21–66 (1998)
4. Pytlinski, J., Skorwider, L., Benedyczak, K., Wroński, M., Bała, P., Huber, V.: Uniform access to the distributed resources for the computational chemistry using UNICORE. In: Sloot, P.M., Abramson, D., Bogdanov, A.V., Gorbachev, Y.E., Dongarra, J., Zomaya, A.Y. (eds.) ICCS 2003, Part II. LNCS, vol. 2658, pp. 307–315. Springer, Heidelberg (2003)
5. Rogers, D., Harvey, I., Huu, T.T., Evans, K., Glatard, T., Kallel, I., Taylor, I., Montagnat, J., Jones, A., Harrison, A.: Bundle and Pool Architecture for Multi-Language, Robust, Scalable Workflow Executions. J. Grid Comput. **11**(3), 457–480 (2013)
6. Balasko, Á., Farkas, Z.K.P.: Building Science Gateway by Utilizing the Generic WS-PGRADE/gUSE Workflow System. Computer Science **14**(2), 307 (2013)
7. Wolstencroft, K., Haines, R., Fellows, D., et al.: The Taverna workflow suite: designing and executing workflows of Web Services on the desktop, web or in the cloud. Nucleic Acids Research **44**, W557–W561 (2013)
8. Bowers, S., Ludäscher, B.: Actor-Oriented Design of Scientific Workflows. In: Delcambre, L.M., Kop, C., Mayr, H.C., Mylopoulos, J., Pastor, Ó. (eds.) ER 2005. LNCS, vol. 3716, pp. 369–384. Springer, Heidelberg (2005)
9. Callaghan, S., Deelman, E., Gunter, D., et al.: Scaling up Workflow-based Applications. Journal of Computer and System **76**, 428–446 (2010)
10. Potapkin, B., Bogdanova, M., et al.: Simulation Platform for Multiscale and Multiphysics Modeling of OLEDs. Procedia Computer Science **29**, 740–753 (2014)
11. Valuev, I.A.: GridMD: program architecture for distributed molecular simulation. In: Hobbs, M., Goscinski, A.M., Zhou, W. (eds.) ICA3PP 2005. LNCS, vol. 3719, pp. 309–314. Springer, Heidelberg (2005)
12. Morozov, I.V., Valuev, I.A.: Distributed Applications from Scratch: Using GridMD Workflow Patterns. In: Shi, Y., van Albada, G.D., Dongarra, J., Sloot, P.M. (eds.) ICCS 2007, Part III. LNCS, vol. 4489, pp. 199–203. Springer, Heidelberg (2007)
13. Ellson, J., Gansner, E.R., Koutsofios, E., North, S.C., Gordon, W.: Graphviz and dynagraph – static and dynamic graph drawing tools. In: Graph Drawing Software (2003)
14. LibSSH project. http://www.libssh.org/
15. PuTTY project. http://www.chiark.greenend.org.uk/~sgtatham/putty/

A Parallel Algorithm for Efficient Solution of Stochastic Collection Equation

Elena N. Stankova[1,2(✉)] and Ilya A. Karpov[2]

[1] Saint-Petersburg State University, Peterhof, Universitetsky pr., 35 198504,
St.-Petersburg, Russia
lena@csa.ru
[2] Saint-Petersburg Electrotechnical University "LETI", ul.Professora Popova 5 197376,
St.-Petersburg, Russia
ilia.karpov@gmail.com

Abstract. A parallel algorithm is presented for the efficient numerical solution of the stochastic collection equation. It is based on Bott flux method for numerical solution of the stochastic collection equation. The algorithm of Bott has been chosen as one of the most popular algorithms intended for calculation of the evolution of cloud particle spectra. The optimized algorithm makes it possible to use multiple CPU cores for computation acceleration without significant accuracy loss and remains free from mass defect. Acceleration is achieved at the cost of reduced accuracy, because the steps of strictly sequential algorithm are executed in parallel. Tests showed a 3.5 time speedup on PC with four CPU cores. Results of the numerical tests show that the parallel algorithm is very promising to be used in the numerical models of convective clouds for calculating the spectra of cloud particles, such as water drops and various types of ice crystals. Stochastic collection equation presents the most computationally expensive part of such models aimed at forecasting thunderstorms, heavy rains and hails. Thus elaborated parallel algorithm can become an efficient instrument in the hardware and software systems designed for operational forecast of dangerous weather phenomena.

Keywords: Stochastic collection equation · Efficient algorithm · Meteorology · Numerical modeling · Multithreading

1 Introduction

Clouds largely affect [3] on the formation of weather and climate on Earth. Power and intensity of precipitation play an important role in people's life. Radiation characteristics of clouds significantly affect the transfer of radiant energy in the atmosphere. The latent heat of phase transitions, released in the process of precipitation formation, is the source of energy for atmospheric effects of different temporal and spatial scales, such as thunderstorms, squalls, powerful tropical cyclones and hurricanes.

Investigation of dangerous convective phenomena such as thunderstorms, hail and rain storms requires consideration of various processes having different nature and

© Springer International Publishing Switzerland 2015
O. Gervasi et al. (Eds.): ICCSA 2015, Part IV, LNCS 9158, pp. 290–298, 2015.
DOI: 10.1007/978-3-319-21410-8_22

scale and presents an extremely complex problem for numerical modeling [1,5]. Cloud model should reproduce both thermodynamical and microphysical processes. The former describe interaction of updraft and downdraft convective flows, turbulence vortexes and temperature variations. The latter describe transformations and interactions of small cloud particles – water drops, aerosols and various kinds if ice particles (ice crystals, snowflakes, graupel and frozen drops). Calculations of the whole set of the processes require a large number of computational resources and time, especially in case of using 2D and 3D cloud models. Numerical simulation of the microphysical processes is the most computationally expensive part in such models especially in case of the so-called "detailed" description of the microphysical process, which demands calculation of cloud particle spectra.

Droplet or ice particle spectrum evolves mainly due to the collection process, numerical description of which is conducted by means of numerical solution of stochastic collection equation (SCE). SCE is the complicated integro-differential equation describing the evolution of the spectrum (distribution function) of particles in dispersed medium through collision and subsequent coalescence. Analytical solution of such equation is possible only in case of the special kernel type; in other cases it should be solved numerically [6-8].

System of stochastic collection equations constitutes microphysical blocks of numerical models of natural convective clouds. They allow calculating evolution of the spectra of water droplets and ice particles through coagulation and thus describe the details of the process of precipitation that are very important for forecasting dangerous convective phenomena such as heavy rain, thunderstorm and hail [4].

Modern numerical models are widely used for rapid predication of such danger convective phenomena as storms, hail, shower and others. These forecasts play an important role in climatic disaster prevention in places like airports. The problem lies in the absence of supercomputers in this kind of places, therefore designing of models with low computing expenditures is the actual subject for scientists.

Usage of common models causes several disadvantages such as mass defect or spectrum widening. Thereby there is a need in model that both eliminates those problems and does not increase calculation time significantly.

The most expensive part of every model is the solution of stochastic collection equation. From this point of view, the optimization of this part is a primary question. The equation is given by

$$\frac{\partial n(x,t)}{\partial t} = \int_{x_0}^{x_1} n(x_c,t)K(x_c,x')n(x',t)dx' - \int_{x_0}^{\infty} n(x,t)K(x,x')n(x',t)dx' \quad (1)$$

where $n(x,t)$ is the drop number distribution function at time t and $K(x_c,x')$ is the collection kernel describing the rate at which a drop of mass $x_c = x - x'$ is collected by a drop of mass x', thus, forming a drop of mass x. Here, x_0 is the mass of the smallest drop being involved in the collection process and $x_1 = x/2$ [1].

One of the most popular schemes for the numerical solution of the SCE has been developed by Berry and Reinhardt (1974). In this approach the SCE is solved at discrete points of the drop spectrum. If necessary, high order Lagrangian polynomials are used to interpolate the drop spectrum at intermediate points. Seeßelberg et al. (1996) presented a stochastic simulation algorithm for the solution to the SCE. The main purpose of their method is to produce benchmark calculations for situations where no analytical solutions of the SCE are available. By comparing their stochastic simulation algorithm with the Berry–Reinhardt scheme, Seeßelberg et al. found a very good agreement between both methods, thus demonstrating the high accuracy of the latter scheme [1].

Sequential algorithms which are used for SCE numerical solution, as a rule, have the computational complexity of $O(N^3)$ order, where N is the total number of spectrum bins or intervals. These algorithms are quite computationally expensive, even for calculation of one type of cloud particles. Computations of the spectra of several particle types in each spatial grid point of 2D and 3D model demand tremendous computational resources and time. This timing does not allow the use of the models in the operational practice for prediction of dangerous convective phenomena such as thunderstorms, hail and heavy rain. The way out is the optimization and parallelization of the sequential algorithms for SCE numerical solution [3].

2   Method of Bott

Not long ago Andreas Bott [2] has introduced new algorithm, which is being the most adequate for the described problem at present time, as long as it has acceptable accuracy and is not vulnerable for mass defect and spectrum widening. Moreover, it fits better to physical world view.

As constructed [2], the Berry–Reinhardt scheme is not mass conservative. In addition, it is computationally relatively time-consuming. These are two serious disadvantages when the scheme is implemented into a dynamic cloud model with detailed microphysics. Furthermore, since the scheme solves the SCE in an integral way, one gets no information on the collision partners being involved in a particular collision–coalescence process. However, such information is necessary if, for instance, one is interested in the redistribution of aerosol mass or other chemical substances that are contained within the colliding droplets. In order to avoid these shortcomings, the new flux method was proposed for the numerical solution of the SCE. The new scheme is exactly mass conservative and is computationally very efficient.

The exact mass conservation as well as the numerical stability and efficiency of the flux method are the most important advantages when the model is compared with other schemes. Therefore, it is very attractive alternative for solving the SCE in dynamic cloud models dealing with explicit microphysics.

The method consists of a two-step procedure [2]. In the first step the mass distribution of drops with mass x' that have been newly formed in a collision process is added to grid box k with $x_k \leq x' \leq x_{k+1}$. By solving an advection equation, in the second step a certain fraction of the cloud water mass is transported from k to $k + 1$. The procedure is schematically illustrated in Fig. 1. Here, the dashed lines indicate the initial mass distributions after the collision process. The stippled area in grid box k represents the mass that will be transported into grid box $k + 1$, while the dark shaded areas are the final mass increase in grid boxes k and $k + 1$.

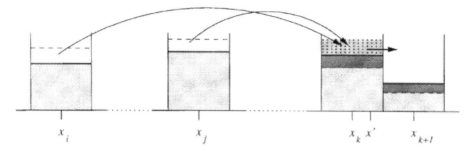

Fig. 1. Schematic illustration of the algorithm

Algorithm could be represented in pseudo code as follows:

```
for (var t = 0; t <= TimeToComputeInSec; t += StepInSec)
  for (var i = 0; i < GridSize - 1; i += 1)
    for (var j = i + 1; j < GridSize; j += 1)
      collide(i, j);
```

3   Modified Algorithm

One approach to accelerate computation is to perform calculations in parallel. The majority of modern personal computers are equipped with multicore CPU, which allows concurrent computations. Despite all advantages of Bott method, it is strongly sequential algorithm. In other words, on each step the previous result is required.

According to algorithm pseudo code the variable "t" is only used for repetition of inner computation, so concurrent processing of main loop is rather useless. As a result, there remains the only way to run code in parallel – calculate collisions on each time step in different threads. This approach will definitely lead to inaccuracy, thus, to achieve acceptable results modified algorithm should be invulnerable to mass defect and spectrum widening along with respectable accuracy and computational cost. Extensive experiments on wide range of input data have shown that the described approach is free from these disadvantages while at the same time improves computation performance.

The main idea of the algorithm is to create a queue of "collisions". Due to constant grid size, this queue is only created at initial phase of the algorithm. That means this is not exactly the "queue", it is more like an array of collision task definitions. Each thread gets next task and performs it. This process repeats while queue has incomplete tasks.

Shared memory access from different threads can significantly reduce computation rate. In order to avoid these locks we have designed our algorithm so that each thread uses its own local memory for storing local grid changes. After completion of all tasks, threads merge its results into global grid (without any locks).

Proposed algorithm consists of three steps:

- Initial phase
- Collisions computation
- Results synchronization

In the first phase (Fig. 2) initial environment setup is performed: initial spectrum extracted from source, local memory is allocated for threads and all drops collisions are queued. All the memory, required for processing, is only allocated in this phase, i.e. there will be no more memory allocations in other phases. This gives additional performance boost, because memory allocation calls are known to be slow.

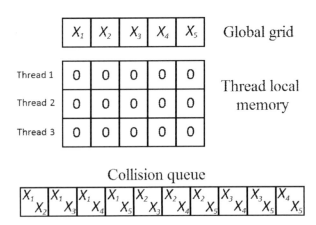

Fig. 2. Initial phase of modified algorithm

In the second phase (Fig. 3) collision queue is equally splitted up between threads. Each thread uses global grid as a read-only source whilst writes computational results into its own local memory. Doing so allows threads not to use shared memory locks and avoids data race conditions.

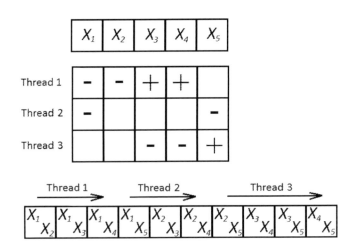

Fig. 3. Second phase of modified algorithm

The computation of each collision occurs in change of the distribution of mass in the four cells. Cells, in which the mass will be decreased, are the parameters of a collision. Cells, in which the mass is increased, calculated according to the rules specified in paragraph 2 of the present paper.

Thread scheduling requires additional CPU time, so the amount of simultaneously running threads should be limited in accordance with the number of the processor cores. The N-core processor can achieve the best performance executing N threads. If CPU supports Hyper-Threading technology, then the amount of threads can be multiplied by 2.

In the third phase (Fig. 4) local data from all threads is merging into global grid in the same manner (in parallel, dividing work between threads).

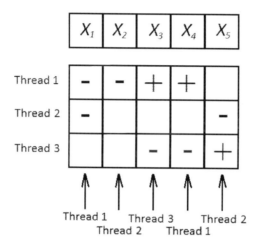

Fig. 4. Third phase of modified algorithm

Phases 2 and 3 are repeated a predetermined number of times in the main loop.

Proposed algorithm was implemented in C# program language with Task Parallel Library used for parallelization.

4   Experiments

In our experiments original algorithm implementation was used as a reference. Simulations were run on different grid sizes: from 50 cells and up to 250.

Testing results are presented in Table 1.

Table 1. Computation time in milliseconds obtained by compared algorithms

Implementation	Grid size				
	50	70	100	150	250
Original, ms	450	757	1534	3276	9824
Optimized, ms	142	304	469	1034	2731

As it can be seen from the table, a 3.5 speedup factor was achieved on CPU with four cores. The speedup value is growing along with grid size, which is good, due to the fact, that in production scenarios the larger grids will be used and performance gains may be even greater.

In the next experiment we study dependency between CPU threads and achieved performance increase. The grid with 250 cells and CPU with four cores were used. Designed algorithm is able to configure the amount of threads in use. Speedup chart is shown in Fig. 5. Here, each value is ratio of computational time of original algorithm to time of modified algorithm with provided amount of threads.

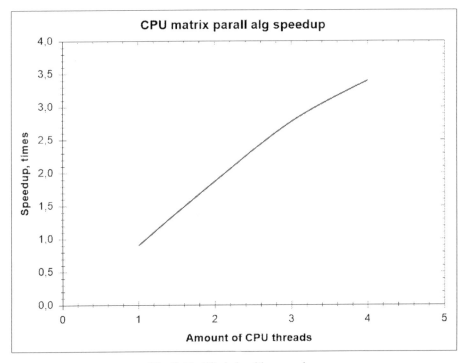

Fig. 5. Modified algorithm speedup

As it can be seen from the plot, the overhead of using multiple CPU threads is not significant. This observation leads us to consider using GPU instead of CPU.

Proposed algorithm accuracy lies in the acceptable range. Chart in Fig. 6 shows correlation of the compared algorithms. Here, the dashed lines indicate the resultant mass distribution after 1800 and 3600 seconds of collision process for modified algorithm, whilst the solid lines demonstrate the same for original method implementation.

For the present time, the error could not be completely removed, because the flow of the original algorithm has been significantly changed. The only way for reducing it is to increase synchronization between CPU threads, which in turn will reduce performance gains. The main idea is to find optimal combination of speedup and accuracy.

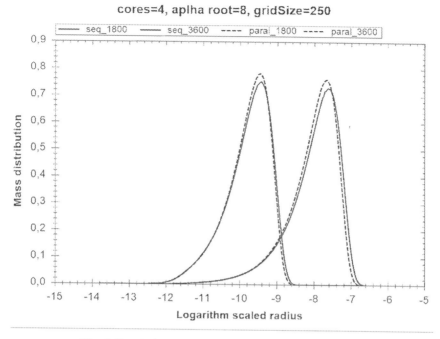

Fig. 6. Correlation of the results of the compared algorithms

The big advantage of method of Bott is in avoidance of mass defect. This is one of the most important reason for using it in modern models and it should be preserved.

Experiments has showed that modified algorithm has no mass defect, regardless of amount of CPU threads in use. Results are summarized in the Table 2.

Table 2. Mass defect value in $g(gm^{-3})$ obtained by compared algorithms

Implementation	Amount of CPU threads			
	1	2	3	4
Original, $g(gm^{-3})$	0			
Optimized, $g(gm^{-3})$	0	0	0	0

5   Conclusions

In this paper we propose a modification of well-known Bott algorithm that allows it to take advantage of modern CPUs with multiple cores by parallelizing algorithm iterations. Modified algorithm is able to achieve 3.5 speedup factor on CPU with four cores. Experiments have shown that algorithm is free from mass defect while having acceptable accuracy.

The proposed modification is able to run on CPU with different amount of computational cores and any grid sizes.

From our experiments, especially from chart on Fig. 5, it can be seen that increasing amount of threads will also increase speedup factor. Therefore, algorithm adaptation for the GPU is an interesting subject.

There are still areas where proposed algorithm can be improved. As a continuation of presented work we are planning to investigate algorithm adaptation for GPU and further study possibilities to decrease error rate.

Acknowledgments. The authors wish to thank Dmitry Karpov for helpful discussions in search of the best performance solution.

References

1. Khain, A., Ovtchinnikov, M., Pinsky, M., Pokrovsky, A., Krugliak, H.: Notes on the state-of-the-art numerical modeling of cloud microphysics Review. Atmospheric Research **55**, 159–224 (2000)
2. Bott, A.: A flux method for the numerical solution of the stochastic collection equation. J. Atmos. Sci. **55**, 2284–2293 (1997)
3. Raba, N.O., Stankova, E.N., et al.: Two parallel algorithms for effective calculation of the precipitation particle spectra in elaborated numerical models of convective clouds. In: Murgante, B. (ed.) ICCSA 2014, Part VI. LNCS, vol. 8584, pp. 289–299. Springer, Heidelberg (2014)
4. Raba, N.O., Stankova, E.N.: On the effectiveness of using the GPU for numerical solution of stochastic collection equation. In: Murgante, B., Misra, S., Carlini, M., Torre, C.M., Nguyen, H.-Q., Taniar, D., Apduhan, B.O., Gervasi, O. (eds.) ICCSA 2013, Part V. LNCS, vol. 7975, pp. 248–258. Springer, Heidelberg (2013). doi:10.1007/978-3-642-39640-3_18
5. Kogan, Y., Mazin, I.P., Sergeev, B.N., Khvorostyanov, V.I.: Numerical cloud modeling. Gidrometeoizdat, 183 p, Moscow (1984).
6. Pruppacher, H.R., Klett, J.D.: Microphysics of Clouds and Precipitation. Kluwer Academic, p. 954 (1997)
7. Tzivion, S., Feingold, G., Levin, Z.: An efficient numerical solution to the stochastic collection equation. J. Atmos. Sci. **44**, 3139–3149 (1987)
8. Prat, O.P., Barros, A.P.: A Robust Numerical Solution of the Stochastic Collection-Breakup Equation for Warm Rain. J. of Applied Meteorology and Climatology **46**, 1480–2014 (2007)

Profiling Scheduler for Efficient Resource Utilization

Alexander Bogdanov[1], Vladimir Gaiduchok[2(✉)],
Nabil Ahmed[2], Amissi Cubahiro[2], and Ivan Gankevich[1]

[1] Saint Petersburg State University, Saint Petersburg, Russia
[2] Saint Petersburg Electrotechnical University "LETI", Saint Petersburg, Russia
gvladimiru@gmail.com

Abstract. Optimal resource utilization is one of the most important and most challenging tasks for computational centers. A typical contemporary center includes several clusters. These clusters are used by many clients. So, administrators should set resource sharing policies that will meet different requirements of different groups of users. Users want to compute their tasks fast while organizations want their resources to be utilized efficiently. Traditional schedulers do not allow administrator to efficiently solve these problems in that way. Dynamic resource reallocation can improve the efficiency of system utilization while profiling running applications can generate important statistical data that can be used in order to optimize future application usage. These are basic advantages of a new scheduler that are discussed in this paper.

Keywords: Computational cluster · Scheduler · HPC · Profiling · Resource sharing · Load balancing · Networking

1 Introduction

Clusters for scientific calculations became widespread during the last decades. Initially, they were specific to government laboratories, large companies or major universities. But nowadays many non-commercial organizations even with a limited budget can afford a computational cluster. Such clusters usually consist of many homogeneous, relatively inexpensive nodes. This approach became quite common for small universities as well as for prominent IT companies: it allows organizations to reduce total cost of ownership and facilitate administrative tasks. Such systems are much cheaper than big SMP nodes. The same can be said about hardware upgrades due to the fact that nodes consist of common, widespread components and many manufacturers produce it. Actually, this fact means that organizations will probably not get vendor lock-in. The next advantage is scalability. SMP systems are known to have limited scalability while clusters can be easily scaled up to thousands of cores.

The same situation can be found in the supercomputer area. The evolution of the TOP-500 list is an illustrative example. If in 2000 (June 2000 list) only 2.2% of the systems have the cluster architecture (providing 3.5% of overall

© Springer International Publishing Switzerland 2015
O. Gervasi et al. (Eds.): ICCSA 2015, Part IV, LNCS 9158, pp. 299–310, 2015.
DOI: 10.1007/978-3-319-21410-8_23

performance of the systems in the list), in 2014 (November 2014 list) 85.8% of systems were clusters (providing 67% of overall performance) [1].

This paper concerns effective computational cluster utilization. It implies that organization will not get resources underloaded or overloaded while users will be able to compute their tasks fast.

2   Computational Infrastructure Problems

Despite the fact that clusters provide users and administrators with many advantages which other architectures have never had, one can face with many issues while creating, maintaining and using a computational cluster. Sometimes it is really difficult to create a reliable computational infrastructure. Some major concerns and problems of the cluster architecture are listed below.

- **Planning.** An HPC organization should pay much attention to infrastructure planning and upgrading. Decisions at this stage determine how flexible and reliable the computational infrastructure will be. There could be many questions. How many cores per node will be acceptable (should we use big SMP nodes)? How much RAM should be installed (a compromise between requirements and cost)? Which vendor offers best price to performance ratio? Should accelerators (e.g. GPGPU) be used (will user programs harness all GPGPU powers)? There is a tendency in the HPC world to use hybrid clusters for scientific computations [2]. Nowadays (the data from the last TOP-500 list, November 2014), 15% of the TOP-500 systems are equipped with some accelerator [1]. And 34.5% of the TOP-500 overall performance is contributed by hybrid systems (one can compare the present situation with the situation in 2009 when systems with accelerators contribute only about 5% of the TOP-500 overall performance). Moreover, some clusters have several accelerators per node (e.g. multi-GPU nodes) as well as nodes with different accelerators (although they usually belong to different clusters). There are several kinds of computational accelerators. It can be a GPGPU, MIC or some other accelerator. Such hardware is used in order to increase the system performance: accelerators can speedup scientific applications; they usually contribute a high percentage of the peak performance of the hybrid system. But sometimes they just can not help. The first possible problem is the algorithm being used: if it can not be parallelized or a program that implements it has a substantial sequential fraction, the manycore device will be underloaded almost all the time of a program run. One should always remember Amdahl's law. But even in case of good algorithm (in terms of parallelization) there can be other issues. For example, data transfer from the RAM to the device memory and vice versa. It can be a possible bottleneck. Despite the fact that contemporary accelerators can give a substantial speedup, one should carefully estimate the possible benefits. Are there enough software that supports GPGPU? Will users run it? Which algorithms does it use? Answering these questions will help to make the decision on the hybrid systems usage.

– **Data Storage.** Should it be a local storage, SAN or NAS? Which protocols to use? Should a parallel file system be used? One should think about many factors while planning the data storage (e.g. data consolidation). The final goal is to alleviate all possible bottleneck which can be caused by slow reading/writing and provide users with comfortable and transparent access to their data. The specific solutions depend on computational center objectives, specialization and user tasks. Different computational centers work with different amounts of data. Data sets differ too. The most challenging case is working with big data, but not every center will face it.

– **Networking.** Cluster architecture offers great scalability, but introduces some new problems, one of which is networking. Networking can become a bottleneck. One of the main factors is the ratio of CPU performance to networking performance. Increasing the number of processes that participate in a computations will not lead to speedup if the cluster network has low performance. Moreover, it could lead to slowdown. In order to alleviate this problem one should use algorithms with as few synchronization as possible. But such requirement could not be suitable taking into account great variety of user applications. Custom interconnect is a big part of the cost of a supercomputer. But even small computational center can consider networking as a very important factor in the effectiveness of its work. It is advisable to use high performance networking for clusters. There can be two separate networks: a network for management and maintenance tasks (e.g. Ethernet 1G) and separate computational network (e.g. Infiniband FDR). Some aspects of networking in computational center are discussed later in this paper.

– **User Friendly Interface.** Not every user is a computer science specialist, they just want to calculate their tasks (e.g. scientific calculations in case of university computational center). Nodes of computational clusters are usually installed with some Linux distribution. Traditional cluster management systems (e.g. some implementation of PBS) are usually have text interface as a default or the only interface, but inexperienced users are not accustomed to work with the console. So, it takes time for them to learn and get accustomed. It is not convenient neither for users nor for organization because users can do a lot of mistakes at that time (while working with the clusters). In order to solve this problem organization can install some additional packages that will ease task submission (e.g. web interface). It can also give users virtual machines. It is more convenient and secure than single server that provide users with an access to computational resource. While virtual machines can be used as computational resources too. Such approach can be very convenient and beneficial for users as well as for organization [3]. There is also another big problem: regardless of the interface, users should choose appropriate resources for their tasks. Making a decision on computational resources can be difficult. Requesting too much resources could lead to poor cluster utilization. But users want to calculate their tasks fast. So, organization should find a compromise solution that is acceptable for different users, provides fair resource sharing and high resource utilization. That is why the following question becomes one of the crucial questions for computational center.

– **Resource Sharing, Resource Monitoring, Resource Utilization.** These are the main questions of this paper, they are discussed below.
– **Security Questions.** Security issues is very important too, but they are out of scope of this article.
– **Other Questions.**

3   Resource Sharing

HPC resource in general can be shared by different groups of users that use open and closed source software packages, that in turn use various standards which makes efficient resource sharing a challenging problem. One should take into account every scientific application separately. Solving this problem seems to be impossible without intelligent software.

One of the possible solutions is to use a profiling scheduler. Such scheduler can collect information about running programs. These information reflects usage efficiency and can be used for future application run.

All in all, we can distinguish the following approaches that are used in order to share computational resources.

– **No Management System.** One can install operating system on computational nodes and grant access to nodes to users. It is the simplest approach from infrastructure point of view, but it is difficult to do resource planning in the described system.
– **Single System Image.** There are several packages that allow administrator to create such system. Several computational nodes will be viewed as a single system. Such systems are quite convenient but they also have drawbacks: computational resources at the physical level are separate, so questions on scaling up emerge. Users do not pay attention to this fact, so they can use some API for SMP systems and get small speedup or even slowdown. Another possible problem is failure of one node, because it could lead to the restart of the whole system.
– **Cloud.** There are several well-known and reliable solutions that can be used in order to create cloud. Many organization uses this approach due to the advantages that it offers (e.g. relatively simple infrastructure management). But such approach implies virtualization overheads. In addition, there are still some questions about load balancing (e.g. virtual machines migration).
– **Classical Management System.** We consider portable batch system (TORQUE, PBS Professional) and similar systems (HTCondor, Univa Grid Engine) as classical because they offer a traditional approach and are widely used in many computational centers [4] [5]. This approach is a base for a new one that will be discussed in this paper. Such system will be described in the next chapter.
– **Other Approaches.**

4   Traditional Approach

Such approach implies reservation of available computational resources: users request some resources by submitting their jobs. We will discuss this approach on the example of PBS system.

The key notion of PBS is a queue. Cluster nodes are assigned to some queue. There are usually several queues in such systems. One node can belong to several queues at the same time. Queues have different parameters: priority, resource limitations, time limitations, and so forth. Cluster administrator declares some policy for job queuing. According to this policy, he creates queues and assign necessary parameters to them. For example, he can create a queue for fast jobs with weak resource limitations and queue for jobs that requires a long time to complete with severe resource limitations or low priority. The common PBS system can be viewed as shown in Figure 1.

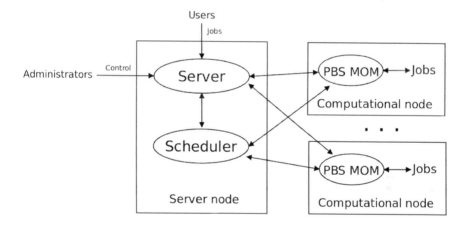

Fig. 1. PBS Scheme

Many organizations across the world use some PBS implementation or similar system to organize resource sharing. PBS provides managers and administrators with relatively simple solution in terms of resource sharing policies as well as administration efforts. It allow users to submit their jobs and specify resource requirements for them.

The PBS scheduler decides at which time and on which resources a job will be run. When it is possible (according to the current resource reservation and policies) the scheduler informs the PBS server. The PBS server, in turn, starts the job on allocated resources. Since that moment the resources that were allocated for the job are reserved. No one else can use those resources (even another job of the same user).

Such approach is quite clear and simple, but it has a drawback. The PBS usually does not monitor dynamic resource load. It monitor available resources only in terms of reservation and requests. The PBS system relies on resource characteristics that were initially assigned by the administrator. Such resources as CPUs, GPUs, the overall amount of RAM will rarely be changed. And when there was some hardware upgrade the system administrator can specify new resource characteristics for PBS. But one can not retrieve the information about the actual resource load. The only thing that a PBS implementation depicts is jobs and resources assigned to it. One can not figure out the current load of nodes (e.g. load average value) using PBS utilities. One can assume that the amount of resources reserved for a job reflects the actual resource usage. But it could not be the case, for example, when a user application that is started within the job is not parallelized very well, or user made a mistake when writing that job. So, when administrator use the PBS he relies on users: users should submit correct jobs and predict the best resource amount in order to reserve only really necessary resources.

The other drawback is the fact that an average PBS implementation provides administrators and managers with scarce accounting information. One can find out only overall resource usage (e.g. CPU time) and only when job is finished. One can not find the detailed information, for example, load on different nodes that participated in computations. Moreover, that scarce information is usually stored simply in log file along with other log messages. If administrator wants to get more user friendly interface he should write it (e.g. some script) or use third-party one. Finally, there is no reliable way to determine particular applications that were executed inside job script. This is the another case when administrator relies on users: they should name their jobs appropriately in order to reflect which application is used within a job. But, of course, users can name their jobs in some different way, the way they think to be the most convenient for them. So, the name of a job is not a reliable source of information. Moreover, users can run several different applications within a single job.

The notion of a job is user-defined anyway. A user determines what will compose a single job. But the organization managers want to know application usage information. For example, they want to know which application is the most frequently used or which one has the worst CPU time to wall clock time ratio. Such information can be very useful for planning the HPC infrastructure (for example, when planning some hardware upgrades or purchase of some new software licenses). The first problem is solved when software package comes with a license server. The administrator can estimate the application usage from the license server log file (calculate the number of license checkouts). But proprietary software may have no license server (for example, some paid closed source scientific packages have no license server), let alone open source software. So, administrator just can not count the application usage in this case. They can write some sophisticated prologue script (in terms of PBS) which will try to find out the job content, but such approach is difficult and not quite reliable, because a job script can call many other scripts and depends on environment variables (for example, a string with application call without absolute path in a script can actually run different versions of the application depending on PATH variable). It is really hard to find out which versions of which applications will be executed in the job in this case. So, standard PBS utilities of an average PBS implementation can not determine which applications are executed within the job while such information can be very helpful for the future HPC infrastructure planning and for the annual reports. Of course, profiling can be done using third-party software (not by a PBS implementation), but such software will not be able to map resource usage to PBS jobs.

The detailed accounting information could be used in another way. One can analyze it in order to improve the overall resource usage. Two similar jobs can show different performance despite the fact that user requested the same resources for them. Despite the fact the CPU cores count was the same the first job, for example, could be executed on a single node while the second on separate nodes in the network. Or different nodes were assigned in the first run and in the second run. And even in case of exactly the same allocated resources it is possible to get different performance because of shared resources like network.

Accounting information can also give the representation about the underlying levels performance (underlying levels that some application uses). As an example, MPI implementation. There are many different MPI implementations that varies greatly in performance: one of them can support Infiniband RDMA and another (probably old version) can not (work only with IP over Infiniband). As a result, the same application run can yield different performance when using different dynamic MPI libraries. Such situation can be figured out by profiling applications.

Figures 2 and 3 illustrate test runs of an MPI application (Intel IMB test PingPong) that were executed on two separate clusters (see Table 1) using TCP/IP (over Ethernet), IPoIB (IP over Infiniband) and RDMA protocol (Infiniband).

Table 1. Cluster characteristics

	Cluster 1	Cluster 2
CPU	2x Intel E5335 2.0 GHz	2x Intel X5650 2.67 GHz
GPGPU	-	3x (8x) NVIDIA Tesla M2050
RAM (GB)	16	96
HDD (GB)	160	120
Network	Ethernet 1G,	2x Ethernet 1G (bond interface),
	Infiniband 4x DDR	Infiniband 4x QDR
Total	768 Tb RAM, 48 nodes,	2.3 Tb RAM, 24 nodes,
characteristics	384 cores	288 cores, 112 GPUs
Peak performance	3.07	59.6
(Tflops)		

As one can see, using RDMA is much more beneficial than IPoIB. Infiniband RDMA is a special protocol that was designed for Infiniband systems (details on working with Infiniband could be found in many sources, e.g. [6]). But RDMA should be supported by MPI implementation. So, planning the computational

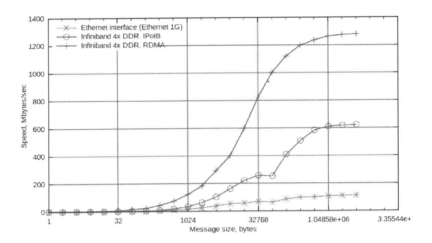

Fig. 2. Network test on the cluster 1

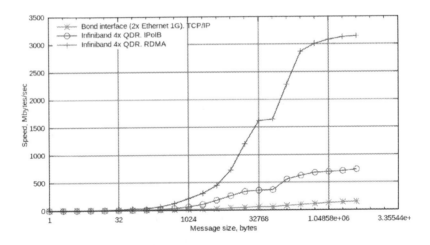

Fig. 3. Network test on the cluster 2

infrastructure is not limited to hardware, it also concerns system software, as well as application software. .

All in all, PBS promotes a way of conventional resource sharing that is quite simple and clear. But such approach works efficiently only under certain assumptions. It does not handle possible user errors efficiently: one can imagine the situation when user requests 128 cores for his job, but the job uses only one CPU core due to some mistake, let alone wrong estimations when users can not predict correctly the necessary amount of resources (for example, inexperienced user requests many cores for a serial job). The opposite situation is one when user application utilize more resources than it was requested. Such situations could be find out using detailed monitoring or profiling which is not specific to the PBS. As a result computational resources can be underloaded or overloaded. The new scheduler was designed in order to solve these problems.

5   New Approach

The new approach is based on profiling. It provides administrator with detailed accounting information about running jobs. The information retrieved from the profiling is used in order to dynamically reallocate the available resources. The main purpose of creating this scheduler is to overcome PBS drawbacks while trying to keep the simplicity of administration and utilization and do not introduce substantial overheads.

This scheduler is designed to be modular. The work is still in progress, we implemented basic functionality and tested it on several nodes at our faculty. Here are the basic features of the scheduler.

- **Profiling, Detailed Accounting and Monitoring.** All jobs are traced by the corresponding module (accounting module). This module starts the job and captures events of new process creation (using fork, vfork or clone system call), stop or execution (when a new binary starts within the process). The main idea is to gather statistical data about each process of the job (binary that was executed, CPU time, memory usage etc.). One can consider it as unnecessary, but such approach allows one to analyze user jobs and find possible bottlenecks. Moreover, statistical data can be used for predictions. Only a few programs perform the analysis of their run. We can collect performance data without the necessity to go deeply into the source code or trace each program step. So, such accounting is suitable for open source programs as well as closed source ones.
- **Dynamic Resource Allocation.** Our system provides flexible reservation of all available resources. This means that the amount of resources user requests will be treated as hints, not precise numbers. Our system will monitor the jobs and check whether jobs meet the requirements specified by users. A proper computational job should correspond to the resources requested for it. But an average job can underload the requested resources or overload them. Using conventional PBS one will have free unused resources in case of underload, and users which jobs are still in a queue. While in case of overload one will have conflict of interests resulting in increase of job completion time. In the both cases one have inefficient resource utilization. Our system will start another job on the underloaded resources (which requirements meet the hardware characteristics) while giving the initial job higher priority (if resources become overloaded by these two jobs, the second one will be suspended). In case of overload our system offers to administrator three possible options: stop the job and send an appropriate message to the user, suspend this job and run another or keep the job running and log this information.

Such flexible approach allows administrators to alleviate user mistakes and improve resource utilization.

- **Rating.** Our system will have a module for keeping user rating. We propose this simple feature which is rarely used when we talk about clusters for scientific computations. User rating will reflect the efficiency of user jobs. Experienced user can estimate the necessary resources for his job, so running such job will not lead to underload or overload. If some job consumes more resources than it was requested, the rating of the job owner will be decreased. And if some user requests more resources than he actually needs, the rating of such user will be decreased too. The goal of such rating is to improve the priority and limitation policies: rating will impact jobs priority and resource limitations for each user. Detailed accounting, once again, can help users to find the possible problems of their jobs.

- **Predictions.** The predictions module can automate resource requests. Such module will offer to user possible amount of the resources that is necessary for his job or even make such requests instead of user (user will have to specify the job only). Of course, the performance of an application depends on the exact task (input data). But one can find out the general tendencies that are specific to the application. For example the fraction of the sequential part will impact the highest speedup that can be achieved (Amdahl's law). The underlying levels can impact the performance too. One can remember the example from the previous section that depicts MPI implementation features. Prediction module can estimate the necessary resources in accordance to the used libraries. And finally it can mark the applications that can use specific hardware (GPGPU) in order to run such applications on the necessary nodes. These can be figured out using statistical data that is gathered by the accounting module. Of course, there should enough statistical data in order to make the correct prediction.

- **Native API.** The accounting module uses ptrace in order to trace and suspend jobs. When this module starts it forks and the newly created process calls ptrace (specifying PTRACE_TRACEME), then it runs the necessary application (by execve system call). The tracer is notified when tracee creates a new process (using fork, vfork or clone), execute another program (using one of exec* functions; it is required in order to get the information about actually running programs) and when one of the tracee processes exits (for retrieving accouting information). The accounting information about process is gathered from the procfs. Another important API that is used in our system is Pthreads. Invocation of the accounting module on a remote node is done by changing the ssh command. Such calls will be traced and the command for the remote node will be changed in order to invoke the accounting module (it is necessary to monitor the jobs resource usage). So, such approach could be considered as a native to Linux systems.

- **Testing.** The dynamic resource allocation is also supported by periodic tests. They are executed in order to reflect the actual resource state (for example, the network load). Using such information the scheduler can allocate the best nodes (in terms of the performance) for new jobs.

– **The Possibility to Use the Modules in Different Ways.** For example, accounting module can be used in a PBS cluster. In this case, administrator should add an invocation of the accounting program before actual user script invocation in the PBS prologue script. Administrator should pass user script with all the options as an arguments to the accounting module.

Testing of our system shows that overheads imposed by profiling are smaller than 1% when compared with PBS implementation on a proper job test set. While an average improper job test set (that underloads and overloads the resources) was computed about 16% faster. The proposed system can be viewed as it depicted in Figure 4.

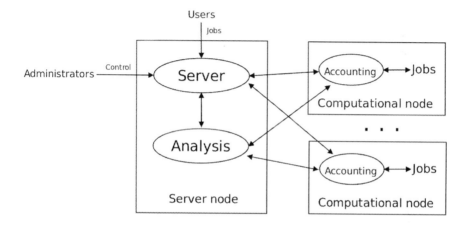

Fig. 4. Scheduler scheme

Such system performs the constant detailed monitoring which generates the accounting information that helps to make real time decisions on resource real-locations (in case of underload or overload) while is also saved for the future use. The rating module will use this information for keeping the user rating that will encourage users to request resources correctly. The prediction module will use this information in order to offer the optimal amount of resources to users and probably automate this task in the future.

Such approach can ease the administration tasks as well as utilization of computational clusters. It could optimize the resource usage and helps to make efficient decisions on computational infrastructure questions.

6   Conclusion

Planning infrastructure for scientific computations sometimes can be challenging. One should take into account many factors. User jobs can underload or overload requested resources, that is why administrator should use systems that will monitor the situation. Statistics generated from such monitoring can be used for future application run.

Scheduling applications can be done using the approach described. Each job submitted to the system is profiled and its optimal resource reservation is determined. For subsequent submissions this reservation is used by default.

Basic modules are implemented using native to Linux APIs and instruments. There can be several modules that perform predictions based on statistics, test actual resource parameters, calculate user rating. When finished, these modules will ease administrative tasks as well as user work.

Acknowledgments. The research was carried out using computational resources of Resource Center Computational Center of Saint-Petersburg State University within frameworks of grants of Russian Foundation for Basic Research (project No. 13-07-00747) and Saint Petersburg State University (projects No. 9.38.674.2013 and 0.37.155.2014).

References

1. The TOP-500 list. http://www.top500.org/statistics/list
2. Owens, J.D., Luebke, D., Govindaraju, N., Harris, M., Kru ger, J., Lefohn, A.E., Purcell, T.: A survey of general-purpose computation on graphics hardware
3. Gayduchok, V.Yu., Bogdanov, A.V., Degtyarev, A.B., Gankevich, I.G., Gayduchok, V.Yu., Zolotarev, V.I.: Virtual workspace as a basis of supercomputer center. In: 5th International Conference on Distributed Computing and Grid Technologies in Science and Education, pp. 60–66. Joint Institute for Nuclear Research, Dubna (2012)
4. TORQUE, Adaptive computing. http://www.adaptivecomputing.com/products/open-source/torque
5. PBS Professional, Altair PBS Works. http://www.pbsworks.com/Product.aspx?id=1
6. Mellanox OFED User Manual. http://www.mellanox.com/page/products_dyn?product_family=26

Storage Database System in the Cloud Data Processing on the Base of Consolidation Technology

Alexander V. Bogdanov[1,2], Thurein Kyaw Lwin[1], and Elena Stankova[1,2(✉)]

[1] Saint-Petersburg State University, Peterhof, Universitetsky pr., 35,
198504, St.-Petersburg, Russia
{alex,trkl.mm}@mail.ru, lena@csa.ru
[2] Saint-Petersburg Electrotechnical University "LETI", ul.Professora Popova 5,
197376, St.-Petersburg, Russia

Abstract. In this article we were studying the types of architectures for cloud processing and storage of data, data consolidation and enterprise storage. Special attention is given to the use of large data sets in computational process. It was shown, that based on the methods of theoretical analysis and experimental study of computer systems architectures, including heterogeneous, special techniques of data processing, large volumes of information models relevant architectures, methods of optimization software for the heterogeneous systems, it is possible to ensure the integration of computer systems to provide computations with very large data sets.

Keywords: Database system · Cloud computing · Cosolidation technology · Hybrid cloud · Distributed system · Centralized databases · Federated databases · IBM DB2 · DBMS · Vmware · Virtual server · SMP · MPP · Virtual processing

1 Introduction

Cloud computing and cloud storage systems have gained popularity as the most convenient way of transferring information and providing functional tools on the Internet. Some cloud services offer a wide range of services and functions to individual consumers (online shopping and online multimedia technologies, social networks, environment for e-commerce and protect critical digital documents), and the other commercial structures, that support the work of small and medium-sized businesses, large corporations, government and other institutions[1].

Some cloud services provide consumers with space for the storage and use of data for free, others charge a particular fee for services provided by subscription. There are also private clouds, owned and operated by organizations. In fact, this secure network is used for storing and sharing critical data and relevant programs [1]. For example, hospitals can use the sharing services for archiving electronic medical records and images of patients or create your own backup storage network. Moreover, it is possible to combine budgets and resources of several hospitals and provide them with a separate private cloud, which the participants of the group will enjoy together[2]. To create a private cloud requires hardware, software, and other tools from

O. Gervasi et al. (Eds.): ICCSA 2015, Part IV, LNCS 9158, pp. 311–320, 2015.
DOI: 10.1007/978-3-319-21410-8_24

different vendors. Management of physical servers at the same time can be both external and internal. Hybrid clouds, as is clear from the title, pool resources of different public and private clouds into a single service or a decision [3]. The basis of all cloud services, products and solutions are software tools that functionality can be divided into three types, means for processing data and running applications (server computing) to move data (network) and for storage (NAS).

The problem of Cloud use for computation stands aside both due to the large overheads in parallel libraries in the Cloud and what is more important serious limitations due to file systems peculiarities in virtual cluster architectures. Thus the way of transferring the data to virtual processes is of a vital importance for large scale computations optimization.

2   Cloud Computing on the Based of Consolidation Technology

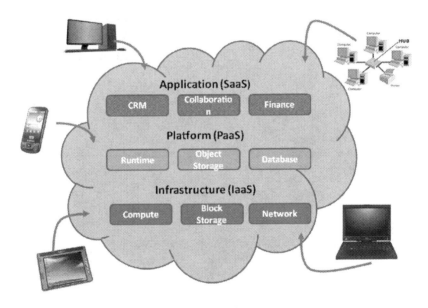

Fig. 1. Cloud Processing

The idea of "cloud computing" is to combine multiple computers and servers into a single environment designed to solve certain kinds of problems, such as scientific problems or complex calculations. Over time, this structure collects a lot of data, distributed computing and storage nodes. Typically, applications, running in a distributed computing environment, address only one of the data sources [4]. However, when the need arises to access simultaneously multiple sources, difficulties arise because these sources may contain different data and tools of heterogeneous access and also are distributed at a distance from each other. In addition, for users performing an analysis of historical data, it is convenient to apply to a single source of information, forming a query and get results in the same format.

Thus, the main problem of the approach to the storage of information in distributed computing systems is the diversity and remote data sources. The solution is to create a centralized point of access, providing a single interface access to all data sources for cloud computing in real time. It is necessary to choose the most appropriate approach and the corresponding platform that provides a consolidation.

3   Data Consolidation Technology

All existing approaches to consolidate distributed data sources can be divided into two types:

1.   The centralized approach

Data from all external sources are transferred to the central repository and are updated periodically. All users work directly with the central repository.

2.   The federated approach

Data is stored directly in the sources, the central link provides transparent redirection of user requests and the formation of the results. In this case, all users can also refer to the central node only which translates requests more data sources.

Each of these approaches has its advantages, it is necessary to consider each and identify the most suitable for data consolidation in the cloud computing.

3.1   Architecture Centralized Databases

Centralized approach to consolidate distributed data sources is duplicating data from all sources in the central database. Such a database called a data store. Typically, the data warehouse using relational databases with advanced tools for integration with external sources. The availability of data combined in a single source, speeds up user access to data and simplifies the normalization and other similar processes compared with data scattered in different systems. However, the integration of information in a centralized source requires that data, which are often in different formats, were brought to a common format, and this process can lead to errors. Also for storage can be difficult to work with new data sources in unfamiliar formats. Moreover, the cost of treatment is often increased because of the need to duplicate the data processing and two sets of data[5].

3.2   Architecture Federated Databases

Federated databases access mechanism and management of heterogeneous data hides the features of the reference to a specific data source, but instead provides a single interface, similar to the classical relational databases [5].

Most applicable approach to creating a platform for the federated database approach is to develop the existing relational database management system and to ensure its interaction with external data sources. This database becomes central to a federal database that stores all the information about data sources, and redirects requests to it [5].

System database directory of the central node must contain all the necessary information about data sources in general and about each object in particular. Such information shall be used by the optimizer SQL-queries to build the most efficient query execution plan.

4 Comparison of Federal and Centralized Approaches

A feature of the federal database is a logical integration of data when the user has a single point of access to all the data, but the data itself physically remain in the original source. This feature is a key differentiator from the centralized federal approach that uses physical integration when data from disparate sources are duplicated at the general assembly that is accessed by all users. Federated approach involves storing data sources themselves, when the central node performs broadcasting requests, taking into account features of a particular source[5].

In the case of cloud computing, the federated database is a better choice for the following reasons:

1. Federated technology less prone to errors with the distortions and integrity because the data will remain in their original locations.

2. In a federated architecture easier to add new sources, this is especially important in dynamic systems.

3. The federated approach, as opposed to a centralized, always guarantees a real-time data from the original source, whereas the centralized approach transferring the data to a central site can become outdated.

It is worth noting that in complex cases that require large amounts of data intersection from different sources, federated database should provide the ability to store the information centrally, providing thus a hybrid approach [6].

5 Software Requirements for Federated Databases

Due to the heterogeneity and distribution of data sources in the cloud, unified information management environment is a challenge. Data sources can be relational databases, business applications, flat files, web services, etc. Each of them has its own storage format, challenges and way of delivery of results. Moreover, the sources may be located at a considerable distance from each other on different networks with different access protocols.

The software manages the federated database, must necessarily meet the following requirements:

1. Transparency
2. Heterogeneity

3. Scalability
4. Support for specific functionality
5. High performance
6. Separation of access rights

6   Existing Platforms of Federated Databases

1.   IBM DB2 Information Integrator

This decision is based on DBMS IBM DB2 Universal Database and initially focused on the creation of distributed systems with federated access. Supported by a lot
of variety of data sources, as well as standard SQL / MED, allowing you to create
your own extensions.

Particular attention is paid to the performance and security platform, as well as
ease of use and management.

2.   Microsoft SQL Server

Integrating Microsoft SQL Server database with external sources is carried out
through the use of Microsoft Integration Services - a platform to build integration
solutions and data transformation at the enterprise level.

The Integration Services can extract and transform data from a number of sources,
such as files XML, flat files, relational databases, etc. It is possible to use graphical
tools of Integration Services to create a ready-made solutions or independent creation
of the object model of Integration Services using the supplied software.

3.   Oracle Streams

Integration of Oracle initially focused on the implementation of the approach with
a centralized access, however, the technology Oracle Streams Transparent Gateways
provides the means to implement the model with federated access. External data
sources can be registered in the Oracle database in the form of links, called DB-links,
and use the data from these sources in distributed queries. Supports access to flat files,
XML-files, ODBC sources, etc[7].

7   Distributed Data Processing

Distributed data processing is an opportunity to integrate fragmented data resources.
One approach to centralizing data is to decommission simply the existing database
system and to build a new integrated database. An alternative approach is to build an
integration layer on top of pre-existing systems. Building an integration layer on top
of existing database systems is a challenge in complexity and performance, but this
option sometimes gives the most effective results in business and engineering sense.
In a data-sharing environment, there is no single best architecture that will solve all

problems. Large installations of database systems may be accessed by hundreds of thousands of times a minute. The irreducible latency present even in a fully optical network is not capable of supporting such a performance requirement. Indeed, local disks are also too slow, and most of this sort of information is cached off disk and into memory. In some organizations, if critical data is unavailable for even a matter of minutes, it could affect millions of dollars of revenue. This is why remote data access is not used in such large-scale situations where high availability is critical. There are many small and medium weight applications with modest performance requirements for data. Often, such applications are designed to work with a copy of data because getting a copy and loading it on a local database seems like the easiest solution. Such design does not factor in the cost of maintaining a separate copy of the data. When the applications are put into production and begin having problems keeping their data in sync, these costs become all too apparent. Such applications would probably do better to remotely reference their data. In such cases, it is a good architecture to remotely reference application databases for shared data. Such "distributed" databases need to incorporate some high availability design, depending on the weight of the applications served and their availability requirements. Each application should be analyzed to determine its performance and reliability requirements [1].

8   Distributed Databases are Working with Data on the Remote Server

Database management system (DBMS) has become universally recognized tool for creating application in software systems. These tools are constantly being improved, and the company database developers are closely monitoring the progress of their competitors, trying quickly to include in their packages the new features implemented in the competition. True internal architecture of the database is not always let to do this successfully. Distributed databases are implemented in a local or global computer network [7]. In this case, one of the logical database is located in different nodes of the network, possibly on different types of computers with different operating systems. Distributed DBMS provides users with access to information, regardless of what equipment and how the application software used in the network nodes. Members are not compelled to know where the data is physically located and how to perform physical access to them [8]. Distributed DBMS allows horizontal and vertical "splitting" of the tables and put the data in one table in different network nodes. Requests to the distributed database are formulated in the local database. Transaction processing operations and backup / restore distributed database integrity is ensured throughout the database.

We have been installing our database system on the server of linux platform in St.Petersburg State University GRID technoligies laboratory. The system includs a DB2 client and DB2 Administration Client, which implements the graphical tools that enable you to select the appropriate performance access to remote servers, to manage all servers from one location, to develop powerful applications and process requests. If the network is working properly, and protocols will function correctly on the workstation, the interaction of the "LAN - Local Area Network" between Database

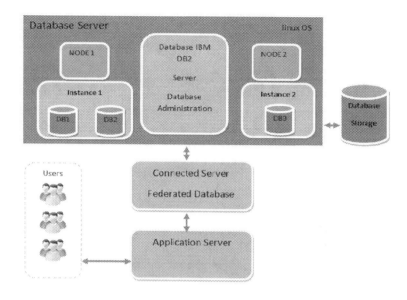

Fig. 2. Architecture, developed in Saint Petersburg State University

servers and clients require no additional software. As long as there is a connection between the local networks of any network client can access any server. Transactions provide access and update data in the databases of both servers while maintaining the integrity of data on both servers. Typically, such a mode of operation is called two-phase (two-phase commit) or access within a distributed unit of work. The first debit account and for the account of the second loan is very important that their update was carried out as a single transaction [8].

We can consider this solution as the DB2 database used for the consolidation of computing components and data storage in a virtual site. DB2 UDB is a completely parallel and support parallelized execution of most operations, including queries, insert, update, and delete data, create indexes, load and export data. Moreover, due to functionality of DB2, the transition of a standard, non-parallel execution environment to a parallel one is not limited by increasing efficiency [2]. DB2 UDB has been specifically designed to work successfully in a number of parallel media systems including MPP, SMP and MPP clusters of SMP nodes. DB2 provides computer data storage solutions for the target problem.

We choose DB2 UDB for our database system. We have installed IBM DB2 in the VMware Infrastructure environment. Then we installed VMware Tools in the guest operating system and created eight identical single-processor virtual machines. We have included only the virtual machines that are used in the particular test and made sure all unused virtual machines on the host ESX Server, were closed [8]. Results show, that one VCPU virtual machines advantages of working with IBM DB2 on a VMware ESX Server to become apparent when we run the test with multiple virtual machines.

We were modeling the virtual processor used in the test run with different load. In other words, the same number of concurrent virtual machines used, we have doubled the number of simulated users within two VCPU SMP virtual machines compared to one VCPU virtual machine. Figure-3 shows that the efficiency is almost doubled in two-VCPU SMP virtual machine. The results verify that the virtual environments can achieve the scale of SMP similar to that seen in the native environment.

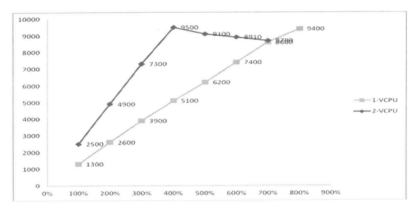

Fig. 3. Scalability compared for a fixed number of virtual machines

9   Analysis of the Database in the Distributed System

We were testing the distributed system in our university and have shown how to improve the performance and extend the range of applications of scientific methods and algorithms for parallel and distributed data processing by optimization of database applications from the point of view of search for promising architectural solutions[9]. The aim is to provide an operating environment for the database and its consolidation in a distributed computing environment, that is some general solution for relatively small networks and can be used in research institutions and commercial enterprises whose resources may be located in the same building and in geographically remote locations To achieve this goal it was necessary to solve fairly complex problem of choosing a prototype system architecture, algorithm development, as well as the problem of creating and adapting existing software products[10]. Such a system is implemented in the form of blocks that make up the distributed virtual computer system and we can call it virtual testbed.

10   Conclusions

Consolidation of data in distributed heterogeneous systems is an important and challenging task. Out of existing approaches to solving this problem, the most appropriate approach is that of federal databases. Creating and managing such a structure requires the use of specialized software, which in turn must meet a number of requirements for transparency, heterogeneity, security, performance, etc. On market integration software there are a number of solutions from major manufacturers, build on industrial relational database, based on which you can organize a federal structure data access. To select a specific solution, the detailed examination for compliance with the requirements for systems of this type must be made. In this research we can point out the following results.

• There is a way to create an operating environment for database consolidation in a distributed computing environment. This environment is some common solution for relatively small networks and can be used in research institutes and commercial enterprises in which resources may be located in the same building as well as in geographically remote locations.

• A special synthetic method of data processing, which allows you to combine the power of SGE and DB2 DBMS for distributed heterogeneous computing resources, was an important step in the consolidation of the global pool of resources.

• The results of testing, which clearly showed that the databases in a distributed computing environment can be an effective means of consolidating software and the practical use of the developed techniques can significantly improve the efficiency of data processing and improve the scalability of distributed systems.

• The proposed methods and products have been tested on a variety of platforms and operating systems, and we hope, that they will be widely used not only in distributed computer systems, but also for cluster computing.

We listed the requirements for the cloud databases and compared the suitability of different database architectures to cloud computing. Based on our performance results, we believe that the Cloud Database vision can be made a reality, and we look forward to demonstrating an integrated prototype of next Big Data Solution. Whether we come to assembling, managing or developing of a cloud computing platform and need a cloud-compatible database is a challenge.

References

1. Bogdanov, A., Lwin, T.K.: (Myanmar)/ Storage database in cloud processing. In: The 6th International Conference "Distributed Computing and Grid-technologies in Science and Education" will be held at the Laboratory of Information Technologies (LIT) of the Joint Institute for Nuclear Research (JINR) on 30 June - 5 July 2014 in Dubna
2. Technical Details Architecture cloud data processing and storage .
   http://www.seagate.com/ru/ru/tech-insights/cloud-compute-and-cloud-storage-architecture-master-ti/
3. Bogdanov, A.: Private cloud vs Personal supercomputer, Distributed computing andGRID technologies in science and education, JINR, Dubna, 2012 Advanced high performance algorithms for data processing, - Computational science ICCS-2004, Poland
4. Bogdanov, A.: The methodology of Application development for hybrid architectures. Computer technologies and applications (4), 543 – 547 (2013)
5. Bogdanov, A.V., Stankova, E.N., Lin, T.K.: Distributed databases. SPb.: ``LETI'', pp. 39–43 (2013)

6. Bogdanov, A.V., Kyaw lwin, T., Stankova, E.: Using Distributed Database System and Consolidation Resources of Data Server Consolidation. In: Computer Science and Information Technology International Conference, pp. 290–294. Armenia, September 23–27, 2013

7. Bogdanov, A.V., Lwin, T.K., Naing, Y.M.: Database Consolidation used for Private Cloud. In: Proceedings of the 5th International Conference GRID 2012, Dubna, July 16–21, 2012

8. Bogdanov, A.V., Lin, T.K.: Database technology for system integration of heterogeneous systems and scientific computing. Proceedings of ETU ``LETI'' (4), 21–24 (2012)

9. Babu, S.: Automated control in cloud computing: challenges and opportunities. In: Babu, S., Chase, J., Parekh, S.(eds.) 1st workshop on Automated control for datacenters and clouds, pp. 13–18 (2009) (DOI:10.1145/1555271.1555275)

10. Armbrust, M., Fox, A., Griffith, R., et al.: A view of cloud computing. Communications of the ACM 53(4), 50–58 (2010)

Integrated Information System for Verification of the Models of Convective Clouds

Dmitry A. Petrov[1([⊠])] and Elena N. Stankova[1,2]

[1] Saint-Petersburg State University, Peterhof, Universitetsky pr.,
35, 198504, St.-Petersburg, Russia
g_q_w_petrov_dm_alex@mail.ru
[2] Saint-Petersburg Electrotechnical University "LETI",
ul.Professora Popova 5, 197376, St.-Petersburg, Russia
lena@csa.ru

Abstract. The paper describes the information system that integrates heterogeneous meteorological information necessary for the verification of the numerical models of convective clouds. Data integration is realized on the base of consolidation technology. The first section of the article describes the implementation of the method of consolidation of meteorological data from heterogeneous sources (using PHP). The design and realization of the relational database (DB) (using MySQL) are described in the second section. The third section is concerned with the development of Web-based applications for verification of 1.5-D convective cloud model [1] using HTML 5, CSS 3 and JavaScript.

Keywords: Data consolidation · Data integration · Meteorological data · Numerical modeling · Weather forecast

1   Introduction

Convective cloud is an aggregation of liquid droplets and / or ice particles formed in the vertical air flow. Numerical simulation is one of the most effective tools for cloud study and forecast of such dangerous convective phenomena as thunderstorms, hail and rain storms.

There are currently a lot of one, two and three-dimensional cloud models, varying by the level of detail in description of microphysical processes [2]. A special place is occupied by the so-called one and half dimensional (1.5-D) models [3], in which the additional half of dimension arises in view of account of involvement of ambient air via lateral edges of a cloud. Such models do not require essential computational resources, but are able to reproduce those cloud characteristics that can be predictors for existing methods of forecasting of dangerous convective phenomena in near real time. Programs based on these models are very relevant, and are used, for example, in the meteorological centers of the airports, where simple models of small dimension are required, as these centers are not usually equipped with powerful computers.

However, verification of these models, and the subsequent "tuning" is complicated because of the lack of free access to integrated data on the vertical distribution of

© Springer International Publishing Switzerland 2015
O. Gervasi et al. (Eds.): ICCSA 2015, Part IV, LNCS 9158, pp. 321–330, 2015.
DOI: 10.1007/978-3-319-21410-8_25

temperature and humidity in the atmosphere in that place and at that time, when there was a dangerous convective phenomenon. These data are used in the model calculations as initial and boundary conditions. In addition, verification involves obtaining statistically reliable data about the applicability of the model for forecasting, and for this it is necessary to carry out a large series of numerical experiments.

It is necessary to develop a software environment that would provide, firstly, the integration of the necessary information from distributed heterogeneous sources, and, secondly, would enable implementation of a series of numerical experiments in automatic mode using the intuitive interface.

Nowadays many business companies and research institutes are engaged in developing of specialized systems for the integration and processing of meteorological information.

RaytheonCompany [4], for example, has developed a system ITWS (Integrated TerminalWeatherSystem). the National Oceanic and Atmospheric Administration (NOAA) has developed meteorological data assimilation system - (MADIS) [5]. LLC "Perspective methods of monitoring» (AdvancedMonitoringMethods) [6], uses SQL for integration and correction of meteorological data and subsequent formation of alerts and reports. Institute for Radar Meteorology (IRAM) [7] provides specialized systems for aviation forecasters, collecting data from meteorological satellites, world forecasting centers, meteorological and aviation station networks.

Major part of these data is provided on the commercial basis (ITWS, DAS, IRAM) and thus does not entertain requests for academic research assistance. Besides, the data is stored in the most cases as the static archives on computer disks or other external storage media. And what is the most important matter; all available meteorological data cannot be directly used for verification of convective cloud models due to the absence of integrated information about the place, date and atmospheric conditions during hazardous convective event.

Thus, it can be argued that there is absence of a free and publicly available software product that could be used for the integration of meteorological information needed to verify the model of convective cloud.

The paper describes the application, including the correction of all deficiencies identified in the above-mentioned software products that perform similar tasks. Software product developed by the authors integrates necessary meteorological information, places it in storage in the required format and submits the information via user friendly interface.

At the same time, this application has the following advantages: no charge of getting all the necessary data, the ability to filter out unnecessary information, the relative independence of the application from a working machine due to the fact that the product is made in the form of Web-based application.

The process of development of such a product can be divided into three stages. Firstly, it is necessary to analyze the sources of weather information and to find out in what form it is stored. Secondly, it is necessary to design a specialized database for storing the integrated data. Third, a software environment should be developed that provides an interface for access to database with the integrated data and allows to provide numerical experiments with a cloud model with the subsequent opportunity to store and to analyze the results of calculations.

The first section of the article describes the implementation of the method of consolidation of meteorological data from heterogeneous sources (using PHP). The design and realization of the relational database (DB) (using MySQL) are described in the second section. The third section is concerned with the development of Web-based applications for verification of 1.5-D convective cloud model [1, 8-11] using HTML 5, CSS 3 and JavaScript.

2   Consolidation of Meteorological Data from Heterogeneous Sources

In order to make a selection of data sources with the required meteorological information, it is necessary to consider the following criteria: data should be provided solely on the basis of free and contain, in the aggregate, full information about time and place of dangerous convective phenomena as well as about vertical profiles of the meteorological data used to set the initial and boundary conditions of convective cloud model [12].

Site «Meteocenter» [13] and the site of the University of Wyoming. [14] were chosen according these conditions. Website «Meteocenter» provides an archive of data which contain information about the type of hazardous event, place and date where it occurred. These data serve as input for the site of the University of Wyoming (UW). The vertical distributions of temperature and humidity can be obtained with the help of these sites. Thus, the required data can be extracted from the second site using the information from the first site as the input parameters.

Sites vary in their structure, the language of the user interface and may contain information that is not needed for verification of the model. Thus, it is necessary to develop an application for extraction the necessary data from the selected heterogeneous sources with subsequent transformation them into a single structure, processing and placement in a specialized database.

Programming language PHP server version 5.5 has been selected for application development. PHP has the ability to use library of functions "libcurl", which allows the use multi-threading technology (multi_curl) [15]. This library both allows to interact with many different servers via a variety of protocols, and allows to run queries of tasks in parallel, speeding up data extraction from sites [16].

The data extracting from the different sources are in different formats (tables and CSV) and so they should be converted into a single format (suitable for the models) before loading into the database.

Of the three main methods of data integration (data propagation, federalization, consolidation [17]), only the consolidation method is suitable for all of the above requirements, so it was decided to use data consolidation.

At the heart of the consolidation process is ETL-process consisting of three consecutive stages: Extraction, Transformation, Loading. Extraction means data extraction from heterogeneous sources. Transformation means transformation of data into a form suitable for storage in a certain structure. Loading means data loading into the data warehouse, which is a special case of a relational database [18].

Consolidation method is implemented by the following algorithm (Fig. 1).

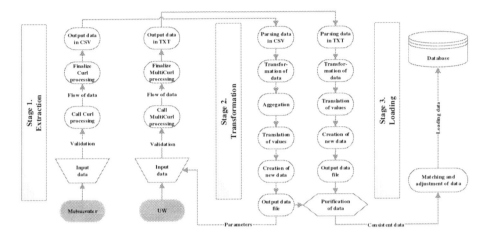

Fig. 1. Brief scheme of realization of consolidation method

At the first stage the data from the archive site «Meteocenter» will be extracted by means of libcurl library, mentioned earlier. As a result, the file in CSV format will be obtained, some of the data from which will be used as input to obtain information from the site UW. Libcurl library will be used also to obtain data from UW site and stored the data in the tables with a specialized structure in a form of the file in TXT format. As a result, there are two temporary files with the data from each site.

At the second stage data transformation takes place. This is necessary, as the data may differ in their structure and format, and it is necessary to transform the data into a form that is required for analysis.

At this stage, conversion of the data from the website «Meteocenter» takes place. The data are analyzed (parsing) and the required parameters are retrieved. This is followed by data restructuring and aggregation over a time interval. Then all the data are converted into «windows-1251» format, suitable for correct operation of the convective cloud model.

If correct model operation requires more data than were obtained, the missing data are created by aggregating already extracted [19].

As a result of the second stage manipulations all the converted data ready to be cleaned are placed in a temporary file. The same processing, except aggregation occurs with the data from the site UW, which are in the format TXT. In this case, the aggregation is not performed due to lack of necessity.

The source data most often needs to be cleaned by filling gaps, eliminating duplicates, etc.. Cleaning is done in a temporary file, which combines data from both sites. As a result of the second stage a file appears with purified, transformed data from the two sources.

The data is transferred from the temporary file to the database structure during Loading, which is the third and last stage of the ETL-process. Since this file already contains all the necessary parameters, it is loaded to the database using SQL-queries without additional manipulations. After that series of post loading operations are provided such as verification and re-indexing of data.

The database was created specifically for the application and meets all the criteria of convective cloud model. The database allows providing search by various parameters for example by the type of convective phenomena.

Figure 2 shows the scheme of database usage for forecasting dangerous convective phenomena.

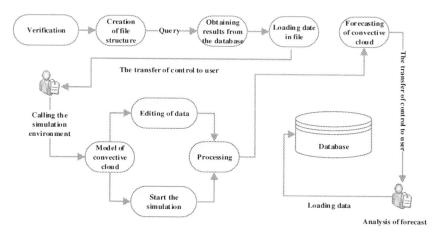

Fig. 2. Database usage for forecasting

3   Design and Realization of the Relational Database

The next section of the paper is concerned with the design and realization of the relational database, which presents a set of related tables (relations), containing information about a certain type of the objects. Each row containing information about one object is called the cortege, and the columns of the table containing the different characteristics of these objects are called the attributes.

Relational model of database has been chosen as it has several advantages:

- displays the information in a simple user-friendly manner;
- allows to create non-procedural languages for manipulating data type;
- is based on the developed mathematical apparatus [20].

Development of a database consisted of several stages:

- conceptual (Infological) design involves data collection, analysis and editing of data requirements;
- logical design involves transformation of the data requirements into the logical structure of the data;
- physical design involves transformation of the logical structure of the database into the physical one, taking into account the aspects of performance of DBMS MySQL [21]

9 relations has been created as a result of database development. The relations present the names of real-world objects: cities, weather stations, convective phemomena, time intervals, the parameters of convective clouds and others. Entity-relationship diagram (ER-diagram) consistent with the data model is shown on Fig. 3. The format of the information stored in the database does not require further transformation, which allows using it directly without additional manipulation for verification of convective cloud model. Database contains information about the time, location and type of dangerous convective phenomena in conjunction with the full range of meteorological data on the state of atmosphere and the earth's surface.

For convenience of data use and storage the database was divided into two large blocks: the block "data" and the block "data management".

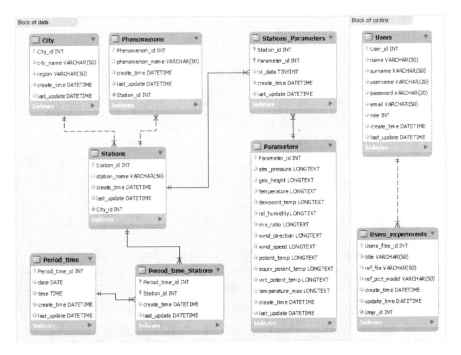

Fig. 3. ER-diagram

4   Web-Based Application Development

Special software environment should be developed in order to provide an interface to the database and the cloud model.

Web-based application and desktop application [22] have been considered as basic types of software products suitable for creating such environment. Web application has been chosen for reason of its mobility. Anyone can use it without downloading any additional software products in any place where there is an access to the Internet and a local machine. This allows to save time that would be required for the downloading and installation of the extra software, and also save web traffic and place on the user's hard drive.

The Web-based application includes six sections. The main ones are: "data extraction", "database" and "verification". In the section "data extraction" a user can specify the parameters based on which data will be extracted from the heterogeneous sources. It is necessary to choose region, city (weather station) and the date (Fig. 4).

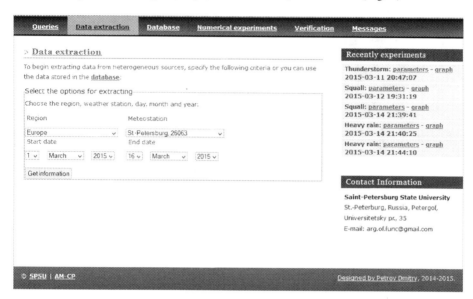

Fig. 4. Specification of parameters needed for data extraction

Section "Database" consists of 9 tables, which contain all the previously received data. A user is able to view them and to manipulate them: to add, to edit and to delete. It is also possible to clean the database completely.

In the "verification" section a user should specify a region, city (weather station), phenomenon and time period of interest. As a result, the system will generate the file with the data that can be used later as an input for the cloud model. To start working with the model it is necessary to save the generated data file to user's local machine and to download the software environment of convective cloud model.

A set of test numerical experiments aimed to the model verification has been conducted with the help of the developed application and the cloud model. The results show adequate behavior of the model against the input data and observed phenomena. In future it will allow obtaining statistically valid data about the effectiveness of using the model for forecasting of dangerous convective phenomena such as thunderstorms, heavy rains, hail and squall.

The fields of the calculated values of dynamical and microphysical cloud characteristics are presented to the left and in the center of the Figure 5. Vertical profiles of the air temperature (black line) and dew point (red line) used as the input data are presented to the right of the figure.

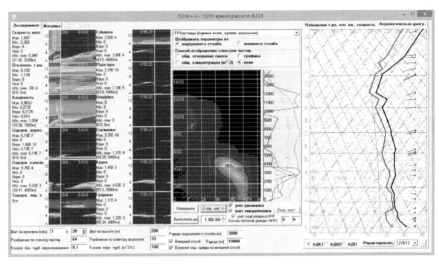

Fig. 5. Interface of the software environment for calculation the characteristics of convective clouds [1, 8-11]. The calculation was performed with the initial data corresponding to the real data of vertical sounding of the atmosphere on a day when there is a strong thunderstorm.

5   Conclusions

Integrated information system for verification of numerical models of convective clouds has been developed.

Meteorological data needed for verification have been analyzed at the first stage of the development. According to this analysis web site «Meteocenter» and the web site of the University of Wyoming have been selected as the sources containing all types of required meteorological information and providing it free of charge.

Different technologies of integration of data from heterogeneous sources have been analyzed. As a result technology of consolidation has been chosen as it provides data extraction, transformation and loading them to data bases. High speed of extraction of the necessary meteorological data has been achieved using multi-threading technology which has been realized by the functions of the "libcurl" library.

Using the MySQL Workbench environment, the relational database has been developed which contains the data about the time and place of dangerous convective phenomena, and the files with vertical profiles of meteorological parameters that can be used as input data for the models of convective clouds.

Files with the input parameters are prepared with the help of the developed Web-based application. The application has been realized in phpDesigner environment using the following tools: PHP, HTML, CSS, JavaScript. The application has an intuitive interface and includes six sections. The main ones are: "data extraction", " database" and "verification". Section "Data extraction" allows to integrate, view and manipulate meteorological data. Section "Verification" allows to prepare and save the data file for later use as an input for cloud model. Software environment has been tested; all known bugs and shortcomings are corrected.

Using the developed system, meteorological data have been integrated from the heterogeneous sources, which have been used as input initial and boundary conditions for the numerical model of convective cloud. The results show adequate behavior of the model against the input data and observed phenomena. In future it will allow obtaining statistically valid data about the effectiveness of using the model for forecasting of dangerous convective phenomena such as thunderstorms, heavy rains, hail and squall.

Acknowledgment. This research was sponsored by the Saint-Petersburg State University under the project 0.37.155.2014 "Research in the field of designing and implementing effective computational simulation for hydrophisical and hydro-meteorological processes of Baltic Sea (and the open Ocean and offshores of Russia)".

References

1. Raba, N.O. Stankova, E.N.: Research of influence of compensating descending flow on cloud's life cycle by means of 1.5-dimensional model with 2 cylinders. In: Proceedings of MGO, V.559, pp. 192–209 (2009). (in Russian)
2. Khain, A., Ovtchinnikov, M., Pinsky, M., Pokrovsky, A., Krugliak, H.: Notes on the state-of-the-art numerical modeling of cloud microphysics Review. Atmospheric Research. **55**, 159–224 (2000)
3. Dovgaluk, Y.A,, Veremey, N.E., Sinkevich, A.A.: Application of 1.5-D model for solution of fundamental and applied problems in the physics of clouds, pp. 32–43. St. Petersburg, Gidrometeoizdat (2007). (in Russian)
4. Integrated Terminal Weather System (ITWS) http://www.raytheon.com/capabilities/products/itws/
5. Meteorological Assimilation Data Ingest System (MADIS) http://madis.noaa.gov/
6. Advanced Monitoring Methods. High Quality Data Collection from any Location. http://advm2.com/
7. Institute for Radar Meteorology http://www.iram.ru/iram/index_en.php
8. Raba, N.O., Stankova, E.N., Ampilova, N.: On Investigation of Parallelization Effectiveness with the Help of Multi-core Processors. Procedia Computer Science 1(1), 2757–2762 (2010)
9. Raba, N., Stankova, E.: On the possibilities of multi-core processor use for real-time forecast of dangerous convective phenomena. In: Taniar, D., Gervasi, O., Murgante, B., Pardede, E., Apduhan, B.O. (eds.) ICCSA 2010, Part II. LNCS, vol. 6017, pp. 130–138. Springer, Heidelberg (2010)
10. Raba, N.O., Stankova, E.N.: On the problem of numerical modeling of dangerous convective phenomena: possibilities of real-time forecast with the help of multi-core processors. In: Murgante, B., Gervasi, O., Iglesias, A., Taniar, D., Apduhan, B.O. (eds.) ICCSA 2011, Part V. LNCS, vol. 6786, pp. 633–642. Springer, Heidelberg (2011)
11. Raba, N.O., Stankova, E.N.: On the effectiveness of using the gpu for numerical solution of stochastic collection equation. In: Murgante, B., Misra, S., Carlini, M., Torre, C.M., Nguyen, H.-Q., Taniar, D., Apduhan, B.O., Gervasi, O. (eds.) ICCSA 2013, Part V. LNCS, vol. 7975, pp. 248–258. Springer, Heidelberg (2013)

12. Petrov, D.A., Stankova, E.N.: Use of consolidation technology for meteorological data processing. In: Murgante, B., Misra, S., Rocha, A.M.A., Torre, C., Rocha, J.G., Falcão, M.I., Taniar, D., Apduhan, B.O., Gervasi, O. (eds.) ICCSA 2014, Part I. LNCS, vol. 8579, pp. 440–451. Springer, Heidelberg (2014)
13. Archive of weather in Russia and in the world. http://meteocenter.net/ussr_fact.htm
14. University of Wyoming. College of Engeneering. Department of Atmospheric Sciences http://weather.uwyo.edu/upperair/sounding.html
15. Khalid, M.I.: PHP/CURL Book. pp. 10–17 (2009)
16. The client library for working with URL. http://www.php.net/manual/ru/book.curl.php
17. Bogdanov, A.V., Stankova, E.N., Lin, T.K.: Distributed databases. SPb.: "LETI", pp. 39–43 (2013). (in Russian)
18. Data consolidation - the key concepts. Corporate management. http://www.cfin.ru/itm/olap/cons.shtml
19. Paklin, N.B., Oreshkov, V.I.: Business Intelligence: from data to knowledge. Tutorial. SPb.: Peter, pp. 61–137 (2013). (in Russian)
20. Evseeva, O.N., Shamshev, A.B.: Working with databases in the language C #. Technology ADO.NET. Tutorial. UlSTU, p. 7–15 (2009). (in Russian)
21. Ershov, G.N.: Information technology in publishing. Tutorial. M.: MGUP, pp. 17–29 (2002)
22. Zamaraev, B.: Types of software. Software & Startups. http://www.softwareandstartups.com/razrabotka-po/tipy-programmnyx-produktov/

Hybrid Approach Perturbed KdVB Equation

Alexander V. Bogdanov, Vladimir V. Mareev, and Elena N. Stankova[✉]

Faculty of Applied Mathematics and Control Processes,
Saint-Petersburg State University, 198504 Petergof, Saint-Petersburg, Russia
{bogdanov,map,lena}@csa.ru

Abstract. The solution of nonintegrable nonlinear equations is very difficult even numerically and practically impossible by standard analytical technic. New view, offered by heterogeneous computational systems, gives some new possibilities, but also need novel approaches for numerical realization of pertinent algorithms. We shall give some examples of such analysis on the base of nonlinear wave's evolution study in multiphase media with chemical reaction.

Keywords: Nonintegrable nonlinear equations

1 Introduction

The solution of nonintegrable nonlinear equations is very difficult even numerically and practically impossible by standard analytical technic. It is particularly hard to get reliable results in asymptotic region (large times).

A lot of useful approaches proposed for vector systems can hardly be ported to current cluster systems. Some new possibilities are offered by modern hybrid technologies, based on GPGPU's. But GPGPU is not yet vector accelerator, so it is difficult to monitor and optimize parallel tasks and onboard memory bottlenecks made it almost impossible to get reliable results for 3D real problems. So it is necessary to make preliminary tests on simple problem to illuminate all possible difficulties and find out optimal numerical approaches for future optimized codes.

Although nonlinear PDEs appear in many applications and there is a lot of work done on their analysis, it is hard to quote any result that can be called conclusive. The only favorable exception is the systems not far from completely integrable. We can cast up perturbation theory, starting from nonlinear integrable system, that can give a direct way to build the unique solution at least for finite time intervals. Some beautiful examples were given in [1,2] on the base of different methods of solution of basic nonlinear integrable problem.

Although those approaches make it possible to study all the details of solution, for practical applications they can be difficult to realize and they are cumbersome for qualitative analysis. Sometimes it is advisable to have a simple approximate approach that gives with small effort possibility to see main features of a solution and obtain the details of evolution of someintegral parameters of it. Our purpose is to attract attention to the possibilities of standard quasi-classical solutions in this problem.

© Springer International Publishing Switzerland 2015
O. Gervasi et al. (Eds.): ICCSA 2015, Part IV, LNCS 9158, pp. 331–341, 2015.
DOI: 10.1007/978-3-319-21410-8_26

For nonlinear equation, the direct rigorous application of this approach [3] is also time consuming, but in many applications it is possible to get a lot of details without evaluation of general solution. As in all asymptotical methods, the use of some art saves a lot of calculations.

2   Governing Equations

We shall give some examples of such analysis on the base of nonlinear waves evolution study in multi-phase media with chemical reaction [4]. It was shown, that for one dimensional gas dynamic problem, described by Navier-Stockes equation, equation of state, and simple linear relaxational equation after expansion up to the second order near the equilibrium state one gets for the velocity nonlinear evolution equation of the form

$$v_t + vv_x + \alpha v_{xx} + \beta v_{xxx} = \gamma I(v) \tag{1}$$

with α being the measure of dissipational effects, β being the measure of dispersion and expressed via transport coefficients and relaxation times, γ is the measure of interphase interactions and is expressed via integral brackets and relaxational times and $I(v)$ is the integral operator, that to the first approximation is linear and can be expressed as

$$I(v) = -\int_0^t G(t, \tau)v_\tau \, d\tau \tag{2}$$

with G for different models of interaction being exponential or inverse power function.

It is useful to introduce the integral over nonlinear wave

$$M(t) = \int_{-\infty}^{\infty} v(x, t) dx)$$

and after some calculations to show, that

$$\dot{M}(t) = -\gamma \int_0^t G(t - \tau)\dot{M}(\tau)d\tau \tag{3}$$

and so interphase interactions are the only source of violation of conservation laws in this problem. After differentiation of (3), integration by parts and some calculations, taking into account properties of $I(v)$ one gets

$$\ddot{M}(t) + [\gamma G(0) + T^{-1}]\dot{M}(t) = 0 \tag{4}$$

with T being the measure of the width of distribution G. Of course(4) can be solved at once and we get that except small initial layer the total intensity of wave is conserved so we can apply simple perturbational consideration for determination of evolution of soliton back parameters. Let's discuss the example of one soliton solution

$$u_s = a\cosh^2[b(z + z_0)], \qquad a = 12b^2$$

with $Z = x - Ut, U = 4b^2$ and ε being maximum between γ and α it is possible to introduce *slow time* $\tau = \varepsilon t$ and obtain equation for evolution of slow variables. For example,

$$\frac{da}{d\tau} = \int_{-\infty}^{\infty} P[u_s]\cosh^{-2}\theta d\theta, \quad \theta - bz,$$

that corresponds to the results of inverse scattering method [1]. In quasi-classical approximation

$$z = x - Q_0(\tau)/\varepsilon - Q_1(\tau)$$

and for $Q_j(\tau)$ we get the equations:

$$\frac{dQ_0}{d\tau} = 4b^2,$$

$$\frac{dQ_1}{d\tau} = \int_{-\infty}^{\infty} P[u_s](\tanh\theta + \theta\cosh^{-2}\theta + \tanh^2\theta)d\theta,$$

again in agreement with [1].

Those results are not very surprising, because the phase and amplitude of soliton are too crude parameters for detail comparison. Let us study the change of form of soliton by quasi-classical methods. So we put $v = u_s + \delta v$ and try to study the equation for δv. The main problem that we have here is that leading term diverges and so we must check if the leading diverging term is zero, that puts some additional conditions on perturbation.

It is not difficult to get the first order equation for perturbation

$$-U\delta v/2 - \alpha(\delta v)_z + u_s\delta v = g(z) \tag{5}$$

with

$$g(z) = -(U/2)\int_0^1 G(\tau)u_s(z + U\tau/2)d\tau$$

and study its solutions for large z, since we are interested only in such features of the wave, that change the form of soliton, i.e. tails and pilot waves.

Our main result comes from the fact, that change dv of the form of soliton can be significant only outside its undistorted pattern because by definition there $|\delta v| \ll |u_s|$. Outside that region δv may become important in forming new structures. Of such structures the tails are most important and are carefully studied by different techniques for KdVB equation. For such equation with the perturbation (2) the pilot wave can appear and we can give here simple quasi-classical conditions for such effect. It can be shown, that change of amplitude in our approximation is given by integral:

$$\Phi(\kappa) = 2\int_0^{\infty} \exp(-\kappa\theta')\int_0^{\infty} \cosh^{-2}(\theta + \gamma\theta')\cosh^{-3}\theta \sinh\theta \, d\theta' d\theta$$

and the evolution of the amplitude is determined by the equation

$$\frac{da}{d\tau} = -\Phi(\kappa)a$$

with κ being the combination of parameters of soliton and perturbation. It is clear, that pilot wave can appear only if the integral is negative and so the condition

$$\Phi(\kappa) < 0$$

is the condition of appearance of pilot wave in (1) with perturbation (2).

For the kernel G of the exponential type the solution of (5) can be obtained in analytical form

$$\delta v = A \exp(-\gamma z +)\mathrm{Ei}(-e^{\gamma z})$$

from which we can extract the qualitative behavior of the solution. As usually with primitive quasi-classical approach we cannot get asymptotical form of the solution for very large negative z from it, nevertheless it might be good for prediction of "physical effects" in the solution. Of course one needs more refined techniques for exact asymptotical behavior of the system.

The last possibility we would like to point out is the use of quasi-classical approximation within nonlinear change of variables for our system. The best known example is Whiner-Hopf transformation [5], that transforms Burgers equation to linear problem. For our equation the corresponding transformation is given by [5]

$$v = (v/\mu)t + f(\xi)$$
$$\xi = x + (v/2\mu)t^2 + (\delta/\mu)t \tag{6}$$

The resulting equation can be represented in the form

$$\mu\beta f''' + \mu\alpha f'' + (\mu f - \delta)f' + \mu = I_r(f) \tag{7}$$

and the sign $"'"$ stands for derivative over ξ and we do not give the cumbersome expression for I_r, that can be obtained by direct substitution of (6) into (2). One can give a beautiful quasi-classical description of (7), even uniform one, so in this approach the only serious problem is to return to initial variables.

3   Numerical Calculations

To check our analytical considerations and to evaluate the optimal approaches for nonlinear waves propagation calculations we shall use two standard models for the kernel of the integral in (2):

$$\text{I. } G(t,\tau) = e^{-k(t-\tau)}, \quad k > 0 \tag{8}$$

$$\text{II. } G(t,\tau) = \left(\frac{\xi}{\pi}\right)^2 \frac{1}{t^2+\xi^2/4} \cdot \frac{1}{\tau^2+\xi^2/4} \tag{9}$$

And for testing of numerical methods we used two initial distributions of the wave, to model smooth and rough waves:

I. $v(x,0) = h(x) = \varepsilon \cosh^{-2}(\delta x)$, at $\varepsilon = 3$, $\delta = 0.5$ (10)

II. $v(x,0) = h \times w$, at $x \in [-w/2, w/2]$ (11)

The initial distribution (11) is a rectangular step with h being the height of the step and w being its width.

The uniform Derichlet conditions are imposed on both ends of the region $v(x_{min}, t) = v(x_{max}, t) = 0$.

Fig. 1 shows two cases of initial distribution.

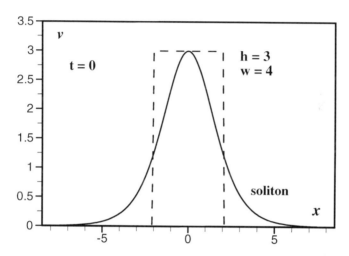

Fig. 1. The initial distributions for soliton (10) and a step function (11) — $h \times w = 3 \times 4$

After numerous tests we have found optimal for solution of (1) explicit MacCormack scheme with flux-corrected procedure ($\Delta x = 0.1$).

Fig. 2 shows the example of the influence of dispersion errors on numerical solution. This influence is quite clear. The solution became non monotone and the causality is violated. Introduction of physical diffusion damps the oscillations but cannot be used for particular case $\alpha = 0$.

One can see, that monotonization procedure helps, otherwise the perturbations cover all numerical region at once.

Detailed numerical modeling of wave's formation process was carried-out for a wide range of α, β, γ parameters. We were more interested in the influence of the source upon waves system formation ($\gamma \neq 0$) dispersion factors impact (β). The analysis of numeric results uncovered an interesting peculiarity of the initial equation (1): we managed to obtain solutions even for "exotic" values of α, β, γ parameters, i.e. for the values that formally exceeded our initial assumptions about coefficients being sufficiently small. A case of negative γ is the least interesting one, as it actually gives the equation and additional smoothening diffusion effect, not unlike a parabolic equation. Initial distribution of parameters was picked in such a way that the "distribution mass" was equal for all the cases (~ 12).

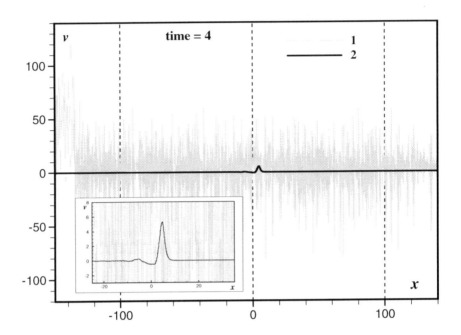

Fig. 2. The comparison of two numerical approaches. Curve 1 is obtained without FCT procedure; curve 2 is shown after FCT procedure; $\alpha = 0$, $\beta = 1$, $\gamma = -0.1$.

This is a regular case, very similar to usual diffusion. But the presence of nonlocal term shows very interesting effect – the solution gets "asymptotic" form much earlier, than in standard case. Of course there is natural explanation of this effect in terms of Fourier components (in linear approximation, of course) interaction.

Here we can face real intramode interaction with ongoing soliton production. As well, as on the next picture, we can see strong influence of nonlocality on the process of soliton birth. That could be very well expected from (4), but calculations make it possible to determine the mechanism of those processes.

On fig.3, 4 the results are shown in presence of source $I(v)$ for different values of $k = 10^{-3} \div 0.1$ and for relatively small times. For such $k's$ the source becomes the principle factor but not dispersion or diffusion. It is clearly seen that specific form of the solution soon produce the profiles, typical for shock waves.

Fig. 5, 6 show the direct influence of the initial distribution on the solution. The main feature was that "mass' of all distributions was kept constant:

$$\int_{-\infty}^{\infty} v_I(x,0)dx = \int_{-\infty}^{\infty} v_{II}(x,0)dx$$

For the case on fig. 5, it is clear, that distributions are similar. For both cases the amplitudes of distributions are equal and physical dispersion is small. For large width of the initial distribution the dispersion becomes key factor and we can see soliton like perturbation but with symmetry violation.

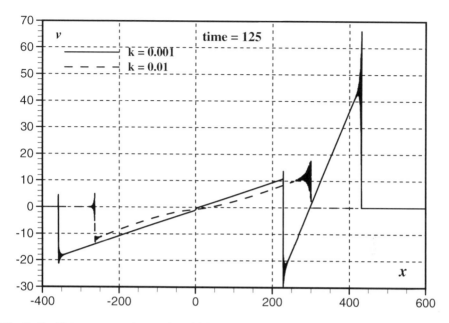

Fig. 3. Coefficients comparison at $k = 0.001$ and $k = 0.01$ for an exponential kernel function (8) for soliton initial distribution (10). $\alpha = -0.01$, $\beta = 0.05$, $\gamma = 0.05$.

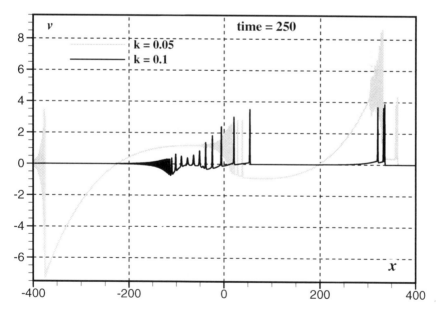

Fig. 4. Coefficients comparison at $k = 0.05$ and $k = 0.1$ for an exponential kernel function (8) for soliton initial distribution (10). $\alpha = -0.01$, $\beta = 0.05$, $\gamma = 0.05$.

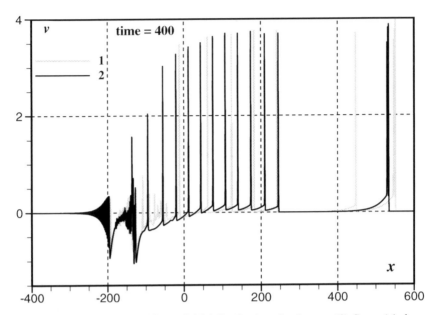

Fig. 5. Results comparison for different initial distributions for the case (8):Curve 1 is $h \times w = 3 \times 4$ (11); curve 2 is a soliton (10). $\alpha = -0.01$, $\beta = 0.05$, $\gamma = 0.05$.

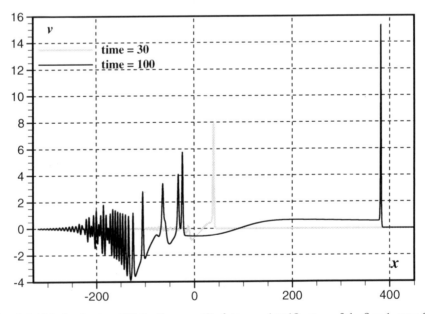

Fig. 6. Initial distribution (11) for the case (8): $h \times w = 1 \times 12$, $\alpha = -0.1$, $\beta = 1$, $\gamma = 0.1$, $k = 0.1$

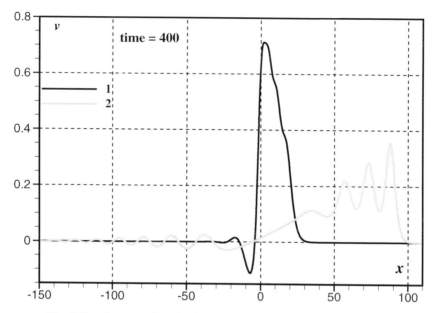

Fig. 7. Results comparison for the kernel (8) and soliton distribution (10);
$\alpha = -0.001, \beta = 0.5, \gamma = -0.05$; curve 1 — $k = 0.1$; curve 2 — $k = 0.001$

This is much more case of disperse waves. Nevertheless we can see again the influence of nonlocal interactions on the production of solitons.

It is surprising; that we cannot detect a pilot waves in this case. Of course they cannot survive in asymptotic region, but the values of parameters in principle make it possible to generate them for intermediate time values.

In that sense it is interesting to make calculations for some nonphysical case, when intramode interaction is stronger, than diffusion and dissipation. As it is seen from the next picture (fig.8) in such situation the solitons expire and we can see irregular structures without any conservation laws visible.

From those and many other examples we can conclude, that

1. Proposed numerical scheme work even for very rough distribution and is optimal for future standard code.
2. Quasiclassical analytical estimations are very effective at least for asymptotical region of solution.
3. There are no substantial bottlenecks for GPGPU onboard memory and the attempts to use heterogeneous systems for 3D computations are justified.

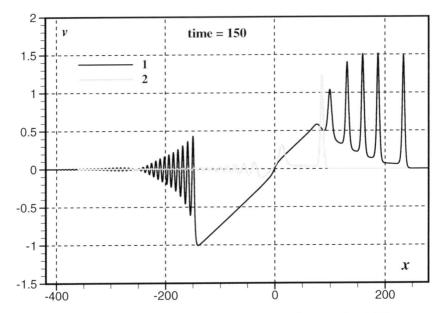

Fig. 8. A case of sub-integral function (9) at $\xi = 2$; $h \times w = 3 \times 4$ (11).
Curve 1 — $\alpha = -0.05, \beta = 1, \gamma = 10$; curve 2 — $\alpha = -0.03, \beta = 1, \gamma = 25$.

4   Conclusion

The above considerations show, that simple quasiclassical approach with some additional effort can make it possible to get a lot of information about the solution of difficult nonlinear problem. This information can be used for optimization of numerical approach for its solution. Some preliminary tests of such approach make the use of powerful heterogeneous systems for its realization very promising. Some ideas to turn to 3D computations effectively were discussed in our presentation [6].

One should clearly understand that such systems are not a panacea: while there are many tasks that can be smoothly mapped on GPGPU there are some classes of algorithms that cannot benefit from implementing them for GPGPU. But hybrid systems are constantly evolving that leads to performance growth while preserving and improving GFLOPS/Watt ratio. Software for hybrid systems is improving too. Software companies develop new libraries for GPGPU and applications that use such libraries, there are already several standards for GPGPU (e.g. OpenCL, OpenACC). And we believe that in near future we will have vector accelerator in hands for solution of more difficult nonlinear problems.

References

1. Karpman, V.I., Maslov, E.M.: JETP **2**(8), 537–548 (1977). (In Russian)
2. Keener, J.R., Mclaughlin, D.W.: Phys. Rev. A **16**(2), 777–790 (1977)
3. Maslov, V.P.: Complex WKB method in nonlinear equations. Moscow, Nauka, (1977). (In Russian)
4. Bogdanov, A.V., Vakulenko, S.A., Strelchenya, V.M.: Tchislennye Metody Mechaniky Sploshnoy Sredy **11**(3), 18–26 (1980). (InRussian)
5. Witham, G.B.: Linear and nonlinear waves. John Wiley & Sons, NewYork–London (1974)
6. Bogdanov, A., Stankova, E., Mareev, V.: High performance algorithms for multiphase and multicomponent media. In: 14th Ship stability workshop, UTMSPACE, Malaysia, pp. 242–245 (2014)

Flexible Configuration of Application-Centric Virtualized Computing Infrastructure

Vladimir Korkhov[✉], Sergey Kobyshev, and Artem Krosheninnikov

St. Petersburg State University,
Universitetskii 35, Petergof, 198504 St. Petersburg, Russia
vkorkhov@apmath.spbu.ru

Abstract. Virtualization technologies enable flexible ways to configure computing environment according to the needs of particular applications. Combined with software defined networking technologies (SDN), operating system-level virtualization of computing resources can be used to model and tune the computing infrastructure to optimize application performance and optimally distribute virtualized physical resources between concurrent applications. We investigate capabilities provided by several modern tools (Docker, Mesos, Mininet) to model and build virtualized computational infrastructure, investigate configuration management in the integrated environment and evaluate performance of the infrastructure tuned to a particular test application.

Keywords: Virtualization · Software defined networks · Virtual cluster · Cloud computing

1 Introduction

Virtualization refers to the act of creating a virtual version of an object, including but not limited to a virtual computer hardware platform, operating system (OS), storage device, or computer network resources [1]. It can be divided onto several types, and each one has its own pros and cons. Generally, hardware virtualization refers to abstraction of functionalities from physical devices. Nowadays, on modern multicore systems with powerful hardware it is possible to run several virtual guest operating systems one physical node. In a usual system a single operating system uses all hardware resources it has (CPU, RAM, etc), whilst virtualized system can use a special layer that spreads low-level resources to several systems or applications. This technique hides the existence of a virtual layer so that it looks like a real machine for launched applications.

In the classical network management, network traffic is distributed over routers ranging between network nodes. With this approach, you must configure each router separately; the failure of one node may be detrimental to the work of a network segment. With a large number of network nodes, this problem is compounded, as the problem is the large number of manufacturers and different software content. A low-level piece of information in networks is a packet. Normally when transmitting a packet in a network segment, a separate route is not

© Springer International Publishing Switzerland 2015
O. Gervasi et al. (Eds.): ICCSA 2015, Part IV, LNCS 9158, pp. 342–353, 2015.
DOI: 10.1007/978-3-319-21410-8_27

created, thus the packets are duplicated in all possible ways. However, there is a recent approach, devoid of these shortcomings – Software Defined Networking (SDN).

The main difference between SDN and traditional networks is that software processing of networking data occurs in the network on the controller, which determines what and where it is necessary to deliver. If necessary, the transfer of information occurs after initial request to the controller, which in turn creates a route for the data flow. Thus, the entire infrastructure network is constructed using simple switches, the main task of which is to store routing tables and forward packets according to this table. Also it simplifies the structure of the package, part of the meta data required for conventional networks. When the output node of the system controller catches this change, it changes in the optimal routes.

OpenFlow [2] is the first standard interface for realizing SDN that can decouple the data and control plane to provide scalable network management. To validate the performance and features of the OpenFlow standard, many researchers have utilized specialized hardware network devices such as NetFPGA. However, these devices are not suitable for implementing a small-scale SDN testbed due to high cost, complexity, and specialized programming languages. One of the options is a well-known SDN emulator, Mininet [17], which is widely utilized. In our research we also use Mininet to represent configurable network infrastructure.

These technologies facilitate creation of a virtual supercomputer or virtual clusters that are adapted to problem being solved and to manage processes running on these clusters (see Figure 1). The work presented in this paper continues our earlier research [9,11].

Virtual clusters

- Collection of virtual machines working together to solve a computational problem
- Can be configured by advanced users; they know exactly what they want

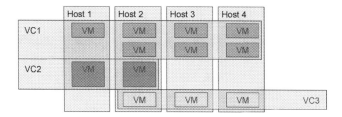

I. Gankevich et al. Constructing Virtual Private Supercomputer Using Virtualization and Cloud Technologies, ICCSA'14, 02.07.2014

Fig. 1. A testbed example with a set of virtual clusters over several physical resources

In this work, we evaluate the capabilities obtained while integrating the OS-level virtualization technology and SDN to build a computational environment with configurable computation (CPU, memory) and network (latency, bandwidth) characteristics. Such configuration enables flexible partitioning of available physical resources between a number of concurrent applications utilizing a single infrastructure. Depending on application requirements and priorities of execution each application can get a customized virtual environment with as much resources as it needs or is allowed to use.

Section 2 gives an overview of related work in the area of software defined networks and virtual testbeds built with their help. Section 3 describes virtualization techniques for computing and networking. Section 4 presents the methods and approaches to build and configure the virtual computing environment along with some results of its experimental evaluation. Section 5 discusses the experience and observed experimental results; and Section 6 concludes the paper.

2   Related Work

In the current paper we investigate the ways to organize virtual clusters based on virtualization technologies both for computing and network resources. Our earlier approach to this issue was presented in [10, 11]. This work also develops the ideas of resource selection and adaptive workload balancing introduced in [12].

One of the first approaches to construct clusters from virtual machines was proposed in [3] and was partially implemented in In-VIGO [4], VMPlants [5], and Virtual Clusters on the Fly [6] projects.

The idea of an adaptive virtual cluster changing its size based on the workload was presented in [8] describing a cluster management software called COD (Cluster On Demand), which can dynamically allocate resources to multiple virtual clusters from a common resource pool.

The work presented in [13] describes an approach to investigate heterogeneous networks based on Linux containers and software emulators. It introduces a modular and flexible testbed NetBoxIT used to study architecture and implementation issues of building virtual environment with several emulators on a single, multi-core hardware platform.

The goal of our research is to create a flexible configurable computing environment that allows us to tune both computation and communication characteristics of the testbed according to requirements of the application. To create a separate environment for a particular application we employ virtualization technologies both for computing and network resources.

Our main concern is using OS-level or container-based virtualization for computational resources as it introduces little overhead compared to other types of virtualization. For network virtualization we address Software-Defined Network (SDN) technology that can allow more control over routing thus optimize network transfers for communication-intensive applications.

Research works on the subject of virtual clusters can be divided into two broad groups: works dealing with provisioning and deploying virtual clusters in high performance environment or Grid and works dealing with overheads of virtualization. Works from the first group typically assume that virtualization overheads are low and acceptable in high performance computing and works from the second group in general assume that virtualization has some benefits for high performance computing, however, the authors are not aware of the work that touches both subjects in aggregate.

In [14] authors evaluate overheads of the system for on-line virtual cluster provisioning (based on QEMU/KVM) and different resource mapping strategies used in this system and show that the main source of deploying overhead is network transfer of virtual machine images. To reduce it they use different caching techniques to reuse already transferred images as well as multicast file transfer to increase network throughput. Simultaneous use of caching and multicasting is concluded to be an efficient way to reduce overhead of virtual machine provisioning.

In [15] authors evaluate general overheads of Xen para-virtualization compared to fully virtualized and physical machines using HPCC benchmarking suite. They conclude that an acceptable level of overheads can be achieved only with para-virtualization due to its efficient inter domain communication (bypassing dom0 kernel) and absence of high L2 cache miss rate when running MPI programs which is common to fully virtualized guest machines.

In contrast to these works the main principles of our approach can be summarized as follows. Do not use full or para-virtualization of the whole machine but use virtualization of selected components so that overheads occur only when they are unavoidable (i.e. do not virtualize processor). Do not transfer opaque file system images but mount standard file systems over the network so that only minimal transfer overhead can occur. Finally, amend standard task schedulers to work with virtual clusters so that no programming is needed to distribute the load efficiently. These principles are paramount to make virtualization lightweight and fast.

3   Virtualization for Computing and Networking

Hardware virtualization provides an abstraction of functionalities from actual hardware components. In modern computer hardware with powerful multi-core processors, virtualization allows multiple virtual guest operating systems (Virtual Machines or GuestOSs) to transparently share the physical machine resources avoiding conflicts. In a traditional non-virtualized system a single operating system distributes available hardware resources among applications; in turn, a virtualized system uses a special software layer (hypervisor or Virtual Machine Monitor) to schedule and multiplex the low-level shared resources among high-level applications and operating systems.

Virtualization has several advantages, for example, increased level of security (applications are launched in an isolated sandbox with limited access outside), easier live backup and migration, more efficient usage of resources, fast deployment of pre-configured application servers and so on.

Speaking about live migration, Linux containers (LXC) [22] support Checkpoint/Restore In Userspace (CRIU) since 1.1.0 release. CRUI gives a possibility to save the state of one or more applications and after some time restore their work from the saved state, even if the system is rebooted or this application is restored on another machine. In terms of operating-system-level virtualization it helps to scale and distribute containers to more powerful hosts in case more resources are needed or reboot is required. However, advantages do not come without disadvantages. The performance can depend on other virtual instances running on the same host; in case of so-called full virtualization, when hardware is completely virtualized, the presence of this layer can dramatically decrease performance. VirtualBox, KVM, VmWare are the examples of full virtualization solutions.

Another approach called para-virtualization, when guest OS has direct access to hardware, removes virtual layer constraint. XEN is a good example of para-virtualization.

The last type of virtualization is a more recent technique called an operating system level (OS-level) or container-based virtualization. This approach allows us to execute applications running in containers of the host operating system. It uses subsystems of OS kernel that provide isolation of applications in virtual environments. Containers provide performance a bit less or equal to native. It is also possible to get direct access to hardware. Moreover, applications running in containers have low start-up costs and effectively use HDD space.

This virtualization method was originally developed as an enhancement of the UNIX 'chroot' mechanism, however it offers much more isolation and better resource control and management. In practice, the kernel of the hosting operating system is modified to allow for multiple isolated user-space instances instead of just one; resource management is available to split the available resources among containers and to limit their impact on each other. This form of virtualization usually brings little overhead, since applications within containers use the host system API interface directly and do not need to rely on hardware emulation or virtual machine monitors. OpenVZ, Linux-VServer, Virtuozzo, FreeBSD Jails, Docker [23] and LXC [22] are the most popular container-based tools.

For network virtualization the following tools can be used: Mininet [17], a system for rapid prototyping of virtual networks on a single laptop: virtual nodes are represented by different namespaces, whilst the network links are based on Virtual Ethernet peers [18]; in-kernel Linux Traffic Control can be employed to apply bandwidth limitations and QoS policies to a link; CORE (Common Open Research Emulator, [19]), where virtual networks are created using the Linux bridging tools, which allow the modulation of bandwidth, propagation delay, packet errors probability etc.; NS-3 [20].

4   Building Flexible Virtual Computing Environment and Its Experimental Evaluation

Experiments show that using lightweight virtualization technologies (para-virtualization and application containers) instead of full virtualization is advantageous in terms of performance [21], hence virtual computing nodes should be created using lightweight virtualization technologies only. However, not every operating system supports these technologies and it should be possible to access virtual supercomputer facilities through fully-virtualized hosts. So, lightweight virtualization is inevitable in achieving balance between good performance and ease of system administration in distributed environment and as a consequence operating system should be UNIX-like for it to work.

Load balance can be achieved using virtual processors with controlled clock rate and process migration. The first technique allows balancing coarse-granularity tasks and the second is suitable for fine-grained parallelism.

The greatest thing containers give to our research is that it allows to quickly prototype networks and test how software that requires some traffic exchange works. In order to create a network, several containers are created and it is possible to attach to them and do anything inside there. But the main idea is to launch distributed application on several nodes. One of the most popular applications for evaluation of computing infrastructure is HPCC (HPC benchmark) [24].

HPCC is a set of tools combined to simultaneously measure and test different aspects of clusters' performance. It consists of basically 7 tests, to name a few: HPL (a well-known Linpack benchmark which measures the floating point rate of execution for solving a linear system of equations; PTRANS (parallel matrix transpose) and others. We are specifically interested in PTRANS benchmark since it measures the rate of transfer for large arrays of data from multiprocessor's memory. Docker – a lightweight and powerful open source container virtualization technology which we use to manage containers – has a resource management system available so it is possible to test different configurations: from "slow network and slow CPUs" to "fast network and fast CPUs".

First of all, a CentOS7 image was used to launch a minimal version of a container, then openssh, openmpi and openblas were installed. Unfortunately, a lot of standard software packages are installed not in usual place so before compilation there is a need to export PATH and libs, but once it was done, this prepared image was saved. This is a great advantage of container-based virtualization, because all steps were done only once. After that, this newly created image was used to run several containers with identical internals. Each container instance has its own ID and IP-address so it is possible to access to them via ssh. That was the way HPCC was launched on 4 containers (see figure 2).

Even though container-based virtualization is easy to run and use, it's not often easy and user-friendly to scale configuration or to limit resources. This is where Apache Mesos [25] and Mesosphere Marathon [26] were used. Apache Mesos abstracts CPU, memory, storage, and other compute resources away from machines (physical or virtual), enabling fault-tolerant and elastic distributed

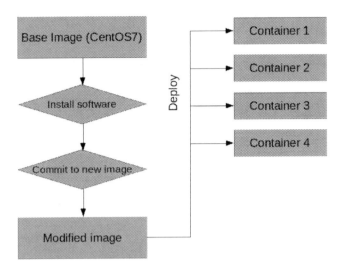

Fig. 2. Deployment of images on containers

systems to easily be built and run effectively. At a high level Mesos is a cluster management platform that combines servers into a shared pool from which applications or frameworks like Hadoop, Jenkins, Cassandra, ElasticSearch, and others can draw. Marathon is a Mesos framework for long-running services such as web applications, long computations and so on.

The goal of our research is to investigate the influence of infrastructure characteristics to the performance of applications. To do this we create a flexible testbed based on OS-level virtualization for computational resources and OpenFlow-enabled networks. This enables fine-grained tuning of both computational power of virtualized computing nodes (based on containers) and network bandwidth/latency.

For the sample applications we use test application kernels with known characteristics on computational and communicational demands. Test executions are performed on a set of infrastructure configurations, ranging from "slow nodes-little memory-slow network" to "fast nodes-much memory-fast network".

Table 1 shows sample results of PTRANS, StarDGEMM, and HPL benchmarks executed on different testbed setups. The experiments were performed on a single multicore machine, with 4 virtual nodes (containers created with Docker) on it. It is not possible to manage network performance though Marathon/Mesos, so only memory limits and CPU limits were set in these tests. We can also say that those tests are heavily dependant on background processes running on the same machine, so the first three containers can get 25 of all CPU available and the last one can get less than 25 due to OS needs. The second thing is that by default the `memory.memsw.limit_in_bytes` value is set to twice as much as the memory parameter we specify while starting a container ($-m$). What does

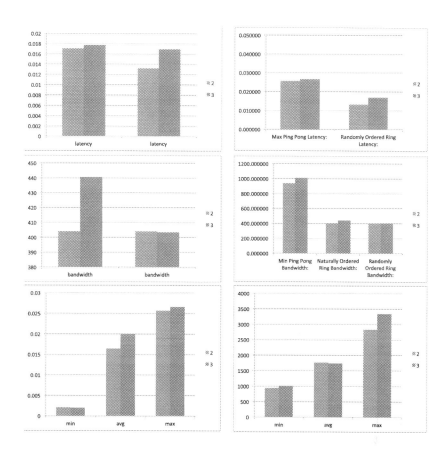

Fig. 3. Sample evaluation of the testbed communications (2 and 3 virtual nodes)

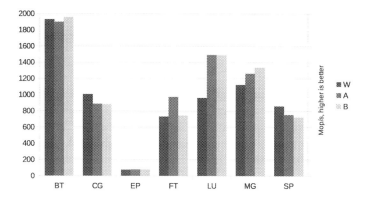

Fig. 4. Performance of different tests from NAS Parallel Benchmarks suite on different configurations

Table 1. Sample execution results for a number of benchmarks on 4 virtual nodes

	PTRANS (GB/s)	StarDGEMM (Gflops)	HPL (tflops)
256mb, 25% CPU, 5000 matrix size	0.937	0.478	0.0055
512mb, 25% CPU, 5000 matrix size	0.736	0.474	0.0063
512mb, 25% CPU, 15000 matrix size	0.724	0.475	0.0026

the `memory.memsw.limit_in_bytes` parameter say? It is a sum of memory and swap. This means that Docker will assign to the container $-m$ amount of memory as well as $-m$ amount of swap. The current Docker interface does not allow us to specify how much (or disable it entirely) swap should be allowed, so we need live with it for now. The last test (15000 matrix size) was too large and did not fit into the memory, so swap was used, and this resulted in decreased performance.

Considering the test results, we observe obvious patterns such as no influence of network capacity to the performance of embarrassingly parallel applications (with low communication demands). On the other hand, we can observe different trends for the applications with different computation/communication ratio on different testbed configurations. The example of experimental results for communication performance is presented in figure 3. Here the virtual network is evaluated with the Latency-Bandwidth-Benchmark R1.5.1 (c) HLRS on different number of virtual nodes.

Figure 4 shows the experimental results for execution of NAS Parallel Benchmarks (NPB) suite on different configurations of virtual testbed. With NPB results are also very different, everything depends on benchmark type. For example, for SP test smaller size system of nonlinear PDEs had better Mop/s than for bigger size. However, for lower matrices sizes in LU test results are worse than for bigger matrices. Table 2 shows detailed results of experimental runs.

5   Discussion

The presented approach can be used as an enabling part of the virtual supercomputer concept [10,11] to ensure proper and efficient distribution of resources between several applications. Knowing the application demands in advance we can create appropriate infrastructure configuration giving just as much resources as needed to each particular instance of a virtual supercomputer running a particular application. In such a way, free resources can be controlled and granted to other applications without negative effect on current executions.

Table 2. NPB execution results

.	W	A	B
BT (Block Tridiagonal)	Size = 24x24x24 Mop/s total = 1928.15	Size = 64x64x64 Mop/s total = 1898.05	Size = 102x102x102 Mop/s total = 1957.95
CG (Conjugate Gradiant)	Size = 7000 Iterations = 15 Mop/s total = 1006.12	Size = 14000 Iterations = 15 Mop/s total = 887.33	Size = 75000 Iterations = 75 Mop/s total = 881.61
EP (Embarrassingly Parallel)	Size = 67108864 Mop/s total = 74.67 Operation type = Random numbers generated	Size = 536870912 Mop/s total = 77.14 Operation type = Random numbers generated	Size = 2147483648 Mop/s total = 78.08 Operation type = Random numbers generated
FT (Fast Fourier Transform)	Size = 128x128x32 Iterations = 6 Mop/s total = 733.28	Size= 256x256x128 Iterations = 6 Mop/s total = 973.63	Size = 512x256x256 Iterations = 20 Mop/s total = 747.78
LU (Lower-Upper symmetric Gauss-Seidel)	Size = 33x33x33 Iterations = 300 Mop/s total = 961.05	Size = 64x64x64 Iterations = 250 Mop/s total = 1491.43	Size = 102x102x102 Iterations = 250 Mop/s total = 1487.62
MG (MultiGrid)	Size = 128x128x128 Iterations = 4 Mop/s total = 1123.20	Size = 256x256x256 Iterations = 4 Mop/s total = 1260.53	Size = 256x256x256 Iterations = 20 Mop/s total = 1333.47
SP (Scalar Pentadiagonal)	Size = 36x36x36 Iterations = 400 Mop/s total = 862.24	Size = 64x64x64 Iterations = 400 Mop/s total = 755.75	Size = 102x102x102 Iterations = 400 Mop/s total = 722.84

6   Conclusions and Future Work

It is known that virtualization improves security, resilience to failures, substantially eases administration due to dynamic load balancing [9] while does not introduce substantial overheads. Moreover, a proper choice of virtualization package can improve CPU utilization.

Usage of standard virtualization technologies can improve overall throughput of a distributed system and adapt it to problems being solved. In that way virtual supercomputer can help people efficiently run applications and focus on domain-specific problems rather than on underlying computer architecture and placement of parallel tasks.

The goal of the current work was to investigate the influence of characteristics of virtualized infrastructure to the performance of applications. To do this we proposed a flexible testbed based on OS-level virtualization for computational resources and OpenFlow emulation for networks. This enabled fine-grained tuning of both computational power of virtualized computing nodes (based on containers) and network bandwidth/latency.

In future, we plan to present an automated solution to create virtualized computing infrastructures based on application demands that are either provided by used or evaluated automatically in a series or preliminary test runs.

Acknowledgments. The research was supported by Russian Foundation for Basic Research (project N 13-07-00747) and Saint Petersburg State University (projects N 9.38.674.2013, 0.37.155.2014).

References

1. Wikipedia. http://en.wikipedia.org/wiki/Virtualization
2. McKeown, N., et al.: OpenFlow: Enabling innovation in campus networks. ACM Communications Review **38**(2), 69–74 (2008)
3. Figueiredo, R.J., Dinda, P.A., Fortes, J.A.B.: A case for grid computing on virtual machines. In: Proceedings of the 23rd International Conference on Distributed Computing Systems (2003)
4. Matsunaga, A.M., Tsugawa, M.O., Adabala, S., Figueiredo, R.J., Lam, H., Fortes, J.A.B.: Science gateways made easy: the In-VIGO approach, Concurrency andComputation: Practice and Experience, vol. 19(6), pp. 905–919 (April 2007)
5. Krsul, I., Ganguly, A., Zhang, J., Fortes, J.A.B., Figueiredo, R.J.: VMPlants:Providing and managing virtual machine execution environments for grid computing. In: Proceedings of the 2004 ACM/IEEE Conference on Supercomputing (2004)
6. Nishimura, H., Maruyama, N., Matsuoka, S.: Virtual clusters on they - fast, scalable, and exible installation. In: CCGRID 2007: Seventh IEEE International Symposium on Cluster Computing and the Grid (May 2007)
7. bibitem Emeneker, W., Stanzione, D.: Dynamic virtual clustering. In: IEEE Cluster 2007, Austin, TX (September 2007)
8. Chase, J.S., Irwin, D.E., Grit, L.E., Moore, J.D., Sprenkle, S.E.: Dynamic virtual-clusters in a grid site manager. In: HPDC 2003: Proceedings of the 12th IEEE International Symposium on High Performance Distributed Computing, p. 90. IEEE Computer Society, Washington, DC (2003)
9. Bogdanov, A.V., Degtyarev, A.B., Gankevich, I.G., Yu. Gayduchok, V., Zolotarev, V.I.: Virtual workspace as a basis of supercomputer center. In: Proceedings of the 5th International Conference Distributed Computing and Grid-Technologies in Science and Education, Dubna, Russia, pp. 60–66 (2012)
10. Gankevich, I., Gaiduchok, V., Gushchanskiy, D., Tipikin, Y., Korkhov, V., Degtyarev, A., Bogdanov, A., Zolotarev, V.: Virtual private supercomputer: Design and evaluation. In: CSIT 2013–9th International Conference on Computer Science and Information Technologies (CSIT) (Revised Selected Papers. 2013, DOI: 10.1109/CSITechnol..6710358)
11. Gankevich, I., Korkhov, V., Balyan, S., Gaiduchok, V., Gushchanskiy, D., Tipikin, Y., Degtyarev, A., Bogdanov, A.: Constructing Virtual Private Supercomputer Using Virtualization and Cloud Technologies. In: Murgante, B., Misra, S., Rocha, A.M.A.C., Torre, C., Rocha, J.G., Falcão, M.I., Taniar, D., Apduhan, B.O., Gervasi, O. (eds.) ICCSA 2014, Part VI. LNCS, vol. 8584, pp. 341–354. Springer, Heidelberg (2014)

12. Korkhov, V.V., Moscicki, J.T., Krzhizhanovskaya, V.V.: User-Level Scheduling of Divisible Load Parallel Applications with Resource Selection and Adaptive Workload Balancing on the Grid. IEEE Systems Journal **3**(1), 121–130 (2009)
13. Calarco, G., Casoni, M.: On the effectiveness of Linux containers for network virtualization. Simulation Modelling Practice and Theory **31**, 169–185 (2013)
14. Chen, Y., Wo, T., Li, J.: An effcient resource management system for on-line virtual cluster provision. In: Proc. of International Conference on Cloud Computing (CLOUD), p. 7279. IEEE (2009)
15. Ye, K., Jiang, X., Chen, S., Huang, D., Wang, B.: Analyzing and modeling the performance in Xen-based virtual cluster environment. In: Proc. of the 12th International Conference on High Performance Computing and Communications (HPCC), pp. 273–280. IEEE (2010)
16. Gupta, D., Yocum, K., Mcnett, M., Snoeren, A.C., Vahdat, A., Voelker, G.M.: To infinity and beyond: time warped network emulation. In: ACM Symposium on Operating Systems Principles (2005)
17. Lantz, B., Heller, B., McKeown, N.: A network in a laptop: rapid prototyping for software-defined networks. In: Proceedings of the 9th ACM SIGCOMM Workshop on Hot Topics in Networks, Hotnets-IX, pp. 19:1–19:6. ACM, New York (2010)
18. VETH. http://code.google.com/p/veth
19. Ahrenholz, J.: Comparison of core network emulation platforms. In: Military Communications Conference, MILCOM 2010, pp. 166–171 (2010)
20. NS-3 Project Homepage. http://www.nsnam.org
21. Barham, P., Dragovic, B., Fraser, K., Hand, S., Harris, T., Ho, A., Neugebauer, R., Pratt, I., Warfield, A.: Xen and the art of virtualization. ACM SIGOPS Operating Systems Review **37**(5), 164–177 (2003)
22. The LXC Linux Containers. http://lxc.sourceforge.net
23. The Docker platform. https://www.docker.com
24. Dongarra, J., Luszczek, P.: HPC Challenge: Design, History, and Implementation Highlights. In: Vetter, J. (ed.) Contemporary High Performance Computing: From Petascale Toward Exascale. CRC Computational Science Series. Taylor and Francis, Boca Raton (2013)
25. Apache Mesos. http://mesos.apache.org/
26. Mesosphere Marathon. https://mesosphere.github.io/marathon/

Distributed Collaboration
Based on Mobile Infrastructure

Suren Abrahamyan[1]([✉]), Serob Balyan[1], Harutyun Ter-Minasyan[2],
Alfred Waizenauer[3], and Vladimir Korkhov[1]

[1] Saint Petersburg State University, Universitetskaia nab., Saint Petersburg 199034, Russia
`suro7@live.com`, `{serob.balyan,vkorkhov}@gmail.com`
[2] RWTH Aachen University, Templergraben 55, Aachen 52062, Germany
`hterminasyan@gmail.com`
[3] Osensus GmbH, Nibelungenstraße 20a, Passau 94032, Germany
`alfred.waizenauer@osensus.com`

Abstract. There are several types of infrastructures to allow people to interact distributed. Together with development of information technologies, the use of such infrastructures is gaining momentum at many organizations and opens new trend in scientific researches. On the other hand, rapid increase of mobile technologies allows individuals to apply solutions without need of traditional office spaces and regardless of location. Hence, realization of certain infrastructures on mobile platforms could be useful not only in daily purposes but also in terms of scientific approach. Implementation of tools based on mobile infrastructures can vary from basic internet-messengers to complex software for online collaboration equipment in large-scaled workgroups. Although growth of mobile infrastructures, applied distributed solutions in healthcare and group decision-making are not widespread. To increase mobility and improve usage of current solutions we propose innovative tools for real-time collaboration on smart devices, which will be described in this article.

Keywords: Scientific mobile infrastructures · Digital collaboration

1 Introduction

Collaboration is not a new term or practice by human beings. Long before technologies, people engaged in a process whereby "individuals and/or groups work together on a practical endeavor". In fact, collaboration has always been "a fundamental feature" of every organizational type [1]. What has changed is that today's information and communication technologies (ICT) provide support for new types of collaboration through "network formation and support". This new ICT enabled "*e-collaboration*" has become an important area of study [2].

E-collaboration gets distributed work done with others through the assistance of network-based services, software, hardware or applications.

E-collaboration is gaining momentum at many organizations, using IT tools such as interactive online meetings, social networks, blogs, twitters, wikis, bookmarks,

© Springer International Publishing Switzerland 2015
O. Gervasi et al. (Eds.): ICCSA 2015, Part IV, LNCS 9158, pp. 354–368, 2015.
DOI: 10.1007/978-3-319-21410-8_28

forums, podcasts, and any other imaginable way of reinventing. People can interact with each other from the application on a smartphone or a browser on a desktop. Participants use online collaboration tools like e-mail, chat, discussion forums, Web conferencing, shared whiteboards, and file sharing to promote collaboration. Chat can be either voice- or text-based, point-to-point or in a conference environment. Conferences can be open, where everyone has the ability to speak and interact, or moderated. People use interactive online meetings to make a sales presentation, demonstrate applications, and review contracts online. People also provide video training or consulting in environments for customers, partners, and employees in any location [2]. In addition, e-collaborations can be used for educational purposes. Examples of such platforms are Google Classroom (https://classroom.google.com/) and Coursera (https://www.coursera.org). Many of universities opened online learning and qualification improvement services for students, and world's top universities created a free collaborative space called EDX (https://www.edx.org). This allows users to attend internet-based courses and get certified by top universities like Berkeley, Harvard, MIT, etc.

We are also seeing more frequent use of webinars and online press briefings. It is not limited to meeting activities. These techniques are also being used to provide IT support for distributed users with remotely controlled desktops, to see and fix the problems in real time.

Organizations or group of individuals, who use e-collaboration for group decision-making purposes, requires infrastructure with special hardware as a server to provide data processing and storage, reliable internet connection with appropriate bandwidth for information transportation, visualization devices like projectors or smart boards. Despite this important hardware, it may need also qualified management and support. Therefore, organizations should have their own resources or buy services from other organizations who own infrastructures like mentioned above. In both cases, users will face with strong financial efforts. However, depending on number of involved people and type of tasks that they should solve with the help of collaboration system, need of costly infrastructures could be not reasonable or acceptable. As people use mobile handheld devices in work process and trend like Bring Your Own Device (BYOD) has forced individuals to reexamine their approach to smartphone and tablet use, we propose realization of cheaper and portable solutions based on mobile infrastructure for user groups.

Despite advantages, which gives e-collaboration to group of people, it can be also extremely important in spheres like healthcare for people with disabilities due to ability to eliminate distance. On the other hand, use of mobile devices will increase their opportunities to communicate.

In this article, we propose innovative tools based on mobile infrastructures for healthcare and real-time group decision-making to increase mobility and improve usage of existing solutions.

2   Challenges in Mobile Development

Worldwide smartphone and mobile devices market grows in large temps and billions gadgets connected to internet every day. Like in any other area of programming, particularly in mobile development, there are several challenges.

- device fragmentation,
- OS fragmentation,
- development approach

Fragmentation is the inability to "write once and run anywhere"[4]. This is the situation, when there is no possibility to compile and get the same behavior of application on different vendor/platform device with the same source code.

Let us see three of more common situations that programmer faces during preparation and coding.

2.1   Device Fragmentation

Different mobile devices have different screen sizes, pixel-densities (pixels per inch), processors and mobile architectures, memory, graphic units, communication capabilities, etc. First generation mobile devices had more or less similarities in terms of processing and screen sizes, but now, in the new era of wearables and connectables, people like to have an application for their all devices, with the same form and functionality. For the best result, developers must take into account and optimize application to fit all possible user's needs. In addition, that is not all problems - with every device and OS updates, developers should consider about new design, guidelines and viewpoints from OS/platform vendors. Aside from that, developing an application for each platform individually will grow the time and cost and adds also maintenance cost of all those different versions of applications. Of course, there are several software solutions to write once for main mobile platforms, which we will see, on the last point, but in general, the problems coming from device fragmentation, are increasing.

2.2   OS Fragmentation

In terms of mobile operating systems, Android is still undisputed and holds largest stake with about 76%. Apple's iOS is in second place, shipping 75 million units in 2014 and taking 20% of market. Other main players are Microsoft's Windows Phone and Blackberry OS accordingly with 2.8% and 0.4% [3]. As we know, it is important to support cross-platform availability for mobile tools. Developing an application or tool for leading platforms is necessary to cover and attract as much users as possible. Multi-platform development can be done simultaneously or one-by-one.

2.3   Development Approach

Traditional of coding way or **native approach** is most stable but not always best approach. In this case, developers use special IDE's and SDK's provided by platform developers and optimized for certain platform. Depending on platform, some capabilities and options may or may not be possible to use on others. By using an SDK, developer targets only on specific mobile OS with its own programming language, like Java for Android and Blackberry OS, Objective C for iOS, C# for Windows Phone, etc. Natively developed applications are not compatible with other mobile operating systems and the main problem after development is in rewriting application on other platforms [4].

In some cases, lack of knowledge in appropriate programming language or time shortage can be a reason for special type of mobile development, called **web-based approach**. This gives users to create programs in simplest way, saving time and money. Given solution sometimes called HTML5 mobile applications, as essentially they are web pages, designed with mobile application's look-and-feel using HTML5, CSS3 and JavaScript. Whereas its use a mobile browser as an engine, they are very limited, highly dependent on device's performance and hardware resources may represent in diverse way in different browsers. Of course it can be very helpful and enough in certain situations, this approach provides weakest performance, usability and potentiality for both developers and users.

Hybrid approach combines advantages of both native- and web-approaches. It is done by enveloping web-based application into native container, which provides access to device's native features and hardware. In this case, developer uses mainly web technologies and scripting languages like JavaScript to create applications, called Hybrid programs. Some cross-platform frameworks, like Cordova, Titanium, Xamarin, etc., make possible to control main capabilities and features of mobile device, and automatically build source code for other platforms. Although it seems very attractive to use such frameworks at first sight, but there are some disadvantages, too. Problems like lack of performance, incompleteness of provided SDK, lack of plugins, not very frequent updates from community or weak graphics support, are main reasons that many of developers try to avoid such frameworks.

3   Key-Points (criterias)

As described before, importance of mobile infrastructures is in providing mobility and increasing electronic communication abilities. Let us see which other key-points are required and mandatory in mobile e-collaboration tools.

3.1   Security

With the development of information technology and software in particular, the protection of the intellectual labor becomes one of the most important directions of science development.[5] To have a secure software in a mobile environment we should consider particular qualities of mobile devices (mobility, limited memory and processing power etc.). Security itself can solve one or complex tasks such as secure data transfer, storing, protection from illegal copy, use, modification etc. [6]. If an application should interact with personal/business data, it would be safer to allow only authorized users to access these data. To have a secure communication between devices encryption should be used, and it is necessary to use key-exchange mechanism (standard solutions like SSL or solutions based on them or unexpected steganography methods) to avoid attacks like man-in-the-middle.

3.2   Price

Another key-point is the price of an application. By determining the right price for mobile application, it would be easier to reach the target audience and have a profit at the same time. Figuring out this price is not so easy. It is all about finding the right balance between what it costs to create and support the application, and how many users will be ready to pay for it.   There are several common pricing types for mobile applications:

- Plans

The idea of this pricing type is to offer users a free version (lite) of a product or a service with limited functionality but charge for more advanced(premium) fully functional version: if users like the lite version they will pay for better, full version. Also, there can be different plans of application with various prices depending on its functionality. For collaboration tools, which allow multiple users to operate with it, it is common to have several pricing plans depending on supported users count.

- In-app purchases

This type assumes that users will be offered to buy additional content, services or subscriptions within the mobile application. Both free and paid applications can offer in-app purchases.

- In-app advertising

In-app advertising can give ongoing revenue to application creators. They will be paid once for an application download, but can continue being paid through advertising. There is a huge range of potential advertisements — everything from small banners to full screen popovers. It is important to manage user expectations for how annoying the advertisements could be. Users in general expect that free versions of paid applications will contain more advertisements. Of course, it is not encouraged to use advertisements in e-collaboration tools, which are created for group decision-making or health care purposes.

3.3   Device-Independency

As we see from above, the main challenges that encounter in every mobile development process are platform/OS fragmentations and the way of programming selected by developers. It is very important to have at least brief notion of work to be done and targeted audience that will mainly use the application. In case when number of developers is limited or development for specific platform is exigent, usually the best solution is to selection the first platform with its appropriate tools, programming language, SDK-s and IDE-s to start immediately. More generic way is to define milestones for each stage of development and separate work to groups based on targeted platform, which will require more investments and people engaged. Using cross-platform tools can be also productive and timesaving, but commonly there are many improvements and need of double coding on top of, so the final decision is up-to developers/managers, which depends on time/money/programmers triangle. Another advantage to the application will bring adding web and PC support (if possible), which in addition with mobile will allow users to put on from any accessible device regardless of location.

3.4   Usability

Usability is a combination of factors and theories, which are targeted to achieve goals like intuitive design and user interface, efficiency of use, error prevention, user satisfaction etc. The more usable application is, the less trainings are necessary for its users. If the application is built around users' needs, they would not have to call a support team for assistance. This reduces the need for online support models and help desk staff, ultimately saving money.

While developing an e-collaboration tool in addition to these goals must be taken into account also other factors, such as network-dependence, number of participants supported by the application. If application requires internet connection, users can work with it from anywhere - where they have access to the internet. But, if collaboration tool is created for cases when group members should be in a same location, there is no need of always-online requirement and participants can interact each other using only local network. This approach will give mobility not the person but the group in general. The choice of network type depends on the tasks, which are going to be solved by a tool. Number of supported participants depends on data type and amount which will be sent simultaneously, network bandwidth, server capabilities etc. Of course, e-collaboration tool will be as useful as many participants will be allowed to use it.

Besides the main points described above, in our opinion, there are several things in e-collaboration application developer's to-do list. From time to time, we can see notifications on our handheld devices about application updates from program stores. This can be annoying sometimes, but one of the important things is to keep users up-to-date with latest version of the applications, where more and more bugs are fixed. Such updates will bring smoother application behavior, security improvements, better integration with OS, etc. Also, users can subscribe to beta channels to receive latest nightly-build versions and see upcoming features of the app. Staying in touch with users via forums, blogs or web-discussions will give a chance to hear from the "other side" of development process, and allow developers to get through dynamic cycle of testing, hearing feedback, new ideas or wishes from users, doing hotfixes, testing again with users. This will increase robustness of a new product and guarantee smarter interaction between developers and final users.

4   Our Solutions

During our research and development in mobile infrastructure sphere, we designed and created two innovative solutions for e-collaboration, first one, for e-collaboration and decision-making, and second, for healthcare, which will be described below.

4.1   Teambrainer

Brainstorming is a group or individual creativity technique by which efforts are made to find a conclusion for a specific problem by gathering a list of ideas spontaneously contributed by its member(s).

The idea and use of e-collaborative tools in brainstorming is not new. Early in 1990's interactive software developed by different tech companies to use in offices and large meetings were first steps to digitize standard way of group collaboration. Even after inventions of interactive whiteboards - analogous to blackboards, which allows rapid marking and erasing of markings on their surface, and smart presenters, people, especially small groups, still prefer to communicate for decision making, generation of ideas or voting on old-school style: using blackboards, brainstorming moderation kits with sticky notes, pen and paper. In some elementary cases it could be the optimal solution, but there are a lot of problems, discomfort and inconvenience during meetings - too much time is wasting on arranging, sorting ideas on sticky notes, not readable ideas, editing possibility is limited, no media (images, videos, presentations) and anonymous voting is possible to use, etc. Although there are programs that partially solve some of problems during e-collaboration process, but currently there is a lack of tool, which combines all these features together.

According to advantages and opportunities from mobile infrastructures and due to narrowness and limitations of applied mobile distributed solutions in group decision-making we decided to create a software for real-time collaboration based on mobile infrastructures to allow people distributively interact each other, increasing mobility and minimizing hardware-dependence. Our tool named Teambrainer (http://teambrainer.com/) encapsulates tools and techniques such as idea generation, voting, analyzing, data storing and visualization, mobility in one smart solution. With Teambrainer, organizing meetings, events, and large discussions can get simpler and smarter.

It is a complex of three different applications each of which has its unique role in whole collaboration process. First, one is called "Cards Application", which should be installed on mobile phones of participants. With the help of this application group members can send their ideas to the centralized mobile server on tablet device ("Board application") and make further interactions. Basically, this application replaces action of creating sticky notes in standard (white board/sticky notes) solution therefore the ideas will appear in the form of cards like sticky notes with written information on them. Second, one is "Board Application" installed on a tablet, which is a main tool of facilitator, a person who plans, guides and organize meetings for group of people. It is a mobile application and a server at the same time. Participants join the same wireless network with facilitator's Boards application, and brainstorming can be done with or without internet connection. All data according to participants' (cards applications') ideas and voting will be stored in this application. Also it will analyze all incoming data, show and save the results. With this application facilitator can control cards applications by doing actions like allowing them to send ideas or to vote for already sent cards. It is a smart version of white board in standard solution. Access to the "Boards Application" has only facilitator, but there are some visual data that need to be shared with other participants also (sent cards' titles, ideas, average voting results, etc.). For this purposes the third application called "Screen Application" is created. It works as a client for "Boards application" and its main goal is to show to participants all data that they need. It can be run in a mobile environment or on a PC as a web application. Perfect choice of device for "Screen Application" would be special cast sticks, like Chrome Cast (http://www.google.com/chrome/devices/chromecast/), which can be connected to any large screens with HDMI input and turn them into "mobile" visualization devices running on Android (https://www.android.com/). After idea generation and voting, analyzing is done, and group decision-making ends with or without consensus by determining standard deviation.

Teambrainer is very simple and intuitive in use and creative groups can focus on their real challenges, the workshop topics, without being distracted. Application enables engaged creativity, collective learning and collaborative decision making processes [7]. In results, we get up to 10 times faster work process for participants. It can be useful in face-to-face real time meetings, workshops, trainings, presentations or conferences for ideation and co-creation, development and management, marketing and sales, voting and decision-making, etc.

Hybrid approach was chosen for implementing all three applications. The reason for this choice is that they all have to be supported by several platforms and Apache Cordova (http://cordova.apache.org/) used as a cross-platform tool. In this approach problems like lack of performance or plugins were solved by writing own plugins with native SDK background.

To give "Board application" centralized server's abilities, we use open-source NanoHTTPd (https://github.com/NanoHttpd/nanohttpd) - a lightweight HTTP server designed for embedding in mobile devices. To connect and exchange data from JavaScript side with NanoHTTPd server, we developed a bridge plugin for Cordova, which allows asynchronous transfer of information to web client and back. By default, NanoHttpd creates one thread per request. It may be not efficient in terms of performance if many devices would send different requests to the server. Also network can be overused if we send multiple requests in short period of time to update data on client devices. That is why we optimized the behavior of NanoHttpd and developed a mechanism that allows keeping alive threads, which can be still useful by keeping them in sleep mode and waking up on action, and kill threads that are not in use as fast as possible.

For testing app performance in strict conditions (in environment of 100 simultaneously active participants), we simulated a network of 100 different devices (IP addresses) and used Apache jMeter performance tester (http://jmeter.apache.org/). App was tested for realistic case: each user sent text message (137 Byte, TEXT) every 10 second, low resolution photo (100KB, *SMALL Image*) every 30 second, high resolution photo (1 MB, *LARGE Image*) every 60 second during a minute. Total average time for samples (request + response times) was 616 ms. while using "OnePlus One" device as a server. The results tests is displayed below.

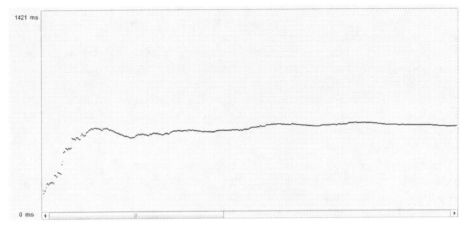

Fig. 1. Samples' send-receive average times during a minute

The average time results for every type of request are shown on the following graph:

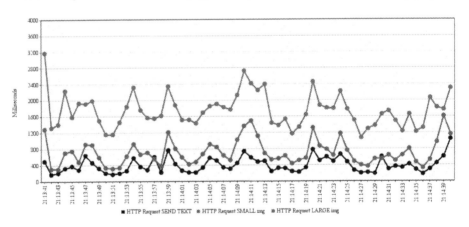

Fig. 2. Samples' send-receive average times for different types of requests

As it can be seen from the graph, sending of larger samples, in general, took more time, which is normal in terms of network transportation. Also, sent files need to be saved on the device's storage, which again affects on total send-receive time of one sample.

It is interesting to see, how much these results depend on the device's performance, especially on number of processor cores, because multithreading is used extensively, and it is interesting to know if different tasks (saving files, showing UI animations, processing of incoming data, sending responses to devices etc.) can be parallelized well.

For testing application on multi-processor devices, we emulated a "Board" device with different number of cores. We deployed server application on GenyMotion (https://www.genymotion.com/) emulating tool, which is great solution for testing application on different devices. It is using modified and Android-optimized version of Oracle's VM VirtualBox (https://www.virtualbox.org/) as an emulator and shows fast performance and ease of use compared to Android's stock emulator. We preconfigured GenyMotion from 1-7 with selected number of cores, 2GB of RAM and deployed under Android 4.4 KitKat (API version 19). Here are detailed results that we got during tests:

Table 1. Average response time depending on device cores

Number of Processors	Results (ms)	Average (ms)
1	2087	2209.5
	1334	
	2564	
	2853	
2	352	325.25
	332	
	323	
	294	
3	238	240.25
	225	
	238	
	260	
4	232	234
	225	
	239	
	240	
5	146	156.25
	163	
	151	
	165	
6	128	139
	128	
	156	
	144	
7	125	120.5
	118	
	119	
	120	

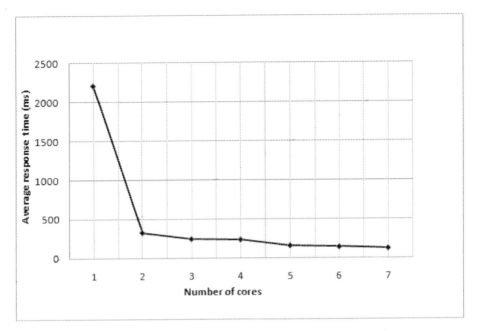

Fig. 3. Samples' send-receive average times for different number of cores

Server's average response time depending on number of cores visualized in graph below:

As we can see, we have a huge drop and response time reduction when moving from single-core server to multi core. This is due by the circumstance that the Android system use the main thread, and also, for every request NanoHTTPd creates a new thread, as we mentioned before. That is why single-core system cannot handle all these threads simultaneously, and threads are moving to que, which grows bigger and bigger, hence the response time increases. Using multi-core device dramatically improves servers' performance, and we can assume that number of cores is the bottleneck of the system. As noted already, we use GenyMotion tool for emulation based on PC with hardware-assisted para-virtualization, where Android system is clear from any other third-party applications and services, which can slow system's performance. That is why we expect weaker tests results on "real" devices, but in general, they should be proportional to initial tests results and it does not make sense in overall outcomes.

We support different price plans for product depending on user's needs, starting from free to business.

Free plan will allow maximum five boards, 100 cards and up to five participants connected to the "Board" application.

Personal plan includes limitations to 15 users, unlimited cards and 100 boards. This is great for small or medium-size groups during events, presentations, discussions and decision-making.

Pro version of Teambrainer supports up to 100 users with unlimited cards and boards. An enterprise-aimed solution could be customized based on client's desire, like supporting more participants, etc.

All versions support capturing ideas, organizing, rating, analyzing, photo taking and screen casting. Video capturing, reporting and sharing, data encryption and email support is available only for Personal and Pro plans.

The teambrainer "Cards" and "Screen" applications for Android, iOS and current browser can be installed and used free.

Offering multiple price-packages will attract more people and try Teambrainer, also "try before buy" is the best approach.

For Teambrainer, we minimized hardware- and device-dependency for facilitators to have all advantages without limitations. For basic usage, only a tablet with preinstalled "Board" app and users with "Cards" application on their smartphone required. For demonstration and visualization, screen or projector with "Screen" application or computer with standard browser is needed. For full experience, screen with HDMI input and cast stick is mandatory.

Further development of product will include completion of iOS version of program, Windows-based version, remote connection of participants, ability of device/cloud backup, restore, and video conferences. We keep in mind mobility of the group, easy of usage and price while developing Teambrainer applications.

4.2   Sezam

Another e-collaboration application based on mobile infrastructures developed by us is a healthcare application called "Sezam" (http://sezamapp.ru/). It is a unique, simple and convenient application of alternative communication for people with speech and/or writing disabilities. Program can be used by adults and children with such disorders as autism Cerebral palsy, Down syndrome etc. and also by people with temporary disorders (such as after a stroke). In the first version of the application there are available about 500 black and white pictograms[8] of international standard, which represent objects, actions, attributes of objects and words necessary in any conversation (such as "yes", "no", "I want", "do not want", "thank you", "please" and so on).

The application allows sending and receiving these pictograms with the help of social networks. All the pictograms by default separated by functional folders: "People", "Time", "Location", "Action" and others, which allows users to find desired pictograms easily. The use of black and white pictograms reduces eyestrain and significantly reduces the time to choose the desired pictogram [9].

Here is the part from conversation between parent and child:

At this moment, there are free and paid programs that allow people with speech or writing disabilities by using special pictograms translate their feelings and needs, but "sender" and "receiver" should be close to each other.

"Sezam" allows to exchange with specialized professional pictograms. From these schematic images with labels is possible to compose complete messages. There is a chance for people with limited communication capabilities and for their relatives to soften a stressful situation caused by problems of communication at a distance.

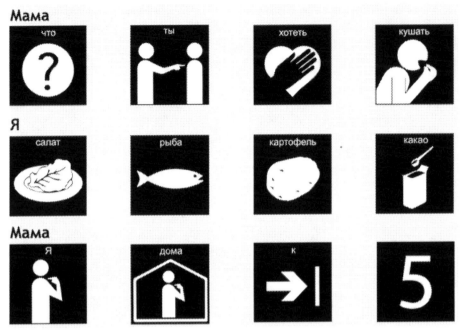

Fig. 4. Pictogram-based conversation between child and parent

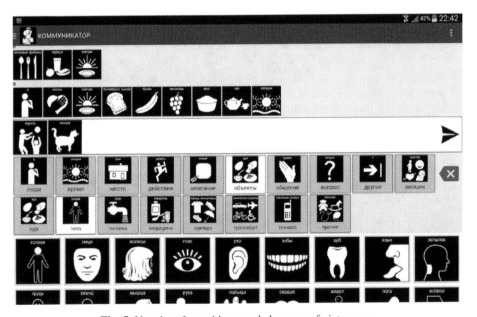

Fig. 5. User interface with expanded groups of pictograms

Added simplified analog keyboard with Russian letters, numbers and basic signs. This allows combining pictograms and texts in message to create a cohesive grammatical structure.

By default, communicator's every pictogram is divided into separate functional folders: "People", "Time", "Place", "Action", etc., allowing the user to easily find necessary pictogram in a given time.

To navigate to the new folder does not need to go back to the beginning, all the folders are constantly visible on the screen, as shown on picture below.

Test Results. We tested beta version of the application with 18 children from 5 to 18 years old, among them: 4 - with EDR, 14 children with complex defect structure (movement disorders, organic disorders GM, secondary autism), three - with additional impairments.

With the help of the program were asked to answer the following questions:

- Question 1 - Do you want a candy? (Question, implying the answers YES and NO).
- Question 2 - What you want to do now? (Question that implies a choice of several actions).
- Question 3 - What did you do yesterday? (Question that implies a choice of several actions, but gives an additional opportunity to make a detailed proposal).
- Question 4 - Check your state? (Implies the choice of the group "emotions" or "adjectives").

Here are the results:

- Question 1 - All succeed.
- Question 2 - All succeed except one child, still not very well familiar with this pictographic system.
- Question 3 - 14 people succeed. With four of them managed to establish an extensive dialogue within the "Sezam" program. These children are experienced users of gadgets and specific pictographic system. These children are then asked to install the software on their tablets.
- Question 4 - Coped without the help from side - 4, coped with minimal help - 6, failed - 8 (mostly due to lack of understanding of emotions as such or insufficient familiarity with specific pictographic system).

Summary of the test: As shown by testing, at an appropriate level of fluency in pictographic signs, work with the program is going very well even in children with severe autism. The gadget as a mediator facilitates communication and provides an additional opportunity to communicate with someone "from the other room" that not only expands the possibilities of communication, but also makes it much "safer" and "safer" from the point of view of the child, because it does not involve any additional contact. This is the main and obvious advantage of the program.

One of our goals is to unify graphics system for people with special communication needs that will enable them to communicate effectively in any situation. A single set of pictograms in different programs will allow, if necessary, to facilitate the transition from one to another.

We assume the following updates and new functionality in "Sezam" project:

- development of program versions for iOS and Windows;
- independent authorization (via through social networks or local registration);
- quick search of users who installed the application;
- connect the speech synthesizer with endings' correction in the set of graphical and combined messages;
- add the function of zoom for the visually impaired;
- the extension of a list of icons with new models from existing database;
- the ability to create their own communication folder with the most frequently used icons;
- individual sorting icons in groups;
- the ability to add your own icons or pictures for the individual learning process;
- setting the "panic button" for emergency messages to pre-selected person;
- customizable calendar/organizer with audible and visual signals;
- interface, adapted to different user groups.

References

1. Scale, E.M.S.: Assessing the Impact of Cloud Computing and Web Collaboration on the Work of Distance Library Services. Journal of Library Administration **50**, 933–950 (2010). United Kingdom
2. Myerson, J.M.: "Collaborate to brainstorm and share projects", developer Works, October 2009. URL: http://www.ibm.com/developerworks/library/wa-aj-social/wa-aj-social-pdf.pdf date
3. http://www.idc.com/prodserv/smartphone-os-market-share.jsp
4. http://www.diva-portal.org/smash/get/diva2:664680/fulltext01.pdf
5. Danielyan, V., "Software Protection Based on IP Steganography". Computer Science and Information Technologies (CSIT-2011), pp. 359–361, Yerevan, Armenia
6. Balyan, S., Hakobyan, R.: "Means of Hidden Protection of Software Against Unauthorized Use" Computer Science And Information Technologies (CSIT-2013), pp. 441–444, Yerevan, Armenia
7. teambrainer - apps for smarter workshops. URL: http://teambrainer.com/en/features# smart-&-secure
8. Pictogram – a language in pictures, URL: http://www.pictogram.se/
9. Frost, L., Bondy, A.: PECS: The Picture Exchange Communication System. Pyramid Educational Consultants, Inc, (2002)

Workshop on Software Engineering Processes and Applications (SEPA 2015)

A Systematic Approach to Develop Mobile Applications from Existing Web Information Systems

Itamir de Morais Barroca Filho[1]([⊠]) and Gibeon Soares de Aquino Jr.[2]

[1] Metropole Digital Institute, Federal University of Rio Grande do Norte,
Natal, Brazil
itamir.filho@imd.ufrn.br
http://www.imd.ufrn.br
[2] Department of Informatics and Applied Mathematics,
Federal University of Rio Grande do Norte, Natal, Brazil
gibeon@dimap.ufrn.br
http://www.dimap.ufrn.br

Abstract. Mobile computing is changing the way society accesses information. People are using mobile devices such as smartphones and tablets to make transactions and consume data more and more each day. Driven by this kind of use, many companies and institutions are developing mobile versions of their information systems. But what should these companies and institutions consider when developing a mobile versions of its information systems? It is important to note that the mobile version is not the redesign or copy of all functionalities from the existing information system. Considering this scenario of mobile application development from existing web information systems, we proposed a process named Metamorphosis. This process provides a set of activities subdivided into four phases - requirements, design, development and deployment -, to assist in the creation of mobile applications from existing web information systems. Thus, this article aims to present a case study of the Metamorphosis process in the development of SIGEventos Mobile, a mobile version based on SIGEventos web information system, approaching the feasibility of the use of this process.

Keywords: Mobile computing · Web information systems · Metamorphosis process · SIGEventos mobile

1 Introduction

In today's world, mobile computing is a reality in people's lives. For example, we use mobile devices to receive and answer our e-mails, like our friends' status, locate a place, order food and much more. Beyond that, we have strategies like *Bring Your Own Device (BYOD)* to access privileged company information and applications, which brings us even closer to the mobile computing context. Considering these devices' hardware, we have processing power with multiples

© Springer International Publishing Switzerland 2015
O. Gervasi et al. (Eds.): ICCSA 2015, Part IV, LNCS 9158, pp. 371–386, 2015.
DOI: 10.1007/978-3-319-21410-8_29

cores, gigabytes of memory and dozens gigabytes of storage. In addition, we have sensors such as infra-red, GPS and NFC. Thus, information can be accessible from mobile devices that are powerful in terms of resources and lower sizes.

Moreover, there is a global trend towards the increasing number of users connected to the network via mobile devices. According to [1], today we have more active mobile devices than people in the world: about 7.2 billion people and 7.3 billion mobile connections. This produces demands for information systems, mobile applications and content for such equipments. As a result of the diversity of features and capabilities offered by such devices, we observe an increase in their sales in 2014. [2] says that smartphone sales grew 20% in the third quarter of 2014, reaching 301 million units. Then, as a result of smartphones and tablets sales, there is also an increase demand for new applications. This can be seen by the growing number of application downloads on mobile application markets such as Google Play and Apple AppStore. About this fact, [3] estimates that by 2017, mobile apps will be downloaded more than 268 billion times, generating a revenue of more than $77 billion and making apps one of the most popular computing tools for users across the globe.

Thus, driven by this mobile computing scenario that is changing the way society accesses information, there is a growing demand for mobile applications. [4] predicts that developers will create apps for virtually every aspect of a mobile user's personal and business lives that will 'appify' just about every interaction between physical and digital worlds. There is a natural tendency for companies that have web information systems to begin to adapt them to fit this computing scenario. It is an essential strategy for such systems to continue attracting and serving its users' needs. According to [5], 90% of 250 IT managers had plans to develop new mobile apps within their company by the end of 2011 and there is a considerable interest in mobile applications and willingness to invest in these technologies.

Therefore, we realize that many information systems are changing to fit this mobile computing context and provide ways to make their data available through mobile applications. However, according to [6], it is important to note that the development of these applications involves several activities, such as:

- Catalog functionalities from the existing information system that should be present in the mobile application;
- Engage of stakeholders in the validation of the selected functionalities;
- Assess the need for offline operations in each selected functionality;
- Evolve the existing information system to enable integration with mobile applications;
- Develop web services on the existing information system to enable obtaining data by the mobile application;
- Develop mobile application considering the target platform (Android, iOS, Windows Phone, and Mobile Web);
- Integrate with exclusive services on these devices, such as GPS, SMS and NFC;
- Publication and publicity of the mobile application.

As said before, this mobile version is not the redesign or copy of all functionalities from the based information system. It's important to review our knowledge of software development, particularly in processes, methods, techniques, patterns and architectural solutions for applications to fit this computing environment.

Finally, this article presents a process named Metamorphosis, which was designed to assist in the creation of mobile applications from existing web information systems. This process provides a set of activities subdivided into four phases: requirements, design, development and deployment, which will be presented in section 3. In Section 2, we outline related work. Section 3 presents the Metamorphosis process, focusing on phases, activities and work products. Section 4, in its turn, presents SIGEventos Mobile, a case study of the Metamorphosis process utilization. In section 5, we present conclusions of this article and future works.

2   Related Works

The creation of the Metamorphosis process started from the development experience of the first version of SIGAA Mobile[1] from SIGAA[2] web information system. This development, described in [6], was conducted using two actions:

- Adaptation of existing features, created specifically for web systems, to be also accessible from mobile devices. Such adaptation involves the development of applications in native platforms, like Android and iOS; the development of Restful Web Services; and the integration with existing enterprise software components which implement the business rules;
- Development of new specific features for mobile devices. These devices offer new possibilities and new technologies embedded in them, therefore an important and promising initiative in this context would be the development of features not existent in the scope of old systems, which is only possible with mobile devices.

In this first experience, we categorized generic approaches related to: business, like choosing the most popular functionalities from the existing web information system for the mobile version; technical, such as the reuse of bussiness components from the exising web information system; and UI (user interface), such as the creation of a logo and the usability review of the mobile version.

Thus, we noted that these approaches could be formalized into a process, and to provide a background for this process we peformed a systematic review to identify strategies, good practices and experiences reported in the literature about the development of mobile applications. With the result of this systematic review and our experience, we defined Metamorphosis, introduced in [7]. Also in [7], we present its utilization for the development of SIGAA Mobile (second version).

---

[1] https://play.google.com/store/apps/details?id=br.ufrn.sigaa.mobile
[2] https://sigaa.ufrn.br/

The following section, Section 3, describes Metamorphosis process with more details, presenting its elements -phases and work products- and phases specifications -requirements, design, development and deployment-.

3   Metamorphosis Process

The Metamorphosis process consists in a set of activities organized in four phases (requirements, design, development and deployment) that should be considered for the creation of mobile applications from existing web-based information systems.

3.1   Process Elements

This section presents the elements of the Metamorphosis process, describing its phases and work products. The activities are described in Section 3.2.

Phases. The Metamorphosis phases, shown in Figure 1, are:

- *Requirements:* phase related to the selection of the information system's functionalities that should also be present in the mobile application. It has activities focused on the definition of the scope of the mobile application, along with elicitation and validation of requirements with stakeholders;
- *Design:* phase related to the architectural design of the mobile application. It has activities focused on its design; and the creation of architectural solutions with technologies, frameworks, design patterns and best practices in development;
- *Development:* phase related to source code implementation and software tests;
- *Deployment:* phase related to application distribution. It has activities focused on publication and distribution of the mobile application.

Fig. 1. Phases of the Metamorphosis process

Work Products. The work products of the Metamorphosis process contains information that is produced throughout the execution of its activities. Thus, these products are:

- *Selected Functionalities Document:* It is a document generated by the activities of the "Requirements" phase. It contains descriptions of selected functionalities, links to access the documentation of these features and indications for adjustments and offline operation;

– *Deployment Document:* It is a document generated by the activities of the "Deployment" phase. It contains a description of the mobile application, the mobile market and a link to download it.

3.2   Metamorphosis Process Specification

The four phases of the Metamorphosis process and its activities will be detailed in this section. These details consider the execution of activities [7] in each of its respective phases: requirements, design, development and deployment.

Requirements Phase. At the beginning of the project, it is necessary to plan which functionalities are relevant in the context of mobile environments. The process of creating a mobile application from an existing web enterprise system is not a direct mapping of functionality-to-functionality. This kind of simplification is a common mistake and must be treated carefully. Mobile devices have some intrinsic restrictions such as screen size, difficulties to type long texts and no guarantee of network access availability. Moreover, it has a different mean of user interaction with touch support, gesture events and rapid actions. According to [8], mobile applications tend to provide relevant advantages to their users in terms of design and usability. For this reason, it is important to know some strategies from the three layers design guideline proposed in [9], which involves the creation of graphical user interfaces: offering shortcuts for functionalities, creating consistent graphical user interfaces for small devices and clearly distinguishing selected items.

Thus, the first activity of the requirements phase **(Identify Functionalities)**, presented on Figure 2, is to analyze which features of the existing web information system would be important in the mobile context. For this, four practices should be considered [6]:

1. Choose popular functionalities;
2. Avoid long-steps functionalities or long-fill forms;
3. Adapt existing functionalities;
4. Create specific functionalities for the mobile application.

The output of this activity is a list of pre-selected functionalities. This list will be validated with the stakeholders, which is the second activity of this phase **(Validate Functionalities)**. If the preselected list is not validated by the stakeholders, there is a need to go back to the previous activity, which is the analysis of the existing system's functionalities. Otherwise, we can move on to the next activity, which is the evaluation of the pre-selected functionalities considering the mobile context restrictions **(Evaluate Mobile Context)**. In this activity, we verify the sizes of the forms and the amount of steps on these functionalities. If we consider a functionality suited for the mobile context, it moves on to the second output of this phase, which is a list of selected functionalities for development. Otherwise, if a pre-selected functionality is not suited for mobile context,

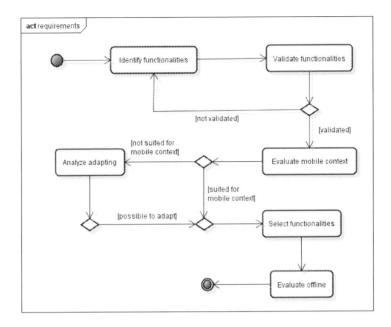

Fig. 2. Metamorphosis requirements phase

we evaluated the possibility of adapting the functionality (**Analyze Adapting**), which can result, for example, in a new design for the operation, reducing the amount of fields and steps. If such a change is possible in the functionality, it is selected for development (**Select Functionalities**).

For each functionality selected for development, we evaluate the need to work offline (**Evaluate Offline**). This is important because mobile devices are often connected to a network using wireless connections, whose availability can be low. Moreover, the connectivity may be unstable while the user is interacting with the system. The necessity to run offline directly impacts on the next phase, which is the design. At the end of this phase, the Selected Functionalities Document, presented in Section 3.1, is created.

Design Phase. The first activity of the Design phase is the choice of the target platform (**Choose Platform**). Nowadays, there are many platforms available, such as Android, iOS, Windows Phone and Web Mobile. After choosing the target platform, the next challenge that needs to be addressed is how to integrate the mobile application with the existing system. Moreover, we need to be concerned about how to reuse its already implemented business components. The use of the layer pattern [10] is particularly common in web information systems. For this reason, a generic approach that can be used to integrate with the existing web system is the definition of a new separated layer providing a set of services that must be used by the mobile application (**Design Services**). This new layer, which integrates with the existing business rules layer using the

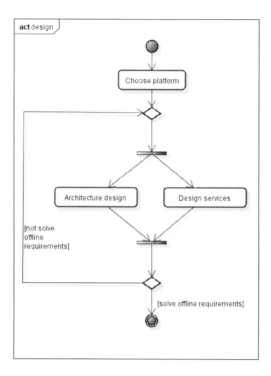

Fig. 3. Metamorphosis design phase

already implemented and stable code, is developed in the Design phase, presented on Figure 3, where there is an activity to design this service layer. It's a strategy to use REST on the service layer. The use of a REST API simplifies the development of clients (mobile apps, web and desktop versions) to consume information [11].

Development Phase. After designing the mobile application's architecture and services, the development phase, which is presented in Figure 4, starts. The first activity of the development phase is the implementation of the mobile application functionalities' source code **(Implementation)**. For this, the *Selected Functionalities Document*, described in Section 3.1, is forwarded to the developers, who use them to obtain details about the functionalities that will be implemented for the mobile application. As soon as the implementations are finalized, developers request their software tests **(Testing)**. If the functionality developed is not validated (which means it does not pass all tests), the developers solve all bugs identified and request new tests. This phase only ends when all functionalities on the *Selected Functionalities Document* are implemented and there aren't any bugs detected on them.

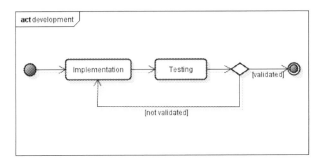

Fig. 4. Metamorphosis development phase

Deployment Phase. After the development of the mobile application, we enter the deployment phase, presented in Figure 5. The first activity of this phase is to evaluate the need for publishing the mobile application on mobile applications markets (**Evaluate Publishing Need**) such as Google Play and Apple App Store. According to [5], there are companies that operate their own mobile app stores in order to distribute the mobile applications to their employees, customers and partners. These app stores are known as "in-house" or "corporate" mobile app stores. It is important to discuss with stakeholders the need to publish the app in in-house or public mobile applications markets. If there is the need to publish the app in a public market (**Publish Application**), one of the most important activities of this phase is the publicity around the mobile application (**Make Publicity**). The potential user must know that this new kind of enterprise system exists and they should be motivated to try this new way to access the system. Only publishing the application on a platform store, e.g. Android Play, Apple App Store, is not enough to make it be well known among users. For this reason, its existence should be well communicated to the

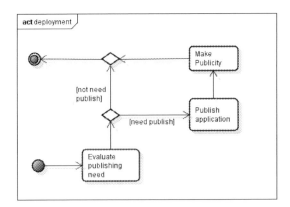

Fig. 5. Metamorphosis deployment phase

target audience. At the end of this phase, the *Deployment Document*, presented in Section 3.1, is created.

4   SIGEventos Mobile: A Case Study

This section describes the case study for the development of the mobile application SIGEventos Mobile. The main objective of this case study was to investigate the feasibility of using the Metamorphosis process to create mobile applications based on existing web information systems. For planning and describing the results, we followed the guidelines proposed in [12], [13] and [14].

4.1   Planning

The case study planning includes a description of research questions, participating subjects, the object, analysis units, evaluated artifacts, evaluation criteria and procedures used for data collection.

Research Questions. In order to reach the proposed objective, the case study should answer the following research questions:

1. Is the Metamorphosis process useful for the development of mobile applications based on existing web information systems?
2. What are the benefits, problems and challenges resulting from the Metamorphosis process utilization?

   To answer these questions, apart from observing and monitoring the development of SIGEventos Mobile, we applied two surveys[3] to the development team.

Subjects. The Metamorphosis process was used by the mobile applications development team of the Informatics Management Office (SINFO) from Federal University of Rio Grande do Norte (UFRN). This team consists of three programmers specialized in Android, iOS and PhoneGap. Therefore, our subjects were the three developers of this team.

Object. The object used for the case study was the Integrated system of conference management (SIGEventos)[4]. This web system was designed to manage the conferences held at Federal University of Rio Grande do Norte (UFRN). Thus, as access profiles on SIGEventos web, we have:

- Evaluation Manager: Performs the final evaluation of the paper, deciding whether it will participate or not in the conference;

---

[3] Links: http://goo.gl/forms/C8cK2qef55 and http://goo.gl/forms/2wqkqHS7PO
[4] Link: http://www.sigeventos.ufrn.br

– Reviewer: Performs evaluation of the paper submitted through an opinion, but is allowed to approve the role of this manuscripts in the conference;
– Participant: It is registered to use the system as a reviewer or to submit papers.

Summarizing the use flow of this system:

1. A conference is created;
2. For each conference, the reviewers must be registered;
3. This conference has a registration period during which the work is submitted;
4. After the submission of the papers, they are distributed to the reviewers;
5. Each reviewer performs his review. The evaluation manager determines, based on the reviewer's opinions, which papers will attend the conference, and sends a notification to users who submitted papers;
6. At the end of the conference, the emission of certificates becomes available to those with approved papers.

SIGEventos was developed using open technologies such as Java, Hibernate, JavaServer Faces, Richfaces, Struts, EJB and Spring. It uses PostgreSQL as DBMS and is available via the JBoss application server.

Analysis Units. The group of programmers was the analysis unit of this case study. They were analyzed during the execution of the Metamorphosis process.

4.2   Metamorphosis Process Utilization

To start, we performed a meeting with the group of programmers to present the Metamorphosis process. At this meeting, all phases and activities of this process were detailed to them. In addition, we provided a document with the specifications of this process for future reference.

After this Metamorphosis process presentation meeting, the programmers began the *Requirements Phase*. In the first activity, known as *Identify Functionalities*, they performed a SQL query in SIGEventos logs trying to group and select the most commonly used functionalities. This technique, which uses the log database, is defined by the process to assist on this activity. The logs of query returned 76 rows with various links of all SIGEventos. This activity took around 1 hour and 30 minutes. After this review of logs and a meeting with the team, developers reached functionalities identified in Table 1.

With the list, developers began the second activity, which is *Validate Functionalities*, and created a survey to be answered by professors and students (stakeholders). This survey[5] listed the functionalities described in Table 1 allowing the stakeholders to evaluate them from 1 to 4, with 1 being the minimum score and 4 the highest score (in increasing order of significance). There were 20 responses. Also through this survey, it was possible to capture the stakeholders' suggestions for other functionalities.

---

[5] Link: http://goo.gl/forms/zQvNe9q7cG

Table 1. Functionalities identified by the activity of *Identify Functionalities* of requirements phase

Functionality	Description
1. Certificates of paper submissions.	List used to issue user's submissions certificates.
2. My paper submissions.	List of papers you have submitted. This list presents: paper title, status of submission, payment and submission period.
3. My conferences registrations.	List of your previous registrations. This list presents: the type of participation, registration status and conference period.
4. Conference schedule.	Calendar with general program of the conference.
5. Conference Location.	Map with conference location and integration with Google Maps for navigation.
6. Feedback of the talks.	List with the conference talks where participants can perform evaluations (bad, good, very good and perfect).

The functionalities evaluated with maximum grade by more than 60% of participants have been selected to be in the first version of SIGEventos Mobile. They are: *Certificates of Paper Submissions, My Conferences Registrations, Conference Schedule and Conference Location.* For a second version, the other functionalities and suggestions will be reviewed by the stakeholders. This activity took around five days since it depended on the application of the survey.

Continuing with requirements phase, after the validation of the functionalities by stakeholders, programmers moved on to the third activity, *Evaluate Mobile Context*, in which the need to adapt each of these functionalities for the mobile environment was verified. The schedule and location functionalities were adapted for the use of GPS in mobile devices (*Analyze Adaptation*). The activities of analyzing and evaluating mobile context adaptation took about 1 hour and 15 minutes to be performed.

The programmers selected these functionalities for development (*Select Functionalities*) and for each functionality, they evaluated the need for offline operation (*Evaluate Offline*). The schedule and registrations functionalities were evaluated as necessary to work offline. These activities, Select Functionalities and Evaluate Offline, took around 30 minutes to be performed. Finally, the programmers created the *Selected Functionalities Document*[6]. After the execution of this phase, the requirements phase survey, described in Section 4.1, was applied to programmers. The answers will be discussed in Section 4.3.

The programmers finalized the requirements phase and started the design phase. In the first activity of this phase, *Choose Platform*, Android was chosen for the development of SIGEventos Mobile. Then they started to *Design Architecture* and *Design Services*. In such activities, they spent about 8 hours. This long period of time was due to the need for refactorings to separate the

---

[6] Link: http://goo.gl/7KZWZY

business components of SIGEventos web and to create of SIGEventos REST. This SIGEventos REST has RESTFul services to enable the integration of SIGEventos Mobile with SIGEventos data. The motivation to separate was to make more independent projects at deployment and runtime level on JBoss application server, and allow any application (desktop, mobile or web) to use the services of SIGEventos REST. The components of SIGEventos and their interactions are shown in Figure 6. As in the requirements phase, at the end of the design phase, the survey described in Section 4.1 was applied. The answers will be discussed in Section 4.3.

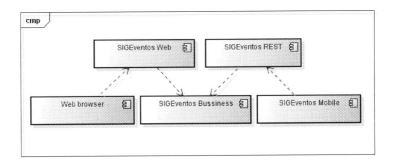

Fig. 6. SIGEventos components

After the architecture and services for SIGEventos Mobile were designed, the programmers started the development phase with the *Implementation* activity. All the functionalities presented on *SIGEventos' Selected Functionalities Document* were implemented. Each programmer spent around 12 hours with it and as the functionalities were implemented, tests were conducted. To perform the test, a tester from the Informatics Management Office Software Test Team used the mobile application for 3 hours. The main SIGEventos Mobile screens are shown in Figure 7 and a video demonstrating the use of the key features is available on the following link: http://goo.gl/OM8SlP.

Thus, the development phase was completed and the deployment phase began. At this phase, only the activity of *Evaluate Publishing Need* was held. It was decided to publish the SIGEventos Mobile on Google Play app market because of the number of potential users for this application, which can be: professors, students, lecturers and participants, in other words, anyone from UFRN or not.

4.3   Answers to Research Questions

As described in Section 2, the subjects of this case study (3 programmers) were observed during the execution of the Metamorphosis process for the creation of SIGEventos Mobile, and answered all questions presented os surveys described on Section 4.1.

Fig. 7. SIGEventos Mobile main screens

1. Is the Metamorphosis Process Useful for the Development of Mobile Applications Based on Existing Web Information Systems? To answer if the Metamorphosis is useful for creating mobile applications based on existing web information systems, these surveys were organized considering the phases of requirements and design of this process. Regarding the requirements phase, all programmers considered that this phase assisted in the selection of SIGEventos' functionalities that should be present in its mobile version. Also concerning to this phase, programmers have not lacked any other activity that is not addressed by this process. This demonstrates that, in these programmers' perspective, the requirements phase of the Metamorphosis process is complete in terms of activities.

At the design phase, all the programmers considered that this phase assisted to create an architecture that facilitated the development of functionalities for SIGEventos Mobile. Also regarding to activities not covered by the design phase, all developers stated that they did not perform any other activity not covered by the design phase of the Metamorphosis process.

Analyzing the answers, it can be considered that the Metamorphosis process is useful for creating mobile applications based on existing web information systems, as it helps in its creation and has complete and well-defined activities. It is important to emphasize that the development and deployment phases were not analyzed by these surveys. All these are avaliable at: http://goo.gl/JfwTEUandhttp://goo.gl/uroh83.

2. What are the Benefits, Problems and Challenges Resulting from the Metamorphosis Process Utilization? To analyze the benefits, problems and challenges resulting from the use of the Metamorphosis process, the questions present in the surveys involved topics such as: difficulty in performing activity, complexity, effort and suggestions for it. Regarding the difficulty of carrying out any activity phase of the requirements, we noted that two programmers struggled to complete the execution of functionalities activity. This was due to communication problems with the stakeholders. This activity took around five days because it depended on the responses of professors and students (stakeholders) to the validation survey about SIGEventos Mobile's functionalities, cited on Section 4.2. Thus, this activity was the longest in the requirements phase.

Regarding the difficulties of the design phase, developers had no difficulty in performing any of its activities. The programmers were also asked about the complexity and effort to perform the activities of the Metamorphosis process. With regard to the effort and complexity, two programmers considered that the execution of the requirements phase has medium complexity and effort. One of the programmers considered that the execution of this phase has easy complexity and effort. In the design phase, the answers about the complexity and effort were inconclusive, because each of the programmers answered with a different level of complexity and effort.

At the end of the surveys, the programmers were asked about suggestions for improvement of the requirements and design phases, and benefits of the use of this process. At the requirements phase, programmers consider that its organization in step-by-step activities facilitates requirements elicitation. Moreover, according to one of the programmers, these well-defined activities led to an optimization of the development time, because it is known what must be done to raise the requirements of mobile application based on an existing web information system. Also in this phase, according to one of the programmers, since such requirements elicitation based on an existing information system becomes simpler, it increases the chance to have a successful mobile application. Finally, according to the programmers, the benefit of the design phase is the conveniently organization of activities, since every activity improves the mobile application.

4.4   Threats to Validity

For the case study, we evaluated four types of validity:

- *Construct Validity:* The data capture of the implementation of this case study was carried out through the application of impersonal surveys (without identifying the programmer) to the programming team. Regarding the interpretation of data, it was peformed by two researchers which avoided errors and tendentious interpretations in the study;
- *Internal Validity:* The case study realization was conducted with three programmers that have the same experience with the development of technologies for mobile devices (Android, iOS and PhoneGap). In addition, since in the first meeting it was held a presentation of the Metamorphosis process and provided a specification document, the trust was raised that the experience inherent factors, in developing applications for mobile devices and knowledge of the process of Metamorphosis programmers, were under control;
- *External Validity:* This case study, performed to solve a real problem, enabled the creation of a mobile application based on a web information system that will be used by thousands of users. Therefore, this allows to generalize the results of this case study and survey responses for industrial practice;

– *Conclusion Validity:* In this case study, a control group did not exist because the goal was not to compare the Metamorphosis process with other software development process. So it was impossible to establish statistical relationships. The quantitative and qualitative data contributed to the assessment of the Metamorphosis process, regarding to its context, which is the development of mobile applications from existing web information systems.

5   Conclusions and Future Works

The implementation of the Metamorphosis process for the development of SIGEventos Mobile, based on SIGEventos web version, allowed the investigation of the feasibility use of this process for creating mobile applications based on existing web information systems. We analyzed the requirements and design phases considering whether they are useful for creating applications for this context, along with its benefits and problems arising from its use. The surveys and observation of process execution were carried out in requirements and design phases and the results were presented in Section 4.3.

As limitations for this case study, we did not monitor development and deployment phases through surveys. In addition, we could not follow the active installations, since the execution of the case study finalized at the first activity of the deployment phase.

Finally, the Metamorphosis process was considered useful for creating mobile applications based on existing information systems. Therefore, the main benefit of the Metamorphosis process is the well defined step-by-step activities that brings to the goal of creating a mobile application based on an existing web information system, a greater chance to be used and successful among its users.

Regarding to future work, we need to perform more cases studies of the use of the Metamorphosis process in creating mobile applications from information systems with different software technologies and architecture. Based on these collected evidence, this process may have its activities enhanced presenting details, for example, of how a particular functionality can be adapted to a mobile application considering the GUI and source code. This evidence collected may also serve for the development of components and architectural patterns to facilitate the integration between web information systems and mobile applications.

References

1. There are now more gadgets on Earth than people. http://www.cnet.com/news/there-are-now-more-gadgets-on-earth-than-people/
2. Gartner Says Sales of Smartphones Grew 20 Percent in Third Quarter of 2014. http://www.gartner.com/newsroom/id/2944819
3. Gartner Says by 2017, Mobile Users Will Provide Personalized Data Streams to More Than 100 Apps and Services Every Day. http://www.gartner.com/newsroom/id/2654115

4. Worldwide and U.S. Mobile Applications, Storefronts, and Developer 2010-2014 Forecast and Year-End 2010 Vendor Shares: The "Appification" of Everything. http://www.idc.com/research/viewdocsynopsis.jsp?containerId=225668

5. Giessmann, A., Stanoevska-Slabeva K., Visser, B.: Mobile enterprise applications - current state and future directions. In: 45th Hawaii International Conference on System Science, pp. 1363–1372. IEEE Press, Hawaii (2012)

6. Aquino Jr., G., Barroca Filho, I.: SIGAA mobile a sucessful experience of constructing a mobile application from an existing web system. In: 25th International Conference on Software Engineering & Knowledge Engineering, pp. 510–516, Boston (2013)

7. de Morais Barroca Filho, I., de Aquino Jr, G.S.: MetamorphosIS: a process for development of mobile applications from existing web-based enterprise systems. In: Murgante, B., et al. (eds.) ICCSA 2014, Part VI. LNCS, vol. 8584, pp. 17–30. Springer, Heidelberg (2014)

8. Nayebi, F., Desharnais, J.-M., Abran, A.: The state of the art of mobile application usability evaluation. In: 25th IEEE Canadian Conference on Electrical & Computer Engineering, Montreal (2012)

9. Ayob, N.Z.B., Hussin, Ab.R.C., Dahlan, H. M.: Three layers design guideline for mobile application. In: Proceedings of the International Conference on Information Management and Engineering (2009)

10. Buschmann, F., Meunier, R., Rohnert, H., Sommerlad, P., Stal, M.: Pattern-Oriented Software Architecture: A System of Patterns. John Wiley & Sons, New York (1996)

11. Zamula, D.; Kolchin, M.: Mnemojno - design and deployment of a semantic web service and a mobile application. In: Open Innovations Association (FRUCT) (2013)

12. Yin, R.K.: Case Study Research: Design and Methods. [S.l.]. SAGE Publications (2003) (Applied Social Research Methods). ISBN 9780761925521

13. Kitchenham, B., Pickard, L., Pfleeger, S.L.: Case Studies for Method and Tool Evaluation. IEEE Softw. **12**(4), 52–62 (1995). http://dx.doi.org/10.1109/52.391832. ISSN 0740-7459

14. Runeson, P., Host, M.: Guidelines for conducting and reporting case study research in software engineering. Empirical Software Engineering **14**(2), 131–164 (2009). http://dx.doi.org/10.1007/s10664-008-9102-8. ISSN 1382–3256

An Ontological Support for Interactions with Experience in Designing the Software Intensive Systems

Petr Sosnin[✉]

Ulyanovsk State Technical University, Severny Venetc str. 32, 432027 Ulyanovsk, Russia
sosnin@ulstu.ru

Abstract. Nowadays, experience bases are widely used by project companies in designing the software intensive systems (SIS). The efficiency of such informational sources is defined by "nature" of modeled experience units and approaches that apply to their systematization. An orientation on a precedent model as a basic type of experience units and an ontological approach to their systematization are defined the specificity of the study described in this paper. Models of precedents are constructed in accordance with the normative schema when the occupational work is fulfilled by a team of designers. In creating the necessary ontology, the team should use a reflection of solved tasks on a specialized memory intended for simulating the applied reasoning of the question-answer type. The used realization of the approach facilitates increasing the efficiency of designing the SIS.

Keywords: Automated designing · Ontology · Precedent · Question-answering · Software intensive system

1 Introduction

Last twenty years, the problem of an extremely low degree of success in designing of SIS is steadily registered in statistics that are presented in reports [1] of the company "Standish Group." The analysis of the successfulness problem indicates that the special attention should be given to the human factor in attempts to increase the degree of success. Designing the SIS is the intensive human-computer activity that is fulfilled in conditions of the high level of complexity.

It is necessary to note that designing (especially conceptual designing) is impossible without researching by designers of numerous and practically unpredictable situations of a task type. In such situations, the designer should behave as a researcher who uses the appropriate type of the research in which the personal and collective experience is creatively applied in the real-time.

In designing of the definite research, the designer would rather work as a scientist who wants to conduct the definite experiment. Results of such experiment will be applied in the creation of the current project and can be useful for the future reuse. For this reason, the search for improved forms of designer interactions with own and collective experience is a promising way of increasing the degree of success in designing of SIS.

© Springer International Publishing Switzerland 2015
O. Gervasi et al. (Eds.): ICCSA 2015, Part IV, LNCS 9158, pp. 387–400, 2015.
DOI: 10.1007/978-3-319-21410-8_30

Four years ago, the group of well-known researchers and developers in software engineering had initiated a process of innovations [2], which have been named SEMAT (Software Engineering Methods And Theory). It is necessary to notice that, in normative documents of SEMAT, a way of working used by a team of designers is marked as a very important essence. There "way-of-working" as a notion is defined as "the tailored set of practices and tools used by the team to guide and support their work." For this reason, the search of effective ways-of-working are perspective and can lead to an increase of success in developments of SISs.

A very promising way-of-working that will facilitate increasing the success in collective designing of SIS can be bound with a creation and use by the team an Experience Base [3]. It is necessary to note that the effectiveness of such version of working essentially depends on kinds of experience models and ways of their systematization and use.

The paper presents an ontological approach to the systematization of the Experience Base and interactions with its units. In addition, such units are formed as experience models corresponding to precedents. According to the Cambridge dictionary, "precedents are actions or decisions that have already happened in the past and which can be referred to and justified as an example that can be followed when the similar situation arises" (http: //dictionary.cambridge.org/ dictionary/ british/ precedent). Methods and means that are offered in the paper are oriented on conceptual experimenting with tasks being solved in an instrumentally-technological environment WIQA (Working In Questions and Answers [4]). The interaction with the toolkit WIQA is based on question-answer reasoning (QA-reasoning) of designers.

2   Preliminary Bases

2.1   Why Ontology

The ontology usually helps to achieve the following positive effects: reaching the coordinated understanding in collective actions; using the controlled vocabulary; systematizing the methods and means used in an occupational activity; specifying the conceptualization; checking the semantics of the built text and applied reasoning; operating with machine-readable and machine-understandable content.

In this paper, the use of the ontology is oriented on all called effects and the systematization of precedents' models embedded into the Experience Base. In addition, it is planned to create such models with the use of conceptual experimenting when designers solve the project tasks in the real time. This way is coordinated with understanding of the ontology as a system of specifications of conceptualization [5].

The called effects are important as for the collective activity so for the activity of one person (individual). Moreover, in a team, the coordinated understanding is achieved as a result of coordinating a personal understanding with a common understanding for all members of the team.

2.2   Why Way-of-Working Oriented on Precedent

As a reusable unit of the occupational work, the precedent should be specified and executed in accordance with an appropriate framework. Such a framework as an artifact should be coordinated with natural reflecting by the individual the environmental conditions of the precedent execution. If the result of reflecting will be bound with understanding the precedent as the behavior that corresponds to the intellectually processed conditioned reflex, then such understanding can be used for its specifying in the precedent framework.

The toolkit WIQA supports the use of the precedent framework that is presented in Fig. 1 where a logical component of the framework demonstrates (without explanations) some details of intellectual processing of the precedent as a behavioral unit.

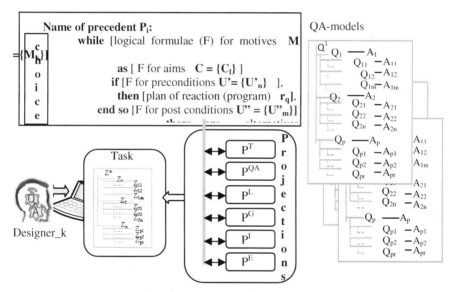

Fig. 1. Framework of precedent

The structure of the framework is coordinated with the process of task-solving and preparing the solution of the task for reuse in designing. This structure includes a textual model P^T of the solved task Z, its model P^{QA} in the form of the registered QA-reasoning about the solution, the logical formulae P^L of the precedent regularity, a graphical (diagram) representation P^G of the precedent on the perceptual level, conceptually algorithmic model P^I in a form of a pseudo-code program and the model which presents its executable code P^E.

2.3   Why Question-Answering

Intellectual processing for natural precedents is being fulfilled by means of natural language that also supports access to units of the experience. Such effects are caused by processes in consciousness which has a dialog nature. Explicit question-answer reasoning helps in controlling of these processes. Thus, question-answer reasoning reflects processes in the consciousness of the designer interacted with own experience. This type of reasoning is expedient for applying in the creation and use of experience models that present the designer's behavior.

In reality, designers usually represent the own professional work by means of a plan to be written in the natural language in its algorithmic usage. Repeatable works fulfilled by individuals are represented by techniques also being written in the natural language in its conceptually algorithmic usage. It prompts to use such way for plans of experimenting. In turn, that was the reason for choosing a pseudo-code language for programming the units of designer's behavior. Interacting with such programmable models of the behavior, the designers will apply the experience of using the natural language.

3   Related Works

A set of typical kinds of ontologies (according to their level of dependence on a particular task or point of view) includes the top-level ontologies, domain ontologies, tasks ontologies and applied ontologies. All these types of ontologies are defined in [5] and [6] as means that are used in different systems.

For the SISs, an adequate type of ontologies is applied type that usually is expanded by means of the other ontologies types. In accordance with the publication [7], the theory and practice of applied ontologies "will require many more experiences yet to be made."

It is necessary to notice that the project ontology as a subtype of applied ontologies is essentially important for SISs. Project ontologies in the greater measure are aimed at the process of designing, but after refining they can be embedded into implemented SISs.

The specificity of project ontologies is indicated in a number of publications. In the technical report [5] the main attention is concentrated on "people, process and product" and collaborative understanding in interactions. Investigating the possibility of the ontology-based project management is discussed in the paper [8].

In developing the program system and ontological problems of program products, the use of ontology possibilities is investigated in the paper [9]. In means suggested in this article, the experience of task ontologies is taking into account also and, first of all, the role of different kinds of knowledge. The place of knowledge into the task ontologies is reflected and discussed in the publication [10]. The role of knowledge connected with problem-solving models is opened in papers [11].

All papers indicated in this section were used as sources of requirements in developing the set of instrumental means provided the creation and use of the proposed ontology. It should be noted that reports of Standish Group were used as "guides" for planning of our research aimed at applications of QA-approach and the development of the toolkit WIQA. Step by step the QA-approach has investigated criteria and factors that facilitate increasing the degree of the success in designing of SISs [12]. The study presented in this paper focuses on the ontological component of QA-approach at the use of the occupational experience.

4 Reflection of Subject Area

4.1 Question-Answer Memory

As told above, the proposed study is aimed at the creation and use of the project ontology by the designers who develop the family of SIS in the definite subject area. In such work, they should use (if they solved) the toolkit WIQA, which helps to include conceptual experimenting in processes (P) of designing.

At the conceptual stage of working, means of WIQA are used by designers for the following aims:

- Registering of the set of created projects each of which is presented by the tree of its tasks in the real time;
- Parallel implementing of the set of workflows $\{W_m\}$ each of which includes subordinated workflows and project tasks $\{Z_n\}$;
- Pseudo-parallel solving of the project tasks on each workplace in the corporate network;
- Simulating the typical units of designers' behavior with using the precedent framework.

Named actions are implemented by designers with using the reflection of processes P on QA-memory some details of which are presented in Fig. 2.

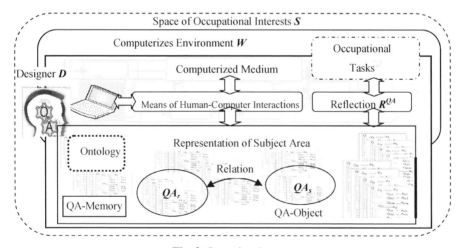

Fig. 2. Operational space

For the process P_l of designing the project $PROJ_l$, all indicated aims are achieved by the following reflections:

$$WW(P_l)=WW(PROJ_l, \{W_m\}, \{Z_n\}, t)\xrightarrow{R^{QA}(P)}ZP_l{}^{QA}(t+\varDelta t) \cup$$

$$\cup \{ZW_m{}^{QA}(t+\varDelta t)\} \cup \{Z_n{}^{QA}(t+\varDelta t)\}, \qquad (1)$$

$$\{Z_n{}^{QA}(t) \xrightarrow{R^{QA}(Z^{QA})} Pr_n{}^{QA}(t+\varDelta t)\},$$

where symbol Z underlines that labeled models have a task type, $Pr^{QA}(t)$ designates a model of a precedent for the corresponding task, $R^{QA}(X)$ indicates that the reflection R^{QA} is applied to the entity or artifact X. For example, $R^{QA}(Z^{QA})$ designates applying this reflection to the model Z^{QA} of the task Z. In WIQA-environment, the model of such kind is called as "QA-model of the task." It is necessary to note that the reflection R^{QA} includes $R^{QA}(P)$ which results are reflected by $R^{QA}(Z^{QA})$ in a set of precedents' models $\{Pr_n{}^{QA}\}$ also located in the QA-memory.

QA-memory is specified and materialized for storing the conceptual descriptions of the operational space S in memory cells in an understandable form. For this reason, cells are oriented on registering the textual units in the form of communicative constructions any of which can be divided on "theme" (predictable starting point) and "rheme" (new information) that should receive additional meaning in answering to the corresponding question.

Such orientation has led to the solution of using two types of corresponding cells. The cell of the first type is intended for registering the simple question (Q) during the cell of the second type registers the corresponding answer (A).

Extracting the theme and rheme from units of the textual description is an important linguistic operation named "actual division". This operation facilitates the process of understanding. Its use corresponds to the dialogic nature of consciousness. Cells of both types are specified equally because any question includes a part of a waited answer, and any answer includes components of the corresponding question. For this reason, a pair of corresponding cells with a question and answer presents the description unit in details. Such pair corresponds to the QA-object in Fig. 2.

QA-memory with its cells are intended for registering the conceptual content of reflected units with taking into account the semantics of their textual descriptions. The necessary semantics is fixed in basic attributes of the cell in additional attributes that can be added by the individual if it will be useful for the simple object stored in the cell. The potential structure of a simple object is presented in Fig. 3.

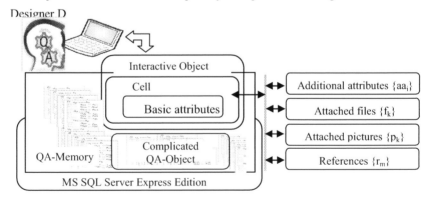

Fig. 3. Specification of the interactive object

The above not only indicates reflections of projects on the QA-memory, but it also demonstrates structures that should find their presentations in such memory. Below, for specifications of question-answer objects (QA-objects) in the QA-memory, a formal grammar GR^{QA} with extended BNF-notations will be used. For example, structures of QA-objects should correspond to the following rules of GR^{QA}:

$$
\left.
\begin{aligned}
&\textit{QA-Memory} \;=\; \{QA\text{-}object\}; \\
&\textit{QA-object} = \textit{Question}, \; \text{``}\leftarrow\text{''}, \textit{Answer}; \\
&\textit{Question} \;=\; Q \,|\, (Q,\text{``}\downarrow\text{''},\{Q\}); \\
&\textit{Answer} \;=\; A \,|\, (A, \text{``}\downarrow\text{''}, \{A\}); \\
&Q = (\{a\}, \{[aa]\}, \{[f]\}, \{[p]\} \{[r]\}); \\
&A = (\{a\}, \{[aa]\}, \{[f]\}) \{[p]\} \{[r]\}; \\
&a \;=\; (\textit{address, type, description, time, the others}); \\
&aa = \{\textit{useful additional attribute}\},
\end{aligned}
\right\} \qquad (2)
$$

where "Q" and "A" are typically visualized objects stored in cells of QA-memory, symbol "$\downarrow$" designates an operation of "subordinating". These objects have the richest attribute descriptions [5]. For example, a set of attributes includes the textual description, index label, type of objects in the QA-memory, the name of a responsible person and the time of last modifying. Any designer can add necessary attributes to the chosen object by the use of the special plug-ins "Additional attributes" (object-relational mapping to C#-classes).

In addition, in QA-memory, any created model is materialized with using the question-answer objects (QA-objects) that are bound by necessary relations. Any QA-object is a corresponding composition uniting the interactive objects "question" (Q) and "answer" (A) types. Such objects are used for saving of ontology concepts in structures the typical scheme of which is presented in Fig. 4.

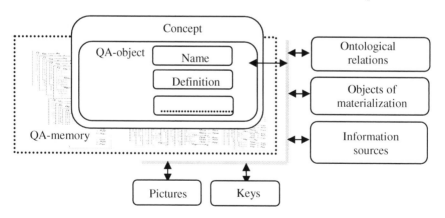

Fig. 4. Framework for specifying the concept

The content embedded in this framework corresponds to the following rules of grammar GR^{QA}:

$$\left.\begin{array}{l} concept = QA \ (*in \ QA\text{-} \ memory*) \\ concept = (concept \ name, \ definition, \ \{list\}); \\ definition = text; \\ list = \{ \ (relation, \ \{name\})\}; \\ relation = is\text{-}a \ | \ part\text{-}of \ | \ association \ | \ synonimity | \\ \qquad | \ picture \ | \ key \ | \ materialization | \ reference. \end{array}\right\} \qquad (3)$$

Concepts can be combined in groups as in frames of project ontologies so as sections in the kernel ontology the content of which is invariant to the specificity of any project fulfilled by designers. Combining is described by the following set of rules:

$$\left.\begin{array}{l} GN = (N, \ "\cap", \ \{N\}); \ (*group \ of \ concepts*, \ "\cap \text{ - } operation \ of \ uniting") \\ PO = \{GN\}; (*project \ ontology*) \\ ontology = (kernel \ ontology, \ \{PO\}). \end{array}\right\} \qquad (4)$$

It is necessary to note that concepts of the project ontology are usually used in definite conditions and that demands to specify variants $\{\Delta N_s\}$ of the concept N into the ontology. This necessity is supported by the following rules:

$$\left.\begin{array}{l} \Delta N = (definition, \ description, \ condition); \ (*concept \ usage*) \\ description = \{attribute \ name\}; \\ condition = programmed \ formulae; \\ N = \quad ("\int", \ \{\Delta N\}); \ (*"\int" - operation \ of \ assembling*) \\ concept = N \ | \ ("\int", \ \{N\}); (*simple \ or \ complicated \ concept*) \end{array}\right\} \qquad (5)$$

Rules also specify the structure of complicated concepts the content of which is defined by a group of the other concepts.

5   The Use of the Ontology

As told above, concepts of any project ontology are used in definite conditions of their uses. Means of the toolkit WIQA support the use of the integrated ontology, its project ontologies and their components till concepts for the following basic aims:

- Systematization of the Experience Base as a whole and its "projections" on any designed project that is implemented in the project organization;
- Access to the necessary assets the models of which are stored in the Experience Base;
- Extraction of requirements, their specifications and evolution in processes of designing;
- Creation of the controlled vocabulary and its applying in the frame of the occupationally natural language used in the project organization;
- Conceptual experimentation with the project solutions created by designers in the real time.

At the bottom level of actions for achieving the called aims, designers use analytical processing the used textual units. First of all they extract usages of concepts $\{\Delta N\}$ from such units. The toolkit WIQA supports three versions for the extraction of $\{\Delta N\}$ from textual units. In all versions, the current state of the ontology is applied.

The first version is based on the automated extraction of the used words with nearest context in the processed text. Extracted phrases are estimated on their conformity to the ontology. Differences can indicate errors or necessity of concepts' evolution.

The second version evolves the first version by visualization of sentences with extracted usages of concepts in forms of "block and line" in the specialized graphical editor embedded to WIQA.

The third version uses more strict rules of processing that is based on transformations of processed into a set of simple sentences. Such transformation is based on the pseudo-physics model of the compound sentence or complex sentence of the other type. In the pseudo-physics model of the sentence, all used words are interpreted as objects that take part in the "force interaction" that is visualized on the monitor screen. Formal expressions of pseudo-physics laws are similar to appropriate laws of the classic physics. In the stable state, (Fig. 5) after finishing dynamic process on the screen each group of words-objects will present the extracted simple sentence.

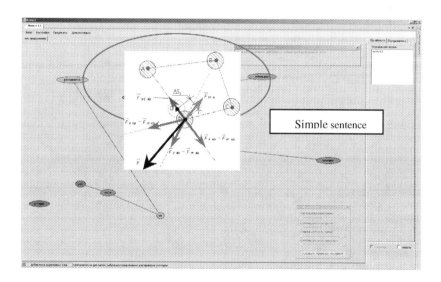

Fig. 5. Extraction of simple sentences

The screenshot in Fig. 5 is used with labels for the generalized demonstrations of the visual forms and objects with which the designers are working. The language of this screenshot is Russian.

Let us notice that in the appointment of attributes (values of mass m_i, charge q_i and others values and parameters) two mechanisms are applied – the automatic morphological analysis and the automated tuning of object parameters. Values are appointed in accordance with the type of part of speech. The suitable, normative values were chosen experimentally.

Extractions of simple sentences are aimed at their transformation in Prolog-like forms in order to assemble logic formulae that correspond to the checked texts. This way is used not only for checking on conformity to the ontology vocabulary. It is also used for controlled creation of logical conditions in models of precedents and the creation of ontology axioms.

It is necessary to note that the concept scheme presented in Fig. 4 and details of the concept reflected in rules (5) indicate that presentations of such units in the project ontology have forms of precedents' models. For example, in the general case, the concept has QA-model presented in Fig. 6.

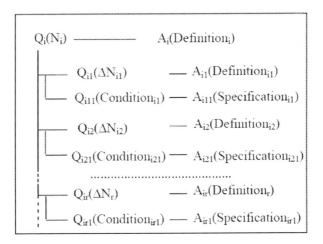

Fig. 6. QA-model of concept

Such presentation of the concept opens the possibility of using the means of pseudo code programming embedded to the toolkit WIQA [12]. In addition, it opens the possibility for:

- Controlling the used concepts;
- Understanding ant its estimation;
- Experiment conducting;
- Conceptual modeling;
- Conceptually-algorithmic modeling;
- Accumulating and improving experience;
- Documenting;
- Systematizing of experience;
- Programmable access to the ontology.

6   Main Features of the Proposed Ontology

As told above, the proposed project ontology is oriented on the experiential way-of-working when designers solve tasks by using the conceptual experimenting. Such way-of-working is based on real-time interactions of designers with the Experience Base the generalized scheme of which is presented in Fig. 7.

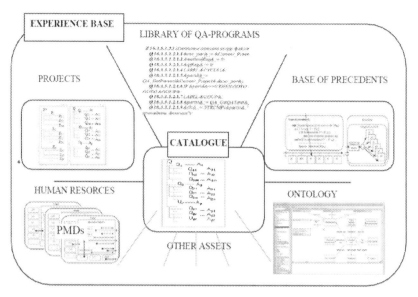

Fig. 7. Structure of the Experience Base

The Experience Base is intended for assembling the assets of a design company that creates the family of SISs. Any asset is presented in this repository as the model of the precedent that presents the inclusion of the corresponding asset in designing. The definite model can be realized in accordance with the full scheme (Fig. 1) or the appropriate form of its projection.

In the current state, the Experience Base is divided into section each of which includes assets of the definite type. The greater part of assets are stored in the corresponding section in forms of their models. For example, the section of "Human resources" saves personified models of designers (PMD). Section "Projects" is intended for registered the information about developed projects of SISs.

There are two versions of the access to models of experience the first of which is provided by the catalog of the experience base. The second version uses the search by keys that are included in the set of the controlled vocabulary.

Only one part of assets is placed in Precedent Base. The greater part of assets is stored in corresponding libraries where they are presented in forms of precedent projections. As told above, components stored in QA-memory can be bound with the attached files that are placed in corresponding libraries too. One of such libraries includes program written in C# that can be used in pseudo code programming.

7   Ontological Support in Designing of Configured Templates

Offered means have been tested in the development of a number of projects the last of which is a system for designing of tooling. In the aircraft industry, for manufacturing the parts of a fuselage, wings and aileron, including details of their covering, technological templates are widely used. Any template for the part is a kind of its machining attachment that supports definite technological operations, for example, manufacturing of the corresponding part. Described means of the ontological support was applied for designing of the template tooling that usually consists of tens of thousands of templates. The built ontology consists of the following sections:

- The system of typical templates as concepts;
- The visualized classification of templates;
- Templates in manufacturing of aviation parts;
- Templates in a production control of parts;
- Templates as models of precedents.

In the section of concepts, the model of the template is described by its type and attributes a number of which are expressed through relations with other concepts. Items in this section are interactively accessible.

The used systematization of concepts and their classifier are shown in Fig. 8 where relations are demonstrated on the example of one of the template. The classifier combines part-of-relations that facilitate the search for appropriate templates.

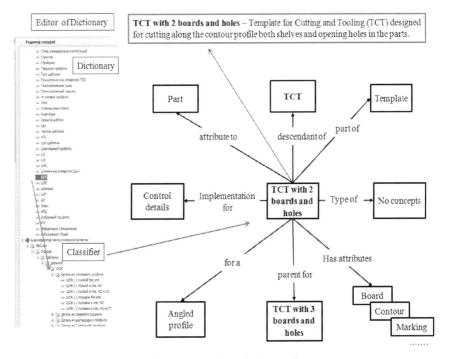

Fig. 8. System relations in the ontology

In the built ontology, any template is defined in the form of the precedent model that has the structure shown in Fig. 9. Since all interfaces in Russian, some fields are marked with labels.

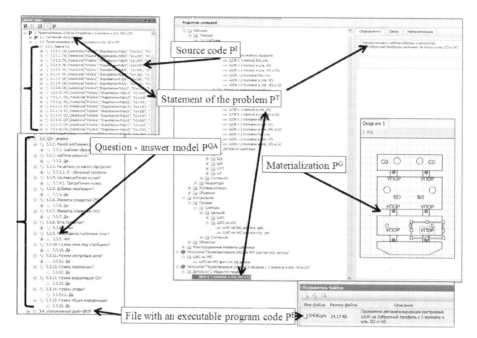

Fig. 9. Structure of the template definition

It should be noted the specificity of the program components used in the precedent model. The life cycle of template manufacturing includes the stages that require the creative-intensive activity of workers that participate in designing of the templates. It is caused by the need to take account of future ways to move the laser beam on all lines that represent the template for its steel billet. Therefore, the template model includes two algorithmic descriptions. The first description imitates the movement of the laser beam in tasks of searching the effective trajectories. The second description is a result of computer numerical control programming that provides the process of laser cutting.

8   Conclusion

This paper presents the system of means for the creation and use of the projects ontology in the development of the family of SISs when enormous quantity of project tasks is being solved by the team of designers in the corporate instrumental network. The success of such activity essentially depends on mutual understanding of designers and means supported conceptual decisions of designers at early stages of them work.

The proposed approach is based on the experiential way-of-working of the designers who interact with the Experience Base accumulating the assets of the design company. All assets are presented in this Base as models of precedents that have been built in accordance with the normative framework. In the real-time work, the designers use the toolkit WIQA intended for the solution of tasks at the conceptual stage of collaborative designing. The toolkit provides the use of question-answer reasoning in conceptual experimenting in solutions of the project task. It also helps to create models of precedents and to achieve positive ontological effects described above.

References

1. Chaos Manifesto (2013).
   http://www.versionone.com/assets/img/files/ChaosManifesto2013.pdf
2. Jacobson, I., Ng, P.-W., McMahon, P., Spence, I., Lidman, S.: The Essence of Software Engineering: The SEMAT Kernel. Queue **10** (2012)
3. Henninger, S.: Tool Support for Experience-based Software Development Methodologies. Advances in Computers **59**, 29–82 (2003)
4. Sosnin, P.: Scientifically Experimental Way-of-Working in Conceptual Designing of Software Intensive Systems. In: Proceedings of the IEEE 12th International Conference on Intelligent Software Methodologies, Tools and Techniques (SoMeT), pp. 43–51 (2013)
5. Garcia, A.C.B., Kunz, J., Ekstrom, M., Kiviniemi, A.: Building a Project Ontology with Extreme Collaboration and Virtual Design & Construction, CIFE Technical Report # 152, Stanford university (2003)
6. Guarino, N.: Formal Ontology and Information Systems in Proceedings of FOIS 1998, Trento, Italy, Amsterdam, IOS Press, pp. 3–15 (1998)
7. Guarino, N., Oberle, D., Staab, S.: What is an ontology? In: Staab, S., Studer, R. (eds.) Handbook on Ontologies, 2nd edn. International handbooks on information systems, pp. 1–17. Springer Verlag (2009)
8. Bullinger, H.-J., Warschat, J., Schumacher, O., Slama, A., Ohlhausen, P.: Ontology-based project management for acceleration of innovation projects. In: Hemmje, M., Niederée, C., Risse, T. (eds.) From Integrated Publication and Information Systems to Information and Knowledge Environments. LNCS, vol. 3379, pp. 280–288. Springer, Heidelberg (2005)
9. Eden, A.H., Turner, R.: Problems in the Ontology of Computer Programs. Applied Ontology **2**(1), 13–36 (2007). Amsterdam, IOS Press
10. Martins, A.F., De Almeida, F.R.: Models for Representing task ontologies. In: Proceeding of the 3rd Workshops on Ontologies and their Application, Brazil (2008)
11. Fitsilis, P., Gerogiannis, V., Anthopoulos, L.: Ontologies for Software Project Management: A Review. Journal of Software Engineering and Applications **7**(13), 1096–1110 (2014)
12. Sosnin, P.: Scientifically experimental way-of-working in conceptual designing of software intensive systems. In: Proceedings of the IEEE 12th International Conference on Intelligent Software Methodologies, Tools and Techniques (SoMeT), pp. 43–51 (2013)

Do We Need New Management Perspectives for Software Research Projects?

Jeong Ah Kim[1], Suntae Kim[2(✉)], Jae-Young Choi[3],
JongWon Ko[3], and YoungWha Cho[3]

[1] Department of Computer Education, CatholicKwandong University,
579 beongil-24, BeomIlRo, KangNeungSi 210-701, Korea
clara@cku.ac.kr
[2] Department of Software Engineering, Jeonbuk National University,
567 Baekje-Daero, DeokjinGu, JeonjuSi, JeollabukDo 561-756, Korea
stkim@jbnu.ac.kr
[3] College of Information and Communication Engineering,
SungKyunKwan Univ., Suwon-si, South Korea
{jaeychoi,jwko0820,choyh2285}@skku.edu

Abstract. Creativity and originality are so important success factors so that many researchers are not interested in formal or official process. But uncertainty and risks should be managed for successful outcome of research. The main goal of our research is to initiate the practical and effective SW engineering techniques and tool which can tracking, monitoring, and managing the SW research life-cycle in R&D projects. Researchers are the knowledge works and research process is very similar with feature of adaptive case management. Also many product data management techniques can be applicable for managing research artifacts. With research descriptor configuration item and flexible process managing environment, SW R&D researchers can define their own creative research process or documents and keep tracking the progress at any time.

Keywords: Software R&D · Monitoring · Process agility · Traceability · Research document configuration management

1 Introduction

Research and development must operate strategically in the organization, becoming a key driver of business success. The drive for growth, the search for new ideas and new ways of doing things and increasing competitive pressures have transformed the R&D function into one of primary strategic importance. There are so many approaches for achieving the software reliability, it is important to perform effective and traceable process. But incorrect application of management might be obstacles for achieving the successes.

R&D development has special features and characteristics in management because of uncertainty and risks [1]. And research phase is difference from the development

© Springer International Publishing Switzerland 2015
O. Gervasi et al. (Eds.): ICCSA 2015, Part IV, LNCS 9158, pp. 401–410, 2015.
DOI: 10.1007/978-3-319-21410-8_31

since there are not stable or fixed requirements at research phase. For these reasons, we need alternative project management for research phase of software R&D projects. A balance needs to be considered between maintaining a suitable level of control whilst at the same time not inhibiting creativity.

Many types of project and process management software for software development projects have developed. But these have assumptions that developer can define their own development process and guideline for performing those activities at the early stage of development. But it is not the real story in research projects. In this paper, we opened some issues for managing the software research projects and suggested complementary solution for these issues.

We identified alternative approach in non-project management domain, adaptive case management and product data management. Researchers are the knowledge works and research process is very similar with feature of adaptive case management. Also many product data management techniques can be applicable for managing research artifacts. We suggested to integrate software engineering practices or principles such as progress management, traceability management, software configuration management and change management, and reuse with adaptive case management and product data management practices.

We suggested flexible process management for research phase and small-grained configuration item as research descriptor for research configuration management. With research descriptor configuration item and flexible process managing environment, SW R&D researchers can define their own creative research process or documents and keep tracking the progress at any time.

The remainder of this paper is organized as follows. Section 2 surveys literature related to R&D project features and new research trends in knowledge-intensive process management. Section 3 surveys previous related research for managing the R&D projects. Section 4 describes the research issues for managing the SW research project. New framework for RLM (Research Lifecycle Management) and RDM (Research Descriptor Management) is described in session 5. Finally, section 6 give the future research directions and plans.

2   Research Context

2.1   Features of SW R&D Projects and R&D Project Management

The uncertainness of R&D projects makes the entire development process is a learning process so that it is difficult to decide the project scopes and concrete architecture in the feasibility analysis stage. Because of the technological uncertainty, R&D project needs to overcome one or more technical problems in the process of research and development [2]. Also, R&D activities can be characterized as complex, interdependent, responsive to research environment or trend and heavily reliant on expert judgment to progress and quality. There are some features of SW R&D projects as followings: (1) projects should follow the variable constant, (2) researchers will to continue the work in spite of the obstacles, (3) projects should employ valid and verified procedures and principles, (4) Finally, researchers should conduct a reliable and complete work product.

The formal disciplines of project management provide a means of planning, organizing and controlling multi-disciplinary projects without stifling innovation. As we identified the features of SW R&D projects, they require some differences in approach from conventional project management. R&D projects should catch up market demands so that many R&D projects can easily lose the overall R&D project strategy. Managing the uncertainties and non-predictable of R&D projects is critical factors for success. It is very difficult to control the research progress since there are always technical problems and technical issues. It does not mean that project does not have a detailed project plan. But the actual implementation of it is difficult to achieve, and unable to timely carry out risk tips and alarm [2].

R&D projects process knowledge information, need a reasonable knowledge management and need a sharing way to make that knowledge can be passed to improve the similar R&D projects. The process of R&D project performance should be monitored and documented. Large-scale, complex construction projects follow clear and validated processes with quarterly milestones and regular review cycles. But, this method is neither appropriate nor meaningful for an R&D program since typically one of a kind that measures progress toward answering a list of key questions through a variety of approaches for periods that may not have a hard fixed end date [3].

2.2 Research Phase .vs. Development Phase

The almost R&D project might start from research phase. At this phase, researchers begin their works with glimmer of interest so they imagine the results and make their concept to be clear and concrete by theory development and idea conceptualization [4]. With this features of research phase, project management might be informal and iterative process. After imagination, idea can be captured as artifact in forms and drawings so idea can transform into substantive issues what should be verified. Argument on these issues can be possible and initial ideas or goal can be concretized. Also researchers can achieve the development scope, strategies, and process.

If something would be clear and fixed, researcher can move to development phase even though all issues are not closed. At development phase, the scope and strategies are much clearer by research phase, process can be stable and robust.

2.3 New Process Management for Knowledge Workers

Knowledge works fundamentally differs from routine process [5] and research is also same. They are less structured and cannot define the exact process. For these kinds of works, new BPM concept Adaptive Case Management(ACM) have been introduced for supporting the knowledge intensive and less structured business process. Dermot McCauley [6] defined the characteristics of processes where Adaptive Case Management is applicable: (1) Goal-driven, (2) knowledge intensive, (3) Highly variable processes, (4) Highly collaborative, (5) Juggling fixed and flexible timescales, (6) Sensitivity to external events, (7) Information complex, (8) Cross-organization visibility, so on. White [7] has formed a set of challenges which a case-handling system should deal with: (1) Striking a balance between Practice and Procedure, (2) Capturing implicit rules and tacit knowledge, (3) Formalizing Experience - supporting learning, (4) Supporting ad hoc change, (5) Involving participant in the design of knowledge processes, (6) Supporting collaboration, (6) Supporting decisions, (7) Managing Artifacts, so on.

3   Previous Research

NTT Software Innovation Center started to resolve the defects and schedule delay by creating standards for software development and clarifying the rules for operation of these standards from 2009. They settled on the new R&D Software Development Standards in April 2013 [8]. This standard has four parts: the main text of the standards, the forms and samples section, operational guidelines, and additional documents that help readers understand the development standards. They define the quality class as 5 levels, A to E depending on the extendibility or availability of R&D software. For example, class A means the R&D software should be directly introduced into a company's package or service so that their quality should meet the level of service agreement with customer. But class E software cannot be introduced in a business in operation and quality factors are not known. They also established checklist based on quality characteristics and sub-characteristics of ISO/IEC 9126 and define the relation between the quality classes.

UK Department of Trade and Industry as part of the National Measurement System Valid Analytical Measurement (VAM) Program suggested quality assurance of R&D [1]. They considered that R&D is compatible with the design element of ISO 9001. However R&D is not formal or non-routine work so that it might not fit easily into a highly documented and formalized quality system. For this reason they suggested the guidance based on good practice rather than compliance with formal standards.

Energy Facility Contractors Group (EFCOG) Project Management Working Group developed a clear and concise value proposition for the use of project management principles in the delivery of R&D projects [3]. This group recognized that change is an integral part of the process so that they suggested more structured approach and framework for project execution by adapting and tailoring project management processes. This group suggested new approaches for 4 issues: (1) progressive baseline management with minimum requirement and goal at early stage but progressively narrowing ranges for scope, (2) Similar concepts for configuration management, change control, and trending have been applied to management of software projects in that their dynamic product environment can be applied to R&D projects, (3) To measure the R&D project, process for application of earned value management techniques was defined.

4   Research Issues of Process Management in SW Research Projects

4.1   Is it Possible to Define Robust Research Process?

Previous researches had assumptions that research and development project defined process at the early stage and performed as planned. Of course, traditional project management was originally applied in the production of hardware. What is the difference of hardware and software? The critical difference is that it is not possible to fixed complete requirements at an early stage of R&D project.

In this paper, we questioned that researchers could catch up the planned schedule or planned process at research phase. The reason not to follow the process because of uncertainty, unfixed software requirements, and risks. At the planning phase, re-searchers defined research goals and schedule which are rough and not consider the risk. Also, almost research projects adopted waterfall process model but waterfall model has many limitations for taking the risks. For handling the risk and uncertainty, iterative or prototyping process model should be adopted for researches. However, not may research projects cannot try to adopt iterative or prototyping model since many software R&D researchers are not expert in software engineering. Waterfall model can be considered as traditional project management process, but we should consider new flexible methods such as spiral model or agile model.

In this paper, we suggested more adaptable approach to these flexible methods as complementary way to not software engineering expertise. ACM can be the comple-mentary approach for research process. Research projects have specific goals and outcomes what are defined in research project plan. Also research projects have rough schedule with milestones of research documents. It means that researchers produce something research materials which are not well-formed and not pattern-based. But, to meet the evaluation criterion, researchers should produce the well-formed docu-ments. If we could provide the software engineering environment to capture the research contents and research process, it is possible to produce several kinds of tem-plate-based research model or document from the captured contents and process. Each research process can be the specific case and has own process. But after com-pleting the research process, its process can be the reference model for other research projects. If then, it is possible to provide the information the researcher needs at the point in a process when he needs it, rather than force him to go searching for it when he hits a problem.

4.2   Is it the Same with Configuration Management in Software Development?

Major configuration baselines known as the functional, allocated, and product baselines as well as the developmental configuration, are associated with milestones in the life cycle of a CI (Configuration Item) [9]. Each of these major configuration baselines is designated when the given level of the CI's configuration documentation is deemed to be complete and correct, and needs to be formally protected from unwar-ranted and uncontrolled change from that point forward in its life cycle. There are 4 types of configuration items can be considered: (1) documentation, (2) software, (3) hardware, (4) data.

The document produced in research phase consists of drawing or narrative sentences for sketching the ideas. It means researchers cannot apply template to their documents. And each part of document can have meaning for evolving the solution so that we should manage each part like data in product data management. Product data management is the use of software or other tools to track and control data related to a particular product [10]. The data in PDM tracked usually involves the technical specifications of the product, specifications for manufacture and development, and the types of materials that will be required to produce goods. The use of product data management allows a company to track the various costs associated with the creation and launch of a product. This feature is very similar with research data management for researchers.

In this paper, we suggest research descriptor item for configuration management in research project. Research descriptor item can be any part in research document. We define the characteristics for research descriptor items: (1) it should be tracked for changing in research phase, (2) it can be reusable or to be aggregated with others to produce the research baseline, (3) it should be traced from the initial project plan to check the research directions.

4.3   What is the Meaning of Monitoring in SW Research Project?

Monitoring is the continuous assessment of project realization according to planned schedules and the goals of research projects. Monitoring provides managers and other stakeholders with continuous feedback on the progress to make a decision to facilitate timely adjustments to project management. For monitoring the progress, metrics should be defined to measure how much the progress meets the cost, schedule, scope, goals and outcomes.

Earned Value Management techniques can be applied for measuring the progress of SW research projects. The benefits of applying earned value management techniques to augment project management has been widely accepted since this technique give an accurate measured of projects progress based on the baseline plan [3]. For this, baseline should be established with resource and time-related information and schedule of activities describing the project scope and deliverables. [3] suggested tailored method for applying EVM techniques but not describe detail guideline how to define scope and deliverables and what unit to estimate for resources.

In this paper, we are interested in measuring the progress in terms of schedule, scope and goals. In development phase, we can have requirement baseline so we can measure functionalities with functional adequacy and functional implementation completeness. But in research phase, it is not easy to define the requirement baseline. What we defined in project plan are the initial scope, goals, and schedule with milestones. It might be the configuration items of requirement baseline. For tailoring the EVM, we need another metric for research projects based on requirement baseline. At first step for this, we suggested the unit of measure for EVM as research descriptor item.

4.4   New Perspective in Software R&D Process Framework

We proposed new Software R&D Process Framework like Fig. 1. In this paper, we introduce research baseline which researcher's idea design and associated research notes which defines the researcher's evolving conceptualization. Several iterative research lifecycles are performed, functional baseline can be established and the development phase can be initiated. Research lifecycle consists of planning, imagination, conceptualization, verification phase and researchers produce their materials asnotes or documents with narrative texts, figures, diagrams, and drawings. With these material, researcher can identify the research descriptors. Research descriptor what was defined in section 4.2 has 3 features: (1) tracking, (2) reusable, (3) traceable.

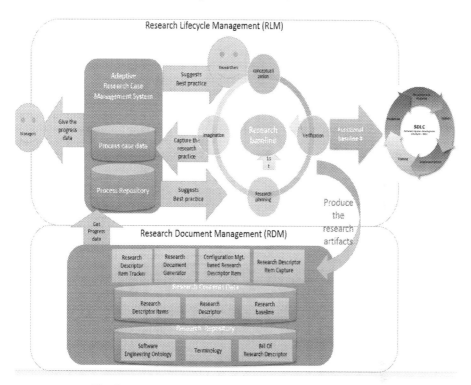

Fig. 1. New Approach for Software R&D Process Framework

Within RDM the focus is on managing and tracking the creation, change and archive of all information related to a research. The information being stored and managed will include research data such as drawing of idea, sketch of architecture, less formal models, benchmark report, weekly report, issues, discussion notes, so on.

With this new framework, the process can be changed from Fig 2-(a) to Fig 2-(b). Fig 2-(a) shows the use case model of project management for software development. Since software development project can establish the WBS (Work Breakdown Structure) for detail plans. Therefore QAO (Quality Assessment Officer) can establish the standard and process model and developer can execute that process based on these standard and templates. And manager can get the progress information of earned value based on the WBS. As mentioned earlier, software research project cannot establish the requirement baseline and detail WBS at the early phase. We suggested new research activities defined in Fig 2-(b). With initial research project plan, researchers execute the imagination, conceptualization, and verification phases. During these phases, researcher can identify the research descriptor items in their research documents which cannot be based on templates or standards.

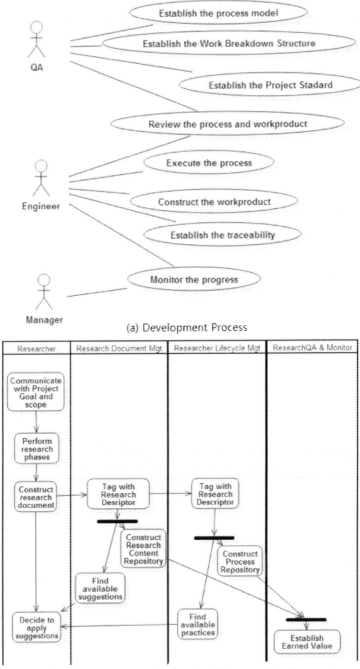

(a) Development Process

(b) New Activity in Research Lifecycle and Document Mgt. Framework

Fig. 2.   Move to new process in Research Projects

With some research descriptor items, RLM searches the best matched process pattern stored in process repository and RDM search the best matched the bill of research descriptor for work product template, software engineering guidelines for performing the research activities. Bill of research descriptor is similar concept of BOM (Bill Of Material) which is also known as the formula, recipe, or ingredients list [11]. Bill of research descriptor define the research document structure. With this, RDM can make a suggestion to fill up more, which leads the researcher can spend more time for complete conceptualization or other research activities. With process case data, RLM make new process model and store it in process repository. Also with process pattern stored in process repository, RLM suggest the research activity to complete the activity such as the process guideline and check list. It can improve the quality of research activity.

5   Future Research Directions

In this paper, we suggested new directions to manage the software research projects in flexible process management, in small-grained research descriptor items, and new earned value measurement based on the research descriptor items. And suggested new research project management framework integrate with project management principles and adaptive case management and product data management. Also we simply check the feasibility of constructing the research descriptor item from the research notes and application of tracing the process and reuse the previous practices. Actually, our research is in the conceptualization of new environment of research project management. We have applied suggested direction to our project to make more detail in conceptualization. From now on, we researched 3 subjects for further conceptualization: (1) How to analyze the semantics of research items for making the meaningful suggestions of process and research document to researchers based on the well-defined practices stored in repository, (2) what structure and techniques are proper for storing and retrieving the research descriptor item, (3) what metrics are available for measuring the progress based on the research baseline and research descriptor items.

Acknowledgements. This research was supported by the Next-Generation Information Computing Development Program through the National Research Foundation of Korea (NRF) funded by the Ministry of Science, ICT & Future Planning(NRF-2014M3C4A7030503)

References

1. Holcombea, D.G., Neidhartb, B., Radvilac, P., Steckd, W., Wegscheidere, W.: Quality assurance good practice for research and development and non-routine analysis. **393**(1–3), 30, June 1999
2. Zhao, X., Yu, K.-C.: Research on ERP and PLM Integration Based on R&D Project Management. In: The 19th International Conference on Industrial Engineering and Engineering Management, pp. 1307–1313 (2013)

3. Energy Facility Contractors Group (EFCOG) Project Management Working Group. "PROJECT MANAGEMENT in Research and Development", White Paper (2010)
4. Wingate, L.M.: Project Management for Research and Development: Guiding Innovation for Positive R&D Outcomes, CRC Press, 2015
5. Herrmann, C., Kurz, M.: Adaptive Case Management: Supporting Knowledge Intensive Processes with IT Systems. In: Schmidt, W. (ed.) S-BPM ONE 2011. CCIS, vol. 213, pp. 80–97. Springer, Heidelberg (2011)
6. Swenson, K.: Mastering the Unpredictable: How Adaptive Case Management Will Revolutionize the Way That Knowledge Workers Get Things Done. Megan-Kiffer Press (2010)
7. White, M.: Case management: Combining Knowledge with Process. A chapter in: BPM & Workflow Handbook, Future Strategies, Lighthouse Point, Florida (2009). Accessed on 08-02-2011,
http://www.bptrends.com/publicationfiles/07-09-WP-CaseMgt-CombiningKnowledgeProcess-White.doc-final.pdf
8. Jinzenji, K., Kasahara, N., Muraki, T.: R&D Software Development Standards and their Operation. NTT Technical Review, **12**(1), January 2014
9. MIL-HDBK-61A, MILITARY HANDBOOK: CONFIGURATION MANAGEMENT GUIDANCE (2001)
10. Crnkovic, I., Asklund, U., Persson Dahlqvi, A.: Implementing and Integrating Product Data Management and Software Configuration Management, Artech House, Inc. (2003)
11. Malakooti, B.: Operations and Production Systems with Multiple Objectives. John Wiley & Sons (2013)

A Binary Fruit Fly Optimization Algorithm to Solve the Set Covering Problem

Broderick Crawford[1,2,3(✉)], Ricardo Soto[1,4,5], Claudio Torres-Rojas[1],
Cristian Peña[1], Marco Riquelme-Leiva[1], Sanjay Misra[6],
Franklin Johnson[1,7], and Fernando Paredes[8]

[1] Pontificia Universidad Católica de Valparaíso, Valparaíso, Chile
{broderick.crawford,ricardo.soto}@ucv.cl,
{claudio.torres.r,cristian.pena.v}@mail.pucv.cl,
marcoriquelmeleiva@gmail.com, franklin.johnson@upla.cl
[2] Universidad San Sebastián, Santiago, Chile
[3] Universidad Central de Chile, Santiago, Chile
[4] Universidad Autónoma de Chile, Santiago, Chile
[5] Universidad Científica del Sur, Lima, Perú
[6] Covenant University, Ota, Nigeria
sanjay.misra@covenantuniversity.edu.ng
[7] Universidad de Playa Ancha, Valparaíso, Chile
[8] Escuela de Ingeniería Industrial, Universidad Diego Portales, Santiago, Chile
fernando.paredes@udp.cl

Abstract. The Set Covering Problem (SCP) is a well known $\mathcal{NP}$-*hard* problem with many practical applications. In this work binary fruit fly optimization algorithms (bFFOA) were used to solve this problem using different binarization methods.

The bFFOA is based on the food finding behavior of the fruit flies using osphresis and vision. The experimental results show the effectiveness of our algorithms producing competitive results when solve the benchmarks of SCP from the OR-Library.

Keywords: Set Covering Problem · Fruit Fly Optimization Algorithm · Metaheuristics · Combinatorial optimization problem

1 Introduction

The Set Covering Problems (SCP) can be formulated as follows: [11]

$$\text{minimize} \quad Z = \sum_{j=1}^{n} c_j x_j \tag{1}$$

Subject to:

$$\sum_{j=1}^{n} a_{ij} x_j \geq 1 \quad \forall i \in I \tag{2}$$

© Springer International Publishing Switzerland 2015
O. Gervasi et al. (Eds.): ICCSA 2015, Part IV, LNCS 9158, pp. 411–420, 2015.
DOI: 10.1007/978-3-319-21410-8_32

$$x_j \in \{0,1\} \quad \forall j \in J \qquad (3)$$

Let $A = (a_{ij})$ be a $m \times n$ 0-1 matrix with $I = \{1, \ldots, m\}$ and $J = \{1, \ldots, n\}$ be the row and column sets respectively. We say that column j can be cover a row i if $a_{ij} = 1$. Where c_j is a nonnegative value that represents the cost of selecting the column j and x_j is a decision variable, it can be 1 if column j is selected or 0 otherwise. The objective is to find a minimum cost subset $S \subseteq J$, such that each row $i \in I$ is covered by at least one column $j \in S$. The SCP was also successfully solved with meta-heuristics such as taboo search [7], simulated annealing [6], artificial bee colony [8], genetic algorithm [12,13,15], ant colony optimization [1,18], swarm optimization particles [9,19] and firefly algorithms [10].

2   Binary Fruit Fly

The Fruit Fly Optimization Algorithm (FFOA) was created by Pan [17] and it is based on the knowledge from the foraging behavior of fruit flies or vinegar flies in finding food represented in figure 1. The traditional FFOA consists of 4 phases. These are initialization, osphresis foraging search, population evaluation, and vision foraging search. In the initialization phase, the fruit flies are created randomly and they have very sensitive osphresis and vision organs which are superior to other species. In osphresis foraging phase, flies use their osphresis organ to feel all kinds of smells in the air and fly towards the corresponding locations. Flies are evaluated to find the best concentration of smell. When they are near food, in the last phase, flying toward it using its vision organ.

The FFOA is used to solve continuous problems such as: the financial distress [17], web auction logistics service [14], power load forecasting [21] and multidimensional knapsack problem [20]. The last, was solved with a new FFOA-based algorithm, which was created by Wang [20], the Binary Fruit Fly Optimization Algorithm (bFFOA).

This algorithm, in contrast from traditional FFOA, the author used: a discrete binary string to represent a solution, a probability vector to generates the population; they adopted change zero to one (or vice versa) to exploit the neighborhood in the smell-based search process; and made a global vision-based search method to improve the exploration ability. The bFFOA was divided in four phases: Initialization, and three search methods: Smell-based, Local-Vision-based and Global-Vision-based. In this paper, the bFFOA was adapted to improve it with other transfer functions, and discrete methods that be will explain in this paper.

The problems that we solve with the algorithm can be downloaded from Beasley OR-Library [1], this files was test in [4,5]. The binary FFOA was divided in the following steps:

---

[1] http://people.brunel.ac.uk/~mastjjb/jeb/info.html

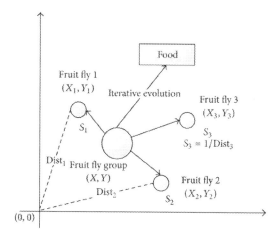

Fig. 1. Figure of the food searching of fruit fly group

2.1 Initialization Population with Probability Vector

In the bFFOA, each fruit fly is a solution and it is represented by a n-bit binary vector. Thus, a fruit fly F_i^d its dth bit is a decision variable $d = \{1, 2, ..., n\}$, in this case is a binary decision (0 or 1). All fruit flies are generated by an n-dimensional probability vector $P(gen) = [p^1(gen), p^2(gen), \ldots p^n(gen)]$ where gen represents the generation (or iteration) and $p^d(gen)$ is the probability that fly F_i^d is equal to 1 in the dth dimension. To generate a uniformly population in the search space, the probability vector is $P(0) = [0.5, 0.5, \ldots, 0.5]$, therefore, all columns have fifty percent chance of being selected. In the next generation, a new population with FN-fruit flies is generated using the probability vector.

2.2 Smell-Based and Local-Vision-Based Search

In this phase, we create randomly FS-neighbors around for each fruit fly F_i, $i \in (1, 2, \ldots, FN)$ using the Smell-based Search, this neighbors are generated using this method: first, we select randomly L-bits, and finally flip these L columns of S, for example, $l \in L$ is a column to change, if the fruit fly F_j^l, $j = \{1, 2, \ldots, FS\}$, is selected in SCP, in this case represents 1, we change to 0, or vice versa.

Before to start Local-Vision-based Search, the population of $(FN \cdot FS)$ fruit flies is evaluated using the objective function, if a solution is infeasible is repaired. When all solutions are feasible, fruit flies find their better local neighbor using vision, and flying together to the best neighbor in the neighborhood and the solution will be replaced with the best neighbor, in otherwise not.

Algorithm 1 bFFOA for SCP Algorithm

1: {Initialization Phase}
2: Initialize parameter values of FN, FS, L and b
3: Initialize Probability vector $P^d(gen = 0)$
4: **for** $i = 0$ to FN **do**
5:     create randomly F_i^d
6: **end for**
7: **repeat**
8:     {Smell-based Search}
9:     **for** $i = 0$ to FN **do**
10:         **for** $j = 0$ to FS **do**
11:             Create a neighbor F_j^d flip L bits around the F_i^d
12:         **end for**
13:     **end for**
14:     Repair and Evaluate
15:     {Local-Vision-based Search}
16:     Find the best neighbor F_{gen}^d
17:     Neighborhood fly towards F_{gen}^d
18:     {Global-Vision-based Search}
19:     Find the best F_{best}^d
20:     Select randomly two flies F_1^d and F_2^d
21:     Update Probability vector $P^d(gen)$
22:     **for** $i = 0$ to FN **do**
23:         Create F_i^d according $P^d(gen)$
24:     **end for**
25: **until** $(gen = gen_{max})$

2.3   Repair Operator

The generated solutions may be not feasible, because some rows are not covered, for to solve this problem this algorithm used a repair operator to make all solutions feasible. To repair the infeasible solutions, first for each solution the operator need identify all *uncovered* rows and the sum of columns [3]. To find these missing columns should calculate following ratio for each column:

$$\frac{\text{cost of a column}}{\text{number of } uncovered \text{ rows which it covers}} \tag{4}$$

Once all columns are covered (i.e. the solution is feasible), any redundant row is removed, a redundant column is the one to be removed the solution remains feasible [3].

2.4   Global-Vision-Based Search

This search was proposed by Wang [20] and it made for improve the exploration ability in the meta-heuristic (in the equations 5 and 6), because the previous phases focused to improving the exploitation.

$$\Delta_{gen}^d = F_{gen}^d + 0.5(F_1^d - F_2^d) \tag{5}$$

$$P^d(gen) = \frac{1}{(1 + \exp^{-b(\Delta^d_{gen} - 0.5)})} \tag{6}$$

A probability vector is used to update the next fruit flies generation using the differential information midst the generation best fruit fly F^d_{gen} and two randomly fruit flies, F^d_1 and F^d_2 and set a coefficient the vision sensitivity b to enhance the exploration.

The algorithm 1 resumes the four phases of bFFOA and explains in detail how are created and updated the fruit flies.

3   Experimental Results

The bFFOA has been implemented in Java in a Common KVM processor 2.66 GHz with 4 GB RAM computer, running Microsoft Windows 7, and the parameter values that we use are a population of 50 fruit flies, (10 neighborhood with 5 individuals for each), 3 bits to flip for generate neighbors and a coefficient the vision sensitivity of 15. For each trial each problems are run 400 iterations.

The algorithm was ran 30 trials for each instance and we report the average value from those 30 trials. We solve currently 65 data files, forty-five of these data files are the instance sets 4, 5, 6 was originally from Balas and Ho [2], the sets A, B, C, D from Beasley [4] and twenty of these data files are the *unknown-solution* problem sets NRE, NRF, NRG, NRH from Beasley [5]. These sixty five files are formatted as: number of rows (m), number of columns (n), the cost of each column $(c_j, j \in J = \{1, 2, \ldots, n\})$, and for each row i, $i \in I = \{1, 2, \ldots, m\}$ the number of columns which cover row i followed by a list of the columns which cover row i.

We solve these problems with the eight transfer functions that was proposed by Mirjalili in [16] and we test the SCP using a modified bFFOA changing the equation (6) for one of these replacing $\rho = b(\Delta^d_i - 0.5)$, where b is the coefficient the vision sensitivity. The transfer functions define a probability to change an element of solution from 1 to 0, or vice versa, and force the fruit flies to move in the interval $[0, 1]$.

After updating the probability vector with *v-shaped* or *s-shaped* transfer function, an element of a fruit fly will be updated using different discrete methods (see Table 1).

Table 1. Transfer Functions [16]

S-Shape	V-Shape
S1 $T(\rho) = \frac{1}{1+e^{-2\rho}}$	**V1** $T(\rho) = \left\lvert erf\left(\frac{\sqrt{\pi}}{2}\rho\right)\right\rvert$
S2 $T(\rho) = \frac{1}{1+e^{-\rho}}$	**V2** $T(\rho) = \lvert tanh(\rho)\rvert$
S3 $T(\rho) = \frac{1}{1+e^{\frac{-\rho}{2}}}$	**V3** $T(\rho) = \left\lvert\frac{\rho}{\sqrt{1+\rho^2}}\right\rvert$
S4 $T(\rho) = \frac{1}{1+e^{\frac{-\rho}{3}}}$	**V4** $T(\rho) = \left\lvert\frac{2}{\pi}arctan\left(\frac{\pi}{2}\rho\right)\right\rvert$

Discrete methods that were used are: the Standard, Elitist, Static Probability, Complement show in the equation (7), (8), (9), (10) respectively and Elitist Roulette. Where F_i^d represents a fruit fly F_i to be updated in the dth position in the solution, F_{best}^d is the best fruit fly in the current generation and α is called static probability.

$$F_i^d = \begin{cases} 1 \text{ if } rand() \leq P^d(gen) \\ 0 \text{ otherwise} \end{cases} \quad (7)$$

$$F_i^d = \begin{cases} F_{best}^d \text{ if } rand() \leq P^d(gen) \\ 0 \quad \text{otherwise} \end{cases} \quad (8)$$

$$F_i^d = \begin{cases} 0 \quad \text{if } P^d(gen) \leq \alpha \\ F_i^d \text{ if } \alpha \leq P^d(gen) \leq \frac{1}{2}(1+\alpha) \\ 1 \quad \text{if } \frac{1}{2}(1+\alpha) \leq P^d(gen) \end{cases} \quad (9)$$

$$F_i^d = \begin{cases} complement(F_i^d) \text{ if } rand() \leq P^d(gen) \\ 0 \quad \text{otherwise} \end{cases} \quad (10)$$

Elitist Roulette. We selected the n best fruit flies and they are assigned with a probability according to their quality of solution, in this case, the fruit fly that has less objective function value (*fitness*) will have more chance of being selected. This selection method is called as roulette-method or Monte Carlo selection-method. According to this selection method, the candidate fruit flies are selected from a population having better quality solution with selection probabilities proportional.

The table 2 and 3 show the detailed results obtained by bFFOA. The second column Z_{BKS} details the best known solution value of each instance evaluated. The next columns Z_{MIN}, Z_{MAX}, and Z_{AVG} represents the minimum, maximum, and average objective function value respectively. The last column reports the relative percentage deviation (*RPD*) value that represents the deviation of the objective value Z (fitness) from Z_{OPT} which in our case is the minimum value obtained for each instance.

$$\text{RPD} = \frac{100(Z_{MIN} - Z_{BKS})}{Z_{BKS}} \quad (11)$$

Reviewing the result tables, we can observe that:

- the binary FFOA is able to obtain optimal results for 45 of 65 problems
- the algorithm could reach a very close value to optimal in the other ten instances
- for *small* problems (4, 5, 6, A, B, C, D) the best combinations were the transfer functions S3 and S4 with the *Standard* discrete method
- for *huge* problems (NRE, NRF, NRG, NRH) the combinations gave better results are the transfer functions S3 and S4 with the *Elitist* method

Table 2. Experimental results of SCP benchmarks (4, 5, 6, A, B and C sets)

Instance	Discrete Method	Transfer Function	Z_{BKS}	Z_{MIN}	Z_{MAX}	Z_{AVG}	RPD
4.1	Standard	S2	429	429	437	431.57	0
4.2	Standard	S4	512	512	512	512	0
4.3	Elitist	S4	516	516	516	516	0
4.4	Elitist	S4	494	494	502	495.53	0
4.5	Standard	S4	512	512	517	514.2	0
4.6	Standard	S3	560	560	571	560.87	0
4.7	Standard	S3	430	430	432	430.67	0
4.8	Standard	S4	492	492	497	494.2	0
4.9	Elitist	V4	641	641	658	646.83	0
4.10	Standard	S3	514	514	517	514.1	0
5.1	Standard	S3	253	253	262	255.6	0
5.2	Standard	S4	302	304	307	305.67	0.66
5.3	Standard	S3	226	226	228	227.73	0
5.4	Complement	S2	242	242	243	242.03	0
5.5	Complement	V4	211	211	211	211	0
5.6	Complement	V4	213	213	216	213.5	0
5.7	Elitist	S4	293	293	297	294.03	0
5.8	Standard	S3	288	288	292	288.87	0
5.9	Standard	S4	279	279	286	279.8	0
5.10	Standard	S4	265	265	267	265.07	0
6.1	Complement	S2	138	138	143	140.07	0
6.2	Standard	S3	146	146	153	148.93	0
6.3	Elitist	S4	145	145	148	146.7	0
6.4	Standard	S3	131	131	131	131	0
6.5	Standard	S3	161	161	165	162.3	0
A.1	Complement	S1	253	253	255	254.8	0
A.2	Standard	S4	252	254	264	258.9	0.79
A.3	Standard	S3	232	233	238	234.8	0.43
A.4	Elitist	S3	234	234	236	234.77	0
A.5	Elitist	V4	236	236	238	236.4	0
B.1	Complement	S2	69	69	72	70.67	0
B.2	Elitist	S3	76	76	80	76.27	0
B.3	Complement	S1	80	80	81	80.17	0
B.4	Complement	V3	79	79	83	80.1	0
B.5	Complement	S1	72	72	72	72	0
C.1	Complement	V3	227	227	233	230.77	0
C.2	Elitist	S4	219	219	224	221.57	0
C.3	Standard	S3	243	247	259	254.27	1.65
C.4	Elitist	S3	219	219	226	223.07	0
C.5	Elitist	V4	215	215	218	216.8	0
D.1	E. Roulette	S1	60	60	60	60	0
D.2	Elitist	S3	66	67	68	67.73	1.52
D.3	Elitist	S4	72	73	78	75.7	1.39
D.4	Complement	S2	62	62	64	63.37	0
D.5	Elitist	S3	61	61	64	62.63	0
Total**							39

** = Total optimal results.

Table 3. Experimental results of SCP benchmarks (NRE, NRF, NRG and NRH sets)

Instance	Discrete Method	Transfer Function	Z_{BKS}	Z_{MIN}	Z_{MAX}	Z_{AVG}	RPD
NRE.1	Elitist	S3	29	29	29	29	0
NRE.2	Elitist	S3	30	30	33	32.13	0
NRE.3	Elitist	S4	27	28	29	28.7	3.7
NRE.4	Elitist	S4	28	29	31	29.63	3.57
NRE.5	Elitist	V4	28	28	30	28.93	0
NRF.1	Elitist	V4	14	14	16	15	0
NRF.2	Elitist	S4	15	15	17	15.9	0
NRF.3	E.Roulette	S4	14	15	17	16.73	7.14
NRF.4	E.Roulette	V1	14	15	16	15.03	7.14
NRF.5	Elitist	V3	13	14	16	15.1	7.69
NRG.1	Elitist	S4	176[*]	178	184	180.3	1.14
NRG.2	Elitist	V4	151[*]	159	162	160.43	3.25
NRG.3	Elitist	S4	166[*]	170	173	171.57	2.41
NRG.4	Elitist	V4	168[*]	170	179	172.2	1.19
NRG.5	Elitist	S4	168[*]	173	179	175	2.98
NRH.1	Elitist	S3	63[*]	66	71	67.47	4.76
NRH.2	Elitist	S3	63[*]	66	66	66	4.76
NRH.3	Elitist	S4	59[*]	61	67	63	3.39
NRH.4	Elitist	S3	59[*]	61	66	63.5	3.39
NRH.5	Elitist	S4	55[*]	55	60	58.07	0
Total[**]							6

[*] = Best known objective function value in the literature.
[**] = Total optimal results.

4   Conclusion

In this paper, a representation of the bFFOA was coded to solve SCP and was improved the algorithm selecting a combination, we used eight different transfer functions, and five discrete methods for transform the continuous algorithm to 0-1 algorithm and reported the best solutions obtained for each benchmark. To be solved the 65 OR-Library's instances, we focused to improve this algorithm and it showed interesting results; in terms of solutions, the experiments had many optimal solutions; this algorithm is robust, because we used the same parameters for different instances gave very good results. Today we will perform the discrete version (row-based representation) of fruit fly algorithm, where we hope to have better or similar results than the binary version.

In the future work we are very interested in realized the hybridization with other meta-heuristics or apply an hyper-heuristics version to enhance the bFFOA, in the close future, we will solve other libraries SCP, like Unicost, Italian railways, American airlines, Random or Euclidean benchmarks. Due to the effectiveness and simplicity of this algorithm, it could be used to solve other combinatorial problems.

Acknowledgments. Broderick Crawford is supported by Grant CONICYT/ FONDECYT/REGU-LAR/1140897 Ricardo Soto is supported by Grant CONI-CYT/FONDECYT/INICIACION/11130459 and Fernando Paredes is supported by Grant CONICYT/FONDECYT/REGULAR/1130455.

References

1. Amini, F., Ghaderi, P.: Hybridization of harmony search and ant colony optimization for optimal locating of structural dampers. Appl. Soft Comput., 2272–2280 (2013)
2. Balas, E., Ho, A.: Set covering algorithms using cutting planes, heuristics, and subgradient optimization: a computational study. In: Padberg, M.W. (ed.) Combinatorial Optimization. Mathematical Programming Studies, vol. 12, pp. 37–60. Elsevier North-Holland, The Netherlands (1980)
3. Beasley, J., Chu, P.: A genetic algorithm for the set covering problem. European Journal of Operational Research **94**(2), 392–404 (1996)
4. Beasley, J.E.: An algorithm for set covering problem. European Journal of Operational Research **31**(1), 85–93 (1987)
5. Beasley, J.E.: A lagrangian heuristic for set-covering problems. Naval Research Logistics (NRL) **37**(1), 151–164 (1990)
6. Brusco, M., Jacobs, L., Thompson, G.: A morphing procedure to supplement a simulated annealing heuristic for cost and coverage correlated setcovering problems. Annals of Operations Research **86**(0), 611–627 (1999)
7. Caserta, M.: Tabu search-based metaheuristic algorithm for large-scale set covering problems. In: Doerner, K., Gendreau, M., Greistorfer, P., Gutjahr, W., Hartl, R., Reimann, M. (eds.) Metaheuristics. Operations Research/Computer Science Interfaces Series, vol. 39, pp. 43–63. Springer, US (2007)
8. Crawford, B., Soto, R., Cuesta, R., Paredes, F.: Application of the artificial bee colony algorithm for solving the set covering problem. The Scientific World Journal **2014** (2014)
9. Crawford, B., Soto, R., Monfroy, E., Palma, W., Castro, C., Paredes, F.: Parameter tuning of a choice-a function based hyperheuristic using particle swarm optimization. Expert Systems with Applications, 1690–1695 (2013)
10. Crawford, B., Soto, R., Olivares-Surez, M., Paredes, F.: A binary firefly algorithm for the set covering problem. In: Silhavy, R., Senkerik, R., Oplatkova, Z.K., Silhavy, P., Prokopova, Z. (eds.) Modern Trends and Techniques in Computer Science. AISC, vol. 285, pp. 65–73. Springer, Heidelberg (2014)
11. Garey, M.R., Johnson, D.S.: Computers and Intractability; A Guide to the Theory of NP-Completeness. W. H. Freeman & Co., New York (1990)
12. Goldberg, D.: Real-coded genetic algorithms, virtual alphabets, and blocking, pp. 139–167. Complex Systems (1990)
13. Han, L., Kendall, G., Cowling, P.: An adaptive length chromosome hyperheuristic genetic algorithm for a trainer scheduling problem. In: Proceedings of the fourth Asia-Pacific Conference on Simulated Evolution And Learning (SEAL 2002), pp. 267–271. Orchid Country Club, Singapore (2002)
14. Lin, S.-M.: Analysis of service satisfaction in web auction logistics service using a combination of fruit fly optimization algorithm and general regression neural network. Neural Computing and Applications **22**(3–4), 783–791 (2013)
15. Michalewicz, Z.: Genetic algorithms + data structures = evolution programs, 3rd edn. Springer-Verlag, London (1996)
16. Mirjalili, S., Lewis, A.: S-shaped versus v-shaped transfer functions for binary particle swarm optimization. Swarm and Evolutionary Computation **9**(0), 1–14 (2013)
17. Pan, W.-T.: A new fruit fly optimization algorithm: Taking the financial distress model as an example. Knowl.-Based Syst. **26**, 69–74 (2012)

18. Ren, Z., Feng, Z., Ke, L., Zhang, Z.: New ideas for applying ant colony optimization to the set covering problem. Computers & Industrial Engineering, 774–784 (2010)
19. Shi, Y., Eberhart, R.: Empirical study of particle swarm optimization. In: Proc. of the Congress on Evolutionary Computation, pp. 1945–1950 (1999)
20. Wang, L., Zheng, X., Wang, S.: A novel binary fruit fly optimization algorithm for solving the multidimensional knapsack problem. Knowl.-Based Syst. **48**, 17–23 (2013)
21. ze Li, H., Guo, S., jie Li, C., qi Sun, J.: A hybrid annual power load forecasting model based on generalized regression neural network with fruit fly optimization algorithm. Knowledge-Based Systems **37**(0), 378–387 (2013)

A Teaching-Learning-Based Optimization Algorithm for Solving Set Covering Problems

Broderick Crawford[1,2,3]([✉]), Ricardo Soto[1,4,5], Felipe Aballay[1], Sanjay Misra[6],
Franklin Johnson[1,7], and Fernando Paredes[8]

[1] Pontificia Universidad Católica de Valparaíso, Valparaíso, Chile
{broderick.crawford,ricardo.soto}@ucv.cl, felipe.aballay.l@mail.pucv.cl,
franklin.johnson@upla.cl
[2] Universidad San Sebastián, Santiago, Chile
[3] Universidad Central de Chile, Santiago, Chile
[4] Universidad Autónoma de Chile, Santiago, Chile
[5] Universidad Científica del Sur, Lima, Perú
[6] Covenant University, Ota, Nigeria
sanjay.misra@covenantuniversity.edu.ng
[7] Universidad de Playa Ancha, Valparaíso, Chile
[8] Escuela de Ingeniería Industrial, Universidad Diego Portales, Santiago, Chile
fernando.paredes@udp.cl

Abstract. The Set Covering Problem (SCP) is a representation of a kind of combinatorial optimization problem which has been applied in several problems in the real world. In this work we used a binary version of Teaching-Learning-Based Optimization (TLBO) algorithm to solve SCP, works with two phases known: teacher and learner; emulating the behavior into a classroom. The proposed algorithm has been tested on 65 benchmark instances. The results show that it has the ability to produce solutions competitively.

Keywords: Set Covering Problem · Teaching-Learning-Based Optimization algorithm · Combinatorial optimization · Metaheuristics

1 Introduction

The Set Covering Problem (SCP) is a popular $\mathcal{NP}$-hard problem [15] that has been used to a wide range of airlines and buses crew scheduling [27], location of emergency facilities [29], railway crew management [8], steel production [30], vehicle scheduling [14] and ship scheduling [13] between others.

The formulation of SCP is as follows: Let $A = (a_{ij})$, a zero-one $m \times n$ matrix, and nonnegative n-dimensional integer vector C, and $I = \{1, \ldots, n\}$ and $J = \{1, \ldots, m\}$ be the row and column set respectively. Given $c_j > 0$ for $(c_j \in C, j \in J)$ the cost of selecting the column j of matrix A. If $a_{ij} \in A$ is equal to 1, we say that row i is covered by column j, otherwise it is not. In SCP, the objective is to find a minimum cost subset of columns of A such that

© Springer International Publishing Switzerland 2015
O. Gervasi et al. (Eds.): ICCSA 2015, Part IV, LNCS 9158, pp. 421–430, 2015.
DOI: 10.1007/978-3-319-21410-8_33

each row is covered by at least one column in the subset S. We can formulated mathematically as:

$$\text{minimize} \quad Z = \sum_{j=1}^{n} c_j x_j \tag{1}$$

Subject to:

$$\sum_{j=1}^{n} a_{ij} x_j \geq 1 \quad \forall i \in I \tag{2}$$

$$x_j \in \{0, 1\} \quad \forall j \in J \tag{3}$$

The SCP was also successfully solved with meta-heuristics such as taboo search [9], simulated annealing [7,28], genetic algorithm [16–18], ant colony optimization [1,18], swarm optimization particles [11], artificial bee colony [10,31] and firefly algorithms [12].

2   Binary Teaching-Learning-Based Optimization Algorithm

Teaching learning based optimization (TLBO) was originally proposed by Rao et al. (2011). The main idea of TLBO is that the teacher is considered as the most knowledgeable person in a class who shares his/her knowledge with the students to improve the output (i.e., grades or marks) of the class. The quality of the learners is evaluated by the mean value of the student's grade in class. In addition learners also can learn from interaction between themselves, which also helps in their results [2].

TLBO is population based method. In this optimization algorithm a group of learners is considered as population and different design variables are considered as different subjects offered to the learners and the results obtained by learners are analogous to the solutions fitness values of the optimization problem. In the entire population the best solution is considered as the teacher.

TLBO is used to solve problems like: global optimization problems [26], unconstrained optimization problems [21].

The working of TLBO is divided into two parts: Teacher phase and Learner phase.

2.1   Teacher Phase

The teacher phase produces a random and ordered state of points called learners within the search space. Then a point is considered as the teacher, who is highly learned person and shares his or her knowledge with the learners, and others learn significant group information from the teacher. It is the first part of the algorithm where the mean of a class increases from M_A to M_B depending upon a good teacher. At this point, assumed a good teacher is one who brings his/her learners up to his/her level in terms of knowledge [20]. However, in practice this is not possible and a teacher can only move the mean of a class up to some extent depending on the capability of the class. This follows a random process depending on many factors [21].

Let M_i be the mean and T_i be the teacher at any iteration i. T_i will try to move mean M_i towards its own level, so now the new mean will be T_i designated as M_{new}. The solution is updated according to the difference between the existing and the new mean given via Eq(4) [22]:

$$DifferenceMean_i = r_i(X_{new} - T_F M_i) \tag{4}$$

where T_F is a teaching factor that decides the value of mean to be changed, and r_i is a random number in the range [0,1]. The value of T_F can be either 1 or 2, which is again a heuristic step and decided randomly with equal probability via Eq (5) [23]:

$$T_F = round[1 + rand(0,1)\{2 - 1\}] \tag{5}$$

This difference modifies the existing solution via Eq (6) [25]:

$$x_{new} = x_{old,i} + DifferenceMean_i \tag{6}$$

2.2   Learner Phase

It is the second part of the algorithm where learners increase their knowledge by interaction among themselves. A solution is randomly interacted to learn something new with other solutions in the population. In order to this statement, a solution will learn new information if the other solutions have more knowledge than him or her. Mathematically the learning phenomenon of this phase is expressed by Eq(7) [24]:

$$\begin{aligned} x_{new}^d = x_i^d + rand()(x_j^d - x_i^d), \quad If \quad f(x_i) > f(x_j) \\ x_{new}^d = x_i^d + rand()(x_i^d - x_j^d), \quad If \quad f(x_i) < f(x_j) \end{aligned} \tag{7}$$

At any iteration i, considering two different learners x_i and x_j, where i $\neq$ j. Consequently, accept x_{new}, if it gives better function value. After a number of sequential teaching learning cycles, where the teacher pass on knowledge among the learners and those level increases toward his or her own level, the distribution of the randomness within the search space becomes smaller and smaller about to point considering as teacher. It means knowledge level of the whole class shows smoothness and the algorithm converges to a solution. Also, a termination criterion can be a predetermined maximum iteration number is reached.

Considering the two key phases described above, the steps of implementing the TLBO algorithm can be summarized as follows [24]:

Algorithm 1. bTLBO for SCP Algorithm

1: Set k=1;
2: Objective Function $f(x)$, $x = (x_1, x_2..., xd)^t$ $d=no.$ of design variables
3: Generate   initial   students   of   the   classroom   randomly x^i, $i=1,2...,n$
   $n=no.$ of students
4: Calculate objective function $f(x)$ for whole students of the classroom
5: **while** the termination conditions are not met **do**
6:    {teacher phase}
7:    Calculate the mean of each design variable x_{mean}
8:    Identify the best solution (teacher)
9: **for** $i = 1 \to N$ **do**
10:      Calculate teaching factor $T_F^i = round[1 + rand(0,1)\{2 - 1\}]$
11:      Modify solution based on best solution(teacher)
12:      $x_{new}^i = X^i + DifferenceMean_i^d$
13:      Calculate objective function for new mapped student $f(x_{new}^i)$
14:    **if** x_{new}^i is better than x^i , i.e $f(x_{new}^i)<f(x^i)$ **then**
15:        $x^i=x_{new}^i$
16:    **end if**
17:      {end of teacher phase}
18:      {student phase}
19:      {Randomly select another learner (x^j), such that j≠i}
20:    **if** x^i is better than x^j , i.e $f(x^i)<f(x^j)$ **then**
21:        $x_{new}^i = x^i + rand(0,1)(x^i - x^j)$
22:    **else**
23:        $x_{new}^i = x^i + rand(0,1)(x^j - x^i)$
24:    **end if**
25:    **if** x_{new}^i is better than x^i , i.e $f(x_{new}^i)<f(x^i)$ **then**
26:        $x^i=x_{new}^i$
27:    **end if**
28:      {end of student phase}
29:    **end for**
30:    k=k+1
31: **end while**

- **Step 1**: Defining the optimization problem, and initializing the optimization parameters.
- **Step 2**: Initializing the population.
- **Step 3**: Starting teacher phase where the main activity is learners learning from their teacher.
- **Step 4**: Starting learner phase where the main activity is learners further tune their knowledge through the interaction with their peers.
- **Step 5**: Evaluating stopping criteria. Terminate the algorithm in the maximum generation number is reached; otherwise return to Step 3 and the algorithm continues.

In teacher and learner phases, the velocity of each student can be calculated according to the following:

$$V_i^d(t+1) = r(X_{teacher}^d - T_F X_n^d) \tag{8}$$

$$V_i^d(t+1) = r(X_i^d - X_j^d) \tag{9}$$

The position in bTLBO is represented by a binary vector and the velocity is still a floating-point vector, however the velocity is used to determine the probability of change from 0 to 1 or vice versa when the position is updating. In the binary version is replaced the equation (6) to equation (8) for teacher phase and the equation (7) is replaced by the equation (9) for learner phase.

2.3   Repair Operator

The solutions generated may be not feasible, because some rows aren't covered, for to solve this problem we need to used a Beasley's Repair Operator [4] to make all solutions feasible. To repair the unfeasible solutions, first for each solution we need identify all uncovered rows and the sum of columns so that all rows are covered. The search for these missing columns is based on the following ratio:

$$\frac{\text{cost of a column}}{\text{number of } uncovered \text{ rows which it covers}} \tag{10}$$

Once all columns are covered or whether the solution is feasible, we begin to eliminate any redundant column. A redundant column is the one to be removed the solution remains feasible.

3   Experimental Results

In the current section we present the experimental results. The algorithm ran 30 trials for each instance and then we got the averages values from these 30 trials. The algorithm solve the 65 data files from the OR-Library, 45 of them are the instance sets 4,5,6 was originally from Balas and Ho [3], the sets A, B, C, D from Beasley [5] and 20 of these data files are the test problem sets E, F, G, H from Beasley [6].

These 65 files are formatted as: number of rows m, number of columns n, the cost of each column c_j, $j = \{1, \ldots, n\}$, and for each row i $i = \{1, ..., m\}$ the number of columns which cover row i followed by a list of the columns which cover rows i. We reduce the SCP with 2 methods created by Beasley [5], called *Column Domination* and *Column Inclusion* where the first consist in deleting (or dominating) columns when another column(s) cover the same row by less cost, and the second is used when a row is covered by a single column in the reduced problem, this column must be in the solution.

The bTLBO was implemented in Java programming language using Eclipse IDE in a computer with the following hardware, Intel i5-3230M 2.60 GHz processor, 4 GB RAM and it ran under windows 7 operating system, the used parameters for execute the algorithm were 20 students and for each trial it was ran 1000 iterations.

We solve these problems with the eight transfer functions that was proposed by Mirjalili in [19]. The transfer functions define a probability to change an element of solution from 1 to 0, or vice versa (Table 1).

Table 1. Transfer Functions [19]

S-Shape	V-Shape
S1 $T(V_i^d) = \frac{1}{1+e^{-2V_i^d}}$	**V1** $T(V_i^d) = \left\| erf\left(\frac{\sqrt{\pi}}{2}V_i^d\right)\right\|$
S2 $T(V_i^d) = \frac{1}{1+e^{-V_i^d}}$	**V2** $T(V_i^d) = \|tanh(V_i^d)\|$
S3 $T(V_i^d) = \frac{1}{1+e^{-\frac{V_i^d}{2}}}$	**V3** $T(V_i^d) = \left\|\frac{V_i^d}{\sqrt{1+(V_i^d)^2}}\right\|$
S4 $T(V_i^d) = \frac{1}{1+e^{-\frac{V_i^d}{3}}}$	**V4** $T(V_i^d) = \left\|\frac{2}{\pi}arctan\left(\frac{\pi}{2}V_i^d\right)\right\|$

Besides the Transfer functions, 5 discretization methods were used, Standard (11), Elitist (12), Complement (13), Static probability (14), Elitist roulette (15) these are showed below:

Standard

$$x_i^d(t+1) = \begin{cases} 1 \text{ if } rand \leq V_i^d(t+1) \\ \\ 0 \text{ otherwise} \end{cases} \tag{11}$$

Elitist

$$x_i^d(t+1) = \begin{cases} x_{best}^k \text{ if } rand \leq V_i^d(t+1) \\ \\ 0 \quad \text{ otherwise} \end{cases} \tag{12}$$

Complement

$$x_i^d(t+1) = \begin{cases} complement(x_i^k) \text{ if } rand \leq V_i^d(t+1) \\ \\ 0 \quad\quad\quad\quad\quad\quad \text{ otherwise} \end{cases} \tag{13}$$

Static Probability

$$x_i^d(t+1) = \begin{cases} x_i^d \quad \text{ if } V_i^d(t+1) \leq \alpha \\ \\ x_{best}^d \text{ if } \alpha \leq V_i^d(t+1) \leq \frac{1}{2}(1+\alpha) \\ \\ x_1^d \quad \text{ if } \frac{1}{2}(1+\alpha) \leq V_i^d(t+1) \end{cases} \tag{14}$$

Table 2. Experimental results of SCP benchmarks (4, 5, 6, A, B, C, and D sets)

Instance	Z_{BKS}	Discrete Method	Transfer Function	Z_{MIN}	Z_{MAX}	Z_{AVG}	RPD
4.1	429	Standard	S1	430	431	430.6	0.23
4.2	512	Standard	S1	524	530	528.2	2.34
4.3	516	Standard	S1	526	528	527.1	1.94
4.4	494	Standard	S1	501	508	504.3	1.42
4.5	512	Standard	S1	518	518	518.0	1.17
4.6	560	Complement	S1	566	587	579.1	1.07
4.7	430	Standard	S1	433	435	433.8	0.7
4.8	492	Standard	S4	507	509	507.5	3.05
4.9	641	Standard	S4	660	676	673.2	2.96
4.10	514	Standard	S3	524	531	526.7	1.95
5.1	253	Standard	S1	257	262	260.3	1.58
5.2	302	Standard	S1	311	311	311.0	2.98
5.3	226	Standard	S1	228	229	228.3	0.88
5.4	242	Complement	S1	244	246	245.6	0.83
5.5	211	Standard	S3	215	220	218.5	1.9
5.6	213	Standard	S1	217	219	217.8	1.88
5.7	293	Standard	S1	293	299	297.0	0
5.8	288	Standard	S2	294	301	297.9	2.08
5.9	279	Standard	S1	281	285	283.5	0.72
5.10	265	Standard	V1	268	275	274.3	1.13
6.1	138	Standard	S4	143	148	144.8	3.62
6.2	146	Complement	S2	148	157	152.5	1.37
6.3	145	Standard	S1	148	150	149.2	2.07
6.4	131	Standard	S1	131	134	132.3	0
6.5	161	Standard	S4	167	173	170.0	3.73
A.1	253	Standard	S3	257	258	257.7	1.58
A.2	252	Standard	S1	263	265	263.6	4.37
A.3	232	Standard	S3	242	245	243.7	4.31
A.4	234	Standard	S3	237	239	238.8	1.28
A.5	236	Standard	S2	239	241	240.1	1.27
B.1	69	Standard	V1	72	79	77.6	4.35
B.2	76	Elite	S3	82	88	86.6	7.89
B.3	80	Complement	V4	80	100	91.4	0
B.4	79	Complement	V2	82	84	83.6	3.8
B.5	72	Standard	V3	72	78	74.2	0
C.1	227	Standard	S1	235	235	235.0	3.52
C.2	219	Complement	V3	226	236	230.0	3.2
C.3	243	Elite	S3	263	269	265.4	8.23
C.4	219	Standard	S4	238	243	241.0	8.68
C.5	215	Standard	S1	220	220	220.0	2.33
D.1	60	Standard	V1	62	62	62.0	3.33
D.2	66	Standard	V1	70	71	70.6	6.06
D.3	72	Complement	V1	77	78	77.4	6.94
D.4	62	Complement	V4	65	68	66.8	4.84
D.5	61	Standard	V3	64	66	65.2	4.92

Elitist Roulette

$$p_i = \frac{f_i}{\sum_{j=1}^{k} f_j} \tag{15}$$

The table 2 and 3 show the results obtained by bTLBO. The second column Z_{BKS} indicates the best known solution value of each instance evaluated. The columns discrete method and transfer functions indicates those that were used. The next columns *Min*, *Max*, and *Avg* represents the minimum, maximum, and average objective function values respectively. The last column reports the relative percentage deviation (RPD) which represents the deviation of the objective

Table 3. Experimental results of SCP benchmarks (NRE, NRF, NRG and NRH sets)

Instance	Z_{BKS}	Discrete Method	Transfer Function	Z_{MIN}	Z_{MAX}	Z_{AVG}	RPD
NRE.1	29	Standard	V1	30	30	30.0	3.45
NRE.2	30	Elite	V1	34	35	34.4	13.33
NRE.3	27	Complement	V4	29	32	30.2	7.41
NRE.4	28	Elite	S2	32	33	32.6	14.29
NRE.5	28	Complement	V1	30	30	30.0	7.14
NRF.1	14	Standard	V1	17	17	17.0	21.43
NRF.2	15	Staticprob	S2	17	18	17.8	13.33
NRF.3	14	Standard	V1	17	17	17.0	21.43
NRF.4	14	Elite	S3	16	18	17.8	14.29
NRF.5	13	Standard	V2	15	16	15.7	15.38
NRG.1	176*	Standard	V3	193	196	194.2	9.66
NRG.2	151*	Standard	V1	164	168	166.4	6.49
NRG.3	166*	Complement	V1	178	180	178.8	7.23
NRG.4	168*	Complement	V4	180	184	181.6	7.14
NRG.5	168*	Complement	V4	183	189	185.4	8.93
NRH.1	63*	Complement	V2	71	73	71.8	12.7
NRH.2	63*	Roulette	V2	67	67	67.0	6.35
NRH.3	59*	Staticprob	V2	68	68	68.0	15.25
NRH.4	59*	Elite	S4	66	68	67.2	11.86
NRH.5	55*	Staticprob	V2	60	61	60.6	9.09

* = Best known objective function value in the literature.

value f (fitness) from f_{opt} which is the minimum value obtained for each instance. RPD is calculated as:

$$\text{RPD} = \frac{100(f - f_{opt})}{f_{opt}} \tag{16}$$

4   Conclusion

In this work was implemented a binary TLBO algorithm to solve the SCP. It uses eight different transfer functions and five discretization methods and it was tested solving 65 benchmarks from OR-Library. We reached only four optimum in instances 5.7, 6.4, B.3 and B.5 and others results are very close to optimum values. But others results are too far: NRG.1 or C.3.

In relation with the *small* problems (4, 5, 6, A, B, C, D) the best combinations observed were transfer functions S2 and S1 with the *Standard* and *Complement* discretization method. Besides, for *huge* problems (NRE, NRF, NRG, NRH) the combinations that gave better results were transfer functions V1 and V2 with the *Complement* and *Elitist* method. In a future work we will solve the other SCP libraries of instances (Italian railways, American airlines and Euclidean benchmarks) and also we are implementing a discrete version of TLBO.

References

1. Amini, F., Ghaderi, P.: Hybridization of harmony search and ant colony optimization for optimal locating of structural dampers. Appl. Soft Comput. **13**(5), 2272–2280 (2013)

2. Bo, X., Gao, W.-J.: Innovative Computational Intelligence: A Rough Guide to 134 Clever Algorithms. Springer (2014)

3. Balas, E., Ho, A.: Set covering algorithms using cutting planes, heuristics, and subgradient optimization: a computational study. In: Padberg, M.W. (ed.) Combinatorial Optimization. Mathematical Programming Studies, vol. 12, pp. 37–60. Elsevier North-Holland, The Netherlands (1980)

4. Beasley, J., Chu, P.: A genetic algorithm for the set covering problem. European Journal of Operational Research **94**(2), 392–404 (1996)

5. Beasley, J.E.: An algorithm for set covering problem. European Journal of Operational Research **31**(1), 85–93 (1987)

6. Beasley, J.E.: A lagrangian heuristic for set-covering problems. Naval Research Logistics (NRL) **37**(1), 151–164 (1990)

7. Brusco, M.J., Jacobs, L.W., Thompson, G.M.: A morphing procedure to supplement a simulated annealing heuristic for cost- and coverage-correlated set-covering problems. Annals of Operations Research **86**, 611–627 (1999)

8. Caprara, A., Fischetti, M., Toth, P., Vigo, D., Guida, P.L.: Algorithms for railway crew management. Math. Program. **79**(1–3), 125–141 (1997)

9. Caserta, M.: Tabu search-based metaheuristic algorithm for large-scale set covering problems. In: Doerner, K., Gendreau, M., Greistorfer, P., Gutjahr, W., Hartl, R., Reimann, M. (eds.) Metaheuristics. Operations Research/Computer Science Interfaces Series, vol. 39, pp. 43–63. Springer, US (2007)

10. Crawford, B., Soto, R., Cuesta, R., Paredes, F.: Application of the artificial bee colony algorithm for solving the set covering problem. The Scientific World Journal (2014)

11. Crawford, B., Soto, R., Monfroy, E., Palma, W., Castro, C., Paredes, F.: Parameter tuning of a choice-function based hyperheuristic using particle swarm optimization. Expert Systems with Applications **40**(5), 1690–1695 (2013)

12. Crawford, B., Soto, R., Olivares-Surez, M., Paredes, F.: A binary firefly algorithm for the set covering problem. In: Silhavy, R., Senkerik, R., Oplatkova, Z.K., Silhavy, P., Prokopova, Z., (eds.) Modern Trends and Techniques in Computer Science. Advances in Intelligent Systems and Computing, vol. 285, pp. 65–73. Springer International Publishing (2014)

13. Fisher, M.L., Rosenwein, M.B.: An interactive optimization system for bulk-cargo ship scheduling. Naval Research Logistics (NRL) **36**(1), 27–42 (1989)

14. Foster, B.A., Ryan, D.: An integer programming approach to the vehicle scheduling problem. Operations Research **27**, 367–384 (1976)

15. Garey, M.R., Johnson, D.S.: Computers and Intractability; A Guide to the Theory of NP-Completeness. W. H. Freeman & Co., New York (1990)

16. Goldberg, D.E.: Real-coded genetic algorithms, virtual alphabets, and blocking. Complex Systems **5**, 139–167 (1990)

17. Han, L., Kendall, G., Cowling, P.: An adaptive length chromosome hyperheuristic genetic algorithm for a trainer scheduling problem. In: Proceedings of the Fourth Asia-Pacific Conference on Simulated Evolution And Learning, (SEAL 2002), Orchid Country Club, Singapore, pp. 267–271 (2002)

18. Michalewicz, Z.: Genetic algorithms + data structures = evolution programs, 3rd edn. Springer-Verlag, London (1996)

19. Mirjalili, S., Lewis, A.: S-shaped versus v-shaped transfer functions for binary particle swarm optimization. Swarm and Evolutionary Computation **9**(0), 1–14 (2013)

20. Rao, R.V., Patel, V.: An elitist teaching-learning-based optimization algorithm for solving complex constrained optimization problems. International Journal of Industrial Engineering Computations **3**, 535–560 (2012)
21. Rao, R.V., Patel, V.: An improved teaching-learning-based optimization algorithm for solving unconstrained optimization problems. Scientia Iranica **20**(3), 710–720 (2013)
22. Rao, R.V., Patel, V.: Multi-objective optimization of heat exchangers using a modified teaching-learning-based optimization algorithm. Applied Mathematical Modelling **37**, 1147–1162 (2013)
23. Rao, R.V., Patel, V.: Multi-objective optimization of two stage thermoelectric cooler using a modified teaching-learning-based optimization algorithm. Engineering Applications of Artificial Intelligence **26**, 430–445 (2013)
24. Rao, R.V., Savsani, V.J., Vakharia, D.P.: Teaching-learning-based optimization: A novel method for constrained mechanical design optimization problems. Computer-Aided Design **43**, 303–315 (2011)
25. Rao, R.V., Savsani, V.J.: Mechanical design optimization using advanced optimization techniques. Springer (2012)
26. Satapathy, S., Naik, A., Parvathi, K.: A teaching learning based optimization based on orthogonal design for solving global optimization problems. SpringerPlus **2**(1), 130 (2013)
27. Smith, B.M.: Impacs - a bus crew scheduling system using integer programming. Math. Program. **42**(1), 181–187 (1988)
28. Thomson, G.: A Simulated Annealing Heuristic for Shift-Scheduling Using Non-Continuously Available Employees. Computers and Operations Research **23**, 275–288 (1996)
29. Toregas, C., Swain, R., ReVelle, C., Bergman, L.: The location of emergency service facilities. Operations Research **19**(6), 1363–1373 (1971)
30. Vasko, F.J., Wolf, F.E., Stott, K.L.: A set covering approach to metallurgical grade assignment. European Journal of Operational Research **38**(1), 27–34 (1989)
31. Zhang, Y., Wu, L., Wang, S., Huo, Y.: Chaotic artificial bee colony used for cluster analysis. In: Chen, R. (ed.) ICICIS 2011 Part I. CCIS, vol. 134, pp. 205–211. Springer, Heidelberg (2011)

A Comparison of Three Recent Nature-Inspired Metaheuristics for the Set Covering Problem

Broderick Crawford[1,2,3]($\boxtimes$), Ricardo Soto[1,4,5], Cristian Peña[1],
Marco Riquelme-Leiva[1], Claudio Torres-Rojas[1], Sanjay Misra[6],
Franklin Johnson[1,7], and Fernando Paredes[8]

[1] Pontificia Universidad Católica de Valparaíso, Valparaíso, Chile
{broderick.crawford,ricardo.soto}@ucv.cl,
{cristian.pena.v,claudio.torres.r}@mail.pucv.cl,
marcoriquelmeleiva@gmail.com, franklin.johnson@upla.cl
[2] Universidad San Sebastián, Santiago, Chile
[3] Universidad Central de Chile, Santiago, Chile
[4] Universidad Autónoma de Chile, Santiago, Chile
[5] Universidad Científica del Sur, Lima, Perú
[6] Covenant University, Ota, Nigeria
sanjay.misra@covenantuniversity.edu.ng
[7] Universidad de Playa Ancha, Valparaíso, Chile
[8] Escuela de Ingeniería Industrial, Universidad Diego Portales, Santiago, Chile
fernando.paredes@udp.cl

Abstract. The Set Covering Problem (SCP) is a classic problem in
combinatorial optimization. SCP has many applications in engineering,
including problems involving routing, scheduling, stock cutting, electoral
redistricting and others important real life situations. Because of its
importance, SCP has attracted attention of many researchers. However,
SCP instances are known as complex and generally NP-hard problems.
Due to the combinatorial nature of this problem, during the last decades,
several metaheuristics have been applied to obtain efficient solutions.
This paper presents a metaheuristics comparison for the SCP. Three
recent nature-inspired metaheuristics are considered: Shuffled Frog Leaping, Firefly and Fruit Fly algorithms. The results show that they can
obtainn optimal or close to optimal solutions at low computational cost.

Keywords: Set Covering Problem · Metaheuristics · Shuffled Frog
Leaping Algorithm · Firefly algorithm · Fruit fly algorithm

1 Introduction

The Set Covering Problem (SCP) is defined as follows, let $A = (a_{ij})$ be an m-row,
n-column, zero-one matrix. We say that a column j covers a row i if $a_{ij} = 1$. Each
column j is associated with a non negative real cost c_j. Let $I = \{1, 2, \ldots, m\}$
and $J = \{1, 2, \ldots, n\}$ be the row set and column set, respectively. The SCP calls

© Springer International Publishing Switzerland 2015
O. Gervasi et al. (Eds.): ICCSA 2015, Part IV, LNCS 9158, pp. 431–443, 2015.
DOI: 10.1007/978-3-319-21410-8_34

for a minimum cost subset $S \subseteq J$, such that each row $i \in I$ is covered by at least one column $j \in S$. A mathematical model for the SCP is:

$$Minimize \quad f(x) = \sum_{j=1}^{n} c_j x_j \tag{1}$$

subject to:

$$\sum_{j=1}^{n} a_{ij} x_j \geq 1, \qquad \forall i \in I \tag{2}$$

$$x_j \in \{0, 1\}, \qquad \forall j \in J \tag{3}$$

The SCP has many practical applications like location of emergency facilities [5], airline and bus crew scheduling [4,15], steel production [11], logical analysis of numerical data [3], ship scheduling [12], vehicle routing [1]. The SCP has been solved using complete techniques and different metaheuristics [6,7,17].

This work proposes to solve the SCP with three recent Nature-inspired metaheuristics: Shuffled Frog Leaping Algorithm (SFLA), Modified Binary FireFly Algorithm (MBFF) and Binary Fruit-Fly Algorithm (bFFOA).

SFLA is based on the observation, the imitation and the modeling of the behavior of a group of frogs searching a location that has the maximum available quantity of food [10].

Firefly algorithm was presented by Yang [19] and its operation, is based on the social behaviour of fireflies.

The Fruit Fly Optimization Algorithm (FFOA) was created by Pan [16] and it is inspired by the knowledge from the foraging behavior of fruit flies in finding food.

2   Shuffled Frog Leaping Algorithm

In SFLA a set of frogs are generated randomly. Then, the population is divided in frog subgroups named *memeplexes*. For each subgroup, a local search is realized to improve the position of the worst frog, which in turn can be influenced by other frogs since each frog has ideas affecting others. This process is named *evolution memetica*, which can repeat up to a certain number of iterations. The ideas generated by every memeplexe are transmitted to other memeplexes in a process of redistribution [14]. The local search, the evolution memetica and the redistribution they continue until the criterion of stop is reached [9].

The initial population of frogs P is created at random. This means that for a problem of n variables, a frog i is represented as a vector $X_i = (x_i^1, x_i^2, \ldots, x_i^n)$. Then, the fitness is calculated for each frog and they are arranged in descending order according to the obtained fitness. m memeplexes are generated from the division of the population, each subgroup contains f frogs (i.e. $P = m \times f$). In this process, the first frog goes to the first memeplexe, the second frog goes to the second memeplexe, frog f goes to the memeplexe m, and the frog $f + 1$ goes back to the first memeplexe, ...

By each memeplexe the best frog is identified as x_b, the frog with the worst fitness as x_w and the best global frog as x_g.

In the local search an adjustment is applied to the worst frog in the following way:

$$d_w^k = rand()(x_b^k - x_w^k), \qquad 1 \le k \le n \tag{4}$$

$$x_{new}^k = x_w^k + d_w^k, \qquad d_{min}^k \le d_w^k \le d_{max}^k \tag{5}$$

Where $rand()$ is a random number ($rand() \sim U(0,1)$) y d_{max}^k is the maximum change allowed in the position of a frog. The result of this process is compared with the fitness of the worst frog, if this one is better than the worst, the worst frog is replaced. Otherwise, the calculation of the equations repeats itself 4 and 5, but with the best global frog (i.e. x_g it replaces to x_b). It turns to compare the obtained result, and if the last fitness calculation is better than the fitness worst frog, the worst frog is replaced. Otherwise, a frog is generated randomly to replace the worst frog. The process is realized by a certain number of iterations.

2.1 A Binary Coded Shuffled Frog Leaping Algorithm

The SCP can not be handled directly by SFLA due to its binary structure. Therefore, to obtain values 0 or 1 a transfer function and a Discretization method are performed. The transfer function is applied to the result of the Eqs. 4 and 5. We tested the eight different functions shown in Table 1. They are separated into two families: S-shape and V-shape (Fig. 1). The result of this operation is a real number between 0 and 1, then a binarization method is required to obtain a value 0 or 1. The algorithm 1 explains the Binary SFLA solving the SCP.

3 Binary FireFly Algorithm

Firefly Algorithm gets its name because its methodology operation is based on the social behaviour of fireflies using three basic rules for its operation [19]:

Rule 1 : All fireflies are unisex, one firefly is attracted to other fireflies regard less of gender.

Rule 2 : The attractiveness of a firefly is proportional to its brightness, this means that in a pair of fireflies the less bright will be attracted by the more bright. In the absence of a brighter firefly will move randomly.

Rule 3 : The brightness of a firefly will be determined by its objective function. in a maximization problem the brightness of each firefly is proportional to the value of the objective function. In the case of minimization (SCP), brightness of fireflies is inversely proportional to the value of its objective function.

These three basic rules of Firefly algorithm can be explained and represented as follows:

Algorithm 1. Binary SFLA for SCP Algorithm.

1: Initialize parameters
2: Generate random population of P solutions(frogs)
3: **for** $f = 1$ to P **do**
4:     Calculate fitness of each solution (f)
5: **end for**
6: **repeat**
7:     Sorting population in ascending order
8:     Partition P into m memeplexes ($P = m \times f$)
9:     **for** $im = 1$ to m **do**
10:         **for** $i = 1$ to it **do**
11:             Determine X_g, X_b and X_w
12:             Apply Eqs. 4 and 5, Transfer function and Discretization method and repair if necessary
13:             **if** $X_{new} < X_w$ **then**
14:                 Replace the worst solution
15:             **else**
16:                 Apply Eqs. 4 and 5), replacing X_b with X_g, Transfer function and Discretization method and repair if necessary
17:                 **if** $X_{new} < X_w$ **then**
18:                     Replace the worst solution
19:                 **else**
20:                     Generate a new feasible solution randomly
21:                 **end if**
22:             **end if**
23:         **end for**
24:     **end for**
25:     Shuffled the m memeplexes
26: **until** (termination = true)
27: return the best solution

Attractive: The formula of attraction $\beta_{(r)}$, can be any monotonically decreasing function as the next.

$$\beta_{(r)} = \beta_0 e^{-\gamma r^m} \tag{6}$$

Where r is the distance between two fireflies, β_0 is the initial appeal of firefly and γ is a light absorption coefficient.

Distance between Fireflies: the distance r_{ij} between any two fireflies i and j at positions x_i and x_j respectively, can be defined as a Cartesian or Euclidean distance as follows:

$$r_{ij} = ||x_i - x_j|| = \sqrt{\sum_{k=1}^{d} (x_i^k - x_j^k)^2} \tag{7}$$

Where x_i^k is the current value of the k_{th} dimension of the ith firefly (a firefly is a solution of the problem) and d is the number of dimensions or variables of the problem.

Movement of the Fireflies: The movement of a firefly i, is attracted to another more attractive (brighter) firefly j, is determined by:

$$d_i^k(t+1) = x_i^k(t) + \beta_0 e^{-\gamma r_{ij}^2}(x_j^k(t) - x_i^k(t)) + \alpha(rand - 1/2) \qquad (8)$$

where the first term $x_i^k(t)$ is the current position of the k_{th} dimension of the firefly i at the iteration t. The second term of the equation corresponds to the attraction and the third term introduces a random value to the equation, where α is a randomization parameter and rand is a random number generated uniformly distributed between 0 and 1.

3.1   A Modified Binary FireFly Algorithm

In this paper, we propose a modification to the classical binary algorithm [7,8], this modification consists in a computational optimization, using the ascending order of the population through the evaluation of the objective function. Then, the brightest firefly occupies the zero position into the population matrix. In lines 3 and 6 of the algorithm, we can appreciate the computational optimization.

Algorithm 2. MBFF for SCP Algorithm

1: Initialize parameter values of α, β_0, γ, Population size, Number of generations.
2: Evaluate the light intensity I determined by $f(x)$, see Eq. 1
3: Sort the population in ascending order according fitness $1/f(x)$ see Eq. 1
4: **while** $t <$**Number of generations do**
5:      **for** $i = 1 : m$ (m fireflies) **do**
6:          **for** $j = i : m$ (m fireflies) **do**
7:              **if** $I_j < I_i$ **then**
8:                  **movement** = calculates value according Table. 1 and Discretization
                    method.
9:              **end if**
10:             Repair solutions using Repair Operator
11:             Update light intensity
12:         **end for**
13:     **end for**
14:     $t = t + 1$
15: **end while**
16: Output result

---

4   Binary Fruit-Fly Algorithm

The Fruit Fly Optimization Algorithm was created by Pan [16] and it is based by the knowledge from the foraging behavior of fruit flies in finding food. The traditional FFOA consists of 4 phases. These are initialization, osphresis foraging search, population evaluation, and vision foraging search. In the initialization phase, the fruit flies are created randomly and they have very sensitive osphresis and vision organs which are superior to other species. In osphresis foraging phase, flies use their osphresis organ to feel all kinds of smells in the air and fly towards the corresponding locations. This flies are evaluated to find the best concentration of smell. When they are near food, in the last phase, flying toward it using its vision organ.

The FFOA is used to solve continuous problems such as: the financial distress [16], web auction logistics service [13], power load forecasting [20] and multidimensional knapsack problem [18]. The last, was solved with a new FFOA-based algorithm, which was created by Wang [18], the Binary Fruit Fly Optimization Algorithm (bFFOA). In this algorithm, in contrast from traditional FFOA, the author used a discrete binary string to represent a solution, a probability vector to generates the population; they adopted change zero to one (or vice versa) to exploit the neighborhood in the smell-based search process; and made a global vision-based search method to improve the exploration ability. The bFFOA was divided in four phases: Initialization, Smell-based search, Local-Vision-based search and Global-Vision-based search. The bFFOA was divided in the following steps:

4.1   Initialization Population with Probability Vector

In the bFFOA, each fruit fly is a solution and it is represented by a n-bit 0-1 vector. Thus, a fruit fly $X_i = \{x_i^1, x_i^2, \ldots, x_i^n\}$ where x_i^k, $k = \{1, 2, ..., n\}$ is a decision variable , in this case is a binary decision (0 or 1). All fruit flies are generated by an n-dimensional probability vector $P(t) = [p^1(t), p^2(t), \ldots p^n(t)]$ where t represents the iteration (or generation) and $p^k(t)$ is the probability that fly x_i^k is equal to 1 in the kth dimension. To generate a uniformly population in the search space, the probability vector is $P(0) = [0.5, 0.5, \ldots, 0.5]$, therefore, all columns have fifty percent chance of being selected. In the next generation, a new population with F_N-fruit flies is generated using the probability vector.

4.2   Smell-Based and Local-Vision-Based Search

In this phase, we create randomly F_S-neighbors around for each fruit fly X_i, $i \in (1, 2, \ldots, F_N)$ using the Smell-based Search, this neighbors are generated using this method: first, we select randomly L-bits, and finally flip these L columns of S, for example, $l \in L$ is a column to change, if the fruit fly x_j^l, $j = \{1, 2, \ldots, F_S\}$, is selected in SCP, in this case represents 1, we change to 0, or vice versa.

Before to start Local-Vision-based Search, the population with $(F_N \cdot F_S)$-fruit flies are evaluated using the objective function, if a solution is infeasible is repaired. When all solutions are feasible, fruit flies find their better local neighbor using vision, and flying together to the best neighbor in the neighborhood and the solution will be replaced with the best neighbor, in otherwise not.

4.3   Global-Vision-Based Search

This search was proposed by Wang [18] and it made for improve the exploration ability in the meta-heuristic (in the equation 9), because the previous phases

Algorithm 3. bFFOA for SCP Algorithm

1: {Initialization Phase}
2: Initialize parameter values of F_N, F_S, L and b
3: Initialize Probability vector $P^k(t = 0)$
4: **for** $i = 0$ to F_N **do**
5:     create randomly x_i^k
6: **end for**
7: **repeat**
8:     {Smell-based Search}
9:     **for** $i = 0$ to F_N **do**
10:        **for** $j = 0$ to F_S **do**
11:            Create a neighbor x_j^k flip L bits around the x_i^k
12:        **end for**
13:    **end for**
14:    Repair and Evaluate
15:    {Local-Vision-based Search}
16:    Find the best neighbor x_g^k
17:    Neighborhood fly towards x_g^k
18:    {Global-Vision-based Search}
19:    Find the best x_{best}^k
20:    Select randomly two flies x_1^k and x_2^k
21:    Update Probability vector $P^k(t)$
22:    **for** $i = 0$ to F_N **do**
23:        Create x_i^k according $P^k(t)$
24:    **end for**
25: **until** $(t = t_{max})$

focused to improving the exploitation.

$$d^k(t+1) = p^k(t+1) = -b\left(x_g^k(t) + \frac{x_1^k(t) - x_2^k(t) - 1}{2}\right) \qquad (9)$$

A probability vector is used to update the next fruit flies generation using the differential information midst the generation best fruit fly x_g^k and two randomly fruit flies, x_1^k and x_2^k and set a coefficient the vision sensitivity b to enhance the exploration.

The algorithm 3 resumes the four phases of bFFOA and explains in detail how are created and updated the fruit flies.

5   Discretization Method (Binarization)

To obtain values 0 or 1 when solving the binary SCP a Transfer function and a Discretization method should be performed.

5.1   Transfer Functions

We tested eight different functions (Table 1) separated into two families: S-shape and V-shape (Fig. 1).

Table 1. Transfer Functions

S-Shape	V-Shape
S1 $T(d_i^k(t+1)) = \dfrac{1}{1+e^{-2d_i^k(t+1)}}$	**V1** $T(d_i^k(t+1)) = \left\lvert erf\left(\dfrac{\sqrt{\pi}}{2}d_i^k(t+1)\right)\right\rvert$
S2 $T(d_i^k(t+1)) = \dfrac{1}{1+e^{-d_i^k(t+1)}}$	**V2** $T(d_i^k(t+1)) = \lvert tanh(d_i^k(t+1))\rvert$
S3 $T(d_i^k(t+1)) = \dfrac{1}{1+e^{\frac{-d_i^k(t+1)}{2}}}$	**V3** $T(d_i^k(t+1)) = \left\lvert \dfrac{x}{\sqrt{1+d_i^k(t+1)^2}}\right\rvert$
S4 $T(d_i^k(t+1)) = \dfrac{1}{1+e^{\frac{-d_i^k(t+1)}{3}}}$	**V4** $T(d_i^k(t+1)) = \left\lvert \dfrac{2}{\pi}arctan\left(\dfrac{\pi}{2}d_i^k(t+1)\right)\right\rvert$

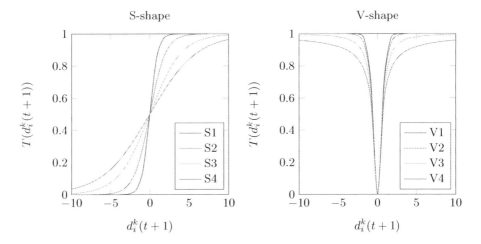

Fig. 1. Transfer functions

5.2   Discretization Methods

The result obatined using a Transfer function is a real number between 0 and 1, then a Discretization method is required to obtain a value 0 or 1.

- Standard.

$$x_{new}^j = \begin{cases} 1 \text{ if } rand \leq T(d_i^k(t+1)) \\[2mm] 0 \text{ } otherwise \end{cases} \tag{10}$$

- Complement.

$$x_{new}^j = \begin{cases} \overline{x_w^j} \text{ if } rand \leq T(d_i^k(t+1)) \\[2mm] 0 \quad otherwise \end{cases} \tag{11}$$

- Static probability.

$$
x_{new}^j = \begin{cases} 0 & \text{if } T(d_i^k(t+1)) \leq \alpha \\ x_w^j & \text{if } \alpha < T(d_i^k(t+1)) \leq \frac{1}{2}(1+\alpha) \\ 1 & \text{if } T(d_i^k(t+1)) \geq \frac{1}{2}(1+\alpha) \end{cases} \tag{12}
$$

- Elitist.

$$
x_{new}^j = \begin{cases} x_b^j & \text{if } rand < T(d_i^k(t+1)) \\ 0 & otherwise \end{cases} \tag{13}
$$

- Elitist Roulette. The Discretization method Elitist Roulette (Monte Carlo) is to select randomly between the best individuals of the population, with a probability proportional its fitness.

6   Avoiding Infeasible Solutions

If a solution does not satisfy the constraints of the problem this is called infeasible. Then a repair operator shoul be applied to the solution becomes feasible. The steps are: identify rows that are not covered by at least one column. Then add at least a column so that all rows are covered. The search for these columns are based on the ratio [2]:

$$
\frac{\text{cost of a column}}{\text{number of uncovered rows which it covers}}
$$

As soon as columns are added converting infeasible solutions to feasibles, it is applied a local optimization phase, which eliminates a set of redundant columns that are in the solution. A redundant column is one that removed from the solution, it continues being feasible [2]. The pseudo-code is the following one:

7   Experimental Evaluation

The table 2 shows the properties of the SCP instances tested: instance set number, number of rows (m), number of columns (n), range of costs, density (percentage of non-zeroes in matrix A) and if the optimal solution is known or unknown.

The results are evaluated using the relative percentage deviation (RPD). The value RPD quantifies the deviation of the objective value Z from Z_{opt} that in our case is the best known value for each instance (BKV in Table 3). The minimum (Min), maximum (Max) and average (Avg) of the solutions obtained using the 40 instances of SFLA, MBFF and bFFOA (different combinations of 8 transfer functions with 5 Discretization methods) are shown. To calculate RPD we use $Z = Min$. This measure is calculated as:

Table 2. SCP instances

Instance set	No. of instance	m	n	Cost range	Density (%)	Optimal solution
4	10	200	1000	[1, 100]	2	Known
5	10	200	2000	[1, 100]	2	Known
6	5	200	1000	[1, 100]	5	Known
A	5	300	3000	[1, 100]	2	Known
B	5	300	3000	[1, 100]	5	Known
C	5	400	4000	[1, 100]	2	Known
D	5	400	4000	[1, 100]	5	Known
NRE	5	500	5000	[1, 100]	10	Unknown
NRF	5	500	5000	[1, 100]	20	Unknown
NRG	5	1000	10000	[1, 100]	2	Unknown
NRH	5	1000	10000	[1, 100]	5	Unknown

$$\text{RPD} = \frac{100(Z - Z_{opt})}{Z_{opt}} \tag{14}$$

In all experiments, each SFLA, MBFF and bFFOA algorithm instance was executed 30 times over each one of the 65 SCP test instances. We test all the

---

Algorithm 4. Repair Operator.

1: I = the set of all rows,
2: J = the set of all columns,
3: α_i = the set of columns that cover row i, $i \in I$,
4: β_j = the set of rows covered by column j, $j \in J$,
5: S = the set of columns in a solution,
6: U = the set of uncovered rows,
7: w_i = the number of columns that cover row i, $i \in I$ in S.
8: Initialise $w_i :=| S \cap \alpha_i |, \forall_i \in I$.
9: Initialise $U := \{i \mid w_i = 0, \forall_i \in I\}$.
10: **for all** row i in U **do** {in increasing order of i}
11:     find the first column j (in increasing order of j) in α_i that minimises $c_j / | U \cap \beta_j |$
12:     $S := S + j$
13:     $w_i := w_i + 1, \forall_i \in \beta_j$
14:     $U := U - \beta_j$
15: **end for**
16: **for all** column j in S **do** {in decreasing order of j}
17:     **if** $w_i \geq 2, \forall i \in \beta_j$ **then**
18:         $S := S - j$;
19:         $w_i := w_i - 1, \forall i \in \beta_j$;
20:     **end if**
21: **end for**
22: S is now a feasible solution for the SCP that contains no redundant columns.

Table 3. Experimental Result of SCP benchmarks

Instances	BKV	SFLA		bFFOA		MBFF	
		MIN	RPD	MIN	RPD	MIN	RPD
4.1	429	430	0,23	429	0	429	0
4.2	512	516	0,78	512	0	517	0,97
4.3	516	520	0,78	516	0	519	0,58
4.4	494	501	1,42	494	0	495	0,20
4.5	512	514	0,39	512	0	514	0,39
4.6	560	563	0,54	560	0	563	0.53
4.7	430	431	0,23	430	0	430	0
4.8	492	497	1,02	492	0	497	1,01
4.9	641	656	2,34	641	0	655	2,18
4.1	514	518	0,78	514	0	519	0,97
5.1	253	254	0,40	253	0	257	1,58
5.2	302	307	1,66	304	0.66	309	2,31
5.3	226	228	0,88	226	0	229	1,32
5.4	242	242	0	242	0	242	0
5.5	211	211	0	211	0	211	0
5.6	213	213	0	213	0	213	0
5.7	293	297	1,37	293	0	298	1,70
5.8	288	291	1,04	288	0	291	1,04
5.9	279	281	0,72	279	0	284	1,79
5.1	265	265	0	265	0	268	1,13
6.1	138	140	1,45	138	0	138	0
6.2	146	147	0,68	146	0	147	0,68
6.3	145	147	1,38	145	0	147	1,37
6.4	131	131	0	131	0	131	0
6.5	161	166	3,11	161	0	164	1,86
A.1	253	255	0,79	253	0	255	0,79
A.2	252	260	3,17	254	0.79	259	2,77
A.3	232	237	2,16	233	0.43	238	2,58
A.4	234	235	0,43	234	0	235	0,42
A.5	236	236	0	236	0	236	0
B.1	69	70	1,45	69	0	71	2,89
B.2	76	76	0	76	0	78	2,63
B.3	80	80	0	80	0	80	0
B.4	79	79	0	79	0	80	1,26
B.5	72	72	0	72	0	72	0
C.1	227	229	0,88	227	0	230	1,32
C.2	219	223	1,83	219	0	223	1,82
C.3	243	253	4,12	247	1.65	253	4,11
C.4	219	227	3,65	219	0	225	2,73
C.5	215	217	0,93	215	0	217	0,93
D.1	60	60	0	60	0	60	0
D.2	66	67	1,52	67	1.52	68	3,03
D.3	72	75	4,17	73	1.39	75	4,16
D.4	62	63	1,61	62	0	62	0
D.5	61	63	3,28	61	0	63	3,27
NRE.1	29	29	0	29	0	29	0
NRE.2	30	31	3,33	30	0	32	6,66
NRE.3	27	28	3,70	28	3.7	29	7,40
NRE.4	28	29	3,57	29	3.57	29	3,57
NRE.5	28	28	0	28	0	29	3,57
NRF.1	14	15	7,14	14	0	15	7,14
NRF.2	15	15	0	15	0	16	6,66
NRF.3	14	16	14,29	15	7.14	16	14,28
NRF.4	14	15	7,14	15	7.14	15	7,14
NRF.5	13	15	15,38	14	7.69	15	15,38
NRG.1	176	182	3,41	178	1.14	185	5,11
NRG.2	154	161	4,55	159	3.25	161	4,54
NRG.3	166	173	4,22	170	2.41	175	5,42
NRG.4	168	173	2,98	170	1.19	176	4,76
NRG.5	168	174	3,57	173	2.98	177	5,35
NRH.1	63	68	7,94	66	4.76	69	9,52
NRH.2	63	66	4,76	66	4.76	66	4,76
NRH.3	59	62	5,08	61	3.39	65	10,16
NRH.4	59	63	8,62	61	3.39	63	6,77
NRH.5	55	59	7,27	55	0	59	7,27

combinations of transfer functions and Discretization methods over all these instances.

The SFLA Metaheuristic used a population of 200 frogs ($P = 200$), 10 memeplexes ($m = 10$), 20 iterations within each memeplex ($it = 20$) and number of iterations as termination condition ($iMax = 100$). In the experiments using the static probability Discretization method, the parameter $\propto$ was set to 0.5.

The MBFF Metaheuristic used a population of 50 fireflies ($m = 50$), randomization parameter ($\alpha = 0.5$), initial appeal of firefly ($\beta_0 = 1.0$), light absorption coefficient ($\gamma = 1.0$).

The bFFOA Metaheuristic used 10 neighborhood ($F_N = 10$), each neighborhood has 5 neighbor ($F_S = 5$), the number of bit to be changed $L = 3$ and set a coefficient the vision sensitivity $b = 15$.

We can see in Table 3 that the bFFOA is much more efficient in finding the global optimums of SCP instances. The bFFOA performs best for all the SCP instances considered. The number of optimal solutions found by this metaheuristic is the greatest: 45. SFLA reached 14 optimums and MBFF found 13. Moreover, the RPD of bFFOA is the best in all cases.

8   Conclusion

In this paper, SFLA, MBFF and bFFOA metaheuristics were presented as a very good alternative to solve the SCP. We used eight transfer function and five methods of Discretization with the aim to solve the binary problem at hand. We have performed experiments through several instances. We presented a comparison of three Nature-Inspired Metaheuristics for Set Covering Problem. Our proposal has demonstrated to be very effective solving the 65 SCP instances obtaining an important number of best known values in a few iterations.

Acknowledgments. Broderick Crawford is supported by Grant CONICYT/ FONDECYT/REGULA-R/1140897, Ricardo Soto is supported by Grant CONICYT/ FONDECYT/INIC-IACION/11130459 and Fernando Paredes is supported by Grant CONICYT/FO-NDECYT/REGULAR/1130455.

References

1. Foster, B.A., Ryan, D.M.: An integer programming approach to the vehicle scheduling problem. Operational Research Quarterly **27**, 367–384 (1976)
2. Beasley, J., Chu, P.: A genetic algorithm for the set covering problem. European Journal of Operational Research **94**(2), 392–404 (1996)
3. Boros, E., Hammer, P.L., Ibaraki, T., Kogan, A.: Logical analysis of numerical data. Math. Program. **79**, 163–190 (1997)
4. Caprara, A., Fischetti, M., Toth, P.: A heuristic method for the set covering problem. Operations Research **47**(5), 730–743 (1999)
5. Constantine, T., Ralph, S., Charles, R., Lawrence, B.: The location of emergency service facilities. Operations Research **19**, 1363–1373 (1971)

6. Crawford, B., Soto, R., Cuesta, R., Paredes, F.: Application of the Artificial Bee Colony Algorithm for Solving the Set Covering Problem. The Scientific World Journal **2014**, 8 (2014)
7. Crawford, B., Soto, R., Olivares-Suárez, M., Paredes, F.: A binary firefly algorithm for the set covering problem. In: Silhavy, R., Senkerik, R., Oplatkova, Z.K., Silhavy, P., Prokopova, Z. (eds.) Modern Trends and Techniques in Computer Science. AISC, vol. 285, pp. 65–73. Springer, Heidelberg (2014)
8. Crawford, B., Soto, R., Olivares-Suárez, M., Paredes, F.: Using the firefly optimization method to solve the weighted set covering problem. In: Stephanidis, C. (ed.) HCI 2014, Part I. CCIS, vol. 434, pp. 509–514. Springer, Heidelberg (2014)
9. Eusuff, M., Lansey, K.: Optimization of water distribution network design usingthe shuffled frog leaping algorithm. Journal of Water Resource Plan Management **129**, 210–225 (2003)
10. Eusuff, M., Lansey, K., Pasha, F.: Shuffled frog-leaping algorithm: a memeticmeta-heuristic for discrete optimization. Engineering Optimization **38**, 129–154 (2006)
11. Vasko, F.J., Wolf, F.E., Stott, K.L.: A set covering approach to metallurgical grade assignment. European Journal of Operational Research **38**(1), 27–34 (1989)
12. Fisher, M.L., Rosenwein, M.B.: An interactive optimization system for bulk-cargo ship scheduling. Naval Research Logistics **36**, 27–42 (1989)
13. Lin, S.-M.: Analysis of service satisfaction in web auction logistics service using a combination of fruit fly optimization algorithm and general regression neural network. Neural Computing and Applications **22**(3–4), 783–791 (2013)
14. Liong, S., Atiquzzaman, M.: Optimal design of water distribution network using-shuffled complex evolution. Journal of Instrumentation Engineering **44**, 93–107 (2004)
15. Smith, B.M.: Impacs a bus crew scheduling system using integer programming. Mathematical Programming **42**(1–3), 181–187 (1998)
16. Pan, W.-T.: A new fruit fly optimization algorithm: Taking the financial distress model as an example. Knowl.-Based Syst. **26**, 69–74 (2012)
17. Valenzuela, C., Crawford, B., Soto, R., Monfroy, E., Paredes, F.: A 2-level meta-heuristic for the set covering problem. International Journal of Computers Communications and Control **7**(2), 377–387 (2012)
18. Wang, L., Zheng, X., Wang, S.: A novel binary fruit fly optimization algorithm for solving the multidimensional knapsack problem. Knowl.-Based Syst. **48**, 17–23 (2013)
19. Yang, X.-S.: Nature-inspired metaheuristic algorithms. Luniver Press (2010)
20. Ze Li, H., Guo, S., Jie Li, C., Qi Sun, J.: A hybrid annual power load forecasting model based on generalized regression neural network with fruit fly optimization algorithm. Knowledge-Based Systems **37**, 378–387 (2013)

Bug Assignee Prediction Using Association Rule Mining

Meera Sharma[1], Madhu Kumari[2], and V.B. Singh[2(✉)]

[1] Department of Computer Science, University of Delhi, Delhi, India
meerakaushik@gmail.com
[2] Delhi College of Arts and Commerce, University of Delhi, Delhi, India
{mesra.madhu,vbsinghdcacdu}@gmail.com

Abstract. In open source software development we have bug repository to which both developers and users can report bugs. Bug triage, deciding what to do with an incoming bug report, takes a large amount of developer resources and time. All newly coming bug reports must be triaged to determine whether the report is correct and requires attention and if it is, which potentially experienced developer/fixer will be assigned the responsibility of resolving the bug report. In this paper, we propose to apply association mining to assist in bug triage by using Apriori algorithm to predict the developer that should work on the bug based on the bug's severity, priority and summary terms. We demonstrate our approach on collection of 1,695 bug reports of Thunderbird, AddOnSDK and Bugzilla products of Mozilla open source project. We have analyzed the association rules for top five assignee of the three products. Association rules can support the managers to improve its process during development and save time and resources.

1   Introduction

The availability of various software repositories namely source code, bugs, attributes of bugs, source code changes, developer communication, mailing list, allows new research areas in software engineering like mining software repositories, empirical software engineering, and machine learning based software engineering. Various machine learning based prediction models have been developed and is currently being used to improve the quality of software in terms of choosing right developer to fix the bugs, predicting bug fix time, predicting the attributes of a bug namely severity and priority, and bugs lying dormant in the software [1-6]. One of the important software repositories is the bug tracking system (BTS) which is used to manage bug reports submitted by users, testers, and developers [7]. Each new reported bug must be triaged to determine if it describes a meaningful new problem or enhancement, and if it does, it must be assigned to an appropriate developer to fix it. A bug is characterized by many attributes shown in table 1[8].

Some of the important bug attributes are severity, priority and summary. The degree of impact of a bug on the functionality of the software is known as its severity. It is defined on seven levels from 1 to 7 namely, Blocker, Critical, Major, Normal, Minor, Enhancement and Trivial, having Blocker as the level 1 and Trivial as the level 7. Bug priority describes the importance and order in which a bug should be fixed compared to other bugs. P1 is considered the highest and P5 is the lowest. The summary attribute of a bug report consists of the brief description (textual description) about the bug.

© Springer International Publishing Switzerland 2015
O. Gervasi et al. (Eds.): ICCSA 2015, Part IV, LNCS 9158, pp. 444–457, 2015.
DOI: 10.1007/978-3-319-21410-8_35

Table 1. Bug Attributes description

Attribute	Short description
Severity	This indicates how severe the problem is. e.g. trivial, critical, etc.
Bug Id	The unique numeric id of a bug.
Priority	This field describes the importance and order in which a bug should be fixed compared to other bugs. P1 is considered the highest and P5 is the lowest.
Resolution	The resolution field indicates what happened to this bug. e.g. FIXED
Status	The Status field indicates the current state of a bug. e.g. NEW, RESOLVED
Number of Comments	Bugs have comments added to them by users. #comments made to a bug report.
Create Date	When the bug was filed.
Dependencies	If this bug cannot be fixed unless other bugs are fixed (depends on), or this bug stops other bugs being fixed (blocks), their numbers are recorded here.
Summary	A one-sentence summary of the problem.
Date of Close	When the bug was closed.
Keywords	The administrator can define keywords which you can use to tag and categorize bugs e.g. the Mozilla project has keywords like crash and regression.
Version	The version field defines the version of the software the bug was found in.
CC List	A list of people who get mail when the bug changes. #people in CC list.
Platform and OS	These indicate the computing environment where the bug was found.
Number of Attachments	Number of attachments for a bug.
Bug Fix Time	Last Resolved time-Opened time. Time to fix a bug.

To the best of our knowledge available in literature no work has been done to discover associations rule among the bug attributes. These rules can support the managers to improve its process during development. In this paper, we have made an attempt to predict the developer that should work on the bug by applying association mining by using Apriori algorithm based on the bug's severity, priority and summary terms. We demonstrate our approach on collection of 1,695 bug reports of Thunderbird, AddOnSDK and Bugzilla products of Mozilla open source project. Our prediction method is based on the association rule mining method which was first explored by [9].

Association rule mining is used to discover the patterns of co-occurrences of the attributes in a database. Associations do not imply causality. An association rule is an expression $A \Rightarrow C$, where A (Antecedent) and C (Consequent) are sets of items. Given a database D of transactions, where each transaction $T \in D$ is a set of items, $A \in C$ expresses that whenever a transaction T contains A, then T also contains C with a specified confidence and support. The rule confidence is defined as the percentage of transactions containing C in addition to A with regard to the overall number of transactions containing A [10]. Support is the number of times the items in a rule appear together in a single entry within the entire set. Association rule mining can successfully be applied to a wide range of business and science problems. Extensive performance studies have also shown that associative classification frequently generates better accuracy than state-of-the-art classification methods [11-20]. The successful use of association rule mining in various fields motivates us to apply it to the open source software bug data set.

The rest of the paper is organized as follows. Section 2 of the paper describes the datasets and preprocessing of data. Results have been presented in section 3. Section 4 presents the related work. Threats to validity have been discussed in section 5 and finally the paper is concluded in section 6.

2   Description of Data Sets and Data Preprocessing

In this paper, an empirical experiment has been conducted on 1,695 bug reports of the Mozilla open source software products namely Thunderbird, AddOnSDK and Bugzilla. We collected bug reports for resolution "fixed" and status "verified", "resolved" and "closed" because only these types of bug reports contain the meaningful information for the experiment. The collected bug reports from Bugzilla have also been compared and validated against general change data (i.e. CVS or SVN records). Table 2 shows the data collection in the observed period.

Table 2. Number of bug reports in each product

Product	Number of bugs	Observation period
Thunderbird	115	Apr. 2000-Mar. 2013
Add-on SDK	616	May 2009-Aug. 2013
Bugzilla	964	Sept. 1994-June 2013

We have used four quantified bug attributes namely severity, priority, summary and assignee.

There is a need to extract terms from bug summary attribute (a textual description of the bug). We pre-processed the bug summary in RapidMiner tool [21] containing the following steps:

Tokenization: Tokenization is the process of breaking a stream of text into words, phrases, symbols, or other meaningful elements called tokens. In this paper a word or a term has been considered as token.

Stop Word Removal: In bug summary prepositions, conjunctions, articles, verbs, nouns, pronouns, adverbs, adjectives, etc. are stop words and have been removed.

Stemming to base stem: The process of converting derived words to their base word is known as stemming. In this paper, we have used Standard Porter stemming algorithm for stemming [22].

Feature Reduction: Tokens of minimum length 3 and maximum length 50 have been considered because most of the data mining algorithm may not be able to handle large feature sets.

As a result of this process we get a set of terms of bug summary attribute for a dataset. In RapidMiner tool for the calculation of summary terms we have set tokenize mode as non-letters. For filter tokens option we have taken min chars parameter value as 3 and max chars parameter value as 50. We have filtered the stop words by English dictionary.

The importance i.e. usefulness and certainty of an association rule is measured by its support and confidence. Rules that discover with high levels of support (or relevance) and high confidence do not necessarily imply causality. Let $Y = \{Y_1, Y_2 \ldots Y_m\}$ be a set of attribute values, called items. A set $A \subseteq Y$ is called an item set. Let a database D be a multiset of Y. Each $T \in D$ is called a transaction. An association rule is an expression $A \Rightarrow C$, where $A \subset Y$, $C \subset Y$, and $A \cap C = \varphi$. We refer to A as the antecedent of the rule, and C as the consequent of the rule. The rule $A \Rightarrow C$ has support $Supp(A \Rightarrow C)$ in D, where the support is defined as $Supp(A \Rightarrow C) = Supp(A \cup C)$. That means $Supp(A \Rightarrow C)$ percent of the transactions in D contain $A \cup C$, and $Supp(A) = |\{T \in D | A \subseteq T\}| / |D|$ is the support of A that is the fraction of transactions T supporting an item set A with respect to database D. The number of transactions required for an item set to satisfy minimum support is referred to as the minimum support count. A transaction $T \in D$ supports an item set $A \subseteq Y$ if $A \subseteq T$ holds. The rule $A \Rightarrow C$ holds in D with confidence $Conf(A \Rightarrow C)$, where the confidence is defined as $Conf(A \Rightarrow C) = Supp(A \cup C) / Supp(A)$. That means $Conf(A \Rightarrow C)$ percent of the transactions in D that contain A also contain C. The confidence is a measure of the rule's strength or certainty while the support corresponds to statistical significance or usefulness. Association rule mining generates all association rules that have a support greater than minimum support $min.Supp(A \Rightarrow C)$, in the database, i.e., the rules are frequent. The rules must also have confidence greater than minimum confidence $min.Conf(A \Rightarrow C)$, i.e., the rules are strong. The process of association rule mining consists of these two steps: 1) Find all frequent item sets, where each $A \cup C$ of these item sets must be at least as frequently supported as the minimum support count. 2) Generate strong rules from the discovered frequent item sets, where each $(A \Rightarrow C)$ of these rules must satisfy $min.Supp(A \Rightarrow C)$ and $min.Conf(A \Rightarrow C)$ [10].

We have carried out following steps for our study:

1. **Data Extraction**
   a. Download the bug reports of different products of Mozilla open source software from the CVS repository: https://bugzilla.mozilla.org/
   b. Save bug reports in excel format.
2. **Data Pre-processing**
   a. Extract individual terms from bug summary attribute.

3.   **Data Preparation**
   a.   Assign severity attribute as numeric values from 1 to 7 and priority levels as 8 to 12.
   b.   Take top 30 terms based on the occurrences of a term in the dataset of summary attribute and assign a numeric value from 13 to 43.
   c.   Assign a unique numeric value to each assignee.
4.   **Modeling**
   a.   Build a model in MATLAB software by using ARMADA tool [23]. ARMADA (Association Rule Miner And Deduction Analysis) is a Data Mining tool that extracts Association Rules from numerical data files using a variety of selectable techniques and criteria. The program integrates several mining methods which allow the efficient extraction of rules, while allowing the thoroughness of the mine to be specified at the user's discretion. We have applied Apriori algorithm to find association rules for assignee as consequent and severity, priority and summary terms as antecedents with minimum confidence 20% and minimum support 7% for AddOnSDK, Bugzilla. We have taken confidence 20% and minimum support 3% for Thunderbird dataset as we are not getting sufficient rules for support 7% because of fewer transactions in the dataset.
5.   **Testing and Validation**
   a.   Assess the association rules in terms of support and confidence.

3   Results and Discussion

We have applied Apriori algorithm for association rule mining using ARMADA tool in MATLAB software to predict bug assignee using bug severity, priority and summary terms. When mining association rules for bug assignee prediction, we have taken minimum confidence 20% and minimum support 7% for AddOnSDK and Bugzilla products. As number of bugs is very less for thunderbird product we have taken minimum support 3% and confidence 20%. As a result we get 3 sets of rules, where each set consist of more than 100 rules. For this reason, we do not list them all, but instead we present top 5 rules for top 5 assignee based on the highest confidence. Table 3 shows the typical forms of the association rules for top five bug assignee of AddOnSDK dataset.

Table 3. Association rules for top five assignee in AddOnSDK

Association Rules with minimum support 7% and minimum confidence 20%

Assignee = Alexandre Poirot

- Priority {P1} ∧ Term {con} ∧ Term {content} ∧ Term {fail}
  ⇒ Assignee {Alexandre Poirot} @ (11%, 79%).
- Severity {Normal} ∧ Priority {P1} ∧ Term {con} ∧ Term {content} ∧ Term {fail} ⇒ Assignee {Alexandre Poirot} @ (7%, 78%)

Table 3. (*Continued*)

- Priority {P1} ∧ Term {con} ∧ Term {test} ∧ Term {content} ∧Term {fail}
  ⇒ Assignee {Alexandre Poirot} @ (10%, 77%)
- Severity{Normal} ∧ Priority {P1} ∧Term {con} ∧Term {content}∧Term {script} ⇒Assignee {Alexandre Poirot} @ (14%, 67%)
- Priority {P1} ∧ Term {con} ∧ Term {test} ∧ Term {fail}
  ⇒ Assignee {Alexandre Poirot} @ (10%, 67%)

Assignee = Will Bamberg

- Severity {Normal} ∧ Term {doc} ∧ Term {document} ∧ Term {page}
  ⇒ Assignee {Will Bamberg} @ (7%, 100%)
- Severity {Normal} ∧ Term {doc} ∧ Term {tab}
  ⇒ Assignee {Will Bamberg} @ (8%, 89%)
- Severity {Normal} ∧ Priority {P1} ∧ Term {doc} ∧ Term {mod} ∧ Term {modul} ⇒ Assignee {Will Bamberg} @ (10%, 83%)
- Severity {Normal} ∧ Priority {P1} ∧ Term {con} ∧ Term {doc}
  ⇒ Assignee {Will Bamberg} @ (9%, 82%)
- Severity {Normal} ∧ Priority {P1} ∧ Term {doc} ∧ Term {mod}
  ⇒ Assignee {Will Bamberg} @ (11%, 79%)

Assignee = Brian Warner

- Severity {Normal} ∧ Priority{P1} ∧ Term {add} ∧ Term {sdk}
  ⇒ Assignee {Nobody} @ (7%, 26%)
- Term {P1} ∧ Term {add} ∧ Term {sdk}
  ⇒ Assignee {Nobody} @ (7%, 24%)
- Severity {Normal} ∧ Term {add} ∧ Term {sdk}
  ⇒ Assignee {Nobody} @ (7%, 21%)
- Severity {Normal} ∧ Term {pack}
  ⇒ Assignee {Nobody} @ (7%, 20%)
- Term {add} ∧ Term {sdk}
  ⇒ Assignee {Nobody} @ (7%, 20%)

Assignee = Erik Vold

- Severity {Normal} ∧ Term {test} ∧ Term {win} ∧ Term {window}
  ⇒ Assignee {Erik Vold} @ (7%, 47%)
- Term {privat} ∧ Term {brows}
  ⇒ Assignee {Erik Vold} @ (11%, 44%)
- Priority {P1} ∧ Term {test} ∧ Term {win} ∧ Term {window}
  ⇒ Assignee {Erik Vold} @ (7%, 44%)
- Severity {Normal} ∧ Term {test} ∧ Term {win}
  ⇒ Assignee {Erik Vold} @ (7%, 44%)
- Term {test} ∧ Term {win} ∧ Term {window}
  ⇒ Assignee {Erik Vold} @ (7%, 42%)

Table 3. (*Continued*)

Assignee = Irakli Gozilalishvili

- Severity {Normal} ∧ Priority {P1} ∧ Term {load}
  ⇒ Assignee {Irakli Gozilalishvili} @ (8%, 32%)
- Priority {P1} ∧ Term {load}
  ⇒ Assignee {Irakli Gozilalishvili} @ (8%, 27%)
- Severity {Normal} ∧ Term {load}
  ⇒ Assignee {Irakli Gozilalishvili} @ (8%, 24%)
- Term{load}
  ⇒ Assignee {Irakli Gozilalishvili} @ (8%, 21%)

The first association rule is a four antecedent rule, which reveals that the assignee *Alexandre Poirot* can be assigned a bug having priority *P1* and summary containing terms *con, content* and *fail* with a significance of 11 percent and a certainty of 79 percent. Second association rule is a five antecedent rule, which means that the assignee *Alexandre Poirot* can be assigned a bug having severity *Normal*, priority *P1* and summary containing terms *con, content* and *fail* with a significance of 7 percent and a certainty of 78 percent. Third rule shows that the assignee *Alexandre Poirot* can be assigned a bug having priority *P1* and summary containing terms *con, test, content* and *fail* with a significance of 10 percent and a certainty of 77 percent. Rule four reveals that 14 percent of the bugs in the bug data set have severity *Normal*, priority *P1* and summary containing terms *con, content, script* and assignee *Alexandre Poirot*. 67 percent of the bugs in the bug data set that have severity *Normal*, priority *P1* and summary containing terms *con, content, script* also have assignee *Alexandre Poirot*. The fifth rule shows that the assignee *Alexandre Poirot* can be assigned a bug having priority *P1* and summary containing terms *con, test* and *fail* with a significance of 10 percent and a certainty of 67 percent. We can similarly interpret the association rules for other assignee.

Table 4 shows top five rules for top five bug assignee of Thuderbird dataset.

Table 4. Association rules for top five assignee in Thuderbird

---

Association Rules with minimum support 3% and minimum confidence 20%

Assignee=David

- Priority {P2} ∧ Term {folder} ∧ Term {mar}
  ⇒ Assignee {David} @ (3%, 100%)
- Term {folder} ∧ Term {mar}
  ⇒ Assignee {David} @ (4%, 100%)
- Term {move} ∧ Term {account}
  ⇒ Assignee {David} @ (3%, 100%)
- Severity {Normal} ∧ Priority {P2} ∧ Term {folder}
  ⇒ Assignee {David} @ (3%, 75%)
- Priority {P2} ∧ Term {folder}
  ⇒ Assignee {David} @ (4%, 67%)

Table 4. (*Continued*)

Assignee=Phil Ringnalda

- Term {show}
  ⇒ Assignee {Phil Ringnalda} @ (3%, 60%)
- Severity {Normal} ∧ Term {text}
  ⇒ Assignee {Phil Ringnalda} @ (3%, 38%)
- Severity {Normal} ∧ Term {remov}
  ⇒ Assignee {Phil Ringnalda} @ (3%, 33%)
- Term {remov}
  ⇒ Assignee {Phil Ringnalda} @ (3%, 30%)
- Severity {Normal} ∧ Priority {P3}
  ⇒ Assignee {Phil Ringnalda} @ (8%, 24%)

Assignee=Mark Banner

- Severity {Normal} ∧ Priority {P3} ∧ Term {thunderbird}
  ⇒ Assignee {Mark Banner} @ (3%, 50%)
- Severity {Normal} ∧ Term {thunderbird}
  ⇒ Assignee {Mark Banner} @ (5%, 33%)
- Term {thunderbird}
  ⇒ Assignee {Mark Banner} @ (5%, 28%)
- Term {move}
  ⇒ Assignee {Mark Banner} @ (4%, 31%)

Assignee=Blake Winton

- Term {config} ∧ Term {auto}
  ⇒ Assignee {Blake Winton} @ (3%, 75%)
- Term {tool} ∧ Term {toolbar}
  ⇒ Assignee {Blake Winton} @ (3%, 60%)
- Term {auto}
  ⇒ Assignee {Blake Winton} @ (3%, 60%)
- Term {tool}
  ⇒ Assignee {Blake Winton} @ (3%, 43%)
- Term {add}
  ⇒ Assignee {Blake Winton} @ (4%, 31%)

Assignee=Andreas Nilsson

- Severity {Normal} ∧ Priority {P3} ∧ Term {icon}
  ⇒ Assignee {Andreas Nilsson} @ (5%, 71%)
- Severity {Normal} ∧ Priority {P3} ∧ Term {window}
  ⇒ Assignee {Andreas Nilsson} @ (3%, 60%)
- Severity {Normal} ∧ Term {icon}
  ⇒ Assignee {Andreas Nilsson} @ (5%, 56%)
- Priority {P3} ∧ Term {button}
  ⇒ Assignee {Andreas Nilsson} @ (3%, 43%)
- Severity {Normal} ∧ Term {button}
  ⇒ Assignee {Andreas Nilsson} @ (3%, 43%)

The first rule is a three antecedent rule, which reveals that the assignee *David* can be assigned a bug having priority *P2* and summary containing terms *folder* and *mar* with a significance of 3 percent and a certainty of 100 percent. Second association rule is a two antecedent rule, which means that the assignee *David* can be assigned a bug having summary containing terms *folder* and *mar* with a significance of 4 percent and a certainty of 100 percent. Third rule shows that the assignee *David* can be assigned a bug having summary containing terms *move* and *account* with a significance of 3 percent and a certainty of 100 percent. Rule four reveals that 3 percent of the bugs in the bug data set have severity *Normal*, priority *P2* and summary containing terms *folder* and assignee *David*. 75 percent of the bugs in the bug data set that have severity *Normal*, priority *P2* and summary containing term *folder* also have assignee *David*. The fifth rule shows that the assignee *David* can be assigned a bug having priority *P2* and summary containing term *folder* with a significance of 4 percent and a certainty of 67 percent. We can similarly interpret the association rules for other assignee.

Table 5 shows top five rules for top five bug assignee of Bugzilla dataset.

Table 5. Association rules for top five assignee in Bugzilla

Association Rules with minimum support 7% and minimum confidence 20%

Assignee = Terry Weissman

- Severity {Normal} ∧ Priority {P3} ∧ Term {mai} ∧ Term {mail}
  ⇒ Assignee {Terry Weissman} @ (11%, 73%).
- Severity {Normal} ∧ Priority {P3} ∧ Term {bug} ∧ Term {bugzilla}
  ⇒ Assignee {Terry Weissman} @ (9%, 41%).
- Severity {Normal} ∧ Priority {P3} ∧ Term {bug}
  ⇒ Assignee {Terry Weissman} @ (21%, 41%).
- Priority {P3} ∧ Term {bug} ∧ Term {bugzilla}
  ⇒ Assignee {Terry Weissman} @ (19%, 42%).
- Priority {P3} ∧ Term {mai} ∧ Term {mail}
  ⇒ Assignee {Terry Weissman} @ (16%, 53%).

Assignee = Max Kanat-Alexander

- Severity {Enhancement} ∧ Term {sql}
  ⇒ Assignee {Max Kanat-Alexander} @ (7%, 78%).
- Severity {Normal} ∧ Priority {P1} ∧ Term {sql}
  ⇒ Assignee {Max Kanat-Alexander} @ (11%, 69%).
- Severity {Enhancement} ∧ Priority {P1}
  ⇒ Assignee {Max Kanat-Alexander} @ (32%, 62%).
- Severity {Enhancement} ∧ Term {bug} ∧ Term {bugzilla}
  ⇒ Assignee {Max Kanat-Alexander} @ (11%, 58%).
- Priority {P1} ∧ Term {sql}
  ⇒ Assignee {Max Kanat-Alexander} @ (21%, 55%).

Table 5. (*Continued*)

Assignee = Joel Peshkin

- Priority {P2} ∧ Term {abl}
  ⇒ Assignee {Joel Peshkin} @ (8%, 38%).
- Priority {P2} ∧ Term {user}
  ⇒ Assignee {Joel Peshkin} @ (7%, 37%).
- Severity {Enhancement} ∧ Priority {P2}
  ⇒ Assignee {Joel Peshkin} @ (10%, 24%).
- Term{abl} ∧ Term{tab}
  ⇒ Assignee {Joel Peshkin} @ (9%, 24%).
- Severity {Major} ∧ Priority {P2}
  ⇒ Assignee {Joel Peshkin} @ (8%, 22%).

Assignee = Gervase Markham

- Severity {Blocker} ∧ Term {cgi} ∧ Term {temp} ∧ Term {templat}
  ⇒ Assignee {Gervase Markham} @ (7%, 100%).
- Priority {P1} ∧ Term {cgi} ∧ Term {temp} ∧ Term {templat}
  ⇒ Assignee {Gervase Markham} @ (7%, 78%).
- Term {cgi} ∧ Term {temp} ∧ Term {templat}
  ⇒ Assignee {Gervase Markham} @ (9%, 75%).
- Severity {Blocker} ∧ Term {temp} ∧ Term {templat}
  ⇒ Assignee {Gervase Markham} @ (11%, 50%).
- Severity {Blocker ∧ Term {temp}
  ⇒ Assignee {Gervase Markham} @ (11%, 48%).

Assignee = Bradley Baetz

- Severity {Blocker} ∧ Priority {P1} ∧ Term {ing}
  ⇒ Assignee {Bradley Baetz} @ (8%, 30%).
- Severity {Blocker} ∧ Term {ing}
  ⇒ Assignee {Bradley Baetz} @ (8%, 26%).

The first association rule is a four antecedent rule, which reveals that the assignee *Terry Weissman* can be assigned a bug having priority *P3* and summary containing terms *mai, content* and *mail* with a significance of 11 percent and a certainty of 73 percent. Second association rule is a four antecedent rule, which means that the assignee *Terry Weissman* can be assigned a bug having severity *Normal*, priority *P3* and summary containing terms *bug* and *bugzilla* with a significance of 9 percent and a certainty of 41 percent. Third rule shows that the assignee *Terry Weissman* can be assigned a bug having severity *Normal*, priority *P3* and summary containing term *bug* with a significance of 21 percent and a certainty of 41 percent. Rule four reveals that 19 percent of the bugs in the bug data set have priority *P3* and summary containing terms *bug, Bugzilla and assignee Terry Weissman*. 42 percent of the bugs in the bug data set that have priority *P3* and summary containing terms *bug* and *bugzilla* also have assignee *Terry Weissman.* The fifth rule shows that the assignee *Terry Weissman* can be assigned a bug having priority *P3* and summary containing terms *mai* and *mail* with a significance of 16 percent and a certainty of 53 percent. We can similarly interpret the association rules for other assignee.

Rules for assignee *Gervase Markham* shows that bugs with severity *blocker* are assigned to him.

We have drawn the distribution of association rules according to their length i.e. number of antecedents for all the three products. Figures 1 to 3 show the number of association rules with different rule length for all products.

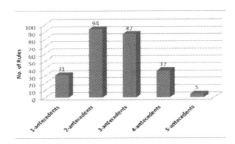

Fig. 1. Distribution of rules by rule length for AddOnSdk data set for min.supp=7% and min.conf=20%

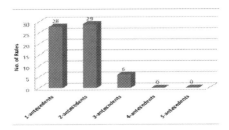

Fig. 2. Distribution of rules by rule length for Thunderbird data set for min.supp=3% and min.conf=20%

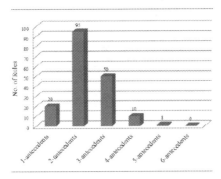

Fig. 3. Distribution of rules by rule length for Bugzilla data set min.supp =7% and min.conf =20%

Figures 1 to 3 show that we are getting maximum number of association rules of length 2 (with two antecedents) across all datasets.

4   Related Work

Each new reported bug must be triaged to determine if it describes a meaningful new problem or enhancement, and if it does, it must be assigned to an appropriate developer to fix it. In recent years, there have been a number of valuable contributions in order to address this problem. An attempt has been made by [7] by using descriptions of fixed bug reports in open bug repositories as machine learning features, and names of developers as class labels in Bayesian classifier. They have achieved the accuracy value of 30% for Eclipse projects. In another approach [24] authors expanded the work of [7] by using additional textual information of bug reports beyond the bug description, to form the machine learning features. The authors have also applied a non-linear Support Vector Machines (SVMs) and C4.5 algorithms in addition to the Naive Bayes classifier which their predecessor work had used and found that SVM is the best one. An approach for assisting human bug triagers in large open source software projects by semi-automating the bug assignment process has been proposed by [25]. This approach employs a simple and efficient n-gram-based algorithm for approximate string matching by collecting the natural language textual information available in the summary and description fields of the previously resolved bug reports and classifying that information in a number of separate inverted lists with respect to the resolver of each issue. In a study [26] authors outlined an approach based on information retrieval in which they report recall levels of around 20% for Mozilla. A new technique which automatically selects the most appropriate developers for fixing the fault represented by a failing test case has been proposed by [27]. Their technique is the first to assign developers to execution failures without the need for textual bug reports. Results reported 81% of accuracy for the top-three developer suggestions.

Extensive performance studies have also shown that associative classification frequently generates better accuracy than state-of-the-art classification methods [11-20]. The successful use of association rule mining in various fields motivates us to apply it to the open source software bug data set.

5   Threats to Validity

Following are the factors that affect the validity of our study:

Construct Validity: The independent attributes taken in our study are not based on any empirical validation.

Internal Validity: We have considered four bug attributes: severity, priority, summary terms and assignee. Developer's reputation attribute can also be considered.

External Validity: We have considered only open source Mozilla products. The study can be extended for other open source and closed source software.

Reliability: RapidMiner and MATLAB software have been used in this paper for model building and testing. The increasing use of these software confirms the reliability of the experiments. But we have not considered and handled any accuracy error for these tools.

6   Conclusion

Bug triaging is a process of deciding what to do with an incoming bug report which takes a large amount of developer resources and time. Triaging all the incoming bugs to determine the assignee to whom the bug should be assigned is a cumbersome task. In literature much work has been done by using classification based on the textual information about bug i.e. bug summary or description. To the best of our knowledge no work has been done till now to find associations between bug attributes to predict the assignee for that bug. We have used association mining to assist in bug triage by using Apriori algorithm to predict the developer that should work on the bug based on the bug's severity, priority and summary terms. We have used 1,695 bug reports of Thunderbird, AddOnSDK and Bugzilla products of Mozilla open source project for result validation. For a minimum confidence of 20% and minimum support of 3 and 7% we have summarized the association rules for top five assignee based on the number of bugs assigned to them. Prediction of assignee will help the managers in software development process by assigning a bug to potentially efficient developer. In future we can extend our study for other association mining algorithms to empirically validate the results.

References

1. Menzies, T., Marcus, A.: Automated severity assessment of software defect reports. In: IEEE Int. Conf. Software Maintenance, pp. 346–355 (2008)
2. Lamkanfi, A., Demeyer, S., Giger, E., Goethals, B.: Predicting the severity of a reported bug. In: Mining Software Repositories, MSR, pp. 1–10 (2010)
3. Lamkanfi, A., Demeyer, S., Soetens, Q.D., Verdonck, T.: Comparing mining algorithms for predicting the severity of a reported bug. In: CSMR, pp. 249–258 (2011)
4. Chaturvedi, K.K., Singh, V.B.: Determining bug severity using machine learning techniques. In: CSI-IEEE Int. Conf. Software Engineering (CONSEG), pp. 378–387 (2012)
5. Chaturvedi, K.K., Singh, V.B.: An empirical Comparison of Machine Learning Techniques in Predicting the Bug Severity of Open and Close Source Projects. Int. J. Open Source Software and Processes 4(2), 32–59 (2013)
6. Sharma, M., Bedi, P., Chaturvedi, K.K., Singh, V. B.: Predicting the priority of a reported bug using machine learning techniques and cross project validation. In: IEEE Int. Conf. Intelligent Systems Design and Applications (ISDA), pp. 27–29 (2012)
7. Cubranic, D., Murphy G.C.: Automatic bug triage using text categorization. In: Int. Conf. Software Engineering. Citeseer, pp. 92–97 (2004)
8. Sharma, M., Kumari, M., Singh, V.B.: Understanding the meaning of bug attributes and prediction models. In: I-CARE 5th IBM Collaborative Academia Research Exchange Workshop, Article No. 15. ACM (2013)

9. Agrawal, R., Imielinski, T., Swami, A.: Mining association rules between sets of items in large databases. In: SIGMOD Conf. Management of Data. ACM, May 1993
10. Song, Q., Shepperd, M., Cartwright, M., Mair, C.: Software defect association mining and defect correction effort prediction. IEEE Transactions on Software Engineering **32**(2), 69–82 (2006)
11. Ali, K., Manganaris, S., Srikant, R.: Partial classification using association rules. In: Int. Conf. Knowledge Discovery and Data Mining, pp. 115–118 (1997)
12. Dong, G., Zhang, X., Wong, L., Li, J.: CAEP: classification by aggregating emerging patterns. In: Arikawa, S., Nakata, I. (eds.) DS 1999. LNCS (LNAI), vol. 1721, pp. 30–42. Springer, Heidelberg (1999)
13. Liu, B., Hsu, W., Ma, Y.: Integrating classification and association rule mining. In: Int. Conf. Knowledge Discovery and Data Mining, pp. 80–86 (1998)
14. She, R., Chen, F., Wang, K., Ester, M., Gardy, J.L., Brinkman, F.L.: Frequent-subsequence-based prediction of outer membrane proteins. In: ACM SIGKDD Int. Conf. Knowledge Discovery and Data Mining (2003)
15. Wang, K., Zhou, S.Q., Liew, S.C.: Building hierarchical classifiers using class proximity. In: Int. Conf. Very Large Data Bases, pp. 363–374 (1999)
16. Wang, K., Zhou, S., He, Y.: Growing decision tree on support-less association rules. In: Int. Conf. Knowledge Discovery and Data Mining (2000)
17. Yang, Q., Zhang, H.H., Li, T.: Mining web logs for prediction models in WWW caching and prefetching. In: ACM SIGKDD Int. Conf. Knowledge Discovery and Data Mining (2001)
18. Yin, X., Han, J.: CPAR: classification based on predictive association rules. In: SIAM Int. Conf. Data Mining (2003)
19. Ying, A.T.T., Murphy, C.G., Ng, R., Chu-Carroll, M.C.: Predicting source code changes by mining revision history. In: Int. Workshop Mining Software Repositories (2004)
20. Zimmermann, T., Weigerber, P., Diehl, S., Zeller, A.: Mining version histories to guide software changes. In: Int. Conf. Software Engineering (2004)
21. Mierswa, I., Wurst, M., Klinkenberg, R., Scholz, M., Euler, T.: YALE: rapid prototyping for complex data mining tasks. In: ACM SIGKDD Int. Conf. Knowledge Discovery and Data Mining (KDD 2006) (2006). http://www.rapid-i.com
22. Porter, M.: An algorithm for suffix stripping. Program. **14**(3), 130–137 (2008)
23. www.mathworks.in/.../3016-armada-data-mining-tool-version-1-4
24. Anvik, J., Hiew, L., Murphy, G.C.: Who should fix this bug? In: Int. Conf. Software Engineering (ICSE) (2006)
25. Amir, H.M., Neumann, G.: Assisting bug triage in large open source projects using approximate string matching. In: Int. Conf. Software Engineering Advances (ICSEA 2012), Lisbon, Portugal (2012)
26. Canfora, G., Cerulo, L.: How software repositories can help in resolving a new change request. In: Workshop on Empirical Studies in Reverse Engineering (2005)
27. Servant, F., Jones, J.A.: Whose fault: automatic developer-to-fault assignment through fault localization. In: Int. Conf. Software Engineering (ICSE 2012), pp. 36–46. IEEE Press, Piscataway (2012)

Zernike Moment-Based Approach for Detecting Duplicated Image Regions by a Modified Method to Reduce Geometrical and Numerical Errors

Thuong Le-Tien[1(✉)], Tan Huynh-Ngoc[2], Tu Huynh-Kha[3], and Luong Marie[4]

[1] Department of Electrical Electronics Engineering, HCM City University of Technology,
Ho Chi Minh City, Vietnam
thuongle@hcmut.edu.vn
[2] Faculty of Information Technique, Mannheim University of Applied Science,
Mannheim, Germany
huynh.ngoctan@yahoo.com
[3] School of Computer Science and Engineering, HCM City International University,
Ho Chi Minh City, Vietnam
hktu@hcmiu.edu.vn
[4] Labo L2TI, Institut Galilee, University of Paris 13, Paris, France
marie.luong@univ-paris13.fr

Abstract. In the paper, the approach is focused on the Zernike Moment-based model of ROI image (Region of Interest) and its parameters for an efficient image processing in the forensic issue. By considering the factors affecting the identification of an duplicated image, the change of ROI's size is determined through the proposed algorithm. The proposed technique has shown a good improvement in reducing significantly Geometrical Errors (G.E) and Numerical Errors (N.E) performed better than that of the Zernike-based traditional technique. The duplicated detection program has been written by C++ and supporting OpenCV and Boost libraries that help to verify the images authentication.

Keywords: Zernike moments · Geometrical Errors · Numerical Errors · Geometric moments · Region of Interest (ROI) · FLANN library-Fast library for approximation nearest neighbors

1 Introduction

A number of proposed algorithms have been developed for detecting tampered image such as Pixel-based, Cloning, Resampling, Color Filter Array [1], Hu Moments (1962), KPCA, PCA, Fourier–Mellin Moments (Sheng and Duvernoy, 1986) [2,3], etc. However, Zernike Moments have proved its superiority in the analysis of invariant points of digital image. It has been becoming an effective tool to detect duplicated regions in such an image due to the excellent capability of features extraction representing, invariant rotations [4]. Moreover, in the cutting process of a tampered image to become much Region of Interests (ROI) by ZMs, it also causes two main errors namely GE and NE [2, 3]. The study of the matching model on reducing G.E and N.E in the paper will help the computation more accurately and performing more

© Springer International Publishing Switzerland 2015
O. Gervasi et al. (Eds.): ICCSA 2015, Part IV, LNCS 9158, pp. 458–475, 2015.
DOI: 10.1007/978-3-319-21410-8_36

highly efficient. The kernel of Zernike Moments is built by the set of orthogonal Zernike polynomials. Those create a marginal boundary being called the unit circle (or unit disc). The equation of Zernike polynomials order n and repetition m in polar coordinates [5, 6] defined as follows,

$$V_{nm}(x, y) = V_{nm}(r, \theta) = R_{nm}(r)\exp(jm\theta)$$
$$= \begin{cases} N_{nm}R_{nm}(r)\cos(m\theta), & for\ m \geq 0 \\ -N_{nm}R_{nm}(r)\sin(m\theta), & for\ m < 0 \end{cases} \quad (1)$$

where $n \in Z^{+}$; m is an integer defining the rotation subject to the conditions $n - |m| =$ even, $|m| \leq n$; r is the length of the vector from origin to pixel coordinate (x, y); $\theta = \angle(r, x)$ in counter clockwise direction, $N_{nm} = \sqrt{\dfrac{2(n+1)}{1+\delta_{m0}}}$ is the normalization factor. The accuracy of ZMs computation is affected by two parameters, including order n and repetition m. The higher order n of ZMs computation gives us more precise values of those pixels computed, it is similar to Taylor's series calculating for higher order. Moreover, the higher repetition m creates more feature points in an image, the visual figure if increasing the repetition m is shown. Since combining higher order n and higher repetition m in Zernike polynomials, there are more feature points having their invariant rotation shown, thus it is very useful to reconstruct exactly the image.

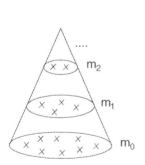

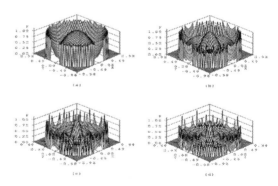

Fig. 1. The model to compute Zernike Moments of repetition m

Fig. 2. The magnitude of some $V_{nm}(x, y)$ polynomials $(a)|V_{2,0}(x, y)|$; $(b)|V_{4,0}(x, y)|$; $(c)|V_{8,0}(x, y)|$; $(d)|V_{12,0}(x, y)|$

According to Fig. 2, the outline is a unit circle if increasing order n and repetition m from Fig. 2(a) to Fig. 2(d), there are not only one main lobe, but also more and more side lobes are resurgent. It means that more feature points are displayed since increasing both the order and repetition of Zernike polynomials. When an object is copied and moved from this area to other one, the invariant points of the copied area are not changed when they located in another area. Therefore, using Zernike Moments, it is able to detect easily the copied areas of a tampered image (Fig. 3).

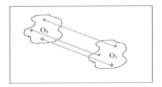

Fig. 3. Forgery model (O_1 original area; O_2 copied area of O_1 area)

To compute the Zernike Moments for a tampered image, a discrete-space image $f(x, y)$, which is combined by its (NxN) sub-images or ROI, is mapped to the unit disc in polar form; the center of each sub-image is the original coordinate of the unit circle, pixels fall outside the circle are not computed. Moreover, there are moment invariances of the Zernike magnitude reflected in the image mapping to the unit circle. To reduce the computation cost, it is necessary having an initial constraint about the side of ROI is not greater than 50 pixels. When f is a complex-valued function on the unit circle, Zernike Moment [1] for f of order n repetition m is

$$Z_{nm}^{(f)} = \frac{n+1}{\pi} \iint_{D^2} f(x, y)V_{nm}^{*}(x, y)dxdy \qquad (2)$$

where V_{nm}^{*} is a complex conjugate of V_{nm}, $D^2 : x^2 + y^2 \leq 1$. Moreover, if F is the digital image of f, the above equation as below

$$Z_{nm}^{(F)} = \frac{n+1}{\pi} \sum_{x} \sum_{y} F(x, y)[V_{nm}(x, y)]^{*} \qquad (3)$$

where $x^2 + y^2 \leq 1$. The Eq. (3) is the discrete form of Green's theorem for computing along the image's object boundary points. The Zernike real-valued radial polynomials

$$R_{nm}(r) = \sum_{s=0}^{(n-|m|)/2} (-1)^s \frac{(n-s)!}{s!\left(\dfrac{n+|m|}{2}-s\right)!\left(\dfrac{n-|m|}{2}-s\right)!} r^{n-2s} \qquad (4)$$

where $n-|m|$ is even, $0 \leq |m| \leq n$ and $n \geq 0$. It implies in Eq. (4) that $R_{n,m}(r) = R_{n,-m}(r)$ and $R_{nm}(r) = 0$ when $n-|m| = 0 \pmod 2$, $|m| \leq n$ not satisfied

[4], obtained $V_{nm}(x, y) = \displaystyle\sum_{k=m}^{n} B_{nmk} r^k e^{jm\theta}$ (5), $\quad B_{nmk} = \dfrac{(-1)^{\frac{n-k}{2}}\left(\dfrac{n+k}{2}\right)!}{\left(\dfrac{n-k}{2}\right)!\left(\dfrac{k+m}{2}\right)!\left(\dfrac{k-m}{2}\right)!}$ (6)

Hence, the traditional equation of Zernike moments in polar form can be rewritten as

$$Z_{nm} = \frac{n+1}{\pi} \int_{-1}^{1} \int_{-\pi}^{\pi} \sum_{k=m}^{n} B_{nmk} r^k e^{-jm\theta} f(r,\theta) r dr d\theta \tag{7}$$

with $dxdy = rdrd\theta$ and $-\pi \leq \theta \leq \pi$. It denotes $f(\rho,\theta)$ and $f^r(\rho,\theta)$, respectively the function of original image and rotated image. The set of Zernike Moments being rotated by α angle in an image and the extracted Zernike rotation invariant obtained as follows,

$$Z_{nm}^r = \frac{n+1}{\pi} \int_{0}^{2\pi} \int_{0}^{1} f(\rho,\theta-\alpha) R_{nm}(\rho) exp(-jm\theta) \rho d\rho d\theta \tag{8}$$

$$= \frac{n+1}{\pi} \int_{0}^{2\pi} \int_{0}^{1} f(\rho,\theta-\alpha) R_{nm}(\rho) \exp(-jm(\theta-\alpha)) \exp(-jm\alpha) \rho d\rho d\theta$$

$$\Rightarrow Z_{nm}^r = Z_{nm} e^{-jm\alpha} \tag{9}$$

$$\left| Z_{nm}^r \right| = \left| Z_{nm} e^{-jm\alpha} \right| = \left| Z_{nm} \right| \tag{10}$$

Hence, the function of piecewise continuous over the unit disc can be expressed as

$$f(x,y) = \sum_{n=0}^{\infty} \sum_{m=0}^{n} \lambda_n Z_{nm} V_{nm}(x,y) \tag{11}$$

where Z_{nm} is calculated over the unit circle; $\lambda_n = (n+1)[\delta(A/\pi)]$ is the normalizing constant and $\delta A = \pi / N^2$ is the ratio between the area of unit disc to the total number of pixels in ROI image. Otherwise, the value of δA plays an important role to provide different values corresponding to different squares. After Zernike computation, the difference between initial image and reconstructed image is [5].

$$E_{nm} = Z_{nm} - \tilde{Z}_{nm} = E_{nm}^{(g)} + E_{nm}^{(n)} \tag{12}$$

where E_{nm} is the total approximate error, Z_{nm} of the continuous original image, $\tilde{Z}_{nm}$ of the discrete approximated, $E_{nm}^{(g)}$ is GE and $E_{nm}^{(n)}$ is Numerical Error.

2 Overview of Geometrical (GE) and Numerical Error (NE)

2.1 Geometrical Error (GE)

Since computing Zernike polynomials in polar coordinate (r,θ) with $|r| \leq 1$, it requires a linear mapping process to map correctly image coordinates (i, j) to the unit circular domain $(r,\theta) \in R^2$. Therefore, the general form of mapping techniques for each ROI image to the unit disc obtained as [4, 6].

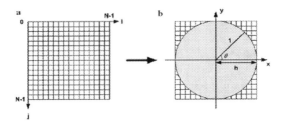

Fig. 4. The traditional square $(N \times N)$ mapped to circular mapping technique

$$\begin{cases} x_i = c_1 i + c_2 \\ y_j = c_1 j + c_2 \end{cases} \quad \text{where} \quad c_1 = \frac{2h}{N-1} \quad \text{and} \quad c_2 = -h \tag{13}$$

Because of the domain of $\{h \in R \mid 0 \le h \le 1\}$, it used ROI to divide the original image into some sub-images having their size of $(N \times N)$ (*denoted* $N \neq n$). In the traditional approach, the value of h is equal the radius of the unit disc ($h = 1$). Then, $c_1 = \frac{2}{N-1}$

and $c_2 = -1$, this approach causes the GE during Zernike Moments computation.

2.2   Numerical Error (NE)

The numerical error, $E_{nm}^{(n)}$, is created by the computation of double integral term of Zernike Moments in Eq. (2) and Eq. (8). Assume that the image coordinates of a $(N \times N)$ square image and $f(x, y)$ is defined as a function of the set of points (x_i, y_j). Hence, Eq. (3) can be rewritten to be

$$Z_{nm} = \frac{n+1}{\pi} \sum_{i=0}^{N-1} \sum_{j=0}^{N-1} f(x_i, y_j) \times \int_{x_i - \frac{\Delta x_i}{2}}^{x_i + \frac{\Delta x_i}{2}} \int_{y_j - \frac{\Delta y_j}{2}}^{y_j + \frac{\Delta y_j}{2}} V_{nm}^*(x, y)dxdy , \quad n \neq N \tag{14}$$

where $\Delta x_i = x_{i+1} - x_i$ and $\Delta y_j = y_{j+1} - y_j$ with $x, y \in$ *Cartesian axis*. In the traditional approach [6], the double integral term is computed by using zero-th order approximation where the values of Zernike polynomials are assumed in two intervals $[x_i - \Delta x_i / 2, x_i + \Delta x_i / 2]$ and $[y_j - \Delta y_j / 2, y_j + \Delta y_j / 2]$. They are obtained through sampling Zernike polynomials at the center points of these intervals.
The approximation for computing zero-th order of Zernike Moments as

$$\tilde{Z}_{nm} = \frac{n+1}{\pi} \sum_{i=0}^{N-1} \sum_{j=0}^{N-1} f(x_i, y_j) V_{nm}^*(x_i, y_j) \Delta x_i \Delta y_j , \quad n \neq N \tag{15}$$

Hence the numerical error in computing the double integral term of Zernike moments is the difference of the initial double term and the sampling term of Zernike Moments

$$E_{nm}^{(n)} = Z_{nm} - \tilde{Z}_{nm} \tag{16}$$

3 Proposed Method for Reducing the GE and NE

3.1 Decreasing of Geometrical Error

According to the traditional model of the ROIs are mapped onto the unit circle, our research focused on changing the size of each ROI by applying the shrinking ratio h_0.

Thus, our proposed technique collects more pixels in order to apply Zernike polynomials (unit circle) computing those. The graphical figure will demonstrate our methodology. From the step of "father", the ROI will be cut arbitrary so that it is as small as possible, then our algorithm will consider a fixed value of N in order to divide the ROI to be a grid having NxN meshes, the position of each pixel located in four vertices of a square but in which having the distance of two contiguous arbitrary vertices is quite large. Hence, in the step of "son", the sum of distance of all contiguous vertices on a side of the ROI is shrinked by the ratio h_0, thus more pixels are located in the unit circle. In the proposed model, it denoted some notations for the computation of the ratio h_0.

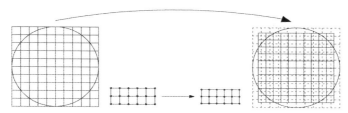

Fig. 5. Two steps - "father-son" in shrinking the size of one ROI

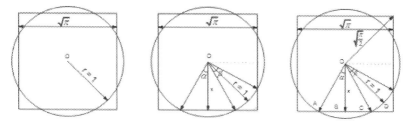

Fig. 6. Analysis the squaring circle

Proved:

$r = $ *radius of the circle*, $r_1 = $ *length of a side of square* .Considering the proposed model,
Four sides of the square are $r_1 = \sqrt{\pi}$ and the diameter of the unit circle is $d = 2r = 2$;
$r_1 = r\sqrt{\pi}$ when the area of the square is equal the area of the circle.

$$r_1 = r\sqrt{\pi} = \sqrt{\pi} \rightarrow x = 1 - \frac{2 - \sqrt{\pi}}{2} = \frac{\sqrt{\pi}}{2} \quad (17), \quad A_{2\Delta AOB} = \frac{1}{2} x 2r\sin\alpha = xr\sin\alpha = \frac{\sqrt{\pi}}{2}\sin\alpha \quad (18)$$

The angle α between x and r obtained by

$$\alpha = \cos^{-1}\left(\frac{x}{r}\right) = \cos^{-1}\left(\frac{\sqrt{\pi}}{2}\right) \rightarrow \beta = \frac{(90 - 2\alpha)\pi}{180} \approx 0.6074753677 \ (rad) \tag{19}$$

The area of $2\angle COD\ (or\ \beta)$ is $A_{2\angle COD} = \frac{1}{2}r^2\beta = \frac{1}{2}\beta$. Finally, the total numerical area

inside the circle and the square is

$$A_{total} = 4A_{2\angle COD} + 4A_{2\Delta AOB} = 2\beta + 4xr\sin\alpha \tag{20}$$

$$\approx 2 * 0.6074753677 + 2\sqrt{\pi}\sin\left(\cos^{-1}\left(\frac{\sqrt{\pi}}{2}\right)\right) \approx 2.857134103$$

Since $A_{total} \approx 2.857134103$, the total pixels of proposed method inside unit disc are

$\left\lfloor 2.857134103 \times \dfrac{N^2}{4} \right\rfloor$ (pixels). Similarly, considering the traditional model, we have the

total numerical area inside the unit circle is $\qquad A'_{total} = \pi$ \hfill (21)

Hence, the value of ratio h_0 (in reducing the G.E) between the proposed model

(A_{total}) and the traditional model (A'_{total}) is $h_0 = \dfrac{A_{total}}{A'_{total}} = \dfrac{2.857134103}{\pi} \approx 0.909$ \hfill (22)

In conclusion, by shrinking the traditional ROI with a ratio $h_0 = 0.909$, our research
will transform a traditional to modified ROI having more pixels within to the unit of
circle. Therefore, there are more feature extraction points compared with other ones in
different ROIs. In the proposed method, mapping a new modified value of

$h = x = \dfrac{\sqrt{\pi}}{2}$, $c_1 = \dfrac{\sqrt{\pi}}{N-1}$ and $c_2 = -\dfrac{\sqrt{\pi}}{2}$ is computed in Eq. (13), then it obtained

$\delta A = \dfrac{A_{total}}{N^2} \approx \dfrac{2.857134103}{N^2}$. The transformed pixel coordinates are located in Fig.7

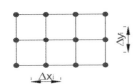

Fig. 7a. Traditional approach

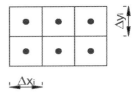

Fig. 7b. Proposed approach

The equation used to compute the transformed coordinate (x_i, y_j) in Eq. (13) is modified to be $x_i = c_1(i + 0.5) + c_2$ and $y_j = c_1(j + 0.5) + c_2$ where $c_1 = \sqrt{\pi}/N$; $c_2 = -\sqrt{\pi}/2$. Although, the area of image ($N \times N$) has been scaled down a bit after mapping interior of the unit disc, but the information of pixels is mostly reserved.

3.2   Decreasing of Numerical Error

There are two kinds of numerical errors (N.E) generated in the computation of the program, including the *overflow error*, due to computers' limitation related to the maximum magnitude which is the numerical value can get and the *finite precision error* due to the finite precision presentation which the number can have. The *overflow error* is the circumstance in which a quantity takes a higher value, from the range of its data type. The second kind of error in NE is *finite precision error*, which easily affects to the power of ZMs computation and it may causes flaws and errors.

The N.E is come about by approximating the zero-th order of the double integral term in Eq. (15). The set of Zernike Moments can be computed by using the geometric moments as [8]

$$Z_{nm} = \frac{n+1}{\pi} \sum_{k=m}^{n} B_{nmk} \sum_{v=0}^{s} \sum_{u=0}^{m} w^u \times \binom{s}{v}\binom{m}{u} M_{k-2v-u,\,2v+u} \qquad (23)$$

where $w = \begin{cases} -i \,, m > 0 \\ +i \,, m \le 0 \end{cases}$ and $s = \frac{1}{2}(k - m); \; i = \sqrt{-1}$

Otherwise, from [1], the function of geometric moments for set of discrete points as

$$M_{nm} = \sum_{i=0}^{N-1} \sum_{j=0}^{N-1} f(x_i, y_j) H_n(x_i) H_m(y_j) \qquad (24)$$

where $H_n(x_i) = \int_{x_i - \frac{\Delta x_i}{2}}^{x_i + \frac{\Delta x_i}{2}} x^n dx$; $H_m(y_j) = \int_{y_j - \frac{\Delta y_j}{2}}^{y_j + \frac{\Delta y_j}{2}} y^m dy$. Thus, from the Eq. (23) and Eq. (24), the set of Zernike moments by computing to order n that restricts NE as,

$$Z_{nm} = \frac{2^s(n+1)}{\pi} \cdot \left(\sum_{u=0}^{m} w^u \binom{m}{u}\right) \sum_{k=m}^{n} B_{nmk} \times \sum_{v=0}^{s} \sum_{u=0}^{m} \left[\sum_{i=0}^{N-1} \sum_{j=0}^{N-1} f(x_i, y_j) H_{k-2v-u}(x_i) H_{2v+u}(y_j)\right] \qquad (25)$$

4   Matlab Results for Zernike Moment Calculation

MATLAB is a computational program running in the structure of top-down illustrated in Fig 8. It can be seen the "Test.m" file playing a coordinating role in each stage of the computation of Zernike Moments. The three main parts of our program to compute the proposed ZMs and show the false pixels of reconstructed image comparing the original one indicated through the following Matlab codes.

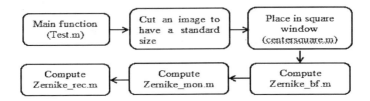

Fig. 8. Flowchart of computing Zernike Moment for binary images

Table 1. The computational codes in MATAB for processing our proposed Zernike Moments

Test.m

```
[x,y]=size(m);
    [x1,y1]=size(m1);
    m=centersquare(m,imsize);    %proposed technique
    m1=centersquare1(m1,imsize); %traditional technique
    m=uint8(m);
    m1=uint8(m1);
    whos
    zs0=zernike1_bf(imsize,i);
    v0=zernike_mom(m,zs0);
    v01=zernike_mom(m1,zs0);
    rec0=zernike_rec(v0,imsize,zs0,3); %proposed technique
    rec01=zernike_rec(v01,imsize,zs0,3); %traditional technique
    err=0;
    tmp=m;
    for q=1:y
        for p=1:x
            if m(p,q)>=125
                tmp(p,q)=1;
            else
                tmp(p,q)=0;
            end
            if (tmp(p,q)~=rec0(p,q))
                err=err+1;
            end
        end
    end
    err1=0;
    tmp1=m1;
    for q=1:y1
        for p=1:x1
        if m1(p,q)>=125
                tmp1(p,q)=1;
            else
                tmp1(p,q)=0;
        end
         if (tmp1(p,q)~=rec01(p,q))
                err1=err1+1;
         end
        end
    end
```

Table 1. (*Continued*)

```
    t_aveg=(err/(x*y))*100;
    t_aveg1=(err1/(x1*y1))*100;
```

Centersquare.m

```
if nargin<2
        OPTSIZE=SQUARE_IMG_SIZE;
    end
    osz=size(I);
    ms=max(osz);
    sqim=zeros(ms);
    y=floor((ms-osz(1))/2);
    x=floor((ms-osz(2))/2);
    sqim((1+y:y+osz(1)),(1+x:x+osz(2)))=I;
    smallersize=floor(OPTSIZE*(sqrt(pi)/2));
    tmp=imresize(sqim,smallersize*ones(1,2),'nearest');
    CS=zeros(OPTSIZE,OPTSIZE);
    offset=floor((OPTSIZE-smallersize)/2);
    range=1+offset:offset+smallersize;
    CS(range,range)=tmp;
```

Zernike_bf.m

```
syms r q
G11=[1  r*sin(q) r*cos(q) r^2*sin(2*q) 2*r^2-1 r^2*cos(2*q)
r^3*sin(3*q) (3*r^3-2*r)*sin(q) (3*r^3-2*r)*cos(q)...
r^3*cos(3*q) r^4*sin(4*q)...
(4*r^4-3*r^2)*sin(2*q) 6*r^4-6*r^2 + 1 (4*r^4-3*r^2)*cos(2*q)
r^4*cos(4*q) r^5*sin(5*q) (5*r^5-4*r^3)*sin(3*q)...
(10*r^5-12*r^3 + 3*r)*sin(q) (10*r^5-12*r^3 + 3*r)*cos(q)...
(5*r^5-4*r^3)*cos(3*q) r^5*cos(5*q) r^6*sin(6*q) (6*r^6-
5*r^4)*sin(4*q) (15*r^6-20*r^4 + 6*r^2)*sin(2*q) (20*r^6-30*r^4
+ 12*r^2-1)...
(15*r^6-20*r^4 + 6*r^2)*cos(2*q) (6*r^6-5*r^4)*cos(4*q)
r^6*cos(6*q)...
r^7*sin(7*q) (7*r^7-6*r^5)*sin(5*q) (21*r^7-30*r^5 +
10*r^3)*sin(3*q) (35*r^7-60*r^5 + 30*r^3-4*r)*sin(q)...
(35*r^7-60*r^5 + 30*r^3-4*r)*cos(q) (21*r^7-
30*r^5+10*r^3)*cos(3*q)...
(7*r^7-6*r^5)*cos(5*q) r^7*cos(7*q)... r^8*sin(8*q) ...
(8*r^8-7*r^6)*sin(6*q) (28*r^8-42*r^6 + 15*r^4)*sin(4*q)...
(56*r^8-105*r^6+60*r^4-10*r^2)*sin(2*q)...
70*r^8-140*r^6+90*r^4-20*r^2+1 ...
(56*r^8-105*r^6+60*r^4-10*r^2)*cos(2*q)...
 (28*r^8-42*r^6+ 15*r^4)*cos(4*q) (8*r^8-7*r^6)*cos(6*q)
r^8*cos(8*q) r^9*sin(9*q) (9*r^9-8*r^7)*sin(7*q) (36*r^9-
56*r^7+21*r^5)*sin(5*q)...
(84*r^9-168*r^7+105*r^5-20*r^3)*sin(3*q) (126*r^9-
280*r^7+210*r^5) -(60*r^3+5*r)*sin(q) (126*r^9-280*r^7 +
210*r^5-60*r^3+5*r)*cos(q)...
```

Table 1. (*Continued*)

```
(84*r^9-168*r^7 + 105*r^5-20*r^3)*cos(3*q) (36*r^9-56*r^7 +
21*r^5)*cos(5*q) (9*r^9-8*r^7)*cos(7*q) r^9*cos(9*q)
r^10*sin(10*q) (10*r^10-9*r^8)*sin(8*q) (45*r^10-72*r^8 +
28*r^6)*sin(6*q)...
(120*r^10-252*r^8 + 168*r^6-35*r^4)*sin(4*q)...
(210*r^10-504*r^8 + 420*r^6-140*r^4 + 15*r^2)*sin(2*q)...
252*r^10-630*r^8 + 560*r^6-210*r^4 + 30*r^2-1 ...
(210*r^10-504*r^8 + 420*r^6-140*r^4 + 15*r^2)*cos(2*q)...
(120*r^10-252*r^8 + 168*r^6-35*r^4)*cos(4*q)...
(45*r^10-72*r^8 + 28*r^6)*cos(6*q) (10*r^10-9*r^8)*cos(8*q)
r^10*cos(10*q)];
    ss=size(G11);
    if nargin<3
        WITHNEG=0;
    end
    limitfastcomp=50;
    F=factorial(0:ORDER);
    pq=zernike_orderlist(ORDER, WITHNEG);
    len=size(pq,1);
    szh=SZ/2;
    pqind=-1*ones(1+2*ORDER,1+2*ORDER);
    src=1+ORDER+pq ;
    pqind(sub2ind(size(pqind),src(:,1),src(:,2)))=(1:len)';
    Rmns=zeros(1+2*ORDER,1+2*ORDER+1,1+2*ORDER);
    ZBF=zeros(SZ,SZ,len);
    for y=1:SZ
    for x=1:SZ
    r=sqrt((szh-x)^2+(szh-y)^2);
    q=atan2(szh-y,szh-x);
    if r>szh
    continue
    end
    r=r/szh;
    ii=0;
    for flat=1:len
    m=pq(flat,1);
    n=pq(flat,2);
    ii=ii+1;
    R=eval (G11(ii));
    ZBF(y,x,flat) = R;
    end
    end
    end
```

In the experiment, MATLAB program (version 2011b) is used, it based on the Zernike table [9], besides $x_i = c_1(i + 0.5) + c_2$ and $y_j = c_1(j + 0.5) + c_2$ where $c_1 = \sqrt{\pi}/N$, $c_2 = -\sqrt{\pi}/2$. The results of four binary images chosen in Fig. 9 and Fig. 10. The reconstruction error can be determined [10] by the equation as follows,

$$\varepsilon = \sum_{i=0}^{N-1} \sum_{j=0}^{N-1} \left\{ \frac{\left[f(i,j) - \hat{f}(i,j) \right]^2}{\left[f(i,j) \right]^2} \right\} \tag{26}$$

where $f(i,j)$ is the sampled original image and $\hat{f}(i,j)$ is the constructed image.

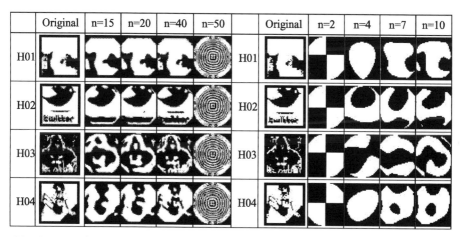

Fig. 9. Reconstructed images using complex Zernike polynomials (n=15-50) (6 left columns)

Fig. 10. Reconstructed images using polar Zernike polynomials (n=2-10) (6 right columns)

If the size of a binary image is $N \times N$ pixels, the maximum order of ZMs is n. In the Eq. 7, when denoting $\chi_{n,m,k} = \int_{-1}^{1} \int_{-\pi}^{\pi} \sum_{k=m}^{n} r^k e^{-jm\theta} f(r,\theta) r\,dr\,d\theta$, the computational complexity

$$\begin{cases} N^2 n \left(\dfrac{n}{2} + 1 \right) \text{ multiplications} \\ 2(N^2 - 1) \left(\dfrac{n}{2} + 1 \right)^2 \text{ additions} \end{cases} \quad \text{without considering } m = 0. \tag{27}$$

According to above figures, the higher order n and more repetition m in Zernike computing, the more feature points extracted by ZMs. If the high orders are exceeded the capability of ZMs, the reconstructed image will be distorted, besides that the magnitudes and geometric distances of extracted feature points defining objects are not changed when copying to other positions. The percentage of total false-pixels in the reconstructed image of proposed technique comparing to the reconstructed image of traditional technique in Zernike Moments computation are clearly illustrated in the Fig11. In the *Fig 11a*, it can be seen from the line graph, the lowest moment order of Zernike polar (0th order) has the total false-pixels' percentage in the *reconstructed image* of proposed technique – *blue line*, and the *original image* (implied as the reconstructed image of traditional technique – *red line*) are 60% and 78%, respectively. However, since increasing the order moment to 10th, the false-pixels between the *reconstructed image* and the *original image* are 18% and 32%, respectively. Hence, it has proved that the proposed ZMs of our study more improved in decreasing G.E and N.E in ZMs computation.

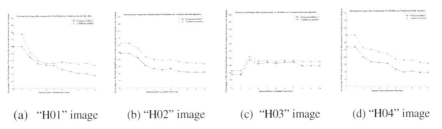

(a) "H01" image   (b) "H02" image   (c) "H03" image   (d) "H04" image

Fig. 11. The percentage false-pixels

In contiguous graphs, *Fig 11b-c-d*, the false-pixels of *reconstructed image* (*blue lines*) for proposed technique and the reconstructed image for traditional technique (*original image – red lines*) are efficiently decreasing since the order of Zernike Moments increasing, besides that the percentage of false-pixels of proposed technique is always lower than the percentage of false-pixels of traditional technique. Hence, by considering seriously on the trend of line graphs, it can be deduced that the percentage of false-pixels of ZMs proposed technique is roundly [20% to 30%] at the moment order 10th.

5   Deployment of Modified Zernike Moments to Detect Duplicated Image on the Hardware BeagleBone Board

The study also used the BeagleBone board (http://beaglenoard.org) for implementing the proposed Zernike Moments in order to detect tampered images. The board is embedded by Ubuntu and then, the environment of OpenCV and Boost libraries (http://theboostlibraries.com) are set up for coding. These libraries support Kdsort and Fastsats. In Kdsort, it used the k-D tree as a balanced binary tree and the block divided the original ROI after Zernike Moment process into many some subparts having specific features in order. To build the k-D tree, it starts from the root-cell and bisect recursively the cells through their longest axis, so that an equal number of particles lie in each sub-volume, and that quadrupole moments of cells are kept to a minimum.

5.1   Flowchart of Proposed Work

This bisection is accomplished using Hoare's median finding algorithm [13], which is an $O(N)$ average time operation per level of the tree. The depth of the tree is chosen so that it end up with a most 8 particles in leaf cells (*buckets*). The use of buckets, only $2[N/8]$ nodes are required, it make the tree structure memory – efficient. Otherwise, the block Fastsats processes in pixels level of a specific characteristic, it creates one feature vector is extracted from the neighborhood of this pixel. This feature vector is matched to its closest neighbor in feature space. Moreover, since the "Make file" of the program is created, then it can detect duplicated region by running command line to execute "Make file" in Ubuntu, Fig 13 as an example for detecting duplicated images.

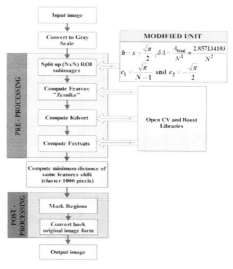

Fig. 12. Flowchart of detecting a tampered image

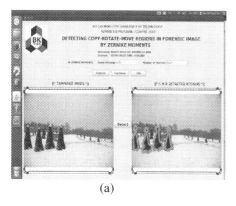 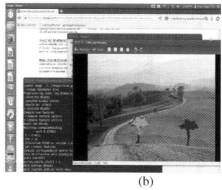

(a)   (b)

Fig. 13. (a) The result using Java GUI,   (b) Deployment of Zernike algorithm in BeagleBone

In the algorithm, it used FLANN library- Fast Library for Approximate Nearest Neighbors, which is as a comparator to calculate Euclidean distances and match the blocks having their similar moment magnitudes in the same ROI [11, 12]. For comparing to other ROIs, the Bhattacharyya distances [14] are used through two different classes having the variance σ_p, mean μ_p of p-th distribution, similarly for q-th have been shown below (Eq. 29)

$$d^2 = \sum_{(n,m)\in D} \sum \left(|Z_{nm}| - |Z'_{nm}| \right)^2 \tag{28}$$

$$D_B(p,\ q) = \frac{1}{4} ln \left(\frac{1}{4} \left(\frac{\sigma_p^2}{\sigma_q^2} + \frac{\sigma_q^2}{\sigma_p^2} + 2 \right) \right) + \frac{1}{4} \left(\frac{\left(\mu_p - \mu_q \right)^2}{\sigma_p^2 + \sigma_q^2} \right) \tag{29}$$

5.2   Comparison of Proposed Zernike Moments and Other Methods

According to some existed previous methods - HU, KPCA and PCA [15]. These methods having their efficiency in detecting duplicated images are not high. It can be easy to realize that the precision (%) of proposed Zernike Moments is the greatest and it means that the average percentage of the true detected duplicated pixels in one ROI divided to total pixels existing in that ROI - Table 2), similarly the order in Table 3. While recall (%) describes the true detected duplicated pixels in one ROI compared with the total samples included true detected duplicated pixels. The correlation of false pixels and true pixels F_ψ (%) in precision and recall gives the efficiency of one method, thus based on the results in the two tables (2,3), it is shown the proposed Zernike Moments to be the best.

Table 2. Results (%) of the methods to detect duplicated copies at pixel level in Fig 14b

Method	Precision (%) - p	Recall (%) - r	F_ψ (%)
HU	86.32	48.05	61.74
KPCA	82.51	63.20	71.57
PCA	88.01	65.47	75.08
ZERNIKE (proposed)	90.47	67.09	77.04
Average	86.83	60.95	71.36

The performance of Zernike Moment algorithm indicated extracted features in tampered region is excellent when comparing with HU, KPCA and PCA [15] as shown in tables (2 and 3)

$$p = \frac{T_P}{T_P + F_P},\ r = \frac{T_P}{T_P + F_N} \text{ and } F_\psi = 2.\frac{p.r}{p + r} \tag{30}$$

where T_P = the number of correctly detected duplicated pixels; F_P = the number of faultily detected duplicated pixels; F_N = the falsely missed pixels; p = *Precision*; r = *Recall* and F_ψ = the combination of precision and recall in a single value [16].

HU and Zernike Moments have quite similar characteristic in detecting copied regions, but Zernike Moments have more advantaged than HU Moments, because Zernike Moments can easily localize the specific features in an object. However, HU is just useful to recognize features generally. Hence, the efficiency of HU Moments is lower than proposed Zernike Moments. While KPCA and PCA use statistic for group of highlight pixels in order to consider which pixels in that group containing the invariant features of a ROI. KPCA and PCA take more time to process one ROI, and then it is good for color classifications and segmentations.

Table 3. Results (%) of some methods to detect duplicated copies at pixel level in Fig 15b

Method	Precision (%) - p	Recall (%) - r	F_ψ (%)
HU	81.71	47.01	59.68
KPCA	84.79	61.27	71.14
PCA	86.02	63.76	73.24
ZERNIKE (proposed)	90.78	62.52	74.06
Average	85.83	58.64	69.53

Moreover, when using ZMs is not only for type *.png, but also *.jpg, the detection also shown duplicated regions well. This means the noise sensitivity of ZMs algorithm is low and it also describe well for multilevel representation in shapes of pattern. The Fig.14 are presented some examples about the results of computing the Zernike Moments based modified algorithm combined with FLANN matching and Bhattacharyya distances [13].

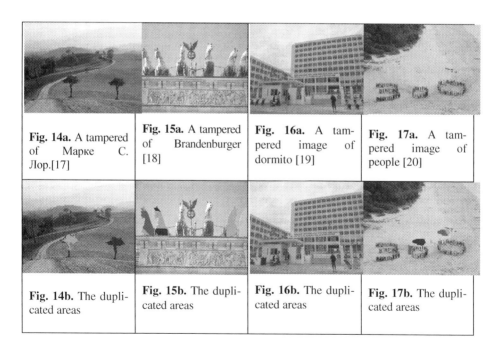

Fig. 14a. A tampered of Марке C. Лор.[17]

Fig. 15a. A tampered of Brandenburger [18]

Fig. 16a. A tampered image of dormito [19]

Fig. 17a. A tampered image of people [20]

Fig. 14b. The duplicated areas

Fig. 15b. The duplicated areas

Fig. 16b. The duplicated areas

Fig. 17b. The duplicated areas

6   Conclusion

In this work, the proposed method to modify the traditional Zernike moment computational of our study is not only enhanced the number of pixels of ROI - Region of Interest or elemental image computed inside the unit circle but also limited the change in size of a reconstructed image comparing with the size of an original image. The Zernike–based modified approach has proven its potential capability in significantly reducing the two main errors of Zernike Moments (ZMs) computation in the reconstructed images. The study is also tested on the BeagleBone hardware for implementing the proposed Zernike moment based approach to detect tampered images with acceptable and comprehensive results.

References

1. Popescu, A.C., Farid, H.: Exposing Digital Forgeries in Color Filter Array Interpolated Images. IEEE Trans. Signal Processing **53**(10), October 2005
2. Sheng, Y., Shen, L.: Orthogonal Fourier-Mellin moments for invariant pattern recognition. J. Opt. Soc. Am. A **11**, 1748–1757 (1994)
3. Farid, H.: Image Forgery Detection- A survey. IEEE Signal Processing Magazine **26**(2), 16–25 (2009)
4. Chong, C.-W.: A formulation of a new class of continuous orthogonal moment invariants, and the analysis of their computational aspects. *Ph.D thesis*, Faculty of Engineering, University of Malaya, 50603 Kuala Lumpur, Malaysia, May 2003
5. Liao, S.X., Pawlak, M.: On the accuracy of Zernike moments for image analysis. IEEE Trans. Pattern Anal. Mach. Intell. **20**(12), 1358–1364 (1998)
6. Chong, C.W., Raveendran, P., Mukundan, R.: A comparative analysis of algorithm for fast computation of Zernike moments. Pattern Recognition. **36**(3), 731–742 (2003)
7. Papakostas, G.A., Boutalisn, Y.S., Papaodysseus, C.N., Fragoulis, D.K.: Numerical error analysis in Zernike moments computation. Image and Vision Computing **24** (2006)
8. Teh, C.H., Chin, R.T.: On image analysis by the methods of moments. Pattern Analysis and Machine Intelligence **10**(4), 496–513 (1988)
9. http://www.lumetrics.com/documents/wavefront/Zernike_table.pdf
10. Chong-Yaw, W., Paramesran, R.: On the computational aspects of Zernike moments. Image Vision Computing **25**(6), 967–980 (2007)
11. Arai, K., Barakbah, A.R.: Hierarchical K-means: an algorithm for centroids initialization for K-means. Reports of the Faculty of Science and Engineering, Saga University, pp. 840–8502 Saga Prefecture, Saga, Japan, 36, (1), (2007)
12. Liao, S.: Image analysis by moments. Ph.D thesis, University of Manitoba, Winnipeg, Manitoba, Canada (1993)
13. Kirschenhofer, P., Prodinger, H., Martınez, C.: Analysis of Hoare's FIND Algorithm with Median-of-Three Partition. Journal Random Structures and Algorithm **10**(1–2), 143–156 (1997)
14. Thacker, N.A., Aherne, F.J., Rockett, P.I.: The Bhattacharyya Metric as an Absolute Similarity Measure for Frequency Coded Data. TIPR 1997, pp. 9–11, Prague, June, 1997
15. Amerini, I., Ballan, L., Caldelli, R., Bimbo, A.D., Serra, G.: A SIFT-based forensic method for copy-move attack detection and transformation recovery. IEEE Trans. Inf. Forensics Security **6**(3), 1099–1110 (2011)

16. Christlein, V., Riess, C., et al.: An Evaluation of Popular Copy-Move Forgery Detection Approaches. IEEE Transactions on Information Forensics and Security **7**(6), 1841–1854 (2012)
17. http://www.italiadavay.ru/public/images/marche/marche_san_lorenzo.jpg
18. http://www.nacht-der-wissenschaften.de/2011/ ansicht.php?id=%20%20%20%20%20%20%20%20101
19. http://www.sggp.org.vn/anninhtrattu/2014/2/340316/
20. http://www.arxiv.org/pdf/1208.3665

Requirements Engineering in an Emerging Market

Emmanuel Okewu[✉]

Centre for Information Technology and Systems, University of Lagos, Lagos, Nigeria
eokewu@unilag.edu.ng

Abstract. The growing importance of requirements engineering (RE) in software development cannot be overemphasized. A faulty requirements gathering exercise and the emergent requirements document could mislead the entire software development drive, resulting in a software product that falls short of user expectation in terms of meeting needs and delivering within budget, time and scope. Achieving the objective of a well articulated and coordinated requirements document in an ideal economic environment is tasking let alone in an emerging market characterized by macro-economic variables such as high cost of doing business, weak institutions, poor infrastructure, lack of skilled and competitive workforce, among others coupled with micro-economic (personal) tendencies like resistance to change, vested interest, technophobia and insider abuse. This paper reports on industrial experience of designing and implementing an n-tier enterprise application in an African university using service oriented software engineering (SOSE) approach. The application is meant to facilitate the actualization of the 25-year strategic plan of the institution. We applied design and software engineering skills: Literature were examined, requirements gathered, the n-tier enterprise solution modeled using unified modeling language (UML), implementation achieved using Microsoft SharePoint and the results evaluated. Though success was recorded, the challenges encountered during the requirements engineering stage were quiet reflective of the challenges of software project management in a typical relatively unstable macroeconomic environment. The outcome of this study is a compendium of lessons learnt and recommendation for successful RE in the context of an emerging economy like Africa in the hope that this will guide would-be software stakeholders in such a business landscape.

Keywords: Africa · Requirements engineering · Service oriented software engineering · Unstable economic environment · Emerging market

1 Introduction

As Africa struggles with its economic, social and military problems, the international market crude oil crash is adding to challenges. Nigeria, Africa's largest economy has a fare share of the current international pressure on emerging economies. As anticipated, policies, programmes and projects are being affected, and software projects are no exception [1]. Software project plans, estimates, schedules, budgets and overall requirements engineering (RE) are under pressure to give precise users specifications

© Springer International Publishing Switzerland 2015
O. Gervasi et al. (Eds.): ICCSA 2015, Part IV, LNCS 9158, pp. 476–491, 2015.
DOI: 10.1007/978-3-319-21410-8_37

for software that will be delivered on tine time, within budget and within scope in such a precarious environment. Needless to say, this is a tall order though achievable. Guiding requirements engineers and software projects managers on the winning way against all odds in an emerging market is the motivation for this work.

In an emerging economy like Africa, government remains the chief spender [2,3]. Though the private sector is growing in importance, it is not yet strong enough to command huge spending like government. Hence, whatever affects government reve- nue impacts seriously on all sectors. The information and communication technology (ICT) sector has grown in strategic importance in recent years as Africa, instructively, joins the rest of the world in the global knowledge economy. Software engineering is a critical component of any ICTs drive and this accounts for the upsurge in software projects in emerging markets. Software project management involves planning, esti- mating, budgeting and scheduling. Project managers meticulously monitor the soft- ware project management cycle and other software stakeholders to ensure that these variables are properly articulated thereby mitigating project risk and business risk associated with software projects. However, while microeconomic factors (personnel, hardware, software) are within the control of the project manager, the same cannot be said of macroeconomic variable (exchange rate, inflation rate, oil price, etc.). Clearly, the costs of software requirements are tied to the macroeconomic environment of the software project.

Requirements can be thought of as belonging to a type for two reasons: as an aid to finding the requirements and secondly, to be able to group the requirements that are relevant to a specific expert specialty [4,5,6] . Pursuant to this assertion, they classi- fied requirements as functional, non-functional, project constraints, design con- straints, project drivers and projects issues.

Whereas functional requirements are the key subject matter of the product, which describes what, the product has to do or what processing actions it must take, non- functional requirements are the properties that the functions must have, such as per- formance and usability. Project constraints are limitations on the product owing to the budget or the time available to build the product while design constraints place restrictions on how the product must be designed such as implementation in the hand- held device being given to major customers, use of existing servers and desktop com- puters, or any other hardware, software, or business practice. Project drivers are the business-related forces such as the purpose of the project, and all stakeholders—each for different reasons. Project issues refer to the conditions under which the project will be done - these present a comprehensive and coherent picture of all factors that contribute to the success or failure of the project.

Meanwhile, the Standish Group and Scientific American have statistically modeled variables that contribute to software project failure based on researches conducted over time as shown in Table 1:

Table 1. Causes of software project failure

SN	Factor	% Contribution
1	Incomplete requirements	13.1
2	Lack of user involvement	12.4
3	Lack of resources	10.6
4	Unrealistic expectations	9.9
5	Lack of executive support	9.3
6	Changing requirements/specification	8.7
7	Lack of planning	8.1
8	Didn't need it any longer	7.5

It is instructive, from the above table, that the combined role of variables like in-complete requirements and lack of user involvement, both of which are requirements engineering concerns, amounted to whooping 25.5% failure rate. This reality provides the underpinning premise for investigating RE, particularly in an emerging market environment.

The environment provided by an emerging market like Africa is less than perfect [7]. This paper reports a study of the challenges of RE in a software engineering project in a Nigerian university - University of Lagos. The study empirically investigated the claims of environmental impact on software architecture and software economics that some authors have referred to in the literature [8,9]. This researcher observed that not many reports on empirical impacts of socio-cultural and socio-economic environment on RE, and by extension, on software development originate from Africa. A number of research-ers and practitioners in RE are of the strong opinion that the RE research community is in need of more industrial experiences and empirical studies. Moreover, the calls for papers of the foremost international conferences in the area of requirements engineering such as Requirements Engineering Conference (RE) have in recent times, underscored the ex-igency for more case studies in requirements engineering. Hence, as a contribution, this works seeks to enrich the body of knowledge of RE by reporting on an exclusive indus-trial experience from Nigeria. Against the backdrop that reports of empirical studies of the impact of socio-cultural environment on RE emanating from Africa are scanty, this is an important development.

The remainder of this paper is made up of the following: section 2 gives the background of study and related work; section 3 presents the methodology and the selected case study; section 4 focuses on results and discussions; and finally, the paper is concluded in section 5.

2   Background and Related Work

2.1   The African Economic Environment

The African economy is largely agrarian and most governments are yet to diversify their economic base. Hence, the ability to compete favorably at the international mar-ket is a mirage. This accounts for the severe economic crisis that African countries experience when there is problem on the international market. Recent international economic crises as articulated by the just concluded World Economic Forum (WEF) in Davous, Switzerland such as oil and gas prices fall, Quantitative Easing, Ukraine crisis, Anti-ISIS war in the Middle East, just to mention a few, all impact on the African economy in one form or the other. Political instability has also been fingered as key contributory factor to the continent's unstable economic/business climate. All these pieced together has left Africa with crumbling infrastructure, macroeconomic instability, weak institutions, low-skilled workforce, among others. Little wonder then that projects in various sectors suffer from uncertainties and surprises. And software projects are not insulated [7], [10].

The fallout of instability and uncertainties is that software project managers and requirements engineers experience difficulty keeping to budget, time and in some cases, scope of projects as the unstable macroeconomic variables affect project planning, estimation, scheduling and budgeting [11,12]. It is safe to say that RE certainly is dealt a lethal blow in such circumstance. Worthy of mention equally is the fact that the challenges posed by political, economic, social and technology (PEST) environments apart, the physical geographical environment may pose challenges that moderate choice of software architecture. The industrial experience gathered from implementing multi-tier enterprise architecture in University of Lagos, Nigeria clearly supports this stance.

2.2   ICTs Strategy and Corporate Objectives of University of Lagos, Nigeria

University of Lagos, Nigeria is a coastal university with the main campus situated by the Lagos Lagoon. Established in 1962 by the Federal Government of Nigeria, it has two campuses - the main campus at Akoka, Yaba and the College of Medicine in Idi-Araba, Surulere. Both sites are in the Mainland of Lagos, Nigeria. The main campus is largely surrounded by the scenic view of the Lagos lagoon and is located on 802 acres (3.25 km2) of land. From a modest intake of 131 students in 1962, enrolment in the university has now grown to over 45,000. It has total staff strength of 3,365 made up of 1,386 Administrative and Technical Staff, 1,164 Junior and 813 Academic Staff. The University is composed of nine Faculties and a College of Medicine. The Faculties offer a total of 117 programmes in Arts, Social Sciences, Environmental Sciences, Pharmacy, Law, Engineering, Sciences, Business Administration and Education.

Its surrounding water implies it has limited space for expansion. This is of strategic consideration to the management as decisions across sectors are guided by the geographic constraint. According to Strategic Plan [13], the University is setting the pace and re-ordering current practices to meet international standards while putting in place the necessary infrastructure to carry on as an international institution. The plan is divided into seven segments: Academic Matters, Student Affairs, Environment, Health, Safety and Security, Physical Infrastructure, Human Resources and Finance. The 25-year strategic plan (2012 -2037) outlined corporate objectives, goals/targets, strategies and initiatives, as well as implementation and monitoring parameters for each of the segments. The ICT infrastructure and strategy are coordinated largely under the Physical Infrastructure segment though it functionalities cut across the entire segment as the life-wire of the strategic plan.

The Strategic Plan outlined the strengths, weakness, opportunities and threats (SWOT) analysis of the university [13]. Threats like under-funding by government, lack of full university autonomy, frequent threats of industrial action by unions, competition from other universities, high cost of living, high labor turnover due to poaching, inadequate spaces for expansion, and high cost of building due to marshy terrain typically illustrate the challenges organizations face in emerging markets. These socio-cultural challenges dictate the tune of project execution, software project management and RE inclusive [10].

Equally worthy of mention is the fact that the strategic plan highlighted the following as major sources of funds to the university: Federal Government subvention, internally generated revenue (IGR), education trust fund (ETF), endowment, and donor.

2.3   Virtualization and Enterprise Software Architecture

Gorton [8] is of the view that enterprise architecture can be analyzed along three lines: architectural requirements, architectural design and validation of architecture. The architectural requirements gathered for the University of Lagos enterprise application include stakeholders (functional) requirements and non-functional requirements. While the functional requirements were basically functionalities expected of the application, non-functional requirements included quality attributes (performance, scalability, etc.) and socio-cultural requirements such as challenges posed by physical environment of the university and attitude of members of the university community [4,5], [14] . This study observed that various departments and units were eager to own their servers courtesy internally generated revenue (IGR). Though the need for computing all these departments and units was apparent and hence the need for ICTs infrastructure procurement, investigation revealed that some had vested interest in spending money and ICT procurement provided the leeway. In the ultimate analysis, the data center already limited in size by virtue of the coastal nature of the university, was struggling to cope with the mounting number of servers by various units. The space issue apart, the manner of procurement and installation of the various systems did not align with the overall corporate goals, objectives and strategy of the university. As an aftermath, the ICT strategy of the university was clearly at variance with its corporate strategy and could not its 25-year strategic plan. This provided the underlining premises for the design and implementation of a web-based enterprise architecture that synchronizes and streamlines the heterogeneous and disparate systems across the university into an integrated and a one-stop information system [15, 75]. A virtualized n-tier web-based enterprise framework would therefore be appropriate.

2.4   Service-Oriented Software Engineering (SOSE)

The development of the University of Lagos enterprise was done using Service-oriented Software Engineering (SOSE). SOSE is a software engineering technique that emphasizes the development of software systems by composition of reusable services (service-orientation) often provided by other service providers. Since it involves composition, it shares many characteristics of component-based software engineering, the composition of software systems from reusable components, with the additional capability to dynamically locate necessary services at run-time [16,17]. Others may provide these services as web services, but the key feature is the dynamic nature of the connection between the service users and the service providers [18]. The three categories of actors in a service-oriented interaction are service providers, service users and service registries. They engage in a dynamic collaboration, which varies from time to time. Service providers are software services that publish their capabilities and availability with service registries. Service users are software systems that perform tasks via the use of services provided by service providers. Service users use service registries to discover and locate the service providers that are available and can be used. This discovery and location occurs dynamically when the service user requests them from a service registry [19, 20].

2.5   Related Work

Oliver [21] opined that in 2013, Africa was the world's fastest-growing continent at 5.6% a year, and GDP is expected to rise by an average of over 6% a year between 2013 and 2023. Growth has been present throughout the continent, with over one-third of Sub-Saharan African countries posting 6% or higher growth rates, and another 40% growing between 4% to 6% per year. In March 2013, Africa was identified as the world's poorest inhabited continent; however, the World Bank expects that most African countries will reach "middle income" status (defined as at least US$1,000 per person a year) by 2025 if current growth rates continue. Though this work did not state how uncertainties in such emerging market impact on requirements engineering, it reflects Africa as a business landscape with all the economic features of an emerging market. Like any other sector, the software sector will have its fair share of the relatively harsh economic climate of such a market. This realization prepares requirements engineers and software project managers in such terrain for the task of software engineering.

Somerville [4] emphasized that a project represents a social system in which people have to interact to achieve results. The design of the project environment relationships is a project management activity. The objective is to determine all relevant environments (human resource management, financial management, communication management, interpersonal relationship, team work, etc.) for the project to succeed. Understanding the appropriate social context that can promote optimal collaboration between project team and the end-user team is critical. Hence it is necessary to consider the relationship of a software project to its own social environment. By implication, the requirements engineers and project manager of the n-tier enterprise application in university of Lagos have to understand the psychology of the staff of university of Lagos as well as the physical environment in which the university operate. Equally, its funding structure, strategic plan, socio-economic environment, among others, have to be understood for well articulated and orchestrated requirements elicitation. However, the study was not specific about requirements gathering in an emerging economy, which is the main motivation for this work.

The quartet of Agyris, Maslow, Festinger and Carlsmith [22,23,24] approached the subject from the standpoint of behavioral theorists. To their minds, the functional and non-functional requirements translate into physiological and psychological needs respectively. They were particular about what motivate people to act the way they act. Though they were not specific about end-users behavior in a software in a software project, their work help the requirements engineer and software project manager to understand human psychology and build tactical skills required to successfully elicit requirements from end-users and gain their support through out the project life cycle. It is an established fact that the level of users cooperation in a project can make or mar the project.

Pressman [5] stressed that the success of an IT project does not depend on meeting functional and non-functional requirements alone, but also depends on tackling constraints inherent in the project environment as they have the potential to sustain or destroy the project. These constraints are technical and business constraints. Though they fell short of mentioning challenges of RE in a relatively fluid and unpredictable emerging market liker Africa, the mere fact that they acknowledged that constraints in the project environment could be a deciding factor simply means that their work is relevant to this discussion. The constraints identified in gathering requirements in University of Lagos include its coastal terrain that limits space for expansion and vested business of individuals that tend to undermine the overall corporate objectives of the institution.

Gorton [8] classified the software architecture process into three: determine architectural requirements, architecture design, and validation. He opined that both functional requirements and stakeholder requirements form the architecture requirements. While functional requirements are considered to be architecturally significant requirements or architectural use cases, stakeholders' requirements are essentially the quality and non-functional requirements of a system. It was further elaborated that some architectural requirements are constraints that impose restrictions on the architecture and non-negotiable. As such, they limit the range of design choices an architect can make. This assertion apparently underscores why some highly sophisticated software architecture cannot be implemented in organizations in emerging economies despite their potential for high value addition. University of Lagos, for example, depends largely on the relatively meager sum it gets from government to fund its activities. Though the work stop short of mentioning the constraints of requirements engineering in a less stable economic climate like Africa, it nonetheless stressed that requirements elicitation should be holistic.

In a nutshell, it was observed from the literature that none of the previous studies had pitched on requirements gathering in an African context with a view to understanding the impact of market instability on RE. This is the main motivation for this work. We are also motivated by the fact that a well articulated requirements document is a veritable input into software project management for quality software delivery.

3   Methodology – RE Activities for N-tier Enterprise Application

University of Lagos is a government-established institution and it relies heavily on the Federal Government for it's funding. The weak economic environment has made it practically impossible for the institution to generate enough revenue to sustain itself. The implication is that macroeconomic downtown like the present low revenue from oil incredibly affects it's funding. This means projects across the seven segments of the university strategic plan would be affected as project plan, estimates, budget, and schedule need to be adjusted in response to the emerging reality. This implies that requirements engineers and software project managers have to be extra vigilant and careful in navigating projects to successful end in such a turbulent business clime.

Specifically, designing and implementing the multi-tier enterprise application to replace existing disparate legacy systems required meticulous planning and execution of RE if goal was be achieved [78], [81]. The paradigm shift entails that end-users have to be carried along apart from eliciting requirements. Hence, people management skills were required. An understanding of the socio-cultural and socio-economic environment was also useful. Thus, the researcher interviewed staff, observed events and read documents during requirements elicitation. Thereafter, a model of the proposed enterprise application was designed using universal modeling language and implemented on Microsoft SharePoint development platform.

Guided by the objective-methodology mapping in the Table 2, the study followed RE activities - requirements elicitation, requirements identification, requirements analysis, requirements specification, system modeling, requirements validation, and requirements management [26] - to actualize the proposed n-tier enterprise system as a replacement for the disparate legacy systems dotting the university landscape.

Table 2. Objective-methodology mapping

SN	Objective	Methodology
1	To determine architectural requirements	Interview staff, observe activities/physical environment and read university documents e.g. 25-year strategic plan
2	To provide an enterprise architectural design	Design an architecture using universal modeling language (UML).
3	Validate architecture	Use test scenarios and prototype

3.1   Requirements Elicitation

The requirements for the proposed enterprise system were gathered by interview, observation and studying existing processes, systems and documents.

3.2   Requirements Identification and Analysis

The cross-cutting functional requirements include add information, access information, edit information, and delete information while the non-functional requirements include quality requirements that span performance, security, usability, aesthetics, availability, reliability, scalability, fault tolerance, modifiability, portability and interoperability. The enterprise system incorporates mechanisms that respond to these requirements.

3.3   Requirements Specification

The study captured the requirements in an elaborated format. Table 3 shows detailed explanation of functional requirements in the software requirements specification. Other non-functional requirements captured include training, coaching and mentoring of end-users as effective change management tools for the transition from disparate legacy systems to the enterprise platform. We equally captured other socio-cultural requirements as enshrined in the 25-year strategic plan document. Of particular mention is the university policy on land space utilization in response to it geographic challenge which stipulates that for the next 25 years, the University has resolved to no longer allow infrastructural facility that is less than a 10-storey building on any of its lands in all campuses because of the constraint of land and spaces. However, donors can be encouraged to pull resources together to achieve this goal in phases. From the requirements engineering point of view, this policy lends credence to the integration of virtualization into the enterprise architecture to conserve space occupied by servers.

Table 3. Functional requirements

Requirement ID	Requirement	Brief Description
R01	Add Information	The system shall allow end-users to add information to the enterprise system
RO2	Access Information	The system shall allow end-users to retrieve and view information from the proposed enterprise system
RO3	Edit Information	The system shall allow end-users to edit information on the enterprise system.
R04	Delete Information	The system shall allow end-users to delete information from the enterprise system.

The above crosscutting functions of the enterprise system are represented using Use Case modeling as shown in Fig. 1. Use case diagram captures the functional aspects of a system by visually representing what transpires when an actor interacts with the system [27].

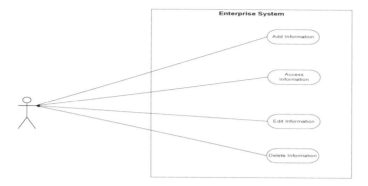

Fig. 1. Use case diagram for the enterprise system

3.4   System Modeling

Both system architecture and software architecture were designed and documented using UML notations such as component diagram, class diagram, sequence diagram, collaboration diagram, among others to highlight components of the proposed solution and their relationships. However, while in this report the system architecture is depicted as a multi-tier structure, the software architecture highlights crosscutting functions using algorithm design. The deployment diagram is shown in Fig. 2 indicating that users can access the system through personal computers, phones, payment systems (ATM/POS) and biometric readers.

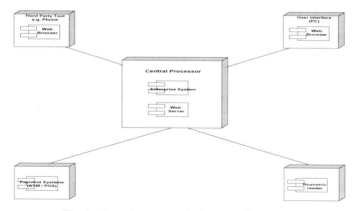

Fig. 2. Enterprise system deployment diagram

We considered that an n-tier web-based enterprise system that supports component reusability and distributed computing would best serve the interest of the university given the challenges associated with current disparate legacy systems found all over the campus [25]. Hence, we designed the n-tier enterprise architecture in Fig. 3 for the proposed solution, which incorporates mechanisms that respond to user requirements.

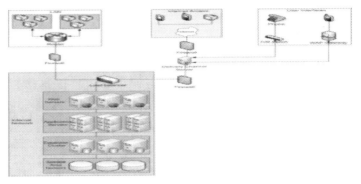

Fig. 3. Enterprise system multi-tier architecture

For optimal efficiency and superior value addition to the 25-year strategic plan of the university that is largely ICT-driven, the study views the above University of Lagos client-server paradigm as processes rather than machines or hosts thus reducing number of physical machines in the Data Centre.

Virtualization [28,29] will facilitate centralized, cost-effective and efficient management of the university's IT infrastructure in the hope of lending greater support to the realization of the 25-year strategic plan. Hardware virtualization and storage virtualization were used to scale up efficiency and effectiveness [30,31].

The software architecture is depicted in using class diagram and algorithm design. Rights (add, access, edit, delete) are assigned to users based on their respective roles in the university. Fig. 4 shows the class diagram.

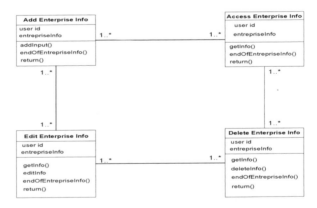

Fig. 4. Class diagram for the enterprise system

The corresponding algorithm designs for the crosscutting functions (addEntrepriseInfo, accessEntrepriseInfo, editEntrepriseInfo, deleteEntrepriseInfo) are as follows:

```
Procedure addEntrepriseInfo()
    entrepriseInfo ← " "
    while (not endOfEntrepriseInfo())
    entrepriseInfo ← addInput()
return(entrepriseInfo)

Procedure accessEntrepriseInfo()
    while (not endOfEntrepriseInfo())
    getInfo(entrepriseInfo)
return

Procedure editEntrepriseInfo()
    while (not endOfEntrepriseInfo ())
    getInfo(entrepriseInfo)
    editInfo(entrepriseInfo)
return(entrepriseInfo)

Procedure deleteEntrepriseInfo ()
    while (not endOfEntrepriseInfo ())
    getInfo(entrepriseInfo)
    deleteEntrepriseInfo()
return(entrepriseInfo)
```

3.5   Requirements Validation

The architectural requirements were validated using test scenarios and prototype with a view to minimizing errors that would surface during implementation [8]. In this light, we concentrated on process-correctness and requirements-compliance of the enterprise architecture by examining the various software representations - requirements documents and design documents. The key concern was to ascertain that user requirements had been well catered for in each software representation at each stage so that the ultimate software product will comply with both operational needs of users and emergent properties.

3.6   Requirements Management

In the course of developing the system, some user requirements changed and were appropriately documented and approved by relevant authorities before affecting them technically. The fluid macro-economic environment also impacted on cost of items required to actualize the task. Specifically, rising costs meant budget had to be adjusted and approval sought from concerned authorities. The end game was to ensure that built system did not deviate significantly from users' expectations even within the context of an unstable environment.

4   Results and Discussions

Amid concerns that operations of the University of Lagos are constrained by limited funding by the Federal Government in the light of its harsh economic realities, every project in its strategic plan has to align with its economic and physical realities. Hence, the university's socio-cultural context cannot be neglected in the execution of any of its projects; the n-tier enterprise application project is no exception. From the experimental design (test bed) set up in University of Lagos, two outcomes emerged: one, a virtualized enterprises architecture and two, a compendium of requirements elicitation lessons learnt which is a vital guiding tool for requirements engineers and software project managers taking up task in an emerging market like Africa [32,33].

4.1   Virtualized Servers in Enterprise Architecture

The constraint posed by the coastal nature of the university means that available space needs to be optimized. With virtually every department/unit eager to have its server and consequent competition for space, virtualization becomes a veritable requirement and space optimization techniques that reduces the amount of physical servers in the Data Centre. So far, 12 technical staff has been trained in Linux, with professional certification obtained. Higher certification training is underway. Linux is strategic to server virtualization and management plan of the university. While we re-organize architecturally, we also reposition the human resources.

4.2   Requirements Elicitation Lessons Learnt

It was realized in the course of the study that requirements elicitation and solution delivery was not only shaped by the demands of the unique physical environment but success was also determined by the way and manner the end-users were handled [4,5]. End-users cooperated more when tactfully communicated to. The study therefore unveiled new end-users handling measures for eliciting requirements. The following findings have been proven to aid users involvement, requirements gathering and maximum co-operation during requirements elicitation.

- End-users want to be respected by solution providers.
- End-users are naturally good but if treated shabbily, will display the other side and not cooperate.
- There are sometimes resistances from end-users owing to perceived pains the project may inflict on them.
- Resistance to change may be due to economic, social or psychological factors.
- Some end-users are technophobes (hate or fear tech.)
- Change management techniques (training, coaching, mentoring, etc.) help in mitigating resistance and technophobic levels.
- End-users appreciate a well-rounded professional with hard skills and emotional intelligence (soft skills).

4.3   Evaluation Threats

It is possible that an expanded review of RE in different emerging markets may unveil new insights. However, the subjects that took part in the experimental survey possess requisite practical knowledge of Nigeria as an emerging market. They likewise had sufficient practical engagements with the n-tier enterprise application. This provided sufficient premise to objectively assess the impact of environmental constraints on requirements gathering. Hence, there is adequate reason to take their views seriously [34], [36,37].

Moreover, only 32 university staff members were interviewed and observed in the requirements elicitation process, which could in a sense limit the statistical significance of the outcome [58],[79]. Nonetheless, the result of the survey clearly indicates socio-cultural constraints and people management impact on requirements engineering. This is considered to be a good result because at this point in the research, the core objective is to gain a first impression of the influence of project environment on successful software delivery. Therefore, despite the limitation of using a limited number of evaluators, there is sufficient ground to infer that there is a positive and preferential disposition to factoring environmental constraints in requirements gathering. We can thus generalize that the environment of an emerging market impact on requirements engineering.

5   Conclusion

In this study, we shared industrial experience of gathering requirements in a software engineering project for implementing an n-tier enterprise application in a Nigerian university. The outcome shows that the environment of a software project can impact on requirements gathering and by extension, software product delivery. In order to succeed in deploying enterprise architecture in this case study, we gathered stakeholder's functional requirements, non-functional quality attributes and the social-cultural requirements of University of Lagos. We secured an understanding of the socio-cultural requirements by studying the behavioral pattern of members of staff towards ICT infrastructure procurement as well study the geographical terrain and its impact on the overall 25-year strategic plan of the university. Ultimately, the study designed and implemented an enterprise application that aligns the ICT strategy of the university to its corporate objectives and strategy. Our recommendation to software project stakeholders in an emerging market like Africa is that in as much as a well articulated RE document is a sine qua non for successful software project management and software delivery, there is need to factor in both microeconomic and macroeconomic variables during the RE stage especially in the context of a relatively unstable economic climate.

References

1. Umble, E.J., Haft, R.R., Umble, M.M.: Enterprise resource planning: Implementation procedures and success factors. European Journal of Operational Research **146**(2), 229–432 (2003)
2. Asiegbu, B.C., Ahaiwe, J.: Software Cost Drivers and Cost Estimation in Nigeria. Interdisciplinary Journal of Contemporary Research in Business **8**(8), 12–14 (2011)
3. Boehm, B.: Software Engineering Economics. Prentice-Hall, Englewood Cliffs, NJ (1981). ISBN ISBN 0-13-822122-7
4. Sommerville, I.: Software Engineering, 9th edn. (2011)
5. Pressman, R.S.: Software Engineering: A Practitioner's Approach, 7th edn. (2009)
6. Aggarwal, K.K., Singh, Y.: Software Engineering. New Age International Publishers (2008)
7. Asiegbu, B.C., Ahaiwe, J.: Software Cost Drivers and Cost Estimation in Nigeria. Interdisciplinary Journal of Contemporary Research in Business **8**(8), 12–14 (2011)
8. Gorton, I.: Essential Software Architecture, Second Edition, Springer (2011)
9. Boehm, B.: Software Engineering Economics. Prentice-Hall, Englewood Cliffs, NJ (1981). ISBN 0-13-822122-7
10. Boehm, B.W.: Software Cost Estimation with COCOMO II (with CD-ROM). Prentice-Hall, Englewood Cliffs, NJ (2000). ISBN ISBN 0-13-026692-2
11. Umble, E.J., Umble, M.M.: Avoiding ERP implementation failure. Industrial Management **44**(1), 25–33 (2002)
12. Umble, E.J., Haft, R.R., Umble, M.M.: Enterprise resource planning: Implementation procedures and success factors. European Journal of Operational Research **146**(2), 229–432 (2003)
13. Strategic Plan: University of Lagos 25-year Strategic Plan. University of Lagos Press, Lagos, Nigeria, ISBN: 978-978-51929-7-1 (2013)
14. Zornada, L., Velkavrh, T.B.: Implementing ERP Systems in Higher Education Institutions (2005)
15. Swartz, D., Orgill, K.: Higher Education ERP: Lessons Learned (2001)
16. Okewu, E.: Component-based software engineering approach to development of a university e-administration system. In: IEEE 6th International Conference on Adaptive Science and Technology, pp. 1–8 (2014)
17. Sirobi, N., Parashar, A.: Component Based System and Testing Techniques. International Journal of Advanced Research in Computer and Communication Engineering **2**(6), 2378–2383 (2013)
18. Cervantes, H., Hall, R.S.: Technical Concepts of Service Orientation. Chapter 1, pp. 1–26 in Stojanović, Zoran and Dahanayake, Ajanthap. Service-oriented software system engineering: challenges and practices. Idea Group Inc. (IGI) (2004). ISBN 978-1-59140-428-6
19. Breivold, H.P., Larsson, M: Component-based and service-oriented software engineering: key concepts and principles. In: Proc. 33rd EUROMICRO Conference on Software Engineering and Advanced Applications, pp. 13–20. IEEE (2007). ISBN 978-0-7695-2977-6
20. Stojanović, Z.: A Method for Component-Based and Service-Oriented Software Systems Engineering. Doctoral Dissertation, Delft University of Technology, The Netherlands (2013). ISBN 90-901910-0-3
21. Oliver, A.: Africa rising A hopeful continent. The Economist. The Economist Newspaper Limited. http://www.economist.com/news/special-report/21572377-african-lives-have-already-greatly-improved-over-past-decade-says-oliver-august (Retrieved15 April 2015)

22. Argyris, C.: Personality and Organization: the Conflict between System and the Individual, p. 291. Harper and Brothers, New York (1957). OCLC 243920
23. Maslow, A.: Motivation and personality, p. 93.. Harper and Brothers, New York, NY (1954)
24. Festinger, L., Carlsmith, J.M.: Cognitive consequences of forced compliance. The Journal of Abnormal and Social Psychology **58**(2), 203–210 (1959)
25. Young, M.: Enterprise Resource Planning (ERP): a review of the literature. Int. J. Management and Enterprise Development **4**(3), 235–264 (2007)
26. Wohlin, C., Runeson, P., Host, M., Ohlsson, M.C., Regnell, B., Wesslen, A.: Experimentation in Software Engineering: An Introduction. Kluwer Academic, Norwell, MA, USA. (2000); Youseff, L., Butrico, M., Da Silva, D.: Towards a Unified Ontology of Cloud Computing. In: Grid Computing Environments Workshop (2008)
27. Aggarwal, K.K., Singh, Y.: Software Engineering. New Age International Publishers (2008)
28. Patterson, D.A., Gibson, G., Katz, R.H.: A Case for Redundant Arrays of Inexpensive Disks (RAID). University of California Berkeley (1988). http://www.cs.cmu.edu/~garth/RAIDpaper/Patterson88.pdf
29. Gupta, M., Sastry, A.C.: Storage Area Network Fundamentals. Cisco Press. ISBN 978-1-58705-065-7
30. Chen, P., Lee, E., Gibson, G., Katz, R., Patterson, D.: RAID: High-Performance, Reliable Secondary Storage. ACM Computing Surveys **26**, 145–185 (1994)
31. Donald, L.: MCSA/MCSE 2006 JumpStart Computer and Network Basics, 2nd edn. SYBEX, Glasgow (2003)
32. Hong, K., Kim, Y.: The Critical success factors for ERP implementation: an organizational fit perspective. Information & Management **40**(2), 25–40 (2002)
33. King, P., Kvavik, R.B., Voloudakis, J.: Enterprise resource planning systems in higher education (ERB0222). EDUCAUSE Center for Applied Research (ECAR), Boulder, CO (2002)
34. Host, M., Regnell, B., Wohlin, C.: Using students as subjects - a comparative study of students and professionals in lead-time impact assessment. Empirical Software Engineering - an International Journal **5**(3), 201–214 (2000)
35. Runeson, P.: Using students as experiment subjects - an analysis on graduate and freshmen student data. In: Linkman, S. (ed.) 7th International Conference on Empirical Assessment & Evaluation in Software Engineering (EASE 2003), pp. 95–102 (2003)
36. Sauro, J., Kindlund, E.: A Method to Standardize Usability Metrics into a Single Score. ACM, CHI (2005)
37. Svahnberg, M., Aurum, A., Wohlin, C.: Using students as subjects -an empirical evaluation. In: Proc. 2nd International Symposium on Empirical Software Engineering and Management ACM, pp. 288–290 (2008)
38. Nielsen, J., Landauer, T.: A mathematical model of the finding of usability problems. In: Proceedings of ACM INTERCHI'93 Conference, pp. 206–213 (1993)
39. Turner, C.W., Lewis, J.R., Nielsen, J.: Determining usability test sample size. In: Karwowski, W. (ed.) International Encyclopedia of Ergonomics and Human Factors, pp. 3084–3088. CRC Press, Boca Raton, FL (2006)

Efficient Utilization of Various Network Coding Techniques in Different Wireless Scenarios

Purnendu Shekhar Pandey[1], Neetesh Purohit[1], Sanjay Mishra[2(✉)],
Broderick Crawford[3,4,5], and Ricardo Soto[3,6,7]

[1] Indian Institute of Information Technology, Allahabad, Uttar Pradesh, India
{purnendu.iiita,neetesh.purohit}@gmail.com
[2] Covenant University, Ota, Nigeria
ssopam@gmail.com
[3] Pontificia Universidad Católica de Valparaíso, Valparaíso, Chile
[4] Universidad Central de Chile, Santiago, Chile
[5] Universidad San Sebastián, Santiago, Chile
[6] Universidad Autónoma de Chile, Santiago, Chile
[7] Universidad Cientifica del Sur, Lima, Peru

Abstract. The nodes of a communication network use network coding technique for generating packets for output links by systematically processing the packets received on its input links such that the original packets should be recovered by the destination nodes. Under low traffic conditions higher bandwidth efficiency and power efficiency can be achieved using network coding but the performance degrades as the traffic in the network increases. Overall performance of the network can be improved if the node is able to take a decision about whether to use or not to use network coding under the current condition of the network. This work presents a scheme for finding out a threshold value below which network coding should be used and above the threshold normal forwarding operation should be adopted by the nodes. The performance of the proposed algorithm on Cross topology under different network coding schemes has been tested under simulation environment i.e. on NS-3. Significant improvement has been observed as compared to only network coded systems as well as the traditional store and forward systems.

Keywords: Maximum Flow Min Cut (MF-MC) · Network Coding (NC) · Coding Gain (CG) · Throughput · XOR · Linear Network Coding (LNC) · Random Linear Network Coding (RLNC)

1 Introduction

Wireless network has become the need of the hour and the sheer reason is that they provide support for mobility and Internet Connectivity at any time and at any place [3]. But there are constrain such as transmission rate achieve by wireless is very less as compared with the wired network [1]. Every time various algorithms tries to improve this constrain but till now it hasnt been achieved

© Springer International Publishing Switzerland 2015
O. Gervasi et al. (Eds.): ICCSA 2015, Part IV, LNCS 9158, pp. 492–502, 2015.
DOI: 10.1007/978-3-319-21410-8_38

effectively [5]. Wireless network still lacks in various areas such as throughput, Security, and robustness [6][9][10] .

NC is indeed a promising way to move a step ahead as compared to previous ways of improving throughput in wireless network [7]. In NC, the intermediate or relay node combines packets from different sources and sends it to various destinations in a single transmission, thus improving the throughput of the network [8].

To understand NC in a more comprehensive ways, consider the diagram

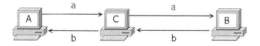

Fig. 1. Scenario without Network Coding

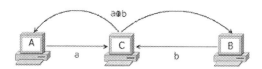

Fig. 2. Scenario with Network Coding

In fig.1, every packet that moves from node A to node B had to go through from relay node C. The packets are transmitted by relay node C as it arrives. For sending a packed named 'a' from node A to Node B takes 2 transmissions similarly for sending a packet 'b' from node B to node A takes 2 transmissions i.e it takes 4 transmissions to send packet 'a' and packet 'b' from node A to node B through node C i.e first transmission will be from Node A which will send packet 'a' to relay node C, second transmission will be from Node C which will forwards the packet 'a' to Node B, third transmission will be from Node B which will send packet 'b' to relay node C, and Fourth transmission will be from Node C which will forwards the packet 'b' to Node A [4]. Whereas, in fig 2. it takes 3 transmissions i.e first transmission will be from Node A which will send packet 'a' to relay node C, second transmission will be from Node B which will send packet 'b' to relay node C, and third transmission will be from Node C which combines (XOR) the packet and send to both Node A and Node B.Thus the CG for fig.1. is 4/4=1 and the CG for fig.2. is 4/3=1.3, which clearly shows that the performance of a network improves as soon as NC approach is applied as it decreases the no of transmission and thus increases the throughput of the network [2][11].

The presented examples are indeed simple scenarios, imagine a complex scenario like big networks topologies containing hundreds of nodes. Now that situation will be altogether different, here in this situation if a congestion in the network increases the throughput gain of NC techniques will change drastically. We will be discussing the changes in the throughput and solution to overcome this problem in the coming sections. This paper is organized in following manner. In section 2, the paper formulates the problems concern with NC in the context of various wireless topologies. Section 3 concern with the algorithm we designed. Simulation Result is presented in Section 4. Finally paper concludes with conclusion in section 5.

2   Problem Description

NC is undoubtedly a better solution for increasing the throughput in various scenarios and in various topologies. The CG in the Table I. clearly shows the difference when NC is applied and when Without-NC is applied in terms of transmission in various topologies.

Table 1. Comparing various Topologies and CG

Topology	W–$NC\,Trans.$	$NC\,Trans.$	$CG = W$–$NC\,Trans./NC\,Trans.$
Chain	4	3	1.33
X	4	3	1.33
Cross	8	5	1.6

The performance degrades as the no of flows increases. As the no of flows increases the no of packet drops also increases in the case of congestion. When the congestion occurs, in Without-NC case, the no of retransmission and the time taken for the retransmission is less as compared to NC scenario. In Without-NC scheme, if a packet is corrupted or dropped, an negative Ack is sent to relay node in order to send that packet again, whereas in NC scheme if something goes wrong within a coded packet the whole of the coded packet is sent again, which is a burden to the system and which accounts for more of congestion and less of throughput. In the next section we have elaborates the technique used in this paper to overcome this problem.

3   Mathematical Solution

To prove above we have taken an arbitrary directed graph X with capacity 'c' of each edge and tried to prove that MF-MC can eventually help us to determine the maximum flow that a network can have. Having that, we can find out the threshold of a network at which we have to stop NC and switch to Without-NC mode, but along with MF-MC algorithm there are various constrains on each node i.e source, receiver and relay (intermediate) nodes that are to be checked continuously for better throughput. We have presented the following iterative algorithm given below to better gauge the network.

A network in a directed graph (digraph) X=(V, A),with a capacity function c: $A \to R$ assigning arcs to non negative real values. V can be partitioned into three sets: the source U, the sink Y and the intermediates I. Where, U and Y must be non empty. For a network, we have associated a flow f:$V \to R$ assigning arcs to non negative real values such that $0 \leqslant f(a) \leqslant c(a)$ for any $a \in A$ and

$$f_{in}(v) = f_{out}(v) \, for \, all \, v \in I, and \tag{1}$$

$$f_{in}(v) = \sum_{uv \in A} f(uv) \, and \, f_{out}(v) = \sum_{uv \in A} f(uv) \tag{2}$$

In other words, the flow over any arc is no more than its capacity, and the inflow is equal to the outflow on the intermediates. The value of a flow f, denoted val f, is defined as

$$val f = \sum_{x \in X} f_{out}(x) - f_{in}(x) \tag{3}$$

To obtain the maximum value attained by any flow: A cut $(S, \hat{S})$ in a network is the set of arcs $\{s\hat{s} \in A / s \in S, \hat{s} \in \hat{S}\}$ where $X \subseteq S \subseteq V - Y$ and $\hat{S} = V - S$. The capacity of a cut K, denoted cap K, is defined as

$$Cap \, K = \sum_{a \in K} c(a) \tag{4}$$

Finally, for each cut $K = (S, \hat{S})$ we can define an anticut $\check{R} = (\hat{S}, S) = \{\hat{s}s \in A / s \in S, \hat{s} \in \hat{S}\}$.

Step 1: Now we have to find out that the maximum of all flow values (i.e the value of the maximum flow), is equal to the minimum of all cut capacities (i.e, capacity of minimum cut). As S is comprised of sources and intermediates, clearly

$$val f = f_{out}(X) - f_{in}(X) = f_{out}(S) - f_{in}(S) \tag{5}$$

Since intermediate contributes nothing to the flow value. Now consider an arc with both endpoints in S: its flow is counted in both $f_{out}(S) \, and \, f_{in}(S)$, and thus makes no net impact on the flow value. Therefore the only arcs flows which positively impact val f are those originating in S and terminating in $\hat{S}$, which are precisely the flow over the cut K. thus we conclude that

$$Val f \leqslant \sum_{a \in K} f(a) \leqslant \sum_{a \in K} c(a) = Cap \, K \tag{6}$$

Step 2: The next step is somewhat more complicated and here we want to prove that if f^* be the maximum flow and K^* is the minimum cut on the network. Then $val f^* \leqslant cap \, K^*$. Consider a u,v-path as a set of intersecting arcs connecting vertex u to vertex v, taken Non- regard to arc direction. Given u,v-path Q and flow f on a network, let the f-augment of Q, shown as lf(Q), be represented as

$$lf(Q) = min_{a \in Q} \, lf(a) \tag{7}$$

$$lf(a) = \{c(a) - f(a)\}, if\ `a'\ points\ towards\ v. \tag{8}$$

$$lf(a) = \{f(a)\}, if\ `a'\ point\ towards\ u. \tag{9}$$

An x,y-path is f-augmenting iff 'x' is a source, 'y' is a sink ,and the 'f' augment will be positive value. Given 'f' augmented path Q, we can construct a new flow $\hat{f}$ with val $\hat{f}$=val f+ lf(Q) if it is a forward arc, or decrease the flow by lf(Q) if it is a reverse arc.

Step 3: For a network there exist flow f and cut K on the network such that val f= cap K. 'f' be a flow such that there are no f-augmenting paths in the network. Let S be a set of containing the network resources and all vertices v such that there exists an x, v-path from some source x to v with positive f-augment. Note, then, that S contain the network sink, and let K=(S, $\hat{S}$). For the sake of contradiction, we have put $f(a) < c(a)$ for some a∈ K. Let a= s$\hat{s}$, and note that an x, s-path from source x to s with positive f-augment may be extended to an x, $\hat{s}$-path also with positive f-augment. Thus $\hat{s} \in S$, which is a contradiction, that f(a)= c(a) for all $a \in K$. Similarly, $s(\bar{a}) = 0$ for all $\bar{a} \in \hat{K}$. Thus combining step1 and step 3 we came to conclusion that for any network, the value of the maximum flow is equal to the capacity of the minimum cut. Let f^* be the minimum flow and K^* the minimum cut on a given network. By step 1 and step 3 there exist some flow f and cut K such that cap K=$valf \leqslant capK^*$, that no cut can have capacity less than the minimum cut, so in fact $valf = capK^*$. Step 2 states $valf^* \leqslant capK^* = valf$, but no flow can have value greater than the maximum flow, so $valf^* = capK^*$.

Step 4: Now in this step the value of max flow i.e Max. val f= Th (Threshold value of network) i.e the maximum flow of a network is equal to the threshold value of the network. Till this Threshold the NC will be applicable but as soon as the value of val f exceeds the threshold the NC will be stopped and a Without-NC steps will be followed i.e store and forward technique will be followed.

If $(valf \leqslant Th)$
Follow NC;
else
Follow Without-NC;

Here step 1-3 clearly shows that MF-MC can clearly proves to be a logical approach to find out the no of flows upto which NC should be applied, step 4 find out when to switch from NC scenario to Without-NC scenario so that throughput should not decrease. For the same graph X, there are various constrains on each node that are needed to be checked continuously before applying the MF-MC algorithm, for achieving considerable throughput and less fluctuation in throughput when we switch from NC to Without-NC. These constrains are as follows:

Constrain 1: Given a node in a graph such as $X = (V, E, V_{out}, V_{in})$, where 'V' represents the set of nodes, 'E' represents set of edges, V_{out} is the upload capacity of a node and V_{in} is the download capacity of a node. The edge rate vector f(e) is said to be achievable if it satisfy the constrains put on upload and download capacity such that sum of all the edge rate at a node should be less or equal to that of V_{in} entering the node and V_{out} leaving the node i.e

$$\sum_{e \in Out(v)} f(e) \leqslant V_{Out}(v) \tag{10}$$

$$\sum_{e \in In(v)} f(e) \leqslant V_{In}(v) \tag{11}$$

Constrain 2: For source x S there are various corresponding receivers, let it be X_r . Any node that is not the part of source set or receiver set is called relay node or intermediate node and is represented by H_r. We have considered as a non-reducible edge rate vector. Then sum of all outgoing edge rate of relay nodes should be equal to Z times sum of all incoming edge rate of relay nodes, where Z= | X_r | i.e

$$\sum_{h \in H_r} \sum_{e \in Out(h)} f_z^*(e) \leqslant Z. \sum_{h \in H_r} \sum_{e \in In(h)} f_z^*(e) \tag{12}$$

$$\sum_{h \in H_r} \sum_{e \in Out(h)} f_z^*(e) \leqslant Z. \sum_{h \in H_r} min[V_{in}(h), \frac{1}{Z}V_{out}(h)] \tag{13}$$

Table 2. Modified MF-MC algorithm

INPUT: Graph X
BEGIN

1	If Constrain 1: TRUE
2	then Constrain 2
3	else exit;
4	If Constrain 2: TRUE
5	then Constrain 3
6	else exit;
7	If constrain 3: TRUE
8	Check: Step 1 to Step 4
9	If (valf ≤ Th)
10	Follow NC;
11	else
12	Follow Without-NC(Non-NC);
13	else exit;
END	

Modified MF-MC algorithm depicted in Table 2. Shows, step by step manner in which the algorithm is applied over the graph. First of all, at the verge of switching from NC scheme to Without-NC scheme the graph is checked for the constrains i.e whether the graph follow all the properties (10-13) before switching or not, if it follows all the constrains, then only the switching takes place from

NC scheme to Without-NC scheme at 'Th' value. If these properties are not satisfied that will prompt us to slow the flow of packet from that particular route.

4   Simulation Results

To gauge the performance of the network, we have deployed a test-bed in NS-3, considering Cross topology. Let's discuss about the five wireless nodes arranged in Cross-topology as shown in fig.4., here all the four nodes send their data through Node n2. Here the node n2 performs NC schemes based on packet send by node n1 and n3 and then broadcast the combined packets to node n1 and node n3 simultaneously. Similarly, it apply same operation of combining the packet received from node n4 and n5. As the node n5 can hear or guess the packet send by node n4 (packet 'a') and as n1 can hear (guess) the packet send by node n3 (packet 'b'). Based on these guessing the node n5 and n1 can decode the packets broadcasted by node n2 e.g based on hearing node n5 knows what packet is send by node n4 (packet 'a') and it also receives the coded packet send by node n2 (packet ('a' NC 'b')), so it will find out the packet send by n4 (packet 'b') by simply negating it i.e ['a'-('a' NC 'b')] = 'b', so at the end node n5 has both the packets 'a', 'b', similarly node n1 can decode both the packets 'a', 'b'.

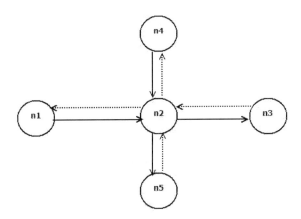

Fig. 3. Cross-Topology

Let see the effects of various NC-schemes on Cross-topology and try to find out the reasons for their changes. Cross-topology Taking the Cross-topology into the consideration for simulation, we have deployed nodes, which are placed in Cross-topology, then we have applied Modified MFMC algorithm shown in Table II. For Cross topology, the threshold value, comes out to be 29 i.e Th=29. After the threshold value, when we have increasing the no of flows, the throughput of the network started decreasing in an unprecedented manner as shown in Fig. 4,

Fig.5 and Fig.6 respectively. It should be noted that, the degree of decrement in the graph after the 'Th' value, not only depends on the type of topology we are using but it also depends on the type of NC scheme that we have used.

4.1   Throughput vs Flow for Cross-Topology Using XOR

Before the threshold has reached i.e at 'Th'=29, there was unprecented increase in the throughput gain of NC-XOR (36.6%), but after the threshold value i.e 'Th' value, throughput value decreased sharply (57.07%) as shown in Fig. 4, the reason lies in inability of NC-XOR to deal with large congestion. But before threshold the NC-XOR scheme out performs the without-NC in terms of throughput, which has throughput gain of (28.8%). So at 'Th'=29 we have applied Modified MF-MC algorithm which has the throughput loss of 4.96% as compared with throughput loss in NC-XOR scheme which turn out to be 57.07% after the threshold value. Thus Modified MF-MC has improved the throughput loss, by switching from NC scheme to without-NC scheme.

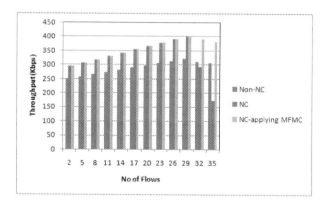

Fig. 4. Throughput Vs No of Flows for XOR scheme in Cross-topolology

4.2   Throughput vs Flow for Cross-Topology Using Linear Network Coding

In the Fig.5, the throughput of NC-LNC surges by 43.17%, before threshold i.e at 'Th'=29 and decreses after the threshold is reached by 59.86%. Cross topology has better guessing capability, so when congestion starts, it has to guess more and more, to find what packets other nodes hold which incurrs more and more time, as a result there is sudden decrease in throughput after the 'Th' value is reached.The second reason which adds to loss of throughput lies in the fact that in LNC, for adding packets various techniques are used such as Galois field theory, once the threshold is reached and the packet starts colliding and receiver starts sending Ack. To generate and send the lost packet this whole process

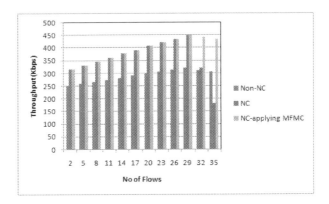

Fig. 5. Throughput Vs No of Flows for LNC scheme in Cross-topolology

of forming packet using Galois theorem starts again which is indeed a time
taking process, which ultimately decreases the throughput of the system drastically. So at 'Th'=29, we have applied the Modified MF-MC algorithm shown in
Table.II.The Modified MF-MC algorithm has throughput loss of 3.32%, as compared with throughput loss in NC-LNC scheme which turn out to be 59.86%
after the threshold value. Thus, Modified MF-MC has considerably improved
the throughput loss, by switching from NC scheme to without-NC scheme.

4.3   Throughput vs Flow for Cross-Topology Using Random Linear Network Coding

As shown in Fig.6, before the threshold is reached i.e Th=29, there is vast difference between the throughput gain achieved by the Without-NC scheme (28.8%)
and throughput gain achieved by NC-RLNC scheme (45.67).The reason for this

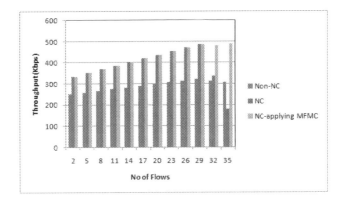

Fig. 6. Throughput Vs No of Flows for RLNC scheme in Cross-topolology

difference lies in the fact that NC-RLNC scheme combines the packets with the random coefficient and at the receivers end this helps in quickly decoding the combined packets efficiently, hence the throughput increases. After the threshold i.e at Th=29, there is sudden decrease in the throughput of NC-RLNC scheme (63.52%), the reason behind this decline in graph lies in the fact that, Galois theorem is applied to generate coefficient randomly and then these coefficient are saved into the header of the combined packet, its indeed a time taking process, so when there is packet loss, the required time to combine packets increases at a drastic rate and hence we can see a drastic decrease in throughput as soon as 'Th' value is reached. So at this point we have applied Modified MF-MC algorithm to improve the throughput. The Modified MF-MC algorithm has throughput loss of 4.09%, as compared with throughput loss in NC-RLNC scheme which turn out to be 63.52% after the threshold value.It can be compared in all the three graphs i.e Fig. 4, Fig. 5, Fig. 6 that, before the 'Th' value, the throughput value of RLNC will be greater than the throughput value of LNC and throughput value of LNC will be greater than throughput value of XOR. But the scenario changes after the threshold value is reached the depreciation of throughput loss value of RLNC will be greater than that of LNC and similarly the depreciation of throughput loss value of LNC is greater than that of XOR. Thus we have attained a comparative increase in throughput in all the three scenarios after we have applied Modified MF-MC.

5   Conclusion

In this paper, we have gauged the efficiency of NC schemes in terms of through-put in Cross topology and come to a conclusion that as the no of flows moves beyond the threshold of Cross topology, the performance in terms of throughput reduce. The sear reason behind this scenario is increases in congestion. Different NC schemes behave differently in Cross topology at the time of congestion. Our result shows that, calculating the threshold of the topology by applying Modi-fied MF-MC algorithm in a network and then applying constrains to check the congestion help the network to determine the threshold before hand and which ultimately helps in switching between NC and without-NC modes .Thus limiting the throughput loss and decreases the packet drop rate in a wireless scenario in a best possible manner.

References

1. Ihsan, J., Iezzi, M., Graziosi, F.: Beyond routing via network coding: an overview of fundamental information theoretic results. In: 21st IEEE International Symposium on Personal Indoor and Mobile Radio Communications, pp. 2745–2750. IEEE Press, New York (2010)
2. Dimakis, A.G., Ramchandran, K., Wu, Y., Suh, C.: A survey on network codes for distributed storage. In: Proc. IEEE, pp. 476–489. IEEE Press (2011)
3. Dougherty, R., Freiling, C., Zeger, K.: Network coding and matroid theory. In: Proc. IEEE, pp. 388–405. IEEE Press (2012)

4. Si, J., Zhuang, B., Cai, A., Li, G.: Unified frameworks for the constructions of variable-rate linear network codes and variable-rate static linear network codes. In: 21st International Symposium on Network Coding, pp. 1–6. NetCod Press (2011)
5. Katti, S., Katabi, D., Hu, W., Medard, M.: The importance of being opportunistic: practical network coding for wireless environments. In: Proc. 43rd Annual Allerton Conference on Communication, pp. 10–17. IEEE Press, New York (2009)
6. Sengupta, S., Rayanchu, S., Banerjee, S.: An analysis of wireless network coding for unicast sessions: the case for coding-aware routing. In: Proc. IEEE INFOCOM, pp. 1028–1036. IEEE Press, New York (2010)
7. Ahlswede, R., Cai, N., Yeung, R.W.: Network information flow Subsequences. IEEE Trans. Inf. Theory **46**, 1204–1216 (2000)
8. Dimakis, A.G., Ramchandran, K., Wu, Y., Suh, C.: A survey on network codes for distributed storage Subsequences. IEEE Proc. Inf. Theory **55**, 2909–2919 (2009)
9. Sanna, M., Izquierdo, E.: A survey of linear network coding and network error correction code constructions and algorithms Subsequences. Int. J. Digt. Mult. Broadcasting **17**, 100–109 (2011)
10. Li, S.Y., Sun, Q.T.: Network coding theory via commutative algebra Subsequences. IEEE Trans. Inf. Theory **57**, 403–415 (2013)
11. Etzion, T., Silberstein, N.: Error-correcting codes in projective spaces via rank-metric codes and Ferrers diagrams Subsequences. IEEE Trans. Inf. Theory **55**, 2909–2919 (2009)

T-Tuple Reallocation: An Algorithm to Create Mixed-Level Covering Arrays to Support Software Test Case Generation

Juliana Marino Balera[✉] and Valdivino Alexandre de Santiago Júnior

Instituto Nacional de Pesquisas Espaciais (INPE), Av. dos Astronautas, 1758
São José dos Campos, São Paulo, Brazil
{juliana.balera,valdivino.santiago}@inpe.br

Abstract. A fact that is known both by researchers and by industry professionals is that exhaustive software testing is impractical. Therefore, one of the most studied activities of software testing process is the generation/selection of test cases. However, selecting test cases that reveal the greatest number of defects within a software is a challenging task, due to the significantly high amount of entries that the system can receive, and even due to the different characteristics of software products in several application domains. This work presents a new algorithm, called T-Tuple Reallocation (TTR), to generate Mixed-Level Covering Array (MCA) which is one of the techniques of combinatorial designs that aims at test case generation. After studying various algorithms/techniques to generate combinatorial designs, starting with pairwise design, TTR was proposed aiming at decreasing the amount of test cases produced to test a software product. The new algorithm was able to create shorter sets of test cases in comparison with classical algorithms/tools proposed in the literature. Although TTR, in general, demanded longer time to generate the sets of test cases, this rise in time can be compensated by a smaller number of test cases so that less time is required for executing them. In the end, this may imply less time for accomplishing the testing process as a whole.

1 Introduction

Exhaustive test case execution is not feasible in practice [11,13]. There is not enough time to execute all test cases if we desire to generate all the possibilities of values of input variables of all functions/methods of an Implementation Under Test (IUT). Thus, generation/selection of test cases comes to address this issue: how to select input values[1] in order to detect the greatest number of defects within the software product?

---

[1] A test case is usually defined as a set of test input data and their respective expected results. However, in the context of combinatorial designs, it is usual to consider a test case only as the test input data generated by the algorithms, being the expected results omitted. Thus, we will take this into account in this work.

© Springer International Publishing Switzerland 2015
O. Gervasi et al. (Eds.): ICCSA 2015, Part IV, LNCS 9158, pp. 503–517, 2015.
DOI: 10.1007/978-3-319-21410-8_39

Software testing community has been used combinatorial designs as a means to generate shorter and more efficient test suites [10]. These techniques have been found to be effective in the discovery of faults (defects) due to the interaction of various input variables. Studies like [6] corroborate this fact where the authors found that a considerable number of software failures are triggered due to the interaction of 6 parameters.

Mixed-Level Covering Array (MCA) is one combinatorial design technique that allows factors (parameters) to assume levels (values) from different sets. MCAs seem to be the most popular choice of combinatorial designs among software testers because they are generally smaller than Mixed-Level Orthogonal Arrays (MOAs) and more adequate for testing [10].

In *t-wise* or *t-way testing*, a set of input data is selected to guarantee a level of interaction, also called the *strength*, between the values of the factors (parameters). Even though there are several studies to generate t-way MCAs [1–4,7,9,12,15], it is interesting to pursue the investigation of new approaches that are able to derive smaller test suites to decrease the amount of time required by the entire software testing process.

In this paper we present a novel algorithm, called T-Tuple Reallocation (TTR), to generate t-way MCAs. After studying various algorithms/techniques to generate combinatorial designs, starting with pairwise design, TTR was proposed aiming at decreasing the amount of test cases produced to test an IUT. The new algorithm was able to create shorter sets of test cases in comparison with classical algorithms/tools proposed in the literature such as the In-Parameter-Order-General-Fast (IPOG-F) [3], the IPO implemented within the TConfig tool [16], and the jenny tool [5]. Although in some cases TTR demanded longer time to generate the sets of test cases, this rise in time can be compensated by a smaller number of test cases so that less time is required for executing them. This may decrease the necessary time to accomplish the testing process as a whole.

This paper is structured as follows. Section 2 provides an overview of combinatorial designs. Section 3 presents in detail our algorithm, TTR. Section 4 shows the application of TTR and compares TTR with some other classical algorithms/tools in the literature. In addition to t-way MCAs, we also present a comparison regarding the generation of pairwise (t = 2) MCAs. Section 5 presents related work. In Section 6, we conclude our work and present future directions to follow.

2   Background

This section presents an overview of combinatorial designs. These are a set of techniques for test case generation that allow the selection of a small set of test cases even when the input domain, and the number of subdomains in its partition, is large and complex [10].

Some basic definitions follow. Consider a program P that takes k inputs corresponding to variables $X_1, X_2, ..., X_k$. These input variables are known as *factors* or *parameters*. Each value assignable to a factor is known as a *level* or simply *value*. A set of levels, one for each factor, is a *factor combination* or *run*.

An Orthogonal Array (OA) is an $N \times k$ matrix in which the entries are from a finite set L of l levels such that any $N \times t$ subarray contains each t-tuple exactly the same number of times. Such an array is denoted by $OA(N, k, l, t)$, where N is the number of runs, k is the number of factors, l is the number of levels, and t is the strength of the OA.

Orthogonal Arrays assume that each factor, f_i, will be assigned a value from the same set of l levels. This is not realistic and thus Mixed-Level Orthogonal Arrays (MOAs) come into picture for situations where factors may be assigned values from different sets. Such an array is denoted by $MOA(N, l_1{}^{k_1} l_2{}^{k_2} ... l_p{}^{k_p}, t)$, meaning N runs where k_1 factors are at l_1 levels, ..., k_p factors are at l_p levels. As before, t is the strength.

Both OA and MOA are examples of balanced designs, i.e. any $N \times t$ subarray contains each t-tuple exactly the same number of times. In software testing, the balance requirement is not always essential. If an IUT has been tested once for a given pair of levels, there is usually no need for testing it again for the same pair unless the IUT is known to behave nondeterministically. For deterministic applications, the balance requirement may be relaxed and unbalanced designs are adequate choices.

A Covering Array (CA) is an example of unbalanced design. A Covering Array $CA(N, k, l, t)$ is an $N \times k$ matrix in which the entries are from a finite set L of l levels such that each $N \times t$ subarray contains each possible t-tuple **at least** a certain number of times. In other words, the number of times the different t-tuples occur in the $N \times t$ subarrays may vary. This difference often leads to combinatorial designs smaller in size than OAs.

A Mixed-Level Covering Array, $MCA(N, l_1{}^{k_1} l_2{}^{k_2} ... l_p{}^{k_p}, t)$, is analogous to an MOA in that both allow factors to assume levels from different sets. MCA is also an instance of unbalanced design. MCAs are generally smaller than MOAs and hence more interesting for testing purposes. Our algorithm, TTR, can generate t-way MCAs.

3   The T-Tuple Reallocation (TTR) Algorithm

TTR allows the generation of combinatorial designs by the t-way testing approach. A high abstract description of TTR is shown in Algorithm 1. The general concept of TTR is as follows. The construction of each row of the solution is made by reallocating t-way tuples from one matrix (α) to other matrix (M) so that each reallocated t-way tuple covers the largest number of the remaining uncovered tuples considering a parameter called a goal. Goals are calculated by the simple combination of strength and value of the number of factors covered by the tuple at certain step of the process (see Section 3.3).

T-way testing is about covering all combinations of factor's levels according to the strength t. Thus, the TTR algorithm follows the following steps: (i) it finds all the possible t-way tuples which have not been covered. The **Constructor**

Algorithm 1. The TTR algorithm

1: $\alpha \leftarrow constructor(f, t)$
2: $M \leftarrow calculateInitialSolution(\alpha)$
3: **while** $\alpha \neq \emptyset$ **do**
4:     $\zeta \leftarrow calculateZeta(M)$
5:     $(M, \alpha) \leftarrow main(M, \alpha)$
6: **end while**

---

procedure thus creates the α matrix; (ii) it generates an initial solution, a matrix M; and (iii) it changes and reallocates t-way tuples aiming at a final solution by means of the **Main** procedure. Then, we have the final set of test cases which is the updated M. These steps are detailed below where we will use the factors and levels presented in Table 1 and $t = 2$.

Table 1. Example of factors and levels

Factor	Level
A	1, 2
B	1, 2
C	1, 2, 3

3.1 The Constructor Procedure

According to the specified input (factors and levels), the Constructor procedure aims to generate all t-way tuples to be covered. Such t-way tuples is represented by a matrix, α^2, whose order is $|C| \times (f + 1)$ where C represents all the $t - way$ combinations of the factors' levels, t is the strength, and f is the number of factors.

Each element of α is a t-way tuple which has not yet been covered. Every $\alpha_{i,j+1}$ element is a value used as a **flag** which aims at helping the process to fit the tuples in the final solution. Figure 1 shows the α matrix for the example presented in Table 1 and $t = 2$. Note that the combinations are due to the levels of the factors "AB", "AC" and "BC". Initially, all flags are false. Algorithm 2 shows the Constructor procedure.

---

[2] In fact, α is implemented as a linked list that decreases. Although the number of columns is already known beforehand $(f + 1)$, the number of rows $(|C|)$ depends on all t-way combinations of factor's levels. During the reallocation process, TTR removes the "rows" of α until it is empty.

A	B	C	Flag
1	1		*false*
1	2		*false*
2	1		*false*
2	2		*false*
1		1	*false*
1		2	*false*
1		3	*false*
2		1	*false*
2		2	*false*
2		3	*false*
	1	1	*false*
	1	2	*false*
	1	3	*false*
	2	1	*false*
	2	2	*false*
	2	3	*false*

Fig. 1. The α matrix

3.2 Initial Solution

The solution is represented by a matrix M of order $ntc \times (f + 1)$ where there are ntc combinations of the factors' levels (i.e. test cases) and f factors[3]. Each element $m_{i,j}$ of M is a level i of a factor j. The $(f + 1)_{th}$ column is not used to represent a factor but rather it is a goal for a given tuple, i.e. every element $m_{i,j+1}$ corresponds to the ζ value associated with a tuple (see Section 3.3). There is an initial solution for the matrix M which is obtained by selecting the combinations of factor's levels C_k ($C_k \subseteq C$) within the α matrix having the largest number of uncovered t-way tuples. All tuples in C_k will be considered in the initial solution, respecting the order of input of factors/levels to TTR as illustrated in Figure 2.

A	B	C	ζ
1		1	2
1		2	2
1		3	2
2		1	2
2		2	2
2		3	2

Fig. 2. The matrix M: initial solution

---

[3] As well as α, M is implemented as a linked list but which may grows. Although the number of columns is already known beforehand $(f + 1)$, the number of rows (ntc) is precisely the number of test cases. Thus, the number of "rows" of M may be greater in the final solution.

Algorithm 2. The Constructor procedure

1: $control \leftarrow false$
2: $Auxiliar[tuples]$
3: $numberLevels \leftarrow 0$
4: **for** $combination\ C\ in\ matrix\ \alpha$ **do**
5:     $control \leftarrow false$
6:     **for** $Factor\ F$ **do**
7:         $numberLevels \leftarrow F.getSizeLevels()$ //$number\ of\ levels\ in\ factor\ F$
8:         **if** $control == false$ **then**
9:             $Auxiliar \leftarrow creatorTuple(F)$ //$create\ an\ tuple\ for\ each\ level\ in\ F$
10:             $control \leftarrow true$
11:         **else**
12:             $size\_tuple \leftarrow auxiliar.size() * numberLevels$
13:             $aux[tuples]$
14:             **for** $tuple\ T\ in\ array\ Auxiliar$ **do**
15:                 **for** $level\ n\ in\ T$ **do**
16:                     $aux.add(l)$
17:                 **end for**
18:             **end for**
19:             $Auxiliar \leftarrow aux$
20:             $cont \leftarrow 0$
21:             **while** $true$ **do**
22:                 **for** $level\ N\ in\ factor\ F$ **do**
23:                     $Auxiliar[cont] \leftarrow N$
24:                     $cont \leftarrow cont + 1$
25:                     **for** $cont == size\_tuple$ **do**
26:                         $stop$
27:                     **end for**
28:                 **end for**
29:                 **for** $cont == size\_tuple$ **do**
30:                     $stop$
31:                 **end for**
32:             **end while**
33:         **end if**
34:     **end for**
35: **end for**

3.3   Goals

The criterion for changing the current solution is the ζ parameter which means a **goal** to be achieved by the solution. The final solution, i.e. the set of test cases, is represented by the matrix M. For each t-way tuple there is an associated goal (ζ).

Considering that the aim is to cover the greatest number of uncovered tuples, ζ is calculated according to the maximum number of uncovered tuples that a certain t-way tuple can address. In order to find ζ we take into account: (i) the disjoint factors covered by the t-way tuple represented by x; (ii) the combinations that have no more tuples to be covered represented by y; (iii) the strength t.

Thus,

$$\zeta = \binom{x}{t} - y.$$

Let us consider Table 1 again and $t = 2$. According to α (Figure 1), the initial solution M (Figure 2) is composed of the tuples from factor combination "AC". This is because from "AB" we have 4 tuples, from "AC" we have 6, and from "BC" there are also 6 tuples as we see in α. As "AC" comes first than "BC", the algorithm picks "AC" as the first factor combination and $|C_k| = 6$.

The amount of disjoint factors, x, is equal to 3. Starting with "AC", the next factor combination that provides the greatest number of uncovered tuples is "BC" (6). Thus, we have all the 3 factors from "AC" and "BC" which explains $x = 3$. As $t = 2$, we have $\binom{3}{2} = 3$. However, one ("AC") out of the three factor combinations has been covered then it is only needed to cover two combinations. Thus, $y = 1$. Therefore, for each t-way tuple in the initial solution M:

$$\zeta = \binom{3}{2} - 1 = 2.$$

This explains the goals (ζ) in Figure 2. It is very important that y is subtracted in order to find ζ. It this is not done, the goal will never be reached because there is no uncovered tuple corresponding to that combination.

Next section we will describe the Main procedure of the TTR algorithm.

3.4   The Main Procedure

The Main procedure takes the initial solution and finds the final one, i.e. the set of combinations of factors' levels which are indeed the definitive set of test cases. Algorithm 3 presents the Main procedure of TTR. After the construction of the α matrix, the initial solution, and the calculation of goals of all t-way tuples, the Main procedure selects the combination with the greatest amount of uncovered tuples (line 3). However, these tuples will not be reallocated from α to the M matrix at once. They will be reallocated gradually, one by one, as the goals are achieved (line 10).

Figure 3 helps understanding the concepts of the Main procedure. All matrices in this figure are snapshots of the matrix M. In part (a) of Figure 3, we have the initial solution. While there exists tuples in α, the Main procedure works. Thus, Main gets from α the highest amount of uncovered tuples. In the example (Table 1), the tuples due to the "BC" factor combination are selected. Every tuple due to "BC" is thus compared with the initial solution M.

When an uncovered tuple "fits" in a tuple already in M in order to define a complete test case, i.e. a row of M with all levels of factors set, this means that by the insertion of the uncovered tuple, the goal for the tuple already in M was reached. Let us consider the first tuple due to "AC" in M, $t_{AC} = (1,1)$, and the first uncovered tuple due to "BC" in α, $t_{BC} = (1,1)$. The insertion of $t_{BC} = (1,1)$ into M is accepted because this reaches $\zeta = 2$. In other words, by the insertion of $t_{BC} = (1,1)$, we have a complete test case $\tau = (1,1,1)$. Thus

Algorithm 3. The Main procedure
```
1: control ← false
2: aux ← 0
3: while !(α.isEmpty()) do
4:      aux ← α.getBigCombination() //combination with highest number of tuples
5:      Aux[ ] ← α.getBigCombination()
6:      while !(Aux[ ].isEmpty()) do
7:          i ← 0
8:          if i == solution.getSize() then              //number of lines in solution
9:              control ← true
10:             for all tuple x in array Aux[ ] do
11:                 if x fit in the test i of solution then
12:                     if solution[i, x.getCombination()].reachesGoals() then
13:                         solution.add(x)
14:                         α.remove(x)
15:                     end if
16:                 else
17:                     if control == true then
18:                         α.markTuple(x) //"mark" the tuple x and keep in matrix α
19:                     end if
20:                     i ← i + 1
21:                 end if
22:             end for
23:         end if
24:     end while
25: end while
```

the other 2 factor combinations ("AB" $\Rightarrow t_{AB} = (1,1)$; "BC" $\Rightarrow t_{BC} = (1,1)$) are covered and the goal is achieved. Part (b) of Figure 3 shows M after these first insertions.

After updating M, the new values of ζ are calculated. Part (c) shows the new values of ζ (see rows 3 and 6). Thus the steps described above are repeated with the insertion/reallocation of tuples in M. Once an uncovered tuple from α is included in M, this t-way tuple is excluded from the α matrix as well as all other tuples in α covered with this (complete) test case. The final set of test cases is the matrix M shown in part (d) of Figure 3.

There is still a possibility that a certain uncovered tuple does not fit in M. Hence, the flag of this row in α is marked as $true$ so that this tuple can be compared with a row of M since Main continues processing while there are uncovered tuples. Figure 4 shows the α matrix after the first interaction. Note that, up to this moment, tuples (1,3) and (2,3) due to "BC" do not fit in any tuple already in M (see values $true$).

This exception related to tuples $t_{BC} = (1,3)$ and $t_{BC} = (2,3)$ occurs because the test cases generated by these tuples, and which are in M, cover tuples already covered in the α matrix ($t_{AC} = (1,1)$ and $t_{AC} = (2,2)$). If we consider the test case created by tuple $t_{BC} = (1,3)$ and row 3 of the M matrix, only one tuple of

(a)

A	B	C	ζ
1		1	2
1		2	2
1		3	2
2		1	2
2		2	2
2		3	2

(b)

A	B	C	ζ
1	1	1	0
1	2	2	0
1		3	2
2	2	1	0
2	1	2	0
2		3	2

(c)

A	B	C	ζ
1	1	1	0
1	2	2	0
1		3	1
2	2	1	0
2	1	2	0
2		3	1

(d)

A	B	C	ζ
1	1	1	0
1	2	2	0
1	1	3	0
2	2	1	0
2	1	2	0
2	2	3	0

Fig. 3. The final solution M: set of test cases

A	B	C	Flag
	1	3	true
	2	3	true

Fig. 4. The α matrix after the first interaction of Main

α $(t_{BC} = (1,3))$ has not already been covered since there are no more tuples for "AC" and "AB". Therefore, the objective for that line ($\zeta = 2$) was not achieved and these tuples can not be found and removed from the α matrix. Thus, we need to recalculate the goal according to a factor combination already covered.

4   Application and Comparison with Other Algorithms

In this section, we present several comparisons between TTR and traditional algorithms/tools for generating test cases based on the creation of combinatorial designs. The algorithms/tools we compared TTR with are:

- **IPO-TConfig**. Version 2.1 of the TConfig tool [16] can generate from 2-way to 6-way MCAs based on IPO. However, it is not evident whether TConfig has implemented IPOG [7], IPOG-F [3] or whether it has a new algorithm for generating MCAs with strength more than 2. Thus, we will call this algorithm IPO-TConfig;

- **Recursive Block Method (RBM)**. It is another algorithm implemented within the TConfig tool. In this case, RBM can generate only 2-way (pairwise) MCAs;
- **IPOG-F**. This refers to our own implementation of the IPOG-F [3] algorithm;
- **jenny tool** [5].

Our results presented in this section were obtained with the following platform: AMD Athlon 64-bit X2 Dual Core Processor 4800+ 2.50 GHz and 3.00 GB RAM, MS Windows 7 Enterprise Operating System.

In Table 2, we compare TTR with IPOG-F and jenny. Note that for both IPOG-F and jenny, we have an interval (a lower and upper limit). This means that IPOG-F and jenny are sensitive to the order in which the factors and levels are input to them. Let us consider the first instance in Table 2: $2^23^24^2$ with $t = 2$. Thus, a test designer may enter factors and levels according to the following orders:

- 2 2 3 3 4 4 (factor 1 → 2 levels, factor 2 → 2 levels, factor 3 → 3 levels, factor 4 → 3 levels, factor 5 → 4 levels, factor 6 → 4 levels),

- 2 3 2 3 4 4 (factor 1 → 2 levels, factor 2 → 3 levels, factor 3 → 2 levels, factor 4 → 3 levels, factor 5 → 4 levels, factor 6 → 4 levels),

- 2 3 3 2 4 4 (factor 1 → 2 levels, factor 2 → 3 levels, factor 3 → 3 levels, factor 4 → 2 levels, factor 5 → 4 levels, factor 6 → 4 levels),

- 4 3 2 3 2 4 (factor 1 → 4 levels, factor 2 → 3 levels, factor 3 → 2 levels, factor 4 → 3 levels, factor 5 → 2 levels, factor 6 → 4 levels),

- 4 4 3 2 3 2 (factor 1 → 4 levels, factor 2 → 4 levels, factor 3 → 3 levels, factor 4 → 2 levels, factor 5 → 3 levels, factor 6 → 2 levels).

In all these cases, TTR generates exactly 16 test cases. However, IPOG-F generates 20, 19, 17, 24, and 18, respectively. On the other hand, the jenny tool creates 18, 18, 17, 20, and 16 test cases. This feature of TTR is very important because it is not interesting that the same instance generates different sizes of test suites in accordance with the order in which factors and levels are input. The test designer does not know beforehand which is the order he/she shall enter factors/levels and it is not interesting, from the practical point of view, to demand that he/she knows this. The test designer only desires to realize what are the test cases generated by the combinatorial design algorithm. In Table 2, we run 5 times the same instance in order to fill the table.

In Table 3, we added IPO-TConfig. Note that TTR generated equal or smaller set of test cases in almost all instances we used comparing it with the IPO-TConfig. There is even two instances, $2^23^24^25^2$ with $t = 5$ and $t = 6$ (Table 3),

Table 2. Comparison between TTR, IPOG-F and jenny: strengths from 2 to 6

Strength	Instance	IPOG-F	Jenny	TTR
2	$2^2 3^2 4^2$	$17 \vdash\dashv 24$	$16 \vdash\dashv 20$	16
3	$2^2 3^2 4^2$	$55 \vdash\dashv 62$	$57 \vdash\dashv 60$	51
4	$2^2 3^2 4^2$	$153 \vdash\dashv 175$	$156 \vdash\dashv 160$	146
5	$2^2 3^2 4^2$	$288 \vdash\dashv 306$	$326 \vdash\dashv 338$	288
6	$2^2 3^2 4^2$	576	576	576
2	$2^2 3^2 4^2 5^2$	$27 \vdash\dashv 29$	$29 \vdash\dashv 32$	25
3	$2^2 3^2 4^2 5^2$	$129 \vdash\dashv 136$	$129 \vdash\dashv 135$	109
4	$2^2 3^2 4^2 5^2$	$479 \vdash\dashv 522$	$488 \vdash\dashv 501$	445
5	$2^2 3^2 4^2 5^2$	-	$1537 \vdash\dashv 1586$	1377
2	$2^2 3^2 4^2 5^2 6^2$	$40 \vdash\dashv 47$	$44 \vdash\dashv 47$	39
3	$2^2 3^2 4^2 5^2 6^2$	$234 \vdash\dashv 263$	$243 \vdash\dashv 258$	217
2	$2^2 3^2 4^2 5^2 6^2 7^2$	$58 \vdash\dashv 62$	$59 \vdash\dashv 62$	56
3	$2^2 3^2 4^2 5^2 6^2 7^2$	$415 \vdash\dashv 464$	$419 \vdash\dashv 428$	372
2	$2^2 3^2 4^2 5^2 6^2 7^2 8^2$	$80 \vdash\dashv 84$	$79 \vdash\dashv 82$	75
2	$2^2 3^2 4^2 5^2 6^2 7^2 8^2 9^2$	$101 \vdash\dashv 113$	$100 \vdash\dashv 111$	100
2	$2^2 3^2 4^2 5^2 6^2 7^2 8^2 9^2 10^2$	$127 \vdash\dashv 140$	$128 \vdash\dashv 137$	134

that IPO-TConfig did not finish to run. This shows the potential of our algorithm to be more scalable especially for critical and more complex applications. Comparing with jenny, TTR had equal but in many instances much more better performance in terms of shorter test suites sizes.

Table 3. Comparison between TTR, IPOG-F, IPO-TConfig and jenny: strengths from 2 to 6

Strength	Instance	IPOG-F	IPO-TConfig	jenny	TTR
2	$2^2 3^2 4^2$	27	20	18	16
3	$2^2 3^2 4^2$	59	61	60	51
4	$2^2 3^2 4^2$	169	175	157	146
5	$2^2 3^2 4^2$	304	304	327	288
6	$2^2 3^2 4^2$	576	576	576	576
2	$2^2 3^2 4^2 5^2$	55	29	31	25
3	$2^2 3^2 4^2 5^2$	163	136	132	109
4	$2^2 3^2 4^2 5^2$	585	522	488	445
5	$2^2 3^2 4^2 5^2$	1785	-	1586	1377
6	$2^2 3^2 4^2 5^2$	4541	-	4133	3694
2	$2^2 3^2 4^2 5^2 6^2$	85	46	45	39
3	$2^2 3^2 4^2 5^2 6^2$	321	263	243	217
2	$2^2 3^2 4^2 5^2 6^2 7^2$	135	60	59	56
3	$2^2 3^2 4^2 5^2 6^2 7^2$	321	263	419	372
2	$2^2 3^2 4^2 5^2 6^2 7^2 8^2$	175	84	82	75
2	$2^2 3^2 4^2 5^2 6^2 7^2 8^2 9^2$	217	113	102	100
2	$2^2 3^2 4^2 5^2 6^2 7^2 8^2 9^2 10^2$	290	140	128	134
2	$2^2 3^2 4^2 5^2 6^2 7^2 8^2 9^2 10^2 11^2$	263	168	160	185

We also accomplished a comparison considering only pairwise testing. In this case, we could compare TTR not only with IPOG-F and IPO-TConfig but also with RBM implemented within TConfig. Table 4 shows the results. There is no instance in which RBM is better than TTR: our algorithm is equal or superior to RBM.

Table 4. Comparing TTR with IPOG-F, IPO-TConfig, RBM: strength = 2 (pairwise testing)

Strength	Instance	IPOG-F	IPO-TConfig	RBM	TTR
2	$3^1 4^1 5^2 6^1 7^1 8^1 9^1$	103	74	72	72
2	$4^1 5^1 6^1 8^2 9^1$	110	79	72	72
2	$2^1 3^1 4^1 6^1$	24	24	24	24
2	$4^1 5^2 6^1 7^1$	55	48	49	43
2	$2^1 4^1 6^1$	24	24	24	24
2	$2^2 3^2 4^2$	27	20	24	16
2	$2^2 3^2 4^2 5^2$	55	29	45	25
2	$2^2 3^2 4^2 5^2 6^2$	85	46	78	39
2	$2^2 3^2 4^2 5^2 6^2 7^2$	135	60	91	56
2	$2^2 3^2 4^2 5^2 6^2 7^2 8^2$	175	84	120	75
2	$2^2 3^2 4^2 5^2 6^2 7^2 8^2 9^2$	217	113	153	100
2	$2^2 3^2 4^2 5^2 6^2 7^2 8^2 9^2 10^2$	290	140	230	134
2	$2^2 3^2 4^2 5^2 6^2 7^2 8^2 9^2 10^2 11^2$	263	168	231	185

The results presented in Tables 2, 3, 4 demonstrated to us the feasibility of our approach by creating shorter sets of test cases. This is important in the current state of the practice because shorter test suites means, in general, less time to execute them.

In Table 5, we compare the time to generate the test suites between TTR, jenny and IPOG-F. The jenny tool had a better performance in relation to the time to generate the test suites. One possible explanation for this fact, although can not be the only reason, is because jenny was developed in C language and TTR has been developed in Java. However, as already pointed out, TTR generated less test cases in all the instances.

IPOG-F has already had better performance than TTR. The explanation about this fact is because IPOG-F changes every test case (row) during its processing while TTR only changes such a test case if certain criterion is reached. But, again in Table 5, TTR generated shorter test suites in all instances. Thus, this issue can be compensated due to the shorter sets of test cases. As we have just pointed out, less test cases means, in general, less time to execute them. Although we have tools and frameworks that allow automated test case execution, in many situations it is necessary the testing team to analyze the report of the test cases especially for system and accepting testing. More test cases means greater reports to analyze.

Table 5. Comparing TTR with jenny and IPOG-F: time to generate the test suites. ms = millisecond

Strength	Instance	jenny	Time[ms]	IPOG-F	Time[ms]	TTR	Time[ms]
2	$2^2 3^2 4^2$	18	2	27	165	16	109
3	$2^2 3^2 4^2$	60	6	59	347	51	438
4	$2^2 3^2 4^2$	157	24	169	486	146	1187
5	$2^2 3^2 4^2$	327	57	304	889	146	953
6	$2^2 3^2 4^2$	576	81	576	265	576	47
2	$2^2 3^2 4^2 5^2$	31	4	55	157	25	328
2	$2^2 3^2 4^2 5^2 6^2$	45	10	85	406	39	969
2	$2^2 3^2 4^2 5^2 6^2 7^2$	59	23	135	545	56	987

5   Related Work

In this section, we present some important studies related to our research. The Tabu Search Algorithm Approach (TSA) makes use of the meta-heuristics Tabu Search (TS). TS has been applied to many problems in the field of Artificial Intelligence. This meta-heuristic provides a solution to a problem by means of neighborhood functions: (i) a routine for setting an initial solution; (ii) the size of the Tabu list; (iii) the use of a mixture of neighborhood functions; (iv) a validation function; (v) a fine-tuning process, the selection of probabilities for each neighborhood function based on a complete test set of discrete probabilities. Like TSA, our algorithm also has an initial solution. But instead of working with neighborhood functions, we use the concept of a goal (ζ) to be achieved in order to change the initial solution and derive the final set of test cases.

In-Parameter-Order (IPO) [8] is one of the most popular pairwise (2-way) test case generation within the combinatorial designs field. In summary, it works as follows. In a system with two or more factors, IPO strategy generates the pairwise testing for the first two factors, extending the set of test cases generated by a set of pairwise testing for the first three factors, and so forth for each additional parameter by means of two steps: (i) the horizontal growth, which consists in adding one more element in the test case; (ii) the vertical growth which adds one more test case in all tests.

Other algorithms have been proposed as an extension of IPO such as the In-Parameter-Order-General (IPOG) [7] and the In-Parameter-Order-General-Fast (IPOG-F) [3]. IPOG-F is an adaptation of the IPO strategy for t-way testing. As IPO strategy, it builds the initial solution by means of horizontal and vertical growths, however, rather than performing comparisons, IPOG-F algorithm makes use of two auxiliary matrices which help in choosing the best value for the modified solution.

The difference between our algorithm, TTR, and the algorithms based on IPO is that TTR does not use auxiliary matrices and defines a more judicious criterion to extend/change the test cases than the IPO-based ones. The IPO-based algorithms alter all elements of a solution corresponding to one column (factor) and they use auxiliary matrices to do so. Thus, TTR demands less memory to run. This can explain why in some instances, IPOG-F (Table 2) and IPO-TConfig (Table 3; see remark below) were not able to finish.

There are several free/open source and commercial tools available for generating MCAs. TConfig [16] is one of such as tools. It has an IPO implementation for 2-way to 6-way[4] testing and also the RBM algorithm. In our implementation of TTR, we can generate t-way MCAs and, as seen in Section 4, TTR generated equal or smaller set of test cases in almost all instances we used comparing it with the IPO-TConfig. There is even two instances, $2^2 3^2 4^2 5^2$ with $t = 5$ and $t = 6$ (Table 3), that IPO-TConfig did not finish to run even that we let the tool running for 2 days. In relation to the RBM algorithm implemented in TConfig, TTR had equal or better performance considering all instances (Table 4).

jenny [5] is a tool initially conceived for regression testing but its reasoning lies with the combinatorial design purpose for test case generation: try to avoid exhaustive testing. Our algorithm TTR had equal but in many instances much more better performance than jenny with respect to the size of the test suites.

6   Conclusions

In this paper, we presented a new contribution to the exhaustive software testing problem. Our new algorithm, TTR, is able to generate t-way MCAs for combinatorial design testing. Different from other classical algorithms proposed in the literature, like the ones based on IPO, TTR does not make use of auxiliary matrices and has a more rigorous criterion to change the test cases. Thus, in comparison with such algorithms, TTR demands less memory to run.

TTR outperformed classical algorithms/tools proposed in the literature such as IPOG-F [3], IPO and RBM implemented within the TConfig tool [16], and the jenny tool [5] with respect the size of the test suites created. For real projects, this is something interesting because less test cases to run means, in general, less time demanded by the entire software testing process. In addition, TTR is case insensitive to the order in which factors/levels are input. This is another advantage over other previous approaches because a test designer must not have to be concerned with the order to input data for a combinatorial designs algorithm.

Future directions follow. We must try to improve the time performance of TTR. In order to do this, we will create a new version of the algorithm with parallel implementation (threads). We must also make an algorithm complexity analysis in order to compare TTR with others algorithms presented in the literature. We will apply TTR to generate test cases for software embedded into computer of space systems such as balloon applications [14].

Acknowledgments. We would like to thank *Financiadora de Estudos e Projetos* (FINEP), project number 01.10.0233.00, *Coordenação de Aperfeiçoamento de Pessoal de Nível Superior (Capes)*, and *Fundação de Amparo à Pesquisa do Estado de São Paulo* (FAPESP), process number 2012/23767-2, to support this research.

---

[4] See remark about IPO and TConfig on Section 4.

References

1. Bracho-Rios, J., Torres-Jimenez, J., Rodriguez-Tello, E.: A new backtracking algorithm for constructing binary covering arrays of variable strength. In: Aguirre, A.H., Borja, R.M., Garciá, C.A.R. (eds.) MICAI 2009. LNCS, vol. 5845, pp. 397–407. Springer, Heidelberg (2009)
2. Cohen, M.B., Gibbons, P.B., Mugridge, W.B., Colbourn, C.J.: Constructing test suites for interaction testing. In: Proceedings of the 25th International Conference on Software Engineering, May 3–10, 2003, Portland, Oregon, USA, pp. 38–48 (2003)
3. Forbes, M., Lawrence, J., Lei, Y., Kacker, R.N., Kuhn, D.R.: Refining the in-parameter-order strategy for constructing covering arrays. Journal of Research of the National Institute of Standards and Technology **113**(5), 287–297 (2008)
4. Gonzalez-Hernandez, L., Rangel-Valdez, N., Torres-Jimenez, J.: Construction of mixed covering arrays of variable strength using a tabu search approach. In: Wu, W., Daescu, O. (eds.) COCOA 2010, Part I. LNCS, vol. 6508, pp. 51–64. Springer, Heidelberg (2010)
5. Jenkins, B.: jenny: a pairwise testing tool (2005). http://burtleburtle.net/bob/math/jenny.html. (accessed February 20, 2015)
6. Kuhn, D.R., Wallace, D.R., Gallo, A.M.: Software fault interactions and implications for software testing. IEEE Trans. Software Eng. **30**(6), 418–421 (2004)
7. Lei, Y., Kacker, R., Kuhn, D.R., Okun, V., Lawrence, J.: IPOG: a general strategy for t-way software testing. In: Proceedings of the Annual IEEE International Conference and Workshops on the Engineering of Computer-Based Systems (ECBS), pp. 549–556. IEEE Computer Society, Washington, DC (2007)
8. Lei, Y., Tai, K.-C.: In-parameter-order: a test generation strategy for pairwise testing. In: Proceedings of the IEEE International Symposium on High-Assurance Systems Engineering (HASE), pp. 254–261. IEEE Computer Society, Washington, DC (1998)
9. Lopez-Escogido, D., Torres-Jimenez, J., Rodriguez-Tello, E., Rangel-Valdez, N.: Strength two covering arrays construction using a SAT representation. In: Gelbukh, A., Morales, E.F. (eds.) MICAI 2008. LNCS (LNAI), vol. 5317, pp. 44–53. Springer, Heidelberg (2008)
10. Mathur, A.P.: Foundations of software testing, 689 p. Dorling Kindersley (India), Pearson Education in South Asia, Delhi, India (2008)
11. Myers, G.J.: The art of software testing, 2nd edn., 234 p. John Wiley & Sons, Hoboken (2004)
12. Nurmela, K.J.: Upper bounds for covering arrays by tabu search. Discrete Applied Mathematics **138**, 143–152 (2004)
13. Santiago, V., Silva, W.P., Vijaykumar, N.L.: Shortening test case execution time for embedded software. In: Proceedings of the 2nd IEEE International Conference SSIRI, pp. 81–88 (2008)
14. Santiago Júnior, V.A., Vijaykumar, N.L.: Generating model-based test cases from natural language requirements for space application software. Software Quality Journal **20**(1), 77–143 (2012). doi:10.1007/s11219-011-9155-6
15. Shiba, T., Tsuchiya, T., Kikuno, T.: Using artificial life techniques to generatetest cases for combinatorial testing. In: Proceedings of 28th Annual International Computer Software and Applications Conference (COMPSAC 2004), pp. 1–6 (2004)
16. Williams, A.W.: Determination of test configurations for pair-wise interaction coverage. In: Ural, H., Probert, R.L., Bochmann, G.V. (eds.) Testing of Communicating Systems. IFIP AICT, pp. 59–74. Springer, Heidelberg (2000)

An Analysis of Techniques and Tools for Requirements Elicitation in Model-Driven Web Engineering Methods

José Alfonso Aguilar[1]([⊠]), Aníbal Zaldívar-Colado[1], Carolina Tripp-Barba[1], Sanjay Misra[2], Roberto Bernal[3], and Abraham Ocegueda[4]

[1] Señales y Sistemas (SESIS) Facultad de Informática Mazatlán,
Universidad Autónoma de Sinaloa, Culiacán, México
{ja.aguilar,azaldivar,ctripp}@uas.edu.mx
[2] Department of Computer and Information Sciences, Covenant University,
Ota, Nigeria
ssopam@gmail.com
[3] Facultad de Informática Culiacán, Universidad Autónoma de Sinaloa,
Culiacán, México
roberto.bernal@uas.edu.mx
[4] ProTech I+D S.A. de C.V., Mazatlán, México
ing.ocegueda@protechid.com

Abstract. Until now, is well-known that Requirements Engineering (RE) is one of the critical factors for success software. In current literature we can find several reasons of this affirmation. One particular phase, which is vital for developing any new software application is the Requirements Elicitation, is spite of this, most of the development of new software fail because of wrong elicitation phase. Several proposals exist for Requirements Elicitation in Software Engineering, but in the current software development market is focusing on the development of Web and mobile applications, specially using Model-Driven methods, that's the reason why we asume that it is necessary to know the Elicitation techniques applied in Model-Driven Web Engineering. To do this, we selected the most representative methods such as NDT, UWE and WebML. We have reviewed 189 publications from ACM, IEEE, Science Direct, DBLP and World Wide Web. Publications from the RE literature were analyzed by means of the strict consideration of the current techniques for Requirements Elicitation.

Keywords: Model-driven web engineering · Requirements · Elicitation

1 Introduction

Recently, it has emphasized the success that Model-Driven Development (MDD) have had for Web application development, the so called Model-Driven Web Engineering (MDWE), this is mainly because with the use of models it is possible

© Springer International Publishing Switzerland 2015
O. Gervasi et al. (Eds.): ICCSA 2015, Part IV, LNCS 9158, pp. 518–527, 2015.
DOI: 10.1007/978-3-319-21410-8_40

to represent (modeling) the user needs (goals) without neglecting the organizational objectives, the software architecture and the business process and from this representation generate the Web application source code. Several MDWE methods [1] have been emerged for the development of Web applications using models to do it, but only some of them strictly complied with the proposal of the Object Management Group (OMG) for Model- Driven Development named Model-Driven Architecture (MDA) [2]. The basic idea of the use of MDA starts from the Computational Independent Model (CIM), in this first level, the application requirements must be elicited and defined, such that we can generate through model-to-model transformations (M2M) the Platform Independent Model (PIM) to finish in the Platform Specific Model (PSM) with the source code. Regrettably, most of the MDWE methods only implements MDA from PIM to PSM leaving aside the requirements phase (CIM level) despite this is a critical phase on which the succes of the development depends directly [3]. This fact can bee seen in their support tools i.e., code generation tools and modeling tools.

Bearing these considerations in mind, in this work is presented a review regarding the current state of the implementation of the Requirements Elicitation techniques in the most well-known MDWE methods according to [1], these are NDT [4], UWE [5] and WebML [6]. Moreover, the tools used or developed for support this phase on each method is analyzed. Requirements Elicitation is a technique to collect the requirements, this is the main movement in the Requirements Engineering process [7]. It try to find out the needs and collecting the appropiate software requirements from the stakeholder's. Therefore, this is a complex process as it decides which techniques to be used in a project, determining, learning, acquiring, discovering the appropriate techniques, so this may negatively influence the development process which results in system failures. Importantly, we do not pretend to stablish a full critic approach since Requirements Engineering is more complex and is sub-divided in another phases like analysis, management and so on. But we emphasize on Requirements Elicitation techniques, because it is the starting point in any development process, and MDWE methods should adopt or develop new techniques for its appliance in Web application development in order to avoid problems like the system may be delivered late, it will be may be more costly than the original estimation, end-user and customer will not be satisfied, system may be unreliable and there may be regular system defects [7].

The rest of the paper is organized as follows: Section 2 presents Requirements Elicitation techniques used for the analysis detailed in this work. Section 3 describes our analysis of MDWE methods and its Requirements Elicitation techniques. Finally, the conclusion and future work is presented in Section 4.

2   Requirements Elicitation Techniques

Requirements Elicitation is used to discover what problems need to be solved [3], and to identify the stakeholders, the objectives that the software system must attain, the tasks that users currently perform and those that they might wish to perform. This phase is often carried out through the application of various techniques, named direct and indirect approach. Direct approach classifies the methods by whom we interact with the domain expert. In direct approach the purpose is to enhance the understanding of the problems of system that is currently in used. Most common techniques used are interviews, case study, prototyping. With these tools a comprehensive analysis of total procedure can been done. Indirect approach are used in order to obtain information that cannot be easily articulated directly. Questioners, documents analysis are its examples. Important thing in this approach is, how things are clarify by using figures and statistics. In it a large quantity of data can be gathering from an alyzing the documents. The results acquire from this type of investigation are easy to measure and an applicable test suggestion can be driven from them [7]. Next are described the techniques used to analyze the MDWE methods selected on this work:

Interviews. The interviews are probably the most traditional and commonly used technique for requirements elicitation due to interviews are essentially human based social activities, they are inherently informal and their effectiveness depends greatly on the quality of interaction between the participants. Interviews provide an efficient way to collect large amounts of data quickly. The results of interviews, such as the usefulness of the information gathered, can vary significantly depending on the skill of the interviewer [8]. There are fundamentally three types of interviews: unstructured, structured, and semi-structured. The latter generally representing a combination of the former two.

Questionnaires. These are mainly used during the early stages of requirements elicitation and may consist of open and/or closed questions. For them to be effective, the terms, concepts, and boundaries of the domain must be well established and understood by the participants and questionnaire designer. Questions must be focused to avoid gathering large amounts of redundant and irrelevant information. They provide an efficient way to collect information from multiple stakeholders quickly, but are limited in the depth of knowledge they are able to elicit.

Observation. In observation methods, the knowledge engineer observes the expert performing a task. This prevents the knowledge engineer from inadvertently interfering in the process, but does not provide any insight into why decisions are made [7].

Prototyping. This technique has been used for elicitation where there is a great deal of uncertainty about the requirements, or where early feedback from stakeholders is required. Actually this is the process to build the model about the system, prototypes help the system designers to build the information system according the requirements and easy to manipulate for end users. Even so, this is an iterative process and it is also part of the analysis phase of system development life cycle [7].

Brainstorming. This is a process where participants from different stakeholder groups engage in informal discussion to rapidly generate as many ideas as possible without focusing on any one in particular. It is important when conducting this type of group work to avoid exploring or critiquing ideas in great detail. It is not usually the intended purpose of brainstorming sessions to resolve major issues or make key decisions. This technique is often used to develop the preliminary mission statement for the project and target system. One of the advantages in using brainstorming is that it promotes freethinking and expression, and allows the discovery of new and innovative solutions to existing problems [8].

3   The Analysis

The purpose of this paper is to provide state of the art regarding Requirements Elicitation techniques used by MDWE methods as well as the tool support their offer. The method we use is described as follows: i) Definition of Research Questions (RQ's), ii) Definition of Search Sources, iii) Elaboration of Search String by means of the keywords extracted from the RQ's, iv) Filtering of the Documents Found and v) The Analysis.

Definition of Research Questions

In general, this study aimed to answer the following RQ's:

1. Which Requirements Elicitation techniques are currently supported for Model-Driven Web Engineering methods?
2. Which methods provide tool support for Requirements Elicitation techniques?

Definition of Search Sources

With regard to the selection of search sources, in our work we used direct search in CONRICYT[1], ACM[2], IEEE[3], Science Direct[4], digital library of scientific literature: DBLP[5] and Google Scholar[6], used in order to obtain grey literature such as Thesis, Technical Reports, etc. In addition, we look into some of the most representative conferences such as ICWE (International Conference on Web Engineering), International Conference on Web Information System Engineering (WISE) and RE Conference (Requirements Engineering Conference). In accordance with the work of Brereton [9], these libraries were chosen because they are some of the most relevant sources in Software Engineering.

---

[1] http://conricyt.mx
[2] http://portal.acm.org
[3] http://ieeexplore.ieee.org
[4] http://sciencedirect.com
[5] http://informatik.uni-trier.de
[6] http://scholar.google.com

Elaboration of Search String

The structure of the RQ's were used as a basis to extract some keywords, which were then used to search for documents. We initially had the following keywords: Web, engineering, requirements, elicitation, MDWE, method and tool. However, in order to obtain more concrete and specific results in the field, we decided to link Web with the keywords engineering and requirements, requirements with the keyword engineering, and the keyword Web with the keywords engineering and methods. In this respect, the choice of concatenating Web with engineering was motivated by our goal, which was to retrieve papers specifically focused on Requirements Elicitation techniques applied in MDWE. Finally, we elaborated three search strings which were used in the search sources previously defined.

Filtering of the Documents Found

The procedure (inclusion and exclusion criteria) for the examination of the documents found was defined as follows: first of all the title and abstract of the documents were read, only those that we considered relevant just in the case if some of the terms used in the search strings appeared in the title or the abstract. Then, the introduction and conclusion to those primary studies dealing with specific Requirements Elicitation issues were read, e.g. those papers concerning elicitation applied in a WE method. Finally, the whole document was then read in order to discover those documents that use Requirements Elicitation techniques (or some of them) along with whether they had any kind of tool support. At the end, only 16 papers were fully-readed.

The Analysis

This section presents and analyses the results obtained after performing the review. Features (derived from RQ's) such as Requirements Engineering techniques and tool support are described for the NDT, UWE and WebML methods. These methods were selected since there are those approaches that have been constantly being improved by research groups and software development enterprises. Moreover, there have been several applications of those methods in real-world projects.

NDT covers the whole life cycle process in the development of Web applications, including testing or quality assurance phases. In the NDT method, the RE process consists of three phases (capture, definition, and validation) by means of use case diagrams and a set of textual templates [10]. For Requirements Elicitation, this approach uses Interviews and Questionnaries. The tool support for the Requirements Engineering [11] is covered by NDT-Suite [12]. NDT hampers the development of a complex Web application, since templates are difficult to complete as they require intensive interviews [13]. Furthermore, the requirements are difficult to maintain in NDT owing to the use of textual templates for their specification.

UWE is a method entirely based on UML [14]. It covers the whole life cycle of the development of Web applications and RIA [7] development too. UWE is based on MDA [15], therefore, requirements are considered in the CIM in order to show what the Web application is expected to do without showing details of how it is implemented. This method proposes the use of interviews, questionnaires and checklists as Requirements Elicitation techniques [13]. A plugin called MagicUWE was developed to be used with the CASE tool MagicDraw. Unfortunately, requirements elicitation support is missing in this tool.

WebML is a visual language for specifying the content structure of a Web application and the organization and presentation of such content in a hypertext [16]. Requirements are specified by using UML use case and activity diagrams. Regarding to Requirements Elicitation techniques, the authors states that this approach uses Interviews. However, in the last years, WebML has worked on implementing the semantics of the Business Process Modeling Notation (BPMN) through Model-Driven Web application generation, the approach transforms BPMN models into Web application models specified according to the WebML notation and then into running Web application [17]. WebML has tool support: WebRatio [18] and WebRatio BPM. The support that WebRatio offers to WebML is mainly focused on providing facilities for the automatic generation of J2EE code.

Discussion

UWE and NDT are those methods which have placed greater importance on Requirements Engineering by defining a set of formal guidelines to be used. In both approaches, the requirements are considered since the early stages of the Web application development process, thus making the development and maintenance of Web applications easier, whilst fulfilling the project budget.

The techniques applied to each approach in the Requirements Elicitation can be seen in Table 1, these are are Interviews, Questionnaires, Observation, Prototyping and Brainstorming. There is a trend towards the use interviews, since this technique is used by WebML, NDT and UWE for Requirements Elicitation.

Table 1. Search results by source

Method/Tech.	Interviews	Questionnaires	Observation	Prototyping	Brainstorming
NDT	+	-	-	-	+
UWE	+	+	-	+	-
WebML	+	-	-	-	-

The techniques presented in this section have advantages and disadvantages, i. e., the use of interviews generates a lot of text, so, in a complex Web application

---

[7] Rich Internet Aplications

development process is a disadvantage because it is difficult to maintain. This technique could, nevertheless, be extremely useful and comprehensible in the development of simple Web applications.

Another important issue that we have seen in this study is the fact that these Model-Driven methods are starting to offer support for RIAs (Rich Internet Applications) which are Web applications emulating desktop software behavior. Moreover, with regard to requirements specification for RIAs some of this methods can easly offer this support due to the use of UML for requirements, for instance UWE and NDT, both of them can easy extends UML to cover special issues regarding RIAs or extending its metamodel. In [19], the authors presents a solution for the treatment of Web Requirements in RIA for NDT method. For this aim they present WebRE+, a requirement metamodel that incorporates RIA features into NDT modelling capabilities. WebML offer limited support for RIAS by means of its Model-Driven tool named WebRatio which actually includes already support for basic RIAs and AJAX behavior. The tool can also be used together with RUX-Tool [20] if very advanced and complex dynamic pages are needed. Furthermore, WebML now evolved into IFML (Interaction Flow Modeling Language)[8], which is since March 2015 an OMG standard. IFML is a more comprehensive language that will natively include custom event management and rich interaction (not bound only to web interfaces). In this respect, IFML can have a great aceptation within the community because in the methods presented in this work is not common the use of standar notation for modelling the behavior of Web applications interfaces.

On the other hand, the dissemination of the different methods is a highly important issue since it assists in the realization of important advances in the standardization of Web application development. The normalization of the way in which requirements (the words/terms used to denominate the Web requirements) are denominated by each method it is necessary because all disciplines need a mutual terminology, which is required to allow researchers to understand and cooperate with each other, thus providing the basis needed to improve the research area. This will help to the methods be well known in both the academic world and the software industry. In this respect, it is worth mentioning the support offered by NDT, UWE and WebML through their websites because all of them provide everyone who visits their websites with examples, published papers and their respective tools, with the exception of WSDM, which only offers the downloading of published papers because the tools license is not free. In the particular cases of NDT, UWE and WebML, they provide guided step-by-step examples with which to study and practice the development of Web systems using their respective support tools. This confirms why these approaches are those most frequently used in the development of Web applications in academic and industry.

We have seen that different techniques and approaches have different and relative strengths and weaknesses, and may be more or less suited to particular types of situations and environments. Likewise some techniques and approaches

---

[8] www.ifml.org

are more appropriate for specific elicitation activities and the types of information that needs to be acquired during those activities. With regard to tool support, all the methods described in the previous section have a tool support for its methodology. NDT its supported by NDT-Suite, WebML is supported by WebRatio and UWE by MagicUWE. With regard to the Requirements Elicitation support tool, only NDT and UWE have it. NDT does this using the NDT-Suite, UWE provides a tool that consists of a MagicDraw plugin.

4   Conclusions and Future Work

Requirements Elicitation is the first stage of Requirements Engineering that requires the use of certain techniques. Selecting the right techniques has a pertinent influence on the quality of a software system. According to [21], the success of this stage does not depend entirely on the selection of techniques, because, due to the heterogeneity of stakeholders, this process must be carefully handled by effectively applying the appropriate techniques towards the people. This paper has presented a review about the MDWE methods that offers Requirements Elicitation support, we analized its techniques and tools. Elicitation techniques (interviews, scenarios, prototypes, etc.) were investigated, followed by representations, models and support tools. The review results suggest that elicitation techniques appear to be relatively mature, although new areas of creative requirements are emerging in order to improve current MDWE methods.

A conclusion that comes out of the analysis performed is that the use of common models and languages is recommended due to this fact will improve the use of the same terminology for requirements, because a topic we found is the different terminolgoy used by the MDWE methods to name their types of requirements. Therefore, if standard concepts are promoted in Web requirements we can stablish a universal forms for modeling requirements. In this respect, although studies have been conducted regarding the benefits of MDD in a development process [22], few refer to the WE domain, among which are the works presented in [23] and [24], thats the reason why it is necessary to conduct more studies with regard to MDD in WE through empirical studies in order to validate and support the potential application of these methods.

Finally, the use of current Web interface techniques for modelling the behavior of the Web application is not common among the methods analyzed. This leads to possible research lines in which mechanisms to represent the behavior are studied to find a form to stadarize the its representation, for instance with the new standar IFML (Interaction Flow Modelling Language).

For future work, we are collaborating with ProTech I+D, which is a software factory, in order to develop a tool for support Requirements Elicitation during its development process. The basic idea is that the tool must be compatible with the WebML approach due to this is the one we detected that does not have a well defined Requirements Engineering phase, basically is only focused on its development tool Webratio (used for Web application development and the letaest version for mobile application development), thus, we think that we can integrate our tool for Requirements Elicitation support withs its approach for MDWE.

Acknowledgments. This work has been partially supported by: Universidad Autónoma de Sinaloa (México) by means of Programa Integral de Fortalecimiento Institucional (PIFI 2014) from Facultad de Informática Culiacán. Special thanks to ProTech I+D for its support and vinculation with our research group.

References

1. Aguilar, J.A., Garrigós, I., Mazón, J.N., Trujillo, J.: Web engineering approaches for requirement analysis- a systematic literature review. In: Web Information Systems and Technologies (WEBIST), vol. 2, pp. 187–190. SciTePress Digital Library, Valencia (2010)
2. Brown, A.: Model driven architecture: Principles and practice. Software and Systems Modeling **3**(4), 314–327 (2004)
3. Nuseibeh, B., Easterbrook, S.: Requirements engineering: a roadmap. In: Proceedings of the Conference on The Future of Software Engineering, ICSE 2000, pp. 35–46. ACM, New York (2000)
4. García-García, J.A., Escalona, M.J., Ravel, E., Rossi, G., Urbieta, M.: NDT-merge: a future tool for conciliating software requirements in mde environments. In: Proceedings of the 14th International Conference on Information Integration and Web-based Applications & Services, IIWAS 2012, pp. 177–186. ACM, New York (2012)
5. Koch, N., Kraus, A., Hennicker, R.: The authoring process of the UML-based web engineering approach. In: First International Workshop on Web-Oriented Software Technology (2001)
6. Brambilla, M., Fraternali, P.: Large-scale model-driven engineering of web user interaction: The WebML and WEBRATIO experience. Science of Computer Programming **89**, 71–87 (2014)
7. Khan, S., Dulloo, A.B., Verma, M.: Systematic review of requirement elicitation techniques. International Journal of Inform ation and Computation Technology **4** (2014)
8. Agarwal, R., Tanniru, M.R.: Knowledge acquisition using structured interviewing: An empirical investigation. J. Manage. Inf. Syst. **7**(1), 123–140 (1990)
9. Ceri, S., Brambilla, M., Fraternali, P.: The history of webml lessons learned from 10 years of model-driven development of web applications. In: Borgida, A.T., Chaudhri, V.K., Giorgini, P., Yu, E.S. (eds.) Mylopoulos Festschrift 2009. LNCS, vol. 5600, pp. 273–292. Springer, Heidelberg (2009)
10. Escalona, M.J., Aragón, G.: Ndt. a model-driven approach for web requirements. IEEE Transactions on Software Engineering **34**(3), 377–390 (2008)
11. Escalona, M.J., Mejías, M., Torres, J.: Developing systems with ndt & ndt-tool. In: 13th International Conference on Information Systems Development: Methods snd Tools, Theory and Practice, Vilna, Lithuania, pp. 149–59 (2004)
12. García-García, J.A., Alba Ortega, M., García-Borgoñon, L., Escalona, M.J.: NDT-suite: a model-based suite for the application of NDT. In: Brambilla, M., Tokuda, T., Tolksdorf, R. (eds.) ICWE 2012. LNCS, vol. 7387, pp. 469–472. Springer, Heidelberg (2012)

13. Escalona, M.J., Koch, N.: Requirements engineering for web applications - a comparative study. J. Web Eng. **2**(3), 193–212 (2004)
14. Koch, N., Kraus, A.: The expressive power of UML-based web engineering. In: Second International Workshop on Web-oriented Software Technology (IWWOST 2002), vol. 16. CYTED (2002)
15. Koch, N.: Transformation techniques in the model-driven development process of UWE. In: Workshop Proceedings of the Sixth International Conference on Web Engineering, p. 3. ACM (2006)
16. Ceri, S., Fraternali, P., Bongio, A.: Web Modeling Language (WebML): a modeling language for designing Web sites. Computer Networks **33**(1), 137–157 (2000)
17. Brambilla, M., Butti, S., Fraternali, P.: WebRatio BPM: a tool for designing and deploying business processes on the web. In: Benatallah, B., Casati, F., Kappel, G., Rossi, G. (eds.) ICWE 2010. LNCS, vol. 6189, pp. 415–429. Springer, Heidelberg (2010)
18. Brambilla, M., Fraternali, P.: Implementing the semantics of BPMN through model-driven web application generation. In: Dijkman, R., Hofstetter, J., Koehler, J. (eds.) BPMN 2011. LNBIP, vol. 95, pp. 124–129. Springer, Heidelberg (2011)
19. Robles Luna, E., Escalona, M.J., Rossi, G.: Modelling the requirements of rich internet applications in webre. In: Cordeiro, J., Virvou, M., Shishkov, B. (eds.) ICSOFT 2010. CCIS, vol. 170, pp. 27–41. Springer, Heidelberg (2013)
20. Linaje, M., Preciado, J.C., Morales-Chaparro, R., Rodríguez-Echeverría, R., Sánchez-Figueroa, F.: Automatic generation of rias using rux-tool and webratio. In: Gaedke, M., Grossniklaus, M., Díaz, O. (eds.) ICWE 2009. LNCS, vol. 5648, pp. 501–504. Springer, Heidelberg (2009)
21. Mishra, D., Mishra, A., Yazici, A.: Successful requirement elicitation by combining requirement engineering techniques. In: First International Conference on the Applications of Digital Information and Web Technologies, ICADIWT 2008, pp. 258–263, August 2008
22. MartíNez, Y., Cachero, C., Meliá, S.: Mdd vs. traditional software development: A practitioner's subjective perspective. Inf. Softw. Technol. **55**(2), 189–200 (2013)
23. Valderas, P., Pelechano, V.: A survey of requirements specification in model-driven development of web applications. ACM Trans. Web 5(2), 10:1–10:51 (2011)
24. Insfran, E., Fernandez, A.: A systematic review of usability evaluation in web development. In: Hartmann, S., Zhou, X., Kirchberg, M. (eds.) WISE 2008. LNCS, vol. 5176, pp. 81–91. Springer, Heidelberg (2008)

Deriving UML Logical Architectures of Traceability Business Processes Based on a GS1 Standard

Rui Neiva[1], Nuno Santos[2(✉)], José C.C. Martins[1], and Ricardo J. Machado[1,2]

[1] Centro de Investigação ALGORITMI, Universidade do Minho, Guimarães, Portugal
[2] CCG - Centro de Computação Gráfica, Campus de Azurém, Guimarães, Portugal
nuno.santos@ccg.pt

Abstract. A good traceability business process (BP) regards a powerful tool for industrial and manufacturing organizations to pursuit effective productivity. However, there is a lack of a common understanding among its key stakeholders on implementing a proper traceability BP. In this paper, we propose the use of software engineering approaches, namely the design of a process-level logical architecture for the traceability BP. This logical architecture captures the main activities, responsibilities, boundaries, and services involved in the traceability BP. The logical architecture was derived based on a use case model that arose from the requirements elicitation of activities as the ones proposed by the GS1 standard.

Keywords: Traceability · Business process · Industrial · Logical architecture · Use case model

1 Introduction

Manufacturing organizations increasingly strive to maximize productivity and reduce costs, which means producing more, faster, more efficiently and with higher quality, in order to increase profits. Traceability business processes (BP) [1-3] play an important role on assisting manufacturing organizations improving their performances regarding those concerns. One of the major goals of traceability BP is to ensure the persistence of relevant information related to the main activities of an organization [4]. Thus, traceability BP allows locating and recalling defective products throughout the supply chain. The fast recall of these products helps reducing the potential negative economic impact of their defects, and preserving consumers' trust on the quality of their favorite brands, as well as their confidence in the systems that are designed to protect their safety [5]. The recalling of defective products is as more important as the physical integrity of their users is com-promised. The contribution of traceability BP to more efficient production processes lies in identifying stages in the production process that cause defects on the products [1-3]. Such identification supports continuous improvement of the production process, responding to demands for more efficiency [4].

This research is sponsored by the Portugal Incentive System for Research and Technological Development PEst-UID/CEC/00319/2013 and by project in co-promotion nº 36265/2013 (Project HMIExcel - 2013-2015).

O. Gervasi et al. (Eds.): ICCSA 2015, Part IV, LNCS 9158, pp. 528–543, 2015.
DOI: 10.1007/978-3-319-21410-8_41

In logistics, traceability data can be used to optimize routes (of traceable items – shipments like truckloads, vessels, certain amounts of pallets of various items, logistic units like pallets or containers, or trade items, which are further on defined in this paper [5], and to improve planning and management, primarily due to improved links with other organizations with which there is collaboration [4]. In accounting, this data can be used for inventory, or by monitoring applications to identify process inefficiencies [4].

According to [5], obtaining an integrated and holistic view of traceability requires all stakeholders to systematically connect the physical flow of traceable items (the physical flow may be a truck with bulk goods or a truck containing big bags of seeds [6] with the flow of information about them (traceability data). Such view is best attained through a common process language.

The GS1 global standard for traceability [6] is thus the starting point for this research and was used as input for the execution of a requirements elicitation approach. The use of the global standard allows that the elicited activities encompass all phases and activities of the traceability BP, where it is assumed that every standard results from consensus and validation from the entities regarding the given industry and that key requirements are addressed.

The main purpose of this research is to design a non-fragmented traceability BP view for software modeling in the industrial context of full supply chain traceability, which means that this view suits the purpose of supporting further development of full supply chain traceability process automation (software) tools. This view also includes the functional requirements of a global traceability BP for full supply chain traceability, which allows contributing with an understanding on the meaning of traceability process concepts Martins and Machado referred in [4] and the common process language mentioned in [5]. This common understanding (*i.e.*, process language) is obtained by creating and transforming a model that captures functional requirements a general traceability process for full industrial supply chain traceability shall meet to satisfy its user needs.

This paper is structured as follows: section two addresses some related work and describes one of the main traceability standards for the industrial context; section three identifies functional requirements of the traceability BP for the industrial context and presents their modeling in a UML use cases diagram; section four illustrates the transformation of standard-based business process requirements for full supply chain traceability into a logical architecture of the traceability BP; and section five presents the conclusions of this research.

2   Related Work

In general, traceability can be understood as the ability to trace the history, application or location of what is under consideration [7]. In manufacture, traceability regards the process to track a product batch and its history throughout the whole or a part of a production chain from raw material to consumer (chain traceability), or internally in one of the steps (non-continuous operations performed at a particular location) in the chain (internal traceability) [8]. Traceability allows: (i) identify the source of all materials and parts of a product (materials, intermediate and finished products) [6]; (ii) processing the entire history of the life cycle of the product [8] ; and (iii) the distribution and location of the product after delivery [7, 8]. Thus, traceability must be ensured throughout all stages of the supply chain, like transport, storage, processing, distribution and sales [8]. According to [8], traceability is only achieved through the unique identification of the products. Traceable items can be located at places of production, handling, storage and sale, each of which represents a traceable item location [5]. Different types of traceability can be found in the literature, as for example: forward and backward traceability, active and passive traceability [9], vertical and horizontal traceability [10], and internal and external traceability [6]. External traceability is particularly relevant for this research.

In [11] it is specified a mapping technique and an algorithm for mapping business process models, using UML activity diagrams, and use cases, so functional requirements specifications support the enterprise's business process. In [12] business process models are derived from object-oriented models. KAOS, a goal-oriented requirement specification method, provides a specification that can be used in order to obtain architecture requirements [13]. On other perspective, Tropos [14] uses notions of actor, goal and (actor) dependency as a foundation to model early and late requirements, architectural and detailed design, and [15] models business goals and derives system requirements, but it outputs a UML state chart. In [16], a methodology is proposed where a business process reference models is used in the first phase of software development. Then, the business model is modified in order to fulfill customer's and organization's needs, and ends with an implemented software system.

The Four-Step-Rule-Set (4SRS) method [17] allows the transformation of user product requirements into an architectural model representation. The method is traditionally applied in a product-level perspective including variability, recursive mechanisms and class diagrams generation. In its process-level perspective, the method is traditionally applied for creating context for product design in ill-defined contexts and specifying participants in service-oriented contexts. In this research, we issue logical architectures for a representation of standard-based activities instead of using them in context of product design. Thus, we use the same rationale as in [17] but the use case model used as input are standard-based activities rather than user process requirements.

Terzi [18, 19] address the interoperability and lack of knowledge problem. Knowledge problem was tackled through the "development of a reference meta model for product traceability", focused on product data structure requirements. He also pursued the interoperability along the diverse enterprise applications, in particular at a manufacturing stage, for managing the product traceability along the entire product life cycle. Jansen-Vullers [9] relies on case studies to create a traceability reference model. As a result an approach to design IS solutions for traceability is proposed. Starting with a model of the goods flow. This model depicts the processes that transforms manufacturing products, through a sequence of operations. The model is then translated into a reference data model that is the basis for designing an information system for tracking and tracing. The intention of Gampl's study is to approach a typology for traceability systems. To achieve that, he surveyed traceability systems, on German food industry, and create a detailed description for traceability systems in terms of dimensions and aspects [20]. Panetto [21] maps the "IEC 62264 standard models to a particular view of Zachman framework in order to make the framework concrete as a guideline for applying the standard and for providing the key players in information systems design with a methodology to use the standard for traceability purposes". On this study Panetto was able to rely on a standard to obtain a abstract model of traceability. Sobrinho on his research modeled an information system aimed at the maintenance of traceability data in the Brazilian wine industry, according to the principles of a service-oriented architectures [22].

The research in [23-25] describes domain-specific implementations of traceability processes in healthcare and in food industry, respectively. Manikas in [26] presents an activity-oriented web application for supply chain traceability for agricultural industries that relies on the Traceability Data Pool (TDP) model. In [27], a framework based on event-driven process chains (EPCs) methodology, the entity-relationship model (ERM) and activity-based costing (ABC) is used to define and analyze the current state of a supply chain and design a future system to develop business process reengineering for a supply chain of fourth range vegetable products.

3   Modeling Traceability with Use Cases

This research was divided in three main phases. This research began by identifying the key requirements of the global traceability BP. The requirements elicitation was based on the GS1 global traceability standard [6] and on literature review. Initially, we used the set of subprocesses from the GS1 global traceability standard previously described. However, the standard is highly generic and does not provide detail about some of the main activities of a generic traceability process. Thus, requirements elicitation was also supported by the domain knowledge and strengthened by the professional experience of the authors namely from the automotive industry. In the end of this phase, the requirements of the traceability BP were gathered and represented in the form of use cases, using the Unified Modeling Language (UML) [28]. UML use cases are mandatory input for the 4SRS method. The 4SRS method has proven successful in the design of Information Systems to represent the requirements for process execution, thus chosen to be executed to the standard-based logical architecture design. The activities were organized by their detail level where, as in [29], functional refinement of use cases was the used strategy to provide different levels of abstraction to modeling (in the case of the present research, the traceability BP modeling). This strategy allows adding requirements incrementally and facilitates their interpretation.

The first phase of our approach was to model the GS1 global traceability standard's subprocesses and steps by using use cases diagrams. This model allows to develop a first draft of the standard subprocesses and steps' representation. Additionally, another main contribution regards the identification of insufficient information to develop an adequate and complete use case model. The identified use cases are depicted in Table 1. These use cases can be used as the starting point for the development of any use case model.

It must be pointed out that in [6] one of the presented subprocesses is called *Recall Product*. In early versions of the global standard, this subprocess was included in the *Use Information* ({UC5}) subprocess. At the time of this research, *Recall Product* was not a subprocess that the interviewed stakeholders wished to address because their concerns regarded other subprocesses. For this reason, *Recall Product* subprocess was not included in the model. However, our proposed approach may easily include *Recall Product* subprocess within the use case model and use it in the logical architecture derivation in future works.

Table 1. Identified use cases from the GS1 global traceability standard

Standard Subprocess	Use Cases with Full Information	Use Cases with Insufficient Information
Plan and Organize	— Manage traceability data — Link traceability phases	— Create traceability plan
Align Master Data	— Identify Traceability Partner — Identify Physical Locations — Identify Assets — Identify Trace Items — Exchange Master Data	— Assign Traceability Partner Data — Assign Physical Locations Data — Assign Assets Data — Assign Trace Items Data — Send Master Data Request — Respond to Master Data Request
Record Traceability Data	— Apply the identification to the identification carrier — Capture the identification of the traceable item — Collect traceability data — Store traceability data	— Assign identification to traceable item — Collect traceability data — Share traceability data
Request Trace	(no use cases were direcly used)	— Send a trace request — Reply to a trace request
Use Information	(no use cases were direcly used)	— Monitor the traceability process
Recall Product	(was not included yet in the approach)	(was not included yet in the approach)

Based in the insufficient information from subprocesses and steps that were identified, new inputs are required that are directly related to the real industrial case study where the approach was adopted. We present the applicability of the approach in a real industrial case study, namely in automotive industry. Thus, it is presented a use case model in Fig. 1 that represents the applicability of our approach to the automotive industry (and ultimately to the organization that regards the presented real industrial case). This use case model captures the functional requirements of all phases and main activities of the global traceability BP. This model also relates these phases with the corresponding stakeholders.

The first-level use cases model (sometimes called level zero) of the traceability BP directly corresponds to the 5 sub-processes proposed by the GS1 global traceability standard (see Fig. 1). All these use cases were decomposed in order to organize them hierarchically in activities at the same level of abstraction. Each use case from this model was textually described as well.

If, for this level of abstraction, modeling of use cases was performed directly from the global standard subprocesses, refining these use cases was not possible to derive directly from standard subprocesses and steps. The refinement was initially performed as directly as proposed by the global standard, but the information from the subprocesses and steps soon was insufficient and too generic to enable a proper use case refinement. The global standard was directly modeled in the first-level use cases, but its steps were able to be directly modeled in second- and in third-level use cases. Thus, there was not a direct correspondence between global standard steps and the lower-level use cases from the model. The use case model mainly included actors from higher hierarchy, and actors from lower hierarchy were only assigned to use cases originated from refinements where their specialization to the given activity was identified. So, such insufficiency was compensated by some literature. This was not however enough, and the next step was to incorporate workflows from the industrial context where this research was conducted, namely the electronics and automotive industry.

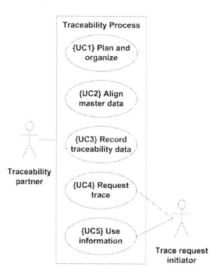

Fig. 1. GS1 global traceability standard subprocesses

{UC1} Plan and Organize

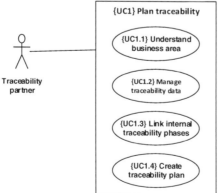

Fig. 2. *{UC1} Plan traceability* use case model

The refinement of this use case gave origin to use cases {UC1.1}, {UC1.2}, {UC1.3} and {UC1.4} (see *Fig. 2*) and all of them were also refined (except for {UC1.3}), *i.e.*, a third-level of abstraction was modeled. Use case {UC1.1} *Understand business area* (included additional lower-level use cases regarding its functional decomposition) was completely modeled and specified based on interviews with some stakeholders from the specific industrial context and literature review, which is to say, no input at all from the global traceability BP subprocess. Some of the use cases from the decomposition of {UC1.2} *Manage traceability data* and {UC1.3} *Link traceability phases* were modeled entirely based on the steps from the subprocess and others from interviews and literature review inputs, which is to say that overall {UC1.2} was partially based on the standard. Use case {UC1.4} *Create traceability plan* was based on the standard, but its specification was strengthened with interviews and literature.

{UC2} Align Master Data

The refinement of this use case gave origin to use cases {UC2.1}, {UC2.2}, {UC2.3}, {UC2.4} and {UC2.5} (see Fig. 3) and all of them were also refined, *i.e.*, a third-level of abstraction was modeled. Use cases {UC2.1}, {UC2.2}, {UC2.3}, {UC2.4} and {UC2.5} were entirely based on the standard subprocess, but refinements from all of them were partially or not at all specified based on the standard steps, thus interviews and literature were input for all of third-level use cases.

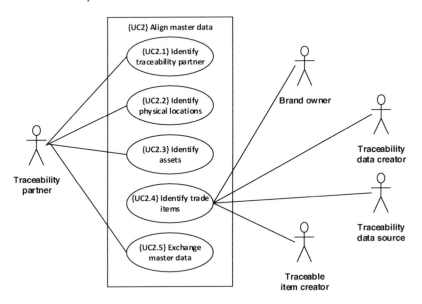

Fig. 3. *{UC2} Align master data* use case model

{UC3} Record Traceability Data

The refinement of this use case gave origin to use cases {UC3.1}, {UC3.2}, {UC3.3}, {UC3.4}, {UC3.5} and {UC3.6} (see Fig. 4) and only {UC3.4} *Collect traceability data* required additional refinements. Use cases {UC3.2} *Apply the identification to the identification carrier,* {UC3.3} *Capture the identification of the traceable item,* {UC3.4} *Collect traceability data* and {UC3.6} *Store traceability data* were entirely based on the standards subprocesses. Use cases {UC3.1} *Assign identification to traceable item* and {UC3.5} *Share traceability data* were based on the standard subprocesses, but its specification was strengthened with interviews and literature. This was the case (but regarding standard steps) for all the use cases from the refinement of {UC3.4} *Collect traceability data* as well. In conclusion, in {UC3} all the use cases were at least partially based on the global standard.

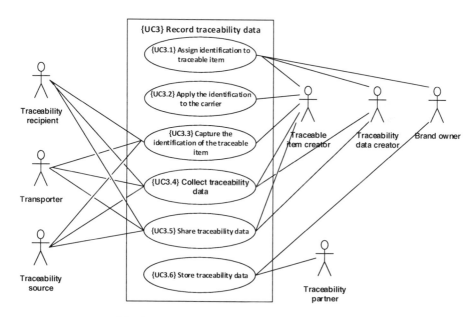

Fig. 4. {UC3} *Record traceability data* use case model

{UC4} Request trace

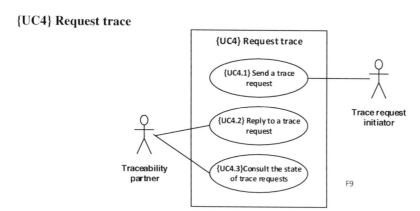

Fig. 5. {UC4} *Request trace* use case model

The refinement of this use case gave origin to use cases {UC4.1}, {UC4.2} and {UC4.3} (see Fig. 5), where in this case no additional refinements were required. Use cases {UC4.1} *Send a trace request* and {UC4.2} *Reply to a trace request* were based on the aggregation of two standard steps, but its specification was strengthened with interviews and literature. This aggregation of steps implied changes on the use case actor, since the assigned role had to be analyzed for addressing both use cases that were aggregated. Use case {UC4.3} *Consult the state of trace requests* was completely modeled and specified based on interviews with some stakeholders from the specific industrial context and their explicit identified need for monitoring traces, which is to say, no input at all from the global standard.

{UC5} Use information
The refinement of this use case gave origin to use cases {UC5.1} and {UC5.2} (see Fig. 6) where in this case no additional refinements were required. Use case {UC5.1} *Monitor the traceability process* was partially based on the standard subprocess and steps, where its specification was strengthened with interviews and literature. Use case {UC5.2} *Consult trace requests with reply* was completely modeled and specified based on interviews with some stakeholders from the specific industrial context and literature review, which is to say, no input at all from the global standard.

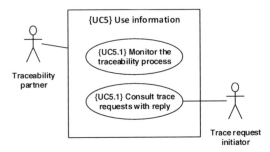

Fig. 6. {UC5} *Use information* use case model

The overall use case model is composed by 64 use cases. Within these use cases, the model included 49 *leaf* use cases and they were used as input in the 4SRS execution. In summary, 18 of the elicited use cases were directly obtained from the standard, 22 were partially obtained from the standard (i.e., with additional inputs from other sources like literature and stakeholder's needs) and 24 were obtained only from literature and meetings with stakeholders. Additionally, it is easily depicted that the standard (in this case, GS1) is too generic to be modeled in use cases and its use is input mainly within higher-level use cases from the model.

4 Deriving a Logical Architecture for Traceability

In this section is presented the logical architecture derived from the use case model from the real case study in industry presented in the previous section. A transformation method, the process-level 4SRS [17], is used in order to derive a logical architecture for the traceability BP. The 4SRS method receives as input a set of functionally decomposed use cases (also called *leaf* use cases) that describe the process requirements. These use cases are refined through successive iterations of the 4SRS method and in the end a logical architecture of that process is obtained. Consider an AE (Architectural Element) as the elementary piece of a logical architecture (a logical architecture is also composed of associations between the AEs and packages, nevertheless associations cannot exist without AEs and packages are not expressive unless they contain AEs). Each AE should have a well-defined set of responsibilities, well defined boundaries and a well-defined set of interfaces, which define the services the architectural element provides to the remaining AEs [30].

The 4SRS is composed of 4 steps (with 10 micro-steps), briefly described as follows [17]: **Step 1 – Architectural Element Creation:** In this step, three types of AEs (interface – *i-type*, control – *c-type*, and data – *d-type*) are created for each use case; **Step 2 – Architectural Element Elimination:** In this step, it is decided which AEs created for each use case are maintained or eliminated, taking into account the entire system, This step is divided into eight micro-steps, *Micro-step 2i – Use Case Classification*, *Micro-step 2ii – Local Elimination*, *Micro-Step 2iii – Architectural Element Naming*, *Micro-step 2iv – Architectural Element Description*, *Micro-step 2v – Architectural Element Representation*, *Micro-step 2vi – Global Elimination*, Micro-step 2vii – Architectural Element Renaming, *Micro-step 2viii – Architectural Element Specification;* **Step 3 – Packaging and Aggregation;** The AEs that were maintained after the execution of step 2 should give origin to packages (or aggregations) semantically consistent; **Step 4 – Architectural Element Association** This last step supports the introduction of associations between AEs. This step is divided into two micro-steps (optional): *Micro-step 4i – Direct Associations* and *Micro-step 4ii – Use Case Model Associations.* The main output of these four steps is a logical architecture, which captures the main responsibilities and boundaries of a traceability process. This model provides a non-fragmented view of the main activities that make up the traceability BP. The specification of traceability process requirements and the logical architecture of this process are described in detail in the following sections.

The execution of the 4SRS over the 49 *leaf use cases* from this real case study resulted in a logical architecture composed of 79 AEs and 390 associations between them. The logical architecture of the standard-based global traceability BP for industrial supply chains comprises a set of abstractions supporting the functional requirements of that process or the activities that comprise that process. AEs were aggregated according to their semantics, which means a package aggregates AEs that cooperate for the execution of the same activity from the traceability BP. For instance, the AEs related to the activity represented by the use case {UC3.4} *Collect traceability data* are contained by the package with the same name ({P4} *Collect traceability data*). If the use cases in Fig. 1 (representing the GS1 global traceability standard subprocesses) correspond to the highest abstraction level of business process functional requirements, the use cases in Figs. 6 (representing the GS1 global traceability standard activities) through 10 correspond to the second highest abstraction level.

Overall, we obtained 41 AEs of *i-type*, 19 AEs of *c-type*, and 19 AEs of *d-type*, which implies that this process is extremely dynamic. Such was expected from the nature and purpose of traceability itself. An interesting fact of this result is that 19 AEs are of *d-type*, and therefore have no computer support. This means that there are plenty of traceability key activities that depend on human decision.

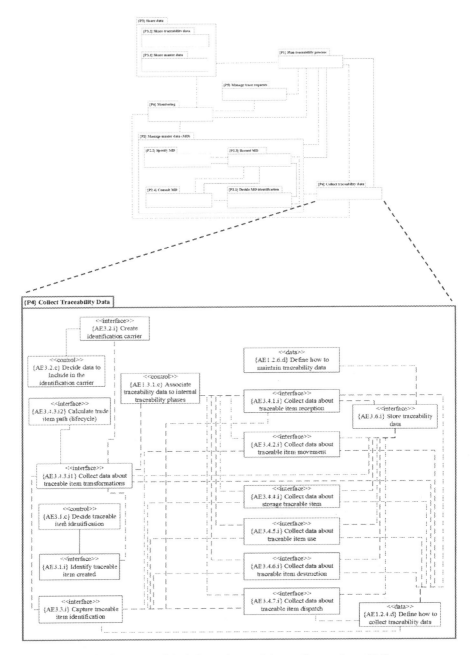

Fig. 7. Subset of the information model regarding package {P4}

The set of AEs contained by each package determines the detailed scope of package {P4} *Collect traceability data*, represented by means of the contained AEs. Hereupon the execution of the 4SRS allows identifying the responsibilities of high abstraction level activities (in this case the activity of collecting traceability data, previously represented by the use case *{UC3.4} Collect traceability data*) and their association with other traceability process activities and subprocesses. For instance, in package {P4}, these responsibilities are: (i) the decision about the traceable item identification; (ii) the identification carrier creation; (iii) the capture of the traceable item identification; (iv) the decision on how maintain the traceability data, among others.

The package {P4} *Collect traceability data* is perhaps the most complex of this architecture, although all are irreplaceable. It consists of 17 AEs. Besides the number of AEs, the complexity of this package lies in the significant number of associations between other packages, which increases their dependency of this package on the proper functioning of the outer AEs. Referring to the premise for the clustering of AEs into packages, the execution of the 4SRS method managed to conceptualize a big picture of the group of activities, and thus identify some challenges related to this abstraction. In the particular example of package {P4} *Collect traceability data*, those challenges are: (i) the need for interoperability between data collection mechanisms; (ii) the need for orchestrating traceability data collection at all stages of the BP; and (iii) the need for determining the life-cycle of the trade item in real-time.

As mentioned in the description of the 4SRS, there are two types of associations between two AEs: direct association and use case model association. The first type (solid line) indicates that two AEs are originated from the same use case. The second type (dashed line) indicates that two AEs are somehow related. These relations are detailed in the textual description of use cases that originated them. Considering the example of package {P4} (see Fig. 7) it is possible to clearly identify the roles of the different associations. An example of these associations is depicted in Fig. 8.

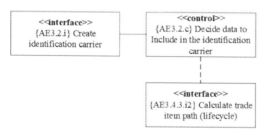

Fig. 8. Direct associations and use case associations

Considering the example presented in Fig. 8, there is an activity deciding which traceability data to include on the identification carrier of the traceable item (performed in {AE3.2.c} *Decide data to Include in the identification carrier*). Additionally, there is an activity regarding the creation of the respective identification carrier of that item (performed in {AE3.2.i} *Create identification carrier*). Both activities originated in the same use case. Another required activity regards collecting data about the transformation of a traceable item (performed in {AE3.4.3.i2} *Calculate trade item path (lifecycle)*). The latter has an association with {AE3.2.i}, requiring the result of the activity of {AE3.2.i} to perform its activity. All associations within the architecture should be thoroughly analyzed, because the existence of anomalies in one AE may imply anomalies in the AE associated with it. Therefore, all identified dependencies must be assured in of the traceability BP implementation, guaranteeing communication between the various parts of the process.

In conclusion, the proposed process modeling approach allows identifying the traceability process key elements. Through the creation of AEs, it is possible to capture a well-defined set of responsibilities, the boundaries of each element of the process, and a set of well-defined interfaces that define the services that this AE provides to other AEs. Thus, it is possible to promote interoperability between the traceability BP subprocesses, including internal and external traceability. This is essential to get a holistic view of the traceability BP. Without such view, any attempt to implement the BP is likely to fail.

5   Conclusions

This paper highlights the potential of traceability to support the sustainability of industrial organizations. Currently, the traceability process implementation presents many challenges particularly in regard to the common understanding among key stakeholders. To address this problem a logical architecture that supports the functional requirements of the traceability BP was proposed. This architecture is the main contribution of this research. This architecture resulted from the application of 4SRS, which is a new and systematic approach in the area of Industrial Engineering. On the one hand, it is aligned with the GS1 global traceability standard. On the other hand, it captures the traceability BP main activities allowing either an alignment among stakeholders about the main activities either a common vocabulary for understanding the scope and the traceability responsibilities. In this research work, an approach was proposed to improve the overall vision and knowledge about the traceability BP in an industrial environment. The literature review (in particular, the works of [4, 5, 31]) allowed realizing that many of the gaps and constraints of the traceability BP reside in the lack of a common understanding among its key stakeholders. Facing such perception, this work proposed the development of a logical architecture that captures the main activities, responsibilities, boundaries, and services involved in the traceability BP. The 4SRS method was used to derive this architecture. This method is typically used in the design of Information Systems to represent the requirements for process execution.

The requirements elicitation involved a detailed analysis of the GS1 global traceability standard, where a set of key use cases that capture the functional requirements of traceability BP were presented. The use of this standard was useful only in a very embryonic moment of the requirements elicitation phase, since its high degree of generalization did not allow a proper functional refinement, so we other sources of information were used (case studies, the authors' experience, among others). The inclusion of an industrial traceability standard promotes the use of best practices that result from consensus and validation of various industry bodies. This is certainly an added-value in our approach. Also during modeling, high complexity was found within the main activities that comprise the traceability BP. The use cases' functional refinement proved to be a useful exercise to minimize this constraint, which includes the detailed description of the 64 use cases to help in this matter. The use of the standard (in this case, GS1) is input mainly within higher-level use cases from the model, although our model was composed by use cases directly obtained from the standard regarding first-, second- and third-level use cases.

The logical architecture inherits the benefits of the proposed use cases. The 4SRS had never been used for this particular context. However, this method has proved very useful since it has formalized and guided the development of our architecture. Some subjectivity can be pointed to the application of 4SRS. To minimize this subjectivity we made some iterations and revisions. A logical architecture of the traceability BP composed by 79 AEs was proposed, promoting an object-oriented vision. These AEs are the fundamental elements in order to enable this architecture addressing the original problem. What distinguish an AE of any other are mainly three aspects: (i) each AE has a well-defined set of responsibilities; (ii) each AE has well-defined boundaries; and (iii) each AE has a set of well-defined interfaces, which define the services that the AE provides to the remaining AEs. By capturing the behavior of the traceability BP, this approach reduces the process ambiguity. Knowing the AEs (and logical architecture as a whole) is useful both for organizations without a formalized traceability BP as for organizations with mature and well established traceability BP. Thus, this architecture could serve as a guide for new traceability BP implementations and/or as an element to assess how well established is a traceability BP. In summary, our systematic approach offers an integrated, comprehensive and holistic overview of the traceability BP that allows to delineate and also detailing the scope of traceability. With the use of the 4SRS we assure a scalable and reusable logical architecture. This method allows adding other activities to the logical architecture.

In the proposed model, *Recall Product* subprocess was not included. However, our proposed approach may easily include *Recall Product* subprocess within the use case model and use it in the logical architecture derivation in future works. In the future we intend to evaluate the logical architecture proposed based on the behavior of a traceability BP already implemented. This study should help align the logical architecture with some traceability organizational practices.

References

1. Cognex, Traceability for the Automotive Industry (2011). http://acrovision.co.uk/wp-content/uploads/2011/07/Expert_Guide__Traceability_for_the_Automotive_Industry.pdf
2. Eckert, W.: Traceability In Electronics Manufacturing. Onboard Technology (2005). http://www.onboard-technology.com/pdf_ottobre2005/100510.pdf
3. Lima, R.C.G.: Indústria Automobilística. Desafios na Cadeia Produtiva (2013). http://www.ibm.com/midmarket/br/pt/articles_general_industry_3Q3.html (retrieved July 22, 2013)
4. Martins, J.C., Machado, R.J.: Ontologies for product and process traceability at manufacturing organizations: a software requirements approach. In: 2012 Eighth International Conference on the Quality of Information and Communications Technology (QUATIC). IEEE (2012)
5. GS1, The GS1 Traceability Standard: What you need to know (2007). http://www.gs1.org/docs/traceability/GS1_tracebility_what_you_need_to_know.pdf
6. GS1, GS1 Standards Document - Business Process and System Requirements for Full Supply Chain Traceability GS1 Global Traceability Standard (2012). http://www.gs1.org/sites/default/files/docs/gsmp/traceability/Global_Traceability_Standard_i1p3p0_November_2012.pdf
7. International Organization for Standardization (ISO), Quality management systems-Fundamentals and vocabulary, vol. 2000 (2005)

8. Moe, T.: Perspectives on traceability in food manufacture. Trends in Food Science & Technology **9**(5), 211–214 (1998)
9. Jansen-Vullers, M.H., van Dorp, C.A., Beulens, A.J.: Managing traceability information in manufacture. International Journal of Information Management **23**(5), 395–413 (2003)
10. Gotel, O., et al.: Traceability fundamentals. In: Software and Systems Traceability, pp. 3–22. Springer (2012)
11. Dijkman, R.M., Joosten, S.M.: Deriving use case diagrams from business process models (2002)
12. Redding, G., et al.: Generating business process models from object behavior models. Information Systems Management **25**(4), 319–331 (2008)
13. Jani, D., Vanderveken, D., Perry, D.E.: Deriving architecture specifications from KAOS specifications: a research case study. In: Morrison, R., Oquendo, F. (eds.) EWSA 2005. LNCS, vol. 3527, pp. 185–202. Springer, Heidelberg (2005)
14. Castro, J., Kolp, M., Mylopoulos, J.: Towards requirements-driven information systems engineering: the Tropos project. Information Systems **27**(6), 365–389 (2002)
15. Ullah, A., Lai, R.: Modeling business goal for business/it alignment using requirements engineering. Journal of Computer Information Systems **51**(3), 21 (2011)
16. Duarte, F.J., Machado, R.J., Fernandes, J.M.: BIM: a methodology to transform business processes into software systems. In: Biffl, S., Winkler, D., Bergsmann, J. (eds.) SWQD 2012. LNBIP, vol. 94, pp. 39–58. Springer, Heidelberg (2012)
17. Santos, N., Machado, R.J., Ferreira, N., Gašević, D.: Derivation of process-oriented logical architectures: an elicitation approach for cloud design. In: Dieste, O., Jedlitschka, A., Juristo, N. (eds.) PROFES 2012. LNCS, vol. 7343, pp. 44–58. Springer, Heidelberg (2012)
18. Terzi, S., Cassina, J., Panetto, H.: Development of a metamodel to foster interoperability along the product lifecycle traceability. In: Konstantas, D., et al. (eds.) Interoperability of Enterprise Software and Applications. Springer Science & Business Media (2006)
19. Terzi, S., et al.: A holonic metamodel for product traceability in product lifecycle management. International Journal of Product Lifecycle Management **2**(3), 253–289 (2007)
20. Gampl, B.: Traceability systems in the German food industry–towards a typology. Department of Agricultural Economics. University of Kiel (2003)
21. Panetto, H., Baïna, S., Morel, G.: Mapping the IEC 62264 models onto the Zachman framework for analysing products information traceability: a case study. Journal of Intelligent Manufacturing **18**(6), 679–698 (2007)
22. Gogliano Sobrinho, O., et al.: Modeling of an information system for wine traceability based on a service oriented architecture. Engenharia Agrícola **30**(1), 100–109 (2010)
23. GS1, GS1 Global Traceability Standard for Healthcare (GTSH) - Implementation Guide (2009). http://www.gs1.org/docs/gsmp/traceability/Global_Traceability_Implementation_Healthcare.pdf
24. GS1, Traceability for Fresh Fruits and Vegetables Implementation Guide (2010). http://www.gs1.org/sites/default/files/docs/gsmp/traceability/Global_Traceability_Implementation_Fresh_Fruit_Veg.pdf
25. Smith, I., Furness, A.: Improving traceability in food processing and distribution. Woodhead Publishing (2006)
26. Manikas, I.: A web application for supply chain traceability. In: Graham, D., Manikas, I., Folinas, D.K. (eds.) E-Logistics and E-Supply Chain Management: Applications for Evolving Business. IGI Global (2013)

27. Bevilacqua, M., Ciarapica, F.E., Giacchetta, G.: Business process reengineering of a supply chain and a traceability system: A case study. Journal of Food Engineering **93**(1), 13–22 (2009)
28. UML, O.: 2.4. 1 superstructure specification. 2011, document formal/2011-08-06. Technical report, OMG
29. Azevedo, S., et al.: The UML «include» relationship and the functional refinement of use cases. In: 2010 36th EUROMICRO Conference on Software Engineering and Advanced Applications (SEAA). IEEE (2010)
30. Rozanski, N., Woods, E.: Software Architecture Systems Working with Stakeholders Using Viewpoints and Perspectives. Addison-Wesley (2005)
31. Ashford, P.: Traceability. Cell and Tissue Banking **11**(4), 329–333 (2010)

Native and Multiple Targeted Mobile Applications

Euler Horta Marinho[1,2]([⊠]) and Rodolfo Ferreira Resende[1]

[1] Computer Science Department, Universidade Federal de Minas Gerais,
Belo Horizonte, Minas Gerais 31270-010, Brazil
{eulerhm,rodolfo}@dcc.ufmg.br
[2] Computer and Systems Department, Universidade Federal de Ouro Preto,
João Monlevade, Minas Gerais 35931-008, Brazil

Abstract. Together with the expansion of the WWW we are seeing the expansion of mobile devices that are becoming more and more pervasive. Mobile application development is becoming more and more complex as users of mobile applications are demanding more high quality software. Our contribution is to frame the positive and negative aspects of native and multiple targeted mobile applications that should be considered by the involved stakeholders more particularly the software organization decision-makers.

Keywords: Mobile application development · Multiple targeted development

1 Introduction

The challenge of mobile application development has several consequences for software professionals [51]. Mobile applications include traditional applications rewritten to run on mobile devices and applications that make use of features related to the mobile context, i.e., information originated in the computing environment allowing behavior adaptation [37]. Behavior adaptation is included in the general class of context-aware applications and inherits the complexity and peculiarities of development related to sensing aspects [4].

Device characteristics, for example screen size, low-power CPU, and network issues such as lost connections and reduced bandwidth influence application requirements. Due to the inherent complexity of these applications, the development should be supported by adequate software engineering methods [52]. Thus, new professional skills are required in mobile application development.

Another defiant matter in mobile application development is related to the diversity of technologies and vendors. The heterogeneity of devices and operating systems, also known as platform fragmentation [8], affects in some extent all mobile platforms. Native applications development is an important development option due, mainly, to the mentioned heterogeneity. However, while there are a

© Springer International Publishing Switzerland 2015
O. Gervasi et al. (Eds.): ICCSA 2015, Part IV, LNCS 9158, pp. 544–558, 2015.
DOI: 10.1007/978-3-319-21410-8_42

few iOS type of devices, Android targets thousands[1]. The platform fragmentation is not always a problem, for example it allows the market expansion giving to users and vendors a broader range of choices and personalization.

The heterogeneity in mobile application domain influences the portability [26] of mobile applications. The attempts to apply solutions to manage portability issues similar to those of the desktop domain are not equally effective due to technological aspects and specific strategies of mobile platform vendors [1]. Quality aspects of mobile applications have been a concern of some industry initiatives [30].

Essentially, the software development can follow a single or multiplatform (cross-platform) strategy [6]. Web applications have the characteristic of cross-platform compatibility [28] and their increased complexity [17] are managed by methodologies in a research area known as Web Engineering [20]. The particularities of Mobile Web applications have created additional demands for development methodologies [23].

Model-driven Engineering (MDE) has been presented as a possible solution to mitigate, among other aspects, software complexity [18] and platform fragmentation problems [43]. MDE includes various model-driven approaches for software development, including model-driven architecture, domain-specific modeling and model-integrated computing [24].

A great number of decision-making factors are involved in software projects. Some of these factors are discussed in the work of Vijayasarathy and Butler [50]. The authors discuss the influence of project, team, and organizational aspects in the choice of software development methodologies. However, other dimensions can be considered when software organizations have to make decisions about which development approach should be adopted. Such examples are team technical skills and technological aspects. As we discussed above and according to the related literature, the mobile application development presents an inherent complexity due to the particularities of this domain. Our contribution in this paper is an analysis of different options for mobile application development in a synthetic context. Our analysis involves solutions based on native and multiple targeted development. Here we loosely refer to (i) "Native" when the development is directed towards a single target e.g. Android and (ii) "Multiple Targeted Development" - MTD when the development effort is directed towards a platform that will redirect the resulting artifacts towards two or more targets, e.g. Android and iOS. We do not discuss the meaning of Java or C++ being more or less "native" [11,16]. Our analysis identified two approaches for MTD: ad-hoc and principled. We consider MDE the only principled MTD option despite the original ambition of the Web, which we consider one of the ad-hoc options. We discuss part of the technological aspects involved in the mobile application development. Among other stakeholders, we expect to help the software development decision-makers of software organizations.

---

[1] opensignal.com/reports/2014/android-fragmentation/

This paper is structured as follows. In Section 2, we give an overview of the role of the Web Engineering in mobile application development. In Section 3, we discuss the different options for mobile application development. Section 4 presents the final considerations.

2   The Role of the Web Engineering in the Mobile Application Development

Our work does not have a focus on the mobile native development options. On the other hand, we have a certain focus on web based mobile software development. Our work wants to compare the native software development with MTD. We refer to the native development options generically by referring to their Software Development Kits (SDKs) (and sometime native APIs or tools). The native development options are not as intricate as the MTD and here we briefly discuss web-based development, unarguably the main MTD option. As we will see on Section 3.2, MDE is not currently such a popular option for mobile platforms.

The advent of new application classes has created new challenges for the Web Engineering. An example is Rich Internet Applications (RIAs). RIAs represent a web application class that has a better responsiveness and a more extended user experience than usual Web applications [32]. RIAs involve a heterogeneous set of solutions aiming to add new characteristics to Web Applications. These characteristics comprise four fundamental aspects [48]: Rich Presentation, Client Data Storage, Client Business Logic, and Client-Server Communication. The peculiarities of RIA development led to the proposal of new methodologies in Web Engineering. Figure 1, adapted from the work of Escalona and Aragon [17], depicts the scenario of web methodologies since 1993, where we included the proposed RIA oriented methodologies since 2006. As we can observe, RIA methodologies (the italic names) basically consist of extensions of existent approaches, providing abstractions to model the structure and the behavior of application components. We refer to the work of Casteleyn and others [7] for a deeper overview of the research involving RIAs.

The adoption of the new characteristics related to the four fundamental aspects listed above did not stop new demands. The need for designing RIAs that could run in multiple devices with screens of different characteristics, such as mobile devices, led to the search of specific web methodologies (for example, UWE + RUX [41], WebML + RUX [29], and AlexandRIA [12]). The integration of cloud computing paradigm in the mobile application development have resulted in specific approaches such as MobiCloUp! [13].

On the other hand, The World Wide Web Consortium (W3C) proposed a specific recommendation[2] for mobile web applications development. Also, the HTML5 standard and several W3C specifications have introduced features that support mobile applications characteristics as presented in Table 1. Some specifications are currently being developed.

---

[2] www.w3.org/TR/mwabp/

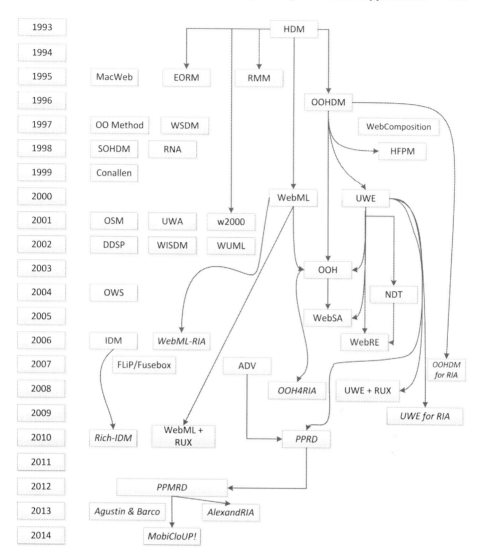

Fig. 1. Web methodologies including those for RIAs (Adapted from the work of Escalona and Aragon [17])

3   Mobile Application Development

There are some works comparing native and web-based development (e.g. [8,45]). Tools supporting multiple targets have also been evaluated for ad-hoc multiple targets (e.g. [22,39]) and for MDE tools (e.g. [43]). Our discussion considers these different options for mobile application development.

Table 1. HTML5 features and W3C specifications related to mobile application development

Feature	Purpose
Offline support [a]	Provide mechanisms to manage application data when it is offline.
Device APIs [b]	A series of APIs to interact with devices services, for example, the Vibration API and the Battery Status API.
Geolocation API [c]	Provide access to geographical location information associated with the hosting device.
Web NFC API [d]	Provide access to the hardware subsystem for Near Field Communication (NFC), an interface and protocol for simple wireless interconnection.
Web Telephony API [e]	Define an API to manage telephone calls.
DeviceOrientation event specification [f]	Define new DOM (Document Object Model) events that provide information about the physical orientation and motion of a hosting device.
Touch events specification [g]	Define a set of low-level events that represent one or more points of contact with a touch-sensitive surface, and changes of those points with respect to the surface and any DOM elements.

[a] http://www.w3.org/TR/offline-webapps/
[b] http://www.w3.org/2009/dap/
[c] http://www.w3.org/TR/geolocation-API/
[d] http://www.w3.org/TR/nfc/
[e] http://www.w3.org/TR/telephony/
[f] http://www.w3.org/TR/orientation-event/
[g] http://www.w3.org/TR/touch-events/

According to a proposed mobile architecture with 6 categories[3], mobile application classes include Native, Web (named *No client*), HTML5, Special, Hybrid, and Message applications. We consider HTML5 based applications a specific case of Web applications. Hybrid applications are those developed combining the use of native and Web technologies. Special is a category that includes applications developed in specific (maybe proprietary) languages and tools (e.g. Adobe Flash applications developed in ActionScript language in Apache Flex or Adobe Flash Builder tools). Message applications are based on legacy frameworks such as SMS and MMS for data exchange. We refer to Hybrid and Special Applications as Alternative Applications.

Several platforms and tools are involved in the mobile domain. In Figure 2, we illustrate the historical scenario of platforms (green lines), cross-platforms

---

[3] www.gartner.com/newsroom/id/2209615

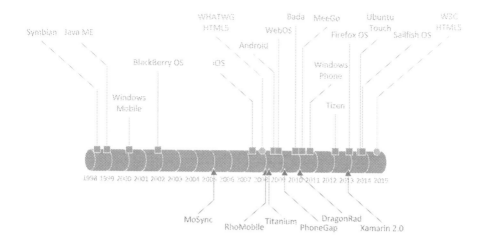

Fig. 2. Platforms and tools for mobile application development

tools (blue lines), and standards (gray lines). We can realize that the platforms proposed in the last years (e.g. Tizen, Firefox OS, and Ubuntu Touch) have openly allowed the use of HTML5 as a development option. The first public working draft of the HTML5 standard was proposed by Web Hypertext Application Technology Working Group (WHATWG[4]) in 2008 and became a W3C recommendation in 2014.

3.1   Native Development

Native applications are generally developed[5] by using a specific SDK provided by a platform vendor. Table 2 depicts in a simplistic way the main characteristics of the SDKs of some mobile platforms.

Table 2. SDK main characteristics

Platform	Programming Language	GUI development	IDE
Android	Java, C, C++	Android UI, Material Design	Eclipse, Android Studio
iOS	Objective-C, Swift	Cocoa Touch	Xcode
Windows Phone	*Multiple*	XAML, DirectX, WinJS, Modern UI	Visual Studio

---

[4] www.whatwg.org
[5] See Section 3.2, for MDE tools that generate native applications

As we can see, each platform has distinguishing characteristics and a specific IDE. From the organizational perspective, this can require particular developer skills and additional efforts in training. Furthermore, the specificity of GUI toolkits is a known issue for MTD [6].

The hardware advances of mobile devices such as multicore CPUs and GPUs are creating new improvement opportunities for native mobile applications. Some types of applications (e.g. 3D games and image processing) can benefit more from multiprocessing and advanced graphical processing features. Moreover, performance of these applications can benefit from the fact that the programming languages used to build native applications are generally compiled or have most of the advantages of a compiled framework.

Concerning the application usability, native applications fits platform look-and-feel. This means that the user interface presents a consistent aspect and behavior across applications of the same platform.

3.2   Multiple Targeted Development

The development of software that is easy to port or deliberately targeted for multiple platforms is known as cross-platform development [6]. Here, we discuss ad-hoc efforts and MDE that we consider a more principled effort.

Ad-hoc Multiple Targeted Development. Wasserman [52] mentions two MTD options for mobile application development: web applications and cross-platform development tools that allow the building and maintenance of multiple platform versions of an application developed from the same code base.

Web applications are developed through web technologies such as HTML, CSS, and JavaScript. HTML allows the specification of content and structure, CSS, the presentation, and JavaScript, the behavior of web applications.

JavaScript engine performance has been part of the focus in the so-called browser wars [42]. The release of JavaScript V8 engine in Google Chrome in the desktop side was accompanied by a similar change in the mobile one. WebKit Web browser engine used in the construction of several mobile browsers is boosting the performance of its JavaScript engine SquirrelFish Extreme (SFX). Both engines use the strategies of compiling JavaScript to native machine code and inline caching.

Web applications have a particular look and feel that can rely on specific UI frameworks (e.g. JQuery and Scriptaculous). On the other side, a particular Web application can present issues such as look and feel inconsistencies or lack of functionalities across different web browsers [33].

One of the early obstacles for mobile web applications was the access to device-specific features. For example, the touch screen is the prevalent interaction method of mobile devices. Nevertheless, due to limitations of HTML language, Web browsers were not able to recognize gestures properly. With the proposal of Touch events specifications for the HTML5 (Table 1), mobile web applications can have their demands of multi touch events supplied.

Alternative mobile applications performance varies considerably depending on the tool adopted [39]. However, the choice criteria of these tools are not so clear and depend on application requirements and development strategies. So benchmark proposals are needed for performance evaluation of mobile applications features such as those adopted in JavaScript evaluation (e.g. SunSpider and V8).

The user interface of Alternative mobile applications relies on web based widgets (e.g. PhoneGap), external libraries implemented in programming languages such as JavaScript to render platform GUI elements (e.g. MoSync) or access to platform GUI widgets through their own APIs (e.g. Titanium). Some tools allow the customization of web-based widgets by Cascading Style Sheets (e.g. Adobe Flash Builder). This permits development of custom user interface widgets but possibly increasing the development effort [39].

The application deployment in online stores encompasses particular packaging and signing formats [14]. Online stores often reject applications that download executable code or interpret code not contained within the application archive [39]. In order to avoid rejection, cross-platform tools generally offer different strategies to package and sign applications to the target platform. One is to compile and package the particular source code to the specific platform code (e.g. the MoSync strategy). Another is to wrap the application in a native format, for example, a mobile web application can be wrapped in a native format where the native portion only invokes an installed web browser or this application embeds a Web View, a kind of component used for the development of functionally limited browsers [47], to access the server side functionalities.

The access to device features in alternative applications is usually performed by the interaction with native APIs. This is often made through external libraries implemented in programming languages such as JavaScript or proprietary frameworks.

Cross-platform tools use different approaches for MTD. Bellow, we briefly present some approaches discussed in the literature [22,39,43]. It is important to note that the same tool can use multiple approaches (for example, Marmalade can be used in the development of Hybrid applications through wrapping or retargeting).

- **Code interpretation:** An interpreter executes instructions directly without previous compilation, in other words not demanding the translation of a language into a format more amenable to execution [3]. The application is often executed in a runtime environment called Virtual Machine (VM) [46] that can require a particular instruction set such as bytecodes [53] for the purpose of increase performance and reduce platform dependence. Sample tools: Titanium, RhoMobile.

- **Retargeting:** Sometimes referred as cross-compilation, it is an approach in which the source code is typically translated to a common format called Intermediate Representation (IR) that is used for generating platform-specific applications [25]. Sample tools: MoSync, Marmalade.

Table 3. Summary of the characteristics of the Ad-hoc MTD approaches

Approach	Possible advantages	Possible drawbacks
Web development	Some standards (such as those from W3C) guide the application development. Improved application portability.	Look and feel inconsistencies. Lack of functionalities. The access to device-specific features can be difficult without proper API support.
Code interpretation	Increase of application portability. The same source-code base can be used in development for multiple platforms.	Performance of pure interpreters is typically lower than static and JIT compilers [44].
Retargeting	The same source-code base is used to generate native applications to each platform. Advantages of native applications.	Increased complexity of the compiler development. Each backend must be optimized to deal with platform specificities when dealing with the same IR [25].
Wrapping	The application can be distributed in online stores.	Similar to those of Web development. The access to device-specific features must be managed by specialized APIs.

- **Wrapping:** Consists on using a wrapper, a native container encapsulating a Web-based application. Sample tools: Phonegap, Marmalade.

Table 3 summarizes possible advantages and drawbacks of the Ad-hoc MTD approaches.

Model-Driven Engineering. MDE defines a set of approaches in which abstract models of software systems are designed and systematically transformed into concrete implementations [18]. It is based on the premise that *"everything is a model"* in software development [5]. Some studies [19,24,36,49] have discussed advantages and drawbacks in the adoption of MDE practices in software industry in general. Suggested advantages of using MDE techniques are improvements in productivity, portability, reusability, and reusability [18]. However, a series of problems can negatively hinder or prevent MDE adoption such as the additional effort required by modelling activities and the lack of appropriate skills, mixing technological and human factors [49].

Kraemer [27] proposes Arctis, an MDE-based tool for Android application development. Arctis uses UML activity diagrams and state machines to generate application code. State machines are used to establish the external contracts of the activity models.

IBM Rational Rhapsody[6] is a commercial tool that allows the development of embedded, real-time and mobile applications through UML or SysML stat-echarts. Models can be simulated and used to generate application code. It is possible to synchronize changes in model and code.

Domain specific languages (DSLs) have been used by different MDE methodologies. The MD^2 approach [21,22] is based on a DSL designed for description of data-driven business applications that are compiled for Android and iOS native code. Mobl [23] uses a declarative DSL to allow the implementation of mobile web applications. Applications built in Mobl are based on HTML5, CSS and Javascript.

Cugola and others [15] present SELFMOTION, an approach for the development of adaptive mobile applications using service compositions. These applications are adaptive with respect to the deployment environment and the external resources and third-party applications they rely upon. The application behavior is described as a set of abstract actions in a declarative language. Abstract actions are then transformed in concrete actions. SELFMOTION applications are executed by a specific middleware.

Miravet and others [35] propose DIMAG, a framework for declarative specification of client-server mobile applications. The workflow of the application is modelled as a state machine whereas the user interface is described in a XML-based specific language. A code generation strategy is used according to the target platform and device characteristics.

Genexus X Evolution[7] is a commercial MDE tool based on objects involving patterns related to user interface widgets and data views described in a proprietary language to define an abstract mobile application. This abstract description can be compiled for native code in Android, iOS and Blackberry or to HTML5.

Table 4 summarizes the characteristics of the MDE tools for mobile application development. Tools such as IBM Rational Rhapsody and Genexus X Evolution are commercial products that have some reports of successfully use in industry[8]. Arctis was initially proposed as a research tool and now it is part of Reactive Blocks Eclipse Plugin[9] that has free and commercial versions. The remaining tools are results of academic research.

Most of the MDE-based tools are based on DSL approaches and support at least the access to device features of geolocation.

In spite of the importance of MDE approaches to control platform fragmentation and complexity problems of the mobile application domain [43], the MDE techniques are considered in early stages or not relevant in general practice [22]. However, technical factors are not the only to be considered in the adoption of MDE approaches. Organizational, managerial and social aspects such as motivation, commitment, and support to process changes are important for successful

---

[6] http://www-03.ibm.com/software/products/en/ratirhapfami

[7] http://www.genexus.com/genexus-versions/genexus-x-evolution-3?en

[8] See for example, http://www-03.ibm.com/software/businesscasestudies/us/en/ andhttp://www.genexus.com/success-stories/success-stories?en

[9] http://www.bitreactive.com

Table 4. Comparison of tool characteristics

Tool	Model	Platform	Device feature supported
Arctis	UML Activity, State Machine	Android	Geolocation
IBM Rational Rhapsody	Statechart	Android	All provided by Android SDK
MD^2	DSL	Android, iOS	Geolocation
Mobl	DSL	Android, iOS	Geolocation
Genexus X Evolution	DSL	Android, iOS, Blackberry OS, HTML5	Geolocation, Camera
DIMAG	DSL, State machine	Android, Java ME, Windows Mobile	Not specified
SELFMOTION	DSL	Android	Geolocation, Camera

adoption of MDE practices [24]. From the reviewed literature, we can observe that the efforts to learn the modelling languages are not clear. Thus, additional empirical studies of MDE tools are needed.

4   Related Work

Some authors have discussed different aspects involved in mobile application development, mainly comparing native and web options.

Charland and Leroux [8] discuss the strengths and weakness of native and Web development options. These options are analyzed according to performance, user experience, and developers skills criteria. The authors mention that, in general, Web applications are cheaper to develop and deploy than native applications. However, the paper does not present MTD as we consider.

Serrano and others [45] evaluate the different technologies for developing mobile web applications. Frameworks are compared through different criteria such as programming languages, platforms, and access to hardware features. Most of the analyzed frameworks converge to the use of HTML5. Our analysis considers other types of mobile applications such as multiple targeted applications.

Orht and Turau [39] and Heitkotter and others [22] compare several features of cross-platform tools for mobile application development. Although the cross-platform approach may become an option for native development, some issues related to development support, hardware resource consumption and User Experience must be solved. Ribeiro e Silva [43] also compare cross-platform tools, but differently from Ohrt and Turau and Heitkotter and others, they included two MDE tools based exclusively on DSL approaches. In our discussion, we include MDE tools based on other modeling approaches.

5   Conclusion

The application requirements are the bottom line in a choice of native or multiple targeted mobile application development. Some application classes, for example games, have particular User Experience demands (e.g. look and feel) that can be decoupled from specific platform considerations. The choice of a strategy for mobile application development and software development in general must be reflected from a specific organizational point of view comprising, the managerial and technical issues. The organization must evaluate, among others, the market options, the development skills, and the project characteristics. This reasoning also applies when existing applications are ported to mobile environments (see for instance, the experience of Facebook[10]). Our contribution with this paper is an analysis of different options for mobile application development in a synthetic context. We assumed a basic knowledge of native mobile approaches and, based on our concepts of MTD options, we discussed some of their advantages and drawbacks.

The characteristics available for the developer in native and web development are converging. Arguably Web and native development will never be equivalent, however considering this convergence, the impact of the choice is diminishing. We are now investigating how to couple some of the tools we already developed with the processes and techniques of Software Product Line (SPL). A definition for SPL is "a set of software-intensive systems sharing a common, managed set of features that satisfy the specific needs of a particular market segment or mission and that are developed from a common set of core assets in a prescribed way" [10]. SPL research involves broad concerns [34] such as Requirements Engineering [2], Variability Management [9] and Testing [38]. In particular, SPL is an important approach for multiple targeted mobile applications and this seems an important direction for mobile application development in general [31,40].

References

1. Agarwal, V., Goval, S., Mittal, S., Mukherjea, S.: MobiVine: a middleware layer to handle fragmentation of platform interfaces for mobile applications. In: ACM/IFIP/USENIX International Conference on Middleware, pp. 1–10. Springer-Verlag, New York (2009)
2. Alves, V., Niu, N., Alves, C., Valenca, G.: Requirements engineering for software product lines: a systematic literature review. Information and Software Techonology **52**(8), 806–820 (2010)
3. Aycock, J.: A brief history of just-in-time. ACM Computing Surveys **35**(2), 97–113 (2003)
4. Bettini, C., Brdiczka, O., Henricksen, K., Indulska, J., Nicklas, D., Ranganathan, A., Riboni, D.: A survey of context modelling and reasoning techniques. Pervasive and Mobile Computing **6**(2), 161–180 (2010)

---

[10] www.techcrunch.com/2012/09/11/mark-zuckerberg-our-biggest-mistake-with-mobile-was-betting-too-much-on-html5

5. Bezivin, J.: On the unification power of models. Software & Systems Modeling **4**(2), 171–188 (2005)
6. Bishop, J., Horspool, N.: Cross-platform development: software that lasts. IEEE Computer **39**(10), 26–35 (2006)
7. Casteleyn, S., Garrigos, I., Mazon, J.N.: Ten years of Rich Internet Applications: a systematic mapping study, and beyond. ACM Transactions on the Web **8**(3), 18:1–18:44 (2014)
8. Charland, A., Leroux, B.: Mobile application development: web vs native. Communications of the ACM **54**(5), 49–53 (2011)
9. Chen, L., Babar, M.A.: A systematic review of evaluation of variability management approaches in software product lines. Information and Software Techonology **53**(4), 344–362 (2011)
10. Clements, P., Northrop, L.: Software Product Lines: Practices and Patterns. Addison-Wesley, Boston (2001)
11. Cohen, R., Wang, T.: NDK and C/C++ optimization. In: Android Application Development for the Intel Platform, pp. 391–444. A press (2014)
12. Colombo-Mendoza, L.O., Alor-Hernandez, G., Rodriguez-Gonzalez, A., Colomo-Palacios, R.: Alexandria: a visual tool for generating multi-device Rich Internet Applications. Journal of Web Engineering **12**(3–4), 317–359 (2013)
13. Colombo-Mendoza, L.O., Alor-Hernandez, G., Rodriguez-Gonzalez, A., Valencia-Garcia, R.: MobiCloUP!: a Paas for cloud services-based mobile applications. Automated Software Engineering **21**(3), 391–437 (2014)
14. Cuadrado, F., Duenas, J.C.: Mobile application stores: sucess factors, existing approaches, and future development. IEEE Transactions on Software Engineering **50**(11), 160–167 (2012)
15. Cugola, G., Ghezzi, C., Pinto, L.S., Tamburrelli, G.: SelfMotion: a declarative approach for adaptive service-oriented mobile applications. The Journal of Systems and Software **92**, 32–44 (2014)
16. Enck, W., Gilbert, P., Han, S., Tendulkar, V., Chun, B.G., Cox, L.P., Jung, J., McDaniel, P., Sheth, A.N.: TaintDroid: An information-flow tracking system for realtime privacy monitoring on smartphones. ACM Transactions on Computer Systems **32**(2), 5:1–5:29 (2014)
17. Escalona, M., Aragon, G.: NDT: A model-driven approach for web requirements. IEEE Transactions on Software Engineering **34**(3), 377–390 (2008)
18. France, R., Rumpe, B.: Model-driven development of complex software: a research roadmap. In: Future of Software Engineering, pp. 37–54. IEEE Computer Society, Washington, DC (2007)
19. Garcia-Diaz, V., Fernandez-Fernandez, H., Palacios-Gonzalez, E., G-Bustelo, B.C.P., Sanjuan-Martinez, O., Lovelle, J.M.C.: TALISMAN MDE: Mixing MDE principles. Journal of Systems and Software **83**(7), 1179–1191 (2010)
20. Gigine, A., Murugesan, S.: Web engineering: an introduction. IEEE Multimedia **8**(1), 14–18 (2001)
21. Heitkotter, H., Kuchen, H., Majchrzak, T.A.: Extending a model-driven cross-platform development approach for business apps. Science of Computer Programming **97**(1), 31–36 (2015)
22. Heitkotter, H., Majchrzak, T.A., Kuchen, H.: Cross-platform model-driven development of mobile applications with MD^2. In: ACM Symposium on Applied Computing, pp. 526–533. ACM Press, New York (2013)
23. Hemel, Z., Visser, E.: Declaratively programming the mobile web with Mobl. In: ACM SIGPLAN Conference on Object Oriented Programming Systems. Languages, and Applications, pp. 695–712. ACM Press, New York (2011)

24. Hutchinson, H., Rouncefield, M., Whittle, J.: Model-driven engineering practices in industry. In: International Conference on Software Engineering, pp. 561–570. ACM Press, New York (2011)
25. Hwang, Y.S., Lin, T.Y., Chang, R.G.: DisIRer: converting a retargetable compiler into a multiplatform binary translator. ACM Transactions on Architecture and Code Optimization **7**(4), 18:1–18:36 (2010)
26. ISO: ISO/IEC 25010: Systems and software engineering - systems and software quality requirements and evaluation (SQuaRE) - system and software quality models (2011)
27. Kraemer, F.A.: Engineering android applications based on uml activities. In: Whittle, J., Clark, T., Kühne, T. (eds.) MODELS 2011. LNCS, vol. 6981, pp. 183–197. Springer, Heidelberg (2011)
28. Li, X., Xue, Y.: A survey on server-side approaches to securing Web Applications. ACM Computing Surveys **54**(4), 54:1–54:29 (2014)
29. Linaje, M., Preciado, J.C., Sanchez-Figueroa, F.: Multi-device context-aware RIAs using a model-driven approach. Journal of Universal Computer Science **16**(15), 2038–2059 (2010)
30. Marinho, E.H., Resende, R.F.: Quality factors in development best practices for mobile applications. In: Murgante, B., Gervasi, O., Misra, S., Nedjah, N., Rocha, A.M.A.C., Taniar, D., Apduhan, B.O. (eds.) ICCSA 2012, Part IV. LNCS, vol. 7336, pp. 632–645. Springer, Heidelberg (2012)
31. Marinho, F.G., Andrade, R.M.C., Werner, C., Viana, W., Maia, M.E.F., Rocha, L.S., Teixeira, E., Filho, J.B.F., Dantas, V.L.L., Lima, F., Aguiar, S.: Mobi-Line: A nested software product line for the domain of mobile and context-aware applications. Science of Computer Programming **78**(12), 2381–2398 (2013)
32. Melia, S., Gomez, J., Perez, S., Diaz, O.: Architectural and technological variability in Rich Internet Applications. IEEE Internet Computing **14**(3), 24–32 (2010)
33. Mesbah, A., Prasad, M.R.: Automated cross-browser compatibility testing. In: International Conference on Software Engineering, pp. 561–570. ACM Press, New York (2011)
34. Metzger, A., Pohl, K.: Software product line engineering and variability management: achievements and challenges. In: Future of Software Engineering, pp. 70–84. ACM Press, New York (2014)
35. Miravet, P., Marin, I., Ortin, F., Rodriguez, J.: Framework for the declarative implementation of native mobile applications. IET Software **8**(1), 19–32 (2014)
36. Mohagheghi, P., Gilani, W., Stefanescu, A., Fernandez, M.: An empirical study of the practice and acceptance of model-driven engineering in four industrial cases. Empirical Software Engineering **18**(1), 89–116 (2013)
37. Muccini, H., Di Francesco, A., Esposito, P.: Software testing of mobile applications: challenges and future research directions. In: International Workshop on Automation of Software Test, pp. 29–35. IEEE Computer Society, Washington, DC (2012)
38. Neto, P.A.M.S., Machado, I.C., McGregor, J.D., Almeida, E.S., Meira, S.R.L.: A systematic mapping study of software product lines testing. Information and Software Techonology **53**(5), 407–423 (2011)
39. Orht, J., Turau, V.: Cross-platform development tools for smartphone applications. IEEE Computer **45**(9), 72–79 (2012)
40. Pascual, G.P., Lopez-Herrejon, R.E., Pinto, M., Fuentes, L., Egyed, A.: Applying multiobjective evolutionary algorithms to dynamic software product lines for reconfiguring mobile applications. Journal of Systems and Software **103**, 392–411 (2015)

41. Preciado, J.C., Linaje, M., Morales-Chaparro, R., Sanchez-Figueroa, F., Zhang, G., Kroib, C., Kock, N.: Designing rich internet applications combining UWE and RUX-method. In: International Conference on Web Engineering, pp. 148–154. IEEE Computer Society, Washington, DC (2008)
42. Ratanaworabhan, P., Livshits, B., Zorn, B.G.: JSMeter: comparing the behavior of JavaScript benchmarks with real web applications. In: USENIX Conference on Web Application Development, pp. 27–38. USENIX Association, Berkeley (2010)
43. Ribeiro, A., Silva, A.R.: Survey on cross-platforms and languages for mobile apps. In: International Conference on the Quality of Information and Communication Technology, pp. 255–260. IEEE Computer Society, Washington, DC (2012)
44. Rohou, E., Williams, K., Yuste, D.: Vectorization technology to improve interpreter performance. ACM Transactions on Architecture and Code Optimization $9(4)$, 26:1–26:22 (2013)
45. Serrano, N., Hernantes, J., Gallardo, G.: Mobile web apps. IEEE Software $30(5)$, 22–27 (2013)
46. Shi, Y., Casey, K., Ertl, M.A., Gregg, D.: Virtual machine showdown: stack versus registers. ACM Transactions on Architecture and Code Optimization $4(4)$, 21:1–21:36 (2008)
47. Shin, D., Yao, H., Rosi, U.: Supporting visual security cues for WebView-based Android apps. In: ACM Symposium on Applied Computing, pp. 1867–1876. ACM Press, New York (2013)
48. Toffetti, G., Comai, S., Preciado, J.C., Trigueros, M.L.: State-of-the-art and trends in the systematic development of Rich Internet Applications. Journal of Web Engineering $10(1)$, 70–86 (2011)
49. Torchiano, M., Tomassetti, F., Ricca, F., Tiso, A., Regio, G.: Relevance, benefits, and problems of software modelling and model driven techniques - a survey in the italian industry. Journal of Systems and Software $86(8)$, 2110–2126 (2013)
50. Vijayasarathy, L., Butler, C.: Choice of software development methodologies - do project, team and organizational characteristics matter? IEEE Software PP 99, 1 (2015)
51. Voas, J., Michael, J.B., van Genuchten, M.: The mobile software app takeover. IEEE Software $24(4)$, 25–27 (2012)
52. Wasserman, A.I.: Software engineering issues for mobile application development. In: FSE/SDP Workshop on the Future of Software Engineering Research, pp. 397–400. ACM Press, New York (2010)
53. Wurthinger, T., Wob, A., Stadler, L., Duboscq, G., Simon, D., Wimmer, C.: Self-optimizing AST interpreters. In: Symposium on Dynamic Languages, pp. 73–82. ACM Press, New York (2012)

Crowdsourcing Based Fuzzy Information Enrichment of Tourist Spot Recommender Systems

Sunita Tiwari[1,3(✉)] and Saroj Kaushik[2]

[1] School of IT, IIT Delhi, New Delhi 110016, India
sunita@cse.iitd.ac.in
[2] Department of Computer Science and Engineering, IIT Delhi, New Delhi 110016, India
saroj@cse.iitd.ac.in
[3] Department of Computer Science and Engineering,
ABES Engineering College, Ghaziabad, India
sunita.tiwari@abes.ac.in

Abstract. Tourist Spot Recommender Systems (TSRS) help users to find the interesting locations/spots in vicinity based on their preferences. Enriching the list of recommended spots with contextual information such as right time to visit, weather conditions, traffic condition, right mode of transport, crowdedness, security alerts etc. may further add value to the systems. This paper proposes the concept of information enrichment for a tourist spot recommender system. Proposed system works in collaboration with a Tourist Spot Recommender System, takes the list of spots to be recommended to the current user and collects the current contextual information for those spots. A new score/rank is computed for each spot to be recommender based on the recommender's rank and current context and sent back to the user. Contextual information may be collected by several techniques such as sensors, collaborative tagging (folksonomy), crowdsourcing etc. This paper proposes an approach for information enrichment using just in time location aware crowdsourcing. Location aware crowdsourcing is used to get current contextual information about a spot from the crowd currently available at that spot. Most of the contextual parameters such as traffic conditions, weather conditions, crowdedness etc. are fuzzy in nature and therefore, fuzzy inference is proposed to compute a new score/rank, with each recommended spot. The proposed system may be used with any spot recommender system, however, in this work a personalized tourist spot recommender system is considered as a case for study and evaluation. A prototype system has been implemented and is evaluated by 104 real users.

Keywords: Recommender systems · Information enrichment · Crowdsourcing · Tourism · Fuzzy inference

1   Introduction

The Location Based Services (LBS) has become quite popular on small computing handheld devices in terms of location aware advertising, security alerts, news updates,

© Springer International Publishing Switzerland 2015
O. Gervasi et al. (Eds.): ICCSA 2015, Part IV, LNCS 9158, pp. 559–574, 2015.
DOI: 10.1007/978-3-319-21410-8_43

disaster management, geo-fencing, location based recommender and so on. A location based recommender system is an important application which generates personalized list of recommendation based on user's position. Location based recommender systems are most beneficial for tourist in recommending interesting locations/spots in vicinity of their current position.

List of recommended spot without considering the current context may not be very beneficial to the user. For example, a user asking for nearby restaurant may be recommended with a popular restaurant in the vicinity without considering the fact that it may be fully reserved for a party or it may be undergoing renovation. In such situations, the current contextual information regarding to be recommended spots may add a lot of value to the spot recommender service. Identifying information suitable for integration with the recommended spot is important to improve the effectiveness of recommendation. Unusual contextual conditions such as construction at a site, a spot is closed on the day of query etc. if provided to user can help user significantly. In addition, most of the contextual information about a spot such as traffic conditions, weather conditions, crowdedness etc. changes regularly, therefore the need to collect information at current time arises.

Contextual information may be gathered using several techniques. Some type of the contextual information such as weather conditions, temperature, traffic conditions etc. may be gathered using sensors. Some information which are available on web and does not change frequently may be crawled from web. Collaborative tagging may also be used to gather contextual information. Some type of contextual information such as a spot is not safe to visit at evening, services at a spots are temporarily unavailable, a restaurant is fully reserved, a road to reach a spot is closed, a spot is not worth visiting etc. are difficult to collect from the traditional sources. However, we believe that human gather such type of information about their vicinity through multiple sensors like vision, touch and audition etc. and use their intelligence to understand them and share them with others. This paper presents a contemporary approach to gather the information about a spot using location aware crowdsourcing approach. Location aware crowdsourcing is an approach to collect information from the people who are present at that spot at current time in order to find the current context. However, some of the information may be captured from sensors or other techniques of information acquisition.

This paper focuses on information enrichment of a push-based personalized tourist spot recommender system (TSRS). This recommender system will recommend the popular spot in the vicinity of a tourist ranked on the based on his/her personal preferences. The additional information may be annotated separately with the recommended spot or they may be integrated with the recommendation rank. However, in this work we choose to integrate the additional contextual information along with personal preferences and reorder the list for recommendation. Integrating the additional information with personal preferences may filter out those spots from the list of recommended spots which are not worth visiting currently due to some or the other reasons.

As mentioned earlier, most of the contextual parameters and personal preferences are fuzzy in nature, therefore the fuzzy inference system is used to get a new score/rank for each spot belonging to the recommended list of spots and then the list is reordered based on this new score/rank.

Rest of the paper is structured as follows- section 2 provides background and related work. Section 3 presents overall design of the proposed system. Section 4 contains experiment settings and results, section 5 discusses evaluation of the system and section 6 concludes the paper.

2   Background and Related Work

In this section we present the related research in location recommender system and crowdsourcing. Also, a brief overview of fuzzy inference system is given.

2.1   Location Based Recommender System

Earlier research in location/spot recommender system includes tourist application such as "COMPASS [18] which is point of interest (POI) recommendation for mobile tourists. The data in this type of approaches is usually gathered and tested on some test users. In such applications, test users are provided with GPS enabled devices and catalogs of POIs are created where users could vote for different POIs, such as restaurants, tourist attractions etc. Authors in [9] created a similar type of system for restaurant recommendation, where test users could vote for restaurants and receive recommendation. For recommendation, collaborative filtering approach is used. First, a list of POIs in the surrounding of the user is computed. All users who had casted a vote on these POIs is taken into consideration for collaborative filtering and the N most similar of them are used to derive a rating.   Various other location recommender systems includes mobile tourist guide system [7], [15] etc. Earlier some work is done in user profile enrichment [5] and geo-tagged images [3], [13] authors clusters a large-scale geo-tagged web photo collection into groups by location and then find the representative images for each group. A user can provide either a photo of the desired scenery or a keyword describing the place of interest, and the system will look into its database for places that share the visual characteristics. Tourist destination recommendations are produced by comparing the query against the representative tags or representative images. To the best of our knowledge, the preliminary work on information enrichment techniques, to ease the selection of a location from the list of recommended nearby interesting location is only discussed in our previous work [17].

2.2   Crowdsourcing

According to Brabham et. al [2], crowdsourcing is an online, distributed model for problem solving and production such that a company's production cost is substantially reduced. Crowdsourcing approach addresses a problem by utilizing the wisdom of crowd. In this approach an organization takes up a job and instead of getting it done by employees, outsource it to an undefined (and generally large) network of people in the form of an open call. The important precondition of crowdsourcing is the use of the open call format and the large network of potential laborers [10]. The problem is broadcasted by the source to a group of unknown people (crowd) for solution and crowd solves the problem and submits their solution back to the source. These solutions are now owned by the source and crowd may be remunerated monitory or with prizes etc. Intellectual satisfaction is one of the most important motivations for crowd in some cases. Crowdsourcing is useful when the task can be divided in subtasks and single person or computer cannot perform the task. Some of the popular crowdsourcing systems are-Wikipedia [19], Yahoo! Suggestion Board [20], Threadless [21], iStockphoto [22], InnoCentive [23], Sheep Market [24], Yahoo's flickr   [25], MobiMission [8], Gopher game [4] and CityExplorer [14] etc. Fig 1. shows the overview of crowdsourcing system. The requestor, platform & workers are main stackholders of the system. Initially, a requester submits the task/job to the platform. All the task management activities are performed by the platform. A popular example of crowdsourcing platform is Mturk- Amazon mechanical turk, which is a highly-available, cheap, programmable prototyping platform for crowd computing [26]. Workers/ crowd resources (sometimes also known as turkers) work on the tasks and return the solution back to the platform.

The use of WWW and the mobile phone with GPS facility has opened opportunities for mobile location aware crowdsourcing in which workers always carry their phones with them to provide them the chance to contribute at any time. In location aware crowdsourcing, location is used as criteria to distribute tasks to the workers. To understand the concept of location aware crowdsourcing, consider a scenario –"Jane is an hour's drive from some great sunset points. She decides to visit a sunset point, but she is not sure which one of these sites to go. She is not sure about mode of transport available, crowdedness and the current weather conditions nearby those sites. She decides to ask the crowd present on these places about the weather conditions, transportation mode and crowdedness. Within a few minutes she gets back information from other people around these sites that help her to decide which of the sunset points to visit". This scenario implies that location aware crowdsourcing may be useful to get real time information.

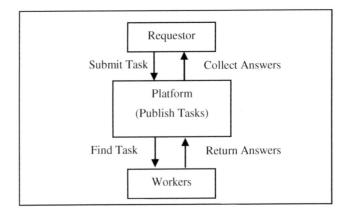

Fig. 1. Crowdsourcing Architecture

Various researchers focused on crowdsourcing based on the use of mobile phones. txteagle [6], is a mobile crowdsourcing system that allows crowd workers to earn money by completing simple tasks such as translation, transcription, and filling out surveys by using their mobile phones. Fashism [28], is a fashion web site allows customers to get comments on their fashion style while doing shopping by sending a dressing-room photo to the community and getting votes and get comments back from the crowds in real time. Google exploits the crowdsourcing idea to collect the road traffic data and provide the traffic conditions. The live traffic data is collected from the user's having Google Map [27] on their phone. Askus [12] is a mobile platform for supporting networked action which allows specifying tasks, which are then matched by the system to specific service person based on geographic location. But none of these examples exploits the user's physical location and context in real sense. Some of the tasks like one described above are location dependent and requires to find people nearby a particular location. Also, such tasks have time constraints which need to be satisfied.

2.3   Fuzzy Inference System

A fuzzy system [11], [16] can model the quantitative aspects of human knowledge and reasoning process without using precise quantitative analysis. Fuzzy inference is a process of transforming a given input to an output using fuzzy logic. This process involves the membership functions (A membership function defines how each point in the input space is mapped to a membership value between 0 and 1), fuzzy if-then rules and fuzzy operators. Fuzzy inference is mainly divided into five parts:

Fuzzyfication: When the crisp inputs are provided to the fuzzy inference system, the system determines the degree to which the given inputs belong to each of the fuzzy set. his process is achieved via membership functions.

Fuzzy Rule Antecedent Matching: Fuzzy inputs and the degree to which each part of the rule antecedent has been satisfied are known by this time. If antecedent comprises of more than one part, the fuzzy operators (min, prod etc.) are applied to get a single result. This single number can now be applied to the output function.

Fuzzy Implication: In this step the consequent part is evaluated. The input to this step is the single value from the previous step and output is a fuzzy value. Implication is evaluated for each rule using methods like min and prod etc.

Aggregation of Consequents: To make the final decision all the matched rules must be aggregated. By this process output is aggregated to a single fuzzy value. Example of aggregation methods are max, prober and sum etc.

De-fuzzyfication of Output: The output of previous step is a single fuzzy value which is now converted to a crisp value.

Fuzzy inference systems are generally of two types - Mamdani type and Sugeno type [16]. These two types differ in the de-fuzzyfication parts. In Mamdani type inference output is a fuzzy value and needs to be converted to crisp value. In Sugeno type output membership function is either linear or constant.

3   Overall System Design

This section presents the overall design of proposed system. The proposed system design is shown in fig 2. The proposed system consists of four sub parts namely mobile client, Spot recommender, crowd resources and Information Enrichment System (IES).

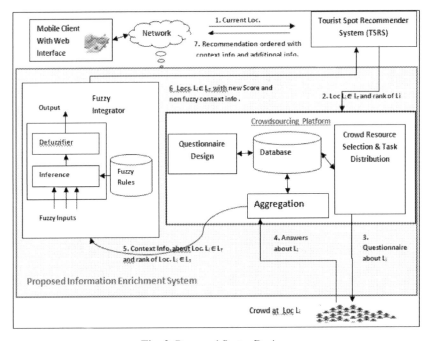

Fig. 2. Proposed Systm Design

The system works as follows-

1. The tourist spot recommender system (TSRS) reads current position of user. Based on user's current position and personal preferences it finds the list of spots (say L_T) in vicinity of the user for recommendations (Assuming a push based service).
2. TSRS sends each spot $L_i \in L_T$ with its rank to Information Enrichment System (IES).
3. TSRS reorder the list L_T of spots and recommend them to user
4. The crowdsourcing platform aggregates the answers from crowd at spot L_i and sends them to fuzzy integrator along with the rank of L_i obtained from TSRS computed in step 1.
5. IES's Crowdsourcing platform having a predefined set of questioner selects crowd currently present at spot L_i and distributes the questioner to them.
6. Crowd resources present at spot L_i answers the questioner with respect to spot L_i.
7. Fuzzy integrator based on the inputs from crowdsourcing platform and rank of spot L_i, computes a new rank of L_i and IES it back to the TSRS.

3.1   Mobile Client with Web Interface

Client is a mobile user who is the target user for pushing the tourist spot recommendation based on his/her personal preferences and current geographical location. Client is a registered user of the system. A new system is developed where crowd resources and requestors may register as fresh users. Current position of the registered user can be automatically obtained by GPS enabled device. The client gets the list of recommended spots ordered on the basis of personal preferences and current contextual information. Client requestor can also see the special information annotated with the rank which includes right time to visit, right transportation mode, security alerts or any other relevant information about the spots [17].

3.2   Tourist Spot Recommender System (TSRS)

Tourist Spot recommender system generates the personalized list of recommended spots L_T nearby client's current position. A personalized spot recommendation strategy similar to one proposed and implemented in [9] is implemented to obtain personalized tourist spot recommendations. However, any other existing spot recommender system or POI recommender system can also be plugged in here. When a mobile client checks in to a new location, the TSRS subsystem generates a ordered list L_T of popular tourist spots in proximity based on the user's personal interest and past preferences. In the proposed work, TSRS generates the list L_T of recommended tourist spots and request the Information Enrichment System (IES) to gather the additional information about each spot in the list L_T at current time. The IES gathers the contextual information of each tourist spot in list L_T using crowdsourcing approach and computes a new rank based on old rank computed by TSRS and gathered contextual information. IES sends back the new rank and some additional information back to TSRS. After getting back the new rank and additional information about each tourist spot in the list, TSRS reorder the list on basis of new score and augment it with the information obtained and push the recommendation to the mobile client.

3.3   Information Enrichment System (IES)

Information enrichment system consists of two sub systems namely crowdsourcing platform and fuzzy integrator. Details of these two sub systems are given in following subsections.

Crowdsourcing Platform

This section discusses the crowdsourcing platform for gathering contextual information about a spot at a specific time. The platform consists of following parts.

Questionnaire Design
The proposed crowdsourcing platform contains a predefined set of general questions about spots. This set of questions is sent to the crowed resources who answer them in context to their respective spot Li. Seven sample questions shown in fig. 3 are

considered for experiment purposes. However, other relevant questions may further
be added or modified based on crowd's feedback.

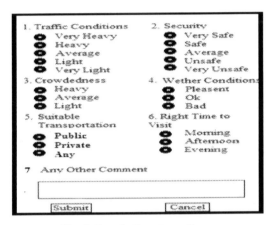

Fig. 3. Sample Questioner Form

Database

The database contains information about the registered users to the system. The proposed system obtains the current position (latitude, longitude, timestamp) of the registered users as soon as they checks-in to the system and keep this updated information
in database. The registered user can play role of client as well as crowd resource. This
database also contains the updated contextual information about popular tourist spots
in a geo-spatial region. This information is proactively gathered from the crowd available at that spot every few minutes. Data of last few hours for every spot is maintained in the system.

Crowd Resource Selection and Task Distribution

To gather contextual information about a spot we need to find the crowd resource at
that time. The selection of crowd resources is made primarily based on their current
geographical location. The registered resource people, who are present in the vicinity
of tourist spot L_i at current time, are selected to answer the questions about spot L_i. A
distance threshold of 300 meters is chosen for selection of crowd i.e. the registered
users present within a radius of 300 meters of a spot L_i are requested to contribute the
information. The crowdsourcing platform distributes the questionnaire to the crowd
resources for each spot L_i on regular interval. The responses from these resources
about spot L_i are then aggregated by result aggregation subsection.

To motivate the crowd some incentive model should be identified. In this work, we
assumed that crowd is willing to answer the question for community service and a
simplest crowd incentive model discussed here is used. However, in future a
crowdsourcing incentive model which can motivate several crowd resources to contribute information will be studied. In proposed system a score is maintained for each
crowd resource person. Initially score for a new resource person is set to zero. Whenever a resource person submits the answer his/her score is incremented by 2 points for

every answer which is considered correct by the aggregator and the score is decremented by 1 as a penalty for every answer which is not considered as correct by the aggregator.

Responses from the crowd resources may come in few minutes depending on the crowd's speed to answer. However, most of the information collected here does takes some time to change and therefore, in order to improve response time information about the popular spots is collected proactively on regular intervals, say every 15 minutes. Since, system always contains the updated information about the spot, it can be returned to TSRS subsystem in real time as soon as the client arrives at some spot and TSRS requests for information.

Result Aggregation

To aggregate the answers sent by crowd regarding spot L_i, some techniques which can ensure the correctness of the results should be applied. In proposed systems, the answers from majority of trustworthy resource persons (one with high reputation score) are considered to be the correct answers for simplicity. For quality control in crowdsourcing, for each resource person we estimate and maintain a trust score based on their past interaction with the crowdsourcing platform and which corresponds to their trustworthiness. The trustworthiness of user u_i is computed as shown in equation (1).

$$trustworthiness(u_i) = \beta * promptness(u_i) + (1 - \beta) * reputation(u_i) \quad (1)$$

Here β is a parameter which can be dynamically adjusted to give precedence to promptness of response or reputation. Initially trust score of each contributor is 10 at system startup and promptness of response may be given higher weightage then the reputation by adjusting the value of parameter β. The initial prompt answers from majority are considered correct and their reputation score is incremented by .5. For each user whose answer is considered correct by aggregator always gets an increment of 0.5 in their reputation score. If the answers are not considered correct their reputation score is decremented by 0.1. The trust score cannot go below -10. After some time, the value of β may be adjusted to give higher weightage to reputation. The aggregated answers from crowdsourcing platform are sent to the fuzzy aggregator. The set of correct answers arrived in last 15 minutes for each tourist spot L_i are aggregated.

Fuzzy Integrator

As traffic conditions, weather conditions, crowdedness, security and the user preference for a spot are all fuzzy in nature, therefore, fuzzy inference system is used for computing the new rating for each tourist spot L_i. The fuzzy integrator computes the new rating of each tourist spot L_i on the basis of fuzzy context parameters (traffic conditions, weather conditions, crowdedness, security) sent from crowdsourcing platform and the rank of spot L_i computed by personalized tourist spot recommender system. The other non fuzzy information about each spot is annotated with newly computed rank and returned back to TSRS. The details of fuzzy integrator are as follows.

Input / Output Variables

We have used five input variables for our fuzzy system namely weather, traffic, security, crowdedness and personalized rank. The fuzzy membership functions for each of these variables are shown in fig. 4 a), 4 b), 4 c), 4 d) and 4 e) respectively. The output of our fuzzy system is a new score/rank and its fuzzy membership functions are shown in fig 4 f).

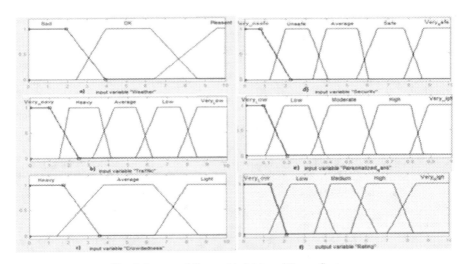

Fig. 4. Input and Output Variables of Fuzzy System

Fuzzy Rules

A set of 200 fuzzy if-then rules is designed intuitively for our system. Some of the sample rules are shown in the fig 5.

• *If (weather is bad) and (traffic is very heavy) and (crowdedness is heavy) and (security is low) and (personalized rank is very low) then (rating is very low)*
• *If (weather is OK) and (traffic is light) and (crowdedness is heavy) and (security is high) and (personalized rank is very high) then (rating is high)*

Fig. 5. Sample Fuzzy if-then Rules

These rules can further be optimized using structured learning techniques such as neural network.

Evaluate Antecedent

For evaluating antecedent, fuzzy min and max operators are used. When the antecedents are connected by "and" operator then min method is used. Whenever the antecedents are connected by "or" operator the max method is used.

Rule Implication and Aggregation Operator

For implication product method is used. The max method is used for aggregation of results.

Other non fuzzy context parameters such as right time to visit, right transportation mode and other comments are annotated directly with the spot L_i.

3.4   Crowd Resources

The crowd resources are registered users on the proposed system. This has already been mentioned that the proposed system can find the current geographical position of the registered user as and when he/she check-in at a spot. The resource person gets the alert for a task anytime anywhere and resource person may choose to perform the task or cancel it. If the resource person chooses to perform the task he/she is provided with a simple interface containing the predefined set of questions and user answers with those questions in context to his/her current location. These responses from crowd resource are stored in the database along with crowd resource person's userid, latitude and longitude of current location and timestamp. The answers submitted by the crowd resources come back to the platform and the platform then aggregates the responses from several crowd resources as discussed earlier [17].

4   Implementation, Experiments and Results

In this Section, we first present the implementation details. Second, we introduce experimental settings and third, some results are reported.

4.1   Implementation

In order to show the feasibility of the proposed concept, a prototype of proposed information enrichment system is implemented. TSRS, crowdsourcing platform and fuzzy integrator modules resides on server side. An Android application is developed to simulate crowd platform. For computing fuzzy crowd-source score, MATLAB fuzzy tool box is used.

Most Android devices allow determining the current geo-location. This can be done via a GPS (Global Positioning System) module, via cell tower triangulation or via WiFi networks. This system has used the GPS receiver in the Android device to determine the best location via satellites. When TSRS receives the request, it generates the list of top ten locations nearby user's current location for recommendation based on user's personal preferences. Personalization is achieved based on the user's profile information and his/her feedback for already visited locations in past. Location similar to the one's user has visited and liked in past are given higher preference. Similarity is computed using Persons's correlation coefficient method. A very basic personalized location based recommender system TSRS is developed using Java. However, any other type of recommender system for locations can be used to generate basic recommendations. The top four out of ten locations for some users are shown in table 1. MYSQL database management system is used for storing the data.

The registered resource person, when checks in at a new location get the questioner interface as a task. A registered user receives questioner interface only once for every location. However, if the user voluntarily wants to resubmit the answer for same location again, he/she may chose to answer the same set of questions after an interval of 30 minutes. The interface is developed in Android Software Development Kit (SDK) in Eclipse IDE. To retrieve user's current location, android.location API has been used. The answers for questioner with respect to their current location of user along with latitude, longitude of the locations, the timestamp and user id are submitted to the crowd platform and stored in the database. The fuzzy integrator is implemented in MATLAB.

The registered resource person, when checks in at a new location get the questioner interface as a task. A registered user receives questioner interface only once for every location. However, if the user voluntarily wants to resubmit the answer for same location again, he/she may chose to answer the same set of questions after an interval of 30 minutes. The interface is developed in Android Software Development Kit (SDK) in Eclipse IDE. To retrieve user's current location, android.location API has been used. The answers for questioner with respect to their current location of user along with latitude, longitude of the locations, the timestamp and user id are submitted to the crowd platform and stored in the database. The fuzzy integrator is implemented in MATLAB.

4.2   Experimental Settings

For assessment of proposed system, 104 volunteer users have been registered (73 males and 31 females) as crowd resource. Each volunteer user carry a GPS enabled android mobile phone along with them. All the users are residents of Delhi or nearby cities. For experiments, volunteers are randomly divided into groups of 10-12 users each and they are requested to visit the top popular locations of Delhi for a span of two weeks in group (so that we can get response from at least some users for every location). Platform fires the job request (consisting of questions shown in fig. 3) for each popular location repeatedly to crowd user present at that location in every 15 minutes. The volunteer users answered the questions sent in context to the current location. The answers submitted by those users are stored in database. The TSRS when wish to send the recommendation to some user say u_x, prepares the personalized list of locations L_T nearby user u_x's current location on the basis of his/her personal preferences.

This list L_T is sent to the information enrichment system. For each location $L_i \in L_T$, crowdsourcing platform's aggregation sub module queries the database and takes the answers received for location L_i in last 15 minutes, aggregate them. The aggregated answers about each location L_i along with the rank of L_i computed by TSRS are forwarded to fuzzy integrator. The fuzzy integrator computes the new rank for each location L_i using fuzzy context parameters and annotates other non fuzzy parameters. The newly computed rank and the additional information is sent back to TSRS for each location $L_i \in L_T$. TSRS reorders the list based on new rank and recommends them to the user u_x.

Table 1. Sample Result Generated by TSRS

S. No.	User Id	Current Location of Mobile Client Requestor	Top 4 Recommended Locations By TSRS			
			Loc-1	Loc-2	Loc-3	Loc-4
1	11	Lotus Temple Delhi	ISKON Temple (0.8)	Tughlaqabad Fort (0.73)	Qutub Minar (0.6)	KalkaJi Temple (0.4)
2	26	Patiala House Court Delhi	Sacred Heart Cathedral (Church) (0.72)	India Gate (0.7)	British Council (Library) (0.4)	National Gallery of Modern Art (0.32)
3	19	TGIP Mall NOIDA	Grate India Place Theme Park (0.8)	Haunted House (0.7)	Tughlaqabad Fort (0.4)	ISKON Temple (0.39)

4.3   Results

The sample result for user id 19 is shown in table 2. The first row in table shows user id of client requestor requesting the recommendation. The second row contains the current location of requestor; third row contains the top four recommended locations with rating generated by TSRS (the rating of a location based on users's preference). Finally, the last row contains new score computed by information enrichment system and the additional non fuzzy information with each recommended location. This result shows that the new rank is different from the original rank computed based on personal preference. The current context may modify the order of list based on the feedback of users currently present at a location.

Table 2. Information Enrichment

User Id-19				
Current Location of user: TGIP Mall NOIDA				
TSRS Rank	Grate India Place Theme Park (0.8)	Haunted House (0.7)	Tughlaqabad Fort Complex (0.4)	ISKON Temple (0.39)
New Rank Based on Current Context	Tughlaqabad Fort Complex (0.75) **Best time to Visit:** Afternoon **Transportation Mode:** Public **Comments:** Click here to see other comments	Grate India Place Theme Park (0.62) **Best time to Visit:** Evening **Transportation Mode:** Personal **Comments:** Click here to see other comments	Haunted House (0.51) **Best time to Visit:** Evening **Transportation Mode:** Private **Comments:** Click here to see other comments	ISKON Temple (0.32) **Best time to Visit:** Morning **Transportation Mode:** Public **Comments:** Click here to see other comments

4.4   Evaluation

With regard to evaluating the generated fuzzy crowd-sourced score, we required the requestor of recommendation to provide his/her feedback. Feedback is taken from 62 different requestors of the service who had used both the TSRS alone and the TSRS with enriched information several times (only those users are requested for feedback that have used both the services at least 2 times each). The first part of survey shows their satisfaction level for information enrichment service. The consolidated results of users are shown below in fig 6.

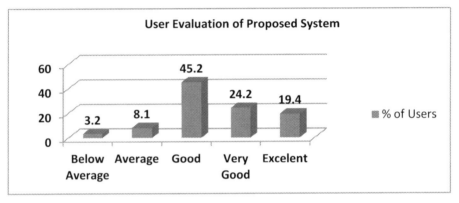

Fig. 6. Consolidate feedback for proposed information enrichment system

The second part of survey shows comparative satisfaction level after using TSRS and Information Enrichment services. The users ranked both the services on a scale of 1-10. The average rating of TSRS is 6.4 whereas for TSRS with information enrichment is 7.7. The result is reported in fig 7 which shows that most of the people more satisfied after contextual information is annotated.

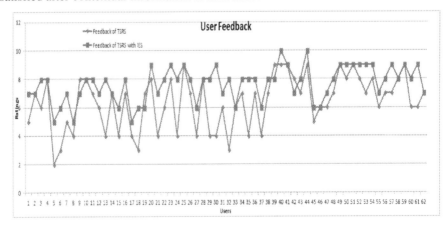

Fig. 7. TSRS VS TSRS with IES

The system is also evaluated over following evaluation metrics-

Accuracy: Accuracy is ratio of correctly recommended items and total number of recommendations made. Accuracy of the implemented recommender system is 0.94.

Response Time: To ensure the fast response time, the context information is collected from the crowd proactively every 15 minutes and therefore the IES system returns the new rank to recommender system within few seconds.

5   Conclusions

In this paper, a novel approach for information enrichment of a recommender system is proposed. The new score/rating for each location in recommendation list is computed based on current context of the location and additional contextual information is also annotated along with new score/rating. This may be helpful to enhance user's travel experiences. This concept when tied up with a location recommendation system helps mobile users (tourists) in selecting which of the popular location in vicinity should be visit. A prototype system is implemented to ensure the feasibility of concept. Also, feedback is obtained for proposed system from 62 real users and more than 89% found it useful. In future, we aim to perform more experiments and evaluation in order to verify the scalability of our system. Also, we aim to use some existing location based social network so that the problem of cold start may be addressed.

Acknowledgments. We express our gratitude to all the volunteers for their contribution towards the evaluation of proposed system.

References

1. Adomavicius, G., Tuzhilin, A.: Toward the next generation of recommender systems: A survey of the state-of-the-art and possible extensions. IEEE Transactions on Knowledge and Data Engineering **17**(6), 734–749 (2005)
2. Brabham, D.C.: Crowdsourcing as a model for problem solving an introduction and cases. Convergence: The International Journal of Research into New Media Technologies **14**(1), 75–90 (2008)
3. Cao, L., Luo, J., Gallagher, A., Jin, X., Han, J., Huang, T.S.: Aworldwide tourism recommendation system based on geotaggedweb photos. In: 2010 IEEE International Conference on Acoustics Speech and Signal Processing (ICASSP), pp. 2274–2277. IEEE (2010)
4. Casey, S., Kirman, B., Rowland, D.: The gopher game: a social, mobile, locative game with user generated content and peer review. In: Proceedings of the International Conference on Advances in Computer Entertainment Technology, pp. 9–16. ACM (2007)
5. De Meo, P., Quattrone, G., Ursino, D.: A query expansion and user profile enrichment approach to improve the performance of recommender systems operating on a folksonomy. User Modeling and User-Adapted Interaction **20**(1), 41–86 (2010)
6. Eagle, N.: txteagle: mobile crowdsourcing. In: Aykin, N. (ed.) IDGD 2009. LNCS, vol. 5623, pp. 447–456. Springer, Heidelberg (2009)

7. Gavalas, D., Kenteris, M.: A web-based pervasive recommendation system for mobile tourist guides. Personal and Ubiquitous Computing **15**(7), 759–770 (2011)
8. Grant, L., Daanen, H., Benford, S., Hampshire, A., Drozd, A., Greenhalgh, C.: MobiMissions: the game of missions for mobile phones, p. 12. ACM (2007)
9. Horozov, T., Narasimhan, N., Vasudevan, V.: Using location for personalized POI recommendations in mobile environments. In: International Symposium on Applications and the Internet, SAINT 2006, pp. 124–129. IEEE (2006)
10. Howe, J.: Crowdsourcing: A definition, crowdsourcing: tracking the rise of the amateur (2006)
11. Jang, J.S.: ANFIS: adaptive-network-based fuzzy inference system. IEEE Transactions on Systems, Man and Cybernetics **23**(3), 665–685 (1993)
12. Konomi, S., Thepvilojanapong, N., Suzuki, R., Pirttikangas, S., Sezaki, K., Tobe, Y.: *Askus:* amplifying mobile actions. In: Tokuda, H., Beigl, M., Friday, A., Brush, A., Tobe, Y. (eds.) Pervasive 2009. LNCS, vol. 5538, pp. 202–219. Springer, Heidelberg (2009)
13. Lu, X., Wang, C., Yang, J.M., Pang, Y., Zhang, L.: Photo2trip: generating travel routes from geo-tagged photos for trip planning. In: Proceedings of the International Conference on Multimedia, pp. 143–152. ACM (2010)
14. Matyas, S., Matyas, C., Schlieder, C., Kiefer, P., Mitarai, H., Kamata, M.: Designing location-based mobile games with a purpose: collecting geospatial data with CityExplorer. In: Proceedings of the 2008 International Conference on Advances in Computer Entertainment Technology, pp. 244–247. ACM (2008)
15. Takeuchi, Y., Sugimoto, M.: An outdoor recommendation system based on user location history. In: Proceedings of the 1st International Workshop on Personalized Context Modeling and Management for UbiComp Applications, pp. 91–100 (2005)
16. Tiwari, S., Kaushik, S.: A non functional properties based web service recommender system. In: 2010 International Conference on Computational Intelligence and Software Engineering (CiSE), pp. 1–4. IEEE (2010)
17. Tiwari, S., Kaushik, S.: Information enrichment for tourist spot recommender system using location aware crowdsourcing. In: 2014 IEEE 15th International Conference on Mobile Data Management (MDM), vol. 2, pp. 11–14. IEEE (2014)
18. van Setten, M., Pokraev, S., Koolwaaij, J.: Context-aware recommendations in the mobile tourist application COMPASS. In: De Bra, P.M., Nejdl, W. (eds.) AH 2004. LNCS, vol. 3137, pp. 235–244. Springer, Heidelberg (2004)
19. www.wikipedia.org/
20. www.suggestions.yahoo.com/
21. www.threadless.com/
22. www.istockphoto.com/
23. www.innocentive.com/
24. www.thesheepmarket.com/
25. www.flickr.com/
26. www.mturk.com/
27. www.maps.google.com/
28. www.fashism.com/

XML Schema Reverse Transformation: A Case Study

Hannani Aman[✉] and Rosziati Ibrahim

University of Tun Hussein Onn Malaysia, Parit Raja, Malaysia
{hanani,rosziati}@uthm.edu.my

Abstract. Reversing XML Schema to conceptual model has been a center of attention. Previous researchers use hierarchy structure of XML Schema component to generate class diagram. However, hierarchy structure involves only primary XML schema components. Our approach involves one more XML schema components, which are group of class and attribute with references. This additional component enriches class diagram results. In this paper, we construct circulation function of Library system to ensure that our formalization method has added a component of dependency relationship in class diagram. This rich class diagram may useful for other research such as evolution, modification and document adaptation in future.

Keywords: Reverse engineering · Formalization · XML schema · UML class diagram

1   Introduction

XML is collectively recognized as a standard format for interchange data among heterogeneous application. It has effectively become the typical storage and exchanging information in heterogeneous online applications. The standard formats are developed and publish in many working group based on respective industry. OASIS and European Union is an example of open community working group that developed their industry based XML rules [1], [2]. User can organize their own data based on XML data model directly developed by working group. XML rules has known as XML schema has been used as the logical schema of XML model. Typically, XML schema is describe by one or more XML files and lots of information is stored in unrelated application such as namespace or forms. For that reason, it is not normal for user to design the schema of XML data themselves.

Conceptual models are tools to capture requirements of an application developed. It helps to interpret the technicality of designer and native understanding of end user. In working group of XML based industry, there is no conceptual model documentation provided [2]–[4]. Many researchers have adapted forward engineering method to develop XML conceptual model [5]–[9]. But, only few publication has concern about reverse method that more reliable to generate XML conceptual model from XML application.

Reverse engineering process results a few kinds of conceptual model which are Class Diagram. [10]–[12], Conceptual XML(C-XML) [13], Entity Relationship

© Springer International Publishing Switzerland 2015
O. Gervasi et al. (Eds.): ICCSA 2015, Part IV, LNCS 9158, pp. 575–586, 2015.
DOI: 10.1007/978-3-319-21410-8_44

Diagram (ERD) [14], ORDB (Object-Relational Databases) [15] and an XML Tree Model(XTM) [16]. These few conceptual models are based on hierarchical structure of XML schema. However, this hierarchical method does not cover every each component of XML schema provided in W3C standard.

In this research, we face several interesting challenges. (1) XML Schema component has not fully transformed to class diagram. (2) Keeping conceptual model easy to generate and enrich the class diagram components. In this work, we extend our previous work to have a complete class diagram generated using XML schema component [17].

2   Related Work

XML Schema is the main resource of XML application used as input in this framework. XML schema used consists of XML construct which is elements, attributes, type definitions, model group and attribute group. A single XML schema may consist of a few pages of text description of a XML document. The text description is hardly read and hardly recognizes if any changes occur. It transform into a graphical form in order to easy understand using graphical diagram. In this discussion, the most accepted conceptual model used in this research is Class Diagram [5], [7], [14], [18]–[21].

Yu et al. [22] state that research on reversing does not yet have optimal standard of transformation. There are trades off in proposing method which is complexity of UML class diagram generated and amount of textual information preserved. Some of XML criteria in XML schema transformed are not incorporate in Class Diagram component. Salim et al. [10] reverse XML schema to class diagram using thirteen building blocks of XML schema without introducing new UML class diagram construct. The result is easy to understand and simple but some XML schema component has not yet been transformed which in her view is not very important. et al. [14] reverse Data Type Definition(DTD) to generate Class Diagram. However, DTD components have less criteria rather than XML Schema.

Fong et al. [16] has use nearly all the XML Schema components in generate conceptual model. Unfortunately the conceptual model generated is not UML as this research focus. Necaksy et al. [11] generate UML class diagram that results a lot of UML criteria which is inheritance, class, attribute, association and cardinality and nesting join. But, the result does only use a few XML Schema components which are element, complex type and nested particle. Table 1 shows xml schema component that has been used by previous researchers.

Earlier transformation method use transformation rules and algorithm [10], [14] , [16] to generate UML diagram. Rules and algorithm has been popular used proposed method. These method is improved with formal method [11] in giving more accurate transformation result with grammar based approach. These grammar based approach are based on XML hierarchy structure. However, these hierarchy method does not fully incorporate every each of XML schema components as stated in W3C [17]. Therefore, component's formal method based on XML Schema components is proposed to generate UML diagram. Based on Table 1, every each researcher did not incorporate Attribute Group, Model Group and Element/Attribute with ref components into their approach. Attribute and Model Group has been proposed in previous paper [18]. Therefore, this paper focuses on Element/Attribute with references component.

Table 1. XML Schema component involved

XML Schema component	[10]	[11]	[14]	[16]
Element	Yes	Yes	Yes	Yes
Attribute	Yes	Yes	Yes	Yes
Complex Type	Yes	Yes	Yes	Yes
Attribute Group	No	No	No	No
Model Group	No	No	No	Yes
Particle	Yes	Yes	Yes	No
Simple Type	Yes	Yes	Yes	Yes
Element/Attribute with references	Yes	No	No	No

Element and attribute are main components of XML schema which generate an XML document element with its characteristics. Meanwhile Element/Attribute with references defines a component that referred from the main component. The functions referred from, explain that any modification of the referred component may impact to the main component. This is due to the fact of dependency between elements.

Element and Attribute has been transforming into class if it is a global element. But if they have a references Element or Attribute nested to global element, can we show the relations within in diagram. Flora et al. [10] suggest artificial naming of this reference in the class generated but it is not shows the relations graphically. It is important to show the relations graphically to maintain the semantic of references element in diagram. Therefore, it is important to show in the class diagram the specific relationship of the referred component.

3   Reverse Method

3.1   Framework of Reverse Method

This section will explain briefly about framework to reverse an XML Schema and generates UML class diagram. An XML Schema consist rules guiding XML data to be use in an application. It describes relationships of XML data. As XML Schema is in textual form, it is hard to see and understand directly XML data relationship by non XML developer. In order to give understanding and keeping conceptual model update of XML Schema that may evolves, a reverse method of generating UML Class Diagram is required.

XML Schema is the source of this process. It transformed to UML Class Diagram component using few Transformation Rule. As informal transformation rules may lead to ambiguous results, a formal method is proposed. The formal method is using semantic matching between component in XML Schema and UML Class Diagram. The formal method used, to ensure the correctness and accurate transformation rules. Formalization of transformation rules has been use to generate a class diagram denoted as Formal Transformation Rules. The informal transformation rules and the formal approach has been discussed in details in [18] where the framework of the rules appeared in [23].

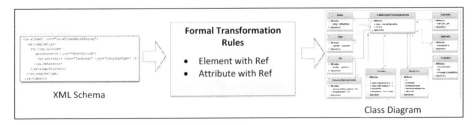

Fig. 1. Reverse Transformation Framework

XML Components consist of Element, Attribute, Complex Type, Attribute, Attribute Group, Model Group, Particle, Simple Type and Element/Attribute with references. Element, Attribute, Complex Type, Attribute, Attribute Group, Model Group, Particle and Simple Type has been explained [18]. Element and attribute with references components show the semantic of relation between global element and the references. Global element that use the references of element or attribute in its declaration show that the referred element or attribute is necessary used in the global element. Therefore, this relationship need to be shown clearly in the class diagram generated. This proposed relationship is explained in section 3.2.

3.2   The Transformation Rules (TR)

In this section, we summarize the transformation rules that are going to be used in this work. Previous work, the transformation rules consist of Definition 1 to Definition 11, Transformation 1 to Transformation 10 that reflect Model Group, Class, Attribute, Particle, Complex and Simple Type and Particle. In this work, Definition 11 and Transformation 10 are added. There are 2 category of XML Schema Class which is class definition and class references. Class definition has been defines in previous work. Class/Attribute reference is a class/attribute that has been referred from a class definition. The function is differs with class definition but the relationship with the references component is semantically same, aggregates with the class referred. Definition 12 explains class/attribute references condition.

Definition 12. *Let "dependency" be an association between classes. Thus it can be denoted as n-ary relationship of class.*

$$\underbrace{CLS \times \dots \times CLS}_{n}$$

where   (1)

$$dependency = (cls_1, \dots \dots, cls_n)$$

where

$$dependency \in ASSOCIATION \text{ and } cls_i \in CLS$$

Definition 13. *Let "Cr" be a group of class or attribute with ref .The group with ref is referring another group from its class. It is a subset of class. The Cr has occurrences to show number of group that may occur in the class referred. Thus it can be denoted as*

$$cr \in e \tag{2}$$

Transformation Rule (TR) 12 defines the dependency relationship between class and it's referred class or attribute.

TR 12. *Let $Trans(e_i, cr_n)$ be a transformation rules for dependency in UML class diagram if e_i is a subset element of cr_n. Thus it can be denoted a*

$$Trans(e_i, cr_n) = dependency \ll uses \gg \quad if \ cr_n \subseteq e_i \tag{3}$$

Table 2 shows the mapping of formalization Transformation Rules between XML Element and Class Diagram.

Table 2. Mapping XML Schema and Class Diagram based on Definition and Transformation Mapping

XML Element	Definition	Transformation Rule	Class Diagram
Global Element	D7, D9	TR1	Class, Attribute
Local Element	D7	TR1, TR8	Class, Attribute
Attribute	D7	TR4, TR9	Attribute
Model Group	D6	TR3,TR10	Class
Attribute Group	D5	TR2,TR11	Class
Simple Type	D2, D3	TR5	Class
Complex Type	D2, D4	TR5, TR6	Class
Nested Element	D10	TR7	Association
Particle	D7, D11	T6, T8, T9	Multiplicity, Ordering
Element/Attribute with references	D12, D13	TR12	Dependency <<uses>>

From this proposed Transformation Rules 12 shows that we have embedded dependency relationship to enrich the class diagram generated. Transformation Rule 1 to 11 have been explained in previous work [18].

4   Modeling Transformation XML Schema

In this section, we introduce a case study that showed the transformation formalization in previous section generates UML diagrams which incorporate one more components, which is Element and Attribute with references. We construct a Library system on circulation function as case study in Figure 2 based on our university library. We extract 3 global elements, a group model and an attribute group as shown in Table 3.

Table 3. XML schema of circulation function when users borrow a book

```
<schema xmlns="http://www.w3c.org/2001/XMLSchema">
<xs:attributeGroup name="MustAttribute">
<xs:attribute name="Id" type="int"/>
<xs:attribute name="Description" type="xs:string"/>
</xs:attributeGroup>

<xs:group name= "UserProfile">
<xs:sequence>
<xs:element name="name" type="string">
<xs:element name="faculty" type="Fac">
<xs:element name="course" type="Course">
</xs:sequence>
</xs:group>

<xs:element name="Borrow">
    <xs:complexType>
      <xs:sequence>
      <xs:group ref="UserProfile"></xs:element>
      <xs:element name="ItemType" type="xs:string" minOccurs="0" maxOc-
curs="unbounded"></xs:element>
<xs:element name="CirculationRule" type="xs:string" minOccurs="1" maxOc-
curs="1"></xs:element>
      </xs:sequence>
<xs:attributeGroup ref="MustAttribute"/>
</xs:complexType>
</xs:element>

<xs:element name="CirculationRule">
   <xs:complexType>
     <xs:sequence>
      <xs:ref name="LoanPeriod" ></xs:element>
      <xs:element name="BillingStructure" type="xs:string"></xs:element>
      <xs:element name="Chargeable" type="xs:string"></xs:element>
      <xs:element name="BookId" type="xs:int"></xs:element>
     </xs:sequence>
<xs:attributeGroup ref="MustAttribute"/>
   </xs:complexType>
</xs:element>

 <xs:element name="LoanPeriod">
    <xs:complexType>
      <xs:sequence>
        <xs:element name="Type" type="xs:string"></xs:element>
        <xs:element name="PeriodCount" type="xs:string"></xs:element>
        <xs:element name="timeDue" type="xs:string"></xs:element>
      </xs:sequence>
<xs:attributeGroup ref="MustAttribute"/>
       </xs:complexType>
</xs:element>

<xs:simpleType name="Fac">
    <xs:restriction base="xs:string">
     <xs:enumeration value="FSKTM"/>
     <xs:enumeration value="FKEE"/>
     <xs:enumeration value="FSTPI"/>
     <xs:enumeration value="FKMP"/>
    </xs:restriction>
</xs:simpleType>

<xs:simpleType name="Course">
    <xs:restriction base="xs:string">
     <xs:enumeration value="BIT"/>
     <xs:enumeration value="BSM"/>
     <xs:enumeration value="BDM"/>
     <xs:enumeration value="BEE"/>
    </xs:restriction>
</xs:simpleType>

</schema>
```

This case study based on Style Mix design schema [24]. This design uses various schema components (attribute/ element declarations, type definitions, model/attribute group definitions) to produce schemas. The design benefits moderate reuse of class. However, this kind of schema is difficult to read and understand. The difficulties make this case study suitable to do the reverse transformation.

4.1 Algorithm Transformation

Circulation Function case study is test in 4 steps algorithm followed:
- Every global XML Schema component transformed into Class and Attributes.
- Relationship generated.
- Particle transformed to Multiplicity.
- Class Diagram generated.

The following subsections discuss in details the steps involved in the transformation algorithm.

4.1.1 Every global XML Schema Component Transformed into Class and Attributes

First, convert all global components to be a class. There are *Borrow*, *CirculationRule*, *LoanPeriod* element (TR1). After the transformation Class *Borrow*, Class *CirculationRule* and Class *LoanPeriod* is generated. There is an attributeGroup named *MustAttribute* transform to Class *MustAttribute* (T2). There is a model group named *UserProfile* transform to Class *UserProfile* (T3). There are 2 simple type named *Fac* and *Course* also transform into stereotype <<type>>Class *Fac* and Class *Course* (TR5). Simple type has restricted the value used. Therefore, a class *String* is generated to show the inheritance function of this type.

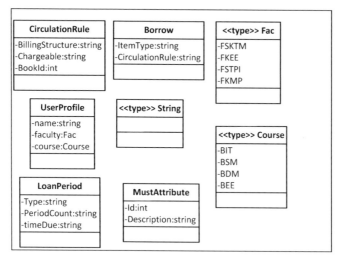

Fig. 2. Class with attributes generated

Second, convert local element that has no attribute be an attribute **(T1)**. Any attribute remain an attribute in class diagram. Figure 2 shows each class transform with their attributes.

4.1.2   Relationship Generated

Global class relates with its nested local class using association relationship. If a class has a nested Model Group or Attribute Group, aggregation relationship is generated (TR10). If the class has nested to other global class, then association relationship is used (TR7). If the class has nested other class with references, relationship dependency <<uses>> is used (TR10). Figure 3 shows relationship generated between classes (Dependency and Association).

Simple type has a class called String which is restricted for Class *Fac* and *Course*. Therefore, Class *Fac* and *Course* is inheriting the attribute of Class *String*. Figure 4 shows inheritance relationship between type class, Class *String*, Class *Fac* and Class *Course*.

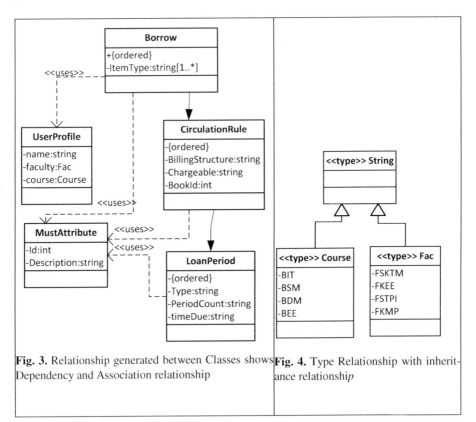

Fig. 3. Relationship generated between Classes shows Dependency and Association relationship

Fig. 4. Type Relationship with inheritance relationship

4.1.3   Particle Transformed to Multiplicity

Particle is use to show the occurrences indicator of an element or group or attribute of relate class. Each class that shows occurrences is transform into multiplicity of class and attribute. *ItemType* and *CirculationRule* has particle that transform into multiplicity. *ItemType* has minimum occurs of 0 and maximum occurs of unbounded transform into [0..*]. *CirculationRule* has one time of minimum and maximum occurrences transform into [1..1] (T8). Figure 5 shows each particle generates to multiplicity.

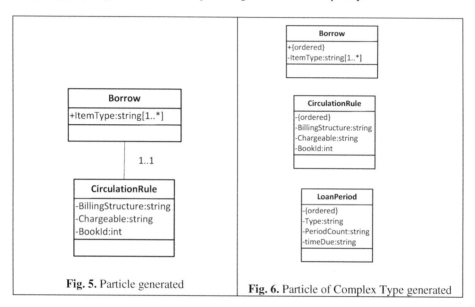

Fig. 5. Particle generated   **Fig. 6.** Particle of Complex Type generated

For particle of complex type, each have sequence particle. The sequence particle is transformed into attribute valued {ordered} (TR6). Figure 6 shows sequence particle of complex type transformed.

4.1.4   Class Diagram Generated

Finally, all these results are combined and generated a complete UML class diagram. Figure 7 shows the complete Class Diagram of function Circulation in Library.

The class diagram generated with our transformation formal method shows that dependency relationship is embedded. Previous method has not yet included element/attribute with references that transformed into dependency relationship in class diagram generated. Therefore, this shows that element/attribute with references are a reuse class may function as dependency with the global element declared. This proposed formal rule enriches the reverse transformation of XML Schema.

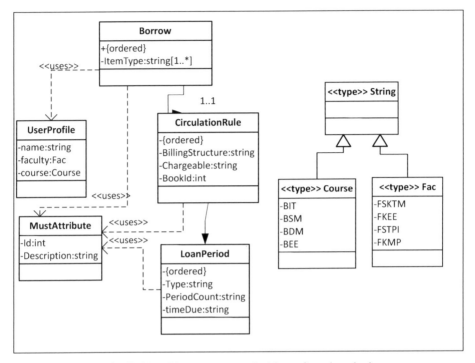

Fig. 7. Class Diagram generated with our formal method

5   Conclusion and Future Work

Though reverse transformation method from XML Schema to UML conceptual model has been developed previously, they include core component of XML Schema. They use hierarchy structure approach that reflect core component of XML Schema.

Our approach is embedded component used in XML Schema but has not yet been included which are model group and attribute group. This component has been introduced new relationship in the UML class diagram according to their semantic. Formal method is the best approach as it guaranteed the correctness of transformation. To ensure correctness of the formal method approach, library case study has been experimented. There are 4 simple steps of transformation algorithm has been shown. UML class diagram generates shows that the transformation is enriches with dependency association.

A complete reverse transformation has been proposed and generate a complete UML class diagram has been established. Using this transformation, any modification, evolution and updating the XML Schema is easier to keep documented and understanding.

Acknowldegement. The authors would like to thank to UTHM, Ministry of Higher Education, Malaysia for supporting this research under the Fundamental Research Grant Schema (FRGS).

References

1. ebXML Business Process Specification Schema Technical Specification v2.0.4. http://docs.oasis-open.org/ebxml-bp/2.0.4/OS/spec/ebxmlbp-v2.0.4-Spec-os-en-html/ ebxmlbp-v2.0.4-Spec-os-en.htm
2. EuroPass Specifications. http://interop.europass.cedefop.europa.eu/data-model/xml-resources/
3. Open Travel Specifications. http://www.opentravel.org/
4. EbXML Specifications. http://www.ebxml.org/specs/index.htm
5. Marchetti, E.: Automatic XML schema generation from UML application profile, vol. 485, pp. 485–487. Springer-Verlag Wien (2005)
6. Conrad, R., Scheffner, D., Freytag, J.-C.: XML conceptual modeling using UML. In: Laender, A.H., Liddle, S.W., Storey, V.C. (eds.) ER 2000. LNCS, vol. 1920, pp. 558–571. Springer, Heidelberg (2000)
7. Domínguez, E., Rubio, Á.L., Lloret, J., Pérez, B., Rodrı, Á.: Evolution of XML schemas and documents from stereotyped UML class models: A traceable approach. In: Information Software Technology, vol. 53, pp. 34–50 (2011)
8. Klettkem M.: Conceptual XML schema evolution. In: BTW Workshop Model Management und Metadaten-Verwaltung, Aachen 2007 (2007)
9. Necasky, M., Mlynkova, I., Klımek, J., Maly, J.: When conceptual model meets grammar. Data Knowledge Engineering **72** (2012)
10. Salim, F.D., Price, R., Indrawan, M., Krishnaswamy, S.: Graphical representation of XML schema. In: Yu, J.X., Lin, X., Lu, H., Zhang, Y. (eds.) APWeb 2004. LNCS, vol. 3007, pp. 234–245. Springer, Heidelberg (2004)
11. Necasky, M.: Reverse engineering of XML schemas to conceptual diagrams. In: Proceedings 6th Asia Pacific Conference on Conceptual Modelling, vol. 96, pp. 117–128 (2009)
12. Maly, J., Necasky, M.: Describing and verifying integrity constraints in XML using OCL. In: 2012 IEEE 19th International Conference on Web Services, pp. 657–658 (2012)
13. Al-Kamha, R.: Conceptual XML for System Analysis. Brigham Young University (2007)
14. Weidong, Y., Ning, G., Baile, S.: Reverse engineering XML. In: Proc. First Int. Multi-Symposiums Comput. Comput. Sci. (2006)
15. Widjaya, N.D., Taniar, D., Rahayu, J.J.: Aggregation transformation of XML schemas to object-relational databases. In: Böhme, T., Heyer, G., Unger, H. (eds.) IICS 2003. LNCS, vol. 2877, pp. 251–262. Springer, Heidelberg (2003)
16. Fong, J., Cheung, S.K., Shiu, H.: The XML Tree Model – toward an XML conceptual schema reversed from XML Schema Definition. Data Knowledge Engineering **64**(3), 624–661 (2008)
17. W3C XML Schema Definition Language (XSD) 1.1 Part 1: Structures. http://www.w3.org/TR/xmlschema11-1/
18. Aman, H., Ibrahim, R.: Formalization of transformation rules from XML schema to UML class diagram. International Journal Software Engineering and its Application **8**(12), 75–90 (2014)
19. Kudrass, T., Krumbein, T.: Rule-Based generation of XML DTDs from UML class diagrams. In: Kalinichenko, L.A., Manthey, R., Thalheim, B., Wloka, U. (eds.) ADBIS 2003. LNCS, vol. 2798, pp. 339–354. Springer, Heidelberg (2003)

20. Bernauer, M., Kappel, G., Kramler, G.: Representing XML Schema in UML – A Comparison of Approaches. In: Koch, N., Fraternali, P., Wirsing, M. (eds.) ICWE 2004. LNCS, vol. 3140, pp. 440–444. Springer, Heidelberg (2004)
21. Huemer, C., Liegl, P.: A UML profile for core components and their transformation to XSD. In: 2007 IEEE 23rd International Conference on Data Engineering Workshop, pp. 298–306 (2007)
22. Yu, Y., Wang, Y., Mylopoulos, J., Liaskos, S., Lapouchnian, A., Cesar, J., Leite, P.: Reverse engineering goal models from legacy code. In: Proceedings of the 2005 13th IEEE International Conference on Requirements Engineering (2005)
23. Aman, H., Ibrahim, R.: Reverse engineering: from XML to Uml for generation of software requirement specification. In: 2013 8th International Conference on Information Technology in Asia - Smart Devices Trend: Technologising Future Lifestyle, Proceedings of CITA 2013 (2013)
24. Quang, N.H., Rahayu, W.: XML Design Approach. Journal Web Information System **3**, 161–177 (2005)

Realisation of Low-Cost Ammonia Breathalyzer for the Identification of Tooth Decay by Neural Simulation

Ima O. Essiet[(✉)]

Department of Electrical Engineering, Bayero University, PMB, Kano 3011, Nigeria
imaessiet82@gmail.com

Abstract. The human mouth contains many kinds of substances both in liquid and gaseous form. The individual concentrations of each of these substances could provide useful insight to the health condition of the entire body. Ammonia is one of such substances whose concentration in the mouth has revealed the presence or absence of diseases in the body. One of such is tooth decay (caries) which occurs when there is insufficient concentration of ammonia in the mouth. This paper proposes an affordable ammonia breathalyzer designed using metal oxide sensor for the detection and prediction of tooth caries in humans with a 87% overall success rate. Selection of appropriate sensor was done via simulation using feed-forward artificial neural network (ANN). The breathalyzer has been designed and constructed to be low-cost such that it can be used for early detection and prevention of tooth decay.

Keywords: Ammonia · Breathalyzer · Caries · Odour sensor · Metal oxide semiconductor

1   Introduction

Tooth decay caused by cavities (caries) is a bacterial infection of the teeth whereby the biofilm on the teeth becomes infected. The biofilm thus becomes acid-producing (low pH) which then leads to cavitation or decay of the teeth [1,2]. A biofilm is a structured community of bacteria embedded within a self-produced matrix and attached to a living or inert surface. Anytime there is a fluid, a surface, and bacteria, there will be a biofilm. A diseased oral biofilm is one in which there is a population shift from normal oral bacteria to the acid-producing (cavity-causing) bacteria that cause dental cavities. Dentists have attributed the increase of acid-producing bacteria to several factors including absence of enough saliva and poor dietary habits. However, an interesting discovery has been the insufficient concentration of ammonia in the mouth [3]. In the past, dentists recommended gaggling with urine as a remedy for tooth caries. This is because they discovered that the bacteria in the mouth acted on the urea in the urine to release ammonia, thereby increasing the concentration of ammonia in the mouth [3]. This theory is still in use today as toothpastes are manufactured which contain compounds with ammonia by-products to increase its concentration in the mouth after brushing.

© Springer International Publishing Switzerland 2015
O. Gervasi et al. (Eds.): ICCSA 2015, Part IV, LNCS 9158, pp. 587–596, 2015.
DOI: 10.1007/978-3-319-21410-8_45

This paper involves the development of an ammonia breathalyzer which was tested on candidates with and without tooth caries in order to compare the ammonia concentrations in their breath. It was also tested on candidates with a high tendency for tooth decay (for instance those who use chewing sticks instead of fluoride-fortified toothpastes and those who ate a lot of candy). This second test was to test the breathalyzer's capability to predict tooth cavities before they actually occured. The breathalyzer is implemented using a metal oxide semiconductor sensor which has been trained to identify ammonia in gaseous form (e.g. in human breath). The sensor's sensitivity to ammonia is in the parts per million (ppm) range [4].

2   Theoretical Background

Metal Oxide semiconductor odour sensors have been around for years, for example see the review by Huang and Wan [5] and also Arshak et al [6]. Semiconducting metal oxides have been known to respond to various gases and other volatile organic compounds. They have found wide application in food industries [7,8], medical applications [9], agriculture [10] as well as the detection of hazardous gases [11].

In spite of their wide application, one of the greatest challenges facing the design and implementation of breathalyzers is that of cost. Most forms of breathalyzers that have been successfully implemented are too expensive for everyday use. As a result, their widespread use has been severely limited. The breathalyzer proposed in this paper has been designed to address this limitation. The components in the circuitry are affordable compared to most of the existing implementations.

Metal oxide semiconductor (MOS) sensors operate using doped semi-conducting metal oxides which employ the principle of resistance change as their method of odour detection. MOS sensors are advantageous because of their relatively low cost, high sensitivity and quick response time. They also have small size and can operate accurately under high temperature and pressure. Thus, the sensor is biased based on the detection of the odorant's *intensity* which is one of the components of an odour [5].

Feed-forward neural networks (as the name implies) allow signal movement in only one direction i.e. from input to output. There is no form of feedback associated with this type of neural network architecture. They are mostly suitable for pattern recognition purposes. There are basically three layers of information processing in a neural network. The first is called the input layer which accepts the information to be processed. The next is the hidden or process layer which acts on the information provided by the input layer along with weights connecting the input and hidden layers. The output layer supplies the results of the process layer along with the weights connecting the hidden and output layers. A neural network with one hidden layer is called a single-layer perceptron, while one with more than one process layer is called a multi-layer perceptron (MLP). Fig. 1 shows an MLP with one input layer (3 nodes), one process layer (4 nodes) and one output layer (2 nodes). For simple cases involving pattern recognition, a single-layer perceptron would suffice. However, for more complex cases of data classification, a multi-layer perceptron would give more accurate results [12].

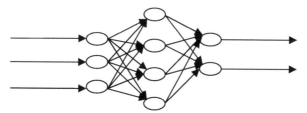

Fig. 1. Feedforward MLP with one input layer, one process layer and one output layer

A number of learning methods are available for training neural networks. The two most commonly used are supervised learning and unsupervised learning. In supervised learning, desired outputs are already known for associated input patterns. Whenever an input is applied to the neural network, the network's parameters are varied based on the difference between the desired and actual output of the neural network. Supervised learning methods include error-correction learning, reinforcement learning and stochastic learning. The aim of supervised learning is to determine a set of weights that minimises the error between the actual and desired outputs. A commonly used method is the least mean square (LMS) convergence method [13]. Due to the fact that the gas sensor's desirable performance is vital to ensuring that the breathalyzer gives accurate results, there is a need to simulate the chosen sensor's response to molecules of ammonia gas. This was achieved by testing the sensor's response to ammonia gas using a neural network. The results of the neural simulation revealed that the sensor identified ammonia in various concentrations with a 92.3% success rate [4]. The sensor being used in the design of the breathalyzer is a TGS 2602 MOS sensor manufactured by Figaro with ammonia sensitivity of 0.033V/1ppm.

3   Neural Network Design and Simulation

The neural network to be designed is to be used for the estimation of varying concentrations of ammonia in human breath for the diagnosis of the onset of tooth decay. Ammonia (NH_3) is one of the compounds whose increasing concentration in the human mouth can be an indication of the onset of tooth decay [2]. In other words, the response of various kinds of gas sensors was analysed using the neural network to determine which one was best suited for measuring oral ammonia in varying concentrations. The sensors being tested have an ammonia sensitivity in the range of parts per million (ppm). NeuroSolutions version 5 was used as the neural network training software.

Supervised learning is a method of training neural networks to automatically perform tasks whose outcome(s) is/are known. One common use of supervised learning is in data classification. The trainer is aware of the correct(desired) results of the classification process. The training process will therefore involve the continuous minimization of the error between the actual outputs of the neural network (OA_n) and the desired outputs (OD_n). The resulting error is given as:

$$\{E_n\} = \{OA_n - OD_n\} \tag{1}$$

The supervised learning paradigm of training neural networks is efficient and offers solutions to several linear and non-linear problems such as classification, plant control, forecasting, prediction and robotics [14]. The two most common forms of supervised learning include error-correction learning and memory-based learning. The method used in this work is the error back-propagation method which is a form of error-correction.

The error back-propagation method involves adjusting network weights in such a way that the error between the actual and desired outputs is minimized as much as possible [15]. This is done by computing the error derivatives of the associated weights. These error dervatives are obtained by error back-propagation. In error back-propagation, the steepest descent minimization method is used [12]. Assume we intend to adjust the weight between two neurons a and b. Weight and threshold coefficents can be adjusted according to the following:

$$w_{ab}^{k+1} = w_{ab}^{k} - \eta \left(\frac{\partial E}{\partial w_{ab}} \right)^{k} \tag{2}$$

$$v_{a}^{k+1} = v_{a}^{k} - \eta \left(\frac{\partial E}{\partial v_{a}} \right)^{k} \tag{3}$$

Where η is the learning rate ($\eta > 0$)

While there is no rule-of-thumb guiding the selection of a particular neural network model for a specific application, there are factors that can provide insight concerning the model that may yield the best results. Some of these include the number of training cases, the amount of noise, the complexity of the function or classification being learned, and also the method of training [16]. There are important aspects of neural network training which must be considered if the network is to yield useful results. These include:

(i) Transfer function: two functions determine the way signals are processed by neurons. The first is the activation function which determines the total signal a neuron receives. For a neuron a connected to a set of neurons b (for b=1.........N), sending signals n_b with the strength of the connections being W_{ab}, the total activation $A_a(b)$ is given as:

$$A_a(b) = \sum_{b=1}^{N} W_{ab} n_b \tag{4}$$

The second function determining the network's processing capability is the output function. The combination of both the activation and output functions determine the transfer function.

(ii) Interconnected Weights: these exist between the various layers of artificial neurons and are used in combination with the transfer function to vary the signals that each neuron receives. Assume two interconnected neurons a and b exchange information through a common axon. The weight associated with both neurons is given as W_{ab} such that $\{W_{ab} \in 0.....1\}$. The closer the value of W_{ab} is to 1, the greater the importance of the associated connection.

(iii) Neuron (node) Biasing: this is a real-valued factor that is used to vary the internal processing operations of a particular node within the network. It is also selected in combination with the transfer function.

(iv) Data Flow: here it has to be determined whether or not data flow is to be recurrent. In other words, the designer should ascertain whether data flow within the network layers should be fed forward continually or fed back at a point.

(v)Input Signals: there are a number of input signals that a neural network can process. These include binary, bipolar and continuous values. Binary input values can be either 0 or 1, while bipolar values are either -1 or 1. Continuous real numbers within a certain range constitute continuous input signals.

The structure of the neural network used is a multilayer perceptron feedforward network with 3 inputs and 2 outputs. The inputs and outputs are normalized values of both tooth decay and non tooth decay breath samples obtained by using a reference of 25ppm and 50ppm ammonia concentration for tooth decay and non-tooth decay samples respectively. These values were obtained according to the following relations:

$$X_{tc} = \frac{\text{Measured ammonia concentration for tooth decay sample (ppm)}}{25\text{ppm}} \tag{5}$$

$$X_{ntc} = \frac{\text{Measured ammonia concentration for non tooth decay sample (ppm)}}{50\text{ppm}} \tag{6}$$

Sigmoid transfer function is used in the hidden layer neurons and the learning is via error back-propagation (EBP) algorithm according to the following expression:

$$E_{12} = (t_{12} - a_{12}) \cdot a_{12} \cdot (1 - a_{12}) \tag{7}$$

Where E_{12} is the associated error for output nodes 1 and 2 respectively, t_{12} is the target activation for nodes 1 and 2 (desired outputs), a_{12} is the activation function for output nodes 1 and 2 respectively. The resulting error is then used to obtain process layer errors using the following relation:

$$E_k = a_i (1 - a_k) \cdot \sum_j E_j w_{ij} \tag{8}$$

Where E_k is error for process node i, a_k is activation for node k, E_j is output error of node j, and w_{ij} is weight connection between nodes i and j.

Due to the fact that the gas sensor's desirable performance is vital to ensuring that the sensor circuit gives accurate results, there is a need to simulate the chosen sensor's response to molecules of ammonia gas. The training cycle was selected as 1000 epochs which provided adequate time for the ANN to adjust to changes in associated synaptic weights. The weights between a hidden node i and output node j are adjusted according to the following expression:

$$\text{new } w_{ij} = \text{old } w_{ij} + \eta \cdot E_k \cdot x_i \tag{9}$$

Where $k = 1, 2$ and $j = 1, 2, \ldots 4$ for each set of 100 sample test points, w_{ij} is the weight between nodes i and j, E_k is error term calculated for output node j, x_i is the output of hidden node j and η is a constant value called the learning rate. The ANN has four hidden nodes. For the simulations carried out the learning rate was fixed at $\eta = 0.01$.

The neural network was used for data classification whereby the sensor's response to ammonia concentration in tooth decay and non-tooth decay breath samples formed the target values for training the neural network using supervised training. Tooth decay samples have an ammonia concentration of between 0 and 25ppm. Non-tooth decay samples have concentrations of 50ppm and above [4]. Two separate data sets were obtained from testing various samples of kidney failure and non-kidney failure samples with each set containing 100 test points. Each set of 100 test points consists of sample measurements being taken using impedance resonance method [17,18]. These two sets of test points were used for the neural network training sessions with a normalized standard of 0 to 0.5 for non-kidney failure breath samples and 0.6 to 1.0 for kidney failure samples respectively. The reason for the non-kidney test samples is to test the classification accuracy of the chosen sensor (via neural simulation). The sample points were normalized according to the following relation:

$$\frac{\text{odor sample concentration}}{\text{overall threshold concentration}} \times 1 \tag{10}$$

4   Neural Simulation Results

The neural network used in this work has three input nodes, four process nodes and two output nodes. Its structure on the NeuroSolutions breadboard is shown in Fig. 2. The normalised values of the hedonic tone were obtained between 0 and 1 with the latter representing extremely bad tooth decay breath samples and the former being non-tooth decay samples. Results obtained were for the two categories of odour analysis i.e. tooth decay and non tooth decay samples. Table 1 shows average overall accuracy of neural classification results using the MLP.

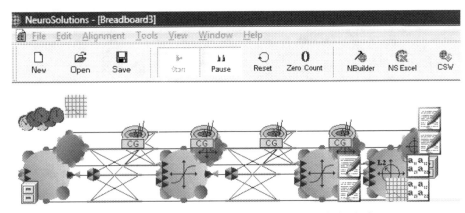

Fig. 2. Structure of MLP Used in Ammonia Breath Analysis

Table 1. Confusion Matrix showing Accuracy of Neural Network Results for Ammonia Breath Test Samples

	Tooth Decay (predicted)	Non tooth decay (predicted)	Accuracy %
Tooth Decay (actual)	85	15	85.0
Non tooth decay (actual)	11	89	89.0
Overall Accuracy			87.0

5   Design and Construction of Ammonia Breathalyzer Sensor Circuit

Based on the electrical characteristics of the chosen gas sensor, the circuit was designed with a DC power supply of 12V but was regulated to 5V in order to supply the sensor circuit and LM324 comparator Integrated Circuit (IC). A red LED was used to indicate proper circuit operation. The illumination (or not) of a green LED indicates whether the concentration of ammonia in a breath sample is below 0.3V or approximately 10ppm concentration of ammonia. A breath sample of below 10ppm ammonia concentration typically reveals that a candidate either has tooth caries or has the tendency to develop the ailment [1]. One kilo Ohm potentiometer was used to adjust the reference voltage so that the circuit can function properly in environments where air quality varies. A toggle switch is used for powering on the circuit. The constructed prototype circuit is housed in a plastic casing and is shown in Fig. 3.

6   Discussions of Circuit Testing and Results

The circuit was left on for 1 to 2 minutes without any sample being introduced to the sensor. This was to allow the sensor's heating element to sufficiently heat up, thereby ensuring rapid and accurate results. The circuit gave an output voltage of between 0.6V and 1.1V (DC) for good breath condition (between 18ppm and 33ppm ammonia), which activated the green Light Emitting Diode (LED). This voltage range is the reference for the comparator's inverting terminal.

Breath samples of five candidates with oral caries were tested using the proposed breathalyzer as well as five candidates without cavities. Each candidate was made to exhale through a rubber tube attached to the ammonia sensor for about 10 seconds while the circuit's DC output voltage was monitored to obtain the variation in ammonia concentration over the time interval. The candidates with cavities were each observed to have breath ammonia concentrations of as low as 9ppm (0.3V). This low ammonia concentration caused the green LED on the breathalyser to go off, thus indicating that oral ammonia concentrations in these candidates was low. The candidates without cavities on the other hand had ammonia concentrations of up to 50ppm (1.65V). Results of breathalyzer tests for both sets of candidates are depicted in graphical form in Figs. 4 and 5.

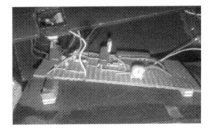

Fig. 3. Constructed Breathalyzer Circuit

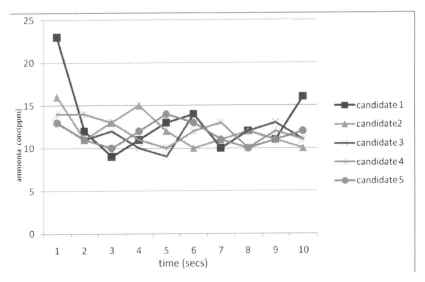

Fig. 4. Oral ammonia concentrations for 5 candidates with oral caries(cavities)

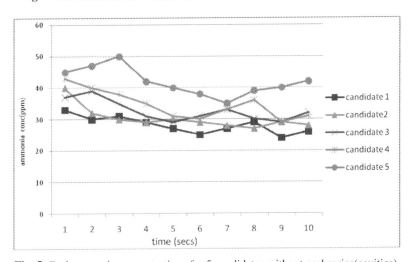

Fig. 5. Oral ammonia concentrations for 5 candidates without oral caries(cavities)

From the Figs. 4 and 5, it can be seen that the candidates without tooth decay have almost twice the concentration of oral ammonia content compared to those with tooth decay, which suggests that a low ammonia concentration in the mouth is a key indicator of tooth decay. Also, candidate 3 (Fig. 4) had the worst case of tooth caries and also the lowest concentration of oral ammonia among those tested. Candidate 5 (Fig. 5) had the highest concentration of oral ammonia and practised proper oral hygiene. From Fig. 5, candidate 1 had the lowest concentration of oral ammonia even though no cavities were noticed. However, the candidate admitted to practicing very poor oral hygiene (frequent eating of candy and consumption of sweet drinks without regular flouride brushing). This discovery suggests that the proposed breathalyzer can be used to also predict the onset of tooth cavities even before they manifest physically.

7   Cost Analysis

The National Science Foundation reports that a research team at Stony Brook University, New York is working on developing a breathalyser that costs 20 dollars (4,000 Naira) [19]. This is currently the cheapest breathalyser model so far. However, the total cost estimate of the ammonia breathalyser developed in this paper is 12 dollars (2,500 Naira) which makes it about 37% cheaper than that being developed at Stony Brook University. This makes the breathalyser affordable even in rural healthcare centers, especially in Africa and other developing nations.

8   Conclusion

Oral ammonia concentration below 18-20 ppm in the human mouth was observed to be an indication of tooth caries. Candidates with concentrations higher than this either had less severe or absence of tooth cavities. Candidates who practiced proper oral hygiene had twice the concentration of ammonia in their breath as those who did not. The proposed breathalyzer was also observed to indicate the onset of tooth decay, thereby making it suitable for the prediction of future tooth decay. As a result, it can be concluded that the breathalyzer can be used both to detect and predict tooth decay to a good degree of accuracy. The choice of sensor for the breathalyzer circuit was selected using simulation results obtained from neural simulations using NeuroSolutions software, thus making the neural network a valuable tool for real-time implementation of functional hardware.

References

1. Alaluusua, S.: Caries in the Primary Teeth and Salivary Streptococcus Mutans and Lactobacillus Levels as Indicators of Caries in Permanent Teeth. Ped. Dent., 126–130 (1987)
2. Dye, B.A.: Trends in Oral Health Status: United States, 1988-1994 and 1999-2004. Vit. Health Stat., 70–92 (2007)

3. Marsh, P.: Microbial Ecology of Dental Plaque and its Significance in Health and Disease. Adv. Dent. Res., 263–271 (1994)
4. Essiet, I.O., Dan-Isa, A.: Practical Discrimination of Good and Bad Cooked Food Using Metal Oxide Semiconductor Odour Sensor. Acta Periodica Technologica (2013)
5. Huang, J., Wan, Q.: Gas Sensors Based on Semiconducting Metal Oxide One-Dimensional Nanostructures. Sensors, 9903–9924 (2009)
6. Arshak, K.: A Review of Gas Sensors Employed in Electronic Nose Applications. Sensor Review, 181–198 (2004)
7. Balasubramanian, S., Panigrahi, S., Logue, C.M., Hu, G., Marchello, M.: Neural Networks-integrated Metal Oxide-based Artificial Olfactory System for Meat Spoilage Identification. Journal of Food Engineering, 91–98 (2009)
8. Berna, A.: Metal Oxide Sensors for Electronic Noses and their Application to Food Analysis. Sensors, 3882–3910 (2010)
9. Kodogiannis, V.S., Lygouras, J.N., Choudrey, H.S.: Artificial Odor Discrmination System Using Electronic Nose and Neural Networks for the Identification of Urinary Tract Infections. IEEE Transactions on Information Technology in Biomedicine, 707–708 (2008)
10. Kizil, U., Lindley, J.A.: Potential Use of Gas Sensors in Beef Manure Nutrient Content Estimations. African Journal of Biotechnology, 2790–2795 (2009)
11. Khawaja, J.: ASIC Gas Sensors Based on Ratiometric Principles. University of Warwick, Warwick (2009)
12. Svozil, D., Kvasnicka, V., Pospichal, J.: Introduction to Multi-layer Feed-forward Neural Networks. Chemometrics and Int. Lab. Sys., 43–62 (1997)
13. Khawaja, J.E.: ASIC Gas Sensors Based on Ratiometric Principles. University of Warwick, Warwick (2009)
14. Alkon, D.L.: Memory Storage and Neural Systems. Chicago. Scientific American (1989)
15. Sathya, R., Abraham, A.: Comparison of Supervised and Unsupervised Learning Algorithms for Pattern Classification. Int. J. Adv. Res. Art. Int., 34–38 (2013)
16. Rumelhart, D.D., Hinton, G.E., Williams, R.J.: Learning Internal Representations by Error Propagation. Glasgow (1986)
17. Hagan, M.T., Menhaj, M.B.: Training Feed-forward Neural Networks with the Marquardt Algorithm. IEEE Trans. Neur. Nets., 989–993 (1994)
18. Yim, H.S, Kibbey, C.E., Ma, S.C., Kliza, D.M., Liu, D., Park, S.B., Torre, C.E., Meyerhoff, M.E.: Polymer Membrane-based ion-, Gas- and Bio-Selective Potentiometric Sensors. Biosensors and Bioelectron, 1–38 (1993)
19. O'Brien, M.: This Breathalyzer Reveals Signs of Disease. National Science Foundation (2012). http://www.nsf.gov

Novel Designs for Memory Checkers Using Semantics and Digital Sequential Circuits

Mohamed A. El-Zawawy[1,2(✉)]

[1] College of Computer and Information Sciences,
Al Imam Mohammad Ibn Saud Islamic University (IMSIU),
Riyadh, Kingdom of Saudi Arabia
[2] Department of Mathematics, Faculty of Science,
Cairo University, Giza 12613, Egypt
maelzawawy@cu.edu.eg

Abstract. Memory safety breaches have been main tools in many of the latest security vulnerabilities. Therefore memory safety is critical and attractive property for any piece of code. Separation logic can be realized as a mathematical tool to reason about memory safety of programs. An important technique for modern parallel programming is multithreading. For a multi-threaded model of programming (*Core-Par-C*), this paper introduces an accurate semantics which is employed to mathematically prove the undecidability of memory-safety of *Core-Par-C* programs. The paper also proposes a design for a hardware to act as an efficient memory checker against memory errors.

Keywords: Operational semantics · Separation logic · Memory-safety · Parallel programs · Digital sequential circuits

1 Introduction

Memory safety breaches were used extensively in many of the latest security vulnerabilities [10,19,31]. This reflects how critical and attractive the property of memory safety is for any piece of code. Therefor not only does the absence of memory safety result in software defect which in turn results in abnormal termination of program executions, but also this absence can be employed maliciously towards security vulnerabilities. Memory safety takes several forms including memory leaks, dangling pointers, and buffer overflows [35].

In presence of shared mutable data structure and to reason about imperative programs, separation logic [26] was designed as enrichment of Hoare logic. Therefore separation logic may be defined as a mathematical tool to reason about memory safety of imperative programs. The enrichment included extending the assertion language with a "separating conjunction" to express that several sub-assertions hold for different regions of the memory. Also a "separating implication" was added to the assertion language of Hoare logic. Defining assertions inductively and the new assertions resulted in flexible and precise depiction of memories with regulated sharing [16].

© Springer International Publishing Switzerland 2015
O. Gervasi et al. (Eds.): ICCSA 2015, Part IV, LNCS 9158, pp. 597–611, 2015.
DOI: 10.1007/978-3-319-21410-8_46

An important technique for modern programming is multithreading [34]. The use of multiply threads is useful in many direction including (a) building interactive servers that are capable of connecting with multiple clients in parallel, (b)utilizing parallelism of multiprocessors that share memory, and (c) building complex user interfaces. Hence studying multithreaded programs and their memory safety are crucial and attract growing interest.

For a multi-threaded model of programming (*Core-Par-C*), this paper presents a formal semantics that is used to mathematically prove the undecidability of memory-safety of *Core-Par-C* programs. The paper also illustrates special cases when the memory-safety problem become decidable.

Shared mutable data structures are used by many areas like artificial intelligence and systems programming. These structures are typically mutable because there are many points for updating and referencing the fields of the data structures. Techniques for reasoning about this approach have been researched for many decades. Either extremely complex or not applicable (even to code of moderate length) techniques are mostly currently used to carry out this reasoning process. Very little research were done to achieve such reasoning process using hardware. However such hardware seems like the convenient solution for the complexity and scalability issues.

This paper also presents a design for a hardware to act as a memory checker against memory errors. The hardware is a digital sequential circuit. Basic operations used in designing the hardware are those used in presenting the separation logic. Also memory states modeled in the hardware design are those considered in separation logic. Therefore this hardware may be realized as a main step towards designing a digital sequential circuit to carry the verifications of separation logic.

More preciously, the second contribution of this paper can be realized as a first attempt to achieve the separation logic as a reasoning tool for shared mutable data structures using digital sequential circuites. Four type of commands are basics for the separation logic: allocation, disposal, mutation, and looking up. Therefore our proposed technique establish codes to these operations (Figure 5). There are four states of memory (an empty heap, a singleton heap, a separation heap, and an error) in separation logic to reason about the memory. The error memory is a memory being treated illegally (may be under attack) and the singleton memory has a single allocated cell. The state where many septated cells are allocated in the memory is denoted by the terminology *separation heap*. The four types of commands and that of memory states are the basics of the design of the proposed digital sequential circuite.

Contributions

This paper has the following contributions:

1. A formal proof that memory safety of multithreaded programs is undecidable.
2. A design for a hardware to act as a memory checker against memory errors.

Paper Outline

The rest of the paper is organized as follows. Section 2 presents the first contribution of this paper; proving the the undecidability of memory safety of multithreaded programs. Section 3 presents the second contribution of this paper; a hardware to carry memory checks. Related and future work are discussed in Section 4. The paper is concluded in Section 5.

2   Memory Checker Using Dynamic Semantics

For a multi-threaded proposed model of programming (*Core-Par-C*), this section presents a formal semantics which is later used to discuss the decidability of memory-safety of *Core-Par-C*. The section also proposes solutions to over come memory-safety difficulties discussed in the section as well. To do so, the memory-safety in *Core-Par-C* is defined formally to express that the safety of a program amounts to being safe through any potential execution of *Core-Par-C*. This also amounts to being safe under any potential effects that the semantics of the parallel command (*par*) and the memory allocation statement (*malloc*) may have. Generally and practically, the concept of memory safety, is undecidable due to the undecidability nature of termination.

For terminating *Core-Par-C* programs, this section proves undecidability of memory-safety. The consequences of this is a believe that any semantics of *Core-Par-C* is not able of statically or dynamically detecting memory-safety problems even for terminating programs.

$$n \in \text{Integers}, \ v \in \textit{Variables} , \ and \otimes \in \{-, +, \times\}$$

$$a \in \textit{A-expressions} ::= v \mid n \mid a_1 \otimes a_2$$

$$o \in \textit{B-expressions} ::= 1 \mid 0 \mid \neg o \mid a_1 \leq a_2 \mid a_1 = a_2 \mid o_1 \wedge o_2 \mid o_1 \vee o_2$$

$$c \in \textit{commands} ::= *v := a \mid v := a \mid \textit{malloc}(n) \mid v_1 := *v_2 \mid \textit{free}(n) \mid c_1; c_2 \mid$$
$$\textit{if } o \textit{ then } c_t \textit{ else } c_f \mid \textit{while } o \textit{ do } c_t \mid \textit{par}\{\{c_1\}, \ldots, \{c_n\}\} \mid$$
$$\textit{par-if}\{(o_1, c_1), \ldots, (o_n, c_n)\} \mid \textit{par-for}\{c\}.$$

Fig. 1. *Core-Par-C*: A Programming Model for Multithreaded Programming with Pointers

---

Figure 1 presents the syntax of our model for multithreaded programming with pointers; *Core-Par-C*. *Variables* is a finite set of program variables. There are three main commands to express the multi-threaded nature of programming. These commands are $par\{\{c_1\}, \ldots, \{c_n\}\}$ for parallel execution of commands, $par\text{-}if\{(o_1, c_1), \ldots, (o_n, c_n)\}$ for conditionally parallel execution of commands, and $par\text{-}for\{c\}$ for executing a randomly-chosen number of copies of c in parallel.

The following definition (Definition 1) presents the states of our proposed semantics.

Definition 1. *1. $E \in Env = Var \rightarrow Integers$.*
2. $M \in Mem = Integers^+ \rightarrow Integers$.
3. $P \in Ptr = Integers^+ \rightarrow Integers^+$.
4. A state is either an abort or a triple (E, M, P).

Boolean and arithmetic expressions semantics are built as usual. However we dot no allow pointers to get involved in Boolean and arithmetic operations. This is given in Figure 2. Semantics of statements of *Core-Par-C* is given in Figure 3.

$$\langle n \rangle E = n \quad \langle v \rangle E = E(v) \quad \langle 1 \rangle E = true \quad \langle 0 \rangle E = false$$

$$\langle a_1 \otimes a_2 \rangle E = \langle a_1 \rangle E \otimes \langle a_2 \rangle E$$

$$\langle \neg o \rangle E = if \ \langle o \rangle E \ then \ false \ else \ true$$

$$\langle a_1 = a_2 \rangle E = if \ (\langle a_1 \rangle E = \langle a_2 \rangle E) \ then \ true \ else \ false$$

$$\langle a_1 \leq a_2 \rangle E = if \ (\langle a_1 \rangle E \leq \langle a_2 \rangle E) \ then \ true \ else \ false$$

$$\langle o_1 \wedge o_2 \rangle (E, M, P) = if \ (\langle o_1 \rangle (E, M, P)) \ then \ \langle o_2 \rangle (E, M, P) \ else \ false$$

$$\langle o_1 \vee o_2 \rangle E = if \ (\langle o_1 \rangle E) \ then \ true \ else \ \langle o_2 \rangle E$$

Fig. 2. Semantics of Boolean and Arithmetic Expressions of *Core-Par-C*

Definition 2 introduces the formal definition of memory safety of *Core-Par-C* programs.

Definition 2. *A program in Core-Par-C is terminating if it has an execution path in our proposed semantics (Figures 2 and 3) that does not lead to an abort. A program in Core-Par-C is memory-safe if for all its possible execution pathes it is terminating.*

Definition 3 presents two programs that paly a vital rule in proving the undecidability of the memory safety of *Core-Par-C* programs.

Definition 3. *We let unsafe and safe be the Core-Par-C programs defined as follows:*

– *unsafe $\equiv par\{x := malloc(1), y := *x\}$, and*
– *safe $\equiv x := malloc(1); par\{z := *x, y := *x\}$.*

Theorem 1[1] uses Definitions 2 and 3 to introduce and formally prove the undecidability of memory safety of terminating and non-terminating *Core-Par-C* programs.

---

[1] This theorem can be realized as a generalization of the work in [28].

$$\frac{}{v := a : (E, M, P) \twoheadrightarrow (E[v \mapsto \ll a \gg E], M, P)}$$

$$\frac{E(v) \in dom(M)}{*v := a : (E, M, P) \twoheadrightarrow (E, M[E(v) \mapsto \ll a \gg E], P)}$$

$$\frac{E(v) \notin dom(M)}{*v := a : (E, M, P) \twoheadrightarrow abort} \qquad \frac{E(v_2) \in dom(M)}{v_1 := *v_2 : (E, M, P) \twoheadrightarrow (E[v_1 \mapsto M(E(v_2))], M, P)}$$

$$\frac{M' = M \otimes M'' \quad dom(M'') = \{p, \ldots, p + n - 1\}}{malloc(n) : (E, M, P) \twoheadrightarrow (E, M', P[p \mapsto n])} \qquad \frac{E(v_2) \notin dom(M)}{v_1 := *v_2 : (E, M, P) \twoheadrightarrow abort}$$

$$\frac{M = M' \oplus M'' \quad dom(M'') = \{p, \ldots, p + n - 1\} \quad P = P' \cup \{(p, n)\}}{free(n) : (E, M, P) \twoheadrightarrow (E, M', P')}$$

$$\frac{c_1 : (E, M, P) \twoheadrightarrow abort}{c_1; c_2 : (E, M, P) \twoheadrightarrow abort}$$

$$\frac{c_1 : (E, M, P) \twoheadrightarrow (E'', M'', P'') \quad c_2 : (E'', M'', P'') \twoheadrightarrow state}{c_1; c_2 : (E, M, P) \twoheadrightarrow state}$$

$$\frac{\ll o \gg E = true \quad c_t : (E, M, P) \twoheadrightarrow state}{if \ o \ then \ c_t \ else \ c_f : (E, M, P) \twoheadrightarrow state} \qquad \frac{\ll b \gg E = false \quad c_f : (E, M, P) \twoheadrightarrow state}{if \ o \ then \ c_t \ else \ c_f : (E, M, P) \twoheadrightarrow state}$$

$$\frac{\ll c \gg E = false}{while \ o \ do \ c_t : (E, M, P) \twoheadrightarrow (E, M, P)} \qquad \frac{\ll o \gg E = true \quad c_t : (E, M, P) \twoheadrightarrow abort}{while \ o \ do \ c_t : (E, M, P) \twoheadrightarrow abort}$$

$$\frac{\ll o \gg E = true \quad c_t : (E, M, P) \twoheadrightarrow (E'', M'', P'') \quad while \ o \ do \ c_t : (E'', M'', P'') \twoheadrightarrow state}{while \ o \ do \ c_t : (E, M, P) \twoheadrightarrow state}$$

$$\frac{(\exists \ \xi : n \to n). \ c_{\xi(1)}; c_{\xi(2)}; \ldots; c_{\xi(n)} : (E, M, P) \twoheadrightarrow state}{par\{\{c_1\}, \ldots, \{c_n\}\} : (E, M, P) \twoheadrightarrow state}$$

$$\frac{par\{\{if \ o_1 \ then \ c_1 \ else \ skip\}, \ldots, \{if \ o_n \ then \ c_n \ else \ skip\}\} : (E, M, P) \twoheadrightarrow state}{par\text{-}if\{(o_1, c_1), \ldots, (o_n, c_n)\} : (E, M, P) \twoheadrightarrow state}$$

$$\frac{\exists k. \ par\{\overbrace{\{c\}, \ldots, \{c\}}^{k - times}\} : (E, M, P) \twoheadrightarrow state}{par\text{-}for\{c\} : (E, M, P) \twoheadrightarrow state}$$

$$\frac{p' = min\{p \mid \{p, \ldots, p + n - 1\} \cap dom(M) = \emptyset\} \quad \begin{array}{l} M' = M \otimes M'' \\ dom(M'') = \{p', \ldots, p' + n - 1\} \end{array}}{malloc(n) : (E, M, P) \twoheadrightarrow (E, M', P[p' \mapsto n])}$$

Fig. 3. Semantics of Statements of *Core-Par-C*

Theorem 1. *The property of memory safety of terminating and non-terminating programs of the language Core-Par-C is undecidable.*

Proof. We suppose that memory-safety is decidable towards a contradiction. Suppose that $\psi(n)$ is a decidable attribute of natural numbers ($n \in \mathbb{N}$). The attribute $\psi(n)$ can be encrypted by a terminating memory-safe program ($P \equiv x := n; S$) of *Core-Par-C* which is Turing complete. A variable r cab be use in P such that $\psi(n)$ is true (does not hold) if and only if the execution of P ends in a state whose environment assigns 1 (0) to r. Now we consider the program

$$P' \equiv x := 0; par\text{-}for\{x := x + 1\}; if \ (r = 1) \ then \ safe \ else \ unsafe.$$

Clearly P' is terminating. Moreover, P' is memory safe if and only if r contains 1 whenever the program terminates. This amounts to the correctness of the attribute ψ for all natural numbers. This is a contradiction because by [15] there exists an attribute ψ such that the attribute $\forall n(\psi(n))$ is proper co-recursively-enumerable.

The details of the proof of Theorem 1 makes it clear that there are several sources of memory un-safety in the *Core-Par-C* programs. These sources are the semantics of the command *malloc* and that of the parallel commands: *par*, *par-for*, and *par-for*. A careful study of the problem reveals that there are restricted versions of these commands semantics that improves the memory-safety characteristics of the Core-Par-C programs. Figure 4 introduces these restricted semantics rules.

Definition 4 introduces the formal definition of *conservatively memory-safety* of *Core-Par-C* programs.

Definition 4. *A program in Core-Par-C is conservatively terminating if it has an execution path in the memory-safe restricted semantics (Figures 2 and 4) that*

$$\frac{c_1; c_2; \ldots; c_n : (E, M, P) \twoheadrightarrow state}{par\{\{c_1\}, \ldots, \{c_n\}\} : (E, M, P) \twoheadrightarrow state}$$

$$\frac{if \ o_1 \ then \ c_1 \ else \ skip; \ldots; if \ o_n \ then \ c_n \ else \ skip : (E, M, P) \twoheadrightarrow state}{par\text{-}if\{(o_1, c_1), \ldots, (o_n, c_n)\} : (E, M, P) \twoheadrightarrow state}$$

$$\frac{par\{\overbrace{\{c\}, \ldots, \{c\}}^{\mu-times}\} : (E, M, P) \twoheadrightarrow state}{par\text{-}for\{c\} : (E, M, P) \twoheadrightarrow state}$$

$$\frac{p' = max(dom(M)) \quad \begin{array}{l} M' = M \otimes M'' \\ dom(M'') = \{p', \ldots, p' + n - 1\} \end{array}}{malloc(n) : (E, M, P) \twoheadrightarrow (E, M', P[p' \mapsto n])}$$

Fig. 4. Memory-Safe Restricted Semantics of Some Statements of *Core-Par-C*

Operation	Binary code
Allocation	00
Disposal	01
Mutation	10
Looking up	11

Fig. 5. Codes of The Main Four Operations of Separation Logic

does not lead to an abort. A program in Core-Par-C is conservatively memory-safe if for all its possible execution pathes, in the memory-safe restricted semantics, it is conservatively terminating.

Theorem 2 uses Definitions 4 to introduce formally the decidability of *conservatively memory-safety* of *conservatively terminating Core-Par-C* programs.

Theorem 2. *The property of conservatively memory safety of conservatively terminating programs of the language Core-Par-C is decidable.*

The proof of Theorem 2 is by contradiction and is similar to that of Theorem 1.

3   Memory Checkers Using Digital Sequential Circuits

Many areas such as artificial intelligence and systems programming use shared mutable data structures. Such structures are typically mutable in the sense that there are many points for referencing and updating the fields of the data structure. For four decades techniques for reasoning about this approach have been researched. Most of the existing software methods that carrying this reasoning process are either extremely complex or not applicable even to code of moderate length. Very little research were done to achieve such reasoning process using hardware. However such hardware seems like the convenient solution for the complexity and salability issues.

More preciously, this section can be realized as a first attempt to achieve the separation logic as a reasoning tool for shared mutable data structures using digital sequential circuites. The separation logic is built on main four type of commands: allocation, disposal, mutation, and looking up. Therefore our technique starts by code these operations as in Figure 5. In separation logic the memory is reasoned about using four states of memory: empty heap, singleton heap, separation heap, error. The singleton memory has a single allocated cell and the error memory is a memory being treated illegally (may be under attack). The separation heap denotes the states where many septated cells are allocated in the memory.

Figure 7 presents a state diagram explaining effects of four main commands on the different memory states. Different memory states are represented by the nodes of the diagram. The the first two digits of the arc labels (denoted later by x and y) represent the command that transfers the memory from the source

State	Symbol	Binary code
Empty heap	S_0	00
Singleton heap	S_1	01
Separation heap	S_2	10
Error	S_3	11

Fig. 6. Main Four States of Memory and Their Codes

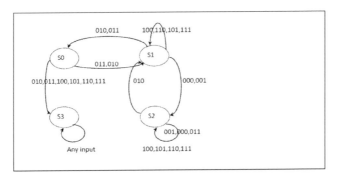

Fig. 7. State Diagram for Effects of Commands on Memory States

state to the target one. The third digit denoted by (MA) is a memory abstraction where 0 means that number of memory cells ≤ 2 and 1 means that number of memory cells > 2.

Figure 8 introduces the truth table of the state diagram of Figure 7.

From the truth table of Figure 8, we can conclude that new states of the memory can be represented as a function of the old states and the inputs which are the command to be executed and the memory abstraction. More precisely, the first column representing the new state can be described by the following equations:

$$A(t+1) = D_A(A, B, x, y, z) =$$

$$\sum(2, 3, 4, 5, 6, 7, 8, 9, 16, 17, 19, 20, 21, 22, 23, 24, 25, 26, 27, 28, 29, 30, 31) =$$

$$\prod(0, 1, 10, 11, 12, 13, 14, 15, 18).$$

This can be represented using the Karnaugh map of Figure 9 which produces the following equation:

$$A(t+1) = D_A(A, B, x, y, z) = AB + Az + Ay' + y'Bx'.$$

The second column representing the new state can be described by the following equations:

Present state		Input command code MA			Next state	
A	B	x	y	z	A	B
0	0	0	0	0	0	1
0	0	0	0	1	0	1
0	0	0	1	0	1	1
0	0	0	1	1	1	1
0	0	1	0	0	1	1
0	0	1	0	1	1	1
0	0	1	1	0	1	1
0	0	1	1	1	1	1
0	1	0	1	0	0	0
0	1	0	1	1	0	0
0	1	1	0	0	0	1
0	1	1	0	1	0	1
0	1	1	1	0	0	1
0	1	1	1	1	0	1
0	1	0	0	0	1	0
0	1	0	0	1	1	0
1	0	0	1	0	0	1
1	0	0	0	0	1	0
1	0	0	0	1	1	0
1	0	0	1	1	1	0
1	0	1	0	0	1	0
1	0	1	0	1	1	0
1	0	1	1	0	1	0
1	0	1	1	1	1	0
1	1	x	x	x	1	1

Fig. 8. The Truth Table of The Sequential Circuit Representing the State Diagram

$$B(t+1) = D_A(A, B, x, y, z) =$$

$$\sum(0, 1, 2, 3, 4, 5, 6, 7, 12, 13, 14, 15, 18, 24, 25, 26, 27, 28, 29, 30, 31) =$$

$$\prod(8, 9, 10, 11, 16, 17, 19, 20, 21, 22, 23).$$

This can be represented using the Karnaugh map of Figure 10 which produces the following equation:

$$B(t+1) = D_A(A, B, x, y, z) = A'B' + A'x + AB + yz'B'x'.$$

Now a corresponding digital sequential circuit that respects the truth table of Figure 8 and the equations $A(t+1)$ and $B(t+1)$ can be built as in Figure 11. This circuit was build in *Logisim* using two JK flip-flops. The circuit was tested for all cases and its correctness was approved.

All in all, what we have presented in this section is a design for a hardware that is capable of acting as a memory checker against some memory errors. The

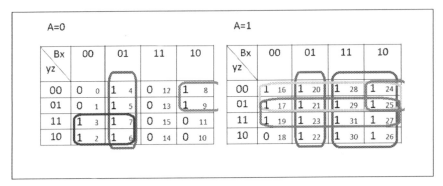

Fig. 9. The First Memory Checker State Diagram

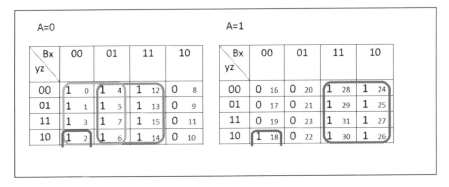

Fig. 10. The Second Memory Checker State diagram

hardware has the form of a digital sequential circuit. Basic operations behind the design of our proposed hardware are that used in presenting the separation logic. Also main memory states modeled in our design are that considered in separation logic. Therefore this hardware is the main step towards designing a digital sequential circuit to carry the verifications of separation logic.

4   Related and Future Work

This sections reviews work most related to our work. The section also discusses directions for future work.

Much research discuss the fact that a main source of unreliability in programs is violations related to memory access [10,19,31]. Research enumerates problems due to such violations. To avoid such problems many programming languages (such as C++) dynamically detect memory errors via software checks augmented to the programming language. Common disadvantages of the software checks include the reliability on inconvenient metadata, focussing on specific errors, execution overheads, and the reliability on manually changing the code [1,6,30].

Robustness is not paid enough attention compared to quick allocation with minimum fragmentation in most runtime systems. Heap corruption and double

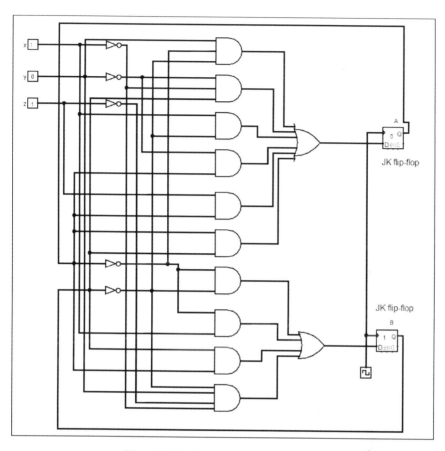

Fig. 11. The Memory Checker circuit

frees due to buffer overflows are caused by most coding applications of malloc; this is true even for the famous GNU C's library. Certain methods are used by some classic garbage collectors and memory controllers [3,27] to support the software robustness. Much more time is needed for reasonably achieving garbage collection than that required by malloc/free [13,36]. Techniques such as DieHard [5] avoids overwrites and invalid and double frees via separating heap from metadata. However DieHard [5] probabilistically (rather than absolutely) avoids dangling references. One more advantage of DieHard [5] over similar techniques is that it discovers unauthorized reads and protects heap content from buffer problems [1,20].

In [29] a static analysis and transformation, *MemSafe*, for guaranteeing protecting the memory safety of C is presented. *MemSafe* can be realized as a technique for casting temporal errors in the form of spatial errors. Merging characteristics of pointer- and object-based procedures, *MemSafe* provides a convenient representation for metadata. *MemSafe* provides a simple and optimal data-flow representation removing unnecessary software checks. However

MemSafe does not treat multi-core programs; it is built for single-core programs [7,32].

Several unsound methods [9,25,27] for preventing memory crashes in programs have been proposed. The idea of automatic pool allocation is to separate regions of memory into pools sharing the same type. This guarantees that dangling references are always replaced only by items of the same type [9]. Unpredictable safely-typed programs result from this method. It is possible to avoid artificial values and illegitimate modifications for manipulation of unprotected regions [27]. Most of these methods significantly increase the performance overhead which may result in faulty program executions. Some methods [25] repair distinguishable errors via using logging and checkpointing together with a file system. This is done via rolling back the software and employees an allocator to avoid defers frees, double frees, pads object requests, and zero-fills buffers [25]. Therefore rolling back techniques [25] are not convenient for softwares with unroll back-based modifications. The fact that rolling back techniques [25] cannot disclose inherent problems resulting in crashes and faulty program executions makes them unsound [11,17].

For C-like programs, a class of memory and type safety approaches [2,18, 22,33,35] ends program execution when discovering an error. Such approaches, like *Cyclone* and *CCured*, are called *fail-stop*. In *Cyclone*, programmers have an explicit and secure control over memory via a revised accurate type system attached to C [14]. *Cyclone* is classified as region-based memory management technique [12]. In *CCured*, the code is protected with dynamic test to guarantee memory safety. This technique also employees static analysis to get rid of tests at programs points that are guaranteed to be error-free [22]. To prevent dangling references and double frees, *CCured* uses a garbage collector. Concerning the underlaying program form, DieHard works with binaries and Cyclone, and CCured work with augmented versions of the source code. These augmentations are typically manually made.

The work in [28] focuses on C as the most popular programming tool to implement imperative systems. The low-level memory access provided by C via high-level abstractions and types makes it a perfect object of treating memory problems. The work in [28] relied on the facts that C enables casting, pointer arithmetic, and memory allocation and deallocation. This is very important to consider as such activities are not easy to use which leads to program bugs and security vulnerabilities such as dangling references and stacks overflows. Typically, memory safety of a program means that memory access errors never occur at runtime. In [28] memory safety is treated as the restrictive strict definition applicable for dynamic verifications. In [28], it is shown that generally checking memory safety is undecidable for C programs, as well for terminating closed programs. However, using a restricted concept of memory safety, [28] shows that dynamic verifications of C programs is decidable.

Many techniques were designed to detect both temporal and spatial memory errors. Protecting safety of heap-allocated objects and working on binaries, [35] is one of such techniques. Although focussing on store operations, the approach proposed in [35] improves the detection cost via the use of static analysis. Although being incompatible as a result of using fat-pointers, the technique presented in [2], *Safe C*, protects complete memory safety. Other techniques, such as [24], approaches the memory safety via establishing, in separate processes, separate performing checks and meta-data. The main disadvantage of such techniques is the need for additional CPU power.

In [21], type systems were used to reduce meta-data recording and eliminate the need to check pointer safety. This technique of [21] relies heavily on the use of fat-pointers which resulted in the need for code modifications and in compatibility problems. On the other hand the similar technique of [33] is more efficient and sound although suffering from issues concerting dealing with down casts. The issue with the yet similar technique of [23] is a serious runtime overhead. However the technique of [23, 29] is conveniently compatible with ANSI C. Utilizing a characteristic hardware, [8] presents a robust procedure to detect memory bugs.

Developing similar techniques to dynamically study the memory safety of other programming techniques like context-oriented programs and quantum programs is an interesting direction for future work. Another direction for future work is to develop a denotational semantics for dynamically checking the memory safety of programs. An important direction for future work is to develop the logic design of Section 3 to carry the memory verifications of separation logic.

5   Summary

This paper presented an accurate semantics which is employed to mathematically prove the undecidability of memory-safety of a multi-threaded model of programming (*Core-Par-C*).

The paper also proposed a design for a hardware that is capable of acting as a memory checker against some memory errors. The hardware is of the form of digital sequential circuit. Basic operations used in presenting the separation logic are that behind the design of our proposed hardware. Also main memory states considered in separation logic are that modeled in our design. Hence this hardware is a main step in the way to design a digital sequential circuit to achieve the separation-logic verifications.

References

1. Abdulla, P.A., Dwarkadas, S., Rezine, A., Shriraman, A., Zhu, Y.: Verifying safety and liveness for the flextm hybrid transactional memory. In: Macii, E. (ed.) Design, Automation and Test in Europe, DATE 13, Grenoble, France, March 18–22, 2013, pp. 785–790. EDA Consortium San Jose, CA, USA / ACM DL (2013)
2. Austin, T.M., Breach, S.E., Sohi, G.S.: Efficient detection of all pointer and array access errors. In: Sarkar, V., Ryder, B.G., Soffa, M.L. (eds.) PLDI, pp. 290–301. ACM (1994)
3. Baker, J., Cunei, A., Kalibera, T., Pizlo, F., Vitek, J.: Accurate garbage collection in uncooperative environments revisited. Concurrency and Computation: Practice and Experience **21**(12), 1572–1606 (2009)

4. Bensalem, S., Peled, D. (eds): Runtime Verification. 9th International Workshop, RV 2009, Grenoble, France, June 26–28, 2009. Selected Papers, volume 5779 of Lecture Notes in Computer Science. Springer (2009)
5. Berger, E.D., Zorn, B.G.: Diehard: probabilistic memory safety for unsafe languages. In: Schwartzbach, M.I., Ball, T. (eds) PLDI, pp. 158–168. ACM (2006)
6. Chatterjee, K., Prabhu, V.S.: Synthesis of memory-efficient, clock-memory free, and non-zeno safety controllers for timed systems. Inf. Comput. **228**, 83–119 (2013)
7. Damm, W., Dierks, H., Oehlerking, J., Pnueli, A.: Towards Component Based Design of Hybrid Systems: Safety and Stability. In: Manna, Z., Peled, D.A. (eds.) Time for Verification. LNCS, vol. 6200, pp. 96–143. Springer, Heidelberg (2010)
8. Dhurjati, D., Adve, V.S.: Efficiently detecting all dangling pointer uses in production servers. In: DSN, pp. 269–280. IEEE Computer Society (2006)
9. Dhurjati, D., Kowshik, S., Adve, V.S., Lattner, C.: Memory safety without runtime checks or garbage collection. In: Mueller, F., Kremer, U. (eds.) LCTES, pp. 69–80. ACM (2003)
10. Dillig, T., Dillig, I., Chaudhuri, S.: Optimal guard synthesis for memory safety. In: Biere, A., Bloem, R. (eds.) CAV 2014. LNCS, vol. 8559, pp. 491–507. Springer, Heidelberg (2014)
11. Godefroid, P., Kinder, J.: Proving memory safety of floating-point computations by combining static and dynamic program analysis. In: Tonella, P., Orso, A. (eds.), Proceedings of the Nineteenth International Symposium on Software Testing and Analysis, ISSTA 2010, Trento, Italy, July 12–16, 2010, pp. 1–12. ACM (2010)
12. Grossman, D., Morrisett, J.G., Jim, T., Hicks, M.W., Wang, Y., Cheney, J.: Region-based memory management in cyclone. In: Knoop, J., Hendren, L.J. (eds.) PLDI, pp. 282–293. ACM (2002)
13. Hertz, M., Berger, E.D.: Quantifying the performance of garbage collection vs. explicit memory management. In: Johnson, R.E., Gabriel, R.P. (eds.) OOPSLA, pp. 313–326. ACM (2005)
14. Jim, T., Morrisett, J.G., Grossman, D., Hicks, M.W., Cheney, J., Wang, Y.: Cyclone: A safe dialect of c. In: Ellis, C.S. (ed.) USENIX Annual Technical Conference, General Track, pages 275–288. USENIX (2002)
15. Rogers Jr, H.: Theory of Recursive Functions and Effective Computability. MIT press, Cambridge, MA (1987)
16. Lamport, L.: The hoare logic' of concurrent programs. Acta Inf. **14**, 21–37 (1980)
17. Lee, H.-C., Seong, P.-H.: A computational model for evaluating the effects of attention, memory, and mental models on situation assessment of nuclear power plant operators. Rel. Eng. & Sys. Safety **94**(11), 1796–1805 (2009)
18. Li, H., Gao, H., Shi, P., Zhao, X.: Fault-tolerant control of markovian jump stochastic systems via the augmented sliding mode observer approach. Automatica **50**(7), 1825–1834 (2014)
19. Marriott, C., Cavalcanti, A.: SCJ: memory-safety checking without annotations. In: Jones, C., Pihlajasaari, P., Sun, J. (eds.) FM 2014. LNCS, vol. 8442, pp. 465–480. Springer, Heidelberg (2014)
20. Nagarakatte, S., Martin, M.M.K., Zdancewic, S.: Watchdog: Hardware for safe and secure manual memory management and full memory safety. In: 39th International Symposium on Computer Architecture (ISCA 2012), June 9–13, 2012, Portland, OR, USA, pp. 189–200. IEEE (2012)
21. Necula, G.C., Condit, J., Harren, M., McPeak, S., Weimer, W.: Ccured: type-safe retrofitting of legacy software. ACM Trans. Program. Lang. Syst. **27**(3), 477–526 (2005)

22. Necula, G.C., McPeak, S., Weimer, W.: Ccured: type-safe retrofitting of legacy code. In: Launchbury, J., Mitchell, J.C. (eds.) POPL, pp. 128–139. ACM (2002)
23. Oiwa, Y.: Implementation of the memory-safe full ansi-c compiler. In: Hind, M., Diwan, A. (eds) PLDI, pp. 259–269. ACM (2009)
24. Patil, H., Fischer, C.N.: Low-cost, concurrent checking of pointer and array accesses in c programs. Softw., Pract. Exper. **27**(1), 87–110 (1997)
25. Qin, F., Tucek, J., Zhou, Y., Sundaresan, J.: Rx: Treating bugs as allergies - a safe method to survive software failures. ACM Trans. Comput. Syst. **25**(3) (2007)
26. Reynolds, J.C.: Separation logic: A logic for shared mutable data structures. In: 17th IEEE Symposium on Logic in Computer Science (LICS 2002), 22–25 July 2002, Copenhagen, Denmark, Proceedings, pp. 55–74. IEEE Computer Society (2002)
27. Rinard, M.C., Cadar, C., Dumitran, D., Roy, D.M., Leu,T., Beebee, W.S.: Enhancing server availability and security through failure-oblivious computing. In: Brewer, E.A., Chen, P. (eds.) OSDI, pp. 303–316. USENIX Association (2004)
28. Rosu, G., Schulte, W., Serbanuta, T.-F.: Runtime verification of c memory safety. In: Bensalem and Peled [4], pp. 132–151
29. Simpson, M.S., Barua, R.: Memsafe: ensuring the spatial and temporal memory safety of c at runtime. Softw., Pract. Exper. **43**(1), 93–128 (2013)
30. Singh, A., Narayanasamy, S., Marino, D., Millstein, T.D., Musuvathi, M.: A safety-first approach to memory models. IEEE Micro **33**(3), 96–104 (2013)
31. Ströder, T., Giesl, J., Brockschmidt, M., Frohn, F., Fuhs, C., Hensel, J., Schneider-Kamp, P.: Proving termination and memory safety for programs with pointer arithmetic. In: Demri, S., Kapur, D., Weidenbach, C. (eds.) IJCAR 2014. LNCS, vol. 8562, pp. 208–223. Springer, Heidelberg (2014)
32. Vazou, N., Papakyriakou, M.A., Papaspyrou, N.: Memory safety and race freedom in concurrent programming languages with linear capabilities. In: Ganzha, M., Maciaszek, L.A., Paprzycki, M. (eds.) Federated Conference on Computer Science and Information Systems - FedCSIS 2011, Szczecin, Poland, 18–21 September 2011, Proceedings, pp. 833–840 (2011)
33. Xu, W., DuVarney, D.C., Sekar, R.: An efficient and backwards-compatible transformation to ensure memory safety of c programs. In: Taylor, R.N., Dwyer, M.B. (eds.) SIGSOFT FSE, pp. 117–126. ACM (2004)
34. Yang, J., Cui, H., Jingyue, W., Tang, Y., Gang, H.: Making parallel programs reliable with stable multithreading. Commun. ACM **57**(3), 58–69 (2014)
35. Yong, S.H., Horwitz, S.: Protecting c programs from attacks via invalid pointer dereferences. In: ESEC / SIGSOFT FSE, pp. 307–316. ACM (2003)
36. Zorn, B.G.: The measured cost of conservative garbage collection. Softw., Pract. Exper. **23**(7), 733–756 (1993)

Towards a Wide Acceptance of Formal Methods to the Design of Safety Critical Software: An Approach Based on UML and Model Checking

Eduardo Rohde Eras[1], Luciana Brasil Rebelo dos Santos[1,2]([⊠]),
Valdivino Alexandre de Santiago Júnior[1],
and Nandamudi Lankalapalli Vijaykumar[1]

[1] Instituto Nacional de Pesquisas Espaciais (INPE), Av. dos Astronautas,
São José dos Campos - São Paulo 1758, Brazil
eduardorohdeeras@gmail.com, lurebelo@ifsp.edu.br,
{valdivino.santiago,vijay.nl}@inpe.br
[2] Instituto Federal de Educação, Ciência e Tecnologia de São Paulo (IFSP),
Av. Rio Grande do Norte, Caraguatatuba - São Paulo 450, Brazil

Abstract. The Unified Modeling Language (UML) is widely used to model systems for object oriented and/or embedded software development, specially by means of its several behavioral diagrams which can provide different points of view of the same software scenario. Model Checking is a formal verification method which has been receiving much attention from the academic community. However, in general, practitioners still avoid using Model Checking in their projects due to several reasons. Based on these facts, we present in this paper a significant improvement of a tool that we have developed which aims to translate several UML behavioral diagrams (sequence, activity, and state machine) into Transition Systems to support software Model Checking. With all the changes, we have applied our tool to a real space software product which is under development for a stratospheric balloon project to show how feasible is our approach in practice.

Keywords: UML · Model Checking · XMITS · Behavioral diagrams

1   Introduction

Almost 30 years ago, Parnas and Clements [21] argued that "... the picture of the software designer deriving his design in a rational, error-free way from a statement of requirements is quite unrealistic. No system has ever been developed in that way, and probably none ever will." After such a long time, we still face the problem of creating high quality software systems and there are many explanations for this fact. Poor Requirements Engineer processes [12,24,28], lack of traceability between requirements and other artifacts created within the software development lifecycle [14], and bad code smells [18,20,25] are just a few of the reasons for faulty software creation.

© Springer International Publishing Switzerland 2015
O. Gervasi et al. (Eds.): ICCSA 2015, Part IV, LNCS 9158, pp. 612–627, 2015.
DOI: 10.1007/978-3-319-21410-8_47

Despite the criticisms against the Unified Modeling Language (UML) [19], recent surveys still show that UML is indeed used in practice, even though its use might be specific to a particular part (e.g. design) of the project in many cases. [26]. Modeling systems for object oriented and/or embedded software development is an approach that has been employed by researchers and practitioners, specially by means of the several UML behavioral diagrams.

Model Checking [2,5,22] is a Formal Verification method which has been receiving much attention from the academic community due to its mathematical foundations. However, in general, practitioners still avoid using Model Checking in their projects due to aspects such as high learning curve and cost, and the lack of commercially supported tools. Thus, the level of automation for Formal Verification methods should be increased so that they can be used as easily as using a compiler. In this line, approaches that translate industry non-formal standards such as UML to Model Checkers notation are a great step towards a wide acceptance of Formal Methods in every day software development.

Considering everything that was exposed, we present in this paper a significant improvement of a tool [9] that we have developed which aims to translate several UML behavioral diagrams (sequence, activity, and state machine) into Transition Systems (TSs) to support software Model Checking. We improved our tool, XMITS, with many important features, such as: the complete development of all individual converters; the implementation of the Unified Transition System (TUTS) which takes all individual converted Transition Systems and unifies them into a unique formal Transition System; the automated translation of the unified Transition System into the NuSMV Model Checker by means of a grammar that we defined; the use of threads to significantly improve the implementation of the tool. In addition, we changed the platform modeling environment to Modelio. With all these changes, we also applied our tool to a real space software product which is under development for a stratospheric balloon project to show how feasible is our approach in practice.

This paper is structured as follows. In Section 2, we present some significant studies related to our work. Section 3 summarizes the methodology that is supported by our tool. All main details of the current version of the XMITS tool are presented in Section 4 as well as in Section 5 we show its application to the real space software system. Conclusions and future directions are presented in Section 6.

2   Related Work

Lima [17] propose a technique to convert a UML sequence diagram into PROMELA, the input language of SPIN verification system. Latella [16] made a rich approach to the conversion of a behavioral subset of UML Statechart diagrams into the SPIN Model Checker with a deep formalism. They aim to guarantee the correctness and efficiency of the conversion process. Lam [15] formally analyzed activity diagrams using NuSMV model checker. The objective was determining the correctness of activity diagrams. Eshuis [13] presented two translations from activity diagrams to NuSMV. The aim was to assess the activity diagrams from the point of view of requirements and also from the point of view of implementation, which represents the current system behavior. Dubrovin [10] implemented a tool that translates UML hierarchical state machine models to the input language of NuSMV. Uchitel [27] proposes translation of scenarios, specified as Message Sequence Charts (MSCs), into a specification in the form of Finite Sequential Processes. This can then be fed to the Labelled Transition System Analyser model checker to support system requirements validation.

All these previous studies deal with a single UML or UML-like diagram to perform Formal Verification. Rather, our tool allows to work with three kinds of UML behavioral diagrams in any number at once. In addition, it is not clear if in the previous studies the authors used specification patterns to formalize the properties. Specification patterns provide clear guidelines to such formalization.

Encarnación Beato [4] presents a tool (TABU - Tool for the Active Behaviour of UML) to convert three UML diagrams into a SMV input for formal verification: class, state and activity diagrams. It seems to see a close approach to our solution, using a tool for convert the XMI inputs into a SMV file. The difference is in the use of the Cadence SMV as a formal verification tool and the need to use all the diagrams to get a output.

Baresi [3] developed a tool to carry out Formal Verification of UML-based models, mainly interested in the timing aspects of systems. It is composed by: static part (class diagrams); dynamic aspects and behavior are rendered through: (a) state diagrams and activity; (b) sequence diagrams; and (c) interaction overview diagrams, used to relate different sequence diagrams; Clocks (and time diagrams) are used to add the time dimension to systems. All these diagrams seem to be required to construct the approach.

Cortellessa [7] proposes a methodology Performance Incremental Validation in UML (PRIMA-UML) aimed at generating a queuing network based performance model from UML diagrams that are usually available early in the software lifecycle (use case, sequence, and deployment).

The goal of our approach is to let the user free to use any number of the three accepted diagrams (with at last one Sequence diagram when working with more than one diagram). Most of the tools we mentioned seems to work with a only diagram at time or multiple required diagrams as input. We keep the idea to use multiple behavioral UML diagrams, representing complementary views of the system behaviors.

3   Approach to Detect Defects Based on UML Artifacts

A prerequisite for Model Checking is a property to be checked and a model of the system under consideration. In Figure 1, we show our approach aiming at Model Checking of software developed in accordance with UML. Activities which are shown in dashed line have been automated by the XMITS tool. In the following, we explain the main activities of the SOLIMVA 3.0 methodology [23].

(i) **identify scenarios** by looking at use case models. A use case can be viewed as a scenario. Each scenario is a set of related subscenarios tied together

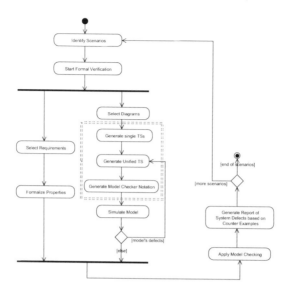

Fig. 1. Workflow of the proposed approach

by a common goal. The mainline sequence ('main success scenario' [6]) and each of the variations ('extensions and sub-variations') are the scenarios identified by our approach; (ii) **formalize properties**. For each selected scenario, we extract requirements from the textual description of use cases. After that, we formalize properties by means of specification patterns [11]; (iii) **generate single TSs**. Based on the available UML behavioral diagrams, we generate a single TS (finite-state model) and then an unified TS **(generate unified TS)**. These are the main activities performed by XMITS. Our approach does not demand that all three UML behavioral diagrams (sequence, activity, behavioral state machines) exist: it is enough to have the sequence diagram and one of the other two to generate the TS; (iv) **generate model checker notation**. Then, we translate the created TS to a model checker. As we will show, our tool accomplishes an automated translation from UML to the notation of the NuSMV Model Checker [1]; (v) Finally, we **apply Model Checking** to realize about defects in the behavioral description of the system represented by the UML diagrams. We repeat activities from (ii) to (v) for each selected scenario. Also note that activities (ii) and (iii)/(iv) may be accomplished in parallel.

4   The XMITS Tool

We have presented a very preliminar version of **XML Metadata Interchange to Transition System** (XMITS) in [9]. As we have already mentioned in Section 3, XMITS is fundamental to support our methodology to improve the design of

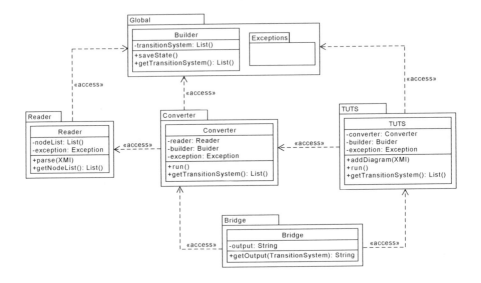

Fig. 2. XMITS 2: software architecture

UML-based software. Currently, XMITS is in version 2 and its architecture is shown in Figure 2.

The first version of XMITS [9] had only two modules: a Reader and a Converter. These two modules were enough to convert two types of UML behavioral diagrams (via their XMI files) into individual TSs: Activity Diagram (AD) and Sequence Diagram (SD). However, not all SD and AD elements were supported by the initial version of XMITS. In this first version, only sequence diagrams with the most simple combined fragments were supported, such as decisions, loops and optional. There was no support to parallel fragment. Both diagrams also got a shallow support to guard conditions. Table 1 summarizes the main differences between version 1 and the current XMITS's version.

As we could have seen in Figure 2 and in Table 1, XMITS 2 not only allows the translation from SD, AD but also from State Machine Diagram (SMD) into individual TSs, provides the unification of all these 3 TSs into a unique TS

Table 1. XMITS comparison: versions 1 and 2 (current)

	Version 1			Version 2		
Resource	SD	AD	SMD	SD	AD	SMD
Decision with Guard Condition	yes	yes	no	yes	yes	yes
Decision without Guard Condition	no	no	no	yes	yes	yes
Parallel	no	yes	no	yes	yes	yes
Transition System unification	no	no	no	yes	yes	yes
Transition System conversion to NuSMV	no	no	no	yes	yes	yes

and converts the TS output into a file in accordance with the NuSMV syntax. In addition, elements such as decision with and without guard condition, and parallel interaction operators are completely supported in XMITS 2.

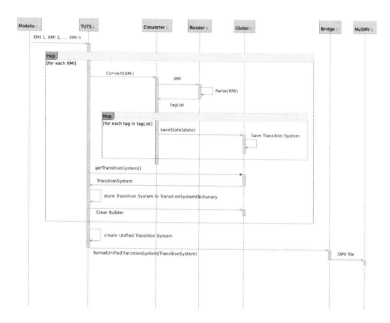

Fig. 3. XMITS 2 workflow

Figure 3 shows the workflow related to XMITS. The modules are called recursively by the system, so it is not necessary to call each module in the right order to get the final output. The user only need to call the last module most of the time. After the TUTS or the Converter, we already have an output, which is a Transition System. It is possible to call the Bridge module to translate the Transition System into the NuSMV notation. But the output provided by the TUTS module (as well as the output provided by the Reader and the Converter modules) are free for use as they are. The Global module also has a Printer tool to visualize the Transition System output directly in the Java terminal, or save as a *txt* file.

All the five modules work together following a flow to produce the final output, but some of then can be called many times or even anytime. The process begins by the XMI files produced by the Modelio software. XMITS can process n XMI files due to each combination of the 3 accepted UML behavioral diagrams. The only restriction is that our methodology, SOLIMVA 3.0 1 demands that at least one SD is available. For n XMI files, XMITS will call n instances of the reader and n instances of the converter. All the converted outputs go into the TUTS (The Unified Transition System) that joins the Transition Systems into a unique TS.

4.1   The Reader Module

To process any XMI input, the first action is to convert the file to a useful format. The reader module is responsible for convert the XMI into a structure for the Converter module. The input file, which is a XML file, is processed by the SAX (Simple API for XML). The reader module uses this API to save all the content of the input into a linked list, storing each tag in a Java class according to its characteristics.

4.2   The Converter Module

The major changes in our tool have been done in Converter module. The actual architecture is: one module for each diagram with a central *Diagram Selector* to define which logic to use. The Sequence Diagrams were processed by a logic based on a *Main Loop* and the Activity diagram logic was based on a *Function Dictionary* structure. The architecture based on dictionaries is significantly better than the previous. There is a logic based on the XMI input with many specific details. Each logic has its own way to process the guard conditions and it only accepts a very limited input format. In XMITS 2 we have a Converter able to process all the three desired UML diagrams with a single logic, generalized to process any diagram. This logic is based on dictionaries as the Activity logic in the first version, but without any specific XMI characteristic, making it significantly smaller and clearer. These facts assure the same resources for any diagram because there is a only logic for all of them.

After the XMI file is processed by the reader, the Converter first parse its content to confirm if its a valid UML diagram and which diagram it is. After the identification the diagram is redirected to its specific collector. The collectors are responsible for reading the file line by line and classifying it's elements into six categories: *State, Fork, Join, Decision, Connection* or *Default*. Those are the basic structures used by the Converter's logic to define all possible elements in a UML diagram, the only not mentioned element is the *Multiple State*, a virtual element used to simulate a parallel processing. For the proposes of this tool, the XMITS accepts only the Sequence, Activity and SMD UML diagrams but, if anytime be necessary to upgrade the system for more diagrams, all we need is to increment a new collector in the *collectors* package. Once the diagram is formatted in this new generic structure its redirected to the logic. This structure is a kind of state machine, so the logic will first get its first state and start to process all the transitions an subsequent states. For each kind of element in the mentioned six generic elements we have a specific function in the *function dictionary*. The functions process the information of the incoming element and call a specific function to process it's transition to next state how will find the specific function in the dictionary to process whatever the next state is. While the elements are been processed the functions call a instance of the *Builder*, a important class of the Global module responsible to create the Transition System output. The *Builder* class always returns its own instance, so we never have more than one *Builder*, no matter how many times it has been used in the system. The Fig.4 shows the flow diagram of the Converter module.

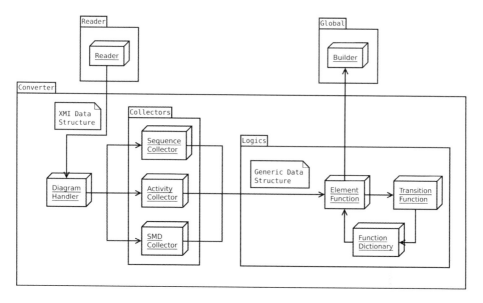

Fig. 4. Flow diagram of the Converter Module

4.3 The TUTS Module

The Unified Transition System (TUTS) module is a implementation of a methodology proposed by Santos [23], and its responsible to unify two or more Transition Systems into a only unified Transitions System output. The TUTS is composed of five packages: *dictionaries, interfaces, facade, logics* and *tools*. A structure based on a *Function Dictionary* is used by this module, just like in the Converter. The *dictionaries* package holds all dictionaries used by the module. The *interfaces* package defines the basic function interfaces to generalize the code. All dictionaries are initialized by a facade class stored in the *facade* package. The core processing happens inside the *logics* package. Finally, the *tools* package provides many important functions for the unification process.

The flow of processing inside the TUTS is based on an iteration on the Transition System inputs, unifying each state of each Transition System according to a rule. Each of those Transition Systems represents an UML diagram transformed by the Converter Module. They are iterate all together, side by side, like cars running in a race track. To walk from a state to another, the iterator needs to obey rules in a specific order. All those rules are stored in a *Rule Dictionary*, in the *dictionaries* package. If the states at the moment fits the first rule, there will be a specific kind of iteration, if not, the system tries to apply the second rule, and so on. The last rule is a "end of race", a function that simply stop all the iterators because there is nothing more to iterate, *i. e.*, all the diagrams reach they last state. This process is observed in the Fig. 5.

The first rule expect find all states in the iterators position with the same guard conditions. This rule means "all states are equal", and is the only time

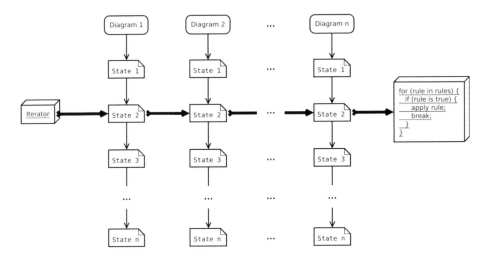

Fig. 5. The iteration over the diagrams in the TUTS module

in the process when the *Builder* class is called to generate a output. A unified state is possible if, and only if all the states are equal. If all states are not equal then the system will try the second rule. The second rule will look for a scenario where some diagrams change they guard conditions but the others didn't. In this hybrid condition the iterator will advance just the diagrams how didn't change to keep in the same position just equal states. If the second rule didn't apply than the system tries the third rule, where all diagrams are in different states. In this case there is no way to create a unified state, so all diagrams start to run in a different thread with no interaction with the others. If even the third rule applies there is a final rule to end all the process.

Every time where the diagrams can not be unified the system divide the process and a new thread is create. This process happens when a division occurs in the diagram or in the third rule when all states are different. In this way the TUTS avoid join different states and simplify the process. While there is more than one thread running the builder class fall into a concurrency of process trying to save a Transition System. To solve this, the save function runs into a critical region.

4.4   The Bridge Module

The Bridge module is responsible to translate the output created by the Converter module or by TUTS into the notation of the NuSMV Model Checker. Bridge accomplishes a model transformation and it uses the context-free grammar shown in Fig. 6. Its possible to visualize the output directly in the Java terminal or save direct in a file.

To make this conversion, the Bridge module iterates over the Transition System several times, one for each section of the output file. First, the *formater()*

Defining a subset of NuSMV input language using Backus–Naur Form.

<nusmv-code> ::=	"MODULE main" <EOL>
	<variable-declaration>
	"ASSIGN" <EOL>
	<statements>
	["CTLSPEC" <formalized-ctl-logic>	"LTLSPEC" <formalized-ltl-logic>]
<variable-declaration> ::=	<variable-declaration>	"VAR" <EOL>
	<identifier> ":" <EOL>
	<enumeration-type> " ;"	<boolean-type> " ;"
<identifier> ::=	<ident-first-charact>	<identifier ident-consecutive-charact>
<enumeration-type> ::=	"{" <integer-numbers> "}"	"{" <symbolic-constants> "}"	"{" <integer-numbers><symbolic-constants> "}"
<boolean-type> ::=	"TRUE"	"FALSE"
<ident-first-charact> ::=	"A"	"B"	"C"	"D"	"E"	"F"	"G"	"H"	"I"	"J"	"K"	"L"	"M"	"N"
		"O"	"P"	"Q"	"R"	"S"	"T"	"U"	"V"	"W"	"X"	"Y"	"Z"	"a"	"b"
		"c"	"d"	"e"	"f"	"g"	"h"	"i"	"j"	"k"	"l"	"m"	"n"	"o"	"o"| "q"| "r"
		"s"	"t"	"u"	"v"	"w"	"x"	"y"	"z"	"_"
<ident-consecutive-charact> ::=	<ident-first-charact>	<digit>	"$"	"#"	"-"
<digit> ::=	"0"	"1"	"2"	"3"	"4"	"5"	"6"	"7"	"8"	"9"
<integer-numbers> ::=	<digit>	"-"<digit>
<symbolic-constants> ::=	"A"	"B"	"C"	"D"	"E"	"F"	"G"	"H"	"I"	"J"	"K"	"L"	"M"	"N"
		"O"	"P"	"Q"	"R"	"S"	"T"	"U"	"V"	"W"	"X"	"Y"	"Z"	"a"	"b"
		"c"	"d"	"e"	"f"	"g"	"h"	"i"	"j"	"k"	"l"	"m"	"n"	"o"	"o"| "q"| "r"
		"s"	"t"	"u"	"v"	"w"	"x"	"y"	"z"	
<statements> ::=	<statements>	<initial-state> <transition-relation>
<initial-state> ::=	<initial-state>	" "	
	"init" "(" <identifier> ")" := " <enumeration-type> " ;"	<boolean-type> " ;"
<transition-relation> ::=	<transition-relation>	" "	
	"next" "(" <identifier> ")" := " <EOL>
	"case" <EOL>
	<transicao> <EOL>
	"TRUE: " <identifier> " ;" <EOL>
	"esac ;"

Fig. 6. Context-Free Grammar to convert the output of the Converter or of the TUTS into the NuSMV Model Checker notation

function iterates over all *String* values looking for non accepted characters and replace then in order to apply the grammar. Then, there is one function dedicated to create each section of the NuSMV file: header, variables, initial state (*Assign*), transitions (*Next*) and guards. The module's main function how call all those process is show below:

```
private String generateInputString(TransitionSystem root){
    formatter.format(root);
    output = writer.createHeader();
    output = writer.createVar(output, root);
    output = writer.createAssign(output, root);
    output = writer.createNext(output, root);
    output = writer.createGuards(output, root);
    return output;
}
```

4.5   The Global Module

The Global module provides some useful resources. The *Builder* and the main data structure, widely used by Converter and TUTS modules are here. The *Builder* is responsible for writing the Transitions System during the the conversion and unification processes. The *Builder* is also responsible for giving access to the Transition System output after all the processes have been finalized. All data traveling through the processes of conversion and unification are encapsulated into a data structure defined in the Global module. The Transition System itself is defined in this module. This module is also responsible for holding all the exception classes used by all other modules. Finally, there are two useful tools for global use: an *ID generator* and a *Printer*. The *ID generator* provides a global unique ID generation for all the inner process. The *Printer* is used for textual visualization of the Transition System tree. It is possible to see this output on a terminal console or save it direct in a *txt* file.

5   Application to a Space Application Software Product

We have applied our tool in a real space software product as our case study. SWPDCpM (Software for the Payload Data Handling Computer protoMIRAX) is an embedded software of the computer of Data Handling Subsystem of protoMIRAX [8]. protoMIRAX is a balloon-borne high energy astrophysics experiment which aims to develop cutting-edge space technology.

We are working with the on-board management system (SGB) that processes the information received from the ground station, as well as gets, generates, formats, stores, and transmits to the ground station via the telecommunications subsystem and flight control, information on the subsystems of the experiment protoMIRAX. There are several internal and external interfaces, among which we can mention: SCA (attitude control) controls the positioning of the protoMIRAX experiment during its operation in azimuth and elevation; CRX (X-ray camera) observes cosmic sources of x-rays to demonstrate innovative techniques for reconstruction of images in x-rays, using shades hardcoded.

The first activity is **identify scenarios**, following the approach proposed in Section 3. In our experiments, we have carried out twelve scenarios of SWPDCpM. (i) Scenario 1: Receive remote commands; (ii) Scenario 2: Forward remote commands; (iii) Scenario 3: Forward telemetry; (iv) Scenario 4: Enable/Disable forwarding of telemetry; (v) Scenario 5: Report current operation mode; (vi) Scenario 6: Change computational operation mode; (vii) Scenario 7: Distribute command turns on and off; (viii) Scenario 8: Download process of memory; (ix) Scenario 9: Report current software version; (x) Scenario 10: Get program charge status; (xi) Scenario 11: Start loading new version; and (xii) Scenario 12: Control interface with SCA.

We detail only one scenario, showing the full cycle of activities proposed in Section 3. With respect to other scenarios, we present only the final results, as we have restriction of space. The scenario we detail is *Scenario 11 - Start loading new version*. This scenario subsumes the function that attempts to remote command (TC) initiation request for loading new version of SWPDCpM.

Then, we start **Formal Verification**. For this, in one side we **select requirements** and on the other side, we define which **diagrams** are related to the scenario we are working with. Only one sequence diagram is related to this scenario. Our tool automatically converts this diagram in a transition system (TS) as can be seen in Figure 7 - the sequence diagram and the TS generated for this diagram.

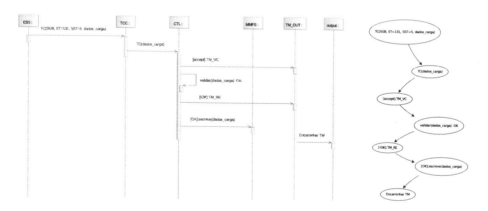

Fig. 7. Sequence Diagram and its respective TS

Each state is characterized by the values of the variables (**generate model checker notation**). We have identified one variable that characterize the TS obtained from sequence diagram:

State = {_tc$_$sgb$coma$__st$equals$131$coma$__sst$equals$5$coma$ __dados_carga$_$, _tc$_$dados_carga$_$,..., _encaminharTM$_$}.

Continuing the approach, we can extract one relevant user-defined property for this scenario, related to requirements. To proceed with the Formal Verification, it is necessary to **formalize the properties** to be checked. We chose

Computation Tree Logic (CTL) to formalize the properties. Note that the properties could be formalized using LTL as well, considering that NuSMV supports such logic.

Requirement: Once the initiation of load (TC_InicicacaoCarga) is received, the system should: accomplish data verification, to persist the data so that it is possible to recover data, and finally to report informing success or anomaly. This property can be formalized using **Precedence Chain Pattern and Scope After Q proposed** by [11], in CTL, as follows:

```
!E[!(State = _tc$_$dados_carga$_$) U
(State = _tc$_$dados_carga$_$ & E[!
(State = _validar$_$dados_carga$_$__$dots$ok)
U State = _#$not$ok#$dots$tm_re]
& E[!(State = _#$not$ok#$dots$tm_re)
U (State = _validar$_$dados_carga$_$__$dots$ok
& !( State = _#$not$ok#$dots$tm_re)
& EX(E[!(State = _#ok#$dots$escreve$_$dados_carga$_$) U
(State = _#$not$ok#$dots$tm_re & !( State =
_#ok#$dots$escreve$_$dados_carga$_$))])))])]
```

When running NuSMV (**apply Model Checking**) for the TS, we have obtained 7 states, and all of them are reachable states. The property we checked is true, that is, the property is satisfied.

We have applied this approach for all twelve scenarios. The results are displayed in Table 2. As we can see, some properties were not satisfied, meaning that some diagrams do not reflect all the requirements. Although this case study has only sequence diagrams, and it is not possible to exercise the whole features (which include unifying the three UML diagrams) of our approach, this case study is very suitable because it is a real project.

Table 2. Final Results after running NuSMV

Scenario	Proprieties		States	
	Verified	Satisfied	Total	Reachable
1	3	3	15	7
2	7	0	7	7
3	2	2	27	11
4	1	1	6	6
5	1	1	5	5
6	2	0	6	6
7	1	0	7	7
8	2	0	36	16
9	1	1	6	6
10	1	1	7	7
11	1	1	7	7
12	2	2	4	4

6 Conclusions

We presented a tool to convert selected UML behavioral diagrams into a NuSMV ready input for formal verification. The tool is able to process Sequence, Activity, and State Machine UML diagrams and unify any number of those diagrams, since we have at least one Sequence diagram. Using more than one diagram provides a rich view of the system by different angles or in different states in time. It also helps to find inconsistencies and incompleteness in the models by confronting multiple views of the same system. The ability of the XMITS tool to unify those diagrams catch this feature. Many other tools were developed to help apply formal verification in UML or UML-like diagrams, but none of them seems to have the freedom of use any number of inputs or unify any number of UML diagrams.

The XMITS tool aims to be modular and ready for upgrade. All its architecture were think to be generic and flexible. One of the most important characteristics of the architecture used in this version of the XMITS tool is the aperture to implement new modules and add new features in the existent modules. Based on this, we propose some future implementations towards new resources and easy to use. One interesting new feature could be the implementation of new UML diagrams compatibility. The Converter module is ready to accept new UML diagrams by adding a new *collector* class. Another important future work is the development of a new module to catch the feedback from the NuSMV and show to the user the results automatically. A addiction of different grammars in the Bridge module can open the possibility to use not only the NuSMV. It could allow the use of the converted and unified diagrams into a different Mode Checker tool.

Acknowledgments. This work is supported in part by Financiadora de Estudos e Projetos (FINEP) under Project Number 01.10.0233.00, by Funda-ção de Amparo à Pesquisa do Estado de São Paulo (FAPESP) under Process Number 2012/23767-2, and by Conselho Nacional de Desenvolvimento Científico e Tecnológico (CNPq) under Institutional Process Number 455097/2013-5, Individual Process Number 313709/2014-9.

References

1. Nusmv home page (2011)
2. Baier, C., Katoen, J.-P.: Principles of model checking. The MIT Press, Cambridge (2008). p. 975
3. Baresi, L., Morzenti, A., Motta, A., Rossi, M.: Towards the UML-based formal verification of timed systems. In: Aichernig, B.K., de Boer, F.S., Bonsangue, M.M. (eds.) Formal Methods for Components and Objects. LNCS, vol. 6957, pp. 267–286. Springer, Heidelberg (2011)
4. Beato, M.E., Barrio-Solórzano, M., Cuesta, C.E., de la Fuente, P.: Uml automatic verification tool with formal methods. Electronic Notes in Theoretical Computer Science **127**(4), 3–16 (2005)

5. Clarke, E.M., Emerson, E.A.: Design and synthesis of synchronization skeletons using branching time temporal logic. In: Grumberg, O., Veith, H. (eds.) 25 Years of Model Checking. LNCS, vol. 5000, pp. 196–215. Springer, Heidelberg (2008)

6. Cockburn, A.: Writing Effective Use Cases, 1st edn. Addison-Wesley Longman Publishing Co. Inc., Boston (2000)

7. Cortellessa, V., Mirandola, R.: Prima-uml: a performance validation incremental methodology on early uml diagrams. Science of Computer Programming **44**(1), 101–129 (2002)

8. Santiago Júnior, V.A., Vijaykumar, N.L.: Vijaykumar. Generating model-based test cases from natural language requirements for space application software. Software Quality Journal **20**(1), 77–143 (2012)

9. dos Santos, L.B.R., Eras, E.R., Santiago Júnior, V.A., Vijaykumar, N.L.: A formal verification tool for UML behavioral diagrams. In: Murgante, B., Misra, S., Rocha, A.M.A.C., Torre, C., Rocha, J.G., Falcão, M.I., Taniar, D., Apduhan, B.O., Gervasi, O. (eds.) ICCSA 2014, Part I. LNCS, vol. 8579, pp. 696–711. Springer, Heidelberg (2014)

10. Dubrovin, J., Junttila, T.: Symbolic model checking of hierarchical uml state machines. In: 8th International Conference on Application of Concurrency to System Design, ACSD 2008, pp. 108–117. IEEE (2008)

11. Dwyer, M.B., Avrunin, G.S., Corbett, J.C.: Patterns in property specifications for finite-state verification. In: Proceedings of the 1999 International Conference on Software Engineering, 1999, pp. 411–420. IEEE (1999)

12. Emam, K.E., Koru, A.G.: A replicated survey of it software project failures. IEEE Software **25**, 84–90 (2008)

13. Eshuis, R., Wieringa, R.: Tool support for verifying uml activity diagrams. IEEE Transactions on Software Engineering **30**(7), 437–447 (2004)

14. Kannenberg, A., Saiedian, H.: Why software requirements traceability remains a challenge. The Journal of Defense Software Engineering, 14–18 (2009)

15. Lam, V.S.W.: A formalism for reasoning about uml activity diagrams. Nordic Journal of Computing **14**(1), 43–64 (2007)

16. Latella, D., Majzik, I., Massink, M.: Automatic verification of a behavioural subset of uml statechart diagrams using the spin model-checker. Formal Aspects of Computing **11**(6), 637–664 (1999)

17. Lima, V., Talhi, C., Mouheb, D., Debbabi, M., Wang, L., Pourzandi, M.: Formal verification and validation of uml 2.0 sequence diagrams using source and destination of messages. Electronic Notes in Theoretical Computer Science **254**, 143–160 (2009)

18. Moha, N., Guéhéneuc, Y.-G., Duchien, L., Le Meur, A.-F.: DECOR: A method for the specification and detection of code and design smells. IEEE Trans. Software Eng. **36**(1), 20–36 (2010)

19. The Object Management Group (OMG), Needham, MA, USA. OMG Unified Modeling Language (OMG UML), Superstructure, V2.1.2 (2007)

20. Palomba, F., Bavota, G., Di Penta, M., Oliveto, R., De Lucia, A.: Do they really smell bad? a study on developers' perception of bad code smells. In: 30th IEEE International Conference on Software Maintenance and Evolution, Victoria, BC, Canada, September 29 - October 3, 2014, pp. 101–110. IEEE (2014)

21. Parnas, D.L., Clements, P.C.: A rational design process: How and why to fake it. IEEE Trans. Software Eng. **12**(2), 251–257 (1986)

22. Queille, J.P., Sifakis, J.: Specification and verification of conurrent systems in CESAR. In: Grumberg, O., Veith, H. (eds.) 25 Years of Model Checking. LNCS, vol. 5000, pp. 216–230. Springer, Heidelberg (2008)

23. dos Santos, L.B.R., de Santiago, Jr., V.A., Vijaykumar, N.L.: Transformation of uml behavioral diagrams to support software model checking (2014). arXiv preprint arXiv:1404.0855
24. Solemon, B., Sahibuddin, S., Ghani, A.A.A.: Requirements engineering problems and practices in software companies: an industrial survey. In: Ślezak, D., Kim, T.-h., Kiumi, A., Jiang, T., Verner, J., Abrahão, S. (eds.) ASEA 2009. CCIS, vol. 59, pp. 70–77. Springer, Heidelberg (2009)
25. Tsantalis, N., Chatzigeorgiou, A.: Identification of move method refactoring opportunities. IEEE Trans. Software Eng. **35**(3), 347–367 (2009)
26. UBM Tech. 2013 embedded market study. DESIGN West 2013 Conference (2013)
27. Uchitel, S., Kramer, J.: A workbench for synthesising behaviour models from scenarios. In: Proceedings of the 23rd International Conference on Software Engineering, pp. 188–197. IEEE Computer Society (2001)
28. van Lamsweerde, A.: Requirements engineering in the year 00: a research perspective. In: Proceedings of International Conference on Software Engineering (ICSE), pp. 5–19. ACM, New York (2000)

A Scheduling Problem for Software Project Solved with ABC Metaheuristic

Broderick Crawford[1,2,3], Ricardo Soto[1,4,5], Franklin Johnson[1,6(✉)],
Melissa Vargas[6], Sanjay Misra[7], and Fernando Paredes[8]

[1] Pontifícia Universidad Católica de Valparaíso, Valparaíso, Chile
{broderick.crawford,ricardo.soto}@ucv.cl, franklin.johnson@upla.cl
[2] Universidad Finis Terrae, Santiago, Chile
[3] Universidad San Sebastián, Santiago, Chile
[4] Universidad Autónoma de Chile, Santiago, Chile
[5] Universidad Central de Chile, Santiago, Chile
[6] Universidad de Playa Ancha, Valparaíso, Chile
melissa.vargas.a@alumnos.upla.cl
[7] Covenant University, Ota, Nigeria
sanjay.misra@covenantuniversity.edu.ng
[8] Universidad Diego Portales, Santiago, Chile
fernando.paredes@udp.cl

Abstract. The scheduling problems are very common in any industry or organization. The software project management is frequently faced with different scheduling problems. We present the Resource-Constrained Project Scheduling problem as a generic problem in which different resources must be assigned to different activities, so that the make span is minimized and a set of precedence constraints between activities and resource allocation to these activities are met. This Problem is a NP-hard combinatorial optimization problem. In this paper we present the model the resolution of the problem through the Artificial Bee Colony algorithm. The Artificial Bee Colony is a metaheuristic that uses foraging behavior of honey bees for solving problems, especially applied to combinatorial optimization. We present an Artificial Bee Colony algorithm able to solve the Resource-Constrained Project Scheduling efficiently.

Keywords: Software project management · Project scheduling · Optimization · Artificial Bee Colony · Metaheuristic

1 Introduction

The software project management is frequently faced with different scheduling problems, such as the assignation of different programmers to different software modules, the assignation of software engineers to different tasks in a software project, so that the whole activities to the software project are completed and the cost and duration of the project are minimized.

The problem of project scheduling with limited resources can be understood as the scheduling of project activities, so that neither resource constraints nor the

© Springer International Publishing Switzerland 2015
O. Gervasi et al. (Eds.): ICCSA 2015, Part IV, LNCS 9158, pp. 628–639, 2015.
DOI: 10.1007/978-3-319-21410-8_48

precedence constraints are violated, with the aim of minimizing the completion time of the last activity, ie, the project makespan. Therefore scarce resources are allocated to competing activities over a given time horizon for the best possible performance [8].

The Resource Constrained Project Scheduling Problem (RCPSP) is considered a NP-hard [3] problem, This is a complex and difficult combinatorial problem to program. Whereby the exact methods have difficulties in solving large RCPSP instances, then other alternatives are needed to solve them. Among these alternatives are heuristics, which have the capacity to lead to a near optimal solution even in scale problems, minimizing the use of resources.

Therefore the use of heuristics is a tool to find good solutions for acceptable quality computation. For this there have been many studies in the literature based on metaheuristics as: Harmony Search, Iterated Local Search, Memetic algorithm, Tabu Search based algorithms and Ant Colony. Worth mentioning that for the general formulation of combinatorial problems has not been identified a method that significantly outperform others.

In the literature, we can find resolution to RCPSP by methods based on linear programming [9, 11] approaches, which proposes a number of solutions where the formulations, the objective function, precedence constraints and resource constraints are specified in a variety of ways to solve. There are also other methods such as branch and bound algorithms [4], Genetic Algorithms (GA) [5,13], Tabu Search (TS) [10], Simulated Annealing (SA) [1], Ant Colony Optimization (ACO) [2,6], Adaptive search method (AS) [12] and Particle Swarm Optimization (PSO) [2,15], which have been well applied to solve the scheduling problem projects with limited resources. Also the Artificial Bee Colony algorithm (ABC) is a new meta-heuristic developed by Karaboga [7], which is a stochastic optimization technique based and inspired by the foraging behavior of the honeybee colonies.

There are other approaches for resolving problems RCPSP which are designed as a hybrid of different approaches, these methods try to exploit the advantages of two or more methods in order to design an algorithm with improved performance in general, algorithms hybrids best performance is obtained by improving the balance between exploration and exploitation by combining the potential of different approaches. Therefore, hybridization can be used as a way to avoid stagnation and premature convergence are known as major deficiencies metaheuristics.

In this paper a mathematical model of RCPSP solved with ABC algorithm is presented. We believe that ABC is a metaheuristic able to solve efficiently RCPSP, capable of competing with other solutions generated by complete and incomplete techniques.

This paper is organized as follows. In Section 2, we describe the RCPSP. Section 3 presents the ABC metaheuristic. The results of the ABC algorithm to RCPSP are presented in section 4. Finally, the concluding remarks of investigation made in the paper are presented in Section 5.

2 Resource Constrained Project Scheduling Problem

RCPSP can be defined as projects with limited resources in an environment which must process a set of activities subject to precedence constraints and resources, the latter being shared by several activities. Thus the problem is to perform such allocation optimizing some objective function.

A project consists of a set of activities, also known as jobs, operations or tasks; to complete the project successfully, each activity can be processed in one of several possible modes, each mode represents a different way of performing the activity being considered. The mode determines the duration of this activity, the requirements of different types of resources, possible flows (positive or negative) of money and other characteristics associated with the activity.

Among the project activities precedence relations are defined when the order to be executed is determined, these relationships are for technological or process design reasons, for which the project is represented as a directed graph where activities are represented in a nodes and the precedence relationships between activities by the arcs.

The programming projects may have different objectives and based on them the goodness or quality of a solution is measured, typically the objectives considered are:

- **Minimize Project duration (makespan):** the measure most applied in the domain of project scheduling. The length is defined as the time interval between the beginning and the end of the project. Since the beginning of the project is usually assumed in time $t = 0$, minimize the length is equivalent to minimizing the maximum completion times of all activities.
- **Minimize Delays of Activities:** other regular performance measure is to minimize the flow time of all project activities, or are given due dates is to minimize delays.
- **Maximize the net present value of the project:** when large projects and long term are present significant amounts of money flow in the form of expenses to initiate activities and payments to complete the project parties; the net present value (NPV) is a suitable criterion to measure whether the project is optimal. This approach generates a critical path cost and not time critical path generated when the duration is minimized.
- **Maximize the Value of the project:** this goal is very important for project managers. The quality of a project is given by the fulfilment of the deadlines planned, budget and meets the customer is satisfied with the products. The formulation of this problem has focused on minimizing the deviation of deadlines and within budget because of which must be pre-processed.
- **Minimize Project Cost:** this has attracted much attention of researchers due to their practical relevance. It delivery the cost of activities as ways of engaging different results in direct costs, which should be minimized (exchange cost-duration). On the other hand may be considered to minimize the cost of resources is determined by the programming of activities, which influence the cost indirectly through resources.

Besides the resources applied to a given project have the following classification:

- **Renewable Resources:** the man hours, machines, tools, space; these resources are available from period to period, ie, the amount available is renewed from one period to another.
- **Non-Renewable Resources:** the money, raw materials or energy resources that are available throughout the project are consumed as activities are implemented.
- **Doubly Constrained Resources:** are resources that are limited both by the period as a whole available in the project, for example, the budget constraint limits not only the cost of the entire project, but also consumption for each period.
- **Resources Limited Partially:** are resources whose availability is renewed at specific intervals. These types of resources can be viewed as a generic term for all types of resources.

In these four classifications, a resource of the same type is equally efficient for another. In general the ideal solution for RCPSP is about considering satisfy the constraints on the ability of different types of resources in each period. The results that may be obtained helping managers decide how many resources can be used for processing an activity and minimize the time of completion of a project.

2.1   Model of RCPSP

We have specified the main components for RCPSP well as the mathematical definition, the mathematical model, the explanation of the mathematical model, an example for RCPSP and PSLIB library.

The scheduling problem with limited resources whose main components:

- Set of activities $A = \{a_0, a_1, a_2, ..., a_n, a_{n+1}\}$
- Renewable resources limited set of $R = \{r_1, r_2, ..., r_m\}$
- Dummy activities that represent the start and end of the project respectively, also consume zero units of time and do not consume any resources $a_0; a_{n+1}$
- Duration for each activity a_i con $i = \{1, .., n\}$ represented by $d_i \geq 0$
- Amount of resources consumed $q_{ij} \geq 0$
- maximum availability $Q_j \geq 0$ each resource $r_j \in R$ at each instant of time.
- Precedence Constraints $P_a \in A = \{a_1, ..., a_{n+1}\}$
- each activity a_i can not be initiated while predecessor activities P_i have not been finalized.
- Set of successor activities S_i for a_i activities being $i = \{1, ..., n+1\}$
- Set of start times for each activity $T = \{t_0, ..., t_{n+1}\}$
- Directed acyclic graph $G = \{A, E\}$ where $E = \{(a_i, a_j)/a_i \in P_a; a \in A\}$

The goal is to find a set of start times for each activity that meets the precedence constraints and availability of resources and minimizes the total duration of the project (makespan).

2.2   Mathematical Model

Mathematically the RCPSP is represented by the following mathematical model containing the minimization of makespan and the constraints of time and resources.

$$Min\ F_{n+1} \tag{1}$$

$$F_h \leq F_j - p_j \tag{2}$$

$$\sum_{j \in A(t)} r_{j,k} \tag{3}$$

$$k \in K; t \geq 0 \tag{4}$$

$$F_j \geq 0 \tag{5}$$

$$where\ j = \{1, ..., n+1\}$$

$$h \in P_j$$

Defined F_j as the completion of the activity j therefore in the mathematical formulation should be minimized F_{n+1} since $n+1$ this is the last activity. This is represented in the objective function presented in Eq. 1, used to define the quality of a solution (Fitness). The Eq. 2 does satisfy the precedence constraint between activities since it shows that the completion of an activity h must be greater or equal to the completion of the activity j unless the predecessor activity. The Eq. 3 and Eq. 4 show the limits for each type of resource k and each time instant t, thus not allowing the demand for activities occurring at present does not exceed its capacity. Finally the Eq. 5 defines the decision variables.

2.3   Example of a Project

An example is shown for a project in which there are 7 activities in the next section, the type of resource and precedence constraints are indicated, as well as also a possible solution should be ordered as activities for the minimum makespan.

In Fig. 1 a example project is observed with activities of 0 to 7 with their respective durations(d) and requirement(r) (d/r)of the resource, where activities 0 and 7 have length 0, only represent start and finish milestones and other activities have resource constraints that are 4 units of resources simultaneously, given that the resource is renewable type so for example, if an activity spends 3 units does not mean that the next activity must spend 1 unit if they can spend between 1 and 4 units of the resource.

In Fig. 2 a possible solution to the shown example, representing resource use R_1 and project duration. 2.4, (1.6), 3.5, which satisfy a possible solution to the problem given that meet the constraints of time and resources thereby minimizing the makespan: the order of tasks is also displayed.

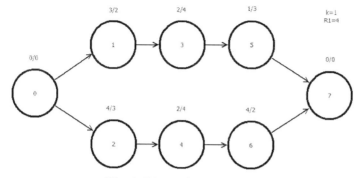

Fig. 1. Directed acyclic graph

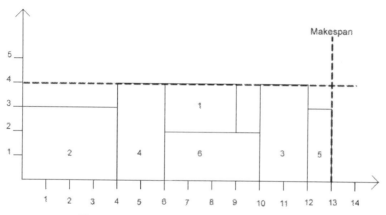

Fig. 2. Representation of a possible solution

3   Description of Artificial Bee Colony

The ABC algorithm is inspired by the behavior of intelligent power of a colony of bees, this is an optimization algorithm based on population introduced by Karaboga [7].

This algorithm is motivated by the intelligent behavior of honey bees, using common parameters such control as a colony size and the maximum number of cycles.

ABC as an optimization tool provides a search procedure based on the population where the individual items of food are modified by artificial bees with time and purpose of the bee is to discover the locations of food sources high amount nectar.

In the ABC system artificial bees fly around in a multidimensional space search and some bees that can be used are those spectators who choose foods sources based on their own experience and their nest mates, and then adjust their positions. Some scout bees fly and food sources are chosen randomly and without the use of experience, if the amount of nectar from a new source is larger than the previous saved in its memory, memorize the new position and forget the above.

Therefore, the ABC system combines local search methods, carried out by employed bees and spectators, with global search methods, managed by spectators and explorers, trying to balance the exploration and exploitation process.

This algorithm comes from the intelligence swarm and this based on the behavior of natural honey bees to find their food is so in the following points is passed on to detail each of these concepts and computational bees are also analyzed.

Initialization phase: In the initialization phase all parameters are set, mainly by the population size m, number of iterations of the algorithm, and the initial values of the solution, which is defined as a vector x with a dimension between 1 and n, where n is the n dimension of the solution. The initial solution may be random or guided.

$$x_i = l_i + rand(0,1) * (u_i - l_i) \tag{6}$$

The Eq. 6 determined as randomly initialize the dimension i of the solution x. l_i and u_i are the minimum and maximum values for x_i.

Bees Employed Phase: The bees take a new neighbor solution to the current solution v, this solution is obtained by applying an operator to change random to one or more of the dimensions of the initial solution x. The fitness of v and x are then calculated, selecting the best.

Bees Spectators Phase: The spectators bees take your probabilistically decision based on the information provided by the Employed bees. Taking the best solution according to the highest probability calculated according to Equation.

$$p_i = \frac{fit(x_i)}{\sum_{j=1}^{m} fit(x_j)} \tag{7}$$

After selecting the best solution x as probability. New neighbor solution v is calculated and the best solution between the two is chosen.

Bees Scout Phase: Bees are choose randomly solution. The worker bees that your solution does not improve over a period of time become explorers and may use the above equation to define a new solution to chance.

3.1   Representation of the Solution

The solution is represented as follows, if the RCPSP is a problem composed of n activities, then the bees will move in solution space n dimensions. ie a solution is a vector x_i, where $i \in \{1, ..., n\}$, the possible values for each dimension are between 0 and 1. Where each bee represents a solution in the solution space. The order of each element of the solution, represents the order in which to perform the activities. This representation is based on list priority represented by Zhang 2006 [14].

We use a change operator based in simple swap. The different types of swap are N, NS, NSC.

Type N: such is the nearest neighbor and is choosing a random value as the image Fig. 3 Activity 3 and exchanged for activity 4.

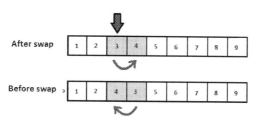

Fig. 3. Swap type N

Type NS: this rate is to choose a random number as in the Fig. 4 can be seen activity 4 is selected and then has 4 spaces to the right and make the change, then exchanged the 4 activity 8.

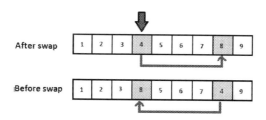

Fig. 4. Swap type NS

Type NSC: in such an activity and randomly select Fig. 5 that activity 7 is selected and then count backwards 4 spaces for putting this activity in the position being located activity in the rest position and ran forward.

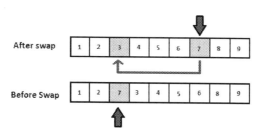

Fig. 5. Swap type NSC

3.2   Representation of the Algorithm

We can see below the pseudocode which shows that through the ABC algorithm to obtain results for RCPSP

Algorithm 1. Pseducode of the ABC algorithm for RCPSP

Initialization

Define (Cycles, Sources, Scouts, Onlooker, Visit)

Initialize food source randomly

While Cycle > 0

(Send Employed Bees)

Calculate $V_{id} = X_{id} + \omega_1 r_{id}(X_{id} - X_{kd})$

(Send Onlooker Bees)

Calculate permutation Fig. 3

Calculate permutation Fig. 4

Calculate permutation Fig. 5

Calculate $V_{id} = X_{id} + \omega_2 r_{id}(X_{id} - X_{kd})$

(Send Scout Bees)

If visit is satisfied

Initialize food source i **randomly**

End While

Return best schedule

---

4   Experimental Results

The algorithm was implemented using Netbeans IDE 8.0 with java programming language. The instances of each bechmark J30, J60, and J90 of PSLIB were taken with the following results which are compared with those of the page http://www.om-db.wi.tum.de/psplib/datasm.html.

Table 1 shows that our algorithm is able to obtain good solutions for instances with less complexity. In some instances the algorithm was possible to obtain

Table 1. Results instances J30

Instance	Makespan	BestSolution
J30_1	38	43
J30_2	49	47
J30_3	35	47
J30_4	61	62
J30_5	31	39
J30_6	36	48
J30_7	43	60
J30_8	45	53

better results than the best found for these instances. In Table 2 and Table 3 we can see the results for more complex instances J60 and J90, where our algorithm was not possible to achieve the best known results. The Fig. 6 shows the convergence of our proposal for the 3 types of tested instances. Proving that the algorithm for J30 instances converges around iteration 100 and for J60 and J90 instances, the algorithm converges near the iteration 1000.

Table 2. Results instances J60

Instance	Makespan	BestSolution
J60_1	85	77
J60_2	99	68

Table 3. Results instances J90

Instance	Makespan	BestSolution
J90_1	148	73
J90_2	140	92

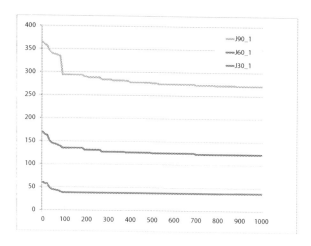

Fig. 6. Convergence of ABC algorithm to RCPSP

5   Conclusion

In conclusion the RCPSP problem, which is a combinatorial NP-hard problem, is possible to represent real problems which can be applied to software project management. We propose to detailing with this problem a bio-inspired meta-heuristic able to find solutions to combinatorial problems using the behavior of natural bee colonies.

The mathematical model of the problem and a description of the behavior of ABC are clearly presented. Also we present the model that solves the problem with ABC. Through this algorithm it is possible generate possible solutions to compete with other solutions proposed by other authors.

As future work is expected to improve the exchange operators of our ABC algorithm so as to increase the exploration of bees, for not converge quickly to a suboptimal solution.

Acknowledgments. Broderick Crawford is supported by Grant CONICYT/ FONDECYT/REGULAR/1140897, Ricardo Soto is supported by Grant CONICYT/ FONDECYT/INICIACION/11130459, Franklin Johnson is supported by Postgradu- ate Grant PUCV 2015, And Fernando Paredes is supported by Grant CONICYT/ FONDECYT/REGULAR/1130455.

References

1. Bouleimen, K., Lecocq, H.: A new efficient simulated annealing algorithm for the resource-constrained project scheduling problem and its multiple mode version. European Journal of Operational Research **149**(2), 268–281 (2003)
2. Chen, R.-M., Wu, C.-L., Wang, C.-M., Lo, S.-T.: Using novel particle swarm opti- mization scheme to solve resource-constrained scheduling problem in psplib. Expert systems with applications **37**(3), 1899–1910 (2010)
3. Chiarandini, M., Di Gaspero, L., Gualandi, S., Schaerf, A.: The balanced academic curriculum problem revisited. Journal of Heuristics **18**(1), 119–148 (2012)
4. Dorndorf, U., Pesch, E., Phan-Huy, T.: A time-oriented branch-and-bound algo- rithm for resource-constrained project scheduling with generalised precedence con- straints. Management Science **46**(10), 1365–1384 (2000)
5. Hartmann, S.: A competitive genetic algorithm for resource-constrained project scheduling. Naval Research Logistics (NRL) **45**(7), 733–750 (1998)
6. Herbots, J., Herroelen, W., Leus, R.: Experimental investigation of the applicability of ant colony optimization algorithms for project scheduling. D'TEW Research Report **0459**, 1–25 (2004)
7. Karaboga, D.: An idea based on honey bee swarm for numerical optimization. Tech- nical report, Technical report-tr06, Erciyes university, engineering faculty, com- puter engineering department (2005)
8. Kempf, K., Uzsoy, R., Smith, S., Gary, K.: Evaluation and comparison of produc- tion schedules. Computers in industry **42**(2), 203–220 (2000)
9. Mingozzi, A., Maniezzo, V., Ricciardelli, S., Bianco, L.: An exact algorithm for the resource-constrained project scheduling problem based on a new mathematical formulation. Management Science **44**(5), 714–729 (1998)

10. Nonobe, K., Ibaraki, T.: Formulation and tabu search algorithm for the resource constrained project scheduling problem. In: Essays and Surveys in Metaheuristics, pp. 557–588. Springer (2002)
11. Pritsker, A.A.B., Waiters, L.J., Wolfe, P.M.: Multiproject scheduling with limited resources: A zero-one programming approach. Management science **16**(1), 93–108 (1969)
12. Schirmer, A.: Case-based reasoning and improved adaptive search for project scheduling. Naval Research Logistics (NRL) **47**(3), 201–222 (2000)
13. Valls, V., Ballestin, F., Quintanilla, S.: Justification and rcpsp: A technique that pays. European Journal of Operational Research **165**(2), 375–386 (2005)
14. Zhang, H., Li, H., Tam, C.: Particle swarm optimization for resource-constrained project scheduling. International Journal of Project Management **24**(1), 83–92 (2006)
15. Zhang, H., Li, X., Li, H., Huang, F.: Particle swarm optimization-based schemes for resource-constrained project scheduling. Automation in Construction **14**(3), 393–404 (2005)

On the Use of a Multiple View Interactive Environment for MATLAB and Octave Program Comprehension

Ivan M. Lessa[1], Glauco de F. Carneiro[1], Miguel P. Monteiro[2], and Fernando Brito e Abreu[3(✉)]

[1] Universidade Salvador (UNIFACS), Salvador/Bahia, Brazil
ivan.lessa@gmail.com, glauco.carneiro@unifacs.br
[2] Universidade Nova de Lisboa (UNL), NOVA LINCS, Lisbon, Portugal
mtpm@fct.unl.pt
[3] Instituto Universitário de Lisboa (ISCTE-IUL), ISTAR-IUL, Lisbon, Portugal
fba@iscte-iul.pt

Abstract. MATLAB or GNU/Octave programs can become very large and complex and therefore difficult to understand and maintain. The objective of this paper is presenting an approach to mitigate this problem, based upon a multiple view interactive environment (MVIE) called *OctMiner*. The latter provides visual resources to support program comprehension, namely the selection and configuration of several views to meet developers' needs. For validation purposes, the authors conducted two case studies to characterize the use of *OctMiner* in the context of software comprehension activities. The results provided initial evidences of its effectiveness to support the comprehension of programs written in the aforementioned languages.

Keywords: Software visualization · MATLAB/octave · Software comprehension

1 Introduction

MATLAB[1] and its open source "clone" Octave[2] are high-level programming languages and development environments that are widely used for rapid prototyping and simulation of scientific applications. As those applications grow in size and complexity, they face the usual maintenance challenges that are common in the so-called "legacy systems" [1]. Maintainability depends on our ability to understand programs, what lead to the creation of a *Program Comprehension* scientific community[3].

By reviewing the available literature, we found evidence of a lack of support for the comprehension of programs coded in MATLAB and Octave, as described in a following section. We tackled this research opportunity by implementing a multiple view interactive environment (MVIE) named *OctMiner*. MVIEs provide resources to

---

[1] A registered trademark of *The MathWorks© (http://www.mathworks.com/products/matlab)*
[2] See *http://www.gnu.org/software/octave/*
[3] See *http://www.program-comprehension.org/*

© Springer International Publishing Switzerland 2015
O. Gervasi et al. (Eds.): ICCSA 2015, Part IV, LNCS 9158, pp. 640–654, 2015.
DOI: 10.1007/978-3-319-21410-8_49

support data analyses and unveiling information that otherwise would remain unnoticed [2] [3]. To validate *OctMiner* effectiveness, we conducted two case studies using the tool to support the comprehension of MATLAB/Octave programs. The first study aimed at characterizing the MVIE support to identify crosscutting concerns following previous research on this issue [4] [5]. The second study focused on analyzing to which extent *OctMiner* can help programmers to understand the solutions proposed in the *StackOverflow* community [4], a popular question-and-answer site for professional programmers, regarding MATLAB and Octave problems.

This paper is structured as follows: section 2 summarizes the main concepts of the MATLAB/Octave programming languages and describes the key functionalities of *OctMiner* and its architecture; section 3 presents two case studies to exemplify how *OctMiner* can support MATLAB/Octave program comprehension; section 4 proposes a set of usage strategies to be performed with *OctMiner* for comprehension purposes; section 5 reviews related work; finally, section 6 presents the final considerations and outlines opportunities for future work.

2   Multiple View Interactive Environments

Visualization provides perceivable cues to several aspects of the data under analysis to reveal patterns and behaviors that would otherwise remain "under the radar" [6]. Card et al. [2] proposed a well-known reference model for information visualization. According to them, the creation of views goes through a sequence of successive steps: pre-processing and data transformations, visual mapping and view creation. Carneiro and Mendonça [7] extended this model to adapt it to the context of MVIEs. The extended model is portrayed in Fig. 1 emphasizing that the visualization process is highly interactive. Moreover, it enables the combined use of resources of a multiple view interactive environment. The process starts with original (raw) data obtained from a repository that undergoes a set of transformations, which is then organized into data structures suitable for information exploration. This process is called *data transformation* [3]. Next, the aforementioned data structures are used to assemble visual data structures. Those structures organize data properties and visual information properties in ways that facilitate the construction of visual metaphors. This step defines the mapping from real attributes – which are derived from the data properties (software attributes, in our case) – to visual attributes such as shapes, colors and positions on the screen. This process is called *visual mapping* [3]. It is important to highlight that these activities do not deal with rendering, but rather with building suitable data structures from which the views can be rendered. The final step, presented in Fig. 1, is the *visual transformation*, aimed at drawing the information on the screen to produce the views. In this step, a specific visual scene is actually rendered on the computer screen [3].

Nunes et al. [8] proposed a toolkit implemented as a Java Eclipse plugin from which MVIEs could be developed. The plugin provides a basic structure that allows the creation and inclusion of new resources and functionalities to develop MVIEs. Fig. 2 presents the way the toolkit was used and extended by other plugins to reify the

---

[4] See *http://www.stackoverflow.com*

SourceMiner MVIE. This MVIE was originally developed to support the comprehension of Java source code. As can be seen in Fig. 2, the extension points of the *toolkit.aimv* plugin enable the inclusion of new plugins to the MVIE. Each of the conveyed extension points provides an interface with methods and their respective signatures. In the case of *OctMiner*, we needed to access and transform raw data – the Abstract Syntax Tree (AST) of MATLAB/Octave programs – to a format compatible with the visual data structure. According to the extended reference model for MVIEs, this is a requirement to feed the views. Fig. 2 presents a set of plugins that comprise the *SourceMiner* MVIE.

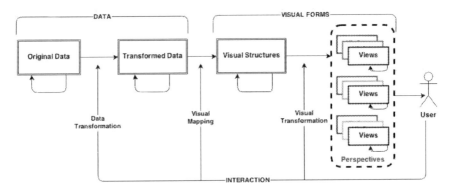

Fig. 1. An extended reference model for MVIEs [3]

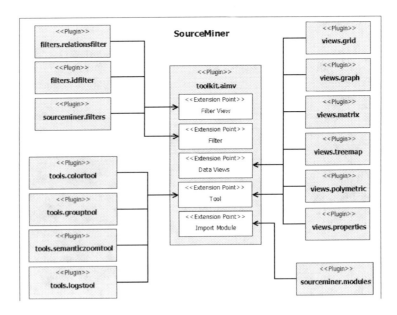

Fig. 2. The *SourceMiner* MVIE [8]

The goal of the toolkit is to provide an infrastructure to develop MVIEs for different domains. The domain targeted in this paper comprises programs written in MATLAB/Octave. The application of the aforementioned toolkit to this domain was reified through *OctMiner*, whose architectural overview is depicted in Fig. 3. The *Grid* and *Treemap* views were provided by the MVIE. On the other hand, the *List* view, the *Filters* and the *Analyzer* were extended from the MVIE specifically for *OctMiner* usage.

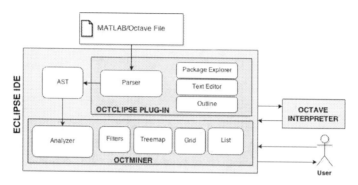

Fig. 3. OctMiner architectural overview [9]

2.1   The MATLAB and OCTAVE Programming Languages

MATLAB is an interpreted language, very popular among students and researchers of physics, biomedical engineering and related areas. It is not uncommon that a young engineer is fluent in using MATLAB, but hardly familiar with C, and even less with Fortran [10] [11].

MATLAB has been used to teach linear algebra, numerical analysis, and statistics. Since the MATLAB language is proprietary, a similar language, named Octave was developed, and is distributed under the terms of the GNU General Public License. It was originally conceived in 1988 to be a companion programming language for an undergraduate-level textbook on chemical reactor design. Due to the similarities between these languages, it is possible to interpret MATLAB programs in the interpreter of the GNU/Octave with no major problems. The main differences between the two languages are the following:

i)   some similar functions have different names in each language;

ii)   comments in MATLAB are written after "%" while in Octave you can use both "%" and "#";

iii)   in MATLAB the control blocks (*while*, *if* and *for*), as well as the functions delimiter all finish with *"end"* while in Octave you can also use *"endwhile"*, *"endif"*, *"endfor"* and *"end-function"* respectively;

iv)   In MATLAB the not equal to operator is *"~="* while in Octave *"!="* is also valid;

v)   MATLAB does not accept increment operators such as "++" and "—", while Octave accepts them.

2.2   The *OctMiner* MVIE

The main motivation for representing concerns manifested in MATLAB/Octave code in a MVIE is the enhancement of the comprehension activities. The plugin structure supporting the MVIE toolkit is the same as presented in Fig. 2. The main difference is that in this case the focus is on MATLAB/Octave rather than Java. Fig. 3 depicts the main four elements of *OctMiner*: the Eclipse IDE RAP/RCP (Remote Application Platform / Rich Client Platform) , the *Octclipse* plugin[5], the Octave interpreter and the MVIE toolkit proposed in [8]. The Eclipse IDE enables its extension through the use of plugins. The MVIE toolkit does this to provide its functionalities, as well as enabling the tailoring of the MVIE for the analysis of data from different domains, e.g., the data gathered from MATLAB/Octave programs. We implemented an *Analyzer* module, as conveyed in Fig. 3, which is analogous to *sourceminer.modules* in Fig. 2. It is an extension of the *Import Module*, whose goal is to import and convert data from the original data repository to be represented in the multiple views. The *Octclipse* plugin also provides an Octave development environment built on top of *Eclipse's Dynamic Languages Toolkit*[6]. This environment enables programmers to create Octave scripts (**.m* files), edit them in a multi featured text editor, run the Octave interpreter and see the results displayed in the IDE's console. *OctMiner* is freely available for download[7].

3   Comprehension Activities with *OctMiner*

This section presents two studies to characterize the use of *OctMiner* in software comprehension activities.

3.1   The First Study

The first study investigated the following question: "to which extent *OctMiner* provides effective support to identify potential symptoms of crosscutting concerns in MATLAB programs?" To answer this question we analyzed 22 MATLAB image-processing routines with *OctMiner*, to identify the presence of scattering and tangling. Scattering [12] is the degree to which a concern is spread over different modules or other units of decomposition. Tangling [13] is the degree to which concerns are intertwined to each other in the same module. Both scattering and tangling are indicators of the presence of crosscutting concerns. The basic units of decomposition (modules) in MATLAB or Octave are *functions* and *files*. For cohesion sake, a file usually contains a set of related functions.

The term "token", to be used hereinafter, represents a function name from the MATLAB/Octave systems. This study considers that the distribution of the occurrence of these tokens can be used as an indicator of scattering and tangling symptoms.

---

[5]   See *http://sourceforge.net/projects/octclipse/*
[6]   See *http://eclipse.org/dltk/*
[7]   See *http://www.sourceminer.org/octminer*

The approach is as follows: sets of tokens can be associated to a given concern, which ideally would be modularized into its own file, with no additional concerns. When the concern is not modularized, its code is scattered across multiple files and its associated tokens are found in such files – an indicator of scattering. Often, such files also betray the presence of tokens categorized under multiple concerns – an indicator of tangling.

To explore the above approach, participants performed the following activities:

i)   Identify tokens most commonly used in the 22 routines;
ii)   Characterize the localization among files of the most commonly used tokens to assess the symptoms of scattering;
iii) Characterize the relationship between the most commonly used tokens and other tokens in the files to assess the symptoms of tangling;
iv)   Determine the category (concern) to which the most commonly used tokens belong;
v)   Using the category of each token, identify the main functionalities (concerns) of the program.

This approach allowed identifying the top most commonly used tokens in the analyzed routines. These tokens presented evidences of scattering. This study was just a pilot-test in using *OctMiner* in comprehension activities and it allowed us to identify a set of improvements, which were added to *OctMiner* before the next study took place.

3.2   The Second Study

The second study had the following research question, based on answers posted at *StackOverflow*: "to which extent *OctMiner* provides effective support to clarify programmer´s doubts"? The main goal of this study was then to show *OctMiner's* effectiveness in supporting the visualization of target functions as the ones reported at *StackOverflow*. In other words, we hypothesized that *OctMiner* can help programmers to understand the context of use of a function in routines that serves as examples supported by the available views. The authors searched for the top questions about the two selected programming languages and their corresponding best answers. For this purpose, the search used the *StackExchange Data Explorer* tool[8]. Applying the following query, using the mentioned tool, we obtained as a result the top 200 questions related with the keywords "MATLAB" and "Octave":

```
SELECT TOP 200 a.creationdate, q.owneruserid, q.title
FROM users u, posts a, (SELECT id, owneruserid, title, tags,
creationdate FROM posts WHERE tags LIKE '%<KEY_WORD>%') q
WHERE q.id = a.parentid and a.owneruserid=u.id
ORDER BY a.creationdate desc
```

---

[8]   Available at http://data.stackexchange.com/

We classified the questions in the following categories:

(a) programming language basic issues – 146 questions;

(b) common mistakes in MATLAB and Octave – 51 questions;

(c) using functions to perform specific work such as numerical calculation and image processing – 98 questions;

(d) using functions to plot data on the screen – 69 questions;

(e) questions that do not fit into any of the previous categories – 56 questions.

As can be seen, category (a) has the greater number of questions, which indicates a lack of basic knowledge of the two languages. We considered this fact as the start point to select the following question: "I want to create a vector without the number 1". The answer with most votes was "I would use *setdiff*". The answer was illustrated as follows "*setdiff(-5:5,1)*".

Configuring *OctMiner* to Answer the Question. The participant configured *OctMiner* according to the goal of the second study. The configuration consists of editing a XML file as follows. *<GroupName>* defines the group to which the function belongs to, whereas *<function>* contains the list of functions to be represented in the views.

```
<group>
  <GroupName title="GroupName" color="color">
    <function>;Function1;Function2;</function>
  </GroupName>
</group>
```

Table 1. Categories and their Colors in *OctMiner*

Category	Color Name	Color
Array and Matrix Creation and Concatenations	Concrete	
Set Operations	Green	
Indexing	MethodBorder	
Parse Strings	Size	
Logical Operations	Blue	
Advanced Software Development	Class	
Mathematics	Abstract	

We selected 22 MATLAB routines to illustrate the use of the *setdiff* function, the target function of the selected doubt. The authors selected these routines searching the *StackOverFlow* repository using the string "MATLAB *setdiff*". The authors also registered *settdiff* function and all the other functions identified in the 22 routines in the *OctMiner* configuration XML file. More details regarding the XML file can be obtained at *OctMiner* page. Table 1 conveys the categories and their respective colors to be presented by *OctMiner*.

Focusing *OctMiner* on the setdiff Function. Based on our experience, we proposed a set of steps in Table 2 focusing on the comprehension of *setdiff* function supported by *OctMiner* to clarify a real doubt registered by a programmer at *StackOverFlow*. In the next section, we explain how these steps were executed.

Table 2. Proposed Steps in *OctMiner*

STEPS
Select a question: to clarify a doubt.
Identify the *setdiff* function in the repository: the programmer should configure *OctMiner* to visually identify occurrences of the *setdiff* function in the repository routines and the way they are used.
Identify the category that the function belongs to: the programmer should configure *OctMiner* to spot other functions that belong to the same category of *setdiff* to help in the comprehension tasks.
Identify similar functions from the repository that can replace *setdiff*: configure *OctMiner* to support the identification of similar functions that can replace the target function.
Verify if the gathered information was enough to clarify the doubt: the user can now be more confident and can agree why the answer was the one with most votes.

Executing the Steps. In this section, we describe how *OctMiner* can help programmers to clarify their doubts about MATLAB and Octave. To this end, we use Fig. 4 through Fig. 7, which apply one of the following two types of configuration. Type I, presented in Fig. 4 and Fig. 5, focuses on files and their respective functions. In these figures, each rectangle from the *Grid view* (part D) represents a file together with the number of function categories found there one. Each rectangle from the *List view* (part E) represents the complete name and path of each file. In the case of the *TreeMap view* – part G of Fig. 4 – all rectangles together convey a panoramic visual representation of the files. In fact, the *TreeMap view* conveys a 2D visualization that maps a hierarchical structure into rectangles with each rectangle representing a file. In that case, files and functions are represented as nested rectangles, where the innermost rectangles are functions and the outermost rectangles are files. In the configuration type I, the size of each innermost rectangle corresponds to the number of functions implemented in each file and the color is associated to the category of the function.

Configuration type II (Fig. 6 and Fig. 7) is focused on the functions. Each rectangle from the *Grid view* (part D) represents a function together with their number of occurrences in the repository, in which multiple occurrences in the same file are counted. Each rectangle from the *List view* (part E) represents the complete name of function from the repository. The rectangles from the *TreeMap view* represent functions and the size of each rectangle is proportional to the number of times a function appears in the repository. The colors represent the category of each function.

When the user executes *OctMiner* and selects the option "*Visualize with OctMiner*", the tool conveys a typical visual scenario like the one presented in Fig. 4. This scenario uses the configuration type I, which focus on the analyzed repository files and their respective functions. In a first step, the user can understand the way the functions are distributed in the repository based on the information provided by the views as described in the following sentences. The *Grid view* (part D of Fig. 4) provides an overview of how many types of functions are implemented in each file (rectangle) of the repository (all rectangles). The *List view* (E) lists each files name and location in the repository. The *TreeMap view* (G) provides an overview of the files from the repository. Using this visual metaphor, functions found in a file are represented in the rectangle representing that file. A single screen shot can show all functions and files in accordance with its position in the file structure. We adapted the *TreeMap* visual paradigm to use colors to represent categories to which each specific function belongs. The difference of Fig. 4 and Fig. 5, is that in Fig. 5 we apply a filter (indicated by a red ellipse in part F) to highlight the files that implement *setdiff*. As a result of the filter, part G of the same figure highlights the functions where *setdiff* occurs by painting the rectangles in green.

The next step is the configuration of *OctMiner* to present the visual scenario of Fig. 6 that applies the configuration type II focusing on functions. The *List view* (part E of Fig. 6) enables checking the exact name of the routine, as well as the category to which the function belongs by looking at the color of the rectangles. The green color indicates that *setdiff* belongs to category "*Set Operations*". The user can access and read the code of specific routines (Part C) and analyze the various ways in which the function is used.

Using the type II configuration, focusing on functions, it is possible to spot the *Treemap view* displaying the largest green rectangle to represent the *setdiff* function. On the other hand, the *Grid view* complements this information by reporting that there are eleven occurrences of this function in the analyzed repository. The second largest rectangle from the *Treemap view* belongs to the same category and refers to function *isMember*. Interestingly, that function can replace *setdiff*: to find out if a vector is a subset of another, we can use *isempty(setdiff(a, b))* where a and b are arrays – but we will get the same result using *all(isMember(a, b))*. Based on information provided by *OctMiner*, we could identify an alternative function to *setdiff* if necessary. Different functions, like those already mentioned from the same category, are indications of the dominant category of the repository under analysis. The function *setdiff* belongs to the group "*set Operations*" but the category with more distinct features is "*Array and Matrix Creation and Concatenations*", which includes a large number of functions to deal with arrays and matrices. In Fig. 7, this category is highlighted to emphasize that a high number of distinct functions in a given category is a possible indicator of auxiliary functions, which may be of interest for the user.

The aforementioned conclusions can be confirmed from the replies registered at *StackOverFlow*. The user now can be more confident to understand the answers provided by the repository, considering both the target and similar functions, their utility, as well as the way they can be used to solve the stated problem.

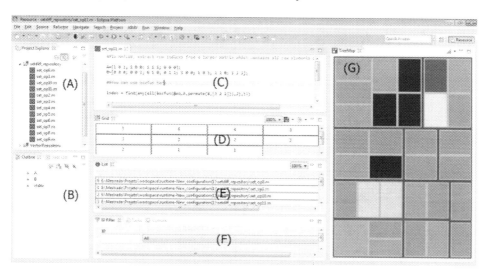

Fig. 4. OctMiner panoramic views for the initial analysis

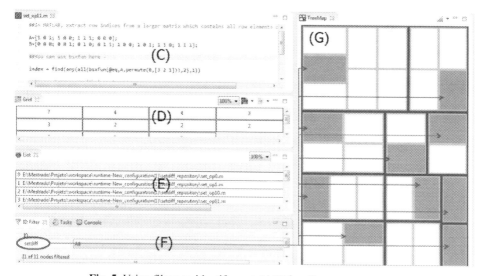

Fig. 5. Using filters to identify `setdiff` function occurrences

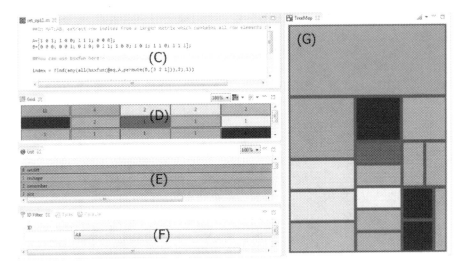

Fig. 6. Visual representation of functions from the repository

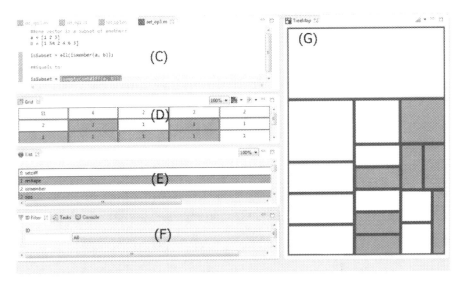

Fig. 7. Visual representation of functions in *OctMiner*

Even though this example is simple, it illustrates the benefits from using *OctMiner* to support comprehension. The combined use of the configuration types I and II can be an effective way for the comprehension of particularities of MATLAB and Octave programming that would be difficult to notice through non-visual, non-interactive approaches.

In the second study, one of the potential threats related to external validity (to which extent results can be generalized) is that just one question from *StackOverflow* was evaluated in *OctMiner*. The environment might not easily fit other issues registered at *StackOverflow*. However, we do not expect that the case studies presented in this paper should to be generalizable to all types of issues and questions from *StackOverflow*. The purpose of both studies was to provide insights about the potential of *OctMiner* as support for the comprehension of MATLAB/Octave programs. The first study had the goal to use *OctMiner* to support the detection and characterization of crosscutting concerns [5], as well as to characterize the use of *OctMiner* and improvement opportunities of its use. The second study explored two configuration types to use *OctMiner* for supporting comprehension of issues posted at *StackOverflow*.

We recognize that *OctMiner* may not be able to provide support for all kinds of comprehension needs. To better characterize and validate its range of applicability, we plan additional studies. Another potential threat to validity is that both the design and the execution of the study were performed by the same person. To overcome this issue, further independent experiments should be carried out to better compare results.

4   A Comprenhension Strategy Based on *OctMiner*

The results from the two studies enabled us to propose a set of usage strategies based on *OctMiner* for comprehension purposes. The set has a comprehension question that drives the strategy steps as a starting point. The question of the first study was related to tangling and scattering, using a set of tokens from programs of a repository as a basis. The second study focused on questions posted at *StackOverflow* by programmers. Table 3 presents the steps proposed from evidences collected from the two case studies presented in this paper.

Table 3. A proposed set of usage strategies

SUGGESTED STEPS
1 - **Select a question:** the programmer needs to identify an issue relevant for his daily activities. Answers to the question should be available considering that the functions used in the code should be registered in the *OctMiner* configuration file. A repository of questions and answers, such as StackOverflow, may be used for this purpose, as illustrated in the second study.
2 - **Identify a target function:** it should be the function that plays a relevant role in the code of the primary solution to the selected question. In repositories such as the StackOverflow, the best ranked answers usually indicate the relevant function to solve the problem.
3 - **Locate repositories that use the target function:** since *OctMiner* aims at assisting the comprehension of a given target function, it is desirable that routines using the target function provide good examples and be the subject of analysis.
4 - **Identify the functions and their respective categories available in the official documentation:** alternative functions used in the repository selected in Item 3 must also be identified. MATLAB and Octave functions are categorized in the official language site of MATLAB and Octave.
5 - **Register the target function as well as other function from the repository in the *OctMiner* configuration file:** the functions should be registered in *OctMiner* configuration file using their specific group, identified according to Item 4.
6 - **Create a To-Do list for identification through visualization:** activities that the user must perform should be described so that the study is conducted as well as possible within *OctMiner*. In the example from the second preliminary study, the user is directed through four comprehension tasks centred on the *setdiff* function.
7 - **Implementation of the proposed activities:** the user must run *OctMiner* according to the activities set out in Item 6.
8 - **Answer the original question:** to prove the effectiveness of the tool, the user should be able to answer the question that started the process.

5   Related Work

Research on MATLAB and Octave program comprehension is in its infancy. A simple proof of this claim can be obtained with *Google Scholar*. While the search string *"Java program comprehension"* returns a considerable number of hits[9], at the time of writing this paper, similar searches with MATLAB[10] or Octave[11] did not match any articles. Therefore, we enlarged our search to include related aspects such as static analysis, code refactoring, reverse engineering or program transformation and optimization.

The oldest reference found was from V. Menon and K. Pingali, where the authors proposed three kinds of source-to-source transformations for optimizing MATLAB programs and show their effectiveness [14]. They claim that transformations yield performance benefits additional to those obtained by (optimizing) compilation, and may be useful for other DSLs that are high-level, untyped, and interpreted.

In spite of MATLAB's popularity, and the need for static analysis (e.g. for program optimization, code smells detection, refactoring), Jesse Doherty claimed that there was no publicly available framework for creating static analyses for that language, until he created the *McLAB Static Analysis Framework (McSAF)* [15] [16]. The goal of this framework was to make new analyses easy to write and to extend to new language features.

Soroush Radpour, also at McGill University [17], developed a tool named *McBench*[12] that is claimed to help compiler writers understand the language better, by giving some insight about how programmers use MATLAB. He also proposed a suite of semantic-preserving refactoring for MATLAB functions and scripts including: function and script inlining, converting scripts to functions, extracting new functions, and converting dynamic *feval* (function evaluation) calls to static function calls.

Last, but not the least, Anton Dubrau and Laurie Hendren, again from the *McLab Project*[13] team at McGill University, claim that MATLAB users often want to convert their programs to a static language such as Fortran [18]. They developed an object-oriented open source toolkit, called *Matlab Tamer*[14], for supporting the generation of static programs from dynamic MALTAB programs.

6   Conclusions and Future Work

MATLAB and Octave are popular languages for numerical computations used by scientists, engineers and students worldwide. As their programs grow in size and complexity, they face the usual maintenance challenges that originated the emergence of the program comprehension domain in Computer Science. Software visualization

---

[9] https://scholar.google.pt/scholar?q="Java+program+comprehension"
[10] https://scholar.google.pt/scholar?q="MATLAB+program+comprehension"
[11] https://scholar.google.pt/scholar?q="Octave+program+comprehension"
[12] See https://github.com/isbadawi/mcbench
[13] See http://www.sable.mcgill.ca/mclab/
[14] See http://www.sable.mcgill.ca/mclab/projects/tamer/

techniques can mitigate those maintenance challenges, but as far as we could devise, their use has not yet been adopted by the MATLAB and Octave communities.

This paper presents the following contributions: a) the provision of an environment called *OctMiner* for the comprehension of MATLAB/Octave routines supported by multiple views; b) Evidences of the effectiveness of *OctMiner* to support the identification of symptoms of code tangling and code scattering as discussed in the first study; c) Evidences of the effectiveness of *OctMiner* to understand the solutions proposed in a popular question-and-answer site for professional programmers, regarding MATLAB and Octave languages as discussed in the second study; d) a set of usage strategies of *OctMiner* for comprehension purposes.

A preliminary version of this paper, describing *OctMiner* architecture, along with an illustrative example of its main functionalities in a real scenario of program comprehension, was presented at ITNG'2015 [9]. A short paper, including a first case study for validating *OctMiner* feasibility, was presented at SEKE'2015 [19]. On the current version, additional details are provided on the validation case studies and additional information on the proposal is provided.

We now plan to conduct a controlled experiment where engineering undergraduate students will perform comprehension activities with and without the support of *OctMiner*. We also plan to include collaborative resources in *OctMiner* to enable programmers to communicate and cooperate among themselves to more effectively achieve software comprehension activities related to MATLAB and Octave software development.

References

1. Seacord, R., Plakosh, D., Lewis, G.: Modernizing legacy systems: software technologies, engineering processes, and business practices. Addison-Wesley Professional (2003)
2. Card, S., Mackinlay, J., Shneiderman, B.: Readings in Information Visualization Using Vision to Think. Morgan Kaufmann, San Francisco (1999)
3. de Figueiredo Carneiro, G., Silva, M., Mara, L., Figueiredo, E., Sant'Anna, C., Garcia, A., Mendonça, M.: Identifying code smells with multiple concern views. In: Proceedings of the XXIV Brazilian Symposium on Software Engineering (SBES 2010), pp. 128–137 (2010)
4. Cardoso, J., Fernandes, J., Monteiro, M., Carvalho, T., Nobre, R.: Enriching MATLAB with aspect-oriented features for developing embedded systems. Journal of Systems Architecture, 412–428 (2013)
5. Monteiro, M., Cardoso, J., Posea, S.: Identification and characterization of crosscutting concerns in MATLAB systems. In: Proceedings of the Conference on Compilers, Programming Languages, Related Technologies and Applications (CoRTA 2010), Braga, Portugal, pp. 9–10 (2010)
6. Spence, R.: Information Visualization: Design for Interaction, 2nd edn. Prentice Hall (2007)
7. de Figueiredo Carneiro, G., de Mendonça Neto, M.G.: SourceMiner: towards an extensible multi-perspective software visualization environment. In: Hammoudi, S., Cordeiro, J., Maciaszek, L.A., Filipe, J. (eds.) ICEIS 2013. LNBIP, vol. 190, pp. 242–263. Springer, Heidelberg (2014)

8. Nunes, A., de Figueiredo Carneiro, G., David, J.: Towards the development of a framework for multiple view interactive environments. In: Proceedings of the International Conference on Information Technology: New Generations (ITNG 2014), Las Vegas, USA, pp. 23–30 (2014)

9. Lessa, I., de Figueiredo Carneiro, G., Monteiro, M., Brito e Abreu, F.: A multiple view interactive environment to support MATLAB and GNU/Octave program comprehension. In: Proceedings of the International Conference on Information Technology: New Generations (ITNG), Las Vegas, USA (2015)

10. Chaves, J., Nehrbass, J., Guilfoos, B., Gardiner, J., Ahalt, S., Krishnamurthy, A., Unpingco, J., Warnock, A., Samsi, S.: Octave and python: high-level scripting languages productivity and performance evaluation. In: Proceedings of the HPCMP Users Group Conference 2006 (2006)

11. Stenroos, M., Mäntynen, V., Nenonen, J.: A MATLAB library for solving quasi-static volume conduction problems using the boundary element method. Computer methods and programs in biomedicine (2007)

12. Robillard, M., Murphy, G.: Representing Concerns in Source Code. ACM TOSEM (2007)

13. Tarr, P., Ossher, H., Harrison, W., Sutton, Jr., S.M.: Degrees of separation: multidimensional separation of concerns. In: Proceedings of the ICSE 1999 (1999)

14. Menon, V., Pingali, K.: A case for source-level transformations in MATLAB. In: Proceedings of the DSL 1999, pp. 53–66 (1999)

15. Doherty, J.: McSAF: An extensible static analysis framework for the MATLAB language. MSc dissertation, McGill University, Montréal, Canada (2011)

16. Doherty, J., Hendren, L.: McSAF: A Static Analysis Framework for MATLAB. Sable Technical Report sable-2011-01, McGill University, Montréal, Canada (2011)

17. Radpour, S.: Understanding and refactoring the MATLAB language. MSc dissertation, McGill University, Montréal, Canada (2012)

18. Dubrau, A., Hendren, L.: Taming MATLAB. Sable Technical Report sable-2011-04, McGill University, Montréal, Canada (2011)

19. Lessa, I., de Figueiredo Carneiro, G., Monteiro, M., Brito e Abreu, F.: Scaffolding MATLAB and octave software comprehension through visualization. In: Proceedings of the 27th International Conference on Software Engineering and Knowledge Engineering (SEKE 2015), Pittsburgh, USA, pp. 552–557 (2015)

Design Phase Consistency: A Tool for Reverse Engineering of UML Activity Diagrams to Their Original Scenarios in the Specification Phase

Jay Pancham and Richard Millham[✉]

Department of IT, Durban University of Technology, Durban, South Africa
{panchamj,richardm1}@dut.ac.za

Abstract. In this paper, we present a tool that preserves phase consistency from specifications to the design phase by reverse engineering UML activity diagrams, designed from scenario specifications, back to scenarios to ensure that all of the original scenarios can be recreated. We use a set of action and action-link rules to specify the activity and scenario diagrams in order to provide consistency and rigor. Given an activity diagram depicting a common telecentre process (es), we present an algorithm that follows this set of action and action-link rules to reverse engineer this activity diagram back to their set of scenarios. The validation of this algorithm is achieved when, given a set of activity diagrams, the algorithm is able to recreate the original set of scenarios. Thus, all original specifications, in the form of scenarios, are ensured to be encapsulated within their activity diagram.

Keywords: Phase consistency · UML · Scenario · Activity diagram · Requirements engineering

1 Introduction

Ensuring that the requirements from the specification phase are consistently and completely represented in the design phase is an ongoing problem in software engineering [9]. Often, traditional software engineering methods are used to first elicit system requirements. After elicitation, these methods are then used to specify these requirements in the specification phase and then develop the system design in the design phase. However system development that relies on traditional methods, without the use of formal notations to specify requirements and automated tools to manage these formalisms, to ensure that the requirements are consistent from the specification to the design phase is difficult [31].

Using a telecentre monitoring system with its requirements represented as UML scenarios and its design represented as UML activity diagrams, this paper presents a tool that reverse engineers, using a set of rules to formally defined scenarios and activity diagrams, activity diagrams back to their corresponding scenarios in order to ensure phase consistency of requirements from the specification to the design phase. Phase consistency is deemed to be achieved when the same set of scenarios, manually

© Springer International Publishing Switzerland 2015
O. Gervasi et al. (Eds.): ICCSA 2015, Part IV, LNCS 9158, pp. 655–670, 2015.
DOI: 10.1007/978-3-319-21410-8_50

derived from the requirements using traditional software engineering methods, matches the set of scenarios generated by the tool from the activity diagram.

2   Literature Review

2.1   Case Study System

The case system used in this paper is a monitoring system for telecentres. Currently, telecentre usage data is gathered using ad-hoc traditional techniques [4]; [21] which is both untimely and subjective.   Many researchers advocate better support processes for systematic monitoring of telecentre usage [4; 13; 7] which led to the development of this telecentre monitoring system [16]. Traditional methods were utilized to draft the system's activity diagram in the design phase but in order to provide both development speed and rigor in the telecentre design model, there was need to automatically verify the activity diagram representation with the original specifications in order to instill greater confidence in the correctness of the design.

2.2   Software Development Models

Although there are many software development models, the classical model is the waterfall software lifecycle. This model consists of several sequential non-overlapping stages which begins with first determining system requirements (specification phase) and then proceeds consecutively to the phases of architectural design, implementation (coding), testing, and maintenance [14]. The specification phase includes requirements elicitation where the functional requirements of the system, which indicates what the client requires and what client business processes are modelled, are captured [24].

2.3   Verification and Validation

When software is being developed through the phases of a software development model, there is a need for verification and validation of this software. Although there are varying definitions of verification and validation, Boehm [2] defines verification as the process to establish the truth of correspondence between a software product/design and its specification. Using static techniques such as expert opinion, walk-throughs, and reviews, documents and processes are examined to determine if they conform to specific requirements and if they satisfy conditions that were imposed at the start of the software phase [25]. Validation, defined by Boehm [2], is the process of determining the appropriateness of a software product/design for its operational goal.

Methods for verification include walkthrough and expert opinion when the requirements have been communicated clearly [24]. Another effective method for both verification and validation is through the use of dynamic analysis of the product/design via a prototype built from its design [11].

2.4   Software Development Methods

As software systems became more widespread in the 1970s and 1980s, a number of different software development methods, including structured analysis and design, emerged. When object oriented programming languages became active, a number of development methods for these object-oriented languages appeared with each method having its own concepts, definitions, notations, terminology, and process. In order unify these diverse methods, in 1995 Grady Booch, James Rumbaugh, and Ivar Jacobson combined many of the concepts within these methods to form the Unified Modelling Language (UML) [10]. UML is now the de-facto standard in requirements engineering [27].

The UML is a general-purpose modelling language that provides multiple system perspectives, a set of semantic and graphical notations for its diagrams, and an interchange format. UML is designed for use in object-oriented software applications. UML was designed to provide documentation of the system, and communicate requirements and design intent [15].

Within UML, a scenario can be defined as a unique set of internal activities within a business process while activity diagrams represent a business process modelled within the system [25]. An activity diagram may also be used to represent a use case (a coherent unit of functionality that depicts transactions between the system and its users) [23]. Although there are varying definitions of scenarios, Stevens and Pooley [26] encase a set of related scenarios into a single use case. Each scenario in this set is a single unique thread through this particular use case. This definition has the advantage of relating scenarios, use case, and consequently activity diagrams to the system core requirements [26].

Despite its advantages, UML is semi-formal, lacks a formal syntax, and does not have automatic verification of its diagrams in various phases of the software lifecycle. Chandra proposed a set of syntactical rules for correctness and consistency and utilized these rules to formalize UML activity diagrams. However, his rules assume a one-to-one correspondence between a formal entity, described by his rules, and an activity in an activity diagram which may not always be the case [3]. Rafe defined formal semantics of UML activity diagrams and then transformed these diagrams to graphs which were automatically verified using AGG, a graphical transformation tool. However, this verification makes several system assumptions such as no queuing of system inputs [20]. Bhattacharjee also formalized activity diagram semantics with constructs of Esterel which simulated behaviour at the implementation, rather than design, phase [1]. Han proposes the use of Petri nets, with methods, to map activity diagrams to their Petri net equivalents [6]. However, these Petri net equivalents retain their activity entities and their high-level view which negates Petri net advantages of identifying non-determinate states [17]. Linzhang also proposed formalization of activity diagrams and the use of activity diagrams to derive test cases and proposed test scenarios. However, Linzhang's work examines the consistency between the implementation and design phases rather than specification and design phase [11].

2.5   Design Science as Development Methodology

Design Science may be defined as a process model that conducts research in Information Systems (IS) that is consistent with previous literature, together with the supply of a mental model and artifact [18]. The process includes six steps: problem definition and motivation, objectives for a solution, design and development, evaluation and communication is shown in Figure 1. Empirical research in relation to artifact design may play two roles, first validation of the designed artifact before it has been deployed to practice and then evaluation of the performance of the implemented design after it has been deployed to practice [28]. Consequently, validation and evaluation can be encompassed within the Design Science model. Both validation and evaluation can be housed in the Design Science process model. In the case of building the tool to check phase consistency, the research role will be one of validation of the artifact and verification of its requirements.

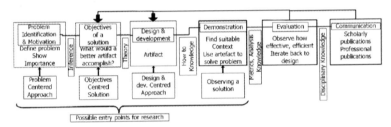

Fig. 1. Design Science Process

3   Methodology

3.1   Traditional Methods: Acquisition and Verification of Scenarios

Because the business processes were not well-defined in this case system, the traditional method of using a series of semi-structured interviews from stakeholders to obtain system requirements were used [30]. These requirements were then analysed and drafted into a scenario diagrams. These scenario diagrams, in turn, were used to derive draft activity diagrams that encompassed the corresponding activities of the telecentre. Scenarios have been advocated as an effective means of acquiring and validating requirements as they capture examples and real world experiences that users can understand [19]. Using expert opinion and walkthroughs (without test data but using experts to verify paths taken as representative of conditions and processes within the requirements), these requirements were verified. To validate these requirements, a triangulation of traditional methods were used in order to ensure that the requirements were verified and draft activity diagrams were accurately validated as the deficiencies of any one method can be overcome by combining methods and by capitalizing on their individual method strengths (see Fig 2) [29]. Expert opinion, in the form of managers who were experts in their domain, was used to ensure that all business scenarios reflected business processes and events in the draft diagrams. Walkthroughs, with typical business test data, were conducted to ensure that all paths of the activity diagrams were traversed and the expected outputs obtained.

Using activity diagrams as the basis of development, a prototype was built to elicit tacit knowledge from stakeholders and to provide both dynamic analysis verification and validation [22]. Given the set of typical business test data as input, the prototype produced the expected results (dynamic analysis). These three methods of expert opinion, walkthroughs, and dynamic analysis through a prototype formed the triangulation of methods for the specification validation.

Fig. 2. Specification Verification and Validation

However, once the activity diagram was validated, there was a need to ensure that all of the specifications from the earlier phase were brought down and incorporated into the activity diagram. To ensure this, another method, a tool to reverse engineer activity diagrams back to their original scenarios, was needed and this tool was consequently developed. Tool Development

3.1.1   Design Science

Figure 3 (DS Verification Process for Phase Consistency) illustrates the process of design science in developing this tool. The problem of phase consistency between the specification and design phases in the software development process was identified, an algorithm to traverse an activity diagram to produce a set of scenarios was developed, action and action link rules were incorporated within this algorithm to provide rigor and consistency, and the tool was evaluated by comparing its generated scenarios from the activity diagram to those original scenarios from which it was derived. As per design science evaluation process, the generated scenarios (results) were reviewed and if any discrepancies with the manually-created scenarios were identified, these discrepancies were resolved by modifying the algorithm or by determining if-there were any unforeseen legitimate scenarios omitted from the original activity diagram. This tool which reverse engineers activity diagrams back to their original sce

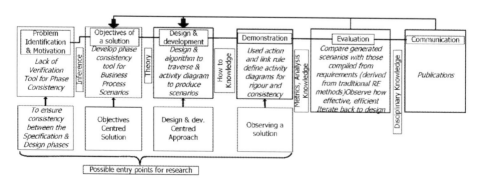

Fig. 3. DS Verification Process for Phase Consistency

narios from which they were designed ensures consistency of requirements encompassed in the design phase and derived from the specification phase. In short, requirements, described as scenarios, in the specification phase could be confirmed as being incorporated within the activity diagrams in the design phase without any loss of requirements between the phases. Inconsistencies between phases in requirements result in greater likelihood of errors and possible different interpretations of the design [32].

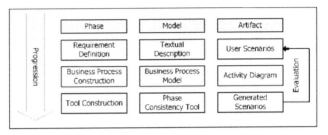

Fig. 4. Phase, Model, and Artifact

Figure 4 illustrates the relationship between this paper's tool and the phases, models, and artifacts of the software development lifecycle. The requirements (or specification) phase produces a textual description of the requirements which are then translated into UML user scenarios. After the specification phase, the business process construction (or design) phase produces a business model which is translated into a UML activity diagram. To ensure that all the requirements from the specification phase are encapsulated with the design phase (phase consistency), a tool is developed to generate scenarios which are compared to the original scenarios of the specification phase.

3.1.2   Action and Action Link Rules

The action rules and action link rules used in both manual specification of and in the tool generation of scenarios are as follows [12]:

1. Action Link Rule Definition

- Strict sequence (A then B): Defines sequential order of actions i.e. action B occurs after the completion of action A
- Alternative (A or B): Defines a choice i.e. action A or action B occurs. This is used in the case of a Branch – Merge condition.
- Concurrent (A and B): Defines a concurrent set of actions where action A and B occur concurrently. This is used in the case of a Fork and Join condition.
- Equal-end (A ends-with B): Define two actions A and B that end together.

2. Rules to formalise an activity diagram

Action rules are defined so that the actions can be linked using the action link rules to formalize the activity diagrams. The following nodes together with transitions will be used in activity diagrams: Start, End, Branch, Merge, Fork, and Join. Guard Condition Each activity diagram will commence at a Start node and finish at the End node and all nodes will be linked via a directed transition.

- Each activity diagram will have one Start node and one End node.
- Each of these actions will be linked to subsequent actions in their paths using the Strict Sequence rule.
- Branch Merge construct rules
  - A Branch – Merge construct will be used in the case of a decision so that a single path can be selected based on a guard condition.
  - An action (generally a question that results in a single guard condition) will link to a decision node using a Strict Sequence rule.
  - A decision will link to branch using a Strict Sequence rule.
  - Each of the subsequent actions following a branch will be linked using the Alternative rule.
  - The Branch will also link to the subsequent actions using the Strict Sequence rule.
  - Each of these actions will be linked to subsequent actions in their paths using the Strict Sequence rule.
  - The last action of each branched path will be linked to the merge using a Strict Sequence rule.
  - Each of these actions will also be linked to each other using the Equal End rule.
- Fork – Join construct Rules
  - An action will link to a Fork node using a Strict Sequence rule.
  - The first set of actions of each path following the Fork will be linked using the Concurrent rule.
  - Each of these actions will also be linked to subsequent actions using the Strict Sequence rule
  - The last action of each forked path will be linked to the join using a Strict Sequence rule.
  - Each of these actions will also be linked to each other using the Equal End rule.

3.1.3   Algorithm for Tool

Table 1 provides a coded description of the activity diagram using the rules to formalize the activity diagram together with the action link rules [12]. This table is used by the algorithm to walk through the activities, flows, and decision/merge/fork/join nodes of the given activity diagram to generate all possible scenarios. The algorithm takes into account decision and parallel activities during its walkthrough.

Table 1. Formalization, by action and action link rules, of the telecentre activity diagram

ID	Action One Name	Rule Name	Action Two Name
30	Start	Strict Sequence	Start of Day
31	Start of Day	Strict Sequence	Request Service
32	Request Service	Strict Sequence	Log User Profile
33	Log User Profile	Strict Sequence	Service Available
34	Service Available	Strict Sequence	Branch-Service Available
35	Service Available-Yes	Alternative	Service Available-No

Table 1. (*Continued*)

36	Branch-Service Available	Strict Sequence	Bill Usage
39	Bill Usage	Strict Sequence	Allocated Service
41	Allocated Service	Strict Sequence	Use Service
43	Use Service	Strict Sequence	Successful Usage
44	Successful Usage	Strict Sequence	Branch-Successful Usage
45	Successful Usage-Yes	Alternative	Successful Usage-No
46	Branch-Successful Usage	Strict Sequence	Rate Service
47	Rate Service	Strict Sequence	Merge-Successful Usage
49	Branch-Service Available	Strict Sequence	Join Queue
50	Join Queue	Strict Sequence	Continue Wait
52	Continue Wait	Strict Sequence	Branch-Continue Wait
53	Continue Wait-Yes	Alternative	Continue Wait-No
54	Branch-Continue Wait	Strict Sequence	Service Available
55	Branch-Continue Wait	Strict Sequence	Merge-Service Available
56	Branch-Successful Usage	Strict Sequence	Reuse Service
57	Reuse Service	Strict Sequence	Branch-Reuse Service
58	Reuse Service-Yes	Alternative	Reuse Service-No
59	Branch-Reuse Service	Strict Sequence	Service Available
60	Branch-Reuse Service	Strict Sequence	Refund Fee
61	Refund Fee	Strict Sequence	Merge-Successful Usage
62	Merge-Successful Usage	Strict Sequence	Merge-Service Available
63	Merge-Service Available	Strict Sequence	End of Day
66	End of Day	Strict Sequence	End

The algorithm iteratively developed for the scenario generation tool is as follows:

```
Create List of Rules
Set Path Traversed
Count Paths Remaining
While there are paths remaining
     Create Scenario Header for a new scenario
     Set Available Paths = 1
     While Scenario is not complete
          Get next Rule
          If ActionOne = 'Branch'
                    IF ActionTwo = 'End' OR Loop Identified
                         Set Scenario Complete
                         Save Action as scenario action
                    ELSE IF ActionTwo = 'Fork'
                         IF AvailablePaths > 0
                           Save BranchPaths traversed
                           Call ProcProcess Concurrent Actions
                         ELSE
                             Find the next available Path
                             IF another path is available
                                  Prepare to Get Next Rule
                    END
```

```
                        ELSE
                            IF AvailablePaths > 0
                                    Save BranchPaths traversed
                                    Save Action as scenario action
                                    Prepare to Get Next Rule
                            ELSE
                                    Find the next available Path
                                    IF another path is available
                                            Prepare to Get Next Rule
                            END
                    END
            ELSE
                    IF ActionTwo = 'End' OR Loop Identified
                        Set Scenario Complete
                        Save Action as scenario action
                    ELSE IF ActionTwo = 'Fork'
                        IF AvailablePaths > 0
                            Call ProcProcess Concurrent Actions
                        ELSE
                                    Find the next available Path
                                    IF another path is available
                                            Prepare to Get Next Rule
                        END
                  ELSE IF ActionTwo = Branch and PathTraversed = 0
                                    Save Action as scenario action
                                    Update BranchPathsUsed
                                    Prepare to Get Next Rule
                        END
                    ELSE
                            IF AvailablePaths > 0
                                    Save Action as scenario action
                                    Prepare to Get Next Rule
                            IF Loop Encountered
                                    Set Scenario Complete
                    END
            END
        END
    END

    Proc ProcessConcurrentActions
    Set PathsRemaining
    While PathsRemaining > 0
        Get Next Rule
        Save Action as scenario action
        IF ActionOne = 'Fork'
            Decrease the available paths
        IF ActionTwo = 'Join'
            Get PathsRemaining
            IF PathsRemaining > 0
                    Set Next action to Fork
                    Save Action as scenario action
        END
    END
```

```
Proc UpdateBranchPaths
Get BranchPaths
Get NoOfBranches in Current Branch
While BranchPaths are available
      Get Next Branch
      Increase BranchPaths = BranchPaths + NoOfBranches -1
END
```

3.1.4   Generated Scenarios

Using the action and action link rules within its algorithm, this tool generates scenarios from the activity diagram. As per Satzinger, the activity diagram represents a use case or coherent unit of functionality [23]. The scenarios follow the definition of Stevens and Pooley [26] which view scenarios as threads within a single use case, or in this example, within an activity diagram.

3.1.5   Verification of Scenarios Using Tool

Once verified using methods in Fig 2, the scenarios were defined manually using action and action link rules as outlined in Tables 2 to 6 [12]. The draft activity diagram was also defined using the same action and action link rules as appropriate. The definition of scenarios and activity diagrams using action and action link rules provided some consistency and rigor to these diagrams. The scenarios, as defined by these rules, followed the action flow steps as outlined in the scenario description.

Using design science, a tool was developed to use activity diagrams to generate the set of scenarios on which they were based. These generated scenarios, specified in terms of actions and action link rules, were compared to the original scenarios to ensure that the original scenarios' flow steps were the same as those generated by the tool. In so doing, this comparison ensured that all of the scenarios derived from the specifications, using traditional methods, had been incorporated into the activity diagram. Consequently, this tool ensured that all scenarios (specification phase) were brought into the activity diagrams (design phase).

4   Results

4.1   Manually-Specified Scenarios of a Given Activity Diagram

Table 2. Activity 2: Telecentre operation   Scenario 1: Successful usage = Yes

	Action One	Action Two
1	Start	Start of Day
2	Start of Day	Request Service
3	Request Service	Log User Profile
4	Log User Profile	Service Available
5	Service Available	Branch-Service Available
6	Branch-Service Available	Bill Usage
7	Bill Usage	Allocated Service
8	Allocated Service	Use Service
9	Use Service	Successful Usage
10	Successful Usage	Branch-Successful Usage
11	Branch-Successful Usage	Rate Service

Table 2. (*Continued*)

12	Rate Service	Merge-Successful Usage
13	Merge-Successful Usage	Merge-Service Available
14	Merge-Service Available	End of Day
15	End of Day	End

Table 3. Activity 2: Telecentre operation   Scenario 2: Reuse Service = Yes

	Action One	**Action Two**
1	Start	Start of Day
2	Start of Day	Request Service
3	Request Service	Log User Profile
4	Log User Profile	Service Available
5	Service Available	Branch-Service Available
6	Branch-Service Available	Bill Usage
7	Bill Usage	Allocated Service
8	Allocated Service	Use Service
9	Use Service	Successful Usage
10	Successful Usage	Branch-Successful Usage
11	Branch-Successful Usage	Reuse Service
12	Reuse Service	Branch-Reuse Service
13	Branch-Reuse Service	Service Available

Table 4. Activity 2: Telecentre operation   Scenario 3: Reuse Service = No

	Action One	**Action Two**
1	Start	Start of Day
2	Start of Day	Request Service
3	Request Service	Log User Profile
4	Log User Profile	Service Available
5	Service Available	Branch-Service Available
6	Branch-Service Available	Bill Usage
7	Bill Usage	Allocated Service
8	Allocated Service	Use Service
9	Use Service	Successful Usage
10	Successful Usage	Branch-Successful Usage
11	Branch-Successful Usage	Reuse Service
12	Reuse Service	Branch-Reuse Service
13	Branch-Reuse Service	Refund Fee
14	Refund Fee	Merge-Successful Usage
15	Merge-Successful Usage	Merge-Service Available
16	Merge-Service Available	End of Day
17	End of Day	End

Table 5. Activity 2: Telecentre operation   Scenario 4: Continue Wait = Yes

	Action One	**Action Two**
1	Start	Start of Day
2	Start of Day	Request Service
3	Request Service	Log User Profile
4	Log User Profile	Service Available
5	Service Available	Branch-Service Available

Table 5. (*Continued*)

6	Branch-Service Available	Join Queue
7	Join Queue	Continue Wait
8	Continue Wait	Branch-Continue Wait
9	Branch-Continue Wait	Service Available

Table 6. Activity 2: Telecentre operation   Scenario 5: Continue Wait = No

	Action One	**Action Two**
1	Start	Start of Day
2	Start of Day	Request Service
3	Request Service	Log User Profile
4	Log User Profile	Service Available
5	Service Available	Branch-Service Available
6	Branch-Service Availa-	Join Queue
7	Join Queue	Continue Wait
8	Continue Wait	Branch-Continue Wait
9	Branch-Continue Wait	Merge-Service Available
10	Merge-Service Available	End of Day
11	End of Day	End

4.2   Activity Diagram

The constructs used in UML activity diagrams are given in Table 5. Activity diagrams used in this paper are subject to the constraints listed by Rafe [20]. The list of rules to

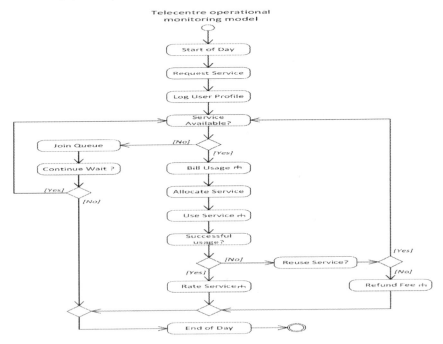

Fig. 5. Activity Diagram of Telecentre Operational Model

formalize an activity diagram together with their link rules is given in section 3.2.2 [12]. Using these rules, the activity diagrams were formally coded so that every activity and every flow in the diagram were linked to their corresponding action using the link rule. The activity diagram represents a common monitoring model of a telecentre.

4.3   Generated Scenarios

These scenarios were generated by the tool from the given activity diagram (Fig. 5) using the set of actions and action link rules.

Table 7. Scenario 1   Successful usage = Yes

	A1 ID	Action One	A2 ID	Action Two
1	1	Start	19	Start of Day
2	19	Start of Day	20	Request Service
3	20	Request Service	21	Log User Profile
4	21	Log User Profile	22	Service Available
5	22	Service Available	3	Branch-Service Available
6	3	Branch-Service Available	23	Bill Usage
7	23	Bill Usage	24	Allocated Service
8	24	Allocated Service	25	Use Service
9	25	Use Service	26	Successful Usage
10	26	Successful Usage	33	Branch-Successful Usage
11	33	Branch-Successful Usage	27	Rate Service
12	27	Rate Service	34	Merge-Successful Usage
13	34	Merge-Successful Usage	4	Merge-Service Available
14	4	Merge-Service Available	28	End of Day
15	28	End of Day	2	End

Table 8. Scenario 2   Reuse Service = Yes

	A1 ID	Action One	A2 ID	Action Two
1	1	Start	19	Start of Day
2	19	Start of Day	20	Request Service
3	20	Request Service	21	Log User Profile
4	21	Log User Profile	22	Service Available
5	22	Service Available	3	Branch-Service Available
6	3	Branch-Service Available	23	Bill Usage
7	23	Bill Usage	24	Allocated Service
8	24	Allocated Service	25	Use Service
9	25	Use Service	26	Successful Usage
10	26	Successful Usage	33	Branch-Successful Usage
11	33	Branch-Successful Usage	31	Reuse Service
12	31	Reuse Service	41	Branch-Reuse Service
13	41	Branch-Reuse Service	22	Service Available

Table 9. Scenario 3   Reuse Service = No

	A1 ID	Action One	A2 ID	Action Two
1	1	Start	19	Start of Day
2	19	Start of Day	20	Request Service
3	20	Request Service	21	Log User Profile
4	21	Log User Profile	22	Service Available
5	22	Service Available	3	Branch-Service Available
6	3	Branch-Service Available	23	Bill Usage
7	23	Bill Usage	24	Allocated Service
8	24	Allocated Service	25	Use Service
9	25	Use Service	26	Successful Usage
10	26	Successful Usage	33	Branch-Successful Usage
11	33	Branch-Successful Usage	31	Reuse Service
12	31	Reuse Service	41	Branch-Reuse Service
13	41	Branch-Reuse Service	32	Refund Fee
14	32	Refund Fee	34	Merge-Successful Usage
15	34	Merge-Successful Usage	4	Merge-Service Available
16	4	Merge-Service Available	28	End of Day
17	28	End of Day	2	End

Table 10. Scenario 4   Continue Wait = Yes

	A1	Action One	A2 ID	Action Two
1	1	Start	19	Start of Day
2	19	Start of Day	20	Request Service
3	20	Request Service	21	Log User Profile
4	21	Log User Profile	22	Service Available
5	22	Service Available	3	Branch-Service Available
6	3	Branch-Service Available	29	Join Queue
7	29	Join Queue	30	Continue Wait
8	30	Continue Wait	37	Branch-Continue Wait
9	37	Branch-Continue Wait	22	Service Available

Table 11. Scenario 5   Continue Wait = No

	A1 ID	Action One	A2 ID	Action Two
1	1	Start	19	Start of Day
2	19	Start of Day	20	Request Service
3	20	Request Service	21	Log User Profile
4	21	Log User Profile	22	Service Available
5	22	Service Available	3	Branch-Service Available
6	3	Branch-Service Availa-	29	Join Queue
7	29	Join Queue	30	Continue Wait
8	30	Continue Wait	37	Branch-Continue Wait
9	37	Branch-Continue Wait	4	Merge-Service Available
10	4	Merge-Service Available	28	End of Day
11	28	End of Day	2	End

5   Conclusion

In order to enable more confidence in a telecentre monitoring model, drafted and developed through traditional software engineering methods, we discovered the need for a formalized tool to provide automated phase consistency of requirements from the software specification to the design phase. In order to first develop this tool, a set of formal semantics (action and action link rules) was chosen to formalize the activity diagram and its related scenarios. Once this formalization was complete, the tool was able to automatically generate scenarios from its activity diagram which were then compared and matched to the original scenarios, generated via traditional software engineering methods. This match of both generated and original scenarios established phase consistency between the specification and design phases of the telecentre monitoring software development cycle.

References

1. Battacharjee, A.K., Shyamasunda, R.K.: Activity diagrams: a formal framework to model business processes and code generation. Journal of Object Technology **8**(1), 189–200 (2009)
2. Boehm, B.W.: Verifying and validating software requirements and design specifications. IEEE software (1984)
3. Chanda, J., et al.: Traceability of requirements and consistency verification of UML use case, activity and Class diagram: a formal approach. In: Proceeding of International Conference on Methods and Models in Computer Science, 2. IEEE (2009)
4. Gomez, R., Pather, S., Dosono, B.: Public access computing in South Africa: old lessons and new challenges. Electronic Journal on Information Systems in Developing Countries **52**(1), 1–16 (2012)
5. Gould, E., Gomez, R.: Community Engagement &Intermediaries: challenges facing libraries, telecentres and cybercafés in developing countries (2010).
https://ideals.illinois.edu/handle/2142/19429
6. Han, K.H.: Qualitative and quantitative analysis of workflows based on the UML activity diagram and Petri net. Information Systems **11**(23) (2009)
7. Harris, R.W.: Telecentre Evaluation in the Malasian Context. In: 5th International conference on IT in Asia, 5–7 July 2007, Sarawak Malaysia (2007)
8. Heitmeyer, C.L., Jeffords, R.D., Labaw, B.G.: Automated consistency checking of requirements specifications. ACM Transactions on Software Engineering and Methodology (TOSEM) **5**(3), 231–261 (1996)
9. Hofmeister, C., et al.: A general model of software architecture design derived from five industrial approaches. Journal of Systems and Software **80**(1), 106–126 (2007)
10. Jacobson, I., et al.: The unified software development process. Addison-Wesley, Reading (1999)
11. Linzhang, W., et al.: Generating test cases from UML activity diagram based on gray-box method. In: Software Engineering Conference, 2004. 11th Asia-Pacific. IEEE (2004)
12. Maiden, N.A.M.: CREWS-SAVRE: Scenarios for acquiring and validating requirements. Automated Software Engineering **5**(4), 419–446 (1998)
13. McConnell, S.: Telecentres around the world: Issues to be considered and lessons learned. ICT Development group for CIDA's Canada-Thai Telecentre Project, Columbia Canada (2001)

14. Munassar, N.M.A., Govardhan, A.: A comparison between five models of software engineering. IJCSI **5**, 95–101 (2010)
15. OMG: Unified Modeling Language TM (OMG UML), Superstructure Version 2.4.1 Information technology-Object Management Group Unified Modeling Language (OMG UML), Infrastructure
16. Pancham, J., R Millham, P Singh: A Validated Model for Operational Monitoring of Telecentres' Activities in a Developing Country. In: 7th International Development and Informatics Association Conference, Bangkok (2013)
17. Peterson, J.L.: Petri nets. ACM Computing Surveys (CSUR) **9**(3), 223–252 (1977)
18. Peffers, K., Tunnanen, T., et al.: The design science research process: a model for producing and presenting information systems research. In: Proceedings of The First International Conference On Design Science Research In Information Systems And Technology (DESRIST 2006), pp. 83–106 (2006)
19. Potts, C., Takahashi, K., Anton, A.I.: Inquiry based Requirements Analysis. IEEE Software **11**(2), 21–32 (1994). http://www4.ncsu.edu/~aianton/pubs/ieeeSW.pdf
20. Rafe, V., et al.: Verification and validation of activity diagrams using graph transformation. In: International Conference on Computer Technology and Development. ICCTD 2009. vol. 1. IEEE (2009)
21. Rajapakse, J: Impact of Telecentres on Sri Lankan Society. In: 8th International Conference on Computing and Networking Technology, pp. 281 – 286 (2012)
22. Rantapuska, T., Millham, R.: 15P. Applying Organisational Learning to User Requirements Elicitation. IRMA. (2010)
23. Satzinger, J.W., Jackson, R.B., Burd, S. D.: Introduction to Systems Analysis and Design: An Agile, Iterative Approach, Course Technology, Cengage Learning (2012)
24. Schach, S.R.: Object-oriented and classical software engineering, vol. 6. McGraw-Hill, New York (2002)
25. Sommerville, I.: Software engineering, Pearson Higher Ed. (2011)
26. Stevens, P., Pooley, R.: Using UML: Software Engineering with Objects and Components, 2nd edn. Addison Wesley, Harlow, UK (2006)
27. Swain, R.K., Panthi, V., Behara, P.K.: Generation of test cases using activity diagram. International Journal of Computer Science and Informatics (2013)
28. Wieringa, R.: Design science methodology: principles and practice. In: Proceedings of the 32nd ACM/IEEE International Conference on Software Engineering-vol. 2, ACM, pp. 493–494 (2010)
29. Yeasmin, S., Rahman, K.F.: Triangulation' Research Method as the Tool of Social Science Research, BUP Journal **1**(1), September 2012
30. Xuping, J.: Modeling and Application of Requirements Engineering Process Metamodel. In: IEEE International Symposium on Knowledge Acquisition and Modeling Workshop. KAM Workshop, 21–22 Dec. 2008. pp. 998–1001 (2008)
31. Zowghi, D., Gervasi, V.: On the interplay between consistency, completeness, and correctness in requirements evolution. Information and Software Technology **45**(14), 993–1009 (2003)
32. Muskens, J., Bril, R.J., Chaudron, M.R.V.: Generalizing consistency checking between software views. In: 5th Working IEEE/IFIP Conference on Software Architecture. IEEE (2005)

Extracting Environmental Constraints in Reactive System Specifications

Yuichi Fukaya and Noriaki Yoshiura[✉]

Department of Information and Computer Sciences, Saitama University,
255, Shimo-ookubo, Sakura-ku, Saitama, Japan
yoshiura@fmx.ics.saitama-u.ac.jp

Abstract. Reactive systems ideally never terminate and maintain some interaction with their environment. Temporal logic is one of the methods for formal specification description of reactive systems. For a reactive system specification, we do not always obtain a program that satisfies it because the reactive system program must satisfy the specification no matter how the environment of the reactive system behaves. This problem is known as realizability or feasibility. The complexity of deciding realizability of specifications that are described in linear temporal logic is double or triple exponential time of the length of specifications and realizability decision is impractical. To check reactive system specifications, Strong satisfiability is one of the necessary conditions of realizability of reactive system specifications. If a reactive system specification is not strong satisfiable, it is necessary to revise the specification. This paper proposes the method of revising reactive system specifications that are not strong satisfiable. This method extracts environmental constraints that are included in reactive system specifications.

1 Introduction

Reactive systems, such as operating systems or elevator control systems, interact with their users or environments, and provide services for their users. Reactive system behaviors depend on the interactions with their environments or users [5]. In reactive systems, events are divided into output events and input events. Output events are controlled or generated by reactive systems and input events are generated by environments such as users of reactive systems. Input events cannot be controlled by the reactive systems [6].

Temporal logic is one of the methods of describing reactive system specifications. One of the advantages of temporal logic is to prove several properties of the specifications. Given a reactive system specification, which is a temporal logic formula, satisfiability of the formula only guarantees possibility of synthesizing the reactive system that satisfies a reactive system specification for some environment behavior. However, it is necessary to synthesize the system that satisfies the specification for all environment behaviors; suppose that there are finite sets I and O of input and output events. A reactive system program is a function $f : (2^I)^* \rightarrow 2^O$ that maps finite sequences of input event sets into an output

© Springer International Publishing Switzerland 2015
O. Gervasi et al. (Eds.): ICCSA 2015, Part IV, LNCS 9158, pp. 671–685, 2015.
DOI: 10.1007/978-3-319-21410-8_51

event set. The function f generates infinite sequences of $2^{I \cup O}$ during interaction between a reactive system and an environment because the environment generates infinite sequences of input event sets. A function f realizes a specification ϕ if and only if ϕ always holds for all infinite sequences that are generated by the function f and an environment. The realizability problem is to determine, given a specification ϕ which is a temporal logic formula, whether there exists a function $f : (2^I)^* \rightarrow 2^O$ that realizes ϕ [1]. The complexity of deciding realizability is 2EXPTIME-complete on CTL and 3EXPTIME-complete on CTL* [4,6]. Therefore, the complexity of deciding realizability on LTL is 2EXPTIME-complete or 3EXPTIME-complete [4,6].

The previous studies on realizability of reactive system specifications that are described by several kinds of temporal logic focus on decidability of realizability but not on speed of deciding realizability. Automata theory has been used in almost all studies to give decision procedures of realizability, but it is insufficient for fast decision of realizability of specifications. There are several researches that are related with implementation of realizability decision [3,4,9]. However, implementation of fast decision procedures is impractical because of complexity of realizability problem.

There are several necessary conditions of realizability of reactive system specifications and strong satisfiability is one of the necessary conditions [10]. Reactive system specifications that do not satisfy strong satisfiability should be revised to develop software from the reactive system specifications. Therefore, it is important to find and modify misfeatures of reactive system specifications. The cause of strong unsatisfiability of reactive system specifications is that the specifications constraint environment of reactive systems; if reactive systems specifications are strong unsatisfiable, all reactive systems that satisfy the specifications require cooperation of environment of reactive systems.

Several researches have focused on extracting environmental constraints in reactive system specifications. Mori proposed the method of extracting environmental constraints from reactive system specifications and the method extracted exact environmental constraints. However, the complexity of this method is high and the environmental constraints that are extracted by the method is difficult to understand. Hagihara proposed another method of extracting environmental constraints. The method of Hagihara extracted environmental constraints whose form is $\square \diamondsuit A$. This form of environmental constraints is not exact environmental constraint, but the complexity of this method is lower then the method that is proposed by Mori and this form of environmental constraints is easy to understand. The logical formula of form $\square \diamondsuit A$ is related with fairness of concurrent processes. This form Environmental constraints requires that environment of reactive systems generate events fairly and this constraint is easy to understand and fairness of events is a basic requirement for reactive systems. The environmental constraints that are extracted by Hagihara method are useful for revision of reactive system specifications. However, the environmental constraint is inexact and too strong; there are some cases that environments do not satisfy this constraint but reactive systems can satisfy reactive system specifications. There-fore, it is desirable to extract the environmental constraints that are more exact and easy to understand.

This paper proposes the new method of extracting environmental constraints from reactive system specifications that are strong unsatisfiable. The environ-mental constraints that are extracted by the new method are of the form $\Box\Diamond(A \lor \bigcirc B)$. The feature of the new method is that extracted environmen-tal constraints express the order of events in environment behavior. This feature is the difference between the new method and Hagihara method. The environ-mental constraints in Hagihara method do not express the order of events in envi-ronment behavior. As a result, the environmental constraints that are extracted by the new method come closer to exact constraints than Hagihara method.

This paper is organized as follows; Section 2 explains formal definition of reactive system and linear temporal logic that is used for reactive system spec-ifications. This section also explains realizability and strong satisfiability that are the properties of reactive system specifications, and environmental con-straints. Section 3 explains the previous methods of extracting environmental constraints. Section 4 proposes the new method of extracting environmental constraints. Section 5 compares the proposed method and the previous method. Section 6 concludes this paper.

2   Reactive System

This section provides a formal definition of reactive systems, based on references [8]. Let A be a finite set. A^+ and A^ω denote the set of *finite sequences* and the set of *infinite sequences* over A respectively. $A^\dagger$ denotes $A^+ \cup A^\omega$. Sequences in $A^\dagger$ are denoted by $\hat{a}, \hat{b}, \cdots$, sequences in A^+ by $\bar{a}, \bar{b}, \cdots$ and sequences in A^ω by $\tilde{a}, \tilde{b}, \cdots$. $|\hat{a}|$ denotes the *length* of $\hat{a}$ and $\hat{a}[i]$ denotes the *i-th element* of $\hat{a}$. Suppose that B is a set whose elements are also sets. '$\sqcup$', which is a composition of sequences, is defined over $B^\dagger \times B^\dagger$ by

$$\hat{a} \sqcup \hat{b} = \hat{a}[0] \cup \hat{b}[0], \ \hat{a}[1] \cup \hat{b}[1], \ \hat{a}[2] \cup \hat{b}[2], \cdots .$$

Definition 1. *A reactive system RS is a triple $RS = \langle X, Y, r \rangle$, where*

- *X is a finite set of input events that are generated by an environment.*
- *Y $(X \cap Y = \emptyset)$ is a finite set of output events that are generated by the reactive system itself.*
- *$r : (2^X)^+ \to 2^Y$ is a reaction function.*

A subset of X is called *input set* and a sequence of input sets is called *input sequence*. Similarly, a subset of Y is called *output set* and a sequence of output sets is called *output sequence*. In this paper, a reaction function corresponds to a reactive system program.

Definition 2. *Let* $RS = \langle X, Y, r \rangle$ *be a reactive system and* $\hat{a} = a_0, a_1, \cdots \in (2^X)^\dagger$ *be an input sequence. The behavior of* RS *for* $\hat{a}$, *denoted* $behave_{RS}(\hat{a})$, *is the following sequence:*

$$behave_{RS}(\hat{a}) = \langle a_0, b_0 \rangle, \langle a_1, b_1 \rangle, \langle a_2, b_2 \rangle, \ldots,$$

where for each i $(0 \le i < |\hat{a}|)$, $b_i = r(a_0, \ldots, a_i) \in 2^Y$.

3   Specification

This paper uses propositional linear-time temporal logic (PLTL) as a specification language for reactive systems. A PLTL *formula* is defined in the usual way: an atomic proposition $p \in \mathcal{P}$ is a formula and if f_1 and f_2 are formulas, $f_1 \wedge f_2$, $\neg f_1$, $\bigcirc f_1$ and $f_1 \mathcal{U} f_2$ are formulas. This paper uses "Weak until operator" $(\mathcal{U})$ and an abbreviation of $\Box f_1 \equiv f_1 \mathcal{U}(f_2 \wedge \neg f_2)$. $\vee$, $\rightarrow$, $\leftrightarrow$ and $\Diamond$ are the usual abbreviations.

The semantics of PLTL formulas is defined on the behaviors. Let P be a set of input and output events and $\mathcal{P}$ be a set of atomic propositions corresponding to each element of P. $\langle \sigma, i \rangle \models f$ denotes that a formula f over $\mathcal{P}$ holds at the i-th state of a behavior $\sigma \in (2^P)^\omega$. $\langle \sigma, i \rangle \models f$ is recursively defined as follows.

– $\langle \sigma, i \rangle \models p$ **iff** $p' \in \sigma[i]$ (p is an atomic proposition corresponding to $p' \in P$)
– $\langle \sigma, i \rangle \models \neg f$ **iff** $\langle \sigma, i \rangle \not\models f$
– $\langle \sigma, i \rangle \models f_1 \wedge f_2$ **iff** $\langle \sigma, i \rangle \models f_1$ and $\langle \sigma, i \rangle \models f_2$
– $\langle \sigma, i \rangle \models \bigcirc f$ **iff** $\langle \sigma, i+1 \rangle \models f$
– $\langle \sigma, i \rangle \models f_1 \mathcal{U} f_2$ **iff** $(\forall j \ge 0)$ $\langle \sigma, i+j \rangle \models f_1$ or
    $(\exists j \ge 0)$ $(\langle \sigma, i+j \rangle \models f_2$ and
      $\forall k (0 \le k < j)$ $\langle \sigma, i+k \rangle \models f_1)$

σ is a model of f if and only if $\langle \sigma, 0 \rangle \models f$. We write $\sigma \models f$ if $\langle \sigma, 0 \rangle \models f$. A formula f is satisfiable if and only there is a model of f. For example, $behave_{RS}(\tilde{a}) \models \varphi$ means that a behavior that is generated by the reactive system RS receiving the input sequence $\tilde{a}$ is a model of φ.

3.1   Specification

A *PLTL-specification* for a reactive system is a triple $Spec = \langle \mathcal{X}, \mathcal{Y}, \varphi \rangle$, where

– $\mathcal{X}$ is a set of *input propositions* that are atomic propositions corresponding to the input events of the reactive system. The truth value of an input proposition represents the occurrence of the corresponding input event.
– $\mathcal{Y}$ is a set of *output propositions* that are atomic propositions corresponding to the output events of the reactive system. The truth value of an output proposition represents the occurrence of the corresponding output event.
– φ is a formula in which all the atomic propositions are elements of $\mathcal{X} \cup \mathcal{Y}$.

This paper writes $Spec = \langle \mathcal{X}, \mathcal{Y}, \varphi \rangle$ just as φ if there is no confusion. Finally, the following defines the realizability of reactive system specifications [8].

Definition 3. *A reactive system RS is an implementation of a specification φ if for every input sequence $\tilde{a}$, behave$_{RS}(\tilde{a}) \models \varphi$. A specification is realizable if it has an implementation.*

3.2   Strong Satisfiability

Reactive system specifications can be divided into several classes. This subsection explains the class of *strong satisfiability*. This property is introduced in [7]. Specifications are strong satisfiable if and and only if the specifications are satisfiable for any truth values of input propositions. Strong satisfiability is a necessary condition of realizability. The formal definition of strong satisfiability is as follows.

Definition 4. *A specifications φ is strong satisfiable if and only if the following condition holds.*

$$\forall \tilde{x} \exists \tilde{y} (\langle \tilde{x}, \tilde{y} \rangle \models \varphi.$$

where $\tilde{x}$ is an infinite sequence of sets of input propositions and $\tilde{y}$ is an infinite sequence of sets of output propositions.

We now explain the meaning of strong satisfiability. In reactive systems, input events are made by the environment and not by the reactive system. If a reactive system is realizable, then for any infinite input event sequence, there is an infinite output event sequence made by the reactive system. Thus, this strong satisfiability is a necessary condition of realizability. We can show that strong satisfiability is not a sufficient condition as follows. We suppose that φ is strong satisfiable. For each infinite input event sequence, we can find an infinite output event sequence from the infinite event sequence such that φ is satisfied. However, the reactive system must produce an output event from the input event sequence until the current time but not from the infinite event sequence. Therefore strong satisfiability does not guarantee that reactive systems behave so that φ is satisfied. Later, we will explain strong satisfiability by using some examples.

Now we show some examples of specifications In each example below, atomic propositions written in bold type denote input propositions.

EXAMPLE 1. *Satisfiable but not strongly satisfiable specification.*

1. $\mathbf{req_1} \to \Diamond res$.
2. $\mathbf{req_2} \to \Box \neg res$.

This example shows that a specification is not strongly satisfiable if it could require conflicting responses at the same time.

EXAMPLE 2. *Strongly satisfiable but not realizable specification.*

1. $(\mathbf{req_1} \, \mathcal{U} \, \mathbf{req_2}) \leftrightarrow res$.

This example shows that a specification is not realizable if it could require a response depending on the future sequences of requests.

4   Extraction of Environmental Constraints

If the reactive system specification is unrealizable, it is revised in order to obtain a program from it. There are several researches for revising reactive system specifications. In [7], Mori and Yonezaki proposed the input condition formula which restricts the environment behavior in order to make a reactive system specification strongly satisfiable. In [14], Cimatti, Roveri, Schuppan and Tchaltsev addressed the problem of providing diagnostic information for the realizability of the specification of an open system.

This paper proposes the new method of extracting environmental constraints. In the following, environmental constraints are defined.

Definition 5 (Environmental Constraints). *Suppose that* φ *is a reactive system specification and that* ψ *is a temporal logic formula that consists of only input propositions.* ψ *is an environmental constraint of* φ *if and only if the following condition holds.*

$$\forall \tilde{a}(\tilde{a} \models \psi \Rightarrow \exists \tilde{b} \langle \tilde{a}, \tilde{b} \rangle \models \varphi)$$

Generally, there are several environmental constraints for one specification. For a specification φ, $EC(\varphi)$ is defined to be a set of environmental constraints of φ. The following defines order relation for $EC(\varphi)$.

Definition 6 (Order of environmental constraints). *Suppose that* ψ_1 *and* ψ_2 *are environmental constraints of a specification* φ. $\psi_1 \leq \psi_2$ *if and only if the following condition holds.*

$$\forall \sigma(\sigma \models \psi_1 \Rightarrow \psi_2)$$

$\psi_1 \leq \psi_2$ *means that* ψ_1 *is stronger than* ψ_2. *"Strong" means that an environmental constraint is strong.* $\psi_1 \equiv \psi_2$ *if and only if* $\psi_1 \leq \psi_2$ *and* $\psi_2 \leq \psi_1$.

The weakest environmental constraint is preferable in a set of environmental constraints. This paper defines the weakest constraint as follows.

Definition 7 (Weakest environmental constraint). *Suppose that* $\mathcal{L}$ *is a set of formulas and that* φ *is a specification.* ψ *is the weakest environmental constraint of* φ *with respect to* $\mathcal{L}$ *if and only if the two following conditions hold.*

– $\psi \in (\mathcal{L} \cap EC(\varphi))$
– $\forall \psi'(\psi' \in \mathcal{L} \cap EC(\varphi) \Rightarrow \psi' \leq \psi)$.

5   Previous Method of Extracting Environmental Constraints

This section explains the previous methods that are proposed in [13]. This paper uses the previous methods to extract environmental constraints. The previous methods extract the environmental constraints that are of the specific forms because the complexity of extracting the environmental constraints is low and the environmental constraints can be extracted within reasonable time. Moreover, specific forms of environmental constraints are easy to understand. The previous methods extract the weakest environmental constraints with respect to specific forms. The specific forms in the previous methods are defined as follows.

Definition 8 (Class of formulas). *$\mathcal{L}_1$ and $\mathcal{L}_2$ are classes of formulas.*

- *$\mathcal{L}_1$ is a set of formulas whose form is $\bigwedge \Box \Diamond f$.*
- *$\mathcal{L}_2$ is a set of formulas whose form is $\bigwedge (\bigvee \Diamond \Box f \vee \Box \Diamond g)$.*

where f and g are classical logic formulas.

The formulas in $\mathcal{L}_1$ or $\mathcal{L}_2$ are used to represent fairness of event occurrence. For example, suppose that users require some computer resources infinitely often in operating systems. This situation is represented by $\Box \Diamond req$ where req represents that users require computer resources. This form of formulas represents that no event is kept from occurring.

The following explains the previous methods. A specification φ is defined as follows.

$$\varphi = \Box((x_1 \wedge x_2) \rightarrow (y \, \mathcal{U} \neg x_1 (\neg x_1 \wedge \bigcirc (\neg x_1 \wedge x_2))))$$

where x_1 and x_2 are input events and y is output events.

This specification is strong unsatisfiable. The previous methods can extract environmental constraints as follows.

$$\psi_{\varphi,1} = \bot$$

$$\psi_{\varphi,2} = \Box \Diamond x_2 \wedge \Diamond \Box \neg x_1$$

where $\psi_{\varphi,1}$ is an environmental constraint in $\mathcal{L}_1$ and $\psi_{\varphi,2}$ is an environmental constraint in $\mathcal{L}_2$.

$\psi_{\varphi,1}$ is the strongest environmental constraint and no environmental behavior does not satisfy this constraint. This constraint is useless because this constraint provides no information to revise specifications. $\psi_{\varphi,2}$ is weaker than $\psi_{\varphi,1}$, but is too strong constraint to revise the specifications. $\psi_{\varphi,2}$ requires that user event x_1 never occurs after some time point so that reactive system satisfies a specification.

The previous methods can extract environmental constraints from a specification within reasonable time, but the extracted environmental constraints do not have expressive power. Therefore, the environmental constraints are useless in some cases.

6   Proposed Method

This section proposes the new method of extracting weakest environmental constraints that are of the form $\Box\Diamond(\bigwedge(f \vee \bigcirc g))$ or $\bigwedge((\bigvee \Diamond\Box f) \vee \Box\Diamond(\bigwedge(g \vee \bigcirc h)))$ where f, g and h are classical logic formulas.

The proposed method is based on previous method. The method consists of constructing nondeterministic Büchi automata from specifications, deleting output propositions from the nondeterministic Büchi automata, constructing complement the nondeterministic Büchi automata and generating logical formulas from the complement the nondeterministic Büchi automata.

The following subsections define the nondeterministic Büchi automata and the methods that are proposed in this paper.

6.1   Nondeterministic Büchi Automata

The method of extracting environmental constraints requires nondeterministic Büchi automata that are constructed from specifications. The method of constructing nondeterministic Büchi automata is proposed by several researches [3,4,9] and several tools are proposed in [12]. This paper uses these tools.

Definition 9 (Nondeterministic Büchi automata). *Suppose that P is a set of atomic propositions. Nondeterministic Büchi automata on alphabet 2^P is defined as $\mathcal{A} = \langle Q, q_0, \delta, F \rangle$ where Q is a set of states, q_0 is a initial state, $\delta \in Q \times B(P) \times Q$ is a transition relation and F is a set of acceptance states. $B(P)$ is a set of classical logic formulas that consist of atomic propositions in P.*

A run on $\mathcal{A}$ is defined for an infinite sequence α of 2^P where $\alpha \in (2^P)^\omega$. gamma is a run for α if and only if the following conditions hold.

- *γ is an infinite sequence of states,*
- *$\gamma[0]$ is initial state of γ,*
- *For any i $(0 \leq i)$, $(\gamma[i], b, \gamma[i+1]) \in \delta$ and $\alpha, i \models b$*

γ is success if and only if $In(\gamma) \cap F \neq \emptyset$ where $In(\gamma)$ is a set of states that occur infinitely often in γ. An infinite sequence α of 2^P is accepted on $\mathcal{A}$ if and only if a run for α is success. $L(\mathcal{A})$ is defined to be a set of accepted infinite sequences of 2^P on $\mathcal{A}$.

6.2   Classes of Environments Constraints

Definition 10 (Classes of environmental constraints). *Classes of formulas $\mathcal{L}_3$ and $\mathcal{L}_4$ are defined as follows.*

- *$\mathcal{L}_3$ is a set of formulas whose form is $\bigwedge\Box\Diamond(\bigwedge(f \vee \bigcirc g))$.*
- *$\mathcal{L}_4$ is a set of formulas whose form is $\bigwedge((\bigvee \Diamond\Box f) \vee \Box\Diamond(\bigwedge(g \vee \bigcirc h)))$.*

where f, g and h are classical logic formulas.

This paper proposes the method of extracting environmental constraints that are included in $\mathcal{L}_3$ are $\mathcal{L}_4$. $\mathcal{L}_3$ is an extension of $\mathcal{L}_1$ and $\mathcal{L}_4$ is an extension of $\mathcal{L}_2$. $\mathcal{L}_1$ and $\mathcal{L}_2$ do not use Next operator ($\bigcirc$) or Until operator ($\mathcal{U}$) and cannot express the order of events. $\mathcal{L}_3$ and $\mathcal{L}_4$ can use Next operator to express the order of events for environmental constraints.

Next, this paper defines the method of extracting environmental constraints for $\mathcal{L}_3$. This method uses nondeterministic Büchi automata and this paper supposes that the method uses several method constructing construction method nondeterministic Büchi automata from specification.

Definition 11 (The method for $\mathcal{L}_3$). *The method of extracting environmental constraints for $\mathcal{L}_3$ is defined as follows.*

Input of the method is nondeterministic Büchi automata $A_{spec} = \langle Q, q_0, \delta, F \rangle$ where $L(A_{spec}) = \{\sigma \mid \sigma \models Spec\}$. Output of the method is a formula $\psi_3 \in \mathcal{L}_3 \cap EC(spec)$. The method obtains A_{spec} by construction method of automata from specifications.

STEP 1

Obtain $A'_{spec} = \langle Q, q_0, \delta', F \rangle$ by deleting output propositions from A_{spec}. δ' is constructed as follows; $\delta' = \{(q, E(b), q') \mid (q, b, q') \in \delta\}$ and $E(b)$ is a formula that is obtained by connecting by disjunction all formulas that are obtained by replacing input propositions by $\bot$ or $\top$ in b. The definition of δ' implies $L(A'_{spec}) = \{\tilde{a} \mid \exists \tilde{b} \langle \tilde{a}, \tilde{b} \rangle \models spec\}$.

STEP 2

Construct the complement nondeterministic Büchi automata $\overline{A'_{spec}}$ of A'_{spec}.

STEP 3

Find a set MSC of maximum strongly connected graphs that include acceptance state in $\overline{A'_{spec}}$.

STEP 4

Generate the following formula ψ_{msc} for each element $msc \in MSC$.
$\psi_{msc} = \Diamond \Box (\bigvee_{s \in q(msc)} ((\bigvee_{c_{in} \in in(s)} c_{in}) \wedge \bigcirc (\bigvee_{c_{out} \in out(s)} c_{out})))$
where $q(msc)$ is a set of states in msc, $in(s)$ is a set of label whose edge goes to state s and $out(s)$ is a set of label whose edge comes out of state s.

STEP 5

This method outputs the following formula as environmental constraint for $\mathcal{L}_3$.
$\psi_3 = \neg(\bigvee_{msc \in MSC} \psi_{msc})$
$= \bigwedge_{msc \in MSC} \Box \Diamond (\bigwedge_{s \in q(msc)} ((\bigwedge_{c_{in} \in in(s)} \neg c_{in}) \vee \bigcirc (\bigwedge_{c_{out} \in out(s)} \neg c_{out})))$

Next, this paper defines the method of extracting environmental constraints for $\mathcal{L}_4$.

Definition 12 (The method for $\mathcal{L}_4$). *The method of extracting environmental constraints for $\mathcal{L}_4$ is defined as follows.*

Input of the method is nondeterministic Büchi automata $A_{spec} = \langle Q, q_0, \delta, F \rangle$ where $L(A_{spec}) = \{\sigma \mid \sigma \models Spec\}$. Output of the method is a formula $\psi_4 \in \mathcal{L}_4 \cap EC(spec)$. The method obtains A_{spec} by construction method of automata from specifications.

STEP 1

Obtain $A'_{spec} = \langle Q, q_0, \delta', F \rangle$ by deleting output propositions from A_{spec}. δ' is constructed as follows; $\delta' = \{(q, E(b), q') \mid (q, b, q') \in \delta\}$ and $E(b)$ is a formula that is obtained by connecting by disjunction all formulas that are obtained by replacing input propositions by $\bot$ or $\top$ in b. The definition of δ' implies $L(A'_{spec}) = \{\tilde{a} \mid \exists \tilde{b}\langle \tilde{a}, \tilde{b} \rangle \models spec\}$.

STEP 2

Construct the complement nondeterministic Büchi automata $\overline{A'_{spec}}$ of A'_{spec}.

STEP 3

Find a set SC of strongly connected graphs that include acceptance state in $\overline{A'_{spec}}$.

STEP 4

Generate the following formula ψ_{sc} for each element $sc \in SC$.
$$\psi_{sc} = (\bigwedge_{c \in l(sc)} \Box \Diamond c) \wedge \Diamond \Box (\bigvee_{s \in q(sc)} ((\bigvee_{c_{in} \in in(s)} c_{in}) \wedge \bigcirc (\bigvee_{c_{out} \in out(s)} c_{out})))$$
where $q(sc)$ is a set of states in sc, $in(s)$ is a set of label whose edge goes to state s and $out(s)$ is a set of label whose edge comes out of state s.

STEP 5

This method outputs the following formula as environmental constraint for $\mathcal{L}_4$.
$$\psi_4 = \neg(\bigvee_{sc \in SC} \psi_{sc})$$
$$= \bigwedge_{sc \in SC}((\bigvee_{c \in l(sc)} \Diamond \Box \neg c) \vee$$
$$\Box \Diamond (\bigwedge_{s \in q(msc)}((\bigwedge_{c_{in} \in in(s)} \neg c_{in}) \vee \bigcirc (\bigwedge_{c_{out} \in out(s)} \neg c_{out}))))$$

The difference between the two methods is that the method for $\mathcal{L}_3$ uses maximum strongly connected graphs in nondeterministic Büchi automata and the method for $\mathcal{L}_4$ uses strongly connected graphs. Therefore, the environmental constraint for $\mathcal{L}_4$ is weaker than that for $\mathcal{L}_3$. The following section proves this fact formally.

6.3   Example of Environmental Constraints

This subsection applies the methods for $\mathcal{L}_3$ and $\mathcal{L}_4$ to the following specification.

$$\varphi = \Box((x_1 \wedge x_2) \rightarrow (y \mathcal{U} \neg x_1 (\neg x_1 \wedge \bigcirc (\neg x_1 \wedge x_2))))$$

where x_1 and x_2 are input propositions and y is an output proposition. The methods require nondeterministic Büchi automata $\overline{A'_\varphi}$, which is shown in Figure 1. In this figure, $s0$ is an initial state and $s2$ and $s3$ are accepted states.

Let $\psi_{\varphi,3}$ be a formula that is obtained by the method for $\mathcal{L}_3$ and $\psi_{\varphi,4}$ be a formula that is obtained by the method for $\mathcal{L}_4$.
$$\psi_{\varphi,3} = \Box \Diamond (\neg x_1 \wedge \bigcirc (\neg x_1 \wedge x_2))$$
$$\psi_{\varphi,4} = (\Diamond \Box (x_1 \vee x_2) \vee \Box \Diamond (x_1 \vee x_2)) \wedge (\Diamond \Box \neg x_1 \vee \Box \Diamond (\neg x_1 \wedge \bigcirc \neg x_1))$$
$$\wedge (\Diamond \Box \neg x_1 \vee \Box (x_1 \vee x_2) \vee \Box \Diamond (\neg x_1 \wedge \bigcirc (\neg x_1 \wedge x_2)))$$
The following relation among $\psi_{\varphi,1}$, $\psi_{\varphi,2}$, $\psi_{\varphi,3}$ and $\psi_{\varphi,4}$ can be proved easily.

$$\psi_{\varphi,1} \leq \psi_{\varphi,2} \leq \psi_{\varphi,3} \leq \psi_{\varphi,4}$$

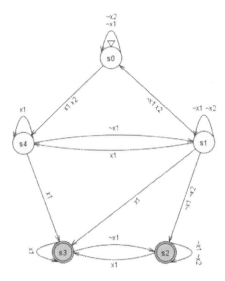

Fig. 1. $\overline{A'_\varphi}$

This result shows that $\psi_{\varphi,3}$ and $\psi_{\varphi,4}$ are more useful than $\psi_{\varphi,1}$ and $\psi_{\varphi,2}$ in this example. The following section shows the general properties of the methods that are proposed in this paper.

7   Properties of the Methods

This section proves soundness and termination of the methods for $\mathcal{L}_3$ and $\mathcal{L}_4$. Moreover, this section proves that the methods extract the weakest environmental constraints for $\mathcal{L}_3$ and $\mathcal{L}_4$. At the beginning, this paper proves termination.

Theorem 1. *Suppose that A_{spec} is nondeterministic Büchi automata where $L(A_{spec}) = \{\sigma \mid \sigma \models spec\}$. Applying the methods for $\mathcal{L}_3$ and $\mathcal{L}_4$ to A_{spec} always terminates.*

Proof: Termination of the method for $\mathcal{L}_3$ is proved. The method consists of five steps. Step 1 terminates because A_{spec} has finite edges. Step 2 terminates because construction of complement nondeterministic Büchi automata terminates [3,4,9]. Step 3, Step 4 and Step 5 terminate because maximum strongly connected graphs are finite in automata [11]. Therefore, the method for $\mathcal{L}_3$ always terminates. Termination of the method for $\mathcal{L}_4$ is proved similarly.   ■

Next, this paper proves soundness of the methods.

Theorem 2. *Suppose that A_{spec} is nondeterministic Büchi automata where $L(A_{spec}) = \{\sigma \mid \sigma \models spec\}$. The environmental constraint that is obtained by the method for $\mathcal{L}_3$ is included in $\mathcal{L}_3 \cap EC(spec)$.*

Proof: Suppose that ψ_3 is the environmental constraint that is obtained by the method for $\mathcal{L}_3$ for the specification $spec$. By the definition of the method for $\mathcal{L}_3$, $\psi_3 \in \mathcal{L}_3$.

Next, $\psi_3 \in EC(spec)$ is proved. Suppose that $\tilde{a}$ is an infinite sequence of input propositions and that $\tilde{a}$ is accepted on $\overline{A'_{spec}}$. In $\overline{A'_{spec}}$, a run for $\tilde{a}$ reaches a maximum strongly connected graph that has a acceptance state. The run includes acceptance states infinite often.

For any state s in the maximum strongly connected graphs, $(\bigvee_{c_{in} \in in(s)} c_{in}) \wedge \bigcirc(\bigvee_{c_{out} \in out(s)} c_{out})$ holds at time point of s. This implies that $\neg\psi_3$ holds. Therefore, the following proves that $\psi_3 \in EC(spec)$.

$$\forall \tilde{a}(\tilde{a} \in L(\overline{A'_{spec}}) \Rightarrow \tilde{a} \models \neg\psi_3) \Leftrightarrow \forall \tilde{a}(\tilde{a} \notin L(A'_{spec}) \Rightarrow \tilde{a} \not\models \psi_3)$$
$$\Leftrightarrow \forall \tilde{a}(\tilde{a} \models \psi_3 \Rightarrow \tilde{a} \in L(A'_{spec})) \Leftrightarrow \forall \tilde{a}(\tilde{a} \models \psi_3 \Rightarrow \exists \tilde{b}\langle \tilde{a}, \tilde{b}\rangle \models spec)$$
$$\Leftrightarrow \psi_3 \in EC(spec) \qquad \blacksquare$$

Theorem 3. *Suppose that A_{spec} is nondeterministic Büchi automata where $L(A_{spec}) = \{\sigma \mid \sigma \models spec\}$. The environmental constraint that is obtained by the method for $\mathcal{L}_4$ is included in $\mathcal{L}_4 \cap EC(spec)$.*

Proof: Suppose that ψ_4 is the environmental constraint that is obtained by the method for $\mathcal{L}_4$ for the specification $spec$. By the definition of the method for $\mathcal{L}_4$, $\psi_4 \in \mathcal{L}_4$.

Next, $\psi_4 \in EC(spec)$ is proved. Suppose that $\tilde{a}$ is an infinite sequence of input propositions and that $\tilde{a}$ is accepted on $\overline{A'_{spec}}$. In $\overline{A'_{spec}}$, a run for $\tilde{a}$ reaches a strongly connected graph that has a acceptance state. The run infinitely often includes all states in the strongly connected graph.

For any state s in the strongly connected graphs, $(\bigvee_{c_{in} \in in(s)} c_{in}) \wedge \bigcirc(\bigvee_{c_{out} \in out(s)} c_{out})$ holds at time point of s. This implies that $\neg\psi_4$ holds. Therefore, the following proves that $\psi_4 \in EC(spec)$.

$$\forall \tilde{a}(\tilde{a} \in L(\overline{A'_{spec}}) \Rightarrow \tilde{a} \models \neg\psi_4) \Leftrightarrow \forall \tilde{a}(\tilde{a} \notin L(A'_{spec}) \Rightarrow \tilde{a} \not\models \psi_4)$$
$$\Leftrightarrow \forall \tilde{a}(\tilde{a} \models \psi_4 \Rightarrow \tilde{a} \in L(A'_{spec})) \Leftrightarrow \forall \tilde{a}(\tilde{a} \models \psi_4 \Rightarrow \exists \tilde{b}\langle \tilde{a}, \tilde{b}\rangle \models spec)$$
$$\Leftrightarrow \psi_4 \in EC(spec) \qquad \blacksquare$$

Next, the weakness of environmental constraints is proved.

Theorem 4. *Suppose that A_{spec} is nondeterministic Büchi automata where $L(A_{spec}) = \{\sigma \mid \sigma \models spec\}$. The environmental constraint that is obtained by the method for $\mathcal{L}_3$ is the weakest environmental constraint in $\mathcal{L}_3$.*

Proof: Suppose that ψ_3 is the environmental constraint that is obtained by the method for $\mathcal{L}_3$ for the specification $spec$. This proof proves the contraposition. Suppose that $\psi' = \bigwedge_{1 \leq i \leq n}(\Box\Diamond(\bigwedge_j(f_{ij} \vee \bigcirc g_{ij})))$ and that $\psi' \not\leq \psi_3$.

Recall the form of ψ_3.

$$\psi_3 = \bigwedge_{msc \in MSC} \Box\Diamond(\bigwedge_{s \in q(msc)}((\bigwedge_{c_{in} \in in(s)} \neg c_{in}) \vee \bigcirc(\bigwedge_{c_{out} \in out(s)} \neg c_{out})))$$

$\psi' \not\leq \psi_3$ implies that there is an infinite sequence $\tilde{a}$ of sets of input propositions such that $\tilde{a} \not\models \psi_3$ and $\tilde{a} \models \psi'$. It follows that there is a maximum strongly connected graph msc such that the following holds.

$$\tilde{a} \models \Diamond\Box \bigvee_{s\in q(msc)}((\bigvee_{c_{in}\in in(s)} c_{in}) \wedge \bigcirc(\bigvee_{c_{out}\in out(s)} c_{out})) \wedge$$
$$\bigwedge_{1\leq i\leq n}(\Box\Diamond(\bigwedge_j(f_{ij} \vee \bigcirc g_{ij})))$$

Therefore, for any i $(1 \leq i \leq n)$, there are a state $s_i \in q(msc)$ and sets a_{in}^i and a_{out}^i of input propositions such that

$$a_{in}^i a_{out}^i \sigma \models ((\bigvee_{c_{in}} c_{in}) \wedge \bigcirc(\bigvee_{c_{out}\in out(s_i)} c_{out})) \wedge (\bigwedge_j(f_{ij} \vee \bigcirc g_{ij}))$$

where σ is an arbitrary sequence of sets of input propositions. For any i $(1 \leq i \leq n)$, there are e_{in}^i, c_{in}^i, e_{out}^i and c_{out} such that the following conditions hold.

- e_{in}^i is an edge that goes into s_i.
- $c_{in}^i \in in(s_i)$ is a label of e_{in}^i.
- e_{out}^i is an edge that goes out of s_i.
- $c_{out}^i \in out(s_i)$ is a label of e_{in}^i.
- $a_{in}^i a_{out}^i \sigma \models (c_{in}^i \wedge \bigcirc c_{out}^i) \wedge (\bigwedge_j(f_{ij} \vee \bigcirc g_{ij}))$

Recall that msc is a maximum strongly connected graph that includes an acceptance state. There is a run from an initial state to msc and let $\overline{x}_{q_0,1}$ be a finite sequence of sets of input propositions where a run of $\overline{x}_{q_0,1}$ is from the initial state to state s_1 that is in msc. For any i $(1 \leq i \leq n)$, there is a run that is from s_i to s_{i+1} via edge e_{out}^i and e_{in}^{i+1} and consists of only states in msc. Let $\overline{x}_{i,i+1}$ be a finite sequence of sets of input propositions where a run of $\overline{x}_{i,i+1}$ is the run from s_i to s_{i+1}. State s_n goes to state s_1 via states in msc. Let $\overline{x}_{n,1}$ be a finite sequence of sets of input propositions where a run of $\overline{x}_{n,1}$ is the run from s_n to s_1. Consider an infinite sequence $\tilde{x}$ of sets of input propositions.

$$\tilde{x} = \overline{x}_{q_0,1} \cdot (\overline{x}_{1,2} \cdot \cdots \overline{x}_{n-1,n} \cdot \overline{x}_{n,1})^\omega$$

The construction of $\tilde{x}$ implies that $\tilde{x} \in L(\overline{A'_{spec}})$. Since $\tilde{x}$ infinitely often includes two long sequence such as $a_{in}^1 a_{out}^1, \cdots$ or $a_{in}^n a_{out}^n$, for any i $(1 \leq i \leq n)$, $a_{in}^i a_{out}^i \sigma \models \bigwedge_j(f_{ij} \vee \bigcirc g_{ij})$. Therefore, $\tilde{x} \models \bigwedge_{1\leq i\leq n}(\Box\Diamond(\bigwedge_j(f_{ij} \vee \bigcirc g_{ij})))$. The following proves $\psi' \notin EC(spec)$.

$$\tilde{x} \models \psi' \text{ and } \tilde{x} \in L(\overline{A'_{spec}}) \Rightarrow \neg(\forall\tilde{a}(\tilde{a} \models \psi' \Rightarrow \tilde{a} \notin L(\overline{A'_{spec}}))$$
$$\Leftrightarrow \neg(\forall\tilde{a}(\tilde{a} \models \psi' \Rightarrow \tilde{a} \in L(A'_{spec})) \Leftrightarrow \neg(\forall\tilde{a}(\tilde{a} \models \psi' \Rightarrow \exists\tilde{b}\langle\tilde{a},\tilde{b}\rangle \models spec))$$
$$\Leftrightarrow \psi' \notin EC(spec)$$

To summarize, $\psi' \in \mathcal{L}_3$ and $\psi' \not\leq \psi_3$ imply $\psi' \notin EC(spec)$. Thus, ψ_3 is the weakest environmental constraint among $\mathcal{L}_3$. ∎

Theorem 5. *Suppose that A_{spec} is nondeterministic Büchi automata where $L(A_{spec}) = \{\sigma \mid \sigma \models spec\}$. The environmental constraint that is obtained by the method for $\mathcal{L}_4$ is the weakest environmental constraint in $\mathcal{L}_4$.*

Proof: Suppose that ψ_4 is the environmental constraint that is obtained by the method for $\mathcal{L}_4$ for the specification $spec$. This proof proves the contraposition. Suppose that $\psi' = \bigwedge_i((\bigvee_j \Diamond\Box f_{ij}) \vee \Box\Diamond(\bigwedge_k(g_{ik} \vee \bigcirc h_{ik})))$ and that $\psi' \not\leq \psi_4$.

Recall the form of ψ_4.

$$\psi_4 = \bigwedge_{sc\in SC}((\bigvee_{c\in l(sc)} \Diamond\Box\neg c) \vee \Box\Diamond(\bigwedge_{s\in q(sc)}((\bigwedge_{c_{in}\in in(s)} \neg c_{in}) \vee$$
$$\bigcirc(\bigwedge_{c_{out}\in out(s)} \neg c_{out}))))$$

$\psi' \not\leq \psi_4$ implies that there is an infinite sequence $\tilde{a}$ of sets of input propositions such that $\tilde{a} \not\models \psi_4$ and $\tilde{a} \models \psi'$. It follows that there is a strongly connected graph sc such that the following holds.

$\tilde{a} \models (\bigwedge_{c \in l(sc)} \Box \Diamond c) \wedge \Diamond \Box (\bigvee_{s \in q(sc)} ((\bigvee_{c_{in} \in in(s)} c_{in}) \wedge \bigcirc (\bigvee_{c_{out} \in out(s)} c_{out})))$

For any i, $\tilde{a} \models (\bigvee_i \Diamond \Box f_{ij}) \vee \Box \Diamond (\bigwedge_k (g_{ik} \vee \bigcirc h_{ik}))$. Therefore, for any i, one of the following two conditions holds.

(1) For some j, $\tilde{a} \models \Diamond \Box f_{ij}$.
(2) $\tilde{a} \models \Box \Diamond (\bigwedge_k (g_{ik} \vee \bigcirc h_{ik}))$

Let V be a set of i that satisfies (1) and W be a set of i that satisfies (2). It follows that $\tilde{a} \models (\bigwedge_{v \in V} \Diamond \Box f_{vj_v}) \wedge (\bigwedge_{w \in W} \Box \Diamond (\bigwedge_k (g_{wk} \vee \bigcirc h_{wk})))$. Therefore, the following holds.

$\tilde{a} \models (\bigwedge_{v \in V} \Diamond \Box f_{vj_v}) \wedge (\bigwedge_{w \in W} \Box \Diamond (\bigwedge_k (g_{wk} \vee \bigcirc h_{wk}))) \wedge (\bigwedge_{c \in l(sc)} \Box \Diamond c) \wedge$
$\Diamond \Box (\bigvee_{s \in q(sc)} ((\bigvee_{c_{in} \in in(s)} c_{in}) \wedge \bigcirc (\bigvee_{c_{out} \in out(c)} c_{out})))$

For any $c \in l(sc)$, there is a set α_c of input propositions where $\alpha_c \models (\bigwedge_{v \in V} f_{vj_v})$.

For any $w \in W$, there are a state $s_w \in q(sc)$, sets a_{in}^w and a_{out}^w of input propositions where $a_{in}^w a_{out}^w \sigma \models (\bigwedge_{v \in V} f_{vj_v}) \wedge (\bigcirc (\bigwedge_{v \in V} f_{vj_v})) \wedge (\bigwedge_k (g_{wk} \vee \bigcirc h_{wk})) \wedge ((\bigvee_{c_{in} \in in(s_w)} c_{in}) \wedge \bigcirc (\bigvee_{c_{out} \in out(s_w)} c_{out}))$. Therefore, for any $w \in W$, there are edges e_{in}^w and e_{out}^w where e_{in}^w goes into s_w and e_{out}^w goes out of s_w. Let c_{in}^w be a label of e_{in}^w and c_{out}^w be a label of e_{out}^w, and $a_{in}^w a_{out}^w \sigma \models (\bigwedge_{v \in V} f_{vj_v}) \wedge (\bigcirc (\bigwedge_{v \in V} f_{vj_v})) \wedge (\bigwedge_k (g_{wk} \vee \bigcirc h_{wk})) \wedge (c_{in}^w \wedge c_{out}^w)$

Now w_m ($1 \le m \le |W|$ is defined to be an element of W. There is a run from an initial state to w_1. Let $\overline{x}_{q_0,1}$ be this run. For any n ($1 \le n < |W|$), there is a run that is from w_n to s_{n+1} via edge e_{out}^n and e_{in}^{n+1} and consists of only states in sc. Let $\overline{x}_{n,n+1}$ be a finite sequence of sets of input propositions where a run of $\overline{x}_{n,n+1}$ is the run from w_n to s_{n+1}. State $s_{|W|}$ goes to state s_1 via states in sc. Let $\overline{x}_{|W|,1}$ be a finite sequence of sets of input propositions where a run of $\overline{x}_{|W|,1}$ is the run from $s_{|W|}$ to s_1. Consider an infinite sequence $\tilde{x}$ of sets of input propositions.

$$\tilde{x} = \overline{x}_{q_0,1} \cdot (\overline{x}_{1,2} \cdot \cdots \cdot \overline{x}_{|W|-1,|W|} \cdot \overline{x}_{|W|,1})^\omega$$

The construction of $\tilde{x}$ implies that $\tilde{x} \in L(\overline{A'_{spec}})$. Since $\tilde{x}$ infinitely often includes two long sequence such as $a_{in}^1 a_{out}^1, \cdots$ or $a_{in}^{|W|} a_{out}^{|W|}$, $\tilde{x} \models \bigwedge_{w \in W} \Box \Diamond (\bigwedge_k (g_{wk} \vee \bigcirc h_{wk}))$. Moreover, for any n ($1 \le n < |W|$), all sets of input propositions between $\overline{x}_{n,n+1}$ and $\overline{x}_{|W|,1}$ satisfy $\bigwedge_{v \in V} f_{vj_v}$. Therefore, $\tilde{x} \models \bigwedge_{v \in V} \Diamond \Box f_{vj_v}$. It follows that $\psi' \notin EC(spec)$ as follows.

$\tilde{x} \models \psi'$ and $\tilde{x} \in L(\overline{A'_{spec}}) \Rightarrow \neg (\forall \tilde{a}(\tilde{a} \models \psi' \Rightarrow \tilde{a} \notin L(\overline{A'_{spec}}))$
$\Leftrightarrow \neg (\forall \tilde{a}(\tilde{a} \models \psi' \Rightarrow \tilde{a} \in L(A'_{spec})) \Leftrightarrow \neg (\forall \tilde{a}(\tilde{a} \models \psi' \Rightarrow \exists \tilde{b} \langle \tilde{a}, \tilde{b} \rangle \models spec))$
$\Leftrightarrow \psi' \notin EC(spec)$

To summarize, $\psi' \in \mathcal{L}_4$ and $\psi' \not\le \psi_4$ imply $\psi' \notin EC(spec)$. Thus, ψ_4 is the weakest environmental constraint among $\mathcal{L}_4$.   ■

8   Conclusion

This paper proposed the methods of extracting environmental constraints from strong unsatisfiable specifications. The constraints that are extracted by the proposed method is the weakest in the classes of formulas. There are several future works. One of them is to implement the methods and to apply the implementation to specifications to revise the specifications.

References

1. Abadi, M., Lamport, L., Wolper, P.: Realizable and unrealizable specifications of reactive systems. In: Ausiello, G., Dezani-Ciancaglini, M., Della Rocca, S.R. (eds.) Automata, Languages and Programming. LNCS, vol. 372, pp. 1–17. Springer, Heidelberg (1989)
2. Bouyer, P., Bozzelli, L., Chevalier, F.: Controller synthesis for MTL specifications. In: Baier, C., Hermanns, H. (eds.) CONCUR 2006. LNCS, vol. 4137, pp. 450–464. Springer, Heidelberg (2006)
3. Duer-Luts, A.: LTL translation improvements in Spot. In: Proceedings of the Fifth International Conference on Verification and Evaluation of Computer and Communication Systems, pp. 72–83 (2011)
4. Filiot, E., Jin, N., Raskin, J.F.: An Antichain Algorithm for LTL Realizaibility. Formal Methods in System Design archive **39**(3), 261–296 (2011)
5. Harel, D., Pnueli, A.: On the development of reactive systems. In: Logics and Models of Concurrent Systems, pp. 477–498 (1985)
6. Kupferman, O., Madhusudan, P., Thiagarajan, P.S., Vardi, M.Y.: Open systems in reactive environments: control and synthesis. In: Palamidessi, C. (ed.) CONCUR 2000. LNCS, vol. 1877, pp. 92–107. Springer, Heidelberg (2000)
7. Mori, R., Yonezaki, N.: Derivation of the input conditional formula from a reactive system specification in temporal logic. In: Langmaack, H., de Roever, W.-P., Vytopil, J. (eds.) FTRTFT 1994 and ProCoS 1994. LNCS, vol. 863, pp. 567–582. Springer, Heidelberg (1994)
8. Pnueli, A., Rosner, R.: On the synthesis of a reactive module. In: Proceedings of the 16th ACM SIGPLAN-SIGACT Symposium on Principles of Programming Languages, pp. 179–190 (1989)
9. Gastin, P., Oddoux, D.: Fast LTL to Büchi automata translation. In: Berry, G., Comon, H., Finkel, A. (eds.) CAV 2001. LNCS, vol. 2102, pp. 53–65. Springer, Heidelberg (2001)
10. Mori, R., Yonezaki, N.: Several Realizability Concepts in Reactive Objects, Information Modeling and Knowledge Bases. IOS Press (1993)
11. Tarjan, R.E.: Depth-First Search and Linear Graph Algorithms. SIAM Journal on Computing **1**(2), 146–160 (1972)
12. Tsay, Y.-K., Chen, Y.-F., Tsai, M.-H., Wu, K.-N., Chan, W.-C.: GOAL: a graphical tool for manipulating Büchi automata and temporal formulae. In: Grumberg, O., Huth, M. (eds.) TACAS 2007. LNCS, vol. 4424, pp. 466–471. Springer, Heidelberg (2007)
13. Hagihara, S, Kitamura, Y., Shimakawa, M., Yonezaki, N.: Extracting environmental constraints to make reactive system specifications realizable In: Proceedings of 16th Asia-Pacific Software Engineering Conference, pp. 61–68 (2009)
14. Cimatti, A., Roveri, M., Schuppan, V., Tchaltsev, A.: Diagnostic information for realizability. In: Logozzo, F., Peled, D.A., Zuck, L.D. (eds.) VMCAI 2008. LNCS, vol. 4905, pp. 52–67. Springer, Heidelberg (2008)

Implementation of Decision Procedure of Stepwise Satisfiability of Reactive System Specifications

Noriaki Yoshiura[✉] and Yuma Hirayanagi

Department of Information and Computer Sciences,
Saitama University, 255, Shimo-ookubo, Sakura-ku, Saitama, Japan
yoshiura@fmx.ics.saitama-u.ac.jp

Abstract. Reactive systems ideally never terminate and maintain some interaction with their environment. Temporal logic is one of the methods for formal specification description of reactive systems. For a reactive system specification, we do not always obtain a program that satisfies it because the reactive system program must satisfy the specification no matter how the environment of the reactive system behaves. This problem is known as realizability or feasibility. The complexity of deciding realizability of specifications that are described in linear temporal logic is double or triple exponential time of the length of specifications and realizability decision is impractical. This paper implements stepwise satisfiability decision procedure with tableau method and proof system. Stepwise satisfiability is one of the necessary conditions of realizability of reactive system specifications. The proposed procedure decides stepwise satisfiability of reactive system specifications.

1 Introduction

Reactive systems, such as operating systems or elevator control systems, interact with their users or environments, and provide services for their users. Reactive system behaviors depend on the interactions with their environments or users[6]. In reactive systems, events are divided into output events and input events. Output events are controlled or generated by reactive systems and input events are generated by environments such as users of reactive systems. Input events cannot be controlled by the reactive systems[7].

Temporal logic is one of the methods of describing reactive system specifications. One of the advantages of temporal logic is to prove several properties of the specifications. Given a reactive system specification, which is a temporal logic formula, satisfiability of the formula only guarantees possibility of synthesizing the reactive system that satisfies a reactive system specification for some environment behavior. However, it is necessary to synthesize the system that satisfies the specification for all environment behaviors; suppose that there are finite sets I and O of input and output events. A reactive system program is a function $f : (2^I)^* \rightarrow 2^O$ that maps finite sequences of input event sets into an output

© Springer International Publishing Switzerland 2015
O. Gervasi et al. (Eds.): ICCSA 2015, Part IV, LNCS 9158, pp. 686–698, 2015.
DOI: 10.1007/978-3-319-21410-8_52

event set. The function f generates infinite sequences of $2^{I \cup O}$ during interaction between a reactive system and an environment because the environment generates infinite sequences of input event sets. A function f realizes a specification ϕ if and only if ϕ always holds for all infinite sequences that are generated by the function f and an environment. The realizability problem is to determine, given a specification ϕ which is a temporal logic formula, whether there exists a function $f : (2^I)^* \rightarrow 2^O$ that realizes ϕ. The complexity of deciding realizability is 2EXPTIME-complete on CTL and 3EXPTIME-complete on CTL*[5,7]. Therefore, the complexity of deciding realizability on LTL is 2EXPTIME-complete or 3EXPTIME-complete[5,7].

The previous studies on realizability of reactive system specifications that are described by several kinds of temporal logic focus on decidability of realizability but not on speed of deciding realizability. Automata theory has been used in almost all studies to give decision procedures of realizability, but it is insufficient for fast decision of realizability of specifications. There are several researches that are related with implementation of realizability decision[3,5]. However, implementation of fast decision procedures is impractical because of complexity of realizability problem. Thus, this paper implements a decision procedure of stepwise satisfiability, which is one of necessity conditions of realizability. Many of actual reactive system specifications are stepwise satisfiable but not realizable[11]. This paper implements stepwise satisfiability decision procedure with tableau method and proof system. This proof system is proposed in [12]. The proof system is not complete but may check stepwise satisfiability fast.

2   Reactive System

This section provides a formal definition of reactive systems, based on references[10]. Let A be a finite set. A^+ and A^ω denote the set of *finite sequences* and the set of *infinite sequences* over A respectively. $A^\dagger$ denotes $A^+ \cup A^\omega$. Sequences in $A^\dagger$ are denoted by $\hat{a}, \hat{b}, \cdots$, sequences in A^+ by $\bar{a}, \bar{b}, \cdots$ and sequences in A^ω by $\tilde{a}, \tilde{b}, \cdots$. $|\hat{a}|$ denotes the *length* of $\hat{a}$ and $\hat{a}[i]$ denotes the *i-th element* of $\hat{a}$. Suppose that B is a set whose elements are also sets. '$\sqcup$', which is a composition of sequences, is defined over $B^\dagger \times B^\dagger$ by

$$\hat{a} \sqcup \hat{b} = \hat{a}[0] \cup \hat{b}[0], \ \hat{a}[1] \cup \hat{b}[1], \ \hat{a}[2] \cup \hat{b}[2], \cdots .$$

Definition 1. *A reactive system RS is a triple* $RS = \langle X, Y, r \rangle$, *where*

- X *is a finite set of input events that are generated by an environment.*
- Y $(X \cap Y = \emptyset)$ *is a finite set of output events that are generated by the the reactive system itself.*
- $r : (2^X)^+ \rightarrow 2^Y$ *is a reaction function.*

A subset of X is called *input set* and a sequence of input sets is called *input sequence*. Similarly, a subset of Y is called *output set* and a sequence of output sets is called *output sequence*. In this paper, a reaction function corresponds to a reactive system program.

Definition 2. *Let* $RS = \langle X, Y, r \rangle$ *be a reactive system and* $\hat{a} = a_0, a_1, \cdots \in$ $(2^X)^\dagger$ *be an input sequence. The behavior of RS for* $\hat{a}$*, denoted* $behave_{RS}(\hat{a})$*, is the following sequence:*

$$behave_{RS}(\hat{a}) = \langle a_0, b_0 \rangle, \langle a_1, b_1 \rangle, \langle a_2, b_2 \rangle, \ldots,$$

where for each i $(0 \le i < |\hat{a}|)$*,* $b_i = r(a_0, \ldots, a_i) \in 2^Y$*.*

3   Specification

This paper uses propositional linear-time temporal logic (PLTL) as a specification language for reactive systems. A PLTL *formula* is defined in the usual way: an atomic proposition $p \in \mathcal{P}$ is a formula and if f_1 and f_2 are formulas, $f_1 \wedge f_2$, $\neg f_1$, $\bigcirc f_1$ and $f_1 \mathcal{W} f_2$ are formulas. This paper uses "Weak until operator" $(\mathcal{W})$ and an abbreviation of $\Box f_1 \equiv f_1 \mathcal{W} (f_2 \wedge \neg f_2)$. $\vee$, $\rightarrow$, $\leftrightarrow$ and $\Diamond$ are the usual abbreviations.

The semantics of PLTL formulas is defined on the behaviors. Let P be a set of input and output events and $\mathcal{P}$ be a set of atomic propositions corresponding to each element of P. $\langle \sigma, i \rangle \models f$ denotes that a formula f over $\mathcal{P}$ holds at the i-th state of a behavior $\sigma \in (2^P)^\omega$. $\langle \sigma, i \rangle \models f$ is recursively defined as follows.

- $\langle \sigma, i \rangle \models p$ **iff** $p' \in \sigma[i]$ (p is an atomic proposition corresponding to $p' \in P$)
- $\langle \sigma, i \rangle \models \neg f$ **iff** $\langle \sigma, i \rangle \not\models f$
- $\langle \sigma, i \rangle \models f_1 \wedge f_2$ **iff** $\langle \sigma, i \rangle \models f_1$ and $\langle \sigma, i \rangle \models f_2$
- $\langle \sigma, i \rangle \models \bigcirc f$ **iff** $\langle \sigma, i+1 \rangle \models f$
- $\langle \sigma, i \rangle \models f_1 \mathcal{W} f_2$ **iff** $(\forall j \ge 0) \langle \sigma, i + j \rangle \models f_1$ or
  $(\exists j \ge 0) (\langle \sigma, i + j \rangle \models f_2$ and
  $\forall k (0 \le k < j) \langle \sigma, i + k \rangle \models f_1)$

σ is a model of f if and only if $\langle \sigma, 0 \rangle \models f$. We write $\sigma \models f$ if $\langle \sigma, 0 \rangle \models f$. A formula f is satisfiable if and only there is a model of f. For example, $behave_{RS}(\tilde{a}) \models \varphi$ means that a behavior that is generated by the reactive system RS receiving the input sequence $\tilde{a}$ is a model of φ.

3.1   Specification

A *PLTL-specification* for a reactive system is a triple $Spec = \langle \mathcal{X}, \mathcal{Y}, \varphi \rangle$, where

- $\mathcal{X}$ is a set of *input propositions* that are atomic propositions corresponding to the input events of the reactive system. The truth value of an input proposition represents the occurrence of the corresponding input event.
- $\mathcal{Y}$ is a set of *output propositions* that are atomic propositions corresponding to the output events of the reactive system. The truth value of an output proposition represents the occurrence of the corresponding output event.
- φ is a formula in which all the atomic propositions are elements of $\mathcal{X} \cup \mathcal{Y}$.

This paper writes $Spec = \langle \mathcal{X}, \mathcal{Y}, \varphi \rangle$ just as φ if there is no confusion. Finally, the following defines the realizability of reactive system specifications[10].

Definition 3. *A reactive system RS is an implementation of a specification φ if for every input sequence $\tilde{a}$, behave$_{RS}(\tilde{a}) \models \varphi$. A specification is realizable if it has an implementation.*

3.2   Stepwise Satisfiability

Reactive system specifications can be divided into several classes. This subsection explains the class of *stepwise satisfiability* specifications. This property is introduced in [8]. Throughout this subsection $\tilde{a}$, $\bar{a}$ and $\tilde{b}$ denote an infinite input sequence, a finite input sequence and an infinite output sequence respectively. Also, r denotes a reaction function. For simplicity the interpretation representing a behavior is denoted by the behavior itself. Stepwise satisfiability is defined as follows:

Definition 4. *A reactive system RS preserves satisfiability of φ if and only if $\forall \bar{a} \exists \tilde{a} \exists \tilde{b}(behave_{RS}(\bar{a})(\tilde{a} \sqcup \tilde{b}) \models \varphi)$. φ is stepwise satisfiable if and only if there is a reactive system that preserves the satisfiability of φ.*

If φ is stepwise satisfiable, a reactive system RS can behave for any input event sequence with keeping a possibility that φ is satisfied even though RS actually does not satisfy φ. The following example explains stepwise satisfiability[8].

EXAMPLE. *Stepwise satisfiable but not realizable specification.*
     This example shows a part of a simple lift specification. In the specification below, an output proposition *Move* is intended to show when the lift can move. An output proposition *Open* means that the lift door is open, and an output proposition *Floor$_i$* means that the lift is on the i-th floor. An input proposition $\boldsymbol{B_{open}}$ represents the request "open the door" and $\boldsymbol{B_i}$ represents the request "come or go to the i-th floor."

1. $\Box(\neg Move \wedge Floor_i \rightarrow Floor_i \ W \ Move)$
   (if *Move* is not true when the lift is at the i-th floor, stay there until *Move* holds)
2. $\Box(Open \rightarrow \neg Move)$
   (if the door is open, do not move)
3. $\Box\big(\neg \boldsymbol{B_{open}} \wedge (\neg Move \ W \ \boldsymbol{B_{open}})$
          $\rightarrow (\neg Open \ W \ (\boldsymbol{B_{open}} \wedge (Open \ W \ \neg \boldsymbol{B_{open}}))))$
   (if *Move* is not true, open the door while $\boldsymbol{B_{open}}$ holds)
4. $\Box(\boldsymbol{B_i} \rightarrow \Diamond Floor_i)$
   (if asked to come or go to the i-th floor, eventually arrive at the floor)

     $\boldsymbol{B_{open}}$ and $\boldsymbol{B_i}$ are input propositions and the other propositions are output propositions. If $\boldsymbol{B_{open}}$ will be true forever after some state where both $\neg Move$ and $Floor_i$ hold, and $\boldsymbol{B_{j(\neq i)}}$ will be true after this state, $\Diamond Floor_j$ could never be satisfied. This example shows that a specification is not realizable if for some infinite input sequence, a $\Diamond$-formula has no opportunity to hold. However, this specification is a typical specification and many specifications of reactive systems are stepwise satisfiable.

4   Decision Procedure

This section gives the decision procedure of stepwise satisfiability, which was given in [11]. This procedure is sound and complete. This procedure is based on the tableau method for PLTL[8].

4.1   Tableau Method

A tableau is a directed graph $T = \langle N, E \rangle$ that is constructed from a given specification. N is a finite set of nodes. Each node is a set of formulas. E is a finite set of edges. Each edge is a pair of nodes. A node n_2 is reachable from a node n_1 in a tableau $\langle N, E \rangle$ if and only if $\langle n_1, a_1 \rangle, \langle a_1, a_2 \rangle, \cdots \langle a_k, n_2 \rangle \in E$.

Definition 5. *A decomposition procedure takes a set S of formulas as input and produces a set Σ of sets of formulas.*

1. *Put $\Sigma = \{S\}$.*
2. *Repeatedly apply one of steps a – e to all the formulas f_{ij} in all the sets $S_i \in \Sigma$ according to the type of the formulas until no step will change Σ. In the following, $f_1 \mathcal{W}^* f_2$ and $\neg(f_1 \mathcal{W}^* f_2)$ are called marked formula. The marks represent that the marked formulae have been applied by the decomposition produce.*
   (a) If f_{ij} is $\neg\neg f$, replace S_i with $(S_i - \{f_{ij}\}) \cup \{f\}$.
   (b) If f_{ij} is $f_1 \wedge f_2$, replace S_i with $(S_i - \{f_{ij}\}) \cup \{f_1, f_2\}$.
   (c) If f_{ij} is $\neg(f_1 \wedge f_2)$, replace S_i with $(S_i - \{f_{ij}\}) \cup \{\neg f_1\}$ and $(S_i - \{f_{ij}\}) \cup \{\neg f_2\}$.
   (d) If f_{ij} is $f_1 \mathcal{W} f_2$, replace S_i with $(S_i - \{f_{ij}\}) \cup \{f_2\}$ and $(S_i - \{f_{ij}\}) \cup \{f_1, \neg f_2, f_1 \mathcal{W}^* f_2\}$.
   (e) If f_{ij} is $\neg(f_1 \mathcal{W} f_2)$, replace S_i with $(S_i - \{f_{ij}\}) \cup \{\neg f_1, \neg f_2\}$ and $(S_i - \{f_{ij}\}) \cup \{f_1, \neg f_2, \neg(f_1 \mathcal{W}^* f_2)\}$.

Definition 6. *A node n of a tableau $\langle N, E \rangle$ is closed if and only if one of the following conditions is satisfied:*

- *n contains both an atomic proposition and its negation.*
- *n contains an eventuality formula[1] and all unclosed nodes that are reachable from n contain the same eventuality formula.*
- *n cannot reach any unclosed node.*

The following describes the tableau construction procedure, which takes a PLTL formula φ as input and produces a tableau $T = \langle N, E \rangle$. In the procedure, a function $temporal(n)$ is used and defined as follows.
$$temporal(n) = \{f_1 \mathcal{W} f_2 \mid f_1 \mathcal{W}^* f_2 \in n\} \cup$$
$$\{\neg(f_1 \mathcal{W} f_2) \mid \neg(f_1 \mathcal{W}^* f_2) \in n\}$$

Definition 7. *The tableau construction procedure takes a formula φ as input and produces a tableau of φ.*

---

[1] An eventuality formula is a formula of the form $\neg(f_1 \mathcal{W} f_2)$.

1. Put $N = \{START, \{\varphi\}\}$ and $E = \{\langle START, \{\varphi\}\rangle\}$ (START is the initial node).
2. Repeatedly apply steps a and b to $T = \langle N, E\rangle$ until T no longer changes.
   (a) (decomposition of states) Apply the following three steps to all the nodes $n_i \in N$ to which these steps have not been applied yet.
      i. Apply the decomposition procedure to n_i (Σ_{n_i} is defined to be the output of the decomposition procedure)
      ii. Replace E with
         $(E \cup \{\langle m, m'\rangle | \langle m, n_i\rangle \in E \text{ and } m' \in \Sigma_{n_i}\}) - \{\langle m, n_i\rangle | m \in N\}$,
      iii. Replace N with $(N - n_i) \cup \Sigma_{n_i}$.
   (b) (transition of states) Apply the following two steps to all the nodes $n_i \in N$ to which these steps have not been applied yet.
      i. Replace E with $E \cup \{\langle n_i, temporal(n_i)\rangle\}$.
      ii. Replace N with $N \cup \{temporal(n_i)\}$.

In [4], it is proved that a formula is satisfiable if and only if the initial node $START$ of the tableau of the formula is unclosed. Thus, the following procedure decides the satisfiability of the formula φ.

Definition 8. *The following procedure decides whether a formula φ is satisfiable.*

1. *By Tableau Construction Procedure, construct the tableau of a formula φ*
2. *If the tableau of the formula φ is unclosed, it is concluded that the formula φ is satisfiable. Otherwise, it is concluded that the formula φ is unsatisfiable.*

4.2 Decision Procedure for Stepwise Satisfiability

This subsection describes the decision procedure of stepwise satisfiability. In the decision procedure it is important to make a tableau deterministic by a set of input and output propositions. At the beginning, some functions are defined for a deterministic tableau.

Definition 9. *Let $T = \langle N, E\rangle$ be a tableau. The function $next(\mathbf{n})$ maps a subset of N to a subset of N, where $\mathbf{n}$ is a subset of N. The function $next(\mathbf{n})$ is defined as follows:*

   $next(\mathbf{n}) \equiv \bigcup_{n \in \mathbf{n}} \{n' \mid \langle n, n'\rangle \in E\}$

The function atm and $\overline{atm}$ map an element of N to a set of atomic propositions. The function atm and $\overline{atm}$ are defined as follows:

   $atm(n) \equiv \{f \mid f \in n \text{ and } f \text{ is atomic formula.}\}$
   $\overline{atm}(n) \equiv \{f \mid \neg f \in n \text{ and } f \text{ is atomic formula.}\}$

For a subset of $\mathcal{N}$, the function atm and $\overline{atm}$ are defined as follows:

   $atm(\mathbf{n}) \equiv \bigcup_{n \in \mathbf{n}} \{f \mid f \in n \text{ and } f \text{ is atomic formula.}\}$
   $\overline{atm}(\mathbf{n}) \equiv \bigcup_{n \in \mathbf{n}} \{f \mid \neg f \in n \text{ and } f \text{ is atomic formula.}\}$

For a subset a of P and a subset of $\mathcal{N}$, $next(\mathbf{n})/a$ is defined as follows:

$next(\mathbf{n})/a \equiv \{n | n \in next(\mathbf{n}), \overline{atm}(n) \cap a = \emptyset \text{ and }$
   $atm(n) \cap (P - a) = \emptyset\}$

Next, a deterministic tableau is defined.

Definition 10. *Let* $\langle N, E \rangle$ *be the tableau of specification* φ. $\langle \mathcal{N}, \mathcal{E} \rangle$ *is the deterministic tableau of* φ *and a set* P *of atomic propositions if and only if the following conditions are satisfied.*

- $\mathcal{N}$ *is a set of tableau node sets, that is* $\mathcal{N} \subseteq 2^N$.
- *Every element of* $\mathcal{N}$ *is not an empty set.*
- $\mathcal{E}$ *is a set of* $\langle \mathbf{n_1}, a, \mathbf{n_2} \rangle$ *such that* $\mathbf{n_1}, \mathbf{n_2} \in \mathcal{N}$ *and* $a \subseteq P$.
- *If* $\langle \mathbf{n'}, a, \mathbf{n} \rangle \in \mathcal{E}$, $a \cap \overline{atm}(\mathbf{n}) = \emptyset$ *and* $(P - a) \cap atm(\mathbf{n}) = \emptyset$
- *If* $\langle \mathbf{n_1}, a, \mathbf{n_2} \rangle \in \mathcal{E}$, *for* $n \in \mathbf{n_2}$, *there is* $n' \in \mathbf{n_1}$ *such that* $\langle n', n \rangle \in E$.

The following defines a procedure for constructing a deterministic tableau that is used in the decision procedure of stepwise satisfiability.

Definition 11. *The following procedure constructs a deterministic tableau* $\mathcal{T} = \langle \mathcal{N}, \mathcal{E} \rangle$ *of specification of* φ *and a set* P *of atomic propositions.*

1. *Construct tableau* $\langle N, E \rangle$ *of* φ *by the tableau construction procedure.*
2. *Set* $\mathcal{N} = \{\{START\}\}$ *and* $\mathcal{E} = \emptyset$.
3. *Repeat the following step until* $\mathcal{T}$ *no longer changes; for* $\mathbf{n} \in \mathcal{N}$ *and* $a \subseteq P$, *if* $next(\mathbf{n})/a \neq \emptyset$, *add* $next(\mathbf{n})/a$ *into* $\mathcal{N}$ *and* $\langle \mathbf{n}, a, next(\mathbf{n})/a \rangle$ *into* $\mathcal{E}$, *where* $next(\mathbf{n})/a \equiv \{n \in next(\mathbf{n}) \mid a \cap \overline{atm}(n) = \emptyset, (P - a) \cap atm(n) = \emptyset\}$.

A deterministic tableau enables to decide whether there is a reactive system RS of φ such that RS behaves for an input event at any time or whether there is an infinite output sequence for any infinite input sequence. However, it is impossible to decide satisfiability of φ using a deterministic tableau. To check the satisfiability of φ requires to examine each part of a tableau that is included in the deterministic tableau.

Definition 12. *The decision procedure of Stepwise Satisfiability is as follows;*

1. *By the tableau deterministic procedure, construct a deterministic tableau* $\langle \mathcal{N}, \mathcal{E} \rangle$ *of specification* φ *and a set* $\mathcal{X} \cup \mathcal{Y}$ *of input and output propositions.*
2. *Repeat the following step until* $\langle \mathcal{N}, \mathcal{E} \rangle$ *no longer changes; for* $\mathbf{n} \in \mathcal{N}$ *and* $a \subseteq \mathcal{X}$, *if there are no* $\mathbf{n'} \in \mathcal{N}$ *and* $b \subseteq \mathcal{Y}$ *such that* $\langle \mathbf{n}, a \cup b, \mathbf{n'} \rangle \in \mathcal{E}$, *delete* $\mathbf{n}$ *from* $\mathcal{N}$ *and elements such as* $\langle \mathbf{n}, c, \mathbf{n'} \rangle$ *or* $\langle \mathbf{n'}, c, \mathbf{n} \rangle$ *from* $\mathcal{E}$.
3. *If* $\mathcal{N}$ *is not an empty set, this procedure determines that* φ *is stepwise satisfiable. Otherwise, this procedure determines that* φ *is not stepwise satisfiable.*

5   Proof System

This paper uses a proof system for checking unrealizability of reactive systems specifications. The proof system is proposed in [12]. This proof system is a sequent-style natural deduction. Before giving several formal definitions, this section informally explains the meanings of sequent and several symbols. This proof system uses sequent $\Gamma, \Delta^\dagger \vdash W$ where Γ is a set of input proposition formula, $\Delta^\dagger$ is a set of mark formulas such $A^\dagger$ and W is a formula. If $\Delta^\dagger$ is an empty set, this sequent denotes that there exists a reactive system RS such that for any input sequence $\tilde{a}$, if $\tilde{a} \models \bigwedge \Gamma$ ($\bigwedge \Gamma$ is a conjunction of all elements of Γ), $behave_{RS}(\tilde{a}) \models W$; if $\Delta^{\dagger}$ is not an empty set, this sequent denotes that there exists a reactive system RS such that for any input sequence $\tilde{a}$, if there exists a formula V such that $\tilde{a} \models V \wedge \bigwedge \Gamma$ and $V^{\dagger} \in \Delta^{\dagger}$, $behave_{RS}(\tilde{a}) \models W$.

To check unrealizability of a specification φ requires to make a proof where $\vdash \varphi$ is the start and the conclusion is in contradiction. While constructing a proof, several input formulas (a formula consisting of input propositions) move from the right side to left side of sequent. If the formulas on the right side of a sequent are in contradiction and the formulas on the left side are satisfiable, it is concluded that φ is unrealizable. If there are several marked formulas on the left side, it is necessary only to check whether all unmarked formulas and one of marked formulas are satisfiable, even if a set of marked formulas are inconsistent. The marked formulas represent the input formulas that can hold in one of possible future states and therefore a set of the marked formulas may be inconsistent. On the other hand, unmarked formulas must be satisfiable in the same state such as current state or future state. This is a reason why the proof system uses mark. The following gives formal definitions.

Definition 13. $A^{\dagger}$ *is defined to be mark formula where A is a formula. $\Delta^{\dagger}$ is defined to a set of marked input formulas where $\Delta^{\dagger} = \{A^{\dagger} \mid A \in \Delta\}$.*

Definition 14. $\Gamma, \Delta^{\dagger} \vdash W$ *is defined to be sequent where Γ is a set of input formulas, $\Delta^{\dagger}$ is a set of marked input formulas and W is a formula. A sequent $\Gamma, \Delta^{\dagger} \vdash W$ holds with respect to a reactive system RS if and only if for any element $V \in \Delta$ and any input sequence $\tilde{a}$, if $\tilde{a} \models V \wedge \bigwedge \Gamma$, $behave_{RS}(\tilde{a}) \models W$.*

Definition 15. *The inference rules are defined in [12]. Figure 1 shows a parts of the inference rules because of page limitation. In the inference rules, $X, X_1, \ldots$ are formulas consisting of only input propositions and A, B, C are arbitrary formulas; a temporal formula is defined to be a formula including temporal operator such as $\Box$, $\Diamond$, W and $\bigcirc$. The proof system in this paper uses "Next operator", which is not defined in the syntax of PLTL. This paper defines next operator ($\bigcirc$) as usual.*

Definition 16. *A proof beginning with formula φ is defined to be a sequence $S_1, S_2 \ldots S_n$ of sequent satisfying the following conditions.*

1. *S_k $(1 \leq k \leq n)$ is one of the following*
   - *$\vdash \varphi$*
   - *an assumption, which must be discarded by inference rule $\vee E$ in Definition 15.*
   - *a conclusion of an inference rule in Definition 15 where the premises of the inference rule must be in $S_l (1 \leq l < k)$.*
2. *S_k $(1 \leq k \leq n)$ is a sequent $\Gamma, \Delta^{\dagger} \vdash A$ satisfying following.*
   - *If Δ is empty, Γ is consistent.*
   - *If Δ is not empty, there exists a formula V such that $V \in \Delta$ and $\Gamma \cup \{V\}$ is consistent.*

$$\frac{\Gamma, \Delta^\dagger \vdash A \wedge B}{\Gamma, \Delta^\dagger \vdash A} \wedge E1 \qquad \frac{\Gamma, \Delta^\dagger \vdash A \wedge B}{\Gamma, \Delta^\dagger \vdash B} \wedge E2 \qquad \frac{\Gamma_1, \Delta_1^\dagger \vdash A \quad \Gamma_2, \Delta_2^\dagger \vdash B}{\Gamma_1, \Gamma_2, \Delta_1^\dagger, \Delta_2^\dagger \vdash A \wedge B} \wedge I$$

$$\frac{\Gamma, \Delta^\dagger \vdash A}{\Gamma, \Delta^\dagger \vdash A \vee B} \vee I1 \qquad \frac{\Gamma, \Delta^\dagger \vdash B}{\Gamma, \Delta^\dagger \vdash A \vee B} \vee I2$$

$$\frac{\Gamma_1, \Delta_1^\dagger \vdash A \quad \Gamma_2, \Delta_2^\dagger \vdash \neg A}{\Gamma_1, \Gamma_2, \Delta_1^\dagger, \Delta_2^\dagger \vdash \bot} \bot I \qquad \frac{\Gamma, \Delta^\dagger \vdash \bot}{\Gamma, \Delta^\dagger \vdash A} \bot E \qquad \frac{\Gamma, \Delta^\dagger \vdash X}{\Gamma, \Delta^\dagger, \neg X \vdash \bot} LM1$$

$$\frac{\Gamma, \Delta^\dagger, \vdash \bigcirc X}{\Gamma, \Delta^\dagger, \bigcirc \neg X^\dagger \vdash \bot} LM2^\star \qquad \frac{\Gamma, \Delta^\dagger \vdash X \vee A}{\Gamma, \Delta^\dagger, \neg X \vdash A} LM3$$

$$\frac{\Gamma, \Delta^\dagger \vdash A \mathcal{W} B}{\Gamma, \Delta^\dagger \vdash B \vee A \wedge \neg B \wedge \bigcirc(A \mathcal{W} B)} W1$$

$$\frac{\Gamma, \Delta^\dagger \vdash \neg(A \mathcal{W} B)}{\Gamma, \Delta^\dagger \vdash (\neg A \wedge \neg B) \vee (A \wedge \neg B \wedge \bigcirc \neg(A \mathcal{W} B))} W2$$

$$\frac{\Gamma, \Delta^\dagger, X_1 \mathcal{W} X_2 \vdash A}{\Gamma, \Delta^\dagger, X_2 \vee X_1 \wedge \neg X_2 \wedge \bigcirc(X_1 \mathcal{W} X_2) \vdash A} WL1$$

$$\frac{\Gamma, \Delta^\dagger, \neg(X_1 \mathcal{W} X_2) \vdash A}{\Gamma, \Delta^\dagger, (\neg X_1 \wedge \neg X_2) \vee (X_1 \wedge \neg X_2 \wedge \bigcirc \neg(X_1 \mathcal{W} X_2)) \vdash A} WL2$$

$\star$ There is no hypothesis or all hypotheses are of the form $\Box A$.

Fig. 1. Parts of Inference rules

S_n is defined to be a conclusion of a proof $S_1, S_2 \ldots S_n$, where all assumptions are discarded.

The definition of the proof uses concept of consistency of a set of formulas; it is undesirable in the definition of proof system. However, this proof system decides realizability but not satisfiability. Realizability is considered to be meta concept rather than satisfiability. It follows that it is allowable to use consistency check in the proof system. Implementation of this proof system does not always have to check consistency.

Definition 17. *Suppose that a sequent* $\Gamma, \Delta^\dagger \vdash \bot$ *is a conclusion of a proof beginning with* φ. *In the case that* $\Delta^\dagger$ *is not an empty set, if* $\Gamma \cup \{V\}$ *is satisfiable for some element* V *of* Δ, *it is concluded that* φ *is unrealizable; in the case that* Δ *is an empty set, if* Γ *is satisfiable, it is concluded that* φ *is unrealizable.*

6 Implementation

This section explains the implementation of stepwise satisfiability checking procedure based on the tableau method and the proof system that is described in the previous section. The tableau based decision procedure of stepwise satisfiability takes too much time and does not always decide stepwise satisfiability of specifications because of time or memory size limit. Thus, this paper implements stepwise satisfiability checking procedure with the proof system. Even if the tableau based decision procedure of stepwise satisfiability does not decide stepwise satisfiability of specifications, the proof system may decide unrealizability of specifications. The checking procedure that is implemented in this paper obtains the results with respect to realizability or stepwise satisfiability within practical realistic duration. The following stepwise satisfiability checking procedure checks realizability of a specification in creating nodes of tableau; proof system checking of unrealizability and creating tableau are performed simultaneously. If the proof system proves unrealizability of a specification, the stepwise satisfiability of checking procedure decide unrealizability of the specification fast without creating tableau. In the following, an element $\mathbf{n}$ of $\mathcal{N}$ of deterministic tableau $\langle \mathcal{N}, \mathcal{E} \rangle$ is interpreted as formula $\bigvee_{n \in \mathbf{n}} \bigwedge_{f \in n} f$.

Definition 18. *This procedure checks stepwise satisfiability of φ.*

1. *Check unrealizability of φ by the proof system within a predefined time or memory.*
2. *If the proof system decides unrealizability of φ, this procedure ends and decides that φ is unrealizable.*
3. *Create tableau $T = \langle N, E \rangle$ of φ.*
4. *Let $\mathcal{T} = \langle \mathcal{N}, \mathcal{E} \rangle$ be a deterministic tableau where $\mathcal{N} = \{\{START\}\}$ and $\mathcal{E} = \emptyset$*
5. *Repeat the following step if $\mathcal{N}$ has the an element $\mathbf{n}$ such that $\langle \mathbf{n}, a, \mathbf{n}' \rangle \notin \mathcal{E}$. Otherwise, this procedure decides that φ is stepwise satisfiable.*
   - *(a) For each $a \subseteq P$, if $next(\mathbf{n})/a \neq \emptyset$, check unrealizability of a formula of $next(\mathbf{n})/a$ by the proof system. If the proof system does not decide that a formula of $next(\mathbf{n})/a$ is unrealizable, add $next(\mathbf{n})/a$ into $\mathcal{N}$ and $\langle \mathbf{n}, a, next(\mathbf{n})/a \rangle$ into $\mathcal{E}$, where $next(\mathbf{n})/a \equiv \{n \in next(\mathbf{n}) \mid a \cap \overline{atm}(n) = \emptyset, (P - a) \cap atm(n) = \emptyset\}$.*
   - *(b) For $\mathbf{n} \in \mathcal{N}$ and $a \subseteq \mathcal{X}$, if there are no $\mathbf{n}' \in \mathcal{N}$ and $b \subseteq \mathcal{Y}$ such that $\langle \mathbf{n}, a \cup b, \mathbf{n}' \rangle \in \mathcal{E}$, delete $\mathbf{n}$ from $\mathcal{N}$ and elements such as $\langle \mathbf{n}, c, \mathbf{n}' \rangle$ or $\langle \mathbf{n}', c, \mathbf{n} \rangle$ from $\mathcal{E}$.*
   - *(c) If $\{START\} \notin \mathcal{N}$, this procedure decides that φ is unrealizable.*

This procedure is different from the tableau based stepwise satisfiability decision procedure. Before adding a new node of deterministic tableau, this procedure checks unrealizability of it by the proof system. The usage of the proof system can omit cost of deterministic tableau and reduce the size of deterministic tableau. This implementation uses MiniSat for satisfiability checking [9].

7 Experiment

This section evaluates the implemented decision procedure of stepwise satisfiability by experiments. This experiment uses a PC running under Fedora 16 Linux, CPU is Core i7 and memory size is 24GB. In the following, $x, x_1, x_2, \ldots$ are input propositions and $y, y_1, y_2, \ldots$ are output propositions.

(1) The following temporal formula is a specification of a three floor elevator control system. The implemented decision procedure cannot decide that specification is stepwise satisfiable because deciding stepwise satisfiability of this specification requires much time and a large size of deterministic tableau. The specification is divided into nine groups of temporal formulas (Some formulas are duplicated) because the groups are used in the following experiments.

(a) $\Box((y_1 \wedge \neg y_2 \wedge \neg y_3) \vee (y_2 \wedge \neg y_1 \wedge \neg y_3) \vee (y_3 \wedge \neg y_1 \wedge \neg y_2))$,
   $\Box(x_1 \rightarrow (\Diamond y_1 \wedge y_7 \mathcal{W}(y_1 \wedge y_7)))$,
   $\Box((y_1 \wedge y_7) \rightarrow (y_5 \wedge y_1 \mathcal{W} y_4))$

(b) $\Box((y_1 \wedge y_4) \rightarrow (\neg y_7 \mathcal{W} x_1))$,
   $\Box((y_1 \wedge \neg y_7) \rightarrow \neg y_5))$,
   $\Box((y_1 \wedge y_9) \rightarrow \neg(\neg y_2 \mathcal{W} y_3))$

(c) $\Box((x_2 \rightarrow \Diamond y_2) \wedge \neg y_8 \mathcal{W}(y_2 \wedge y_8))$,
   $\Box((y_2 \wedge y_8) \rightarrow (y_5 \wedge y_2 \mathcal{W} y_4))$,
   $\Box((y_2 \wedge \neg y_8) \rightarrow \neg y_5)$

(d) $\Box((y_2 \wedge y_4) \rightarrow (\neg y_8 \mathcal{W} x_2))$,
   $\Box(x_3 \rightarrow (\Diamond y_3 \wedge (y_9 \mathcal{W}(y_3 \wedge y_9))))$,
   $\Box((y_3 \wedge y_9) \rightarrow (y_5 \wedge y_3 \mathcal{W} y_4))$

(e) $\Box((y_3 \wedge \neg y_9) \rightarrow \neg y_5)$,
   $\Box((y_3 \wedge y_4) \rightarrow (\neg y_9 \mathcal{W} x_3))$,
   $\Box((y_3 \wedge y_7) \rightarrow \neg(\neg y_2 \mathcal{W} y_1))$

(f) $\Box(y_5 \rightarrow (\neg y_4 \mathcal{W} \neg y_5))$,
   $\Box(\neg y_5 \rightarrow (y_4 \mathcal{W} y_5))$,
   $\Box(y_5 \rightarrow \Diamond y_{10})$

(g) $\Box((x_4 \wedge \neg y_{10}) \rightarrow y_6)$,
   $\Box(y_{10} \rightarrow \neg y_5)$,
   $\Box((x_5 \wedge \neg y_6) \rightarrow \neg y_5)$

(h) $\Box((y_6 \wedge \neg y_4) \rightarrow y_5)$,
   $\Box(y_{10} \rightarrow \neg y_5)$,
   $\Box((x_5 \wedge \neg y_6) \rightarrow \neg y_5)$

(i) $\Box((y_6 \wedge \neg y_4) \rightarrow y_5)$,
   $\Box(y_5 \rightarrow \Diamond y_{10})$,
   $\Box((x_4 \wedge \neg y_{10}) \rightarrow y_6)$

(2) Within one second, this procedure decides stepwise unsatisfiability of the specification that is obtained by exchanging input and output propositions in the previous specification. This fast decision depends on the proof system. Although this result seems inconsistent with the result of (1), the result of (2) does not deduce the property of the specification in (1).

(3) The specification of (1) is difficult for the implemented procedure. The experiment checks each group in the specification of (1). The result is that all groups are stepwise satisfiable. The time of deciding stepwise satisfiability is as follows.

Group	(a)	(b)	(c)	(d)	(e)	(f)	(g)	(h)	(i)
Time (sec)	13.98	12323.03	17.24	273.34	12290.13	0.39	1.99	0.83	9.73

The implemented procedure can check a small number of temporal logic formulas. All groups of formulas are stepwise satisfiable and therefore the experiment show the efficiency of the proof system.

8   Conclusion

This paper proposed a proof system of stepwise satisfiability of reactive system specifications. This paper implemented the decision procedure of stepwise satisfiability, which consists of the tableau method and the proof system. The experiment showed that the procedure can sometimes decide very fast that several specifications are stepwise unsatisfiable because of the proof system.

References

1. Abadi, M., Lamport, L., Wolper, P.: Realizable and unrealizable specifications of reactive systems. In: Ausiello, A., Dezani-Ciancaglini, M., Della Rocca, S.R. (eds.) Automata, Languages and Programming. LNCS, vol. 372, pp. 1–17. Springer, Heidelberg (1989)
2. Bouyer, P., Bozzelli, L., Chevalier, F.: Controller synthesis for MTL specifications. In: Baier, C., Hermanns, H. (eds.) CONCUR 2006. LNCS, vol. 4137, pp. 450–464. Springer, Heidelberg (2006)
3. Duer-Luts, A.: LTL translation improvements in Spot. In: Proceedings of the Fifth International Conference on Verification and Evaluation of Computer and Communication Systems, pp. 72–83 (2011)
4. Emerson, E.A.: Temporal and modal logic. In: Handbook of Theoretical Computer Science, vol. B, pp. 995–1072 (1990)
5. Filiot, E., Jin, N., Raskin, J.F.: An Antichain Algorithm for LTL Realizaibility. Formal Methods in System Design archive **39**(3), 261–296 (2011)
6. Harel, D., Pnueli, A.: On the development of reactive systems. In: Logics and Models of Concurrent Systems, pp. 477–498 (1985)
7. Kupferman, O., Madhusudan, P., Thiagarajan, P.S., Vardi, M.Y.: Open systems in reactive environments: control and synthesis. In: Palamidessi, C. (ed.) CONCUR 2000. LNCS, vol. 1877, pp. 92–107. Springer, Heidelberg (2000)
8. Mori, R., Yonezaki, N.: Derivation of the input conditional formula from a reactive system specification in temporal logic. In: Langmaack, H., de Roever, W.-P., Vytopil, J. (eds.) FTRTFT 1994 and ProCoS 1994. LNCS, vol. 863, pp. 567–582. Springer, Heidelberg (1994)
9. Eén, N., Mishchenko, A., Sörensson, N.: Applying logic synthesis for speeding up SAT. In: Marques-Silva, J., Sakallah, K.A. (eds.) SAT 2007. LNCS, vol. 4501, pp. 272–286. Springer, Heidelberg (2007)

10. Pnueli, A., Rosner, R.: On the synthesis of a reactive module. In: Proceedings of the 16th ACM SIGPLAN-SIGACT Symposium on Principles of Programming Languages, pp. 179–190 (1989)
11. Yoshiura, N.: Decision procedures for several properties of reactive system specifications. In: Futatsugi, K., Mizoguchi, F., Yonezaki, N. (eds.) ISSS 2003. LNCS, vol. 3233, pp. 154–173. Springer, Heidelberg (2004)
12. Neya, Y., Yoshiura, N.: Stepwise satisfiability checking procedure for reactive system specifications by tableau method and proof System. In: Aoki, T., Taguchi, K. (eds.) ICFEM 2012. LNCS, vol. 7635, pp. 283–298. Springer, Heidelberg (2012)

Cryptic-Mining: Association Rules Extractions Using Session Log

Shaligram Prajapat[1(✉)] and Ramjeevan Singh Thakur[2]

[1] MANIT Bhopal and DAVV, Indore, India
shaligram.prajapat@iips.edu.in
[2] MANIT, Bhopal, India
ramthakur2000@yahoo.com

Abstract. Security of gargantuan sized data has always posed as a challenging issue. This domain has witnessed a number of approaches being introduced to counter such issues. This paper first reviews approaches for investigation of mining algorithms in cryptography domain and sheds light on application of mining techniques and machine learning algorithms in cryptography. The paper presents key computation using parameters-only scheme for automatic variable key (AVK) based symmetric key cryptosystem. A cryptanalysis based on association rule mining for key and parameter prediction has been discussed using both analytical method and WEKA tool. The paper also presents some research questions regarding the design issues associated with the implementation of parameter based symmetric Automatic Variable Key (AVK) based cryptosystem.

Keywords: Mining algorithms · Symmetric key cryptography · AVK · WEKA

1 Introduction

The term "CRYPTIC MINING" is derived by merging two well established disciplines: Cryptography and Data Mining. Both of these fields have their own style of Algorithms and applications. One of the important characteristic of Data Mining techniques is to discover large database in order to find novel and useful patterns, which are still unknown[1]. These techniques are data dependant and provide capabilities to predict the outcome in future. So, they are applied for market prediction and analyzing business related data. On the other hand, field of cryptography (the science of hiding information) has a very long history for secret sharing. Cryptanalyst attempts to recover plaintext from the encrypted text, partially or completely without knowing the secret key [2]. Author determine any one element from: {Plaintext, Key, any weakness in the Encrypted Text that results in finding plaintext or key}. So according to principle, Cryptography is to hide properties or patterns of data, and Data Mining reveals patterns in data. On one hand, Data Mining is useful for bulk (large volume) data; on the other hand, Cryptography is inefficient in

© Springer International Publishing Switzerland 2015
O. Gervasi et al. (Eds.): ICCSA 2015, Part IV, LNCS 9158, pp. 699–711, 2015.
DOI: 10.1007/978-3-319-21410-8_53

handling large amount of data. Since, both fields are different and contradictory to one another; it is very difficult to merge them to exploit benefits of one field over another. Recently, Machine Learning has gained pace in the field of computer science and data analysis. According to A. Samuel, "It is the field of study that enables computers to learn without being explicitly programmed." So, efficient algorithms of Data Mining domain can be used to teach (computer) classification of packets on the server: like we get classified messages in our inbox and spam folders. Such algorithms can train computers on email messages, so that they can learn to classify SPAM and HAM. Once algorithm is learned then it can be used to classify new email messages into SPAM and non-SPAM folders.

In traditional cryptographic algorithms key plays central role. A user chooses key in the form of a string of characters (digits, numbers, special symbols etc. depending upon the type of implementation) which is checked by source or destination computer. If the supplied key matches with the one which is associated with the actual user's resource (files, databases, etc), access is granted to all the resources of authorized user. To enhance the level of security, the length of key is recommended to be large enough.

A simple straightforward and inexpensive Key design scheme would be choosing a relatively short string of characters. The user may be allowed to choose the key and in such a casethe chosen key would require little effort to remember but it is easy for an intruder to guess. In case of a difficult key less obvious key is selected, then the user has to write or record the key somewhere on paper or may be in the system also, and making it equally vulnerable. Fixing the key length but varying it from session to session forms basis of automatic variable key (AVK) in the system where both sender (say Alice) and receiver (say Bob) uses similar key. This time variant key or automatic variable key (AVK) approach has been investigated from the perspective of data mining in subsequent sections. In the last section, research questions have been raised for investigating the applicability of data mining techniques, and machine learning for Symmetric Key Cryptographic techniques of information exchange.

2   Related Work

For security domain, standard literature of Data mining insinuates about classification of credit card transactions into fraudulent or genuine based upon the credit history and patterns of earlier transactions [4]. These techniques are also useful for decision making of new transaction that falls into one of the given categories: Authorized transaction state, State of asking for further identification before authorization , State of unauthorized transaction, Critical State- with not to authorize but contact police stock market price graphs.

Table 1. Types of cryptanalysis methods

Cryptanalysis Method Based on	Information assumed to be available		
	Encryption Algorithm	Cipher text (to be decoded)	Addition Requirements
Cipher text only	✓	✓	--
Known plain-text	✓	✓	One or more (P, C) pairs, where P = plain text , C= Cipher text
Chosen plain-text	✓	✓	Cipher text chosen by cryptanalyst and corresponding decrypted plain text i.e.(C, p = (C))
Chosen text	✓	✓	Plain text chosen by cryptanalyst & corresponding cipher text, Cipher text chosen by cryptanalyst & corresponding decrypted plain text

Online analysis with limited amount of memory and continuously changing data creates various difficulty levels [1]. In early 21st century, data stream mining work was in inception phase. In 2000, Domingos and Hulten started their work on classification for mining high speed data streams. For association rule, team of Giannella in 2003 contributed by exploring frequent patterns in data streams in multiple time granularities. Similar detailed work of Guha et.al. was found on Clustering data streams-Theory and practices (2003). For change detection, Kifer team worked for detecting changes in data streams in 2004. Dimensionality reduction principle came first in 2004 by law et al. for adaptive unsupervised stream mining. In the same year, time series analysis work proposed by Papadimitriou and colleagues for Nonlinear Manifold learning for Data. Similar to the grand decade's history of Databases and Cryptography, evolution of machine learning started since 1956 from A. Samuel to the recent age. Now, it has been well established for Data Analytics, Artificial Intelligence and Computational Intelligence. The classical example of learning can be seen in email accounts i.e. automatic classification of emails into useful mails and SPAMS.

In literature, the landmark article of Rivest [3] delineates a survey of the relationship between the fields of cryptography and machine learning. Along with an emphasis on how each field has contributed ideas and techniques to the other. He gives directions for future cross-fertilization. He established relations as follows: {Secret Keys ≡ Target Functions, Attack Types ≡ Learning Protocols, Exact Inferences ≡ Approximate Inference, Unicity Distance ≡ Sample Complexity}.These available literatures lay foundation for making assumptions with application of Mining techniques for development of effective Cryptosystems. In [2, 5, 6, 7, 8] towards classification of cipher some work at IIT-K has been done. Ideally, there should not be any pattern in packets after receiving cipher texts. Theoretically, all generated ciphers must be uniform, but, their classification algorithms observe some pattern in the ciphers. These patterns can be used to identify or guess about cipher text and receive knowledge of enciphering algorithms. Though, cipher text can be obtained easily, but to identify which encryption algorithm is used for generating that cipher is difficult task for cryptanalysts. In [2] "Cipher text only" experiments was carried out to identify the encryption algorithm for producing the given cipher text. For their experiment Blowfish, RC4 ciphers were taken for classification using SVM. Automatic Variable Key concept is taking shape for symmetric key encryption decryption techniques. Professor C. T. Bhunia, a pioneer researcher in this domain proposed key variability concept. Later P. Chakrabarti, introduced some concepts and extended work. Fibonacci-Q matrix [9,10], sparse approach[12] are alternative variants for realization of time variant key towards symmetric key for AVK based methods. A novel approach of parameter based key computation has been proposed and analyzed for adding more security in key exchange [14]. In section 2.1, a new model for symmetric key has been presented .As the keys of both are parties are functionally related with each other, so this model adds one level of security by exchanging parameters instead of key exchange.

2.1   The Parameters Only Scheme for Automatic Variable Key

In order to enhance the security level over insecure public network, instead of exchanging key it is better to exchange parameters, both sender and receiver can compute key using these parameters. This scheme can be well understood by following algorithms:

Algorithm parameters4Key-Alice (parameters p_1, p_2)

{

1. Sense   parameters p_1,p_2;
2. Compute the key for information exchange   by: $key_i = (p_1 * p_2)^{1/2}$;
3. Sense the information to exchange=D_i ;
4. If (mode==transmit)

> Generate Cipher text C_i = Encrypt(D_i, key_i);
> Transmit C_i;

5. else

> Receive Plain text P_i = Decrypt(D_i, key_i);
> Use P_i ;

}

Algorithm Parameters4Key-Bob (parameters p_1, p_2)

{

1. Sense   parameters p_1,p_2;
2. Compute the Arithmetic Mean A.M.= ($p_{1+}p_2$)/2;
3. Compute the Harmonic Mean H.M.=2* $p_1 * p_2$ /($p_1 + p_2$)
4. Compute the $Key_i = (A.M.. * H.M)^{1/2}$
5. If (mode==transmit)

> Generate Cipher text C_i =Encrypt(D_i, key_i);
> Transmit C_i;

6. else

Receive Plain text P_i = Decrypt(D_i, key$_i$);
Use P_i ;

}

Advantage of parameter based AVK cryptosystem will be as under:

1. Without exchanging entire key Alice and Bob will securely communicate with each other.
2. In this model keys are computed using different functions which also enhances the level of security.

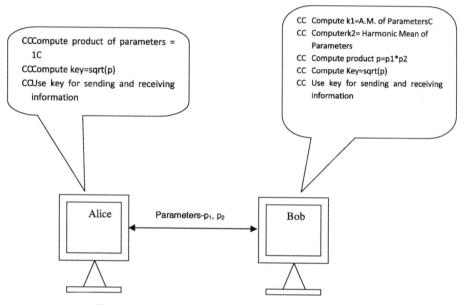

Fig. 1. Model for key computation using parameter only

After the successful key computation at both ends (i.e. sender Alice and receiver Bob), Table-2 demonstrates the exchange of data with variable parameters in successive sessions in AVK based cryptosystem:

Table 2, illustrates that sender (Alice) has successfully sent first data 00000011 and second data 00000100 in session S1 and S2 respectively and similarly Bob has exchanged his first information 00000111 and second information 00001000 respectively using automatic variable key model.

Table 2. AVK Scheme in Symmetric Key Cryptography

Session ID	Alice (T_x)	Bob (R_x)	Bob(T_x)	Alice(R_x)	Remarks
Initial agreement	00000010 (K_a)	00000010	00000110 (K_b)	00000110	For next slot Alice and Bob will use 00000110, 00000010 respectively
S_1	0000001= D_1 Compute C_1=D$\oplus$ K_b and Transmit C_1	Compute $C_1\oplus$ K_a and Get plaintext D_1=00000011	00000111=D_2 Compute C_2=$D_2\oplus$ K_b and Transmit C_2	Compute $C_2\oplus$ K_b and Get plaintext D_2=00000111	Alice and Bob will Compute new keys by Compute $D_2\oplus$ K_b and D_1 $\oplus$ K_a Respectively for new session
S_2	00000100=D_1 Compute C_1=$D_1\oplus$ K_b and Transmit C_1	Compute $C_1\oplus$ K_a and Get plaintext D_1=00000100	00001000=D_2 Compute C_2=$D_2\oplus$ K_b and Transmit C_2	Compute $C_2\oplus$ K_b and Get plaintext D_2=00001000	Compute new session keys similar to previous step. i.e. Alice and Bob will Compute new keys by Compute $D_2\oplus$ K_b and D_1 $\oplus$ K_a Respectively for new session

Generation of automatic variable key under various approaches in cryptographic system had been investigated for random input parameters recently [13].

In [9] intelligent techniques have been pointed out for shared key generation. In case of multiparty communication, the concept of shared key is used to enhance the security level. Chakrabarti [9] has proposed techniques that are based on minimal frequent set, candidate generation, partition scheme, intersection of item-set count. These methods are based on feature analysis, centroid analysis, inter-centroid distance, extraction scheme of vowel, index position of character, support analysis and confidence rule. The minimal frequent set is formed by minimum probability of combination of items. The shared key is the $\oplus$ of $\oplus$ of each of the pairs of elements of the set. Among the combination of the keys only (p_1, p_5) and (p_2, p_4) have least probability and it is zero. So, Minimal frequent set = { p_1, p_5 , p_2, p_4}. And Shared key = ($p_1 \oplus p_5$) $\oplus$ ($p_2 \oplus p_4$).

Table 3. Transaction log of parameter usage for computation of AVK

Data	Shared Key	Key with parameters
D_1	SK_1	$SK_1 = f(p_1, p_3, p_4, p_6)$
D_2	SK_2	$SK_2 = f(p_3, p_5)$
D_3	SK_3	$SK_3 = f(p_4, p_5, p_6)$
D_4	SK_4	$SK_4 = f(p_2, p_3, p_5)$
D_5	SK_5	$SK_5 = f(p_1, p_2)$
D_6	SK_6	$SK_6 = f(p_1, p_2, p_3, p_6)$

Table 4. Single parameter frequent sets

C_1		
Parameter	1-Frequent	Minimum frequent parameter
p_1	3	
p_2	3	
p_3	4	Parameter with minimum
p_4	2	frequency = p_4
p_5	3	
p_6	3	

Table 5. paired parameter frequent sets

Key evaluation based on candidate generation, shared key = $p_4 \oplus (p \oplus p_5) \oplus (p_2 \oplus p_4)$

C_2		
Parameter pair	Frequent Set	Minimum frequent parameters
$p_1.p_2$	2	
p_1, p_3	2	
$p_1.p_4$	1	
$p_1.p_5$	0	
$p_1.p_6$	2	
$p_2.p_3$	2	
$p_2.p_4$	0	
$p_2.p_5$	1	(p_1, p_5) and (p_2, p_4) =
$p_2.p_6$	1	$((p_1 \oplus p_5) \oplus (p_2 \oplus p_4))$
$p_3.p_4$	1	
$p_3.p_5$	2	
$p_3.p_6$	2	
$p_4.p_4$	1	
$p_4.p_5$	2	
$p_4.p_6$	1	

Table 6. Session wise parameter used for key computation in AVK

Data ↓	Parameters for Key→					
	P_1	P_2	P_3	P_4	P_5	P_6
D_1	1	0	1	1	0	1
D_2	0	0	1	0	1	0
D_3	0	0	0	1	1	1
D_4	0	1	1	0	1	0
D_5	1	1	0	0	0	0
D_6	1	1	1	0	0	1

During the first pass support court are: $\{p_1 :3, p_2 :3, p_3 :4, p_4 :2, p_5 :3, p_6 :3\}$, we get most frequent = p_3. In second pass support counts are: $\{(p_1, p_2) :2, (p_1, p_3) :2, (p_1, p_4) :1, (p_1, p_5) :0, (p_2, p_3) :2, (p_2, p_4) :0, (p_2, p_5) :1, (p_3, p_5) :2, (p_4, p_5) :1\}$, we get most frequent=$\{(p_1, p_2), (p_2, p_3), (p_3, p_5)\}$ or $\{p_1, p_2, p_3, p_5\}$ so shared key may be intersection of two element set = p_3 .

3   Association Rule Extractions

For cryptanalysis of parameters-only based AVK cryptosystem, using association rule mining, consider session log information of Table 7 with parameters for key construction. Here we have only four parameter space for key computations $\{p_1, p_2, p_4, p_5\}$ within four sessions $\{s_1, s_2, s_3, s_4\}$. For investigation of association rules we assume minimum support of 0.5 and minimum confidence level of 0. 75.

Table 7. Session wise log of used parameters

Session ID	Parameters for Key
s_1	p_1, p_2
s_2	p_1, p_2, p_4
s_3	p_1, p_5
s_4	p_2, p_4, p_5

Using straightforward approach (analytical method) one may compute all the combination of parameters from Domain space and will identify frequent combinations as in Table 8.

Table 8. Parameters sets and corresponding frequencies

Parameter sets	Frequency
p_1	3
p_2	3
p_4	2
p_5	2
$\{p_1, p_2\}$	2
$\{p_1, p_4\}$	1
$\{p_1, p_5\}$	1
$\{p_2, p_4\}$	2
$\{p_2, p_5\}$	1
$\{p_4, p_5\}$	1
$\{p_1, p_2, p_4\}$	1
$\{p_1, p_2, p_5\}$	0
$\{p_1, p_4, p_5\}$	0
$\{p_2, p_4, p_5\}$	1
$\{p_1, p_2, p_4, p_5\}$	0

With minimum support of 0.5, the parameter set that occur in at least two transactions are explored, and would be frequent for this case is given in Table 9.

Table 9. Parameters groups and corresponding decisions

Parameter Group	Decisions for frequent	Result Set
Single	All individual Frequent	p_1, p_2, p_4, p_5
2-parameter sets	Only 2 parameters out of 6	(p_1, p_2) and (p_2, p_4)
3-parameter sets	None	Null Set
4-parameter sets	None	Null Set

Exclusion of non frequent parameters will result in table 10:

Table 10. The set of all frequent parameter sets

Parameters	Frequency
p_1	3
p_2	3
p_4	2
p_5	2
p_1, p_2	2
p_2, p_4	2

Table 11 illustrates determination (identification) process of useful association rules from all possible combinations using desirable confidence. Only the rule $p_5 \rightarrow p_2$ qualifies the criteria and insinuates the possibility of next parameters for the computation of key in subsequent sessions.

Table 11. Association rules from Possible parameters

Possible Rule	Confidence	Desirable confidence
$p_1 \rightarrow p_2$	2/3	< 0.75
$p_2 \rightarrow p_1$	2/3	<0.75
$p_2 \rightarrow p_4$	2/3	<0.75
$p_5 \rightarrow p_2$	3/3	>0.75

From Table 11, it is obvious that for only 4 parameters detection of frequent parameters requires scanning of transaction logs, and construction of intermediate tables for frequent set computation. This means that for computation of large frequent sets more memory is needed with increase in number of parameters sets and size of transaction logs.

WEKA-64 bit tool can be used to verify the association rule generated from analytical method. This can be extended and verified for variable number of parameters and session logs. For Table no. 7 the Run information is given below:

Scheme: weka .associations .Apriori -N 10 -T 0 -C 0.9 -D 0.05 -U 1.0 -M 0.1 -S -1.0 -c -1
Relation: iris
Instances: 4
Attributes: 4
              p1
              p2
              p4
              p5
=== Associator model (full training set) ===

Using Apriori method the generated rule base are:
===

Minimum support: 0.25 (1 instances)
Minimum metric <confidence>: 0.9
Number of cycles performed: 15

Generated sets of large itemsets:

Size of set of large itemsets L(1): 4

Size of set of large itemsets L(2): 6

Size of set of large itemsets L(3): 2

Best rules found:

1. p4=t 2 ==> p2=t 2   conf:(1)
2. p1=t p4=t 1 ==> p2=t 1   conf:(1)
3. p4=t p5=t 1 ==> p2=t 1   conf:(1)
4. p2=t p5=t 1 ==> p4=t 1   conf:(1)

It can be seen that the output of association process has been presented in the format antecedent => Consequent. The count associated with the antecedent = absolute coverage in the dataset. The number next to the consequent = absolute number of instances that match the antecedent and the consequent. The number in brackets at the end is the support for the rule (number of antecedent divided by the number of matching consequents).

So one can say that a cutoff of 90% was used in selecting rules, of "Association rule generation" window and indicated in that no rule has coverage less than 0.90.

This aligns with analytical result. So the association rules can be extracted for large number of parameter set and cryptanalyst may generate rule based on key computation.

4   Result Discussion

In addition, together with issues of association rules mining in Cryptography domain, other mining algorithms like classification, clustering for AVK based cryptosystem should be taken care. These issues also include:

1. How classification, association rule mining, clustering algorithms would add learning to generate   new   session keys, for secure information transmission by varying the key?
2. Do mining algorithms compromise with the security of the information on key variation?
3. How AVK based mining algorithm work for multiple sessions and parties?
4. Is there any efficient choice instead of reversible function like $\oplus$?
5. Can the cryptic mining, cryptic clustering and cryptic classification algorithm be improved further?
6. What are the time, power, and space efficiency of that alternative algorithm?
7. How concepts of Distributed Computing, Parallel Processing would be exploited to achieve gain?
8. What are the perspectives of cryptanalyst andhacker?

These issues are key challenges and need to be sorted out in near future.

5   Future Enhancement

Once Cryptic Mining algorithm has been implemented, tested and deployed, that can be deployed on the server for disfiguring the nature of information being transferred from servers and it would be possible to know which information going is genuine that is sent by some malicious application. Theft of such information could be analyzed and prevented. Honey Pot System can save their information and can prevent stealing. Similarly, clustering of information request and reply pattern can provide useful insights.

6   Conclusion

Parameter-only communication presented in this paper highlights key and data exchange using secure parameter exchange mechanism. Parameters of the key is used to compute symmetric key by both transmission and receiver end. Using association rules mining cryptanalyst may attempt to predict and form rules for parameter prediction and ultimately future keys. Thus, Cryptic mining opens a platform to explore the mining algorithm at bit level, and may be useful developing intelligent AVK based cryptosystem in near future for efficiently securing industry and business related information over the noisy communication channel. The paper poses some research question also. After resolving raised issues, it would provide higher gain in terms of efficiency for low power devices.

Acknowledgement. The work is supported by Research grants from MPCST, Bhopal, India Project ref. No. 1080/CST/R&D/2012, dated 30/06/2012 and also under Fast Track Scheme for Young Scientist from DST, New Delhi, India. Scheme 2011-12, No. SR/FTP/ETA-121/ 2011 (SERB), dated 18/12/2012.

References

1. Tan, P.-N., Steinbach, M., Kumar, V.: Introduction to data mining. Pearson education, p. 14 (2008)
2. Saxena, G., Karnik, H., Agrawal, M.: Classification of Ciphers using Machine learning. IIT (2008)
3. Rivest, R.L.: Cryptography and machine learning. In: Matsumoto, T., Imai, H., Rivest, R.L. (eds.) ASIACRYPT 1991. LNCS, vol. 739, pp. 427–439. Springer, Heidelberg (1993)
4. Dunham, H.: Data Mining: Introductory and Advanced Topics. Pearson education (2008)
5. Anoop, J.: Classification among DES, AES, and IDEA Ciphers. IIT Kanpur (2003)
6. Rao, B.M.: Classification of RSA and IDEA Ciphers. IIT Kanpur (2003)
7. Chandra, G.: Classification of Modern Ciphers. IIT Kanpur (2002)
8. Pooja, M.: The Classification of Ciphers. IIT Kanpur (2000)
9. Chakrabarti, P., Choudhary, A., Naik, N., Bhunia, C.: Key Generation in the Light of Mining and Fuzzy Rule. IJCSNS **8**, 332–337 (2008)
10. Prajapat, S., Jain, A., Thakur, R.S.: A Novel Approach For Information Security With Automatic Variable Key Using Fibonacci Q-Matrix. IJCCT **3**(3), 54–57 (2012)

11. Prajapat, S, Thakur, R.S.: Time variant approach towards symmetric key. In: SAI 2013. IEEE, pp. 398–405 (2013)
12. Prajapat, S., Thakur, R.S.: Sparse approach for realizing AVK for Symmetric Key Encryption. IJRDET & proceeding of International Research Conference on Engineering, Science and Management (IRCESM 2014) **2**, 15–18 (2014)
13. Goswami, R., Chakrabarti, S., Bhunia, A., Bhunia, C.: Generation of automatic variable key under various approaches in cryptographic system. Journal of The Institution of Engineers (India): Series B **94**(4), 215–220 (2013)
14. Prajapat, S., Thakur, R.S.: Association rules for parameter prediction of AVK. In: proceedings of International Conference on Intelligent, Computational and Informative Systems (ICICIS-2014) and published in International Journal of Electronics Communication and Computer Engineering, October 2014

Feature Based Encryption Technique for Securing Digital Image Data Based on FCA-Image Attributes and Visual Cryptography

Quist-Aphetsi Kester[1,2,3(✉)], Anca Christine Pascu[2], Laurent Nana[2], Sophie Gire[2], Jojo M. Eghan[3], and Nii Narku Quaynor[3]

[1] Faculty of Informatics, Ghana Technology University College, Accra, Ghana
Kester.quist-aphetsi@univ-brest.fr, kquist@ieee.org
[2] Lab-STICC (UMR CNRS 6285), European University of Brittany,
University of Brest, Brest, France
[3] Department of Computer Science and Information Technology,
University of Cape Coast, Cape Coast, Ghana

Abstract. Lossless pixel value encrypted images still maintains the some properties of their respective original plain images. Most of these cryptographic approaches consist of visual cryptographic techniques and pixel displacement approaches. These methods of cryptography are useful in cases as medical image security where pixel expansion is avoided in both the encryption and decryption processes. In this paper we propose a hybrid cryptographic encryption approach by using features generated from digital images based on Galois lattice theory and a visual cryptographic technique based on RGB pixel displacement. The features extracted from a plain image and a lattice was generated which was then used to generate a key used to encrypt the plain image. At the end of the process, there was no pixel expansion and the arithmetic mean, the entropy as well as the Galois lattice of both ciphered and plain image remained the same. The features extracted from the plain image were the same as that of the ciphered image irrespective of pixel displacement that occurred, this makes our approach a suit-able basis for image encryption and storage as well as encrypted image indexing and searching based on pixel values. The implementation was done using Galicia, Lattice Miner and MATLAB.

1   Introduction

Cyber security is a major concern and continues to be a debate in our ever growing and data driven society. With more organizations migrating to cloud computing in order to reduce cost, centralize data easily for access etc. there are corresponding growth of issues on the safety of such data. The emergence of social networks has contributed to the rapid growth of multimedia on the internet especially with digital

This work was supported by Lab-STICC (UMR CNRS 6285) Research Laboratory, UBO France, AWBC Canada, Ambassade de France-Institut Français-Ghana and the DCSIT-UCC.

© Springer International Publishing Switzerland 2015
O. Gervasi et al. (Eds.): ICCSA 2015, Part IV, LNCS 9158, pp. 712–724, 2015.
DOI: 10.1007/978-3-319-21410-8_54

images [1]. Cloud storage also has provided easy backup and easy access to stored data by individuals, organizations such as medical institutions, etc [2]

2   Related Works

The need of symmetric and asymmetric encryption approaches that generate keys based on data sensing techniques is required to make systems to automatically generate distinct keys for encryption processes. An algorithm should also have the capability to feed on certain features of the images to aid in determining their behavior during encryption and decryption processes. This means that if an algorithm picks two frames from a given spot, the keys generated from these two images should differ as well as how they are encrypted. This can only be true based on varying properties of pixel values (sampled space geometrical mean values, arithmetic mean, entropy etc) of the image due to varying physical conditions (humidity, temperature, luminous intensity etc) from nature that will affect the pixel values for each shot. Based on extracted features from the image, we proposed a cryptographic encryption technique of digital images. Social network data have biometric capabilities of auto detection of individuals for self tagging etc [3]. This capability can also be exploited for surveillance purposes and image searching search engines (Zhang, C et al 2009) by both authorized and unauthorized users. Henceforth security of digital images is a major issue. Cryptographic approaches to securing images must provide adequate security for images as well as maintain the quality of the image after decryption. This led to the adoption of lossless pixel value approaches to securing images such as medical images, biometric images etc [4]. The application of Formal concept analysis can extensively be seen in software engineering and applied to applied to several software engineering problems, such as: restructuring the code into more cohesive components, identifying class candidates, locating features in the code by means of dynamic analysis, reengineering class hierarchies [5][6]. It has also finds its way in security analysis like crime patterns, criminal behavior analysis etc.[7][8]. And based on the fact that a change in concept due to a change in property or an object can affect can cause a change in the lattice, we constructed a lattice on the image to generate a unique key as well as using the lattice for image authentication in our approach. The paper has the following structure: methodology, results and analysis, and conclusion.

3   Methodology

Our method employed a hybrid cryptographic encryption approach by using features generated from a digital images based on Galois lattice theory and a visual cryptographic technique based on RGB pixel displacement to encrypt the plain digital image. At the end of the process, there was no pixel loss and the arithmetic mean, the entropy as well as the Galois lattice of both ciphered and plain image remained the same. The overall process is indicated in figure 1 below.

From figure 1:
PI=Plain image
CI=Ciphered image

g(PI)=function that operated on the plain image to pro-
        duce the features
n(f,a,s,e)=function of the features
fe= the feature results
f=sum of all frequency of each pixel in the image
a=arithmetic mean of all the pixel values in the image
s=standard deviation all the pixel value in the image
e=entropy of all the pixel value the image
x=a distinct chosen pixel value number
x'=frequency of x
f'=x'/f, a'=x'/a, s'=x'/s and e'=x'/e
G =set objects extracted from the image
M=sets attributes obtained from the image

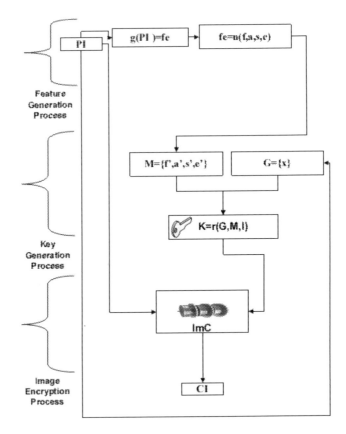

Fig. 1. Summary of the Encryption process

K=generated key from r(G,M,I)
r(G,M,I)=the function that operated on G,M an d I concept to produce K
ImC=the image encryption algorithm that operated on K and Pi to produce CI

3.1   The Feature Extraction

Let I= an image=f (R, G, B)
I is a color image of m x n x 3 arrays

$$
\begin{pmatrix}
R & G & B \\
r_{i1} & g_{i2} & b_{i3} \\
\vdots & \vdots & \vdots \\
\vdots & \vdots & \vdots \\
r_{n1} & g_{n2} & b_{n3}
\end{pmatrix}
\tag{1}
$$

(R, G, B) = m x n
Where R, G, B $\in$ I
(R o G) i j = (R) ij. (G) ij
Where r_11 = first value of R
       r= [ri1] (i=1, 2... m)
       x $\in$ r_i1 : [a, b]= {x $\in$ I: a $\leq$ x $\leq$ b}
       a=0 and b=255
      R= r= I (m, n, 1)
Where g_12 = first value of G
       g= [gi2] (i=1, 2... m)
       x $\in$ g : [a, b]= {x $\in$ I: a $\leq$ x $\leq$ b}
       a=0 and b=255
       G= g= I (m, n, 1)
And   b_13 = first value of B
       g= [bi3] (i=1, 2... m)
       x $\in$ b_i1 : [a, b]= {x $\in$ I: a $\leq$ x $\leq$ b}
       a=0 and b=255
       B=b= I (m, n, 1
Such that   R= r= I (m, n, 1)
Step3 Extraction of the red component as 'r'
Let size of R be m x n   [row, column] = size (R) = R (m x n)

Let X=freq(x) which is the number of times x occurred in r,g and b

$$
f= \sum_{i=m}^{n} X_i
\tag{2}
$$

$$
a= \frac{\sum_{n=1}^{k} x_n}{k}
\tag{3}
$$

Where $x \in b\_i1 : [a, b] = \{x \in I : a \leq x \geq b\}$

$$s: \sqrt{\frac{1}{N} \sum_{i=1}^{N} (x_i - \mu)^2}, \quad \text{where} \quad \mu = \frac{1}{N} \sum_{i=1}^{N} x_i. \tag{4}$$

Entropy is defined as

$$e = -\sum_{\eta=0}^{\varepsilon-1} \Psi(xi) . \log2 (\Psi(xi)) \tag{5}$$

Where:
δ = Entropy of image
ε = Gray value of an input image (0-255).
$\Psi(\eta)$ = Probability of the occurrence of symbol η

3.2   The Feature Classification Using FCA

Formal concept analysis (FCA) is a method of data analysis with growing popularity across various domains. FCA analyzes data which describe relationship between a particular set of objects and a particular set of attributes [10][11]. In FCA a formal context consists of a set of objects, G, a set of attributes, M, and a relation between G and M, $I \subseteq G \times M$. A formal concept is a pair (A, B) where $A \subseteq G$ and $B \subseteq M$. Every object in A has at least an attribute in B. For every object in G that is not in A, there is an attribute in B that that objects does not have. For every attribute in M that is not in B there is an object in A that does not have that attribute. A is called the extent of the concept and B is called the intent of the concept [12].
If $g \in A$ and $m \in B$ then $(g,m) \in I$,or gIm.
A formal context is a triple (G,M,I), where
•G is a set of objects,
• M is a set of attributes
•and I is a relation between G and M.
• $(g,m) \in I$ is read as „object g has attribute m.
For $A \subseteq G$, we define
$A' := \{m \in M \mid \forall g \in A : (g,m) \in I \}$.
For $B \subseteq M$, we define dually
$B' := \{g \in G \mid \forall m \in B : (g,m) \in I \}$.
For A, A1, A2 $\subseteq$ G holds:
• $A1 \subseteq A2 \Rightarrow A`2 \subseteq A`1$
• $A1 \subseteq A``$
• $A` = A```$
For B, B1, B2 $\subseteq$ M holds:
• $B1 \subseteq B2 \Rightarrow B\_2 \subseteq B\_1$
• $B \subseteq B``$
• $B` = B```$
A formal concept is a pair (A, B) where
• A is a set of objects (the extent of the concept),

• B is a set of attributes (the intent of the concept),

•A`= B and B`= A.

The concept lattice of a formal context (G, M, I) is the set of all formal concepts of (G, M, I), together with the partial

Order (A1, B1) ≤ (A2, B2): ⟺ A1 ⊆ A2 (⟺ B1 ⊇ B2) [13].

The concept lattice is denoted by (G,M,I) .

• Theorem: The concept lattice is a lattice, i.e. for two concepts

(A1, B1) and (A2, B2), there is always

•a greatest common sub-concept: (A1∩A2, (B1∪ B2) ´´)

•and a least common super-concept: ((A1 ∪ A2) ´´, B1∩B2)

More general, it is even a complete lattice, i.e. the greatest common sub-concept and the least common super-concept exist for all (finite and infinite) sets of concepts.

Corollary: The set of all concept intents of a formal context is a closure system. The corresponding closure operator is

h(X):= X``.

An implication X→Y holds in a context, if every object having all attributes in X also has all attributes in Y.

Def.: Let X ⊆M. The attributes in X are independent, if there are no trivial dependencies between them

	y_1	y_2	y_3	- - -
x_1	✕	✕	✕	
x_2	✕	✕		⋮
x_3		✕	✕	
⋮		- - -		

Fig. 2. A table of attributes and properties

The table above represents logical attributes represented by a triplet (X, Y, I), where I is a binary relation between X and Y. The elements of X are called objects and correspond to table rows, elements of Y are called attributes and correspond to table columns, and for $x \in X$ and $y \in Y$, (x, y) ∈ I indicates that object x has attribute y while (x, y) ∈ I.

From the image we chose our objects as a classified range of values of x, where x ∈ b_i1 : [a, b]= {x ∈ I: a ≤ x ≥ b} and a=0, b=255. G= {0-25, 26-50, 51-75, 76-100,101-125, 126-150,151-175, 176-200,201-225, 256-255}.

$$f'=j/f$$
$$a'=j/a$$
$$s'=j/s \qquad (6)$$
$$e'=j/e$$

where j is the sum of all the frequencies of all numbers that fall within the range of each object.

Where B= {f,,a',s',e'} Major attributes and f' has {B_1, B_2, B_3 and B_4} as sub attributes. Therefore, a', s' and e' have the same sub attributes as f'. But {B_1, B_2, B_3, B_4} maps directly and exactly on at least one element of {0-0.25, 0.26-0.50, 0.51-0.75, 0.76-1.0}.

3.3   The Key Generation Process

The key is generated by from the table of attributes and properties.

Table 1. Table Objects X and attributes

I	Y 1	Y 2	.	.	.	.	Y n
X 1	x						
X 2		x					
.			x				
.				x			
.					x		
.						x	
X n							x

$$K=ABS(((\textstyle\sum(xi,yi))*f)+f^2)*f)mod(m*n) \tag{7}$$

Where m is the height of the image and n is the width of te image.

3.4   The Image Encryption Process

The below process was use to perform the pixel displacement based on the generated key from the features [5]

Step1. Start
Step2. Extraction of data from a plain image,
Let I= an image=f (R, G, B)
I is a color image of m x n x 3 arrays

$$
\begin{pmatrix}
R & G & B \\
r_{i1} & g_{i2} & b_{i3} \\
\vdots & \vdots & \vdots \\
\vdots & \vdots & \vdots \\
r_{n1} & g_{n2} & b_{n3}
\end{pmatrix}
\tag{8}
$$

$(R, G, B) =$ m x n

Where R, G, B $\in$ I

$(R \ o \ G) \ i \ j = (R) \ ij. \ (G) \ ij$

Where r_11 = first value of R

r= [ri1] (i=1, 2... m)

x $\in$ r_il : [a, b]= {x $\in$ I: a $\leq$ x $\leq$ b}

a=0 and b=255

R= r= I (m, n, 1)

Where g_12 = first value of G

g= [gi2] (i=1, 2... m)

x $\in$ g : [a, b]= {x $\in$ I: a $\leq$ x $\leq$ b}

a=0 and b=255

G= g= I (m, n, 1)

And   b_13 = first value of B

g= [bi3] (i=1, 2... m)

x $\in$ b_il : [a, b]= {x $\in$ I: a $\leq$ x $\leq$ b}

a=0 and b=255

B=b= I (m, n, 1

Such that   R= r= I (m, n, 1)

Step3. Extraction of the red component as 'r'

Let size of R be m x n   [row, column] = size (R) = R (m x n)

$$
rij= r= I \ (m, n, 1) =
\begin{pmatrix}
R \\
r_{i1} \\
\vdots \\
\vdots \\
r_{in}
\end{pmatrix}
\tag{9}
$$

Step4. Extraction of the green component as 'g'

Let size of G be m x n   [row, column] = size (G)

$$
gij= g= I \ (m, n, 1) =
\begin{pmatrix}
G \\
g_{i2} \\
\vdots \\
\vdots \\
g_{n2}
\end{pmatrix}
\tag{10}
$$

Step5. Extraction of the blue component as 'b'
Let size of B be m x n   [row, column] = size (B) = B (m x n)

$$bij = b = I\,(m, n, 1) = \begin{pmatrix} B \\ b_{i3} \\ \vdots \\ \vdots \\ b_{n3} \end{pmatrix} \tag{11}$$

Step6. Getting the size of r as [c, p]
Let size of R be [row, column] = size (r) = r (c x p)
Step7. Engagement of K which is the symmetric secret key generated. The key is then engaged to iterate the step 8 to 14.
Step 8. Let r = Transpose of rij

$$r = \begin{pmatrix} R & & & & \\ r_{11} & \cdots & \cdots & \cdots & r_{n1} \end{pmatrix} \tag{12}$$

Step 9. Let g = Transpose of gij

$$g = \begin{pmatrix} G & & & & \\ g_{i3} & \cdots & \cdots & \cdots & g_{n3} \end{pmatrix} \tag{13}$$

Step10. Let b = Transpose of bij

$$b = \begin{pmatrix} B & & & & \\ b_{i2} & \cdots & \cdots & \cdots & b_{n2} \end{pmatrix} \tag{14}$$

Step11. Reshaping of r into (r, c, p)

$$r = reshape\,(r, c, p) = \begin{pmatrix} R \\ r_{i1} \\ \vdots \\ \vdots \\ r_{in} \end{pmatrix} \tag{15}$$

Step12. Reshaping of g into (g, c, p)

$$g = reshape\,(g, c, p) = \begin{pmatrix} G \\ g_{i2} \\ \vdots \\ \vdots \\ g_{n2} \end{pmatrix} \tag{16}$$

Step13. Reshaping of b into (b, c, p)

$$b = \text{reshape (b, c, p)} = \begin{pmatrix} B \\ b_{i3} \\ \vdots \\ \vdots \\ b_{n3} \end{pmatrix} \qquad (17)$$

Step14. Concatenation of the arrays r, g, b into the same dimension of 'r' or 'g' or 'b' of the original image

$$= \begin{pmatrix} R & G & B \\ r_{i1} & g_{i2} & b_{i3} \\ \vdots & \vdots & \vdots \\ \vdots & \vdots & \vdots \\ r_{n1} & g_{n2} & b_{n3} \end{pmatrix} \qquad (18)$$

Step15. Finally the data will be converted into an image format to get the encrypted image.

We chose a 24 bit depth image jpg of dimension 960 pixels by 720 pixels with a horizontal resolution of 96 dpi and a vertical resolution of 96 dpi.

Fig. 3. Plain image

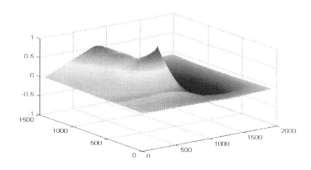

Fig. 4. The graph of the normalized cross-correlation of the matrices of the plain image

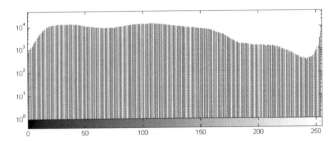

Fig. 5. A graph of frequency of pixel values

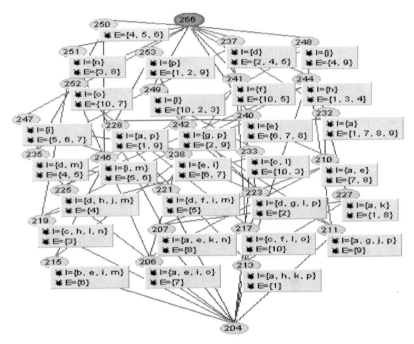

Fig. 6. A Galois lattice generated from the plain image

Fig. 7. The ciphered image

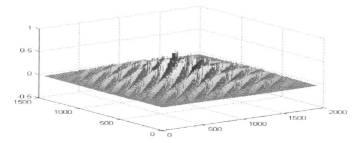

Fig. 8. The graph of the normalized cross-correlation of the matrices of the ciphered image

Table 2. Table Objects X and attributes

G	J	f'	a'	s'
0-25	169938	0.081953	1502.998	2343.901
26-50	316517	0.152641	2799.4	4365.619
51-75	235764	0.113698	2085.189	3251.819
76-100	289149	0.139443	2557.347	3988.141
101-125	319446	0.154054	2825.306	4406.018
126-150	235758	0.113695	2085.136	3251.736
151-175	146001	0.070409	1291.29	2013.746
176-200	46113	0.022238	407.8414	636.0221
201-225	31294	0.015092	276.7764	431.6283
226-255	283620	0.136777	2508.446	3911.881

The graph of the normalized cross-correlation of the matrices of the plain image in figure 3 was plotted as shown in figure 4. The features f=2073600, a=113.066, s=72.5022 and e =7.1945 were extracted from the plain images. The frequencies of the pixel values of the plain image were plotted as shown in figure 5 and a Galois lattice was generated from the features extracted from the plain image based on table 2. A key K was then generated from the plain image and used to encrypt the ciphered image to yield the results as shown in figure 6 and 7.

4   Conclusions

From our work we have realized that the Gallois lattice generated from the plain image as well as the features extracted from the plain image was the same as that of the ciphered image irrespective of pixel displacement that occurred. This means that the image maintains its pixel properties after encryption as a result of conservation of pixel values. This makes our approach a suitable basis for image encryption and storage as well as image indexing and searching based on pixel values, since the properties of the plain image and the ciphered image remain the same. This means that a plain image's property can be used to search for its encrypted one in the cloud or in a n encrypted video frame without encrypting the plain image first. Based on the fact that a change in concept due to a change in property or an object can affect can cause a change in the lattice, we constructed a lattice on the image to generate a unique key as well as using the lattice for image authentication in our approach.

References

1. Cameron, J.J., Leung, C.K.-S., Tanbeer, S.K.: Finding strong groups of friends among friends in social networks. In: 2011 IEEE Ninth International Conference on Dependable, Autonomic and Secure Computing (DASC), pp. 824–831, December 12–14, 2011
2. Löhr, H., Sadeghi, A.-R., Winandy, M.: Securing the e-health cloud. In: Veinot, T. (ed.) Proceedings of the 1st ACM International Health Informatics Symposium (IHI 2010), pp. 220–229. ACM, New York (2010)
3. Kester, Q.-A., Nana, L., Pascu, A.C., Gire, S., Eghan, J.M., Quaynor, N.N.: Feature based encryption technique for securing forensic biometric image data using AES and visual cryptography (AIMS). In: 2014 IEEE International Conference on Artificial Intelligence, Modelling and Simulation, pp. 191–205. IEEE, November 2014
4. Zhang, C., Li, H., Guo, Q., Jia, J., Shen, I.F.: Fast Active Tabu Search and its Application to Image Retrieval. IJCAI 9, 1333–1338 (2009)
5. Tonella, P.: Formal concept analysis in software engineering. In: Proceedings of 26th International Conference on Software Engineering, ICSE 2004, pp. 743–744, May 23–28, 2004
6. Tilley, T., Cole, R., Becker, P., Eklund, P.: A survey of formal concept analysis support for software engineering activities. In: Ganter, B., Stumme, G., Wille, R. (eds.) Formal Concept Analysis. LNCS, vol. 3626, pp. 250–271. Springer, Heidelberg (2005)
7. Faria, A.M., Cintra, M.E., de Castro, A.F., Lopes, D.C.: Criminal Hot Spot Detection using Formal Concept Analysis and Clustering Algorithms
8. Kester, Q.A.: Computer Aided Investigation: Visualization and Analysis of Data from Mobile Communication Devices Using Formal Concept Analysis. arXiv preprint arXiv:1307.7788 (2013)
9. Kester, Q.A., Nana, L., Pascu, A.C., Gire, S., Eghan, J.M., Quaynnor, N.N.: A hybrid encryption technique for securing biometric image data based on feistel network and RGB pixel displacement. In: Pérez, G.M., Thampi, S.M., Ko, R., Shu, L. (eds.) Recent Trends in Computer Networks and Distributed Systems Security. CCIS, vol. 420, pp. 530–539. Springer, Heidelberg (2014)
10. Ganter, B., Wille, R.: Formal Concept Analysis: Mathematical Foundations, p. 1. Springer, Berlin. ISBN 3-540-62771-5
11. Pascu, A., Desclés, J.P.: Attribute-value formalization in the framework of the logic of determination of objects (LDO) and categorization. In: FLAIRS Conference, pp. 506–511 (2008)
12. Chaudron, L., Maille, N.: Generalized formal concept analysis. In: Ganter, B., Mineau, G.W. (eds.) Conceptual Structures: Logical, Linguistic, and Computational Issues. LNCS, vol. 1867, pp. 357–370. Springer, Heidelberg (2000)
13. Priss, U.: A graphical interface for document retrieval based on formal concept analysis. In: Proceedings of the 8th Midwest Artificial Intelligence and Cognitive Science Conference, pp. 66–70, March 1997

Empirical Studies of Cloud Computing in Education: A Systematic Literature Review

Mohamud Sheikh Ibrahim[1], Norsaremah Salleh[1], and Sanjay Misra[2(✉)]

[1] International Islamic University Malaysia, Kuala Lumpur, Malaysia
faaqir1@gmail.com, norsaremah@iium.edu.my
[2] Covenant University, Ota, Nigeria
sanjay.misra@covenantuniversity.edu.ng

Abstract. The purpose of this paper is to present the evidence about adoption of cloud computing in the education system in universities or higher education institutions. We performed a systematic literature review (SLR) of empirical studies that investigated the current level of adoption of cloud computing in the education systems and motivations for using cloud computing in the institution. Twenty-Seven papers were included in our synthesis of evidence. It has been found that several universities are interested in using cloud computing in their education systems, and they have utilized different types of cloud computing service models (IaaS, PaaS, SaaS). The results of this SLR show that a clear gap exists in this research field: a lack of empirical studies focusing on utilizing cloud computing within educational institutions.

Keywords: Systematic review · Cloud computing · Education system · Universities

1   Introduction

U.S National Institute of Standards and Technology (NIST), department of commerce defines cloud computing as a model for enabling ubiquitous, convenient, on-demand network access to a shared pool of configurable computing resources (e.g., networks, servers, storage, applications, and services) that can be rapidly provisioned and released with minimal management effort or service provider interaction [19]. Gartner explains cloud computing as a style of computing in which scalable and elastic IT-enabled capabilities are delivered as a service using Internet technologies [15] [4]. According to [15], for IT professionals cloud computing is an innovative business form and a new technology platform for developing and deploying applications and for end-users is a new tendency and cheaper way to use applications. By providing many applications and services in the cloud to the learners and teachers which can be used for educational purposes cloud computing allows greater flexibility and mobility in the use of resources for teaching purposes. The fact that cloud computing is based on the internet brings many advantages and some limitations as well to the education systems. Some of the benefits of cloud computing for academic institutions are economics, elasticity, enhanced availability, lower environmental impact, concentration

O. Gervasi et al. (Eds.): ICCSA 2015, Part IV, LNCS 9158, pp. 725–737, 2015.
DOI: 10.1007/978-3-319-21410-8_55

on core basics and end user satisfaction [15], [32], [40] but it has some risks such as security [37], data security, un-solicited advertisement, lock-in, functionality, platform, and technical issues [32]. Furthermore, cloud computing is usable in various activities of everyday life, including in education. In addition to providing students and teachers access to many applications and services in the cloud, which can be used in formal and informal education, cloud computing allows for greater flexibility and mobility in the use of resources for teaching and learning, greater degree of collaboration, communication and sharing of resources, and creates a customized learning environment or virtual communities of learning and teaching [10], [15].

Current research on the use of cloud computing in education mainly focused on cloud computing frameworks, security, and implementation and has not mainly touched the use and the adoption of cloud computing in education [9-12].

The aim of this paper is to identify and analyze empirical evidence related to the use or adoption of cloud computing in education context using a systematic review method. This study will contribute to identify the challenges and gaps in current research and suggest areas for further investigation.

2   Related Work

Cloud computing is referred to as a networked infrastructure software which has the capacity of providing resources to users in on-demand environment. Furthermore, researchers have reported that information is stored in centralized servers and cached temporally on clients' devices like desktops, computers and other devices [9], [19]. Cloud computing infrastructure can be in the company's Datacenters (internal cloud) or off-premise of the company and can be used through service providers (external cloud). The subscription of cloud computing internally or externally is based on the company's pay-per-use or financial standards [9], [19], [32]. Moreover, NIST views cloud computing as having three service models, four deployment models and five essential characteristics. NIST's three service models for cloud computing are: Software as a service (SaaS), Platform as a Service (PaaS) and Infrastructure as a service (IaaS). NIST also considered four deployments to the cloud computing which are: Public cloud, Private Cloud, Community Cloud and Hybrid cloud. Lastly, NIST proposed five essential characteristics for the cloud computing which are: on-demand self-service, a broad network access (mobiles, tablets and notebooks) resource pooling, rapid elasticity and a service measured and billed according to the customer utilization[9], [19], [32].

Masud and Huang (2012) [16] have found out that traditional education systems construction and maintenance are located inside the educational institution which will cost universities to pay a lot. In addition, cloud based education systems model introduces scale efficiency mechanism the construction is entrusted to the cloud computing suppliers. Cloud computing is becoming adoptable for many organizations because of it scalability and usage of virtualized resources as a service through the internet [16] [9]. It is believed to have a significant impact on the future education. Cloud based education management systems will be an alternative for many universities which are under budget constraint. In his research, Ercan [9] has summarized a survey done by Gartner in 2009 which resulted that universities and educational institutions are not among top users of cloud computing (i.e. only 4%). Higher usage of cloud computing come from industrial sector such as financial services (12%), business and management services (10%) and manufacturing services (10%).

In addition, some other researchers [10], [18] have summarized the current situation of cloud computing implementation in education institutions especially in the developing countries as follows:

- There is a huge imbalance between developed regions and undeveloped regions in which some schools cannot afford any education systems.
- Some of the institutions do not have the technical expertise to support, operate and maintain their cloud infrastructure
- Some of the institutions perceived that available systems in the market are not customizable for their use or they cannot afford application system that they want.

Looking at this scenario, we are motivated to investigate the current state of research on cloud computing usage or adoption within education sector.

3   Research Method

Systematic Literature Review (SLR) is defined as a procedure of identifying, assessing and interpreting all available research evidence with the goal to provide answers for particular research questions. In this SLR we have followed the procedures of Kitchenham and Charters (2007) [14], [31]. The search process involves the use of five online databases: IEEExplore, ACM, Scopus, Springer, and ScienceDirect.

3.1   Research Questions

In our SLR, we included all empirical studies that investigated cloud computing adoption in higher education. We have selected research papers that have investigated the usage and/or implementation of cloud computing in education as well as researches that have come out with specific outcome/product implementation for cloud computing in education sector. Therefore, we did not include researches that merely describe a proposed framework or design without any empirical evaluation. In other words, we only consider studies that have investigated and analyzed data about cloud computing adoption in education or studies that have produced a product that shows the implementation of cloud computing in education. The primary focus of our SLR was to understand and identify whether universities and higher education institutions adopted cloud computing in their education management and E-learning systems. While the primary reason for using cloud computing in education is to gain benefits in terms of economic advantage (e.g. pay per usage, maintenance) [19], [32], students also can benefit it by not needing high processor computers because they can do their tasks in the cloud. We organized the measurement of cloud computing adoption into two categories: whether universities adopted cloud computing in their education systems and research's specific outcome/product implementation in cloud computing in education. Therefore, our SLR aimed to answer the following primary research question (RQ):

Primary question: *What evidence is there about cloud computing adoption in higher education institutions?*

Our SLR also aimed to answer the following secondary sub questions:

Sub question 1: *Which cloud computing service models (IaaS, PaaS, SaaS) are most commonly investigated by researchers and/or used by the universities?*

Sub question 2: *What is the specific outcome/product/framework or implementation of cloud computing model or technology?*

3.2   Identification of Relevant Literature

The strategy we used to create the search strings was as follows [14] [31]:

- Finding papers about cloud computing in education sector.
- Listing keywords mentioned in primary studies which we knew about.
- Use synonyms word (usage) and sub subjects of cloud computing in education such as (E-learning, management systems in education).
- Use the Boolean OR to incorporate alternative spellings and synonyms.
- Use the Boolean AND to link the major terms from population, intervention, and outcome.

The complete search string initially used for the searching of the literature was as follows:

Cloud computing AND education. It has been highlighted in [26] [31] that there are two main issues on conducting an SLR search which are the sensitivity and specificity of the search. The sensitivity is when a search retrieves a high number of relevant studies. Specificity leads the search to retrieve a minimum number of irrelevant studies. In our preliminary search, when we used the complete search string defined above we retrieved a very high number of articles. For instance, IEEEXplore, ProQuest education, Science Direct, Scopus and Springer Link retrieved more than two hundred results. Therefore, we have deepened our search and used this search string: (*Cloud OR "cloud computing"*) AND (*Adoption OR Usage*) AND (*Education OR E-learning*). The revised search string has given us a reasonable number of studies and we finally selected relevant empirical studies

3.3   Selection of Studies

In this SLR, we applied the following inclusion criteria: 1) studies that investigated and collected data about cloud computing adoption or usage in education, 2) studies that have come out with specific outcome/ implemented a cloud-based product for use in education institution, 3) Papers written in English. We excluded studies that did not provide empirical data or evidence about cloud usage or adoption and studies that presented authors' opinion or proposed framework about cloud computing implementation without supporting evidence.

3.4   Data Extraction

To ease the data extraction process, a form was designed that was used to gather evidence relating to our research questions and to measure the quality of the primary studies. The data extraction form consists of the following items: paper ID, aim of study, research method, hypothesis, research question, analysis method, sample size, summary of findings, limitations and specific outcome or cloud implementation. The quality assessment was performed during data extraction phase and we refer to the quality checklist reported in [14] and [31].

4   Results and Discussion

In this section we present the results of our SLR, starting with analysis from the literature search results. We have chosen the IEEExplore database as the baseline database. In addition, each article retrieved from the other databases was compared with the existing list of papers accumulated from IEEExplore screening process in order to avoid duplication. The initial phase of our search process identified 569 primary studies. We have considered 27 relevant papers for inclusion in this SLR based on the screening of titles and abstracts and removing duplicates. Each of these studies was filtered according to the inclusion and exclusion criteria before being accepted for the synthesis of evidence. We also carefully checked if there were any duplicate studies or if very similar studies were published in more than one paper. Based on the primary searches, 27 studies were accepted for the synthesis of evidence after a detailed assessment of abstracts and full text and exclusion of duplicates. Note that this SLR only included published studies indexed in online databases including conference papers and journal articles. List of studies included in this SLR is shown in Table 1.

Table 1. List of Included Studies

Author(s)	Title	Country
Morgado & Schmidt, R. [23]	Increasing Moodle Resources Through Cloud	Brazil
Meske et al. [20]	Cloud storage services in higher education - Results of a preliminary study in the context of the Sync&Share-project in Germany	Germany
Tantatsanawong et al. [36]	Enabling Future Education with Smart Services	Thailand
Vaquero [38]	EduCloud: PaaS versus IaaS Cloud Usage for an Advanced Computer Science Course	Spain
Nguyen et al. [24]	Acceptance and Use of E-Learning Based on Cloud Computing: The Role of Consumer Innovativeness	Vietnam
Selviandro et al. [33]	Open Learning Optimization Based on Cloud Technology: Case Study Implementation in Personalization E-learning	Indonesia
Ratten [28]	Cloud computing: A social cognitive perspective of ethics, entrepreneurship, technology marketing, computer self-efficacy and outcome expectancy on behavioural intentions	Australia

Table 1. (*Continued*)

Chandran & Kempegowda [6]	Hybrid E-learning Platform based on Cloud Architecture Model: A Proposal	Australia
Pandian & Kasiviswanathan [25]	Effective use of cloud computing concepts in engineering college	India
Masud & Huang [17]	A Novel Approach for Adopting Cloud-based E-learning System	Australia
Alabbadi [2]	Cloud computing for education and learning: Education and Learning as a Service (ELaaS)	Saudi Arabia
Chang et al. [7]	Implications of learning cloud for education: From the perspective of Learners	Taiwan
Mokhtar et al. [22]	Organizational Factors in the Adoption of Cloud Computing in E-learning	Malaysia
Smith et al. [34]	Cloud computing: adoption considerations for business and education	UK
Masud et al. [18]	Cloud Computing for Higher Education: A Roadmap	Australia
Kihara & Gichoya [13]	Use of Cloud Computing Platform for E-Learning in Institutions of Higher Learning in Kenya	Kenya
Akande & Belle [1]	Cloud Computing in Higher Education: A snapshot of Software as a Service	South Africa
Rodrick & Mwangoka [30]	Road Map towards Eco-efficient Cloud Computing Adoption in Higher Learning Institutions in Tanzania	Tanzania
Ewuzie & Usoro [10]	Exploration of cloud computing adoption for e-learning in higher education	UK
Misevicien et al. [21]	Application of Cloud Computing at KTU: MS Live@Edu Case	Kaunas, Lithuania
Wu et al. [41]	Factors hindering acceptance of using cloud services in university: a case study	China
Brandabur [5]	Cloud Computing in Romanian Educational Environment- A Qualitative Research	Romania
Velicanu et al. [39]	Cloud E-learning	Romania
Despotović-Zrakić et al. [8]	Scaffolding Environment for e-Learning through Cloud Computing	Serbia
Surya & Surendro [35]	E-Readiness Framework for Cloud Computing Adoption in Higher Education	Indonesia
Ramachandran et al. [27]	Selecting a suitable Cloud Computing technology deployment model for an academic institute- A case study	India
Atchariyachanvanich et al. [3]	What Makes University Students Use Cloud-based E-Learning?: Case Study of KMITL Students	Thailand

4.1 Answering the Research Questions

Primary Question: *What evidence is there about cloud computing adoption in higher education institutions?*

From the synthesis of evidence, we found there are fourteen (14) studies that have investigated various aspects relating to adoption of cloud computing in educational sector [5], [6], [10], [13], [17], [18], [20], [22], [23], [24], [28], [30], [35], [36]. In general, studies have shown that there is a growing interest from students and universities to move to the cloud. In [23], the authors mentioned that universities are concerning using Moodle resources and moving the Moodle to the cloud. They reported that there are 719 downloads in less than 14 months for virtualization of Moodle resources which indicates that there are a lot of interest in this cloud technology.

In [36], the authors described about the infrastructure services for Thai education and they pointed out four principles that the cloud infrastructure services must provide which are: *availability, accessibility, affordability,* and *administratively.* They also reported about the education information services which have to support and provide data integrity to the schools, universities, parents and the ministry of education. Finally, the paper has given preview about the learning services which provides cloud computing facilities for cyber learning system, digital library and educational broadcasting services called E-TV and Teacher TV. The services allow teachers to provide contents anywhere, anytime, easy to access and use, create and reuse. In addition, learning cloud services has several integrated applications which include: personal website, interactive application, picture and video sharing, social network application, e-book and e-learning, peer-to-peer sharing application, collaborative tools such as email, messaging, blog, Skype and video conferencing system. The services also provide reference database service, union catalog, digital collection, E-journal and e-book directory.

In [24], the study has found out that adoption of cloud computing is significantly influenced by four factors: performance expectancy, social influence, hedonic motivation and habit. Using the UTAUT2 model, they have conducted a survey responded by 282 participants which aimed to empirically validated the factors that influence consumer intention and use of cloud-based E-learning system. The results also showed that demographic characteristics such as age, gender, education and experience moderate the effects of UTAUT2 predictors on cloud-based e-learning intention and cloud-based e-learning usage.

Masud & Huang [17] proposed an interoperable framework known as *EducationCloud* for academic institutions to adopt cloud computing. The framework has an open structure and divided into management subsystem and service subsystem. The management subsystem uses virtualization technology to coordinate actions between the multilevel frameworks and facilitate the sharing of computing resources. On the other hands, the service subsystem comprised the execution cloud and the storage cloud. Some of the expected functionalities of the framework include: ESaaS, Digital Library, Online Storage, Collaboration, Access, Interoperability, Provisioning, and Security/Privacy.

In [22], the study aims to investigate the organizational factors influencing the adoption of cloud computing in education focusing on the e-learning system. The authors proposed four (4) organizational elements that should be considered for adopting cloud computing in education institutions: a) needs assessment, b) readiness assessment, c) organisational change, d) budgeting and return on investment (ROI).

Masud et al. [18] explore the salient features of the nature and educational potential of cloud computing in order to exploit the affordance of cloud computing in teaching and learning in a higher education context. A roadmap for cloud computing in higher education with a focus on the adoption strategies is presented in the previous work of same authors. Following the steps described in the roadmap, successful cloud computing adoption can be achieved in higher education.

Kihara & Gichoya [13] investigate the significant issues related to the Strengths, Weaknesses, Opportunities and Threats (SWOT) of cloud computing adoption in Institutions of Higher Learning in Kenya for hosting their e-learning resources, e-library services and digital repositories. They adopted a SWOT analysis to assess the implementation level of this technology and presented a framework for adoption and realization of this technology in institutions of Higher Learning. They conclude that institutions of Higher Learning that use cloud computing to facilitate e-learning have students that perform better and therefore have relevant knowledge and skills that the market needs.

Rodrick & Mwangoka [30] investigates the traditional computing environment in higher learning institutions in Tanzania (HLIT) in order to identify requirements for establishing eco-efficient cloud computing solutions. They proposed necessary stages to create eco-cloud awareness for HLIT to effectively move into cloud computing. The stages start with strategic preparation, planning and design, implementation and finally optimization of cloud resources.

In Romania, the perceived usage of cloud computing paradigm in academic environment has been studied by Brandabur [5]. The study conducted sixteen interviews with the vice-rectors from main universities in Romania. The study found out that limited knowledge about cloud computing solutions, lack of vision, huge bureaucratic decisional process and lack of interest in cloud computing are the causes that hinder Romanian universities to adopt cloud computing solutions.

Ewuzie & Usoro [10] performed an early exploration of cloud computing adoption for E-learning in higher education. They highlighted the concepts of E-learning on the cloud, research gaps in adopting cloud computing in higher education and outlined several proposed research objectives.

Surya & Surendro [35] presented empirical evidence on the implementation of eReadiness framework to measure the degree of IT readiness level prior to adopting cloud computing technology. By using IT governance COBIT 5, the assessment indicates that the higher education readiness level of cloud computing adoption in West of Java is still at a low level. They noted important aspects to be considered for adopting cloud computing in higher education such as the need to understand Cloud computing, how it should be developed and evaluated, and conduct analysis on how cloud computing will influence the strategy of the organization.

In [28], the main purpose of this study is to examine behavioral intentions towards cloud computing in an educational setting through the use of social cognitive theory. The research was conducted by using qualitative (in-depth interviews, focus groups) and quantitative (Survey questionnaire) methods and had participated by 183 respondents. The results showed that social cognitive theory can improve understanding the main internal and external drivers of increasing an individual's intention to adopt cloud computing as a learning instrument. Finally, in [20], the paper describes the Sync&Share NRW-project in North Rhine-Westphalia in Germany with a target audience of up to 500,000 users and presents the main results of a preliminary large-scale survey at the University of Muenster with more than 3,000 participants. The result of the study shows that German higher education demands internal cloud service solution with mobile access, with big storage, collaborative features such as simultaneous work on text documents and high data processing standards.

Sub question 1. *Which cloud computing service models (IaaS, PaaS, SaaS) are commonly investigated by researchers and/or used by the universities?*

In [38], the paper investigates whether the use of IaaS or PaaS (as infrastructure for the assignment) increases or decreases the performance of the students and how they perceive the use of IaaS and PaaS. This paper which reported the results from controlled experiment with 84 participants found out that the introduction of cloud computing helps to keep students focus. A preliminary comparison of students subject to the first scenario in their sequence revealed minor differences between these technologies (control, IaaS, or PaaS). This study has also suggested that universities are more interested to use software as a service (SaaS).

Akande & Belle [1] report that *Software as a Service* (SaaS) is the most common service model used by higher educational institutions. In their study, they have explored the use of SaaS in South African higher educational institutions with the purpose to determine whether SaaS is a viable option for higher educational institutions and to identify the benefits and limitations of SaaS. Despite the high level of awareness rate on SaaS, the implementation of SaaS is still lacking due to the lack of fast and affordable internet services. They also found that the two most commonly SaaS ERP system being used among South Africans higher education institutions are Oracle PeopleSoft and Office 365.

Sub question 2. *What is the specific outcome/product/framework or implementation of cloud computing model or technology?*

Some researches has produced specific products or implemented specific product related to adopting cloud computing in their education system. In [33], the article discusses the learning architecture and the basic concept of open educational resources, the proposed open learning, the approach of the implementation, experiment in personalization learning and the evaluation. The study has come out with an architecture called IOER. The term open learning is used in order to encouraging the development of the concept of Indonesia Open Educational Resources (IOER) and as well as the adoption of concept of cloud computing. It was found out in their evaluation that the use of cloud computing in open learning meets the user needs. In accessibility, cloud-Based Open Learning has provided easy access to the users. The results of the evaluation shows that by implementing the cloud based open learning portal could decrease the investment cost up to 59% when compared to non-cloud E-learning system. In [20], they produced Sync&Share NRW-project in North Rhine-Westphalia (Germany) with a target audience of up to 500,000 users.

Chandran & Kempegowda [6] proposed hybrid e-learning platform based on existing E-learning architecture models. The authors also analyzed issues in current e-learning applications and discuss advantages of cloud computing adoption, e-learning platforms, cloud delivery models and proposed three scenarios of using hybrid cloud-computing. The first scenario involves migrating the stand alone system to virtual environment, the second is adoption of e-learning application to be introduced for an institution and thirdly introduction of new imitative program similar to The National Assessment Program for Literacy and Numeracy (NAPLAN).

Pandian & Kasiviswanathan [25] proposed a strategy to implement cloud-computing project at Tirunelveli, a city located in the South of Tamilnadu, India where a number of engineering colleges are available in the area. Cloud computing implementation will enable remote sharing of resources in particular high-end software connected through the use of high speed networks such as NKN (National Knowledge Network). The study proposed setting up the new centralized server for providing easy access of software to colleges students.

4.2   Implications for Research

Our SLR found out that there is considerably less empirical evidence about cloud computing adoption in education context. Most papers we found had focused on frameworks and implementations of cloud services. Some studies have reported factors that influence cloud computing in education[23], [28], [33] while others produced and implemented specific systems for cloud computing in education [6] ,[38]. Finally some studies suggested that universities are interested in all cloud computing service models (IaaS, PaaS, SaaS) while others reported that universities are more into software as service (SaaS) [20] [38].

5   Conclusions

Cloud computing stands as a new paradigm of providing IT services which includes rental of resources located in the cloud and in the future of IT development .Today we are increasingly doing our work online, from checking emails, using other forms of communication, writing and editing documents and collaboration, through watching movies and videos up to storing personal documents and images online. Cloud computing usability in education is very wide and it is recognized by many educational institutions around the world. Financial nature is the main reason for the worldwide introduction of cloud computing for educational establishment[15], [29] but it should be noted that cloud has also creative potentials because it enables that ideas, thoughts and knowledge to be created, used and shared easily. However, to achieve this, academic institutions, students and teachers should be willing to use services in cloud and be familiar with their advantages and limitations. In this SLR paper, we have presented current studies about usage of cloud computing in education.

The results from our SLR showed that there is a growing interest in utilizing cloud computing technology in various universities. In addition, some researchers have come out with specific systems or cloud implementation such as IOER in Indonesia. Finally, we found less empirical evidence have been reported about cloud computing adoption in education and future researchers should give attention to cloud computing usage in education and find out the current systems that universities are using prior to developing a cloud framework or implement a cloud based educational system.

References

1. Akande, A.O., Van Belle, J.-P.: Cloud computing in higher education: a snapshot of software as a service. In: IEEE 6th International Conference on Adaptive Science & Technology (ICAST), pp. 1–5 (2014)
2. Alabbadi, M.M.: Cloud computing for education and learning: education and learning as a service (ELaaS). In: 14th International Conference on Interactive Collaborative Learning (ICL2011), pp. 589–594 (2011)
3. Atchariyachanvanich, K., Siripujaka, N., Jaiwong, N.: What makes university students use cloud-based e-learning?: case study of KMITL students. In: International Conference on Information Society (i-Society), pp. 112–116 (2014)
4. Bittman, T.: Cloud Computing Inquiries at Gartner (2009). http://blogs.gartner.com/thomas_bittman/2009/10/29/cloud-computing-inquiries-at-gartner/
5. Brandabur, R.E.: Cloud computing in romanian educational environment- a qualitative research. In: 9th International Scientific Conference eLearning and software for Education Burchares, pp. 290–296 (2013)
6. Chandran, D., Kempegowda, S.: Hybrid E-learning platform based on cloud architecture model: a proposal. In: 2010 International Conference on Signal and Image Processing (ICSIP), pp. 534–537 (2010)
7. Chang, C.-S., Chen, T.-S., Hsu, H.L.: Implications of learning cloud for education: from the perspective of learners. In: 2012 7th IEEE International Conference on Wireless, Mobile and Ubiquitous Technology in Education, pp. 157–161 (2012)
8. Despotović-Zrakić, M., Simić, K., Labus, A., Milić, A., Jovanić, B.: Scaffolding Environment for e-Learning through Cloud Computting. Educational Technology & Society **16**(3), 301–314 (2013)
9. Ercan, T.: Effective use of cloud computing in educational institutions, ScienceDirect, vol. II, pp. 938–994 (2012)
10. Ewuzie, I., Usoro, A.: Exploration of cloud computing adoption for e-learning in higher education. In: IEEE Second Symposium on Network Cloud Computing and Applications, pp. 151–154 (2012)
11. Fernandez, A., Peralta, D., Herrera, F., Benitez, J.M.: An overview of E-learning in cloud computing. In: Proc. Workshop on Learning Technology for Education Cloud (LTEC 2012), pp. 35–46 (2012)
12. Hussain, R.M.R.: eLearning in Higher Education Institutions in Malaysia, International, E-Mentor **5**(7) (2004)
13. Kihara, T., Gichoya, D.: Use of cloud computing platform for E-Learning in institutions of higher learning in kenya. In: IST-Africa 2014 Conference Proceedings (2014)
14. Kitchenham, B.A., Charters, S.: Procedures for Performing Systematic Literature Review in Software Engineering, Software Eng. Group, Keele Univ., Univ. of Durham, Durham, EBSE Technical Report version 2.3 EBSE-2007-01 (2007)
15. Kurelović, E.K., Rako, S., Tomljanović, J.: IT Glossary. In: MIPRO 2013, Opatija, pp. 726–731 (2013). http://www.gartner.com/itglossary/
16. Masud, M.A.H., Huang, X.: An E-learning System Architecture based on Cloud Computing. World Academy of Science, Engineering and Technology, 74–78 (2012)

17. Masud, M.A.H, Huang, X.: A novel approach for adopting cloud-based E-learning system. In: 2012 IEEE/ACIS 11th International Conference on Computer and Information Science, pp 37–42 (2012)
18. Masud, M.A.H, Yong, J, Huang, X.: Cloud computing for higher education: a roadmap. In: Proceedings of the IEEE 16th International Conference on Computer Supported Cooperative Work in Design, pp. 552–557 (2012)
19. Mel, P., Grance, T.: The NIST Definition of Cloud Computing, Computer Security Division, National Institute of Standards and Technology (2011)
20. Meske, C.A., Stieglitz, S.A., Vogl, R.B., Rudolph, D.B.,Öksüz, A.A.: Cloud storage services in higher education - results of a preliminary study in the context of the sync & share-project in Germany. In: 1st International Conference on Learning and Collaboration Technologies, LCT 2014, Greece, pp. 161–171 (2014)
21. Misevicien, R., Budnikas, G., Ambrazien, D.: Application of Cloud Computing at KTU: MS Live@Edu Case. Informatics in Education 10(2), 259–270 (2011)
22. Mokhtar, S.A., Ali, S.H.S., Al-Sharafi,A., Aborujilah, A.: Organizational factors in the adoption of cloud computing. In: E-learning, 3rd International Conference on Advanced Computer Science Applications and Technologies, pp. 188–191 (2014)
23. Morgado, E.M., Schmidt, R.: Increasing moodle resources through cloud. In: Proc. Iberian Conference on Information Systems and Technologies (CISTI), Madrid, pp. 1–4 (2012)
24. Nguyen, T.D., Nguyen, T.M., Pham, Q.-T., Misra, S.: Acceptance and use of E-Learning based on cloud computing: the role of consumer innovativeness. In: Murgante, B., Misra, S., Rocha, A.M.A., Torre, C., Rocha, J.G., Falcão, M.I., Taniar, D., Apduhan, B.O., Gervasi, O. (eds.) ICCSA 2014, Part V. LNCS, vol. 8583, pp. 159–174. Springer, Heidelberg (2014)
25. Pandian, R.S.R., Kasiviswanathan, K.S.: Effective use of cloud computing concepts in engineering college. In: IEEE International Conference on Technology for Education, pp. 233–236 (2011)
26. Petticrew, M., Roberts, H.: Systematic Review in the Social, A Practical Guide. Blackwell Publishing (2006)
27. Ramachandran, N., Sivaprakasam, P., Thangamani, G., Anand, G.: Selecting a suitable Cloud Computing technology deployment model for an academic institute- A case study. Campus-Wide Information Systems 31(5), 319–345 (2014)
28. Ratten, V.: Cloud computing: A social cognitive perspective of ethics, entrepreneurship, technology marketing, computer self-efficacy and outcome expectancy on behavioural intentions. Australasian Marketing Journal 21, 137–146 (2013)
29. Razak, S.F.A.: Cloud computing in malaysia universities. In: Conference on Innovative Technologies in Intelligent Systems and Industrial Applications (CITISIA 2009), Kuala Lumpur (2009)
30. Rodrick, M., Mwangoka, J.: Road map towards eco-efficient cloud computing adoption in higher learning institutions in tanzania. In: Pan African International Conference on Science, Computing and Telecommunications, pp. 154–159 (2014)
31. Salleh, N., Mendes, E., Grundy, J.G.: Empirical Studies of Pair Programming for CS/SE Teaching in Higher Education: A Systematic Literature Review. IEEE Transactions on Software Engineering 37(4), 509–523 (2011)
32. Sclater, N.: Cloud Computing in Education, UNESCO Institute (2010)
33. Selviandro, N., Suryani, M., Hasibuan, Z.A.: Open learning optimization based on cloud technology: case study implementation in personalization E-learning. In: Proc. 16th International Conference on Advanced Communication Technology (ICACT), Pyeongchang, pp. 541–546 (2014)

34. Smith, A., Bhogal, J., Sharma, M.: Cloud computing: adoption considerations for business and education. In: International Conference on Future Internet of Things and Cloud, pp. 302–307 (2014)
35. Surya, S.F.G., Surendro, K.: E-Readiness framework for cloud computing adoption in higher education. In: International Conference of Advanced Informatics: Concept, Theory and Application (ICAICTA), pp 278–282 (2014)
36. Tantatsanawong, P., Kawtrakul,A., Lertwipatrakul, W.: Enabling future education with smart services. In: Proc. SRII Global Conference, San Jose, pp. 550–556 (2011)
37. Tribhuwan, M.R., Bhuyar, V.A., Pirzade, S.: Ensuring cloud computing security through two-way handshake on token management. In: Proc. International Conference on Advances in Recent Technologies in Communication and Computing (ARTCom), pp. 386–389 (2010)
38. Vaquero, L.M.: EduCloud: PaaS versus IaaS Cloud Usage for an Advanced Computer Science Course. IEEE Transactions on Education **54**(4), 590–598 (2011)
39. Velicanu, A., Lungu, I., Diaconita, V., Nisioiu, C.: Cloud E-learning. In: Conference proceedings of eLearning and Software for Education (eLSE), pp. 380–385 (2013)
40. Verma, K., Rizvi, M.A.: Impact of cloud on E-learning. In: 5th International Conference on Computational Intelligence and Communication Networks, pp. 480–485 (2013)
41. Wu, W.-W., Lan, L.W., Lee, Y.-T.: Factors hindering acceptance of using cloud services in university: a case study. The Electronic Library **31**(1), 84–98 (2013)

A Review of Student Attendance System Using Near-Field Communication (NFC) Technology

Mohd Ameer Hakim bin Mohd Nasir[1], Muhammad Hazimuddin bin Asmuni[1],
Norsaremah Salleh[1], and Sanjay Misra[2(✉)]

[1] International Islamic University Malaysia, Selangor, Malaysia
{amer.hkm,hazim.asmuni}@gmail.com, norsaremah@iium.edu.my
[2] Covenant University, Ota, Nigeria
sanjay.misra@covenantuniversity.edu.ng

Abstract. The rapid growth of system development is no longer subtle and continuously improving today's system. In education sector, the student attendance system is able to be applied by Near-Field Communication (NFC) technology. NFC can be referred to as a device that can detect information and/or command from a tag by bringing them together in a close proximity or even by touching together. Traditionally, the manual attendance system would require a lecturer to pass around an attendance sheet for students to sign beside their names and another method would require the lecturer to call out the students' names one by one and register their attendance. The attendance system based on NFC is meant to improve the manual attendance system and therefore the aim of this paper is to review the existing research on this topic.

Keywords: NFC · RFID · Attendance system · Systematic review

1   Introduction

Near-Field Communication (NFC) has been growing and developing endlessly throughout the globe. It has been introduced and used in several modern educational institutions where its services are proved beneficial [8], [9], [12]. The intention of applying NFC in an educational institution is for the purpose of observing and improving student's attendance. Applying NFC could function as a platform that drives institutions towards a modern environment through innovative solutions.

There are various inventions that allowed NFC to become a powerful tool which only needs proper requirements and installation. In this regard, related work around the globe has innovatively provided direct and indirect realization in creating NFC. Biometrics, web-based attendance, and bar scanning are some of the earliest inventions that functions almost similar as NFC[1], [3], [4],

The traditional process of taking student's attendance is time consuming and takes too much effort especially to record and to maintain it. Lecturers are required to call out the names of each student one by one and this process would consume time especially for a large class. Another method is to distribute an attendance sheet where students can check their credentials and sign it. However, the only drawback is that

© Springer International Publishing Switzerland 2015
O. Gervasi et al. (Eds.): ICCSA 2015, Part IV, LNCS 9158, pp. 738–749, 2015.
DOI: 10.1007/978-3-319-21410-8_56

students can be dishonest by signing on their friends' behalf. In general, the manual or traditional way in recording attendance is usually time consuming and may distract the teaching process [5].

In creating NFC-based system, the two main components required are "reader/ writer" and a "tag" [5]. NFC covers standard communication protocol and data exchange formats that is based on Radio Frequency Identification, also known as RFID [5], [6]. The introduction of NFC technology in students' attendance system will likely have a significant impact on the teaching and learning environment [21]. The expectation of using NFC based attendance system is also to utilize the two items that students always have and should always bring; i.e. their smart phone and matric cards.

This paper review previous researches made for a variety of purposes related to the use of NFC in registering attendance. To achieve this objective, we have carried out a systematic literature review to identify all existing research evidence that is relevant to this topic. This study can contribute to the body of knowledge of the use or adoption of NFC in education institution in particular for managing attendance or participants in an event/class. Section 2 describes the related work contributing to NFC research and Section 3 describes the review method that we have followed in our research. Our review and analysis on existing research are available in Section 4. Finally, Section 5 concludes our study.

2   Related Work

NFC technology and student attendance system has existed for quite some time. We refer to NFC as a system that transfers data in a short distance wirelessly and the standards communication protocol are based on existing RFID standards [5]. The introduction of NFC technology allows developers to write applications such as attendance system, ticketing, cashless payment, and membership authentication [6].

Biometric is another contactless technology that has been successfully commercialized worldwide where it uses the physical and behavioral aspect of a person to secure authentication. Using retina, voice, and/or thumbprint allows the biometric system to uniquely distinguished one person from the other [4]. However, one of the issues with implementing biometric in student attendance system is that its high-tech nature requires quite an expensive tool; therefore their implementation might not be cost-effective for an institution.

One of the oldest technological break-through was the barcode system [4]. It is far more affordable than the biometric system and is still effective in certain aspects. Barcode uses unique identification of retrieving data and make use of symbols generally, like a bar, vertical, space, square and dots which contains different widths. Some institutions use this method but it was reported to be less reliable [4].

The most promising research conducted by several different researchers is about Radio Frequency Identification (RFID) [4], [1]. It has been reported that RFID is one of the most suitable technology for a student environment particularly for an attendance system. For instance, Mohamed et al. [1] report that the innovation of a system that records data automatically in universities using RFID is a suitable solution to their problem in recording students' attendance.

One of the examples from existing projects related to NFC is the TouchIn NFC supported attendance system [5]. The system focuses on using NFC for improving the manual approach of recording attendance. The system is equipped with NFC tagging that keeps the attendance record using web-based method. The system components included the reader unit and server unit; reader unit reads the tag (student's credentials) and the server unit records it.

Another project that consists of using NFC for attendance is the *Attendance System using NFC Technology and Embedded Camera Device on Mobile Phone* reported by Subpratatsavee et al. [6]. In their study, they have increases the level of security of the previous research by making use of the camera located on the back of every smartphone. To avoid students passing around their NFC tag for attendance, the camera is used for verification that whoever tags it must be the same person as in the photo, which of course would be monitored by the lecturer or administrator. This project uses the concept of one centralized server to store its data efficiently while all related mobile phones connected to the server. Users can also import and export data effortlessly.

Bucicoiu & Tapus (2013) [7] propose a location-based authentication for attendance system using NFC technology integrated with Moodle. The system exploits both NFC and pictures to ensure double verification on student attendance. The unique part about this project is that it uses a Moodle, which is one of the most popular e-learning platforms. By applying Moodle, lecturers can verify their students more effectively and a lot faster than verifying them individually.

Enabling various functions and creative uses of NFC will further improve and enhance previous research in terms of what is lacking in certain fields for instance in the security and performance aspects [2]. Extracting information on what these researches are providing and what they are lacking are the key to pushing ourselves towards innovatively implementing and improving NFC-based student attendance system. In the next Section, we describe about the method we used in identifying studies related to the use of NFC in managing students' attendance.

3   Research Method

We used a systematic literature method to systematically identify, evaluate, and analyze all available studies on NFC technology on attendance system in the education sector. The research question addressed in this paper is *"What studies are available regarding the use of NFC technology in university/school attendance system?"* As suggested by Petticrew & Roberts [23], the formulation of the research question was based on the five elements known as PICOC – *Population, Intervention, Comparison, Outcome* and *Context* (see Table 1).

We followed the guidelines by Kitchenham & Charters (2007) in conducting this SLR [10]. The search process involves the use of five online databases: IEEExplore, ACM, Scopus, Springer, and ScienceDirect. The search terms used were *"NFC AND attendance"*. The breakdown of literature searches from online databases is shown in Table 2.

Table 1. PICOC Elements

Population	Studies reporting about attendance system
Intervention	Use of NFC technology
Comparison	N/A
Outcome	Implementation or proposed attendance system
Context	Research conducted within academic setting

Table 2. Breakdown of Literature Search

Database Name	# Studies for screening (A)	# Excluded Studies (B)*	# Relevant Studies (A – B)
IEEEXplore	10	0	10
ACM	10	6	4
Scopus	5	2	3
SpringerLink	50	49	1
ScienceDirect	66	63	3
TOTAL	141	120	21

(* *after screening of Titles and Abstracts*)

Our major inclusion criteria is to include studies that have included NFC technology in their research paper relating to monitoring students' attendance or participants. We excluded studies that were not conducted in academic setting or commercial based attendance system. Additionally, we excluded research that appears as work-in-progress, posters, and papers not written in English.

During the initial phase of search, we found a total of 21 studies related to NFC and attendance system. However, after reading the full text articles and removing duplicates, only 17 studies can be included for analysis. Note that this review only included published studies indexed in online databases including conference papers and journal articles. List of studies included in this review is shown in Table 3.

Table 3. List of included studies

Author(s) & year	Title	Country
Chavira et al. (2007) [8]	Spontaneous Interaction on Context-Aware Public Display: An NFC and Infrared Sensor approach	Spain
Ervasti et al. (2009) [9]	Bringing Technology into School – NFC-enabled School Attendance Supervision	Finland
Rahnama et al., (2010) [11]	Securing RFID-Based Authentication Systems Using ParseKey+	Turkey
Ervasti et al. (2009) [12]	Experiences from NFC Supported School Attendance Supervision for Children	Finland
Isomursu et al. (2010) [13]	Evaluating Human Values in the Adoption of New Technology in School Environment	Finland

Table 3. (*Continued*)

Ninomiya et al. (2012) [14]	Bridging SNS ID and User Using NFC and SNS Design of NFC and SNS based event attendance management system	Japan
Ninomiya et al. (2012 [15]	Near Friends Communication Encouragement System Using NFC And SNS	Japan
Benyo et al. (2012) [16]	Student attendance monitoring at the university using NFC	Hungary
Bueno-Delgado et al. (2012) [17]	The Smart University experience: A NFC-based ubiquitous environment ()	Spain
Benyo et al. (2012) [18]	University life in contactless way - NFC use cases in academic environment	Hungary
Ichimura & Kamada, (2013) [19]	Early Discovery of Chronic Non-attenders by Using NFC Attendance Management System	Japan
Bucicoiu & Tapus (2013) [7]	Easy Attendance: Location-based authentication for students integrated with Moodle	Romania
Subpratatsavee et al. (2014) [6]	Attendance System using NFC Technology and Embedded Camera Device on Mobile Phone	Thailand
Ayu & Ahmad (2014) [5]	TouchIn: An NFC Supported Attendance System in a University Environment	Malaysia
Isomursu et al. (2011) [20]	Understanding human values in adopting new technology—A case study and methodological discussion	Finland
Fernandez et al. (2013) [21]	Control of attendance applied in higher education through mobile NFC technologies	Spain
Shen et al. (2014) [22]	Developing a NFC-equipped smart classroom: Effects on attitudes toward computer science	Taiwan

We reused some of the questions available in the literature to identify the study quality assessment. Table 4 shows the criteria used to evaluate the quality of the included studies. Each question was rated based on the following ratio scales: 1=partially; 0.5=Partially; 0= No. We used a score scale of 0 to 6: Very Poor (Score < 2), Poor (Score of 2 to <3), Fair (Score of 3 to <4), Good (Score of 4 to <5), and Very Good (Score of 5 to 6).

Table 4. Quality criteria

Item	Answer
1. Are the aims clearly stated? [10]	Yes/No/Partially
2. Is there an adequate description of the context in which the research was carried out? [10]	Yes/No/Partially
3. Was the data collection done very well? [24]	Yes/No/Partially
4. If the study involves assessment of a technology, is the technology clearly defined? [10]	Yes/No/Partially
5. How well has the approach to, and formulation of, analysis been conveyed? [10]	Yes/No/Partially
6. How has knowledge or understanding been extended by the research? [10]	Yes/No/Partially

4   Review Findings

Our analysis discovered that NFC technology has been implemented not only in higher education institutions, but it is also used in primary schools mainly in Finland [9], [12], [13], [20]. From the literature we found that NFC technology has been equipped in controlling attendance system in various countries including Spain [8], [17], [21], Japan [14], [15], [19], Hungary [16], [18], Romania [7], Turkey [11], Thailand [6], Malaysia [5], and Taiwan [22] (see Figure 1).

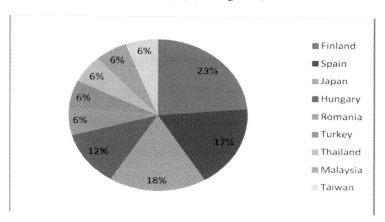

Fig. 1. Breakdown of Countries

There have been various kinds of system design based on existing researches. In each application, the design aims to interact with other components of the system effectively; based on multiple background study of previous designs. The architecture was design to ensure that each component is used optimally. General overview of the students' attendance system using NFC technology can be seen in the Figure 2. In the following, we discuss the findings from our systematic review regarding the use of NFC in student attendance system.

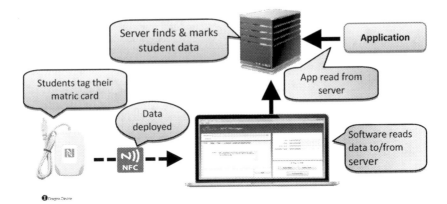

Fig. 2. Overview of NFC-based Students' Attendance System

Ayu & Ahmad (2014) [5] reported their research work that utilize NFC-enabled device to allow students to register courses that have been set up by the lecturers. The system known as *TouchIn* is equipped with two main components: the *reader unit* and the *server unit*. The reader unit is responsible for reading the NFC-enabled smartphone or a tag/card that can store information. If a student is using an NFC-enabled device, then the student ID will be stored in a file stored in the device; tagging it will allow the NFC reader to retrieve that particular file. The web server unit acts as a dedicated server for storing the students' credentials who tagged and display them on a website for recording and verification purposes.

In Thailand, Subpratatsavee et al. [6] described the implementation of NFC technology in the attendance system by incorporating a smartphone's camera capabilities. The system consists of a dedicated server for mobile devices and database. The smartphone/NFC-enabled devices can be connected to one particular server at all times, thus allowing unlimited access and mobility. The system allows lecturers to place their smartphone (acting as NFC reader) and the students can tag it with their NFC-enabled smartphone or NFC tag/card to register attendance for the current class. While tagging, the lecturer's NFC reader will quickly verify the tag and capture a picture of the person who is tagging. The reader will then send both the student's ID from their NFC-enabled device/tag/card along with the captured photo to the server. The server will be able to be accessed by the user depending on their access level (i.e. student, lecturer, or admin). The data stored on the server can be used for various purposes such as record-keeping and verification of student attendance.

Bucicoiu & Tapus (2013) [7] have proposed a location-based authentication method (based on NFC) and using a photo as an authentication in the attendance system. They utilized the existing learning management platform known as Moodle as their backend server. Moodle is an educational platform that enables services like learning on the web. It already provides a stable platform that lecturers could use for online assessment, projects, or even assignments. Combining it with the student attendance system will coherently bring online education to a whole new level. Monitoring becomes effortless, synchronization is on-the-go, and the mobility is limitless. When the first contact of tagging has been made, a session will be created between the lecturer and their students. Moodle will act as their platform for the sessions. Such a system easily facilitates interaction between students and the lecturers. The researchers have implemented the design using a prototype; Samsung Galaxy S3 with Android 4.0.4 against Moodle 2.3.1 plug-in that uses PHP language. Communication between phone and server uses XMLRPC over HTTPS. Moodle uses two-way authentication; either token or username-password in order to ensure the security and integrity of data. The GUI on client side was programmed using Java programming.

Chavira et al. [8] propose two approaches in interacting with context-aware public display: an NFC-enabled phone and infrared sensor. They have implemented a ViMOS system to test both approaches and they suggested that NFC-enable device is a more viable and effective option.

In 2008, an NFC-based attendance system was implemented at primary school in Finland involving 23 pupils [9]. The system allows parent to receive real-time information on children's attendance. The results of the study show that NFC based attendance system bring positive experience to the end-user groups (i.e. children, parents, and teachers).

Rahmana et al. [11] proposed RFID authentication system utilizing ParseKey++ multi-way authentication scheme for the purpose to strengthen the security of the system. The algorithm was implemented on RFID based attendance control system. Such a secure authentication system is needed for highly secure environment for example to access confidential resources on computer network or authentication to access high security locals.

In 2009, Ervasti et al. [12] reported a study that describes user experience in adopting attendance supervision system at primary school in Finland. The same authors have reported their work in [9]. Data were collected through classrooms observation, interviews with children and teachers, and phone interviews with parents. The findings showed that adoption of such system clearly benefits parents, children and teachers.

A similar study was reported by Isomursu et al. [13]. In this study the authors presented results from a case study of elementary school children in using attendance control system based on NFC technology. Using the same model of human values from social psychology, and based on their analysis of subjective perceived values, the authors found that children were the most satisfied user group of the system compared to parents and teachers.

While in Japan, Ninomiya et al. [14] have developed a system that matches real ID of a person with online SNS ID using NFC-enabled smart phone. The aim was to bridge the gap between an online identity on SNS and the real identity. To achieve this aim, the authors have designed an event attendance management system based on NFC and SNS using Twitter. Another study that is also conducted in Japan and reported by the same authors [15] have proposed a participant managerial system that makes use of human relation in SNS to an actual event site. The approach is to use NFC technology and linked with SNS ID and during the event day, the organizer as well as the participants can understand background and interest of each participant based on SNS information. This could aid in promoting exchange between participants in an actual event site.

Benyo et al. (2012) [16] described the design and implementation of student attendance monitoring at Budapest University of Technology and Economics in Hungary utilizing the NFC technology. The system was implemented in a highly autonomous distributed environment comprising NFC enabled contactless terminals and a scalable back-office. The terminals also support biometric identification by fingerprint reading to enhance the security aspect. In the pilot project, more than 1000 students had tested the system and they become accustomed to use the system properly.

Bueno-Delgado et al. (2012) [17] presented the overview of implementation of an ubiquitous computing platform based on NFC known as *Smart University* in Technical University of Cartagena, Spain. The two major projects development were NFC-attendance registering system and NFC administrative fee payment system. Prior to the development, data collection through opinion poll was conducted to study the impact of the use of NFC technology in a university environment. The results showed promising use of NFC applications among the university community.

Benyo et al. (2012) [18] have summarized uses of NFC applications in a contactless infrastructure at the Budapest University of Technology and Economics. These include enrolment at the University, registering attendance during lecture, registration of end-of-semester exam and access to University resources. These contactless services have shown to be beneficial for students and lecturers. The same authors have presented similar works in [16].

A more recent work was reported by Ichimura & Kamada (2013) [19]. In this study, the authors describe the functionality of Attendance Management System (AMS) build upon NFC technology. They have developed the AMS for the University and tested using Nexus 7 devices connected to each other via peer-to-peer network. From the implementation, they found out that the number of absentees in nearly all classes was decreased and this could be due to students' conscientiousness that they are being observed.

Isomursu et al. (2011) [20] described a case study that explores the adoption of attendance control system in the school environment. The system was implemented using networked technology components, including smart cards, NFC enabled mobile phones and card readers, a web portal, and SMS messaging. Using Schwartz's value model adopted from social psychology as a framework, the study analyze the technology adoption from the viewpoint of three end user groups, namely children, parents and teachers. The study presented value analysis to help understand user experience perceived by different user groups when adopting technology-supported attendance system.

Fernández et al. (2013) [21] provide description of an attendance control system based on NFC technology developed and implemented at Pontifical University of Salamanca, Madrid, Spain. The project was carried out as Final Degree Project and developed in collaboration with Samsung Electronics. The aim was to ensure continuous assessment so that lecturers' teaching time will not be affected by the manual way in recording attendance. Results from the survey of the pilot project indicate that students and lecturers perceived high level of satisfaction and usefulness of the project.

Shen et al. (2014) [22] presented the most recent work in the area of NFC usage in classroom. In this study, the authors proposed the NFC-equipped smart classroom system to automate attendance management, to locate students, and to provide real-time feedback to students. The positioning feature of the system shown to be very useful to users; particularly for large classes. The study also evaluates the effect of the proposed system by measuring the students' attitude towards computer science education. Using a 5-point Likert scale questionnaire, students' attitude was measured based on five aspects: i) self-concept in computer science, ii) learning computer science at school, iii) learning computer science outside of school, iv)future participation in computer science, and v) importance of computer science. The results showed that students' attitude toward computer science is generally improved and the students perceived computer science as interesting, exciting, beneficial and helpful after experiencing the proposed system.

5   Conclusions

The innovation and perception of using Near-Field Communication technology for student attendance has been and always will be supported by many researchers as shown in our systematic review [3]. As can be seen from our review, we found a total of 17 studies that have presented various system design and architecture in implementing NFC-based student attendance system for University and school children. The earliest study found was published in 2007 and the latest was in 2014. Hence, we believe that research will continue to bloom and fork on various paths in this area of research.

The developing idea and prototype of Student's Attendance System using NFC is on the verge of introducing worldwide institutions into an automated way of recording student's attendance. It will, by far, improve the current manual process of tracking and recording student's attendance. This system promotes a way for students to 'sign' their attendance on a digital form and in contactless mode. For lecturers, implementation of such system would ease their effort in tracking students who are missing from class as well as reducing effort to verify students' attendance.

References

1. Mohamed, A.A., Kameswari, J.: Web-Server based Student Attendance System using RFID Technology. International Journal of Engineering Trends and Technology 4(5), 1559–1563 (2013)
2. Farrow, R.: Smart Card and Security Future. University of Michigan (2002). http://delivery.acm.org/10.1245/960000/951226/p159basu.pdf?key1=959896&key2=5275 191611&coll=portal&dl=ACM&CFID=2974836&CFTOKEN=65539847
3. Sivalingam, M.: Smart Card Application For Campus E-Services, pp. 5–20. University Malaya, Kuala Lumpur (2009)
4. Wahab, A.H., Kadir, H.A., Yusof, M.N.M., Sanudin, R., Tomari, M.R.: Class Attendance System using Active RFID: A Review, FKKE Compilation of Papers. Pusat Penyelidikan dan Inovasi, University Tun Hussein Onn (2009)
5. Ayu, M.A., Ahmad, B.I.: TouchIn: An NFC Supported Attendance System in a University Environment. International Journal of Information and Education Technology 4(5), 448–453 (2014)

6. Subpratatsavee, P., Promjun, T., Siriprom, W., Sriboon, W.: Attendance system using NFC technology and embedded camera device on mobile phone. In: Proceedings of 2014 International Conference on Information Science and Applications (ICISA), pp. 1–4, IEEE, Seoul (2014)
7. Bucicoiu, M., Tapus, N.: Easy attendance: location-based authentication for students integrated with Moodle. In: Proceedings: 11th Roedunet International Conference, Siania, pp. 1–4 (2013)
8. Chavira, G., Nava, S., Hervás, S., Bravo, J., Sánchez, C.: Spontaneous interaction on context-aware public display: an NFC and infrared sensor approach. In: Proceedings : First International Conference on Immersive Telecommunications (ImmersCom 2007), Brussels, Belgium (2007)
9. Ervasti, M., Isomursu,M., Kinnula, M.: Bringing technology into school–NFC-enabled school attendance supervision. In: Proceedings of the 8th International Conference on Mobile and Ubiquitous Multimedia (MUM 2009), pp. 1–10. ACM (2009)
10. Kitchenham, B., Charters, S.: Procedures for performing SLRs in Software Engineering. Keele University and University of Durham, UK (2007)
11. Rahnama, B., Elci, A., Celik, S.: Securing RFID-based authentication systems using ParseKey+. In: Proc. 3rd International Conference on Security of Information and Networks (SIN 2010), pp 212–217. ACM (2010)
12. Ervasti, M., Isomursu, M., Kinnula, M.: Experiences from NFC supported school attendance supervision for children. In: Proc. Third International Conference on Mobile Ubiquitous Computing, Systems, Services and Technologies (UBICOMM 2009). IEEE (2009)
13. Isomursu, M., Ervasti, M., Isomursu, P., Kinnula, M.: Evaluating human values in the adoption of new technology in school environment. In: Proc. 43rd Hawaii International Conference on System Sciences, pp. 1–10 (2010)
14. Ninomiya, H., Ito, E., Flanagan, B., Hirokawa, S.: Bridging SNS ID and user using NFC and SNS design of NFC and SNS based event attendance management system. In: International Conference on Anti-Counterfeiting, Security and Identification (ASID), pp 1–5. IEEE, Taipei (2012)
15. Ninomiya,H., Ito, E., Flanagan, B., Hirokawa, S.: Near friends communication encouragement system using NFC And SNS. In: Proc. International Conference On Advanced Applied Informatics (IIAIAAI), pp 145–148. IEEE, Fukuoka (2012)
16. Benyo, B., Sodor, B., Doktor, T., Fordos, G.: Student attendance monitoring at the university using NFC. In: Proceedings of the Wireless Telecommunications Symposium (WTS), pp. 1–5. IEEE, London (2012)
17. Bueno-Delgado, M.V., Pavón-Marino, P., De-Gea-García, A., Dolón-García, A.: The smart university experience: a NFC-based ubiquitous environment. In: Proceedings Sixth International Conference on Innovative Mobile and Internet Services in Ubiquitous Computing, pp. 799–804. IEEE, Palermo (2012)
18. Benyo, B., Sodor, B., Doktor, T., Fordos, G.: University life in contactless way - NFC use cases in academic environment. In: Proceedings of the 16th International Conference on Intelligent Engineering Systems (INES 2012), pp. 511–514. IEEE, Lisbon (2012)
19. Ichimura, T., Kamada, S.: Early discovery of chronic non-attenders by using NFC attendance management system. In: IEEE 6th International Workshop on Computational Intelligence and Applications, pp. 191–196. IEEE, Hiroshima (2013)
20. Isomursu, M., Ervasti, M., Kinnula, M., Isomursu, P.: Understanding human values in adopting new technology—A case study and methodological discussion. International Journal of Human Computer Studies **69**, 183–200 (2011)

21. Fernández, M.J.L., Fernández, J.G., Aguilar, S.R., Selvi, B.S., Crespo, R.G.: Control of attendance applied in higher education through mobile NFC technologies. Expert Systems with Applications **40**(1), 4478–4489 (2013)
22. Shen, C.W.: Jim Wub, Y.C., Lee, T.C. : Developing a NFC-equipped smart classroom: Effects on attitudes toward computer science. Computers in Human Behavior **30**, 731–738 (2014)
23. Petticrew, M., Roberts, H.: Systematic Reviews in the Social Sciences: A Practical Guide. Blackwell Publishing, USA (2006)
24. Salleh, N., Mendes, E., Grundy, J.: Empirical Studies of Pair Programming for CS/SE Teaching in Higher Education: A Systematic Literature Review. IEEE Trans. Softw. Eng. **37**(4), 509–525 (2011)

A Decision Support Map for Security Patterns Application

Rahma Bouaziz[1]([✉]) and Slim Kammoun[2]

[1] ReDCAD Laboratory, Sfax University, Sfax, Tunisia
rahma.bouazizkammoun@redcad.org
[2] LaTICE Laboratory, Tunis University, Tunis, Tunisia
slim.kammoun@esstt.rnu.tn

Abstract. In software engineering, security concerns should be addressed at every phase of the development process. To do that, patterns based security engineering approach has been proposed and investigated becoming a very active area of research. Security patterns capture the experience of experts in order to solve a security problem in a more structured and reusable way. With the proliferation of security patterns, thus it is becoming harder to select which ones should be applied and in each case. In this paper, our main contribution consists in the proposition of a map layered security patterns. This map allows software engineer to select and apply patterns in a systematic manner in order to guide the security decisions.

Keywords: Security · Pattern · Software development · Patterns application · Cartography

1 Introduction

To build secure component based systems it is not enough to compose pre-secured atomic component. Security has to be thought during the entire component composition process. It has been argued that security concerns should be addressed at every phase of the development process [1]. Several approaches and methodologies have been proposed to help non-expert in security to implement secure software systems. One of them is Security Patterns approach [2], which is defined as an adaptation of design patterns to security.

Security engineering with patterns is currently a very active area of research since security patterns capture the experience of experts in order to solve a security problem in a more structured and reusable way. A security pattern presents a well-understood generic solution to a particular recurring security problem that arises in specific contexts [2]. Moreover, the advantages of security patterns are that they provide novice security developers with security guidelines so that they will be able to solve security problems in a more systematic and structured way.

A pattern can be defined as a generative tool which is helpful to the designer, but, another problem arises: there is a proliferation of security patterns, thus it is becoming harder to select which ones should be applied. In this work we will propose a systematic

O. Gervasi et al. (Eds.): ICCSA 2015, Part IV, LNCS 9158, pp. 750–759, 2015.
DOI: 10.1007/978-3-319-21410-8_57

manner in order to guide engineers in their decisions to select and use adequate security patterns. To do that, we formalize the concept of pattern cartography or pattern map. We make it navigable, so that it makes it easier for a designer to make security decisions.

The remainder of this paper is organized as follows. Section 2 presents the motivations of this work. Section 3 details the proposed decision making map of security pattern in which we detail relationships among security patterns used to enforce a security policy. A use case of GPS system is presented in Section 4 to illustrate the use of the proposed map. Section 5 present the proposed framework for securing Component based applications using the proposed pattern Map. Section 6 ends the paper with a conclusion and an outlook.

2   Motivations

Making security decisions is a complex task that cannot be done automatically. In the literature, we can note the lack of systematic rules helping designers to secure their applications or systems so patterns help by gathering the knowledge of security experts. However, security patterns exist and propose a set of security knowledge.

During last decades, and with the development of engineering patterns in general and security patterns in particular, we can note that many of these patterns are proposed in both academic and industrial context. Although these patterns have efficient solutions to specific problems, the choice of a pattern over other remains a major problem for developers not expert in security [13]. For example we can find two or three patterns with the same purpose, but using different solutions to solve the same problem. So, according to the proposed design, one of these patterns will be more suitable to use than others. To facilitate this choice, we will try to offer a clear mechanism to help engineers in their decisions.

In this paper, we propose to organize the security patterns as cartography to help developer making security decision. The proposed map will be navigable, in order to help the designer to choose suitable security pattern within the current context and the development phase. In other words, according to a decision tree, the developer is guided to select the appropriate pattern depending on the problem faced.

3   Security Pattern Cartography

In Software development process the problem is first analyzed and then the solution is iteratively refined. During each phase of the software life cycle, the solution is treated with a different level of abstraction. Security is a non-functional requirement that must be thought from the requirements analysis phase. In this context, the security patterns must be defined and used at different levels of abstraction [10].

The higher layer gathers security patterns that describe security policies, that will be applied regardless of the technology or platform used. At the next layer, security patterns that implement these security policies are defined. Then the third layer proposes rules for implementing those patterns. Finally at the lower and end layer aspects are proposed to implement these security solutions.

3.1   Relationships Between Security Patterns

In this section we will focus on different relationships between security patterns that appear in the proposed map. To do that we have already study classification of the relationship between design patterns which are proposed in [3], [4] and [5]. We note that relations between patterns can exist at the same layer or on two different layers.

The relationship that connects one or more patterns from one layer to another pattern of a lower-level layer (rules or aspects), is a "**realization**" relationship.

In the same time, we have identified three basic relationships offered by most known formalisms describing patterns [3], [6], [7]. Those relationships are "**use**", "**refinement**" and "**alternative**". In the following we will describe those relationships with more details.

- **Use** relationship: This is the most common relationship in patterns systems. It allows the decomposition of a complex problem into sub-problems of less complexity. We can define this relationship as follows: Pattern X uses a pattern Y if and only if some problems of X can be solved by Y.

-**Refinement / Extension** Relationship: These are two reciprocal relations. The relationship "Extension" tends to generalize the intention of a pattern while the relationship "refinement "restricts it. So if a pattern X is an extension of a pattern Y, X can solve the Y pattern's problem. Conversely, we say that Y is a refinement of X.

- **Alternative** Relationship: A pattern X is an alternative of a pattern Y, if Y has the same intention of pattern X, but proposes a different solution.

3.2   The Decision Guiding Cartography

Using previous described relationship, we propose a decision map as shown in figure 1 for structuring security patterns. In this first tentative map, all patterns used and listed have been proposed in the literature by various authors as find in [8], [9] and in different areas. However, this map is not exhaustive, new patterns can be proposed and incorporated in a simple way.

In the proposed map, the higher layer, is called "security policies" and contains all security policies that can be defined by the literature such as access control, authentication, etc. These policies define needs in terms of application security.

The second layer, which is a less abstract level and we called it "Security Patterns". This level describes security mechanisms that present several solutions that can be used to implement security policies mentioned above. For example, the "single access point" pattern and the "Authenticator" pattern of one layer can be used together to achieve the "access control" policy.

We note that we can also add other layers to implement these patterns as shown in Figure 1. This figure shows one layer containing the rules for integration of different security patterns and another layer grouping aspects templates enabling aspects code generation according to the selected security solutions. We keep the same semantics relationships for all layers of this mapping.

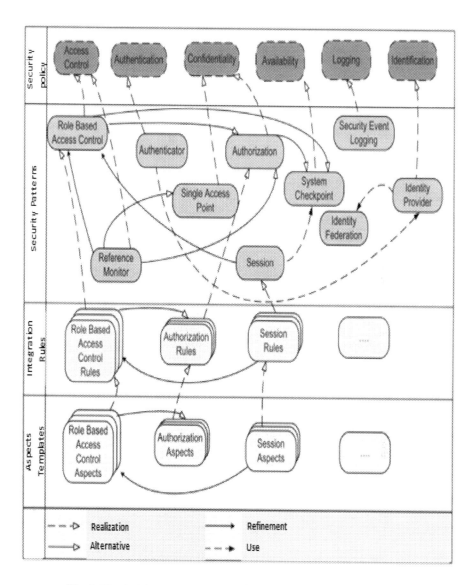

Fig. 1. The decision support cartography for security patterns application

3.3   The Usage of the Map

Since developers must add security to their models during the software development lifecycle phases, they have to make security decisions. In our approach, security decisions are actually about choosing appropriate security patterns that should be applied to the model.

A decision tree is a decision analysis tool that uses a graph of nodes to model all the choices and their possible consequences. The proposed map can be considered as a decision tree forest used to select the appropriate security patterns (in what follows we present an example of a decision tree). We can say that:

- All nodes in the top layer are the following goal: "How can we implement the security policy represented by this node?"
- Arcs from decision trees are relations between the patrons of the map.
- The roots of the trees are selected based on security needs i.e. security policies to be applied.

The security decisions are ultimately taken by developers (humans) from reading the textual parts of patterns templates. In particular, the "consequences" candidates patterns sections because they describe the results following the implementation of the selected patterns.

In addition, the use of the map also provides good traceability of decisions throughout the development cycle. Security decisions are recorded as a sequence of nodes (applied security patterns). The process can be iterative, so, a flashback in the same tree is possible.

4   Application to the Simple GPS Case Study

In order to illustrate the use of the proposed map to select the appropriate security pattern, we present a case study inspired from the use of GPS (Global Positioning System) based systems, illustrated by Figure 2. Those systems allow determining the geographical coordinates of any point on the globe. GPS is used in combination with a map to locate and position.

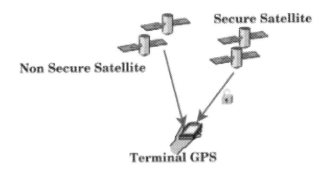

Fig. 2. Basic GPS example

In this example, we consider mainly the management of access control to various services offered by phone operator's especially downloading geographic maps in real-time and manage secure access to satellites. The Basic GPS system described above works as follows:

(1) The GPS Terminal receives continuously the signal of Satellite as well as that of SecureSatellite. The SecureSatellite is active if the user has access rights to it.

(2) The GPS Terminal sends a request to download map to the Phone Operator.

(3) The Phone Operator allows the user, depending on it's access rights, to download the requested geographic map.

With regard to the Basic GPS system, we have identified the following use case:

Access to secure satellites. To implement this use case, we have identified the security policy to be applied: "access control".

Therefore, we consider only decision trees with root as the access control policy. As shown in Figure 1, we can see that starting from this root and following the links of "realization" of "refinement" and "Alternative", we select the "RBAC" pattern, the "Session" pattern, the "System check Point" pattern, the "authorization" pattern and the "reference monitor" pattern. For design phase, the designer can choose to apply one or more patterns belonging to the same family, for example, the "RBAC" pattern. Thus the tree leads us to the selection integration rules of the selected pattern.

Once and for the implementation stage, the decision tree will also give us the aspects templates to generate the code of the chosen the security solution offered by the selected pattern, in our case it is the "RBAC" pattern.

5   A Framework for Securing Component Based Applications Using the Proposed Pattern Map

In order to use the proposed patterns cartography, we propose a framework as shown in Figure 2. This framework will be used to secure component based applications using security patterns. The proposed framework called SCRI-PRO for SeCurity Pattern Integration apPROach will describe how models are derived from one another.

SCRI-PRO covers all phases of a system development process, namely the design phase, the integration of non-functional properties phase (i.e. security), and finally their integration into the functional code of the application.

SCRI-PRO combines the advantages of the use of engineering patterns (especially security patterns), the strengths of the model-driven engineering and the flexibility of aspect-oriented programming. This gives the user the ability to integrate the solutions proposed by the patterns in a way both semi-automatic and modular way.

The proposed framework presents two dimensions:

Vertically: two MDA views are represented, platform independent view and platform specific view. We can clearly see two phases that constitute the proposed framework: the modeling phase, which is independent form a given platform and the implementation phase, which is the specific view of a platform.

The modeling phase comprises:

1. Propose the components application model to secure

2. Annotate the defined application model by the patterns security solutions through:

- Firstly, the definition of UML profiles to extend the components meta-model with all necessary concepts for the specification of non-functional properties (e.g. Security).

- In a second step, the specification of integration rules adapted to the solutions from engineering patterns. These rules once specified may be applied to the model of the application to generate a model that supports non-functional properties.

The implementation phase covers the generation of the functional code of the application.

This code contains the business logic of the application and does not cover non-functional properties. These properties will be specified through the aspects templates. The generated aspects are automatically integrated, in a modular manner, to the functional code defined in the same phase.

Horizontally, we present the separation between the different concerns.

This separation is maintained in all phases of the proposed approach. In Our approach, we think that it is necessary to separate the expertise of the application domain (functional concerns) from expertise encapsulated by specific patterns like security patterns.

The transformations between models are represented by large arrows in Figure 2. These transformations are of two different types:

- Patterns are integrated into a model using UML profile(s) that extend(s) the UML metamodel with security concepts. Then in a second step, this integration will be semi-automatic, requiring that designer make security decisions, through the execution of transformation rules. The implementation of these rules allows the annotation of the components based model with the pattern's concepts.

- To transform a platform independent model (PIM) to a platform dependent (specific) model (PSM) (principle MDA), we will use aspect templates that will be woven in the functional code.

The large arrow between security patterns and aspects templates illustrates the use of aspect paradigm in the implementation of the solutions proposed by the security patterns. Specifically, when selecting security patterns (according to security requirements for a particular application) in the abstract level (PIM), templates corresponding security aspects are generated.

The second vertical red arrow illustrates the use of the proposed decision tree: when security patterns are selected dynamically (according to the security requirements for the particular application eg. security policy) at the abstract level, then the corresponding concrete security patterns (corresponding security rules and corresponding aspects template) can be implied.

The development process is iterative and incremental: activities are repeated through successive refinements, which allow the reuse of proposed security patterns available in the repository and organized as a map, as proposed in this paper. The structure of the proposed process follows the classical life cycle, in which we have an elicitation phase, a modeling phase and finally an implementation phase.

In the elicitation phase, the designer identifies and models the basic functionality of the system. Security concepts are not introduced.

The modeling phase consists first in identifying and analyzing the security requirements from the application component model. Those security requirements define which security policies are necessary for the analysis model. After that, security patterns are selected to enforce security policies and UML profiles are defined according to the selected security patterns. These patterns are integrated into the application component model in order to obtain a secure Application Component Model.

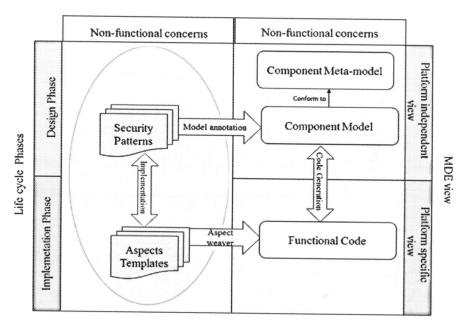

Fig. 3. A framework for securing component based applications using the proposed pattern Map

In the implementation phase, a component-based platform must be selected (CCM, EJB, etc.) and the secure application component model is refined into security aspects code together with the functional code for producing the secure application code.

As one can note with this phase, some activities are collaborative, in the sense that several participants working together towards a common goal should perform them. In the following, we put the focus on the collaborative aspects of this phase.

To describe the collaborative aspect of our process, we use CMSPEM, an extension of the SPEM standard, proposed by Kedji et al. [11]. CMSPEM introduces new concepts to represent collaborative processes, and relationships among them. For describing collaborative activities, CMSPEM introduces the concept of Actor (human actor), a specific human participant in a project, associated with a role and provides relations to specify what is done by each actor.

Our work investigates how non-security experts can take profits from security patterns to easily implement secure component-based applications. In previous work [10, 12], we proposed an engineering process, called SCRIP (SeCurity patteRn Integration

Process), which provides guidelines for integrating security patterns into component-based models. SCRIP defines activities and products to integrate security patterns in the whole development process, from UML component modeling until aspect code generation.

6   Conclusion

Security is one of the most important properties to consider in systems development and the goal of developing secure software system has remained an area of active research. A promising way to address this issue is the application of solutions proposed by security patterns. Because of the proliferation of security patterns, we proposed a structured decision making map for security patterns so that it makes it easier for a designer to make security decisions. The proposed framework has the advantage of separating the application domain expertise and expertise in security. The integration of security in the software development process becomes easier for the architects/designers. Furthermore, it is relatively simple and suitable for use by non-security experts. Understanding security patterns from their description and having knowledge on applications-based components are sufficient skills to use this process. The implementation and the experimentation of the framework were done in a prototype as a partial validation of our approach because further work is still needed to get a true validation.

As future work, we aim to provide a complete development environment to design secure component based application using the proposed engineering process and guide developer in the selection of security patterns by navigating the proposed decision map and suggesting what reachable patterns could be applied.

References

1. Devanbu, P.T., Stubblebin, S.: Software engineering for security: A roadmap. In: Proceedings of the Conference of The Future of Software Engineering (2000)
2. Yoder, J., Barcalow, J.: Architectural patterns for enabling application security. In: Fourth Conference on Patterns Languages of Programs (1997)
3. Zimmer, W.: Relationships between design patterns. In: Pattern Languages for Programming Design, pp. 345–364. Addison-Wesley (1994)
4. Fernandez, E.B., Washizaki, H., Yoshioka, N., Kubo, A., Fukazawa, Y.: Classifying security patterns. In: Zhang, Y., Yu, G., Bertino, E., Xu, G. (eds.) APWeb 2008. LNCS, vol. 4976, pp. 342–347. Springer, Heidelberg (2008)
5. Hafiz, M., Adamczyk, P., Johnson, R.E.: Organizing Security Patterns. IEEE Softw. **24**, 52–60, juillet 2007. URL http://dl.acm.org/citation.cfm?id=1435593.1435870
6. Noble, J.: Classifying relationships between object-oriented design patterns. In: Proceedings of the Dans Software Engineering Conference, 1998, pp. 98–107. Australian (1998)
7. Conte, A., Fredj, M., Giraudin, J.P., Rieu, D.: P-sigma : un formalisme pour une représentation unifiée de patrons. In: Dans INFORSID 2001, pp. 67–86 (2001)

8. Yoder, J., Baracalow, J.: Architectural patterns for enabling application security. In: Dans Proc. Fourth Conf. Pattern Languages of Programming (PLoP) (1997)
9. Schumaher, E., Fernandez-Buglioni, M., Hyberston, D., Buschmann, F., Sommerlad, P.: Security Patterns : Integrating Security and Systems Engineering (Wiley Software Patterns Series). John Wiley and Sons, mars 2006. ISBN 0470858842. URL http://www.worldcat.org/isbn/0470858842
10. Bouaziz, R., Coulette, B.: Applying security patterns for component based applications using UML profile. In: Proceedings of the International Conference on Computational Science and Engineering, Paphos, Cyprus, pp. 186–193 (2012)
11. Akpédjé Kedji, K., Lbath, R., Coulette, B., Nassar, M. Baresse, L.: Florin Racaru Supporting collaborative development using process models: a Tooled Integration-focused Approach. Journal of Software : Evolution and Process (JSEP), February 2014. Wiley
12. Bouaziz, R., Kallel, S., Coulette, B.: An engineering process for security patterns application in component based models. In: Proceedings of the International Conference on Enabling Technologies: Infrastructure for Collaborative Enterprises (WETICE), pp. 231–236. IEEE Computer Society, June 17–20, 2013
13. Schumacher, M.: Security Engineering with Patterns: Origins, Theoretical Models, and New Applications. Springer-Verlag (2003)

Author Index